AF757229

Silicon Devices and Process Integration

Deep Submicron and Nano-Scale Technologies

Badih El-Kareh

Silicon Devices and Process Integration

Deep Submicron and Nano-Scale Technologies

Springer

Badih El-Kareh
Independent Consultant
Cedar Park, TX
USA
bek@ieee.org

ISBN: 978-0-387-36798-9 e-ISBN: 978-0-387-69010-0
DOI: 10.1007/978-0-387-69010-0

Library of Congress Control Number: 2008936300

© Springer Science+Business Media, LLC 2009
All rights reserved. This work may not be translated or copied in whole or in part without the written permission of the publisher (Springer Science+Business Media, LLC, 233 Spring Street, New York, NY 10013, USA), except for brief excerpts in connection with reviews or scholarly analysis. Use in connection with any form of information storage and retrieval, electronic adaptation, computer software, or by similar or dissimilar methodology now known or hereafter developed is forbidden.
The use in this publication of trade names, trademarks, service marks, and similar terms, even if they are not identified as such, is not to be taken as an expression of opinion as to whether or not they are subject to proprietary rights.

Printed on acid-free paper

springer.com

Preface

State-of-the-art silicon devices and integrated process technologies are covered in this book. The eight chapters represent a comprehensive discussion of modern silicon devices, their characteristics, and the relationship between their electrical properties and processing conditions. The material is compiled from industrial and academic lecture-notes and reflects years of experience in the development of silicon devices.

The book is prepared specifically for engineers and scientists in semiconductor research, development and manufacturing. It is also suitable for a one-semester course in electrical engineering and materials science at the upper undergraduate or lower graduate level.

The chapters are arranged logically, beginning with a review of silicon properties that lays the groundwork for the discussion of device properties, including mobility-enhancement by straining silicon.

Junctions and contacts are inherent to practically all semiconductor devices. Chapter 2 covers junctions under forward and reverse characteristics, including high-level injection and high-field effects. Understanding the properties of contacts has become increasingly important as the contact size is reduced to deep submicron and nanoscale dimensions. The last part of Chap. 2 discusses ohmic and rectifying contacts.

Chapter 3 begins with bipolar fundamentals and moves to an advanced treatment of bipolar enhancements with silicon–germanium (SiGe). This chapter is particularly important to analog and mixed-signal applications where complementary metal-oxide semiconductor (CMOS) and bipolar transistors are integrated in a BiCMOS process. It also benefits engineers in understanding important bipolar effects in CMOS-only applications, such as subthreshold current and parasitic latch-up.

The metal-oxide silicon (MOS) capacitor is a key part of a metal-oxide semiconductor field-effect transistor (MOSFET) and a powerful process and device characterization tool. The physics and characterization of MOS structures are detailed in Chap. 4, beginning with an ideal stack of a conductor, an insulator and silicon, and gradually moving to real structures and quantum effects.

Chapter 5 deals with the insulated-gate field-effect transistor. It begins with a description of the modes of transistor operation and the different transistor types. Transistor current–voltage characteristics are detailed, followed by a discussion of scaling the structure to smaller dimensions, scaling limitations, short-channel, reverse short-channel, narrow-channel, and reverse narrow-channel effects. Mobility enhancement techniques are described, including strained silicon and optimization of crystal orientation. The discussion extends to ultra-thin gate-oxide, high-K dielectrics, advanced gate-stacks, and three-dimensional structures.

Analog devices and passive components are introduced in Chap. 6. As an extension of bipolar transistors detailed in Chap. 3, the properties of junction field-effect transistors are described, followed by optimization of MOSFETs for analog applications. The design and properties of integrated precision resistors, capacitors, and varactors are then detailed. The chapter concludes with the important topics of component matching and noise.

Chapter 7 covers advanced enabling processes and process integration. It begins with integrated CMOS and BiCMOS processes to illustrate typical sequences of processing steps. Crystal growth and wafer parameters, including properties of silicon-on-insulator (SOI), relevant to modern integrated processes are discussed. Front-end of the line unit processes include short-duration thermal processes, atomic-lay deposition (ALD), ionized physical-vapor deposition (IPVD), optical proximity correction (OPC), double exposure and patterning, immersion lithography, and new silicides. Back-end of the line processes include copper interconnects and low-K dielectrics.

The last chapter reviews selected CMOS and BiCMOS digital and memory applications. The inverter is used to analyze the important parasitic latch-up effect and methods to suppress it. The second part covers memory cells, including dynamic random-access memory (DRAM), static random-access memory (SRAM), and nonvolatile memory (NVM).

It would not have been possible for me to complete this book in its present form without the continuous invaluable help with corrections and suggestions for improvement and encouragement from Dr. Wendell Noble, independent consultant, retired IBM semiconductor physicist, Professor Carlton Osburn of the North Carolina State University, and Dr. Albert Puttlitz, IEEE-Components, Packaging and Manufacturing Technology Society, VP of Education. I also thank my former colleagues at IBM, Russell Houghton and Ashwin Ghatalia, for their reviews and inputs. My special thanks to the personnel of the University of Texas library for their kind support in my research.

August 18, 2008 *Badih El-Kareh*

Contents

List of Symbols

a	acceleration (cm/s^2)
a	JFET metallurgical channel width (cm)
a	lattice constant (cm)
a	voltage ramp-rate (V/s)
A	geometry-dependent factor
A	area (cm^2)
A^*	effective Richardson constant ($A/cm^2 \cdot K^2$)
A_E	emitter area (cm^2)
A_C	cross-sectional area (cm^2)
$A_{\Delta R}$	process-related resistor mismatch factor (cm)
$A_{\Delta VT}$	process-related threshold voltage-mismatch factor (cm)
A_G	gate area (cm^2)
A_S	surface area (cm^2)
b	mobility ratio (μ_n/μ_p)
BL	bit-line
BV	breakdown voltage (V)
BV_{CBO}	collector-base breakdown voltage, emitter open (V)
BV_{CBS}	collector-base breakdown voltage, emitter-base shorted (V)
BV_{CEO}	collector-emitter breakdown voltage, base open (V)
BV_{DGO}	drain-gate breakdown voltage, source open (V)
BV_{DGS}	drain-gate breakdown voltage, source-gate shorted (V)
BV_{EBO}	emitter-base breakdown voltage, collector open (V)
BV_{EBS}	emitter-base breakdown voltage, collector-base shorted (V)
C	capacitance per unit area (F/cm^2)
c	velocity of light (2.998×10^{10} cm/s)
C_{BL}	bit-line capacitance (C)
C_D	diffusion capacitance per unit area (F/cm^2)
C_{decap}	decoupling capacitance (F)
C_{deep}	deep deletion capacitance per unit area, CV plot (F/cm^2)
$C_{FG\text{-}CG}$	capacitance between floating and control gate (F)

C_{GCh}	gate to channel capacitance per unit area (F/cm^2)
C_{GD}	gate to drain capacitance (F)
C_{GS}	gate to source capacitance (F)
C_{HF}	high-frequency capacitance per unit area, CV plot (F/cm^2)
C_i	intrinsic capacitance, varactor (F)
C_{ILD}	inter-level dielectric capacitance (F)
C_{inv}	silicon inversion capacitance per unit area, CV plot (F/cm^2)
C_j	junction capacitance (F)
C_{jE}	emitter-base junction capacitance (F)
C_{jC}	collector-base junction capacitance (F)
C_L	atomic concentration in liquid state (cm^{-3})
C_L	load capacitance (F)
C_{LF}	low-frequency capacitance per unit area, CV plot (F/cm^2)
C_{max}	maximum capacitance per unit area, CV plot (F/cm^2)
C_{min}	minimum capacitance per unit area, CV plot (F/cm^2)
C_{ox}	oxide capacitance per unit area, CV plot (F/cm^2)
C_{par}	parasitic capacitance (F)
C_{PMD}	pre-metal dielectric capacitance (F)
C_{poly}	polysilicon, e.g., depletion capacitance per unit area, (F/cm^2)
C_S	atomic concentration in solid state (cm^{-3})
C_S	storage node capacitance (F)
C_{Si}	silicon capacitance per unit area, CV plot (F/cm^2)
C_{Sidep}	silicon depletion capacitance per unit area, CV plot (F/cm^2)
C_{SiFB}	silicon capacitance at flatband per unit area, CV plot (F/cm^2)
C_{Simin}	silicon minimum capacitance per unit area, CV plot (F/cm^2)
C_{STI}	shallow-trench capacitance (F)
d	distance (cm)
D	diffusion constant (cm^2/s)
$\tilde{D}$	effective diffusion constant (cm^2/s)
D_n	electron diffusion constant (cm^2/s)
D_p	hole diffusion constant (cm^2/s)
E	energy (eV)
E	electric field (V/cm)
e	tensile strain (Pa)
E_C	critical field (V/cm)
$\mathrm{E_C}$	bottom of conduction band energy level (eV)
$\mathrm{E_{CNL}}$	charge neutrality level (eV)
$\mathrm{E_D}$	donor energy level (eV)
$\mathrm{E_F}$	Fermi level (eV)
$\mathrm{E_{Fn}}$	electron quasi-Fermi level (eV)
$\mathrm{E_{Fp}}$	hole quasi-Fermi level (eV)
$\mathrm{E_g}$	energy gap (eV)
E_{grad}	field induced by grading Ge profile (V/cm)
Ei	intrinsic silicon energy level (eV)

E_i	ionization energy (eV)
$E_{i(A)}$	acceptor ionization energy (eV)
$E_{i(D)}$	donor ionization energy (eV)
E_n	nitride field $Q_n \approx 0$ (V/cm)
E_{ox}	oxide field (V/cm)
E_{OO}	characteristic tunneling energy (eV)
E_P	phonon energy (eV)
E_{peak}	peak electric field (V/cm)
E_s	surface field (V/cm)
E_{Si}	field in silicon (V/cm)
E_T	trap energy level (eV)
E_V	top of valence band energy level (eV)
E_x	field in silicon normal to surface (V/cm)
E_y	surface field parallel to silicon surface (V/cm)
F	force (N)
F	dimensionless electric field (F-function)
f	frequency (Hz)
$f(E)$	Fermi-function
f_T	gain-bandwidth product, cut-off frequency (Hz)
f_{max}	maximum frequency of operation (Hz)
G	constant
G	bulk generation rate $(cm^{-3} \cdot s^{-1})$
g_D	channel (drain) conductance (S)
$g_{D\text{-lin}}$	linear channel (drain) conductance (S)
$g_{D\text{-sat}}$	saturated channel (drain) conductance (S)
g_m	transconductance (S)
$g_{m\text{-lin}}$	linear transconductance (S)
$g_{m\text{-sat}}$	saturated transconductance (S)
GR	generation-recombination
G_0	lumped JFET parameter
I	current (A)
I_B	base current (A)
I_B	body current (A)
I_{BC}	base-collector current (A)
I_{BE}	base-emitter current (A)
I_C	collector current (A)
I_{CBO}	collector-base current, emitter open (A)
I_{CEO}	collector-emitter current, base open (A)
I_{Csat}	collector saturation current (A)
I_D	drain current (A)
I_{Diff}	diffusion current (A)
I_{Dsat}	saturated drain current (A)
I_{D0}	drain current per channel-square at threshold (A)
I_E	emitter current (A)
I_{EBO}	emitter-base current, collector open (A)

I_{Esat}	emitter saturation current (A)
I_{F}	forward-bias current (A)
I_{G}	gate current (A)
I_{gen}	generation current (A)
$I_{\mathrm{gen\text{-}bulk}}$	bulk generation current (A)
$I_{\mathrm{gen\text{-}surf}}$	surface generation current (A)
I_{H}	holding current, latch-up (A)
I_{leak}	total leakage current (A)
I_{n}	electron current (A)
i_{n}	noise current (A)
I_{NW}	current in *n*-well (A)
I_{off}	MOSFET off-current (A)
I_{p}	hole current (A)
I_{PT}	punch-through current (A)
I_{PT0}	current at onset of punch-through (A)
I_{PW}	current in *p*-well (A)
I_{R}	reverse-bias current at $V_G = V_T$ (A)
I_{r}	surface recombination current (A)
I_{S}	source current (A)
I_{S}	saturation current (A)
I_{s}	surface current (A)
I_{sB}	base saturation current (A)
I_{sC}	collector saturation current (A)
I_0	drain current at $V_G = V_T$ (A)
j	current density (A/cm^2)
j_{direct}	direct tunneling current density (A/cm^2)
j_{F}	forward current density (A/cm^2)
j_{FN}	Fowler-Nordheim tunneling current density (A/cm^2)
j_{G}	gate current density (A/cm^2)
j_{n}	electron current density (A/cm^2)
$j_{\mathrm{n(dif)}}$	diffusion electron current density (A/cm^2)
j_{p}	hole current density (A/cm^2)
$j_{\mathrm{p(dif)}}$	diffusion hole current density (A/cm^2)
j_{R}	reverse current density (A/cm^2)
j_{s}	saturation current density (A/cm^2)
j_{T}	total current density (A/cm^2)
K	dielectric constant $= \varepsilon/\varepsilon_0$
k	Boltzmann constant $\approx 8.618 \times 10^{-5}$ eV/K
k	wave number (cm^{-1})
k_l	dimensionless resolution factor
k_{seg}	segregation coefficient
kT	thermal energy (=0.0258 eV at 300 K)
kT/q	thermal voltage (=0.0258 V at 300 K)
L	length (cm)

L	inductance (H)
L_D	Debye length (cm)
L_D	drawn length (cm)
L_E	emitter length (cm)
L_E	electrical length, e.g., resistor (cm)
L_e	extrinsic Debye length (cm)
L_eff	effective channel length (cm)
l_eff	effective mean-free path (cm)
L_I	impact-ionization mean-free path (cm)
L_i	intrinsic Debye length (cm)
L_met	metallurgical channel length (cm)
L_n	electron diffusion length (cm)
L_nB	electron diffusion length in base (cm)
L_p	hole diffusion length (cm)
L_pE	hole diffusion length in emitter (cm)
L_poly	polysilicon line-width, channel length (cm)
L_r	optical phonon mean-free path (cm)
L_T	contact transfer length (cm)
M	multiplication factor
m_D	density of states effective mass (kg)
m_0	electron mass ($\approx 9.1 \times 10^{-31}$ kg)
m_n^*	electron effective mass (kg)
m_ox	oxide electron effective mass (kg)
m_p^*	hole effective mass (kg)
N	number of electrons per unit area (cm^{-2})
n	electron concentration (cm^{-3})
$\bar{n}$	thermal equilibrium electron concentration (cm^{-3})
n	number of squares
n	ideality factor
n	index of refraction
n	number of circuits
NA	numerical aperture
N_A	acceptor concentration (cm^{-3})
N_A^-	ionized acceptor concentration (cm^{-3})
N_A0	acceptor concentration at $x = 0$ (cm^{-3})
$N_\mathrm{A}(x)$	acceptor concentration as function of depth (cm^{-3})
N_B	background dopant concentration (cm^{-3})
N_B	concentration of lightly-doped region (cm^{-3})
n_b	electron concentration in bulk (cm^{-3})
N_C	effective density of states at conduction band edge (cm^{-3})
N_D	donor concentration (cm^{-3})
N_D^+	ionized donor concentration (cm^{-3})
$N_\mathrm{D}(x)$	donor concentration as function of depth (cm^{-3})
N_D0	donor concentration at $x = 0$ (cm^{-3})
N_f	number of fixed oxide charges per unit area ($=Q_\mathrm{f}/q$ cm^{-2})

N_I	number of mobile ions per unit area (cm^{-2})
n_i	intrinsic carrier concentration (cm^{-3})
n_{i0}	intrinsic carrier concentration without energy-gap lowering (cm^{-3})
N_{inv}	number of inversion electrons per unit area (cm^{-2})
N_{it}	number interface traps per unit area (cm^{-2})
NM	noise margin
NM_H	high noise margin
NM_L	low noise margin
n_n	majority electron concentration in n-region (cm^{-3})
$\bar{n}_n$	thermal equilibrium electron concentration in n-region (cm^{-3})
n_{n0}	electron concentration in n-region at $x = 0$ (cm^{-3})
n_p	minority electron concentration in p-region (cm^{-3})
$\bar{n}_p$	thermal equilibrium electron concentration in p-region (cm^{-3})
n_{p0}	electron concentration in p-region at $x = 0$ (cm^{-3})
n_s	surface electron concentration (cm^{-3})
N_s	number of secondary carrier pairs
n_{sL}	surface electron concentration at drain (cm^{-3})
n_{so}	surface electron concentration at source (cm^{-3})
N_t	density of generation-recombination centers (cm^{-3})
N_{teff}	effective density of generation-recombination centers (cm^{-3})
N_V	effective density of states at valence band edge (cm^{-3})
N_0	fixed diffusion-source concentration (cm^{-3})
n_0	electron concentration at $x = 0$ (cm^{-3})
O_i	concentration of interstitial oxygen (cm^{-3})
P	parameter
P	perimeter (cm)
P	power (W)
P	probability
p	hole concentration (cm^{-3})
p	momentum (N · s)
$\bar{p}$	thermal equilibrium hole concentration (cm^{-3})
p_b	bulk hole concentration (cm^{-3})
P_j	junction perimeter (cm)
p_n	concentration of minority holes in n-region (cm^{-3})
$\bar{p}_n$	thermal equilibrium hole concentration in n-region (cm^{-3})
p_{n0}	hole concentration in n-region at $x = 0$ (cm^{-3})
$\bar{p}_{n0}$	equilibrium minority hole concentration at $x = 0$ (cm^{-3})
p_p	concentration of majority holes in p-region (cm^{-3})
$\bar{p}_p$	thermal equilibrium hole concentration in p-region (cm^{-3})
$\bar{p}_{p0}$	equilibrium majority hole concentration at $x = 0$ (cm^{-3})
p_{p0}	hole concentration in p-region at $x = 0$ (cm^{-3})
p_s	surface hole concentration (cm^{-3})
$P_{standby}$	standby power (W)
Q	charge per unit area (C/cm^2)
Q	quality factor

q electron charge ($\approx 1.6 \times 10^{-19}$ C)
Q_B minority-carrier charge in base (C)
Q_b bulk depletion charge per unit area (C/cm^2)
Q_{bmax} maximum bulk depletion charge per unit area (C/cm^2)
$Q_{b\text{-deep}}$ bulk charge in deep depletion per unit area (C/cm^2)
Q_{eff} effective dielectric charge per unit area (C/cm^2)
Q_f oxide fixed charge per unit area (C/cm^2)
Q_{it} silicon-oxide interface trap charge per unit area (C/cm^2)
Q_{itm} gate-insulator interface trap charge per unit area (C/cm^2)
Q_m charge induced at gate-oxide interface per unit area (C/cm^2)
Q_m mobile charge per unit area (C/cm^2)
Q_{max} maximum quality factor
Q_n surface electron charge per unit area (C/cm^2)
Q_{ot} oxide trap charge per unit area (C/cm^2)
Q_p surface hole charge per unit area (C/cm^2)
Q_S stored charge per unit area (C/cm^2)
Q_s surface charge per unit area (C/cm^2)
R resistance (Ω)
r radius of curvature (cm)
r correlation coefficient
r fraction
r_A Auger recombination rate (cm^6/s)
R_B base resistance (Ω)
R_{Bext} extrinsic base resistance (Ω)
R_{Bint} intrinsic base resistance (Ω)
R_{B0} base resistance without applied bias (Ω)
R_C collector resistance (Ω)
R_C contact resistance (Ω)
R_{Ch} channel resistance (Ω)
R_D drain resistance (Ω)
R_E emitter resistance (Ω)
r_E emitter dynamic resistance ($=kT/qI_C$, Ω)
R_{ext} extrinsic resistance (Ω)
$R_{ext\text{-}S}$ source extrinsic resistance (Ω)
$R_{ext\text{-}D}$ drain extrinsic resistance (Ω)
R_G gate resistance (Ω)
R_L load resistance (Ω)
R_{LDD} resistance of lightly-doped drain region (Ω)
R_N noise resistance (Ω)
R_{NBL} n-buried layer resistance (Ω)
R_p parallel resistance (Ω)
R_p projected range (cm)
R_{PBL} p-buried layer resistance (Ω)
R_{pinch} intrinsic-base (pinched) resistance (Ω)

R_S	source resistance (Ω)
R_S	sheet resistance (Ω/Square)
R_{SD}	source-drain resistance (Ω)
R_{Sp}	spreading resistance (Ω)
R_{S0}	sheet resistance at $T = T_0$ (Ω/Square)
R_{wire}	wiring resistance (Ω)
r_0	output wiring resistance (Ω)
S	subthreshold swing (V/decade)
s, s_0	surface recombination velocity (cm/s)
s_i	current noise power spectral density $(A/\sqrt{Hz})$
s_v	voltage noise power spectral density $(V/\sqrt{Hz})$
t	time (s)
t	thickness (cm)
T	temperature (K)
T_0	reference temperature (K)
t_{eq}	equivalent oxide thickness (cm)
t_{metal}	metal thickness (cm)
T_N	noise temperature (K)
t_n	nitride thickness (cm)
t_{ox}	oxide thickness (cm)
$t_{ox\text{-}phys}$	physical oxide thickness (cm)
t_{poly}	polysilicon thickness (cm)
t_{Si}	path-length in silicon (cm)
$t_{silicide}$	silicide thickness (cm)
t_{STI}	shallow-trench isolation thickness (cm)
U	generation-recombination rate $(cm^{-3} \cdot s^{-1})$
U_s	surface generation rate $(cm^{-3} \cdot s^{-1})$
u	dimensionless Fermi potential $(=q\phi/kT)$
u_b	dimensionless bulk Fermi potential $(=q\phi_b/kT)$
u_s	dimensionless surface Fermi potential $(=q\phi_s/kT)$
$u(x)$	dimensionless Fermi potential versus depth x $[=q\phi(x)/kT]$
v	velocity (cm/s)
v	dimensionless potential $(=q\psi/kT)$
V_A	Early voltage (V)
V_a	applied voltage (V)
V_B, V_{BS}	body to source voltage (V)
V_b	barrier height (V)
V_b	built-in voltage (V)
V_{BC}	collector-base voltage (V)
V_{BE}	base-emitter voltage (V)
V_{CBO}	collector-base voltage, emitter open (V)
V_{CBS}	collector-base voltage, emitter shorted to base (V)
V_{CC}	power supply voltage, bipolar transistor (V)
V_{CE}	collector-emitter voltage (V)
V_{CEO}	collector-emitter voltage, base open (V)

V_{CEsat}	collector saturation voltage (V)
V_{Ch}	channel to source voltage, FET (V)
V_D, V_{DS}	drain to source voltage
v_d	drift velocity (cm/s)
V_{DA}	measured dielectric absorption (V)
V_{DD}	power supply voltage, MOSFET (V)
V_{Dsat}	saturation drain voltage, MOSFET (V)
v_{dy}	drift velocity along surface, in y-direction (cm/s)
V_{EBS}	emitter-base voltage, collector shorted to base (V)
V_F	forward voltage (V)
V_{FB}	flatband voltage (V)
V_G, V_{GS}	gate to source voltage (V)
V_H	holding voltage, latch-up (V)
V_{IH}	high input voltage (V)
V_{IHmax}	maximum high input voltage (V)
V_{IHmin}	minimum high input voltage (V)
V_{IL}	low input voltage (V)
V_{ILmax}	maximum low input voltage (V)
V_{ILmin}	minimum low input voltage (V)
V_j	junction voltage (V)
V_{jG}	junction to gate voltage, gated diode (V)
v_n	electron velocity (cm/s)
V_{OH}	high output voltage
V_{OHmax}	maximum high output voltage
V_{OHmin}	minimum high input voltage
V_{ox}	voltage across oxide (V)
V_P	pinch-off voltage (V)
v_p	hole velocity (cm/s)
V_{PT}	punch-through voltage (V)
V_R	reverse voltage (V)
V_{RD}	read voltage (V)
V_{SS}	typically ground potential, MOSFET (0 V)
v_s	dimensionless surface potential $(=q\psi_s/kT)$
v_{sat}	saturation velocity (cm/s)
V_T	threshold voltage (V)
$V_{T\text{-drain}}$	threshold voltage at drain (V)
v_{th}	thermal velocity ($\approx 10^7$ cm/s at 300 K)
$V_{T\text{-source}}$	threshold voltage at source (V)
V_{T0}	threshold voltage for zero floating-gate charge (V)
$v(x)$	dimensionless potential versus depth $[=q\psi(x)/kT]$
v_0	initial velocity (cm/s)
W	width (cm)
W_b	neutral base width (cm)
$W_{Contact}$	contact width (cm)
W_D	drawn width (cm)

W_E	emitter width (cm)
W_E	electrical width (cm)
W_{eff}	effective width (cm)
WL	word-line
W_{Metal}	metal width (cm)
W_n	width of neutral n-region (cm)
W_p	width of neutral p-region (cm)
W_{Total}	sum of all on-chip MOSFET width (cm)
W_{Via}	via width (cm)
x	depth normal to silicon surface (cm)
x_{Ch}	channel depth below the surface, thickness (cm)
x_d	depletion width (cm)
x_{dD}	depletion width at MOSFET drain (cm)
$x_{d\text{-}deep}$	depletion depth in deep depletion (cm)
$x_{d\text{-}field}$	depletion width under field oxide (cm)
$x_{d\text{-}inv}$	steady-state depletion depth in inversion (cm)
$x_{d\text{-}lat}$	lateral junction depletion width at surface (cm)
x_{dmax}	maximum depletion depth below surface (cm)
x_{dn}	depletion width in n-side of pn junction (cm)
x_{dp}	depletion width in p-side of pn junction (cm)
x_{dS}	depletion width at MOSFET source (cm)
x_{ds}, x_{dsurf}	junction depletion width at surface intercept (cm)
x_i	depth below surface where $n = p = n_i$ (cm)
x_J	junction depth (cm)
x_{jC}	collector-base junction depth (cm)
x_{JE}	emitter-base junction depth (cm)
x_{jlat}, x_{jl}	lateral extent of junction at surface (cm)
x_m	depth of potential peak, image-force barrier lowering (cm)
y	direction from source to drain
α	grounded base current gain $(=I_C/I_E)$
α	temperature coefficient of resistance (K^{-1})
α_F	forward grounded base current gain $(=I_C/I_E)$
α_i	impact ionization rate (cm^{-1})
α_R	reverse grounded base current gain $(=I_C/I_E)$
α_T	base transport factor
α_T	pre-tunneling factor
β	grounded emitter current gain $(= I_C/I_B)$
β	$= \mu_{eff} \cdot C_{ox} \cdot W/L$ (MOSFET)
β_F	forward grounded emitter current gain $(=I_C/I_B)$
β_R	reverse grounded emitter current gain $(=I_C/I_B)$
β_R	ratio of NMOS β to PMOS β
γ	injection efficiency
γ_n	electron injection efficiency, NPN $\gamma_n = I_n/(I_n + I_p)$
γ_p	hole injection efficiency, PNP $\gamma_p = I_p/(I_n + I_p)$
δ	thickness of interface barrier gap (cm)

δ_L	length of pinch-off region (cm)
Δ	width of gap layer (cm)
Δ	voltage drop across interface oxide (V)
ΔE_C	change (offset) in minimum conduction band edge (eV)
ΔE_g	change in energy bandgap (eV)
ΔE_V	change (offset) in maximum valence band edge (eV)
ΔI_B	increment in base current (A)
ΔI_C	increment in collector current (A)
ΔI_E	increment in emitter current (A)
ΔL	change in channel length (cm)
Δn	change in electron concentration (cm^{-3})
Δn_{p0}	change in minority-electron concentration at $x = 0$ (cm^{-3})
Δp	change in hole concentration (cm^{-3})
Δp_{n0}	change in minority-hole concentration at $x = 0$ (cm^{-3})
ΔV_{BE}	increment in emitter-base forward voltage (V)
ΔV_G	increment in gate voltage (V)
ΔV_R	increment in reverse voltage (V)
ΔV_T	change in threshold voltage (V)
ΔW	change in channel width (cm)
$\Delta\phi$	barrier lowering (eV)
ε_0	permittivity of free space ($\approx 8.86 \times 10^{-14}$ F/cm)
ε_{gap}	dielectric constant of interface gap
ε_{ox}	oxide dielectric constant (≈ 3.9)
ε_n	nitride dielectric constant (≈ 7.0)
ε_{Si}	silicon dielectric constant (≈ 11.7)
η	multiplier of inversion-layer concentration to calculate field
θ	aperture
θ	mobility degradation factor (V^{-1})
κ	same as $K = \varepsilon/\varepsilon_0$
λ	wavelength (cm)
λ	channel length modulation factor ($= -1/V_A$, V^{-1})
λ	mean-free path (cm)
λ	De Broglie wavelength (cm)
λ_i	average interstitial point-defect diffusion length (cm)
λ_0	wavelength in vacuum (cm)
μ	mobility ($cm^2/V \cdot s$)
μ_{eff}	effective mobility ($cm^2/V \cdot s$)
μ_h	high mobility, normal to crystallographic axis ($cm^2/V \cdot s$)
μ_I	ionized-impurity scattering limited mobility ($cm^2/V \cdot s$)
μ_l	lattice-scattering limited mobility ($cm^2/V \cdot s$)
μ_l	low mobility, along crystallographic axis ($cm^2/V \cdot s$)
μ_{ln}	electron lattice mobility ($cm^2/V \cdot s$)
μ_{lp}	hole lattice mobility ($cm^2/V \cdot s$)
μ_n	electron mobility ($cm^2/V \cdot s$)
$\tilde{\mu}_n$	effective electron mobility ($cm^2/V \cdot s$)

μ_p hole mobility ($cm^2/V \cdot s$)
$\tilde{\mu}_p$ effective hole mobility ($cm^2/V \cdot s$)
μ_0 surface mobility at $V_G = V_T$ ($cm^2/V \cdot s$)
ρ resistivity (Ω-cm)
ρ volume charge concentration (C/cm^3)
ρ_C specific contact resistance (Ω-cm^2)
ρ_0 resistivity at $T = T_0$ (Ω-cm)
σ conductivity ($\Omega^{-1} cm^{-1}$ or S/cm)
σ charge sheet (C/cm^2)
σ surface recombination velocity in buried plane (cm/s)
σ capture cross-section (cm^2)
σ_{VT} variance in threshold voltage (V)
$\sigma_{\Delta P}$ variance in parameter P
τ lifetime (s)
τ mean-time between collisions (s)
τ time constant (s)
τ_B base transit time (s)
τ_C collector transit time (s)
τ_E emitter transit time (s)
τ_n electron transit time, lifetime (s)
τ_{nB} electron transit time, lifetime in base (s)
τ_p hole transit time, lifetime (s)
τ_{pE} hole transit time, lifetime in emitter (s)
τ_{SRH} Shockley-Read-Hall recombination lifetime (s)
τ_0 assumed same lifetime for electrons and holes (s)
ϕ potential (V)
ϕ dose (cm^{-2})
ϕ_I pulsed-shaped implant dose (cm^{-2})
ϕ_B barrier height (V)
ϕ_b bulk Fermi potential (V)
ϕ_{bn} bulk electron Fermi potential (V)
ϕ_{bp} bulk hole Fermi potential (V)
ϕ_{Fn} electron quasi Fermi potential (V)
ϕ_{Fp} hole quasi Fermi potential (V)
ϕ_m metal (gate) workfunction (V)
ϕ_{ms} workfunction difference between metal (gate) and Si (V)
$\phi_{m\text{-}app}$ apparent metal (gate) workfunction (V)
ϕ_n electron Fermi potential (V)
ϕ_p hole Fermi potential (V)
ϕ_s surface Fermi potential (V)
ϕ_{Si} silicon workfunction (V)
ϕ_0 surface neutrality level (V)
χ electron affinity (V)
χ stress (Pa)
χ_{Si} silicon electron affinity (V)

χ_{ox}	silicon-dioxide electron affinity (V)
ψ	band-bending, potential (V)
ψ_s	surface potential (V)
$\psi_{s\text{-field}}$	surface potential under field oxide (V)
ω	angular frequency (s^{-1})

Chapter 1
Silicon Properties

1.1 Introduction

A review of silicon properties is important to understanding silicon components, in particular modern components such as strained-silicon *MOSFETs* and hetero-junction bipolar transistors. Several books cover this subject in detail. The objective of this chapter is to highlight those features that are most important to silicon device operation and characteristics.

1.2 Valence-Bond and Two-Carrier Concept

The valence-bond model is frequently used to qualitatively describe the properties of semiconductors [1]. In this model, the covalent bond between two adjacent atoms formed by two valence electrons, one from each atom contributing to the bond, is visualized as localized bars along which electrons shuttle back and forth with opposite spins (Fig. 1.1).

When a pure silicon crystal is near 0 K, all valence electrons remain locally bound to their covalent bonds since they do not have sufficient energy to break loose. In this case, no quasi-free electrons are generated and the crystal behaves like a perfect insulator. As the temperature is increased, the amplitude of vibration of lattice atoms increases around their equilibrium positions. A fraction of the vibrational energy is transferred to valence electrons. Some electrons can acquire sufficient energy to break loose from their bonds and move quasi freely in the crystal. Hence, the number of quasi free electrons and holes (missing bond electrons) in the crystal increases as the temperature is increased.

The energy required to break a silicon bond is an ionization energy ($\sim$1.1 eV) which differs from the ionization energy of an isolated silicon atom ($\sim$8 eV) because

B. El-Kareh, *Silicon Devices and Process Integration: Deep Submicron and Nano-Scale Technologies*,
© Springer Science+Business Media, LLC 2009

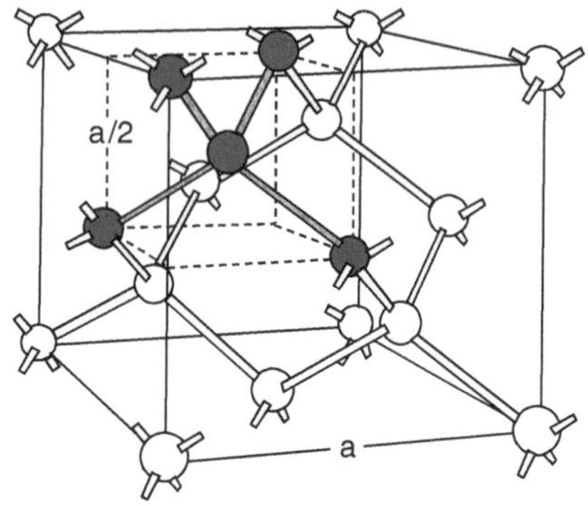

Fig. 1.1 Three dimensional representation of the silicon crystal. Dark atoms define the unit cell. Lattice constant $a = 0.54307\,\text{nm}$

it is influenced by other forces in the crystal.[1] When ionization occurs, the crystal as a whole remains neutral, although locally the ion becomes positively charged. A vacancy is left where an electron breaks loose from a bond. It behaves as a positive free carrier that is referred to as a hole (or "defect electron") with a mass comparable to that of the electron. The hole can move while, under normal operating conditions, the positive ion remains fixed.

To visualize the motion of holes, imagine the lattice sites to be occupied by only bound electrons and disregard the ions [2]. Suppose that one lattice point is void of an electron, that is, a hole is created. If now a field is applied to such an "electron crystal," the electron that is on the negative side of the hole will move into the hole, thus creating another hole. The hole moves as if it were a positively charged particle, although it is the bound electron that has moved in the opposite direction. Imagine now that in this "thought electron crystal" an electron is set free from its lattice site by some external force, for example, thermal agitation, and wanders around independently along interstitial sites. This quasi-free electron will be repulsed by all other electrons but not by the hole. The hole will appear to the interstitial electron as a positive charge. During its random motion, the interstitial electron may fill the hole, thus annihilating a positive and a negative charge simultaneously. In pure silicon, the concentration of electrons and holes are equal since they are generated and annihilated in pairs. In this case, silicon is said to be intrinsic and

$$n = p = n_i \quad \text{cm}^{-3}. \tag{1.1}$$

n and p are the electron and hole concentrations, respectively, and n_i the intrinsic carrier concentration ($\sim 1.4 \times 10^{10}\,\text{cm}^{-3}$ at 300 K).

[1] One electron-Volt (eV) is the energy dissipated or acquired by one electron that goes through a potential difference of one Volt. Since the charge of one electron is 1.6×10^{-19} Coulomb, $1\,\text{eV} = 1.6 \times 10^{-19}$ Joule. In this book, eV and cm are frequently used in place of J and m, as a convenient departure from SI units.

1.2.1 Doping

Intrinsic silicon has very limited use in device applications since the conductivity is very low and conduction of electrons and holes essentially occurs in pairs. One can, however, modify the type and magnitude of conductivity by adding small and controlled amounts of certain elements to the otherwise pure silicon. The crystal can be doped to have more conduction electrons than holes and vice versa. To be active, the dopants must occupy substitutional sites, that is, occupy a lattice site normally occupied by silicon. The doping process is described in more detail in [3]. Of particular importance to silicon devices is doping with elements from the third and fifth columns of the periodic table (Fig. 1.2).

1.2.1.1 Dopants from the Fifth Column: Donors

Phosphorus, arsenic, antimony and bismuth are elements of the fifth column. They differ mainly in their diffusivity and solid solubility in silicon [3]. These elements have five electrons in their outer shell, that is, five valence electrons. When they occupy a substitutional site in silicon, only four of the valence electrons are needed to complete the covalent bonding in the crystal. The fifth electron does not contribute

III	IV	V
5 10.811 B Boron	6 12.011 C Carbon	7 14.007 N Nitrogen
13 26.982 Al Aluminum	14 28.086 Si **Silicon**	15 30.974 P Phosphorus
31 69.723 Ga Gallium	32 72.610 Ge **Germanium**	33 74.922 As Arsenic
49 114.820 In Indium	50 118.710 Sn Tin	51 121.757 Sb Antimony
81 204.383 Ti Titanium	82 207.200 Pb Lead	83 208.980 Bi Bismuth

Typical acceptors (B, Al, Ga, In)

Typical donors (P, As, Sb, Bi)

Fig. 1.2 Important semiconductor elements

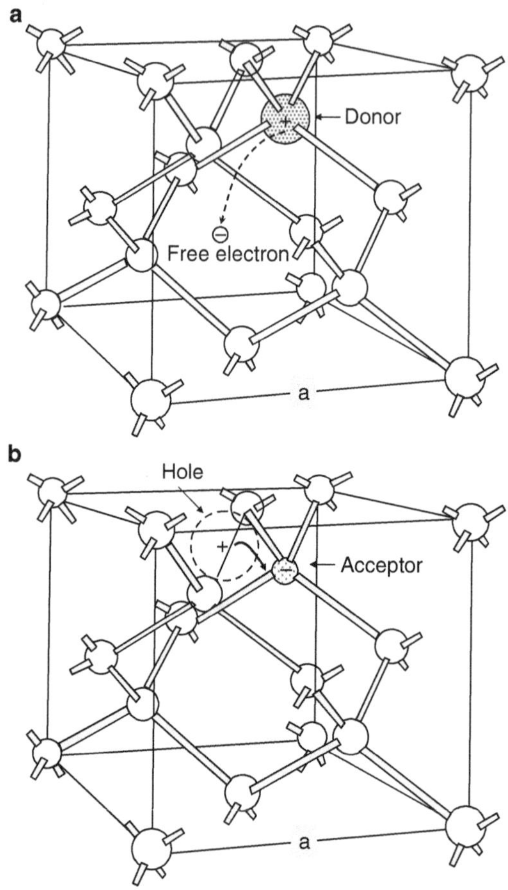

Fig. 1.3 Simplified model for dopants in silicon. **a** Quasi-free electron liberated from donor. **b** Nearby bound electron transferred to acceptor to complete bond, hole created

to the bonding and is set free, or "donated" (Fig. 1.3). The ionization energy required to set the fifth electron free is considerably smaller than the energy required to break a silicon bond.

Most of donor atoms remain ionized even at temperatures as low as 100 K. Each positive ion left behind has four bound valence electrons, the same as the original silicon atom. Note that by ionizing the donor, a free electron is generated without creating a hole. Thus, for a donor concentration N_D, the free electron concentration is $n \cong N_D^+$, where $N_D^+ \cong N_D$ is the ionized donor concentration. Ionized donors are fixed positive charges that do not contribute to conduction. The crystal becomes rich in electrons, n-type, but as a whole remains neutral. At temperatures below ~100 K, the probability for electrons to break loose from donors begins to decrease and an increasing fraction of electrons remains "frozen" to donors.

One can estimate the energy required to ionize a donor by treating the fifth electron as the electron in a hydrogen atom [4]. The electron moves around the central field of the donor core with a net charge of q. The central force is

$$\boldsymbol{F} = -\frac{\boldsymbol{q}^2}{4\pi\varepsilon_0\varepsilon_{Si}\boldsymbol{r}^2},$$

and the ionization energy of hydrogen is

$$\boldsymbol{E}_{i(\mathrm{H})} = -\frac{\boldsymbol{m}_0\boldsymbol{q}^4}{32\pi^2\varepsilon_0^2\hbar^2} = -13.6 \quad \mathrm{eV}, \tag{1.2}$$

where q is the electronic charge $(1.60218 \times 10^{-19}\,\mathrm{C})$, r the radius of the impurity atom, ε_0 the permittivity of free space $(8.85418 \times 10^{-14}\,\mathrm{F/cm})$, $\varepsilon_{\mathrm{Si}}$ the relative dielectric constant of silicon (11.7), and m_0 the free electron mass $(9.1095 \times 10^{-31}\,\mathrm{Kg})$.

The ionization energy of the donor can be estimated from the ratio

$$\frac{E_{i(\mathrm{D})}}{E_{i(\mathrm{H})}} = \frac{\varepsilon_0^2}{\varepsilon_{Si}^2}\frac{m^*}{m_0},$$

where $E_{i(\mathrm{D})}$ is the donor ionization energy and m^* the effective mass of electron in silicon. The effective mass is a quantum-mechanical value that takes into account internal forces exerted on the electron by the various atomic cores and other carriers in the crystal. It is a measure of the ease with which an external field can accelerate electrons and holes along an axis in the crystal and allows the use of a relation between force and momentum that is similar to the classical Newton's law. The effective mass is further discussed in Sect. 1.3.5. Assuming an electron effective mass of $0.26m_0$ and substituting the values for $E_{i(\mathrm{H})}$, ε_0, and $\varepsilon_{\mathrm{Si}}$ into (1.2) gives $E_i(\mathrm{D}) = 0.025\,\mathrm{eV}$ [5]. The actual measured donor ionization energy ranges from 0.044 to 0.067 eV for different group V elements in silicon [1].

1.2.1.2 Dopants from the Third Column: Acceptors

Boron and indium are acceptor elements from group III in the periodic table. When substituted for silicon, the three available valence electrons in the outer shell take part in the bond structure, leaving one vacancy since the fourth bond is not filled. The vacancy is "attractive" to an adjacent bound electron that easily moves to fill it. Boron or indium "accepts" the fourth electron, creating a hole where the filling electron came from without producing a free electron (Fig. 1.3b). When silicon is doped with boron, holes become the majority carriers and silicon is said to be p-type. For an acceptor concentration $\mathrm{N_A}$, the hole concentration at room temperature is $p \cong N_A^-$, where $N_A^- \cong N_A$ is the ionized acceptor concentration. Acceptors are fixed, negatively charged ions that do not contribute to conduction. As for donors, the ionization energy for acceptors can be estimated in terms of the ionization energy for hydrogen by using the conductivity effective mass of holes in silicon and assuming that the hole is "liberated" from the acceptor. The acceptor ionization energy is estimated as $\mathrm{Ei(A)} \approx 0.05\,\mathrm{eV}$, compared to the actual measured ionization energy of 0.045 eV for boron [4].

1.3 Energy Bands in Silicon

The properties of semiconductors are more accurately described with the quantum-mechanical energy-band model than with the over-localized valence-bond model discussed in the previous section. Energy bands in semiconductors are discussed extensively in reference books on solid-state physics [4–6]. The sole objective of this section is to highlight, in simple terms, the energy-band model concepts that are most pertinent to understanding the properties of silicon devices.

1.3.1 Energy Band Model

A free electron is allowed to occupy a continuum of energy levels, similar to the classical case of molecules in an atmospheric column that can occupy any energy level without restrictions. From quantum mechanics it is known that when an electron is bound to, for example, an isolated hydrogen atom, it is allowed only discrete energy levels separated by energy gaps. When two hydrogen atoms are far from each other, they behave as two isolated entities with independent, identical sets of discrete energy levels. As the atoms are brought close to each other, their wave functions begin to overlap so that the electrons of the two atoms begin to interact. Electrons in the first atom can also occupy energy levels in the second and vice versa. A study of energy levels shows that, in the limit, when a hydrogen molecule is formed, each energy level of the isolated hydrogen atom splits into a pair of levels when the atoms are bound to form the molecule. The total number of energy levels in the molecule is the same as in the system of two isolated atoms. A mechanical analogy may help illustrate this situation [2]. Consider, for example, the coupling of two identical pendulums that are connected by an elastic band and can oscillate with negligible friction in planes normal to the paper (Fig. 1.4a). When not coupled by the band, each pendulum can be treated as a harmonic oscillator of constant amplitude. Coupled with the band, when one pendulum is made to oscillate while the other starts at rest, the oscillation energy of the first pendulum is gradually transferred to the second pendulum. The second pendulum in turn gradually transfers the oscillation energy to the first, and so on. This results in an oscillation pattern similar to that of beats (Fig. 1.4b). When one pendulum loses all its energy to the other, it comes to rest while the other reaches its maximum amplitude. Beats are thus produced by the coupling of two oscillating systems that have exactly the same natural frequency when they are isolated from each other.

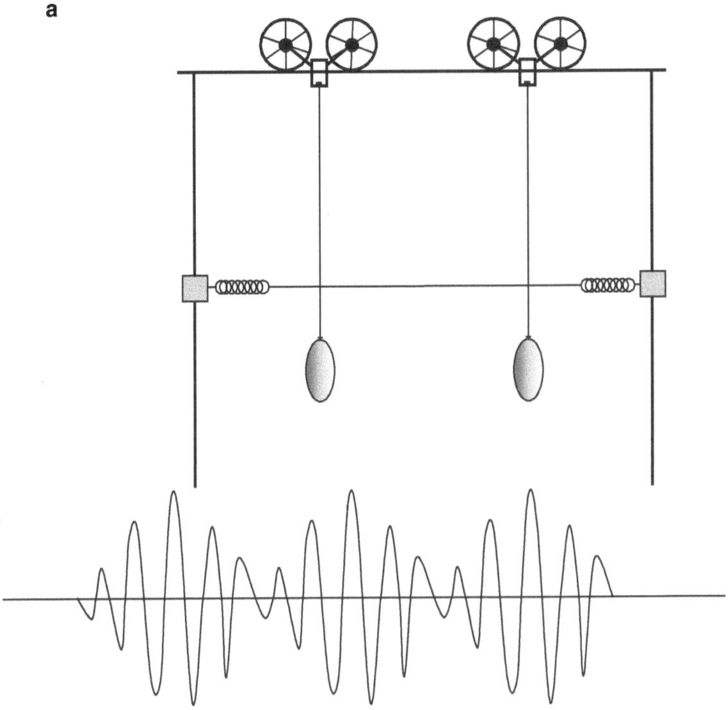

Fig. 1.4 Mechanical analogy of energy splitting. **a** Model of coupled pendulums to visualize the splitting of energy levels in coupled atomic systems. **b** Schematic representation of resulting beats [2]

The result is similar to beats caused by tuning two separate forks of equal amplitude but slightly different natural frequencies. Thus, when two oscillating systems of equal natural frequencies are brought to coupling, the result is similar to the resonance of two oscillating systems of equal amplitudes but frequencies that are slightly different [2]. This suggests that coupling two oscillating systems of the same natural frequency results in the splitting of the initially undisturbed frequency into two slightly different frequencies, one higher and one lower than the natural frequency. The difference between the two new frequencies increases as the coupling-strength increases, that is, as the frequency of amplitude exchange between the two pendulums increases. This is not surprising since the oscillation frequency of the coupled system is lower when the pendulums are in phase and higher when they are out of phase. The pendulums move toward each other when they oscillate in phase and away from each other when they are out of phase.

One can project the above observations to atomic systems. When two atomic systems are brought in proximity of each other, coupling of the Ψ-waves in Schrödinger's wave function results in an amplitude-exchange of Ψ-oscillations between two atomic systems, with an exchange frequency $\Delta\nu$. Since the square of Ψ represents the probability of finding an electron (electron density), $\Delta\nu$ represents the electron–exchange frequency between the two systems. This exchange results in a split of energy levels $\Delta E = h\Delta\nu$, where h is Planck's constant.

In a crystal of N identical lattice atoms, there exists the possibility of exchange of every valence electron with valence electrons of the remaining $N-1$ atoms. One expects then that each energy level of the isolated atom would split in the crystal into N energy levels that can be distinguished by a quantum number k, and each of which can be occupied by two electrons of opposite spins. Since N is a very large number ($\sim 10^{23}\,\mathrm{cm}^{-3}$), the levels are too close to each other to be distinguished. They are thus described by a band of energy levels, bounded by a maximum and minimum level. The difference between maximum and minimum energy levels (the width of the band) depends on the degree of coupling of wave-functions between atoms, that is, on the distance between atoms and probability of electron exchange which is a function of temperature, pressure and stress, and independent of N. One expects therefore that the energy levels of innermost electrons remain sharp because the probability for them to interact is very small, and that the band-width increases as the principal quantum number increases, as shown schematically in Fig. 1.5. The number of quantum states in the energy band is the same as the number of states from which the band was formed.

Of primary interest for conduction are the uppermost two bands, the conduction band of quasi-free electron energy levels, and the band just below the conduction band, referred to as the valence band of bound electrons. It will be shown that for silicon (and germanium) at crystal temperatures near absolute zero, the valence band

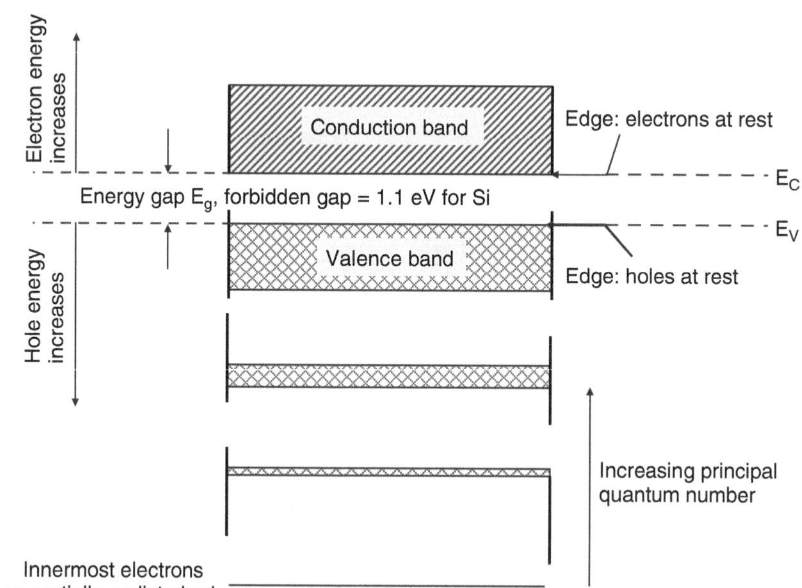

Fig. 1.5 Simplified energy-band diagram illustrating the increase in band-width with increasing principal quantum number

is full (all bound electrons in place) and the conduction band is empty; the crystal behaves like a perfect insulator. As the temperature increases, some electrons acquire sufficient energy from crystal vibrations to overcome the energy gap E_g and are elevated from the valence band to the conduction band where they are quasi-free to move. Holes are created in the valance band where electrons are missing.

A simplified one-dimensional representation of the energy bands relative to the periodic potential in the crystal is shown in Fig. 1.6. It will be shown later that most of the transitions occur between the upper edge of the valence band and the lower edge of the conduction band. In such situations only the edges of the conduction and valence bands are drawn, as indicated with dashed lines in Fig. 1.5.

A theoretical analysis of energy levels as a function of atomic space in the diamond structure, to which C, Si and Ge belong, was made by varying the atomic spacing, in a thought experiment, from infinity to below the actual spacing in the crystal, as illustrated for carbon in Fig. 1.7 [7,8]. At large spaces, the $s(l = 0)$ and

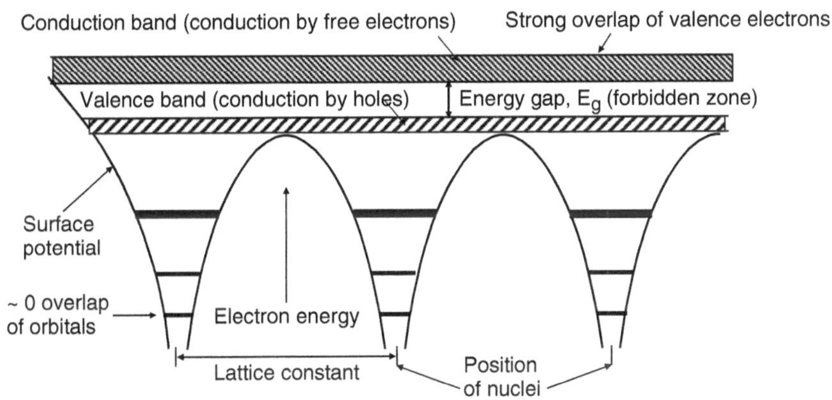

Fig. 1.6 Schematic one-dimensional representation of energy bands relative to the periodic potential in the crystal

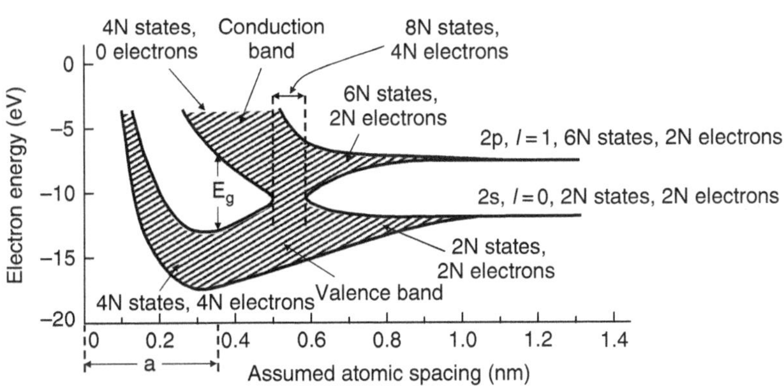

Fig. 1.7 Schematic representation of theoretical energy levels for carbon versus assumed interatomic spacing. a: actual spacing (Adapted from [1,6])

p($l = 1$) levels of the valence shell ($n = 2$) in carbon are sharp. For N atoms, the 2s level contains $2N$ allowed states and is fully occupied by $2N$ electrons while the 2p level has $6N$ states and is partially occupied by the remaining 2N valence electrons. As the atomic space is theoretically reduced below ~1.2 nm, the atoms begin to interact with each other and the energy levels split in bands that contain the same number of energy levels as for the isolated atoms.

As the space is further reduced, the bands merge. The energy gap disappears and there is no distinction between the two levels. This merger is not predicted from the discussion in the preceding section and cannot be explained in simple terms. The total number of available states remains as the sum of states in both levels ($8N$), and the total number of occupied states remains $4N$. At the actual space of about 0.37 nm, however, the bands split again, exhibiting an energy gap E_g and a repartitioning of energy levels into $4N$ in the lower band (the valence band) and $4N$ in the upper band (the conduction band). The lower band is now filled with the $4N$ electrons and the upper band completely empty.

1.3.2 Metals, Semiconductors and Insulators

The simplified energy-band model is now used to distinguish between metals, semiconductors and insulators. A solid conducts electricity only if carriers are free to move under the influence of an electric field, that is, if the carriers can acquire kinetic energy from the field and be accelerated in the solid. When an electric field is applied to the crystal, electrons can gain electric energy only if they can be placed at a higher energy level in the band. If the band is completely filled with electrons, the carriers cannot gain energy from the field since there is no "place" for higher-energy electrons to be placed. Therefore, if we assume that electrons do not get enough thermal or optical energy to make the transition from the completely filled band to a high-level empty or partially-filled band, the solid behaves like an insulator. A crystal can therefore conduct electricity only if its highest energy band is not completely filled.

It was shown in the preceding section that in a metal crystal of N atoms, the bands consist of N levels that can each be occupied by two electrons of opposite spins. The band, therefore, has $2N$ available states. For monovalent metals only half the band is filled since the metal can only provide one electron per atom (Fig. 1.8a). The metal, therefore, exhibits good conductivity. One would expect divalent metals to be insulators since the $2N$ available states in the upper band would be completely filled with $2N$ electrons. The strong coupling between valence electrons, however, results in an overlap of the upper bands, so that the net is a band that is not completely full and the metal behaves as a good conductor (Fig. 1.8b). A solid in which the upper band is not completely filled exhibits metallic character.

It can be shown that in silicon, germanium and carbon the valence band contains $4N$ states that are completely filled by $4N$ electrons when the temperature is near

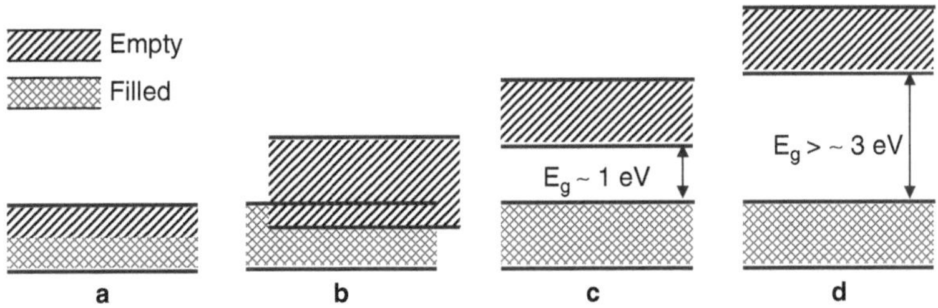

Fig. 1.8 Energy-band model for **a** A monovalent metal, for example, Na **b** Divalent metal, for example, Beryllium, **c** Semiconductor, for example, Si, Ge **d** Insulator, for example, Carbon

0 K, so the crystal behaves like an insulator. What distinguishes silicon and germanium from carbon is the magnitude of the energy gap. At near 0 K, the band-gap of germanium, silicon and carbon is, respectively, ~0.74, ~1.17 eV, and ~5.48 eV. When the bandgap is small, of the order of 1 eV, as for silicon and germanium, the crystal exhibits semiconductor properties (Fig. 1.8c). In this case, as the temperature is raised above 0 K, an increasing number of electrons can gain sufficient energy from crystal vibrations to be excited from the valence band into the conduction band, increasing the crystal conductivity. If, however, the energy gap is larger than ~3 eV, as for carbon, the probability of raising one electron from the valence band to the conduction band becomes negligible and the crystal behaves like an insulator at normal temperatures (Fig. 1.8d).

1.3.3 Band Model for Impurities in Silicon

When shallow donors such as arsenic (As), phosphorus (P), antimony (Sb) or bismuth (Bi) are incorporated into substitutional sites in the silicon crystal, their "fifth electron" is not bound sufficiently tight to be in the valence band. It is almost free to move in the conduction band. At low to moderated concentrations ($< \sim 10^{17}\,\mathrm{cm}^{-3}$), donor levels are represented by short bars, indicating localized energy states E_D that do not interfere with each other, just below the conduction band (Fig. 1.9). The numbers next to the bars are ionization energies measured from the conduction band edge. It will be shown by statistical analysis in Sect. 1.4.4 that for low to moderate concentrations at a temperature not too far below ~100 K, practically all donor atoms are positively ionized. Thus, the probability that their energy levels are not occupied by electrons is almost 100%.

Similarly, when an electron occupies an acceptor level to complete its bond structure, it is only a little less tightly held from being free than in a normal bond. Acceptor levels are thus represented by short bars, indicating discrete energy states E_A just above the valence band. The numbers below the bars are ionization energies measured from the valence band edge. Statistical analysis shows that for lightly doped silicon with shallow acceptors, such as boron and indium, the probability for their

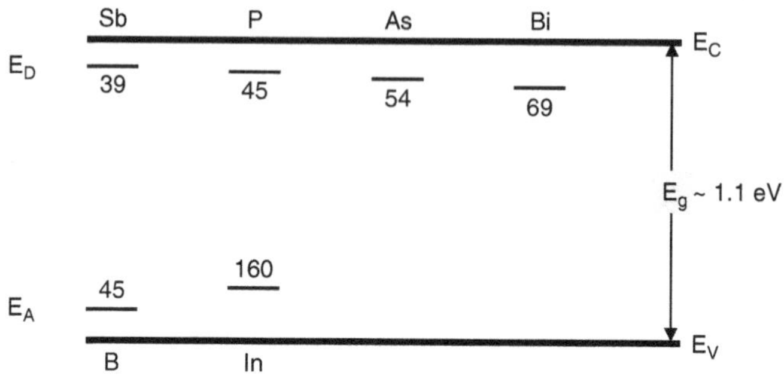

Fig. 1.9 Measured ionization energies (meV) of donors and acceptors in silicon (Adapted from [9]). Diagram is not to scale

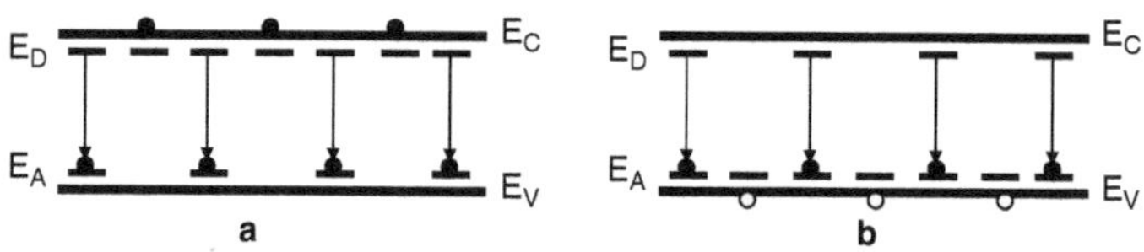

Fig. 1.10 Schematic representation of partially compensated impurities. **a** $N_D > N_A$, $n \approx N_D - N_A$; **b** $N_A > N_D$, $p = N_A - N_D$

energy levels to be occupied by electrons from the valence band is almost 100% at temperatures above ~100 K. Thus, practically all acceptors are negatively ionized, each creating a hole in the valence band.

In most cases, both donors and acceptors are present in the same region of the crystal. Since acceptor levels are below both the conduction band and the donor levels, any donor or conduction band electron will tend to fill an acceptor level. Both donors and acceptors will then be ionized, but only the difference between their concentrations will be available for conduction (Fig. 1.10). The crystal is said to be compensated. In case donors and acceptors are equal in concentration, all donors would be positively ionized without contributing free electrons, and all acceptors negatively ionized without contributing holes for conduction.

As the donor or acceptor concentration increases above $\sim 10^{18}\,\text{cm}^{-3}$, the impurity atoms come closer to each other, their wave functions begin to overlap and they begin to share each other's electrons. Because of this coupling, the levels begin to split and form bands. The donor ionization energy decreases because the field that binds the fifth electron to a particular donor atom is reduced by the charged donor neighbors [1]. Similarly, the acceptor ionization energy decreases at high concentration. Properties of heavily doped silicon are further discussed in Sect. 1.4.4.

1.3.4 Energy Band Theory

The characteristics of most silicon devices can be adequately described with the simplified energy band model presented in the preceding section. There are, however, several situations where a more in-depth discussion of the band theory would be beneficial. For example, the dependence of carrier mobility on crystallographic directions and the modulation of mobility by mechanical stress in silicon can be best understood with a more detailed energy band diagram than shown in Figs. 1.6–1.8.

The potential energy of a free electron is arbitrary within a constant which is set to zero for convenience. Thus, the time-independent one-dimensional Schrödinger wave equation simplifies to

$$\frac{d^2\psi}{dx^2} + \frac{8\pi^2 m}{h^2} E\psi = 0, \tag{1.3}$$

where E is the fixed total energy. Solutions to the above differential equation are periodic traveling plane waves of the form

$$\psi(x) = e^{ikx}. \tag{1.4}$$

Substituting for ψ gives

$$k = \frac{2\pi}{h}\sqrt{2m_0 E} = \frac{2\pi p}{h} = \frac{2\pi}{\lambda}, \tag{1.5}$$

where m_0 is the electron mass, p the electron momentum, h Planck's constant, λ the electron De Broglie wavelength, and $2\pi/\lambda$ the wave number. k is therefore proportional to the electron momentum and inversely proportional to the electron wavelength. The relation between E and k is then

$$E = \frac{\hbar^2}{2m}k^2. \tag{1.6}$$

The E–k diagram should therefore be a parabola, as shown in Fig. 1.11. The free electron can acquire a continuum of energy levels without restrictions.

In a crystal, electrons are not quite free but move in the periodic potential of lattice ions and the E–k plot of Fig. 1.11 does not quite apply. One can, however, make the approximation of a free electron in which the electron wave functions are

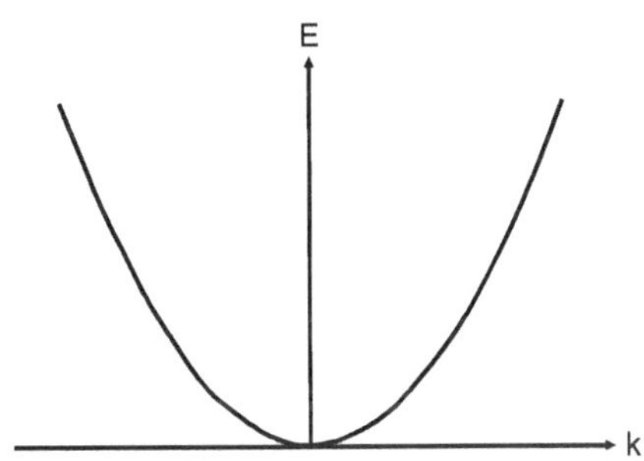

Fig. 1.11 E–k diagram for a free electron

traveling plane waves, and treat their interaction with the crystal as a diffraction of waves. The crystal behaves then as a three-dimensional diffraction grating. This approximation is only valid for metals. It can be seen, for example, that when the electron De Broglie wavelength λ approaches the critical value $2\boldsymbol{a}/\text{n}$ and $k = n\,\pi/\boldsymbol{a}$, where $\boldsymbol{a}$ is the lattice constant (Fig. 1.1) and $n = 1, 2, 3, \ldots$, the waves are reflected in phase by all lattice points, with 180° phase-shift relative to the incident wave. Because of this reflection, referred to as Bragg reflection, standing waves are produced for critical values of λ and hence k, where electrons will be Bragg-reflected and cannot propagate in the crystal [2, 4, 5, 7, 9]. The standing waves can be represented as a combination of two waves traveling in opposite directions, in one dimension as

$$e^{ikx} = \cos(kx) + i\,\sin(kx) \quad \text{and} \quad e^{-ikx} = \cos(kx) - i\,\sin(kx) \tag{1.7}$$

The sum of the above equations is $2\cos(kx)$ and the difference $2i\,\sin(kx)$, both representing standing waves. Either $\cos(kx)$ or $\sin(kx)$ can represent the electron at the critical De Broglie wavelength of $2a/n$. The probability of finding an electron is in one dimension

$$\psi_1^2 = \cos^2\frac{\pi\,x}{a} \quad \text{and} \quad \psi_2^2 = \sin^2\frac{\pi\,x}{a}. \tag{1.8}$$

The cos-function shows a maximum of electron density over the positive lattice ions while the sin-function shows the maximum mid-way between the ions. The energy has therefore a large negative value q^2/x in the first case, since x is small and a smaller value in the second case, that is, a higher potential energy, since x is larger. The difference between the two potential energies is the energy gap (Fig. 1.12).

The E–k diagram in Fig. 1.12 is plotted for the region between $k = -\pi/a$ and $+\pi/a$. The plots are shown flat at the critical values $k = \pi/a$ and $-\pi/a$ because the electron velocity, and hence the slope $\mathrm{d}E/\mathrm{d}k$, is zero at those points.

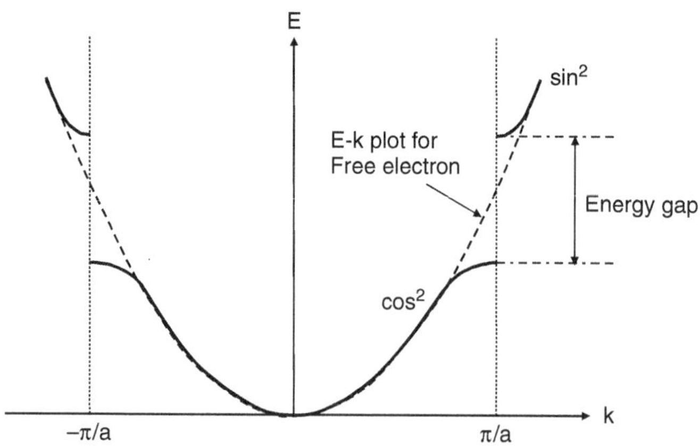

Fig. 1.12 E–k diagram for a free electron (dashed line) and an electron in the periodic crystal (solid line) showing the energy gap for critical values of k [10]

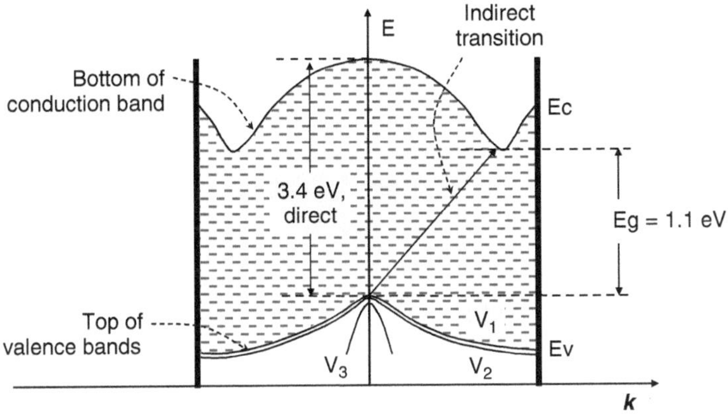

Fig. 1.13 E–k diagram for silicon along the $\langle 100 \rangle$ direction [6]

The actual E–k diagram for a silicon crystal is, however, more complicated than in Fig. 1.12. For example, the E–k plot obtained for silicon along the $\langle 100 \rangle$ axis exhibits a multi-valley band diagram as shown in Fig. 1.13. From this figure, several important observations can be made:

(a) There exist three valence bands, V_1, V_2, and V_3 whose band maxima occur at $k = 0$. V_1 and V_2 meet at $k = 0$ while the maximum of V_3 is about 0.04 eV below that of the other two bands. Their relative population depends on temperature.
(b) There are two equivalent minima along the k_x, equally spaced on opposite sides of $k = 0$. Similarly, there are equivalent minima along the k_y and k_z axes.
(c) The actual energy gap E_g is measured from the maximum of the valence band to the minimum of the conduction band. E_g decreases with increasing temperature because the crystal expands and the average distance between atoms increases.

The temperature dependence of E_g is approximated by [11]

$$E_g = E_g(0) - \frac{\alpha T^2}{(T+\beta)} \approx 1.17 - \frac{4.73 \times 10^{-4} T^2}{T + 636} \quad \text{eV}, \tag{1.9}$$

where $E_g(0)$ is the energy gap at near 0 K and T the absolute temperature. Another useful approximation is [12]

$$E_g = 1.187 - 3.6 \times 10^{-4} T \quad \text{eV}. \tag{1.10}$$

Electron transitions between bands are governed by the laws of conservation of energy and momentum. Since photons have practically zero momentum, a direct, vertical transition from the valence band to the conduction band occurs without a change in momentum when an electron absorbs a photon larger than 3.4 eV. Since the minimum of the conduction band does not occur at $k = 0$, but at a distance Δk from $k = 0$, the generation of an electron–hole pair by an indirect transition from the valence-band maximum to the conduction-band minimum must be accompanied by

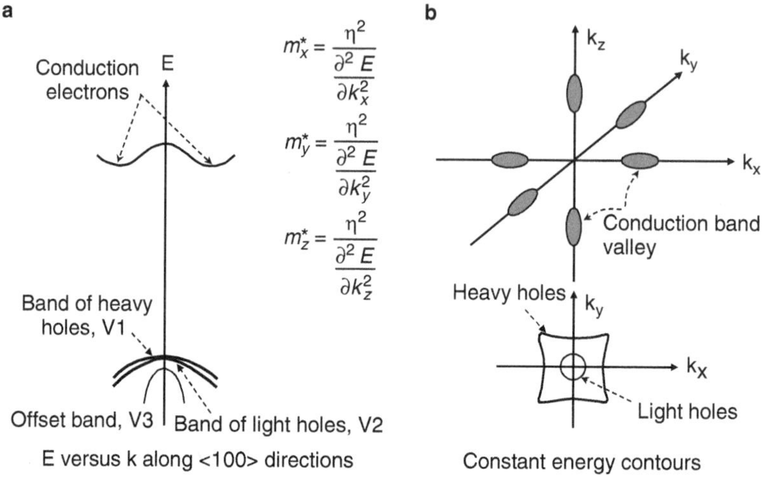

Fig. 1.14 Schematic representation of the conduction band and valence band structure in silicon [13]

the gain of momentum Δk. Similarly, the electron–hole recombination process must involve the loss of momentum Δk. Since a phonon (crystal vibration modes) has a large momentum, the transitions can occur by absorption or emission of phonons. At very low temperature only few phonons are present and indirect transitions are improbable. As the temperature increases, transitions become possible through an increase in the number of phonons. The constant energy surfaces near the band minima and the valence band maxima are shown in Fig. 1.14.

In Si there are six conduction band valleys, two along each of the $\langle 100 \rangle$ crystallographic directions. In the vicinity of the minima, the surfaces of the six valleys can be approximated by ellipsoids of revolution. The constant energy surfaces of valence bands are more complicated because of the existence of multiple bands. They can be approximated as spheres although they appear somehow "warped" [13].

1.3.5 Effective Mass

When an electric field is applied to a crystal, carriers do not only feel the external electric force but also forces that originate at the atomic cores and other carriers in the periodic crystal array. The effect of all these internal forces can be represented by a quantum-mechanical parameter that has the unit mass and is known as the effective mass, m^*. The effective mass is inversely proportional to the curvature of the E–k curve and defined as [6]:

$$m^* = \frac{\hbar^2}{\partial^2 E/\partial k^2}. \tag{1.11}$$

The motion of electrons in a perfect crystal can then be treated in a manner similar to the free electrons provided m^* is used rather than the actual free-electron mass, m_o. The acceleration a of a free electron by the external electric field is expressed by Newton's law as

$$a = \frac{F}{m} = \frac{qE}{m}. \tag{1.12}$$

In a perfect crystal, the acceleration of the electron can now be expressed by a similar relation

$$a = \frac{qE}{m^*}. \tag{1.13}$$

Since the curvature of the E–k plot depends on the crystallographic orientation, the effective mass should be treated as a tensor. It can be seen from Figs. 1.11 and 1.12 that the curvature of the first valence band near $k = 0$ is smaller than that of the other two bands. Holes near that point have therefore a larger effective mass than holes created in V_2 and V_3. Holes in V_1 are referred to as "heavy holes," and holes in V_2 as "light holes." The density of holes in V_3 is negligible because of the large energy offset.

Choosing one of the conduction minima in Fig. 1.11 as the origin, the surface of constant energy can be represented by an expression of the form [6]:

$$E(k) = \frac{\hbar^2}{2}\left(\frac{k_l^2}{m_l^*} + \frac{k_t^2}{m_t^*}\right) \tag{1.14}$$

where the subscripts l and t represent the longitudinal and transverse components of the wave vector k and effective mass m^*, that is, the components parallel and perpendicular to the major axis of the constant-energy ellipsoid. For silicon at 4.2 K, the longitudinal and transverse effective masses for electrons near the bottom conduction band edge are found in term of the electron rest mass m_o as [14]:

$$m_{nl}^* = 0.9163 m_o; \quad m_{nt}^* = 0.1905 m_o \tag{1.15}$$

The effective mass of heavy and light holes near the top of the valence bands is reported for $T = 4.2$ K as [15]:

$$m_{p(V_1,V_2)}^* = m_{\text{heavy}}^* = 0.537 m_o; \quad m_{p(V_3)}^* = m_{\text{light}}^* = 0.153 m_o \tag{1.16}$$

1.4 Thermal Equilibrium Statistics

Thermal equilibrium is a dynamic state whereby every detail of one process is balanced by its own inverse, on the average at exactly the same rate. This is known as the principle of detailed balance. This section discusses the thermal equilibrium distribution of free electrons and holes in silicon.

1.4.1 The Boltzmann Distribution Function

From the kinetic theory of gases, it is known that the mean kinetic energy associated with the motion of one particle is proportional to temperature. At a given temperature, the average energy of each particle is $3kT/2$, where T is the absolute temperature and k the Boltzmann constant $= 8.625 \times 10^{-5}\,\text{eV/K}$. From statistical mechanics, the probability P_i of finding a particle in an energy state E_i relative to the probability P_0 being in an energy state E_0 is found for a given uniform temperature T as

$$\frac{P_i}{P_0} = \frac{e^{-E_i/kT}}{e^{-E_0/kT}}, \tag{1.17}$$

which for a large number of particles is the same as

$$N_i = N_0 e^{-(E_i - E_0)/kT}, \tag{1.18}$$

where N_i is the number of particles at energy E_i, N_0 is the number of particles at energy E_0, E_0 is the energy level chosen as reference, and E_i the energy at a level i above or below the reference energy.

The inverse exponential dependence indicates that at a given temperature the number of particles at energy E_i decreases as E_i increases above E_0. This means that it is more probable to find particles at low energies than at high energies. The relation also indicates that the probability that a particle "occupies" an energy level increases as the temperature is increased. One classical example is the distribution of molecules in a column of an idealized atmosphere at a uniform temperature. The number of molecules at a height Δh above a reference height h is

$$N_{h+\Delta h} = N_h e^{-mg\Delta h/kT} = N_h e^{-\Delta E/kT} \tag{1.19}$$

where N_h is the number of molecules at reference height h, m is the molecule mass, and g is the acceleration of the earth gravity, assumed constant. A relation of the form of (1.19) is referred to as the Boltzmann distribution function.

1.4.2 Fermi-Dirac Distribution and Density of States

The Boltzmann distribution function applies only to the classical case of noninteracting, indistinguishable particles. It does not apply to particles that obey Pauli's exclusion principle, such as electrons and holes. If, for example, we apply Boltzmann's distribution law to quasi-free electrons in a metal, we would find that at near 0 K all electrons would occupy the single lowest energy state. This, however, violates the exclusion principle. As the result of Pauli's principle, each of the *4N* states in the valence band or conduction band in Fig. 1.6 can be occupied by only one electron. At a given temperature, the number of electrons in a small energy interval of a band depends on two factors: the probability that an electron will have this energy, and the number of available energy states in that interval. At thermal equilibrium, the probability that an energy E be occupied by an electron is given by the Fermi-Dirac distribution function

$$f(E) = \frac{1}{1+e^{(E-E_F)kT}}, \tag{1.20}$$

where E_F, referred to as the Fermi level, is a normalizing parameter that is determined by the requirement that the total expectation number of electrons is equal to the actual number of electrons involved [6]:

$$\int_0^\infty n(E)\mathrm{d}E = \int_0^\infty f(E)N(E)\mathrm{d}E = n, \tag{1.21}$$

where $N(E)\mathrm{d}E$ is the density of allowed states in the energy interval $\mathrm{d}E$. The probability that a level E is not occupied by an electron is then given by

$$1-f(E) = \frac{1}{1+e^{(E_F-E)kT}}. \tag{1.22}$$

At $E = E_F$, $f(E) = 0.5$ and there is an equal probability that the level is occupied or vacant. This is sometimes used to define the Fermi level in a system. At thermal equilibrium, the Fermi level must be constant throughout a system even if the system consists of different materials in contact with each other. In a metal, E_F has a straightforward physical meaning: it is the energy of the highest occupied state near absolute zero (Fig. 1.15).

Near absolute zero, all levels below E_F are filled since the probability f(E) of finding an electron below E_F is 100% (Fig. 1.16). At energies above E_F, f(E) is ~0%. $q\phi_m$ is the minimum energy required to free an electron at the Fermi level from the metal.[2] It is referred to as the work function of the metal. Near 0K, the probability function shows an abrupt discontinuity between the values *0* and *1*. It becomes more and more gradual as the temperature increases. For energies above or

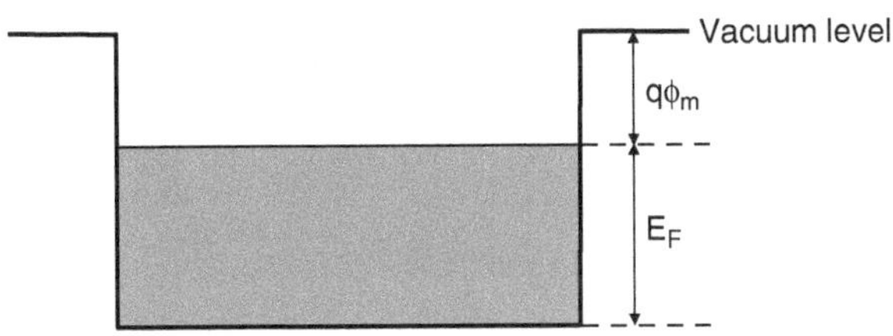

Fig. 1.15 Fermi level in a metal at absolute zero

[2] Unless otherwise stated, E, E_F, E_i, E_C, E_V, and Eg, are energies expressed in eV; ϕ, ψ, χ, are potentials expressed in V. Thus, energies and potentials have the same numerical value but different units.

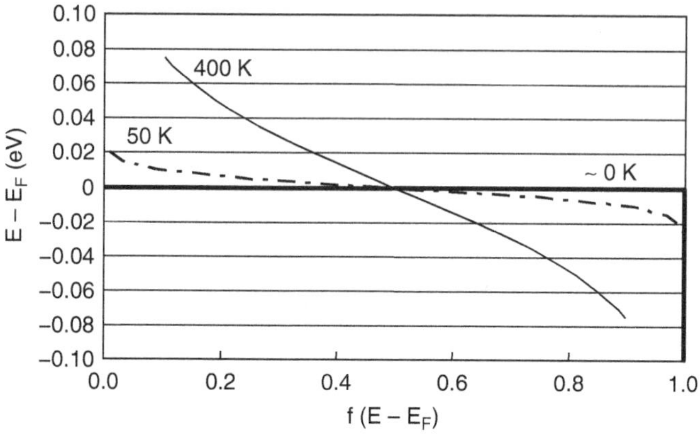

Fig. 1.16 Probability function $f(\mathrm{E}-\mathrm{E_F})$ at two temperatures. As the temperature approaches 0 K, the function becomes a discontinuous step

below the Fermi level $\mathrm{E_F}$ such that $|\mathrm{E}-\mathrm{E_F}| > 3kT$, the Boltzmann approximation can be made by neglecting the "1" in the denominator of (1.22) and the relation becomes similar to (1.19).

1.4.3 Density of States and Carrier Distribution in Silicon

In a semiconductor, the interpretation of the Fermi level is not as simple as for a metal. In most cases, $\mathrm{E_F}$ will be located within the forbidden gap. In this section, several simplifying assumptions are made to determine the position of $\mathrm{E_F}$ and calculate the concentration of electrons and holes in intrinsic silicon.

From spectral analysis, it is found that the distribution of energy states $N(E)$ within a band can be approximated by two half-parabolas peaking near the center of the band (Fig. 1.17) [2].

Of primary interest is the effective density of states (states per unit volume) near the band edges since most transitions occur between these two regions. Near the lower edge E_C of the conduction band, the density of states is approximated by

$$N_\mathrm{C}(E) \propto \sqrt{(E-E_\mathrm{C})} \quad \mathrm{cm}^{-3}, \tag{1.23}$$

and near the upper edge E_V of the valence band

$$N_\mathrm{V}(E) \propto \sqrt{(E_\mathrm{V}-E)} \quad \mathrm{cm}^{-3}. \tag{1.24}$$

The effective density of states is found as [4–7]

$$N_\mathrm{C} \quad \text{or} \quad N_\mathrm{V} = 2\left(\frac{2\pi m_o kT}{h^2}\right)^{3/2}\left(\frac{m_D}{m_o}\right)^{3/2} \quad \mathrm{cm}^{-3}, \tag{1.25a}$$

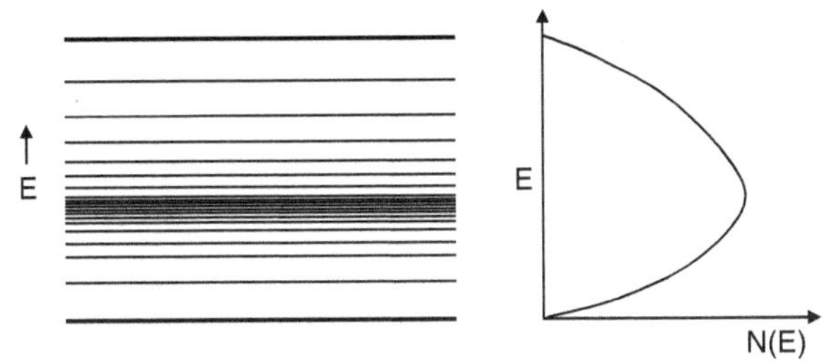

Fig. 1.17 Schematic representation of the energy-level distribution within a band and density of states N(E) [2]

or

$$N_{\mathrm{C}} \text{ or } N_{\mathrm{V}} = 4.82 \times 10^{15} T^{3/2} \left(\frac{m_{\mathrm{D}}}{m_o}\right)^{3/2} \quad \mathrm{cm}^{-3}, \tag{1.25b}$$

where m_o is the electron rest mass and m_D the density-of-state effective mass. For the silicon conduction band with six minima, m_D is given in terms of the longitudinal and transversal effective masses as

$$\frac{m_{De}}{m_o} = 6^{2/3}[m_l^*(m_t^*)^2]^{1/3} \cong 1.06 \text{ (at 4.2 K)}. \tag{1.26}$$

At 300 K, the ratio increases to ~1.18 [15]. For the valence band, m_D is given in terms of the heavy- and light-hole effective masses

$$\frac{m_{Dp}}{m_o} = \frac{[(m_{\text{light}}^*)^{3/2} + (m_{\text{heavy}}^*)^{3/2}]^{2/3}}{m_o} \cong 0.59 \text{ (at 4.2 K)}. \tag{1.27}$$

At 300 K, the ratio increases to ~0.80 [15]. In the temperature range 200–400 K, the effective density of states for electrons and holes can be approximated as

$$N_{\mathrm{C}} \cong 2.80 \times 10^{19} \left(\frac{T}{300}\right)^{1.5} \quad \mathrm{cm}^{-3}, \tag{1.28}$$

$$N_{\mathrm{V}} \cong 1.02 \times 10^{19} \left(\frac{T}{300}\right)^{1.5} \quad \mathrm{cm}^{-3}. \tag{1.29}$$

N_{C} and N_{V} in (1.28) and (1.29) are convenient values to estimate the density of conduction electrons and holes. It should be emphasized, however, that the values are only approximations.

The equilibrium number of conduction-band states that are occupied by electrons, that is, the number of conduction electrons $\bar{n}$ near E_C, is found by multiplying the probability function $f(E)$ by the number of available states N_{C}

$$\bar{n} = \frac{N_{\mathrm{C}}}{1 + e^{(E_{\mathrm{C}} - E_{\mathrm{F}})/kT}} \quad \mathrm{cm}^{-3}. \tag{1.30}$$

Similarly, the equilibrium number of unoccupied levels in the valence band, that is, the number of holes $\bar{p}$ near E_V, is

$$\bar{p} = \frac{N_V}{1 + e^{(E_F - E_V)/kT}} \quad \text{cm}^{-3}. \tag{1.31}$$

The bars over n and p identify them as equilibrium levels. For moderately to low dopant concentrations, the Fermi level E_F is several kT away from the band edges so that the Boltzmann approximation can be made for $f(E)$ and $1 - f(E)$. The expressions for carrier concentration can then be simplified to:

$$\bar{n} = N_C e^{-(E_C - E_F)/kT}, \tag{1.32}$$

$$\bar{p} = N_V e^{-(E_F - E_V)/kT}. \tag{1.33}$$

Employing the fact that in intrinsic silicon at thermal equilibrium, $\bar{n} = \bar{p} = n_i$, where n_i is the intrinsic carrier concentration, it follows that in intrinsic silicon

$$E_F = \frac{E_V + E_C}{2} - \frac{1}{2} kT \ln \frac{N_C}{N_V} \quad \text{eV}. \tag{1.34}$$

At near 0 K, the Fermi level is exactly half-way between E_C and E_V, that is, in the middle of the energy gap since the second term in (1.34) goes to zero. Also, the valence band is completely full and the conduction band completely empty ($n = p = n_i = 0$). As the temperature increases, E_F moves slightly toward the valence band because of the difference in the effective mass and hence effective density of states. At room temperature, for example, E_F is about 8 meV below mid-gap. For all practical purposes, however, E_F is assumed to be at mid-gap in intrinsic silicon. It is also referred to as the intrinsic energy level, E_i. The intrinsic carrier concentration is then

$$n_i = N_C e^{-(E_C - E_i)/kT} \quad \text{cm}^{-3}. \tag{1.35}$$

$$n_i = N_V e^{-(E_i - E_V)/kT} \quad \text{cm}^{-3}. \tag{1.36}$$

From (1.32)–(1.36) it is found that, at thermal equilibrium, the product of electron and hole concentrations depends only on the energy gap

$$\overline{pn} = n_i^2 = N_C N_V e^{-(E_C - E_V)/kT} = N_C N_V e^{-E_g/kT} \quad \text{cm}^{-6}. \tag{1.37}$$

At room temperature $n_i^2 \cong 2 \times 10^{20}\,\text{cm}^{-6}$. Equations (1.10), (1.28), and (1.29) can be combined to give a practical relation for n_i

$$n_i \cong 2.63 \times 10^{16} T^{1.5} e^{-6885/T}. \tag{1.38}$$

A plot of n_i as a function of temperature is shown in Fig. 1.18.

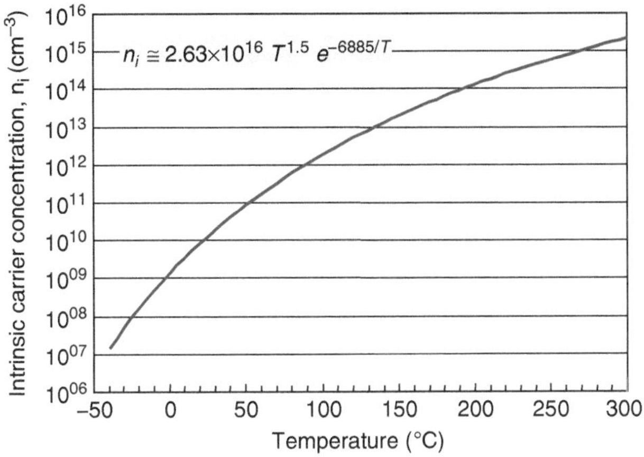

Fig. 1.18 Temperature dependence of intrinsic carrier concentration n_i

1.4.4 Doped Silicon

The thermal-equilibrium $\overline{pn}$ product in (1.37) is extremely useful in device analysis. While derived for intrinsic silicon, it also applies to doped silicon since it depends only on temperature and energy gap. The principle of mass action shows that when the concentration of one type of carriers increases, the concentration of other type must decrease so that the product remains constant.

As discussed in Sect. 1.2.1, the concentration of free electrons can be increased by doping silicon with donors, such as arsenic, phosphorus, antimony or bismuth. Donors become positively ionized when they create free electrons. Similarly, the concentration of holes can be increased by doping silicon with acceptors, such as boron or indium. Acceptors become negatively ionized when they create holes.

At low concentrations, practically all dopants are ionized at room temperature. Consider, for example, silicon doped with 10^{16} phosphorus atoms per cm^3. Near absolute zero, all donor levels would be filled with electrons, that is, all phosphorus atoms are *not* ionized. As the temperature increases slightly above 0 K, a fraction of donors becomes ionized as some "fifth" donor electrons gain sufficient energy to "make it" to the conduction band. The temperatures is, however, still too low for electrons to be excited from the valence band to the conduction band, so the presence of the valence band can be disregarded in this analysis. Assuming the Boltzmann approximation, the free electron concentration is estimated by (1.32) as

$$\bar{n} = N_C\, e^{-(E_C - E_F)/kT}.$$

Since all these electrons come from donors, the neutrality principle requires that their density be equal to the density of positively ionized donors, that is, to the density of donors not occupied by electrons defined by

$$N_D^+ = N_D\, e^{-(E_F - E_D)/kT}. \tag{1.39}$$

Combining (1.31) and (1.38) gives a relation for E_F that is similar to (1.34)

$$E_F = \frac{E_C + E_D}{2} - \frac{1}{2} kT \ln \frac{N_C}{N_D}. \tag{1.40}$$

Near 0 K, E_F lies exactly halfway between the donor level and the conduction band edge. As the temperature is increased, E_F moves downward toward mid-gap. At 300 K, $N_C \sim 2.8 \times 10^{19}\,\text{cm}^{-3}$ and E_F lies about 80 meV below the phosphorus level. Therefore, for $N_D = 10^{16}\,\text{cm}^{-3}$, the probability for phosphorus to be ionized is ~96%. One can then assume

$$\bar{n} = N_D^+ \cong N_D. \tag{1.41}$$

A similar analysis for lightly doped p-type silicon gives

$$\bar{p} = N_A^- \cong N_A. \tag{1.42}$$

When the Boltzmann approximation is valid, the Fermi level can also be determined from (1.32) and (1.35). Taking the ratio, one finds

$$E_F - E_i = kT \ln \frac{\bar{n}}{n_i} \quad \text{eV}. \tag{1.43}$$

Similarly, (1.33) and (1.36) give

$$E_i - E_F = kT \ln \frac{\bar{p}}{n_i} \quad \text{eV}. \tag{1.44}$$

Neglecting the contribution to carriers by thermal generation and utilizing the neutrality principle, (1.43) and (1.44) can be written as

$$\frac{E_F - E_i}{q} = \phi_b = \frac{kT}{q} \ln \frac{N_D^+}{n_i} \quad \text{V}, \tag{1.45}$$

$$\frac{E_i - E_F}{q} = -\phi_b = \frac{kT}{q} \ln \frac{N_A^-}{n_i} \quad \text{V}. \tag{1.46}$$

Noting that kT and kT/q are numerically equal, E_F and E_i in (1.45) and (1.46) are divided by q to convert the values from energy (eV) to Volt. ϕ_b is referred to as the Fermi potential. For n-type silicon, E_F lies above the intrinsic level E_i and ϕ_b is positive. For p-type silicon, E_F lies below E_i and ϕ_b is negative (Fig. 1.19).

For temperatures above ~100 K and low dopant concentrations (N_D or N_A less than $\sim 10^{17}\,\text{cm}^{-3}$), impurities are assumed to be fully ionized and (1.44) and (1.45) can be approximated as

$$\phi_b \cong \frac{kT}{q} \ln \frac{N_D}{n_i} \cong -\frac{kT}{q} \ln \frac{N_A}{n_i} \quad \text{V}. \tag{1.47}$$

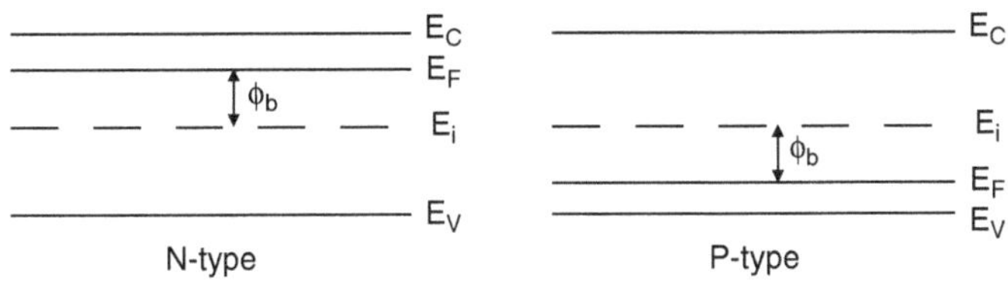

Fig. 1.19 Definition of Fermi potential

For concentrations or temperatures where the Boltzmann approximation is not valid, the Fermi-Dirac distribution function must be used to give a better approximation of carrier concentrations. At thermal equilibrium, the Fermi level will "settle down" to a value that satisfies two conditions, namely the neutrality of the system as a whole and the statistical distribution for all energy levels involved.

For the general case where the crystal is doped with both donors and acceptors, the neutrality principle requires that

$$\bar{p}+N_D^+ = \bar{n}+N_A^-. \tag{1.48}$$

The Fermi-Dirac distribution for all energy levels involved is

$$\bar{p} = \frac{N_V}{1+e^{(E_F-E_V)/kT}}, \tag{1.49}$$

$$N_D^+ = \frac{N_D}{1+2e^{(E_F-E_D)/kT}}, \tag{1.50}$$

where it is noted that the donor is ionized when it is not occupied by an electron and the factor of 2 in the denominator is due to the fact that when the donor is positively charged, an electron can be replaced in the level in two ways, spin up and spin down. After the electron occupies the donor level, it is not possible to place another electron there. For electron and negatively charged acceptors, the distributions are:

$$\bar{n} = \frac{N_C}{1+e^{(E_C-E_F)/kT}}, \tag{1.51}$$

$$N_A^- = \frac{N_A}{1+4e^{(E_A-E_F)/kT}}. \tag{1.52}$$

The 4 in the denominator comes from the fact that the valence band is two-degenerate and the impurity can accept one hole of either spin. The third valence band, V_3 in Fig. 1.14 is sparsely populated and neglected in this analysis.

The distribution functions in (1.49)–(1.52) are substituted in the neutrality expression 1.47. The Fermi level is then calculated by iteration or determined graphically. When both donors and acceptors are present, silicon is said to be "compensated" and the net concentration is the difference between N_D and N_A. An example of graphical determination of E_F is shown in Fig. 1.20. The concentrations of electrons, holes and ionized impurities are plotted as a function of E_F for

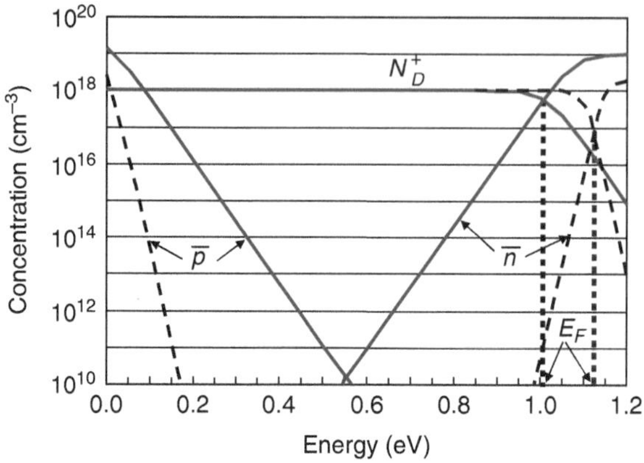

Fig. 1.20 Graphical determination of the Fermi level for $N_D = 10^{18}\,\text{cm}^{-3}$, $T = 300\,\text{K}$ (*solid line*), and $T = 100\,\text{K}$ (*dashed line*)

phosphorus (ionization energy 45 meV at low concentrations). The plot is made for a phosphorus concentration of $10^{18}\,\text{cm}^{-3}$ at 300 K (solid lines) and 100 K (dashed lines). The reference energy level is chosen as $E_V = 0$. Since it is assumed that only donors are present and the hole concentration is practically zero, E_F is defined at the point where $\bar{n} = N_D^+$.

1.4.4.1 Heavily Doped Silicon

As the dopant concentration increases above $\sim 10^{18}\,\text{cm}^{-3}$, several phenomena occur that require extensive modifications to the above relations. This section is not intended, however, to discuss high-doping effects in detail but to briefly discuss the mechanisms involved and summarize empirical relations for energy-gap lowering at high concentrations. Specifically:

1. As the Fermi level approaches the impurity levels and the probability for nonoccupancy of donor levels or occupancy of acceptor levels decreases, an increasing fraction of impurities become de-ionized.
2. At a concentration of $\sim 3 \times 10^{18}\,\text{cm}^{-3}$, the impurity sites can no longer be considered as discrete because their electron wave functions overlap. This increases the probability for electrons to be shared between dopant sites and their levels E_D or E_A split into a set of allowed energy levels [16].
3. As the dopant concentration is further increased, the energy levels broaden into bands, as schematically shown in Fig. 1.21 for compensated silicon (both N_A and N_D present) [17, 18]. As a consequence of band broadening of impurities, their ionization energy decreases until it vanishes completely. This occurs at a concentration of $\sim 3 \times 10^{19}\,\text{cm}^{-3}$ [16]. At this point, it is assumed that all impurities are ionized.

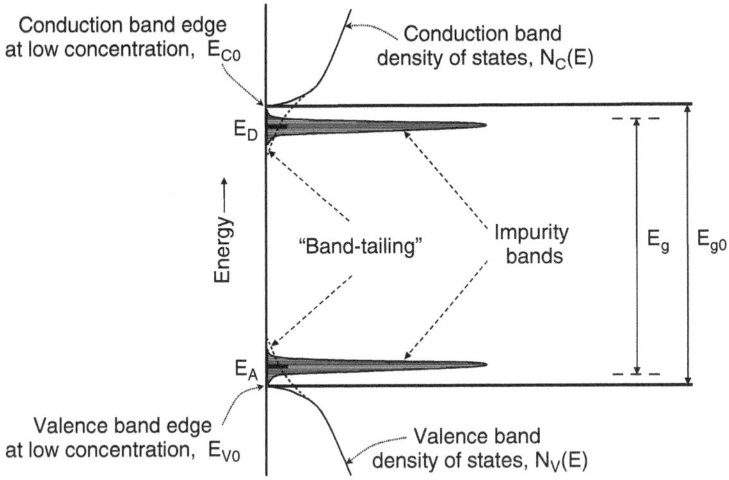

Fig. 1.21 Broadening of impurity levels and "band-tailing" shown for compensated silicon at high concentrations. E_{g0} is the low-concentration energy gap [17]

4. Interactions between majority carriers themselves, for example, electron–electron interactions in the conduction band and between majority carriers and impurity levels, can no longer be disregarded. These interactions distort the band structure and create "band-tails" extending into the band-gap, which reduce the energy gap, as shown schematically in Fig. 1.21 [17]. At still higher concentrations, the impurity bands merge with the conduction or valence band, forming single broad bands.

Bandgap Lowering, ΔEg

An increase in the intrinsic carrier concentration n_i above the value given in Fig. 1.18 is observed at high dopant concentrations. This is attributed to the lowering of the energy gap as impurity concentrations are increased (Fig. 1.21). Although extensive theoretical work has been done on heavily doped silicon, there is no agreement yet on a unified bandgap narrowing model [18–23]. There is, however, considerable data on ΔE_g reported in the literature that can be utilized in device analysis, in particular bipolar structures, where ΔE_g becomes important [24–31]. A typical method to extract bandgap narrowing is to derive the actual intrinsic-carrier concentration from minority-carrier injection parameters in the base and emitter of the bipolar structure (Chap. 3). The bandgap narrowing, ΔE_g, is then derived by assuming that the increase in n_i can be fully attributed to ΔE_g. Modifying (1.35) and (1.36) yields

$$n_i^2 = n_{io}^2 e^{\Delta E_g/kT} \quad \text{cm}^{-6}. \qquad (1.53)$$

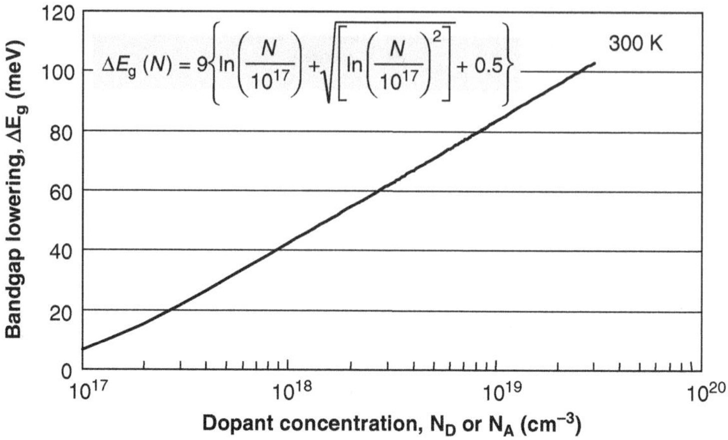

Fig. 1.22 Bandgap narrowing versus dopant concentration in the range 10^{17} to $3 \times 10^{19}\,\text{cm}^{-3}$ [24,25]

In (1.53) n_i is the actual intrinsic-carrier concentration and n_{io} is the value for lightly to moderately doped silicon. The most widely used expression for bandgap lowering was obtained by fitting the measured ΔE_g as a function of acceptor or donor concentration N into an empirical relation given by [24,25]

$$\Delta E_g(N) = 9\left\{\ln\left(\frac{N}{10^{17}}\right) + \sqrt{\left[\ln\left(\frac{N}{10^{17}}\right)^2\right] + 0.5}\right\} \quad \text{meV}. \tag{1.54}$$

The above relation is valid for dopant concentrations below $\sim 3 \times 10^{19}\,\text{cm}^{-3}$. A plot of ΔE_g versus dopant concentration is shown in Fig. 1.22.

Experimental values for ΔE_g obtained from other sources are reproduced in Fig. 1.23 [27]. The "apparent bandgap narrowing" in the figure is the value that would account for an increase in n_i if no other heavy doping effects occurred. A fit that describes the collection of points is given in [27] as

$$\Delta E_g^{\text{app}} = 18.7 \ln \frac{N_D}{7 \times 10^{17}} \quad \text{meV}. \tag{1.55}$$

Considerable scattering in the extracted values is, however, still present.

1.5 Carrier Transport

Electric current is defined as the number of charged carriers transported per unit time across a given surface in a direction normal to it. The transport of carriers can occur in two ways:

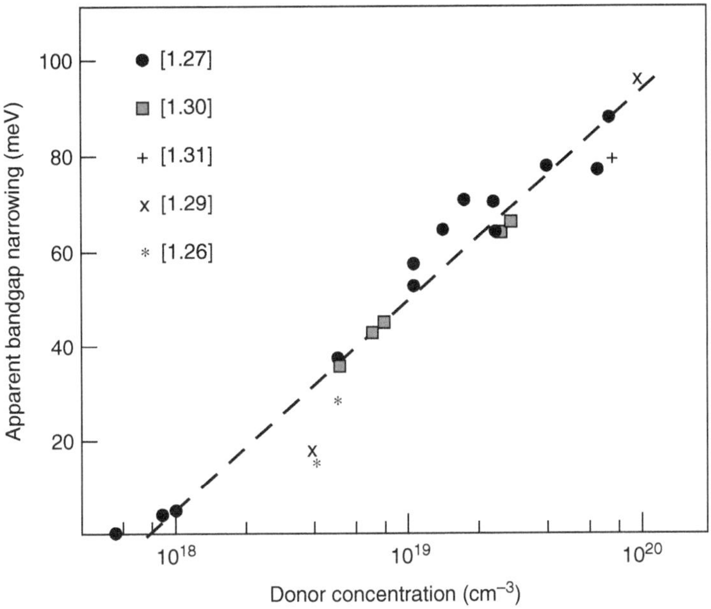

Fig. 1.23 Apparent bandgap narrowing versus phosphorus concentration (Adapted from [27])

(a) Under the influence of an electric field. The field forces the otherwise randomly moving carriers to drift in the direction of the field.
(b) Under the influence of the gradient of carrier concentration. In this case carriers diffuse from regions of high concentration to regions of low concentration.

Both drift and diffusion are involved in different degrees in the transport of carriers. In both cases, the electric current density depends on the concentration of charged carriers free to move and their velocities in the direction of current. For electrons

$$j = qn\langle v\rangle \quad \mathrm{A/cm^2}, \tag{1.56}$$

where: j = current density in $\mathrm{A/cm^2}$, n = electron concentration ($\mathrm{cm^{-3}}$), q = electron charge, $1.60218 \times 10^{-19}\,\mathrm{C}$, $\langle \mathrm{v}\rangle$ = average velocity in the direction of current (cm/s).

1.5.1 Carrier Transport by Drift: Low Field

At a given temperature, both the free carriers and crystal atoms are in random motion. The free carriers fly in all directions at their thermal velocity making random collisions with lattice atoms. Without external disturbances, electrons and holes share their thermal motion with the crystal and the carriers, as a group, do not carry a net current in any direction because, on the average, as many move in one direction

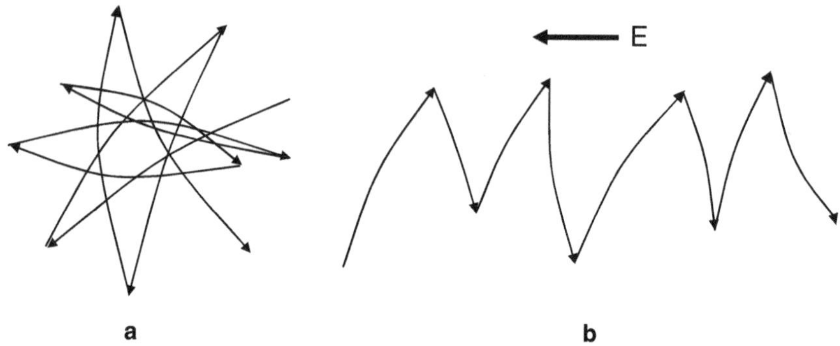

Fig. 1.24 Exaggerated representation of random carrier motion. **a** No field: random motion at thermal velocity, average current is zero. **b** Applied field: organized velocity component (drift velocity) superimposed upon thermal velocity component

as in the other. This is illustrated in Fig. 1.24a. Each carrier acquires a thermal energy of *3kT/2* that is lost after collisions. At thermal equilibrium

$$\frac{1}{2}m * v_{\text{th}}^2 = \frac{3}{2}kT, \tag{1.57}$$

where: $m^* =$ carrier effective mass, $v_{\text{th}} =$ carrier thermal velocity, $k =$ Boltzmann constant, $T =$ Absolute temperature.

Substituting the values for m^* in (1.57), the room temperature thermal velocities are found as:

$$\begin{matrix} \text{Electrons} & v_{\text{th}} \cong 1.0 \times 10^7 \\ \text{Holes} & v_{\text{th}} \cong 1.3 \times 10^7 \end{matrix} \quad \text{cm/s.} \tag{1.58}$$

The lattice atoms vibrate about their mean positions without contributing to the current. The vibrations, however, cause collisions to carriers, randomly deflecting their motion.

A field applied to the crystal causes a departure from thermal equilibrium. For small electric fields ($<5 \times 10^3\,\text{V/cm}$), however, the disturbance is very small so that the thermal equilibrium carrier distribution can be assumed. Holes organize themselves in the direction of the field and electrons in a direction opposite to the field (Fig. 1.24b). Since the electron charge is negative, the total current is the sum of electron and hole current. An exaggerated two-dimensional representation of the influence of a small electric field on the carrier velocity is shown in Fig. 1.25 which illustrates the superposition of a drift velocity component upon the thermal velocity.

The force exerted by the field on an electron is $F = q\text{E}$, where q is the charge of the electron. For a given field, this force is constant. From mechanics, it is known that when a constant force is applied to a particle, the particle is uniformly accelerated. Consequently, the drift velocity and hence the current density in (1.56) would be expected to increase indefinitely under the influence of a constant electric field. This is in disagreement with Ohm's law that specifies a constant current at a given field:

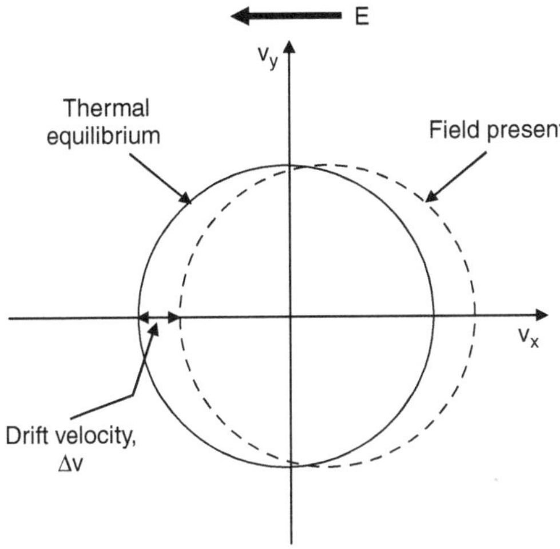

Fig. 1.25 Exaggerated two-dimensional representation of the influence of an electric field on the electron velocity distribution

$$j = \sigma E \quad \mathrm{A/cm^2}, \tag{1.59}$$

where σ is the conductivity of the material, assumed to be constant.

Obviously, there must be frictional forces in a direction opposite to the accelerating force and a balance between forces must be reached rapidly so that the drift velocity settles at a constant average value. The motion of carriers in the crystal under the influence of a constant field can be compared with the fall of a steel ball in a viscous medium such as heavy oil. While the constant force of gravity accelerates the steel ball, the frictional force that the viscous medium exerts on it slows it down. The friction increases as the velocity of the steel ball increases. Eventually, a point is reached where the force of gravity and the frictional forces cancel each other. At this point there will be no net force exerted on the ball that now falls at a constant limiting velocity. The work done by frictional forces produces a small rise in the temperature of the medium.

The average electron drift velocity can be expressed by combining (1.56) and (1.59) as

$$\langle v_{dn} \rangle = \frac{\sigma_n}{qn} E \quad \mathrm{cm/s}. \tag{1.60}$$

The term σ_n/qn is a material constant that is defined as the electron mobility, μ_n, in the crystal. Equation (1.60) can be written as

$$v_{dn} = \mu_n E, \tag{1.61}$$

where, for simplicity, the average velocity is defined as v_{dn} instead of $\langle v_{dn} \rangle$. Similarly, the drift velocity of holes is expressed as

$$v_{dp} = \mu_p E, \tag{1.62}$$

where μ_p is the hole drift mobility. From (1.61) and (1.62) it can be seen that the unit for mobility is cm^2/Vs. Electron and hole mobilities are a measure of the ease with which the carriers move in the crystal. They are very important parameters that play a key role in device analysis.

It should be noted that electrons and holes move in opposite directions but, since their charges are opposite their sum is the total current

$$j_n + j_p = q(v_{dn}n + v_{dp}p) = qE(\mu_n n + \mu_p p) \quad \text{A/cm}^2. \tag{1.63}$$

The conductivity of the crystal is then[3]

$$\sigma = \sigma_n + \sigma_p = q(\mu_n n + \mu_p p) \quad \text{S/cm}. \tag{1.64}$$

The resistivity is defined as:

$$\rho = \frac{1}{\sigma} = \frac{1}{q(\mu_n n + \mu_p p)} \ \text{Ohm} - \text{cm}. \tag{1.65}$$

If only donors at a concentration N_D are present, then ρ can be approximated as

$$\rho \cong \frac{1}{q\mu_n N_D}. \tag{1.66}$$

Similarly, for only N_A

$$\rho \cong \frac{1}{q\mu_p N_A}. \tag{1.67}$$

In the case where both donors and acceptors are present at different concentrations, one type is totally compensated by the other. The carrier concentration is approximated as the net difference in concentration between the two impurities. The mobility, however, is affected by the sum of both impurities. For $N_D = N_A$, silicon becomes intrinsic.

1.5.1.1 Scattering Mechanisms

If the crystal were perfect and the atoms not vibrating, there would be no scattering of carriers by the lattice because the presence of the atoms in the periodic lattice is already taken into account in the effective mass [6]. Only a departure from ideality and periodicity acts as a scattering mechanism that obstructs the motion of carriers, reducing their mobilities. The most important scattering mechanisms are lattice (phonon) scattering and ionized impurity (Coulomb) scattering. Other mechanisms, such as scattering by neutral atoms at low temperature, crystal-defect scattering, carrier-carrier scattering, are less important and will be discussed only where applicable. Stress-induced mobility enhancement or degradation is covered in Chap. 5.

---

[3] The unit Siemens (S) is used for 1/Ohm.

The carrier mobility is limited by the rate of scattering events along its path. The scattering events are referred to as collisions. Between collisions, carriers gain energy from the electric field and transfer this energy to the atoms of the crystal upon collision. The energy transfer increases the oscillations of crystal atoms and causes Joule heating. It is also assumed that upon collision a carrier loses all the energy it gained from the external field and starts over again with a random velocity after collision. In other words, after collision, the carrier has no memory of what happened before collision.

Let the probability for a carrier to collide during a small time dt be dt/τ, where τ is the mean-free time between collisions, assumed to be constant for simplicity. The gain and loss of energy can now be expressed in one dimension as

$$\frac{\partial v_{dx}}{\partial t} = \frac{qE_x}{m*} \quad \text{(Gain from field)} \tag{1.68}$$

$$\frac{\partial v_{dx}}{\partial t} = -\frac{v_{dx}}{\tau} \quad \text{(Loss by collisions)} \tag{1.69}$$

In steady state, there is a balance between loss and gain and the net change in drift velocity is zero

$$\frac{dv_{dx}}{dt} = \frac{\partial v_{dx}}{\partial t}(\text{Field}) + \frac{\partial v_{dx}}{\partial t}(\text{Collisions}) = 0. \tag{1.70}$$

The drift velocity is obtained from (1.68) and (1.70) as

$$v_{dx} = \frac{q\tau}{m^*}E_x = \mu E_x \quad \text{cm/s}, \tag{1.71}$$

where the mobility is now defined in terms of the mean-free time between collisions τ and effective mass m^*. The drift velocity can be increased by increasing τ and reducing m^* (Chap. 5).

Suppose that for a given field E the carriers have a certain drift velocity v_{dx} and that, at time $t = 0$, the field is instantaneously turned off. Because of collisions, the carriers will lose their kinetic energy and v_{dx} will rapidly approach zero. The rate of change by collisions alone is given in (1.68) and the decay follows the relation

$$\langle v_{dx}(t)\rangle = \langle v_{dx}(0)\rangle e^{-t/\tau}, \tag{1.72}$$

where $v_{dx}(0)$ is the average drift velocity at $t = 0$. The time constant τ is then defined as the relaxation time for v_{dx} to return to zero when the electric field is turned off, or to reach its steady state value when the field is turned on. It can be shown that τ in (1.72) is the same as the mean free time between collisions in (1.68).

Lattice Scattering

In the presence of thermal vibrations, lattice atoms are compressed or pulled apart over small regions. Such displacements cause a disturbance to the periodic potential, creating additional local electric fields that modify the momentum and direction of carriers. This scattering mechanism is called lattice scattering. The probability of collisions with the lattice increases with temperature because the number of vibrational modes and the amplitude of oscillations increase with temperature. As the oscillation amplitude increases, the cross-sectional area for collisions increases. By treating the lattice atoms as harmonic oscillators, it can be shown that their collision cross-section, A_c, increases linearly with temperature

$$A_c \propto T. \tag{1.73}$$

For a collision event to be completed, the carrier must find a vacant energy level to be placed after collision. Therefore, the collision probability will not only depend on the collision cross-section but also on the number of vacant states available for the carrier to occupy after collision. In Sect. 1.4.3 it was found that the density of states slightly above the conduction band edge, E_C, or below the valence band edge, E_V, is approximated to be proportional to the square root of energy (1.23 and 1.24). Since energy is proportional to temperature, it is concluded that the probability of finding a vacant state in which to place the carrier after collision increases with the square root of temperature. The collision probability can therefore be assumed to be proportional to $T^{3/2}$. Since the mobility is inversely proportional to the scattering probability, one can approximate the temperature dependence of lattice-limited mobility as

$$\mu \propto T^{-3/2} \quad \text{cm}^2/\text{V.s.} \tag{1.74}$$

The measured temperature dependence of mobility does not exactly follow a $T^{-3/2}$ rule. Instead, careful measurements show the following dependence for electron and hole mobilities [32]

$$\mu_{\text{ln}} \cong 2.10 \times 10^9 T^{-2.5} \quad \text{cm}^2/\text{Vs}, \tag{1.75}$$

$$\mu_{lp} \cong 2.34 \times 10^9 T^{-2.7} \quad \text{cm}^2/\text{Vs}, \tag{1.76}$$

where μ_{ln} and μ_{lp} are the lattice limited electron and hole mobilities, respectively. The results do not fully agree with the temperature dependence in (1.74) but this is not surprising since several simplifying assumptions were made to obtain the $T^{3/2}$ model.

Ionized Impurity Scattering

A carrier moving in the vicinity of an ionized impurity is deflected by the Coulomb field of the ion. When the crystal is doped with both donors and acceptors, the sum of all ionized impurities contribute to scattering. In their path near the ion, electrons are attracted by donors and repelled by acceptor while holes are attracted by acceptors and repelled by donors. The scattering probability can be approximated from a simplified Rutherford model where it is found that the scattering cross-section A_c is inversely proportional to the square of temperature [33]

$$A_c \propto T^{-2}. \tag{1.77}$$

The mean free path λ between collisions is defined as

$$\lambda = \frac{1}{A_c N_I}, \tag{1.78}$$

where N_I is the concentration of ionized impurities. Combining the above two relations gives

$$\lambda \propto \frac{T^2}{N_I}. \tag{1.79}$$

Since the average time between collisions, τ, is the ratio of mean free path to average velocity and the average velocity is proportional to the square root of temperature (1.57), it follows that

$$\tau \propto \frac{\lambda}{v} \propto \frac{T^{3/2}}{N_I}. \tag{1.80}$$

Also, since mobility is proportional to mean free time between collisions, then

$$\mu_I \propto \frac{T^{3/2}}{N_I}, \tag{1.81}$$

where μ_I is the ionized-impurity limited mobility. From (1.74) and (1.81) one would expect the ionized impurity scattering to dominate at low temperature and the lattice scattering to dominate at high temperature. An analysis based on the Rutherford scattering model gives the following relation for the ionized-impurity scattering-limited mobility as a function of temperature and ionized impurity concentration [33]

$$\mu_I = \frac{(8/\pi)(\sqrt{2/\pi})((\varepsilon_0^2 \varepsilon_{Si}^2 (\boldsymbol{kT})^{3/2})/\boldsymbol{q}^3 \sqrt{\boldsymbol{m*}})}{N_I \ln[1 + (3\varepsilon_0 \varepsilon_{Si} \boldsymbol{kT}/\boldsymbol{q}^2 N_I^{1/3})^2]} \quad \text{cm}^2/\text{Vs}, \tag{1.82}$$

where: μ_I = ionized – impurity limited mobility (cm^2/Vs), ε_Si = relative dielectric constant of silicon, 11.7, ε_0 = permittivity of free space, 8.85418×10^{-14} F/cm, k = Boltzmann constant = 8.618×10^{-5} eV/K = 1.38066×10^{-23} J/K, T = absolute temperature (K), N_I = ionized impurity concentration (cm^{-3}), $q = 1.60218 \times 10^{-19}$ C, m^* = effective mass (Kg).

Substituting all the numerical values for the physical and material constants, (1.82) can be written in the simple form

$$\mu_I = \frac{AT^{3/2}}{N_I \ln[(1+B(T^2/N_I^{2/3}))]} \quad \text{cm}^2/\text{Vs}, \tag{1.83}$$

With, A $\cong$ 1.5 $\times$ 10^{16} for electrons, A $\cong$ 7.3 $\times$ 10^{15} for holes, B $\cong$ 2.81 $\times$ 10^6 for electrons and holes.

The dependence of ionized-impurity mobility on temperature can be intuitively understood by considering that elevating the temperature increases the carrier thermal velocity. This reduces the time a carrier spends in the vicinity of the ionized impurity and hence its deflection when it passes the ion, which increases the mobility.

1.5.2 Matthiesson's Rule

In a simple model, the probabilities for scattering are additive and proportional to the reciprocals of the relaxation times. The lattice and ionized impurity scattering can therefore be combined statistically to estimate the effective relaxation time (mean free-time between collisions)

$$\frac{1}{\tau_{eff}} = \frac{1}{\tau_l} + \frac{1}{\tau_I}, \tag{1.84}$$

where τ_l and τ_I are the lattice- and impurity-limited mean free time, respectively. Since μ_{eff} is proportional to relaxation time (1.71),

$$\frac{1}{\mu_{eff}} = \frac{1}{\mu_l} + \frac{1}{\mu_I}. \tag{1.85}$$

The combination of different mobilities in the form of (1.85) is referred to as the Matthiessen's rule. The electron and hole mobilities obtained from (1.75), (1.76), (1.83), and (1.85) are plotted in Fig. 1.26 for 25 °C as a function of dopant concentration N. It should be noted, however, that there is appreciable spread in mobility data in the literature, particularly at higher concentrations [34–39]. Figure 1.27 compares room-temperature mobilities as function of N using the following empirical expression derived from experimental data by [34], and modified slightly by [35–38]

$$\mu = \frac{\mu_{\text{max}} - \mu_{\text{min}}}{1 + (N/N_{\text{ref}})^{\alpha}} + \mu_{\text{min}} \quad \text{cm}^2/\text{Vs}. \tag{1.86}$$

The values for μ_{max}, μ_{min}, (cm^2/Vs), N_{ref}(cm^{-3}) and α are shown in Table 1.1.

The temperature dependence of mobilities is shown in Fig. 1.28 for $N = 10^{15}$, $10^{16}, 10^{17}$ and 10^{18} cm^{-3}. The plots agree fairly closely with those reported in [39]. When combined with the mobility, (1.66) and (1.67) provide a good approximation

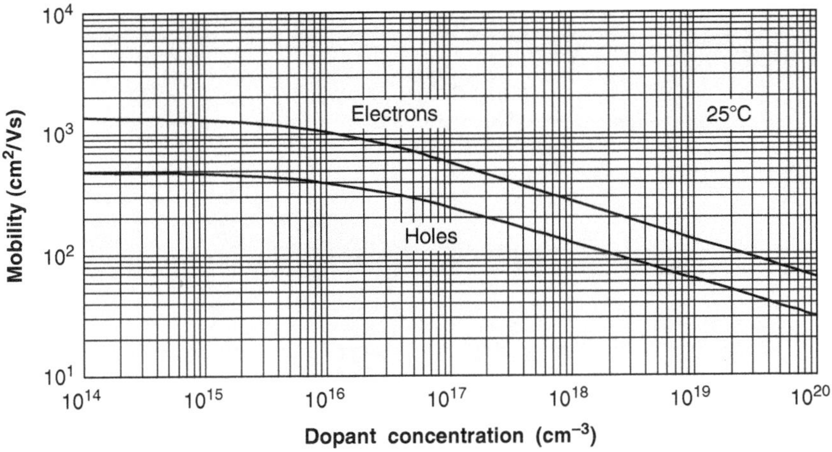

Fig. 1.26 Electron and hole mobilities as a function of dopant concentration at 25 °C

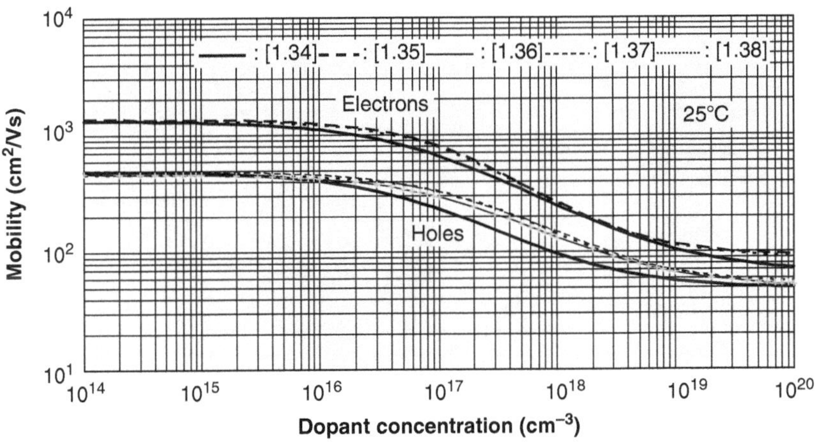

Fig. 1.27 Comparison of electron and hole mobilities at 25 °C

of resistivity as a function of impurity concentration and temperature. Figure 1.29 compares the resistivity obtained from the above equations (dashed lines) with measured results from [40, 41] (solid lines).

1.5.3 Carrier Transport by Drift: High Field

The drift velocity of electrons and holes in silicon is shown as a function of electric field in Fig. 1.30. The plots are constructed from the following empirical relation that constitutes a best fit to measured data at 27 °C [34]:

Table 1.1 Values for parameters in (1.86), μ in cm^2/Vs, N in cm^{-3}

Reference	Electrons				Holes			
	μ_{max}	μ_{min}	N_{Ref}	α	μ_{max}	μ_{min}	N_{Ref}	α
[34]	1330	65	8.5×10^{16}	0.72	495	47.7	6.3×10^{16}	0.76
[35]	1360	92	1.3×10^{17}	0.91				
[36]					495	47.7	1.9×10^{17}	0.76
[37]					468	49.7	1.6×10^{17}	0.70
[38]	1330.1	88.3	1.295×10^{17}	0.891	461.2	54.3	2.35×10^{17}	0.88

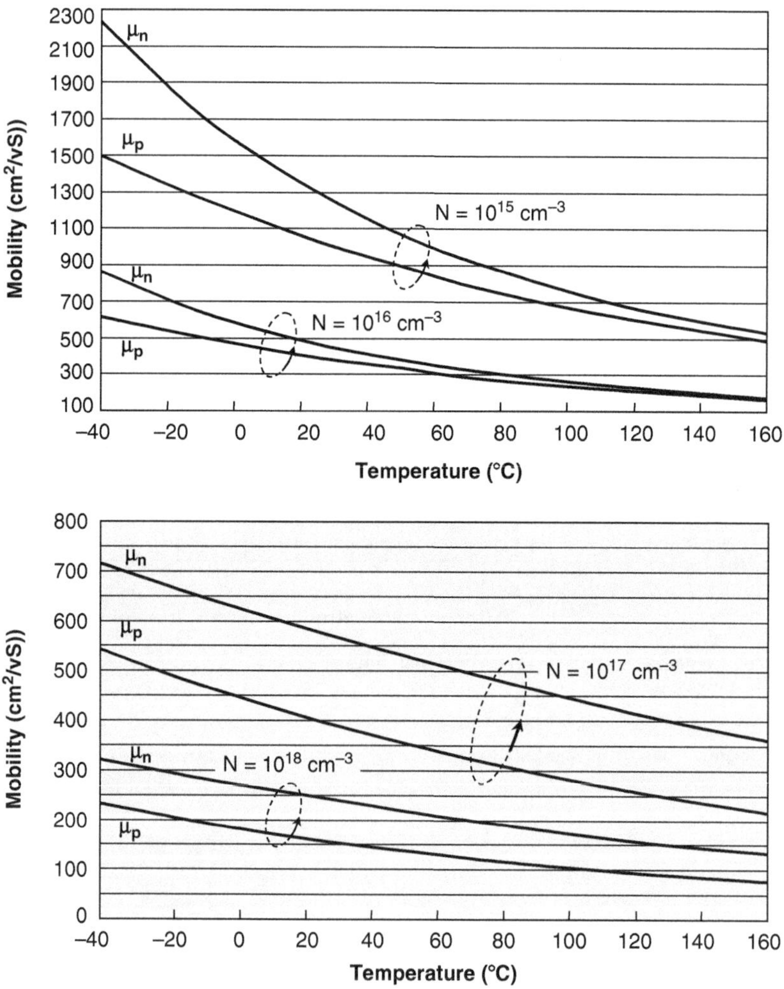

Fig. 1.28 Temperature dependence of electron and hole mobilities for four different concentrations

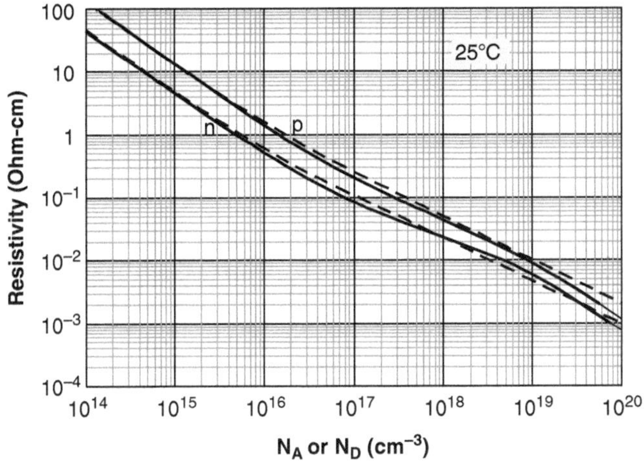

Fig. 1.29 Silicon resistivity versus impurity concentration at 25 °C. Solid lines: measured resistivity [40,41]; dashed lines: calculated resistivity from 1.66, 1.67, 1.74, 1.75, and 1.82

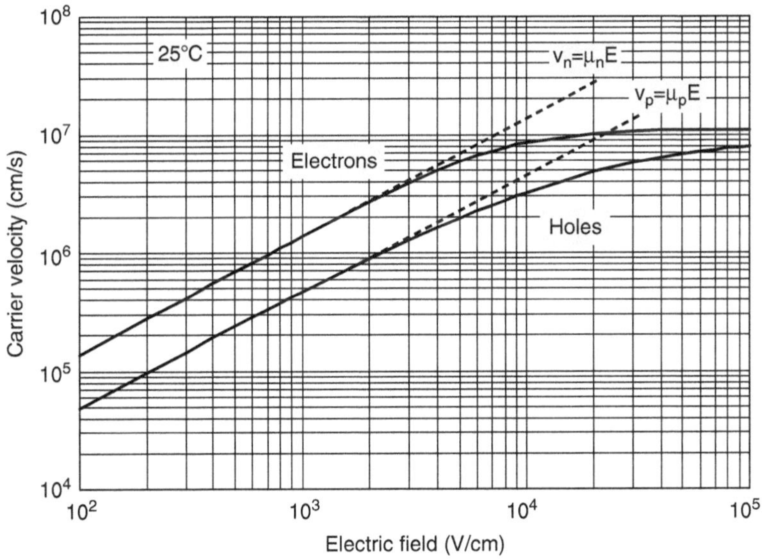

Fig. 1.30 Carrier drift-velocity versus electric field

$$v_d = v_m \frac{E/E_c}{[1+(E/E_c)^{\beta}]^{1/\beta}} \quad \text{cm/s}. \tag{1.87}$$

The values for the fitting parameters are summarized in Table 1.2 [34].

At fields below $\sim 5 \times 10^3$ V/cm, the drift velocity is proportional to electric field. The proportionality factor is the mobility, as defined in (1.61) and (1.62). In the

Table 1.2 Values for fitting parameters in (1.87)

	$E_c(V/\text{cm})$	$v_m(\text{cm/s})$	β
Electrons	8×10^3	1.1×10^7	2
Holes	1.95×10^4	9.5×10^6	1

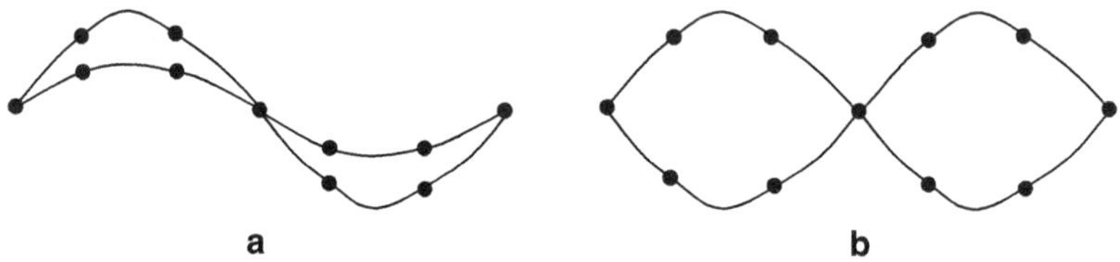

Fig. 1.31 Schematic representation of oscillation modes in a crystal. **a** Atoms in phase, acoustical phonons. **b** Atoms out of phase, optical phonons

range between 5×10^3 and $5 \times 10^4\,\text{V/cm}$, the drift velocity increases approximately with the square root of electric field, that is, the mobility decreases. As the field increases above $\sim 5 \times 10^4\,\text{V/cm}$, the drift velocity begins to saturate to a value close to the thermal velocity.

The departure from linearity and saturation of velocity can be qualitatively explained in terms of different vibration mechanisms in the crystal.

Oscillations in the crystal can be visualized by comparing the crystal lattice with a three-dimensional periodic set of mass-points, spaced at distance a apart and each attached by identical springs to six adjacent springs. In such a system, a multitude of standing waves of different wavelengths can be generated at a frequency that depends on the force-constant of the springs. The minimum meaningful oscillation wavelength is $\lambda = 2a$. Standing waves of any wavelength between $2a$ and the total length of the crystal are allowed, with nodes appearing at the system boundaries. In one dimensional set of N mass-points, for example, one would expect the possible wavelengths to be

$$\lambda = \frac{Na}{n} \quad n = 1,2,3,\ldots N/2.$$

In a crystal, there are two modes of oscillation of different frequencies associated with each wavelength, one in which the two atoms in a unit cell oscillate in phase and the other where they move out of phase (Fig. 1.31). As in the case of electrons and photons, an oscillation in the crystal has a frequency ν and an associated quantum of energy $\text{E} = h\nu$ called phonon. When atoms oscillate approximately in phase, the relative displacement between atoms is very small and the restoring force is small. Therefore, the frequency associated with such oscillations is small and the phonon energy is also small. The wave propagation corresponds to ordinary sound waves in the crystal so they are referred to as acoustical phonons. When the atoms move in approximately opposite directions, they are characterized by high frequencies, hence high energies, because of the large restoring force associated with the opposite motion of atoms. This mode of oscillation is called optical because, if the atoms were oppositely charged, such as in NaCl crystals, the vibrations give rise to oscillating electric dipole moments and electromagnetic radiation is emitted or absorbed during the process.

When a field smaller than $\sim 5 \times 10^3\,\mathrm{V/cm}$ is applied to the crystal, the carriers interact predominantly with acoustical phonons because their kinetic energy is not sufficiently large to generate optical phonons. When the field is varied in this range, the carriers rapidly exchange energy with acoustical phonons and reach a new steady-state average velocity that is proportional to the field, that is, $dv_d/dE = \text{constant} = \mu_o$, where μ_o is the low-field mobility (1.71). In this region, the carrier temperature does not increase appreciably over the crystal temperature. As the field increases above $\sim 5 \times 10^3\,\mathrm{V/cm}$, the carrier energy increases to a level where the probability for generation of optical phonons becomes significant. An increasing fraction of energy gained from the electric field is now lost by generation of optical phonons and the rate of change in drift velocity decreases: $dv_d/dE \neq \text{constant} = \mu < \mu_o$. Therefore, as the field increases above a "critical level" the carrier mobility begins to decrease below its low-field value. As the field is increased above $\sim 5 \times 10^4\,\mathrm{V/cm}$, a point is reached where the carriers lose to the crystal all the kinetic energy gained by increasing the field. The drift velocity does not further increase but saturates to its maximum value, v_{sat}. This means that dv_d/dE approaches zero. In this range, the energy imparted by the electric field is transferred to the crystal and increases the number of carriers at velocity saturation instead of increasing the carrier drift velocity. In the range above $\sim 5 \times 10^4\,\mathrm{V/cm}$, the carrier temperature increases above the lattice temperature and carriers are said to be hot.

The effective mobility as a function of electric field can be extracted as the ratio of velocity to field from (1.87) with the parameter constants defined in Table 1.2 [34]

$$\mu_{eff} = \frac{\mu_o}{[1 + (E/E_c)^{\beta}]^{1/\beta}} \quad \mathrm{cm^2/Vs}, \tag{1.88}$$

where μ_o is defined as v_m/E_c. From the values in Table 1.2, $\mu_o = 1375\,\mathrm{cm^2/Vs}$ for electrons and $487\,\mathrm{cm^2/Vs}$ for holes.

When a field is applied to the crystal, there is also the probability that carriers gain sufficient energy from the field to generate electron–hole pairs by breaking valence bonds. This process is referred to as impact ionization. When the field is low, the probability for breaking bonds is extremely small and the generation of electron–hole pairs by impact ionization can be neglected. As the field is increased, the probability for impact ionization increases and eventually results in avalanche breakdown (Chap. 2). Depending on impurity concentration, impact ionization becomes significant in the field range $8 \times 10^4 - 9 \times 10^5\,\mathrm{V/cm}$. As the field approaches $\sim 10^6\,\mathrm{V/cm}$, the probability for direct band-to-band tunneling increases (Chaps. 2 and 5). Hot-carriers are key device parameters that have become increasingly important as device dimensions are reduced in deep-submicron and nano-scale technologies. They will be discussed in more detail in subsequent chapters.

1.5.4 *Carrier Transport by Diffusion*

Free particles are in constant random thermal motion. When the particle concentration is uniform and no external disturbance is present, on the average, as many particles move in one direction as in the other and the net number of particles passing through a unit area per unit time, the carrier flux in any direction, is zero. In the presence of a concentration gradient, however, the probability per carrier to move in one direction can be the same everywhere but since there are more carriers in a region of high concentration than outside that region, there will be a net flux of carriers moving outward from the high-concentration region. The flow of particles in response to a concentration gradient is referred to as diffusion. It is the same process that spreads smoke in air. In a semiconductor, electrons and holes are carriers of charge and their transport by diffusion gives rise to diffusion current. A thorough analysis shows that the electron diffusion current density in one dimension is

$$j_{n(dif)} = qD_n \frac{\mathrm{d}n}{\mathrm{d}x} \quad \mathrm{A/cm^2}. \tag{1.89}$$

The gradient dn/dx is negative since it points in the direction of decreasing concentration. Since for electrons q is negative, j_n is positive. Similarly, the hole current density in one dimension is

$$j_{p(dif)} = -qD_p \frac{\mathrm{d}p}{\mathrm{d}x} \quad \mathrm{A/cm^2}, \tag{1.90}$$

where D_n and D_p are, respectively, the electron and hole diffusion constants. The total current density in three dimensions is

$$j_{(dif)} = j_{n(dif)} + j_{p(dif)} = q(D_n \nabla n - D_p \nabla p) \quad \mathrm{A/cm^2}, \tag{1.91}$$

where

$$\nabla = \frac{\partial}{\partial x} + \frac{\partial}{\partial y} + \frac{\partial}{\partial z}.$$

The average carrier velocity in one direction can be obtained from (1.56), (1.90) and (1.91) as

$$v_{n(dif)} = D_n \frac{\mathrm{d}n}{\mathrm{d}x}\frac{1}{n} \quad \mathrm{cm/s}, \tag{1.92}$$

$$v_{p(dif)} = -D_p \frac{\mathrm{d}p}{\mathrm{d}x}\frac{1}{p} \quad \mathrm{cm/s}. \tag{1.93}$$

1.5.4.1 Total Current Density

Both drift and diffusion contribute in varying degrees to the total current density. In most cases the two mechanisms may be assumed to be independent events and can be combined to give the total current density. In one dimension this gives

$$j_n = j_{n(\text{drift})} + j_{n(\text{dif})} = q\mu_n nE + qD_n\frac{dn}{dx} \quad \text{A/cm}^2, \tag{1.94}$$

$$j_p = j_{p(\text{drift})} + j_{p(\text{dif})} = q\mu_p pE - qD_p\frac{dp}{dx} \quad \text{A/cm}^2. \tag{1.95}$$

The drift terms in the above equations have the same algebraic sign because the electric field produces conventional current in the same direction for opposite charged particles.

1.5.4.2 Nonuniform Profile

When the crystal is not uniformly doped, which is the typical case in silicon devices, the bulk Fermi potential ϕ_b in (1.47) becomes a function of position, as illustrated for n-type silicon at thermal equilibrium in Fig. 1.32. Since the crystal is at thermal equilibrium, there is no net current and the Fermi level is shown flat with respect to a reference level, E_{ref} that is chosen to be any convenient value. The energy gap $E_C - E_V$ remains constant, but the bands are bent.

The electron electrostatic potential is

$$V(x) = -\frac{E_C(x) - E_{\text{ref}}}{q} \quad \text{V}. \tag{1.96}$$

Because of dopant nonuniformity, there is a built-in electric field

$$E = -\frac{dV}{dx} = \frac{1}{q}\frac{dE_C}{dx} = \frac{1}{q}\frac{dE_V}{dx} = \frac{1}{q}\frac{dE_i}{dx} = -\frac{d\phi}{dx} \quad \text{V/cm}. \tag{1.97}$$

This built-in field induces a drift current component that, at thermal equilibrium, is balanced by a diffusion current component in the opposite direction. This can be

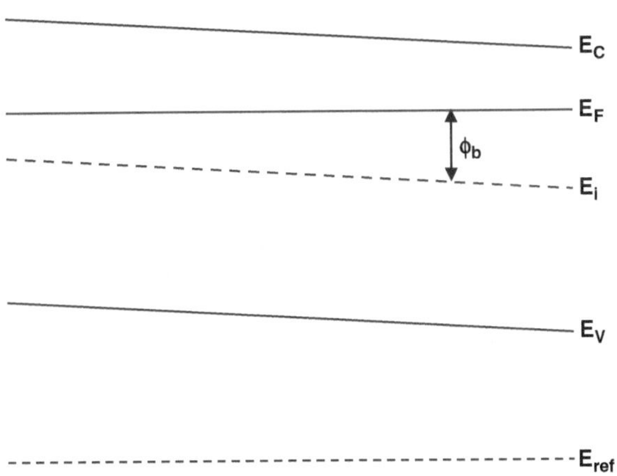

Fig. 1.32 Simplified band diagram for nonuniformly doped silicon at thermal equilibrium

visualized by considering that the concentration gradient in Fig. 1.32 causes electrons to diffuse from right to left but as they diffuse, electrons leave uncompensated positively charged donor ions behind creating a field that forces electrons to drift from left to right. At equilibrium, drift is exactly balanced by diffusion for each type of carrier and the net current is zero. In this case, (1.94) and (1.95) can be expressed as

$$q\mu_n\bar{n}E = -qD_n\frac{\mathrm{d}\bar{n}}{\mathrm{d}x}, \tag{1.98}$$

$$q\mu_p\bar{p}E = qD_p\frac{\mathrm{d}\bar{p}}{\mathrm{d}x}. \tag{1.99}$$

where $\bar{n}$ and $\bar{p}$ are the equilibrium electron and hole concentrations.

1.5.4.3 Einstein Relation

In the above equations, the carrier mobility and diffusion constant are both controlled by scattering mechanisms in the crystal. They are independent of each other and in most cases connected by an equation known as Einstein's relation [1, 4, 42]

$$D_n = \mu_n\frac{kT}{q} \quad \mathrm{cm}^2/\mathrm{s}, \tag{1.100}$$

$$D_p = \mu_p\frac{kT}{q} \quad \mathrm{cm}^2/\mathrm{s}. \tag{1.101}$$

Substituting the above expressions for D_n and D_p in (1.98) and (1.99), respectively, gives important relations between the built-in field and gradient in dopant concentration as

$$E = -\frac{1}{\bar{n}}\frac{kT}{q}\frac{\mathrm{d}\bar{n}}{\mathrm{d}x} = -\frac{1}{N_D}\frac{kT}{q}\frac{\mathrm{d}N_D}{\mathrm{d}x} \quad \mathrm{V/cm}, \tag{1.102}$$

$$E = \frac{1}{\bar{p}}\frac{kT}{q}\frac{\mathrm{d}\bar{p}}{\mathrm{d}x} = \frac{1}{N_A}\frac{kT}{q}\frac{\mathrm{d}N_A}{\mathrm{d}x} \quad \mathrm{V/cm}. \tag{1.103}$$

1.6 Nonequilibrium Conditions

When the crystal is at thermal equilibrium, the rate of thermal electron–hole pair generation in a volume element is balanced by a statistically equal rate of electron–hole pair recombination. In this case (1.37) is valid

$$\bar{p}\bar{n} = n_i^2 \quad \mathrm{cm}^{-6}. \tag{1.104}$$

A disturbance to this condition occurs when a change in the carrier concentration is induced by an external stimulus, such as light or applied voltage. When the equilibrium concentration is upset, the new concentrations can be larger or smaller than in (1.104). When the carrier concentrations are reduced

$$pn < n_i^2. \tag{1.105}$$

If excess carriers are created

$$pn > n_i^2. \tag{1.106}$$

When the stimulus is removed, the system always tends to restore equilibrium, in the first case by an excess in thermal generation over recombination and in the second case by an excess recombination over generation. The next section discusses the mechanisms involved in the generation-recombination processes.

1.6.1 Carrier Lifetime

Excess carriers can be created, for example, by illuminating the crystal with photons of energy above ~1.1 eV. If the intensity of light is kept constant, a nonequilibrium steady-state condition is achieved where the electron and hole concentrations are larger than their thermal equilibrium values and a new rate of generation is balanced by a new rate of recombination. Let Δp and Δn be the excess carrier concentrations. The steady-state concentrations are

$$p = \bar{p} + \Delta p, \tag{1.107}$$

$$n = \bar{n} + \Delta n. \tag{1.108}$$

Since the crystal as a whole must remain neutral, $\Delta p = \Delta n$. In p-type silicon, holes are the majority carriers and electrons the minority carriers. In this case, Δp is negligible when compared to p but can be considerably larger than n. Similarly, in n-type silicon Δn is negligible when compared to n but can be much larger than the minority p. Assume, for example, p-type silicon with $N_A = 10^{16}\,\mathrm{cm}^{-3}$. At room temperature, the hole concentration is $\sim 10^{16}\,\mathrm{cm}^{-3}$ and, from (1.104), the electron concentration is $\sim 2 \times 10^4\,\mathrm{cm}^{-3}$. If the sample is illuminated with light of intensity such that $\Delta n = \Delta p = 10^{12}\,\mathrm{cm}^{-3}$, the percentage increase in the hole concentration is 0.01% while the minority electron concentration increases by eight orders of magnitude. Excess increases in carrier concentrations at such levels are referred to as low-level minority-carrier injection. The word "injection" describes the increase in minority carrier concentration above its thermal equilibrium value. Injection parameters are central to the operation of bipolar structures, as discussed in Chap. 3.

Excess majority carriers will remain present to neutralize excess minority carriers. If, by some means, excess majority carriers could be created in silicon without a simultaneous increase in minority carriers (such as in accumulated surfaces, Chap. 4) and the stimulus instantaneously removed, the excess carriers would dissipate in an average time, the dielectric relaxation time, that is extremely small when compared to the time it takes minority carriers to dissipate (See also Problem 8). It is the minority-carrier lifetime that plays a key role in device applications.

At low level injection, one can assume that the rate of recombination of excess minority carriers is proportional to their concentration

$$\frac{d\Delta n}{dt} = -C\Delta n, \tag{1.109}$$

where C is a constant. The decay of excess carriers can then be expressed as

$$\Delta n = \Delta n_0 e^{-t/\tau} \quad \text{cm}^{-3}. \tag{1.110}$$

In (1.110) Δn_0 is the excess minority concentration at the time the stimulus was stopped and τ is the average minority carrier lifetime. This is the time when the excess minority carrier concentration has decayed to 1/e of its value at the time $t = 0$. The minority carrier lifetime is determined by the rate at which excess electrons and holes recombine, or annihilate. Recombination processes are discussed next.

1.6.1.1 Direct Band-to-Band Transitions

Band to band recombination is a direct transition of an electron from the bottom of the conduction band to the top of the valence band whereby an electron–hole pair is annihilated. The rate of this recombination process is expected to be proportional to the electron and hole concentrations. Such a transition not only involves a loss in energy but also a loss in momentum (Fig. 1.13). When it does occur, a photon of energy ~1.1 eV is emitted. The photon, however, cannot handle the momentum change and the transition becomes very unlikely without assistance of other mechanisms that absorb the excess momentum.

1.6.1.2 Transitions Involving Intermediate States

The minority carrier lifetime in silicon crystals is found to be considerably smaller than the lifetime expected from the direct band-to-band transition probability. In low-level injection and for low to moderately doped silicon (10^{14}–$10^{17}\,\text{cm}^{-3}$), this can be explained by the presence of recombination centers in the crystal. These are energy states distributed within the bandgap and created by impurities other than dopants and faults in the crystal. The minority carrier can orbit for a short time around the center with an energy intermediate between the valence and conduction band edges and give its energy to the crystal in a two-step process, transferring to the crystal a smaller amount of energy in each step. The different transitions through intermediate steps are shown schematically in Fig. 1.33 [42]. The centers are shown at mid-gap for simplicity, but in real crystals this is not the case as can be seen from Fig. 1.34.

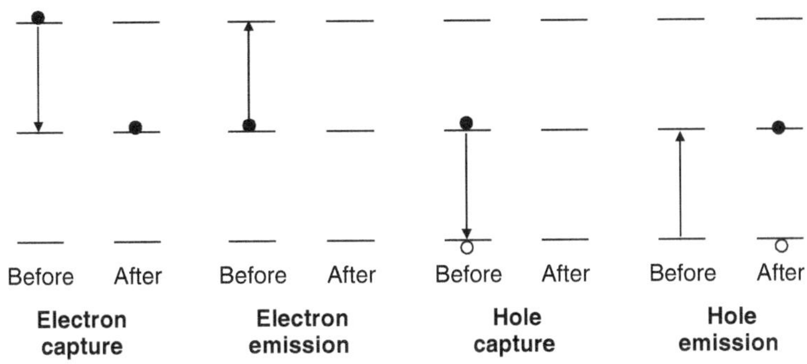

Fig. 1.33 Schematic model for transitions through intermediate states. Arrows represent direction of electron transition (Adapted from [42])

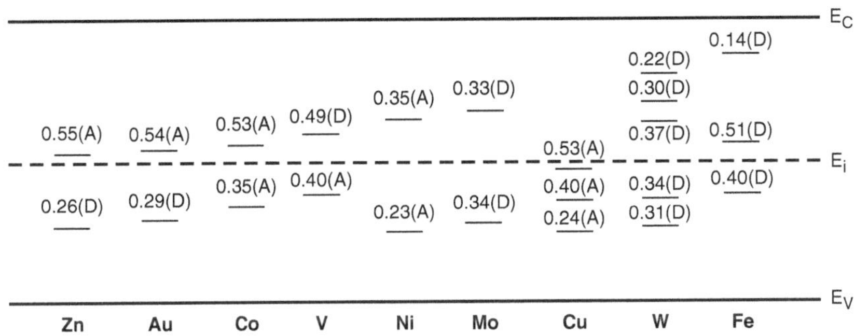

Fig. 1.34 Measured ionization energies for selected deep-level impurities in silicon. Levels above E_i are referenced to E_C and levels below E_i are referenced to E_V (Adapted from [43])

From statistical analysis it can be shown that the efficiency of a center to act as a recombination or generation site is at a maximum when its energy level is exactly at mid-gap [44, 45].

By assuming an effective density of centers, N_t, located at mid-gap, the rate of recombination (and generation) of minority carriers can be approximated as

$$U = \sigma v_{th} N_t \Delta n \text{ (electrons)} \quad \text{cm}^{-3}/\text{s}, \tag{1.111}$$

$$U = \sigma v_{th} N_t \Delta p \text{ (holes)} \quad \text{cm}^{-3}/\text{s}. \tag{1.112}$$

In (1.111) and (1.112), U = recombination rate (cm^{-3}/s), σ = capture cross section, assumed same for electrons and holes (cm^2), v_{th} = thermal velocity ($\cong 10^7$ cm/s), N_t = effective density of recombination-generation centers (cm^{-3}).

The capture cross section is typically in the range of 10^{-15} cm^2 and assumed for simplicity to be the same for electrons and holes. The density of intermediate states depends on the crystal properties and processing conditions. In good crystals N_t ranges typically from 10^{11}–10^{12} cm^{-3}. In case of n-type silicon, the hole lifetime can now be approximated as

$$\tau = \frac{\Delta p}{U} = \frac{1}{\sigma v_{\text{th}} N_t} \quad \text{s}. \tag{1.113}$$

A similar relation is found for minority-carrier electrons. Generation-recombination through intermediate states is referred to as the Shockley-Read-Hall (SRH) process [44, 45].

1.6.1.3 Auger Recombination

Another mechanism for minority carriers to recombine is to transfer its excess energy and momentum to a third carrier available in its vicinity. This recombination process is known as Auger recombination and illustrated in Fig. 1.35. Auger recombination requires a cluster of free carriers and becomes significant, even dominates, in heavily doped regions ($\sim 10^{19}$–$10^{21}\,\text{cm}^{-3}$) where free carriers are abundant. Auger recombination plays a key role in bipolar current gain, as discussed in Chap. 3. In the presence of Auger recombination the effective carrier lifetime is defined as

$$\frac{1}{\tau_{n-\text{eff}}} = \frac{1}{\tau_{SRH}} + r_{A(n)} n^2 \quad \text{s}^{-1}, \tag{1.114}$$

$$\frac{1}{\tau_{p-\text{eff}}} = \frac{1}{\tau_{SRH}} + r_{A(p)} p^2 \quad \text{s}^{-1}, \tag{1.115}$$

where, τ_{SRH} = Shockley-Read-Hall lifetime as approximated by (1.114), r_{A} = Auger recombination rate (cm^6/s), n, p = majority carrier electrons and hole concentration (cm^{-3}).

The room-temperature electron and hole Auger recombination rates are found as [46]

$$r_{A(n)} = 2.8 \times 10^{-31} \quad \text{cm}^6/\text{s}, \tag{1.116}$$

$$r_{A(p)} = 9.9 \times 10^{-32} \quad \text{cm}^6/\text{s}. \tag{1.117}$$

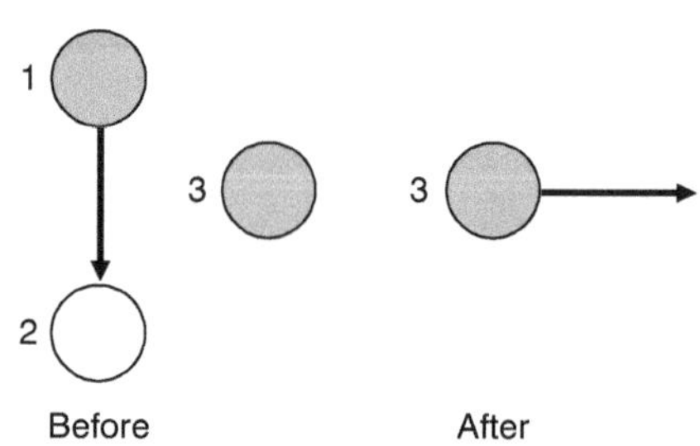

Fig. 1.35 Illustration of Auger recombination. A third carrier in the vicinity absorbs the excess energy and momentum

1.6.2 Diffusion Length

The equation of continuity for each type of carrier relates the net time-rate of change of the number of carriers in a volume element to the flow of carriers into and out of the volume element. It is defined for electrons and holes as

$$\frac{\mathrm{d}n}{\mathrm{d}t} = G - \frac{\Delta n}{\tau_n} + \frac{1}{q}\nabla . j_n, \tag{1.118}$$

$$\frac{\mathrm{d}p}{\mathrm{d}t} = G - \frac{\Delta p}{\tau_p} - \frac{1}{q}\nabla . j_p, \tag{1.119}$$

where G is the generation rate of carriers per unit volume and time. The second term in the equations describes the recombination rate and the last term defines the flow of carriers in and out of the volume element.

Consider, for example, n-type silicon in which electron–hole pairs are generated, e.g., by light in an elemental region of the crystal [7]. Under steady-state conditions, the rate of change in the minority carrier concentration is

$$\frac{\mathrm{d}p}{\mathrm{d}t} = \frac{\mathrm{d}\Delta p}{\mathrm{d}t} = 0. \tag{1.120}$$

The one-dimensional form of (1.119) can be written in the form

$$\frac{1}{q}\frac{\mathrm{d}j_p}{\mathrm{d}x} = G - \frac{\Delta p}{\tau_p} \quad \mathrm{cm}^{-3}\mathrm{s}^{-1}. \tag{1.121}$$

If a point is chosen far away from the source of generation, $G = 0$ and

$$\frac{1}{q}\frac{\mathrm{d}j_p}{\mathrm{d}x} = -\frac{\Delta p}{\tau_p} \quad \mathrm{cm}^{-3}\ \mathrm{s}^{-1}. \tag{1.122}$$

The current density j_p has a drift and a diffusion component. Since holes are minority carriers, they do not contribute appreciably to the total drift current. In this case, only the diffusion component of hole current is significant and can be approximated as

$$j_p = -qD_p\frac{\mathrm{d}\Delta p}{\mathrm{d}x} \quad \mathrm{A/cm}^2. \tag{1.123}$$

Substituting the above expression in (1.122) gives

$$\frac{d^2\Delta p}{dx^2} = \frac{\Delta p}{D_p\tau_p}. \tag{1.124}$$

For regions very far away from the source of generation, one can set $x = \infty$ and $\Delta p = 0$. The solution of (1.124) with this boundary condition has the form

$$\Delta p = \Delta p_0 e^{-x/\sqrt{D_p\tau_p}} \quad \mathrm{cm}^{-3}, \tag{1.125}$$

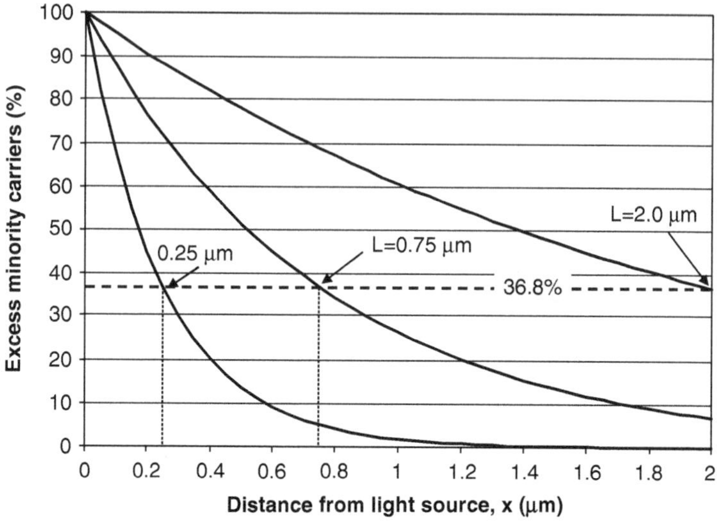

Fig. 1.36 Excess minority-carrier concentration versus distance for three diffusion lengths

where Δp_0 is the excess hole concentration at the point of reference $x = x_0$. The square root term in the exponent is the average distance an excess hole travels before it recombines. It is the distance at which the excess carrier concentration has decayed to 1/e of its value at the reference point $x = 0$ (Fig. 1.36). This value is referred to as the diffusion length, L_p:

$$L_p = \sqrt{D_p \tau_p} \quad \text{cm.} \tag{1.126}$$

Similarly, for minority carrier electrons

$$L_n = \sqrt{D_n \tau_n} \quad \text{cm.} \tag{1.127}$$

From (1.125) and (1.126), it is concluded that a shorter minority-carrier lifetime reduces the average distance carriers travel before they recombine.

1.7 Problems

1. The region between source and drain of a MOSFET is uniformly doped with 10^{18} boron atoms/cm^3. Assume a distance of 50 nm between the two regions and find the average number of boron atoms along a straight line between source and drain.

2. What is the percentage of covalent bonds broken in pure silicon at 100°C?

3. Silicon is doped with 10^{16} phosphorus atoms/cm^3. At what temperature would the hole concentration be equal to 10% of the ionized impurity concentration?

4. Silicon is doped with 10^{15} phosphorus atoms/cm^3. At what temperature is $n = 0.9N_D$?

5. Calculate the conductivity of pure silicon at 25 °C and for silicon doped with 10^{16} boron atoms/cm^3 plus 10^{16} arsenic atoms/cm^3.

6. The sheet resistance R_S of a film is defined as the resistance measured between two opposite sides of a square of the film. Show that for a film thickness x

$$R_S = \frac{\bar{\rho}}{x}$$

where $\bar{\rho}$ is the average film resistivity.

7. The boron profile in the base of a bipolar npn transistor can be approximated by an exponential function of the form

$$N_A(x) = 5 \times 10^{18} e^{-90x}$$

where x is the depth in µm.

Assume a base width of 0.1 µm and 25 °C and

(a) Calculate the average base sheet resistance at 25 °C.
(b) Show that the built-in field is constant in the base.
(c) Calculate the electron current density for a steady-state injected electron concentration of $10^{15}\,\mathrm{cm}^{-3}$ at $x = 0$.

8. In a homogeneous conductor of conductivity σ and dielectric permittivity ε, the mobile charge concentration at time $t = 0$ is $\rho(x,y,z)$. From electromagnetism we have

$$\nabla.D = \rho; D = \varepsilon E; J = \sigma E; \nabla.J = -\frac{\mathrm{d}\rho}{\mathrm{d}t}.$$

(a) Show from the above relations that

$$\rho(x,y,z,t) = \rho(x,y,z,t=0)e^{-t/(\varepsilon/\sigma)}$$

(b) Use this result to show that mobile charge cannot remain in the bulk of uniform material, but must accumulate at surfaces of discontinuity.
(c) Calculate the "dielectric relaxation time" ε/σ for a typical metal and for p-type silicon with $N_A = 10^{15}\,\mathrm{cm}^{-3}$ and n-type silicon with $N_D = 10^{17}\,\mathrm{cm}^{-3}$ [Adapted from 1].

9. Assume that the n-type emitter of a bipolar transistor is uniformly doped with $10^{20}\,\mathrm{cm}^{-3}$ with all impurities ionized. Estimate the hole diffusion length in the emitter at 25 °C.

References

1. R. B. Adler, A. C. Smith, and R. L. Longini, Introduction to Semiconductor Physics, Semiconductor Electronics Education Committee, Vol. 1, J. Wiley & Sons, New York, 1964.
2. W. Finkelburg, Einfuehrung in die Atomphysik, Springer Verlag, Berlin, 1958.
3. B. El-Kareh, Fundamentals of Semiconductor Processing Technologies, Kluwer Academic Press, Boston, 1995.
4. J. L. Moll, Physics of Semiconductors, McGraw-Hill, New York, 1964.
5. C. Kittel, Introduction to Solid State Physics, John Wiley, New York, 1956.
6. A. J. Dekker, Solid State Physics, Prentice-Hall, New Jersey (USA), 1965.
7. W. Shockley, Electrons and Holes in Semiconductors, D. Van Nostrand Company, Princeton, New Jersey, 1950.
8. G. E. Kimball, "The electronic structure of diamond," J. Chem. Phys., 3 (9), 560–564, 1935.
9. E. M. Conwell, "Properties of silicon and germanium," Part II, Proc. IRE, 46, 1281, 1958.
10. R. A. Levy, Principles of Solid State Physics, Academic Press, 1968.
11. C. D. Thurmond, "The standard thermodynamic functions for the formation of electrons and holes in Ge, Si, GaAs, and GaP", J. Electrochem. Soc.: Solid State Sci. Technol., 122, 1133, 1975.
12. F. J. Morin and J. P. Maita, "Electrical properties of silicon containing arsenic and boron," Phys. Rev., 96, 28–35, 1954.
13. A. Many, Y. Goldstein, and N. B. Grover, Semiconductor Surfaces, North-Holland Publishing Co., 1971.
14. J. C. Hensel, H. Hasegawa, and M. Nakayama, "Cyclotron resonance in uniaxially stressed silicon," Phys. Rev., 138 (1A), A132, 1965.
15. H. D. Barber, "Effective mass and intrinsic concentration in silicon," Solid-State Electron., 10, 1039, 1967.
16. G. D. Mahan, "Energy gap in Si and Ge: Impurity dependence," J. Appl. Phys., 51, 2634–2646, 1980.
17. T. N. Morgan, "Broadening of impurity bands in heavily doped silicon," Phys. Rev., 139, 343–348, 1965.
18. R. J. Van Overstraeten, H. J. DeMan, and R. P. Mertens, "Transport equation in heavily doped silicon," IEEE Trans. Electron. Devices, ED-20, 290–298, 1973.
19. S. R. Dhariwal, V. N. Ojha, and G. P. Srisvastava, "On the shifting and broadening of impurity bands and their contribution to the effective electric bandgap narrowing in moderately doped semiconductors," IEEE Trans. Electron. Device, ED-32 (1), 44–48, 1985.
20. H. P. D. Lanyon and R. A. Tuft, "Bandgap narrowing in moderately to heavily doped silicon," IEEE Trans. Electron. Devices, ED-26 (7), 1014–1018, 1979.
21. C. M. Van Vliet, "Bandgap narrowing and emitter efficiency in heavily doped emitter structures revisited," IEEE Trans. Electron. Devices, 40 (6), 1040–1047, 1993.
22. D. S. Lee and J. G. Fossum, "Energy-band distortion in highly doped silicon," IEEE Trans. Electron. Devices, ED-30 (6), 626–634, 1983.
23. N. Shigyo, N. Konishi, and H. Satake, "An improved bandgap narrowing model based on corrected intrinsic carrier concentration," IEIC Trans. Electron., E-75-C (2), 156–160, 1992.
24. J. W. Slotboom and H. C. De Graaff, "Measurement of bandgap narrowing in Si bipolar transistors," Solid State Electron., 19, 857–862, 1976.
25. J. W. Slotboom and H. C. De Graaff, "Bandgap narrowing in silicon bipolar transistors," IEEE Trans. Electron. Devices, ED-24 (8), 1123–1125, 1977.
26. R. P. Mertens, J. L. Van Meerbergen, J. F. Nus, and R. J. Van Overstraeten, "Measurement of the minority-carrier transport parameters in heavily doped silicon," IEEE Trans. Electron. Devices, ED-27 (5), 949–955, 1980.
27. J. del Alamo, S. Swirhun, and R. M. Swanson, "Simultaneous measurement of hole lifetime, hole mobility and bandgap narrowing in heavily doped n-type silicon," IEDM Technol. Digest, 290–293, 1985.

28. H. S. Bennett and C. L. Wilson, "Statistical comparisons of data on band-gap narrowing in heavily doped silicon: electrical and optical measurements," J. Appl. Phys., 55 (10), 3582–3587, 1984.
29. A. Neugroschel, S. C. Pao, and F. A. Lindholm, "A method for determining energy gap lowering in highly doped semiconductors," IEEE Trans. Electron. Devices, ED-29 (5), 894–902, 1982.
30. A. W. Wieder, "Emitter effects in shallow bipolar devices" Measurements and consequences," IEEE Trans. Electron. Devices, ED-27 (8), 1402–1408, 1980.
31. G. E. Possin, M. S. Adler, and B. J. Baliga, "Measurement of the p-n product in heavily doped epitaxial emitters," IEEE Trans. Electron. Devices, ED-31 (1), 3–17, 1984.
32. G. W. Ludwig and R. L Watters, "Drift and conductivity mobility in silicon," Phys. Rev., 101(6), 1699–1701, 1956.
33. E. M. Conwell and V. F. Weisskopf, "Theory of impurity scattering in semiconductors," Phys. Rev. 77(3), 388–390, 1950.
34. D. M. Caughy and R. E. Thomas, "Carrier mobilities in silicon empirically related to doping and field," Proc. IEEE, 2192–2193, 1967.
35. G. Baccarani and P. Ostoja, "Electron mobility empirically related to phosphorus concentration in silicon," Solid State Electron., 18 (6), 579–580, 1975.
36. D. A. Antoniadis, A. G. Gonzalez, and R. W. Dutton, "Boron in near intrinsic ⟨100⟩ and ⟨111⟩ silicon under inert and oxidizing ambients – diffusion and segregation," J. Electrochem. Soc.: Solid-State Sci. Technol., 125 (5), 813–819, 1978.
37. S. Wagner, "Diffusion of boron from shallow ion implants in silicon," J. Electrochem. Soc.: Solid-State Sci. Technol., 119 (1), 1570–1576, 1972.
38. N. D. Arora, J. R. Hauser, and D. J. Roulston, "Electron and hole mobilities in silicon as a function of concentration and temperature," IEEE Trans. Electron. Dev. ED-29, 292–295, 1982.
39. W. W. Gartner, "Temperature dependence of junction transistor parameters," Proc. IRE, 45 (5), 667, 1957.
40. J. C. Irvin, "Resistivity of bulk silicon and of diffused layers in silicon," Bell Syst. Tech. J. 41, 387, March 1962.
41. W. R. Thurber, R. L. Mattis, Y. M. Liu, and J. J. Filliban, "Resistivity-dopant density relationship for phosphorus-doped silicon," J. Electrochem. Soc.: Solid State Sci. Technol., 12 (8), 1807, 1980.
42. A. S. Grove, Physics and Technology of Semiconductor Devices, John Wiley and Sons, 1967.
43. S. M. Sze, Physics of Semiconductor Devices, John Wiley and Sons, 1981.
44. R. N. Hall, "Electron-hole recombination in germanium," Phys. Rev., 87 (2), 387, 1952.
45. W. Shockley and W. T. Read, "Statistics of the recombination of holes and electrons," Phys. Rev. 87 (5), 835–842, 1952.
46. J. Dziewier and W. Schmid, "Auger recombination coefficients for highly doped and highly excited silicon," Appl. Phys. Lett., 31, 346 (1977).

Chapter 2
Junctions and Contacts

2.1 Introduction

A junction is formed when two dissimilar materials come in contact with each other. The junction between a p-type and n-type semiconductor is called a pn junction. A heterojunction is formed when the semiconductors on both sides of a pn junction are not the same. An example of a heterojunction is when one side is made of silicon and the other of a silicon–germanium alloy. A junction formed between a metal, or a material of metallic character, and a semiconductor is called a contact. The contact is ohmic if it exhibits no barrier to majority carriers in either direction, resulting in a symmetrical current–voltage characteristic with respect to the zero origin. A rectifying contact is asymmetrical, the resistance to current being much larger in one direction than in the other.

The pn junction is the fundamental building block for other silicon devices. The junction shape, profile and characteristics have a direct impact on device parameters. A thorough understanding of the properties of pn junctions is therefore essential to the understanding of the operation of transistors and integrated circuits.

Semiconductor device terminals are brought to the "outside world" by means of contacts and wires. The most important feature of an ohmic contact is its resistance. Reducing the contact resistance becomes increasingly important as the contact size is scaled down. Rectifying contacts, in particular Schottky-barrier diodes are used in specific applications where switching speed is important.

This chapter discusses the physics, technology and characterization of pn junctions and contacts and the relation of their properties to the integrated process.

2.2 PN Junction

Consider two uniformly doped n-type and p-type silicon regions and assume for simplicity that $N_A = N_D$ (Fig. 2.1a). Since the regions are uniformly doped, the

B. El-Kareh, *Silicon Devices and Process Integration: Deep Submicron and Nano-Scale Technologies*,
© Springer Science+Business Media, LLC 2009

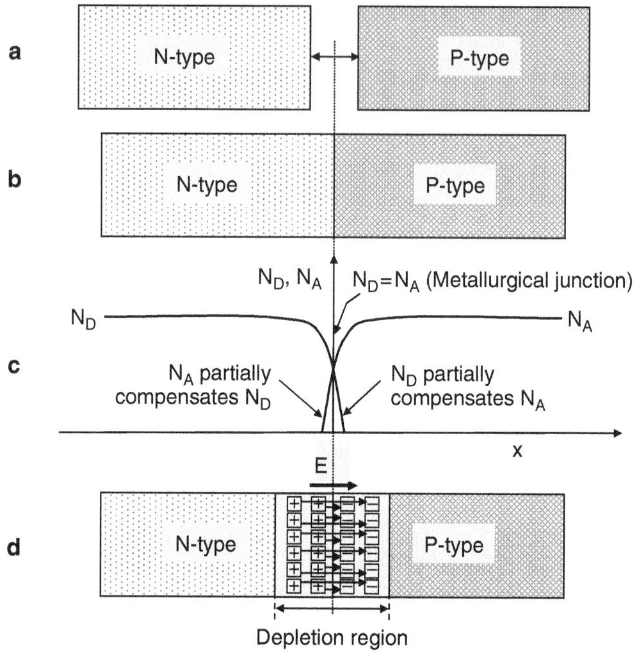

Fig. 2.1 Idealized pn junction. **a** Separated n-type and p-type crystals. **b** Regions brought in contact. **c** Donor and acceptor profiles near metallurgical junction. **d** Space charge (depletion) region at equilibrium; E: Electric field

electric field inside each region is zero; electrons neutralize positively charged donors in the n-regions and holes neutralize negatively charged acceptors in the p-region.

In a thought experiment, let the two regions be instantaneously brought in contact (Fig. 2.1b). Initially, one would expect to find an infinitesimally thin transition region from p-type to n-type material. In reality, however, such abrupt transitions do not exist because, as the junction is formed, the large acceptor and donor concentration gradients at the boundary will cause impurities to diffuse across the boundary. The transition will then have a finite width within which donors and acceptors are partially compensated (Fig. 2.1c). The locus of points where donors and acceptors exactly cancel each other is called the metallurgical junction. Because of the large concentration gradients in electrons and holes, electrons diffuse away from the n-region, where they are majority carriers, into the p-region where they become minority carriers. Similarly, holes diffuse from the p-region to the n-region. The diffusion of electrons leaves behind a space charge of uncompensated, positively charged, fixed donor ions in the n-region. Similarly, the diffusion of holes leaves a negative space charge of uncompensated, fixed acceptor ions in the p-region.

The charges face each other across the boundary. An electric field is thus created causing a drift current in a direction opposite to the diffusion current. At equilibrium, drift and diffusion current components cancel each other for each carrier type.

Although there is a constant traffic of carriers in both directions, the net current is zero. A space charge region is left straddling the metallurgical junction as shown schematically in Fig. 2.1d. This region is also referred to as the depletion region since it contains a negligible number of free carriers compared to the neutral regions. Since the crystal as a whole must remain neutral, the positive space charge must be neutralized by an equal space charge of opposite polarity. This is illustrated in Fig. 2.1d with field lines drawn crossing the metallurgical junction, emanating from positive ions and ending on negative ions.

2.2.1 *Junction Profiles and Shapes*

Junctions of the form in Fig. 2.1 are only found is special cases such as growing a uniformly doped n-type epitaxial layer on a uniformly doped p-type substrate. Most modern junctions, however, are formed by diffusion or implantation followed by an anneal cycle to activate or redistribute impurities and are hence non-uniformly doped. The resulting junction profile and geometrical shape can be complex, requiring sophisticated two-dimensional or even three-dimensional analytical techniques and numerical simulation tools to predict their impurity profiles and electrical properties. These tools must take into account important effects such as concentration-dependent diffusivities, interaction of dopant species with each other, impact of process-induced point defects on diffusivity, and segregation effects near surfaces. There are, however, approximations that can be made on the shape of the impurity profile near the metallurgical junction, depending on the method the junction is formed. These approximations provide a fast and excellent insight into junction properties.

2.2.2 *Step-Junction Approximation*

A step-junction approximation can be made for abrupt transition from p-type to n-type and uniform concentrations in both regions, as shown in Fig. 2.1. A transition is considered abrupt if it occurs in a region that is considerably narrower than the space charge region. When the concentration on one side of the junction is much larger than on the other (e.g., $N_D > 10N_A$), the junction is called a one-side step junction.

2.2.2.1 Linearly Graded Approximation

A junction is graded when the impurity concentration changes gradually from the p-side to the n-side. A linearly-graded approximation can be made when the net impurity concentration changes linearly with distance (Fig. 2.2).

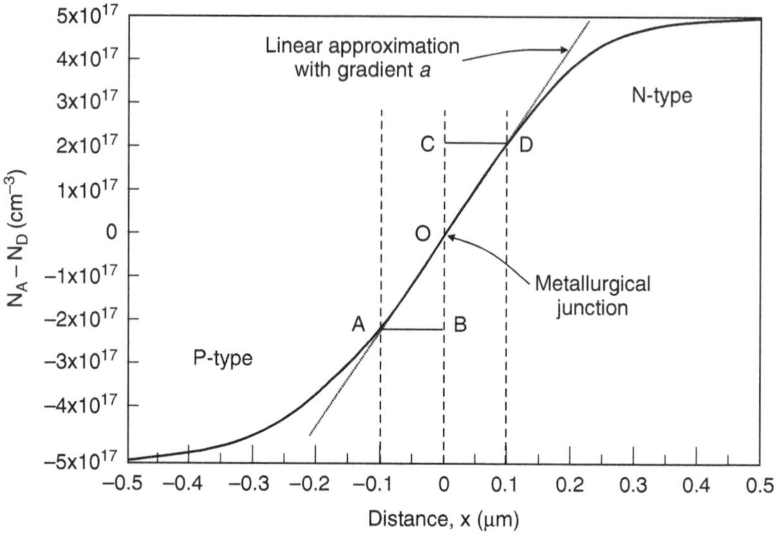

Fig. 2.2 Linearly-graded approximation, assumed symmetrical for simplicity. Triangles ABO and CDO have equal areas

The profile can be described by

$$N_D - N_A = ax \quad cm^{-3}, \tag{2.1}$$

where a is the concentration gradient in cm^{-4} and x the distance from the metallurgical junction. This approximation can be made, for example, for a junction between the *PMOS* drain and the retrograde n-well (Chap. 5), or between the p-type base and the selectively implanted n-type collector of an *NPN* transistor (Chaps. 3 and 7).

2.2.2.2 Gaussian, Two-Sided Gaussian Approximation

A Gaussian approximation can be made when the diffusion source is instantaneous, that is, when a fixed amount of dopants (per cm^2) is deposited and subjected to an anneal temperature whereby it redistributes as a function of time keeping the integrated amount of dopants constant. The resulting one-dimensional Gaussian distribution is of the form

$$N(x,t) = \frac{\phi}{\sqrt{\pi Dt}} e^{-x^2/4Dt} - N_B \tag{2.2}$$

where

$N(x,t)$ = net concentration versus depth as a function of time (cm^{-3}),
ϕ = total amount of dopants, or dose (cm^2),
D = temperature-dependent dopant diffusivity (cm^2/s),
t = time (s),

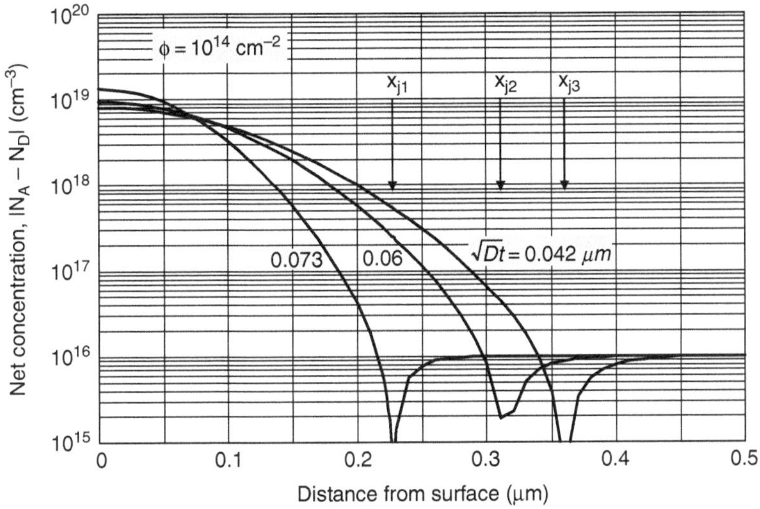

Fig. 2.3 Boron diffusion profiles for a fixed surface dose of $10^{14}\,\text{cm}^{-2}$ and three different thermal cycles. The background concentration is $N_B = 10^{16}\,\text{cm}^{-3}$. The surface concentration drops as Dt increases, but the total amount of dopants remains constant

$x =$ depth (cm),
$N_B =$ background concentration of opposite polarity (cm^{-3}).

Figure 2.3 illustrates the profile of boron diffusion from a fixed surface source into a uniformly doped n-type background. The metallurgical junction is the junction depth x_j.

A Gaussian approximation is also made for, for example, phosphorus implanted at energy E and a dose ϕ, characterized by a projected range and a standard deviation, or straggle. The concentration versus depth is

$$N(x) = N_{peak}e^{-(x-R_p)^2/2[\Delta R_p]^2} - N_B = \frac{\varphi}{\sqrt{2\pi}\Delta R_p}e^{-(x-R_p)^2/2[\Delta R_p]^2} - N_B \tag{2.3}$$

where

$R_\text{p} =$ project range (cm),
$\Delta R_\text{p} =$ standard deviation, or straggle (cm),
$N_\text{peak} =$ peak concentration at R_p (cm^{-3}).

Figure 2.4 illustrates the distribution of a phosphorus n-well implanted into 10 Ohm-cm p-type silicon at a dose of $10^{14}\,\text{cm}^{-2}$ and energy of 500 keV.

A Gaussian distribution with a single straggle as given by (2.3) is applicable to only a few applications. A two-sided Gaussian profile with two straggles ΔR_{p1} and ΔR_{p2} given by

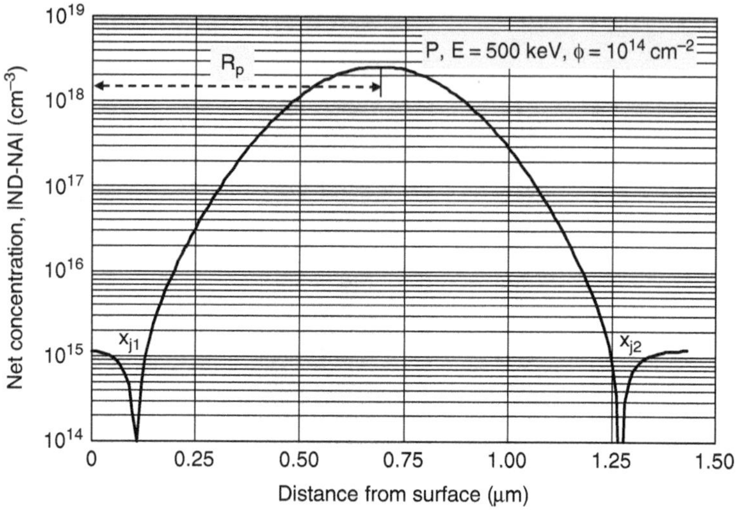

Fig. 2.4 Profile of a phosphorus-implanted N-well into p-type silicon, approximated by a Gaussian distribution

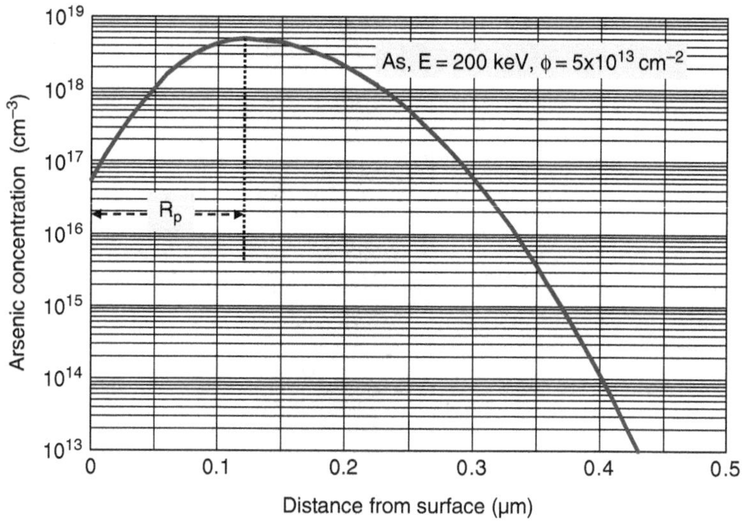

Fig. 2.5 Two-sided Gaussian profile for implanted arsenic

$$N(x) = N_{peak}e^{-(x-R_p)^2/2[\Delta R_{p1}]^2} - N_B \quad \text{cm}^{-3}, \tag{2.4a}$$

$$N(x) = N_{peak}e^{-(x-R_p)^2/2[\Delta R_{p2}]^2} - N_B \tag{2.4b}$$

is a more adequate approximation for a broader range of implanted and annealed profiles [1]. Figure 2.5 illustrates a two-sided Gaussian profile for arsenic implanted at 200 keV at a dose of 5×10^{13} cm^{-2}. The profile is shown skewed to the right.

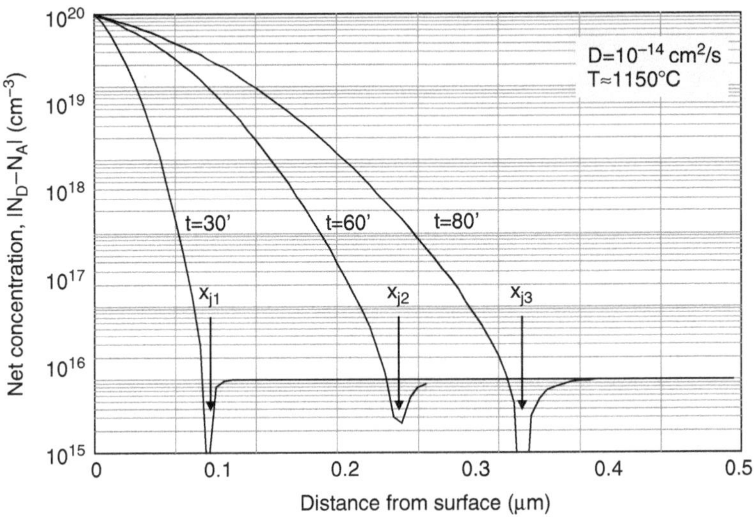

Fig. 2.6 Complementary error-function (erfc) approximation for diffused arsenic at a constant surface concentration of $10^{20}\,\mathrm{cm}^{-3}$, shown for 30, 60 and 80 min

2.2.2.3 Complementary Error-Function (erfc) Approximation

The complementary error function (erfc) approximation is made for the diffusion profile from a constant-concentration source, typically located at the silicon surface. The function is defined as:

$$N(x,t) = N_0 erfc\left(\frac{x}{2\sqrt{Dt}}\right) - N_B \tag{2.5}$$

where N_0 is the fixed source concentration at $x = 0$. The source is continuously replenished to keep its concentration constant. Figure 2.6 illustrates the diffusion profile of arsenic at a fixed temperature of $\sim 1150\,^\circ\mathrm{C}$ and a constant surface concentration of $10^{20}\,\mathrm{cm}^{-3}$. As the diffusion time increases, the junction depth, x_j increases while the surface concentration remains constant.

2.2.2.4 Exponential Profile

Sections of profiles such as the emitter and base of an NPN transistor can be approximated by a simple exponential function as shown in Fig. 2.7. The figure is a replica of a one-dimensional Secondary Ion Mass Spectroscopy (*SIMS*) profile taken through the polysilicon emitter and underlying base. The boron distribution within the polysilicon is irrelevant. Both approximations are of the form

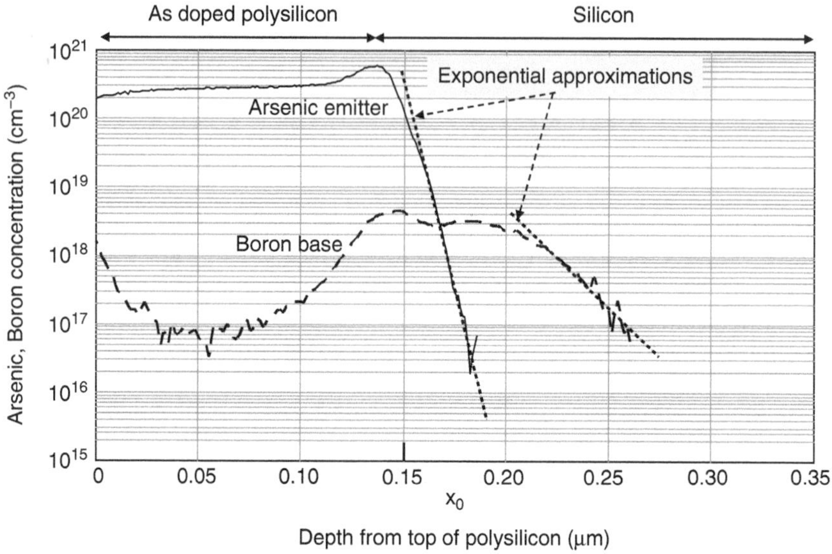

Fig. 2.7 Exponential approximation for sections of the SIMS emitter and base profiles for an NPN transistor

$$N(x) = N_0\, e^{-kx} \quad \text{cm}^{-3}, \tag{2.6}$$

where N_0 is a reference value taken at a reference point labeled as x_0, for example, $3 \times 10^{20}\,\text{cm}^{-3}$ at $x = 0.15\,\mu\text{m}$ for the arsenic-doped emitter in Fig. 2.7. k is a constant that can be extracted from the profile: for the emitter $k \cong 252.2\,\mu\text{m}^{-1}$ and for the base, $k \cong 20.4\,\mu\text{m}^{-1}$.

2.2.2.5 Cylindrical and Spherical Shape Approximations

When dopants are introduced through an opening to form a junction of depth x_j below the surface, the species also distribute laterally. The lateral extent of the metallurgical junction, x_{jl}, is the distance along the silicon surface from the mask edge where impurities are introduced into the crystal (Fig. 2.8a). The junction depth and lateral extent depend on several factors, including temperature and thermal budget, impurity species and concentration, junction size and doping method. The lateral extent is typically smaller than the junction depth, ranging from $x_{jl} \cong 0.5x_j$ to $x_{jl} \cong x_j$. The actual two-dimensional lateral profile can be approximated from electrical measurements. The profile can also be measured directly with sophisticated analytical tools, such as 2-D *SIMS*, or predicted by 2-D or 3-D simulation tools. Figure 2.8b illustrates a junction formed through a rectangular opening and bounded by silicon on all sides.

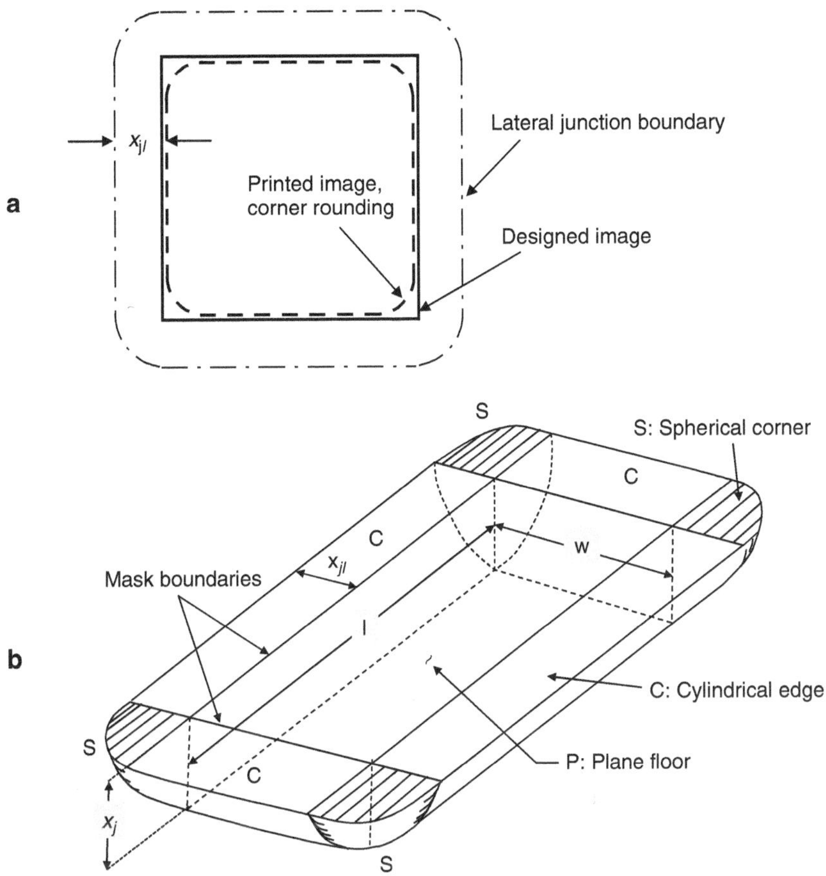

Fig. 2.8 **a** Top view of a junction formed through a square opening illustrating corner rounding and lateral extent. **b** Junction formed through an opening of length l and width w. The junction depth is x_j. Junction edges are approximated by cylindrical shapes, C, and corners by spherical shapes, S [2]

The junction edges are approximated by cylindrical shapes and the corners by spherical shapes [2]. This approximation allows a quick analysis of junction electrical properties in cylindrical or spherical coordinates, as discussed in the next section.

In *MOSFETs* (Metal-Oxide-Silicon-Field-Effect-Transistors, Chap. 5), the source and drain junctions are typically bounded on three sides by dielectric-filled shallow-trench isolation (*STI*). The side facing the channel is approximated by a cylindrical shape. When formed on thin-film silicon-on-insulator (*SOI*, Chap. 7), the junction floor is also bounded by the buried oxide (Fig. 2.9).

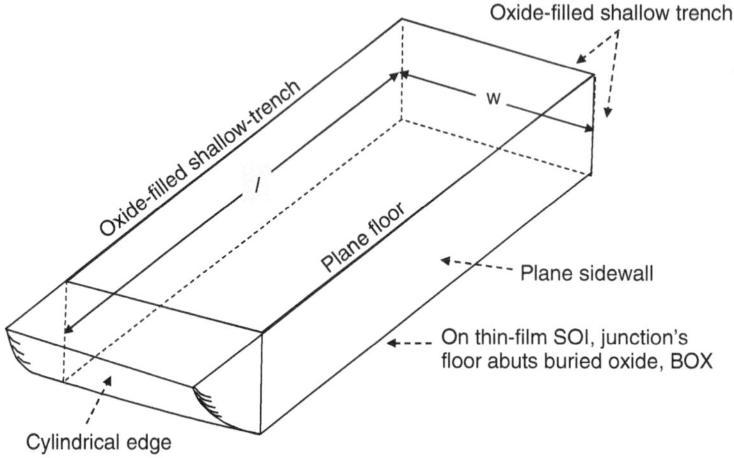

Fig. 2.9 Junction dielectrically bounded on three sides

2.2.3 PN Junction at Thermal Equilibrium

Figure 2.1 is a schematic of a pn junction at thermal equilibrium showing the depletion region consisting of only fixed ionized impurities. It also assumes that the transition from depletion to neutral regions to be abrupt. In reality, the depletion region is not totally void of free carriers (Fig. 2.10), and the transition from neutral to depletion is gradual (Fig. 2.11). An exact analysis of the junction must take these two factors into account. One can, however, make a depletion approximation by defining a depletion region of width x_d that is totally void of free carriers and assume that transitions to the neutral regions are abrupt. The latter assumption is justified by considering that the transition occurs within a region that is very narrow compared to x_d. The depletion approximation brings with it an artificial discontinuity in electric field at the boundaries, but it simplifies the theory without introducing an appreciable error in calculations.

The space charge creates a barrier for electron diffusion from the n-side to the p-side and for hole diffusion from the p-side to the n-side.

The carrier concentrations at the depletion boundaries are assumed to obey the Boltzmann approximation

$$\bar{n}_{p0} = \bar{n}_{n0} e^{-qV_b/kT} \quad \text{cm}^{-3}, \tag{2.7a}$$

$$\bar{p}_{n0} = \bar{p}_{p0} e^{-qV_b/kT} \quad \text{cm}^{-3}, \tag{2.7b}$$

where V_b is the barrier height, also called built-in voltage, $\overline{n}_{p0}, \overline{p}_{n0}$ are the minority electron and hole equilibrium concentrations at depletion boundaries, and $\overline{n}_{n0}, \overline{p}_{p0}$ the majority electron and hole equilibrium concentrations at depletion boundaries. For the junction described by the carrier concentrations in Fig. 2.10 and (2.7) one finds $V_b \simeq 0.7\,\text{V}$ at 300 K.

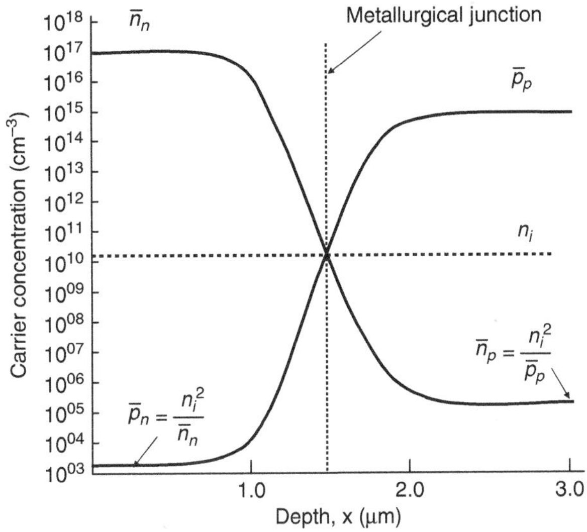

Fig. 2.10 Electron and hole distribution in an abrupt pn junction at thermal equilibrium, with $N_D = 10^{17}\,\text{cm}^{-3}$, $N_A = 10^{15}\,\text{cm}^{-3}$, 300 K

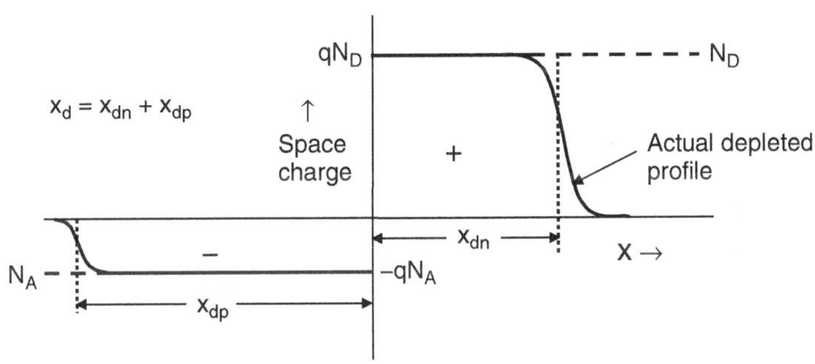

Fig. 2.11 Depletion approximation and depletion width

2.2.3.1 Energy-Band Diagram

The energy-band diagram of a pn junction at thermal equilibrium is drawn by first considering that the Fermi levels on both sides of the junction must align (Fig. 2.12). The position of band edges with respect to the Fermi level is then determined as a function of carrier concentration.

Within the depletion region, the position of band-edge with respect to the Fermi level varies from point to point. For example, the variation of the conduction band edge describes the varying potential energy with respect to an arbitrary reference level that an electron sees as it moves between the points and remains at the band edge. For uniform n-type and p-type regions, the bands are flat outside the depletion

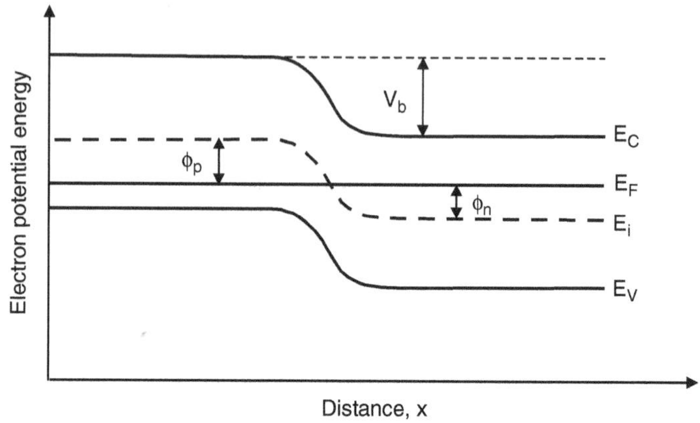

Fig. 2.12 Band-diagram for a pn junction at thermal equilibrium

boundaries. Of primary interest is the potential difference, V_b, between the boundaries of the depletion regions, which is the built-in voltage. The Fermi potential in the n- and p-region is

$$\phi_n = \frac{E_F - E_i}{q} = \frac{kT}{q} \ln \frac{\bar{n}}{n_i} \quad V, \tag{2.8}$$

$$\phi_p = \frac{E_F - E_i}{q} = -\frac{kT}{q} \ln \frac{\bar{p}}{n_i} \quad V. \tag{2.9}$$

The potential difference between the depletion boundaries is

$$V_b = \phi_n - \phi_p = \frac{kT}{q} \ln \frac{\bar{n}.\bar{p}}{n_i^2} \cong \frac{kT}{q} \ln \frac{N_D N_A}{n_i^2}, \tag{2.10}$$

where $\bar{n},\bar{p}$ are the thermal-equilibrium carrier concentrations at the depletion boundaries and N_D,N_A the dopant concentrations at the depletion boundaries. For non-uniformly doped n- and p-regions, the built-in field and resulting potential differences in those regions must be taken into account. In this case, the regions outside the depletion boundaries are described as quasi-neutral. The built-in voltage in (2.10) does not include potential differences outside the depletion boundaries. Depending on dopant concentrations and temperature, the built-in potential can reach values between 0 and approximately 1V. As the temperature increases, V_b decreases because of the rapid increase in n_i.

The existence of a built-in voltage without externally applied bias is sometimes difficult to visualize because the voltage cannot be measured directly with a voltmeter. This difficulty may be overcome by analyzing all contacts in the loop that includes the pn junction and voltmeter. The theory of metal-to-silicon and metal-to-metal contacts shows that a built-in voltage is created at each of the contacts. The loop-sum of all built-in voltages must be 0, otherwise one would draw current without using energy.

2.2.3.2 Poisson's Equation

Poisson's equation states that the divergence of electric field in a volume element is proportional to the space-charge density within that volume[1]

$$div\, E = \frac{\rho}{\varepsilon_0 \varepsilon_{Si}} \tag{2.11}$$

where ρ is the net charge density defined as

$$\rho = q(p - n + N_D^+ - N_A^-) \quad \mathrm{C/cm^3}. \tag{2.12}$$

ε_0 is the permittivity of free space ($\varepsilon_0 = 8.86 \times 10^{-14}\,\mathrm{F/cm}$) and $\varepsilon_{\mathrm{Si}}$ is the dielectric constant of silicon ($\varepsilon_{\mathrm{Si}} = 11.7$). In one dimension, (2.11) is written as

$$\frac{dE}{dx} = \frac{\rho}{\varepsilon_0 \varepsilon_{Si}}. \tag{2.13}$$

The electric field is defined as the force F exerted on a unit charge. For an electron, $E = -F/q$. Since force is also defined as the negative gradient of potential energy and the potential energy of an electron in the conduction band is $\mathrm{E_C}$, the electric field can be defined as

$$\mathrm{E} = \frac{1}{q}\, grad\, \mathrm{E}_C \quad V/\mathrm{cm}. \tag{2.14}$$

In one dimension

$$\boldsymbol{E} = \frac{1}{q}\frac{d\mathrm{E}_C}{dx}. \tag{2.15}$$

The gradient is the same for the conduction band, the valence band and the intrinsic level since the energy gap is assumed to remain constant. In practice, it is convenient to use the gradient of the intrinsic level:

$$\boldsymbol{E} = \frac{1}{q}\frac{d\mathrm{E}_i}{dx}. \tag{2.16}$$

The quantity whose gradient is the negative of electric field is called the electrostatic potential, ϕ. It is defined as

$$\mathrm{E} = -\frac{d\phi}{dx};\ \phi = -\frac{\mathrm{E}_i}{q}. \tag{2.17}$$

[1] Italic E is used for electric field, normal E for energy.

The one-dimensional Poisson's relation can now be defined as

$$\frac{d\mathrm{E}}{dx} = \frac{d^2\mathrm{E}_i}{dx^2} = -\frac{d^2\phi}{dx^2} = \frac{\rho}{\varepsilon_0\varepsilon_{Si}}. \tag{2.18}$$

2.2.3.3 Depletion Region

Poisson's relation and the Boltzmann approximations are now applied to establish the field and voltage distributions within the depletion region, and the depletion width.

Step-Junction

The electric field distribution can be visualized by considering field lines emanating on positive ions in the n-sided depletion region and ending on negative ions in the p-sided depletion region, and imagining a plane parallel to the metallurgical junction moving from the n-side to the p-side (Fig. 2.13).

Within the neutral n-region the number of field-lines crossing the plane is, on the average, zero. As the plane moves into the depletion region, the density of field lines that cross it, and hence the field intensity, increases linearly. The field peaks when the plane reaches the metallurgical junction and then begins to decrease linearly to zero as the plane moves toward the neutral p-region (Fig. 2.14).

The space charge density in the four regions is:

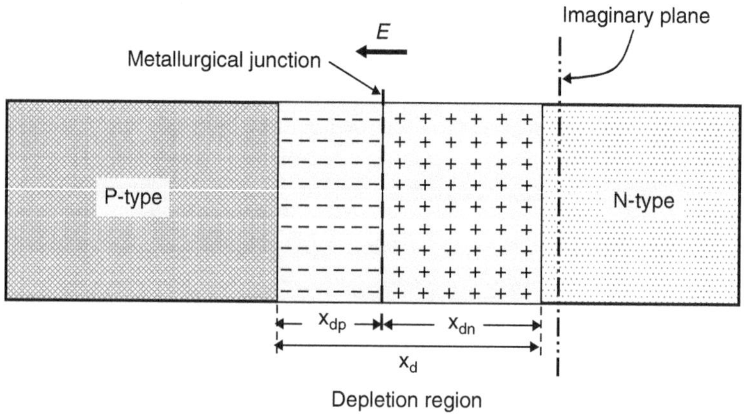

Fig. 2.13 Schematic of an abrupt pn junction

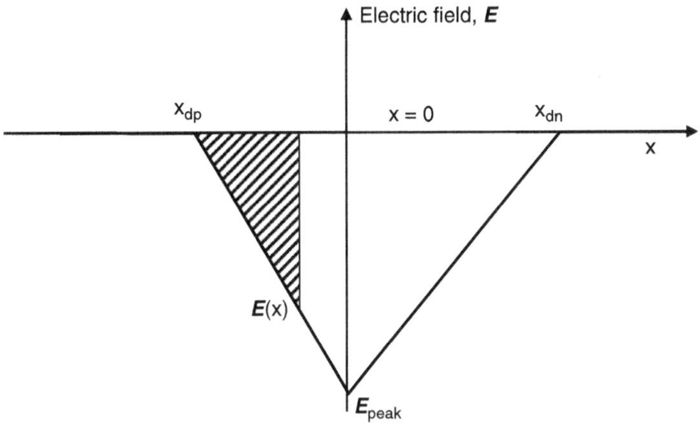

Fig. 2.14 Field distribution in an abrupt pn junction

$$x \geq x_{dn} \text{ and } x \leq x_{dp} \quad \rho(x) = 0, \tag{2.19a}$$
$$0 < x < x_{dn} \quad \rho(x) = qN_D > 0, \tag{2.19b}$$
$$x_{dp} < x < 0 \quad \rho(x) = -qN_A < 0. \tag{2.19c}$$

The boundary conditions for Poisson's equation are:

a) The fields on both sides of the 0-plane must be equal to ensure voltage continuity at $x = 0$,
b) The fields must vanish at $x = x_{dn}$ and $x = x_{dp}$.

From the first condition

$$\left|\frac{dV}{dx}\right|_{0-} = \left|\int_{x_{dp}}^{0} \frac{\rho(x)}{\varepsilon_0 \varepsilon_{Si}} dx\right| = \left|\frac{dV}{dx}\right|_{0+} = \left|\int_{0}^{x_{dn}} \frac{\rho(x)}{\varepsilon_0 \varepsilon_{Si}} dx\right|. \tag{2.20}$$

This simply means that the total charge in the depletion layers on either side of the junction must be equal and of opposite polarity. For a step junction, the integrals simplify to

$$\frac{\int_{x_{dp}}^{0} \rho(x)dx}{\varepsilon_0 \varepsilon_{Si}} = -\frac{qN_A x_{dp}}{\varepsilon_0 \varepsilon_{Si}} = \frac{\int_{0}^{x_{dn}} \rho(x)dx}{\varepsilon_0 \varepsilon_{Si}} = \frac{qN_D x_{dn}}{\varepsilon_0 \varepsilon_{Si}}. \tag{2.21}$$

Poisson's equation is integrated twice to obtain the voltage. The first integration gives the electric field

$$\frac{dV}{dx} = -\frac{qN_D}{\varepsilon_0 \varepsilon_{Si}} x + C_1, \tag{2.22a}$$
$$\frac{dV}{dx} = \frac{qN_A}{\varepsilon_0 \varepsilon_{Si}} x + C_2, \tag{2.22b}$$

where C_1 and C_2 are constants. The second boundary condition defines the constants as

$$C_1 = -\frac{qN_A}{\varepsilon_0 \varepsilon_{Si}} x_{dp}; \quad C_2 = \frac{qN_D}{\varepsilon_0 \varepsilon_{Si}} x_{dn}.$$

Substituting in (2.22) gives

$$\frac{dV}{dx} = \frac{qN_A}{\varepsilon_0 \varepsilon_{Si}}(x - x_{dp}) \quad p\text{–}side, \tag{2.23a}$$

$$\frac{dV}{dx} = -\frac{qN_D}{\varepsilon_0 \varepsilon_{Si}}(x - x_{dn}) \quad n\text{–}side. \tag{2.23b}$$

Integrating once more gives the voltage distribution within the depletion regions as

$$V = \frac{qN_A}{\varepsilon_0 \varepsilon_{Si}} \left[\frac{x^2}{2} - x_{dp}.x\right] + C \quad p\text{–}side, \tag{2.24a}$$

$$V = -\frac{qN_D}{\varepsilon_0 \varepsilon_{Si}} \left[\frac{x^2}{2} - x_{dn}.x\right] + C \quad n\text{–}side. \tag{2.24b}$$

The voltage distribution in the space charge region consists of two parabolas. The integration constants must be equal because of the boundary condition that the voltage must be the same coming from either side of $x = 0$. The voltages at $x \leq x_{dp}$ and $x \geq x_{dn}$ are

$$V_p = -\frac{qN_A}{2\phi_0 \varepsilon_{Si}} x_{dp}^2 + C, \tag{2.25a}$$

$$V_n = \frac{qN_D}{2\varepsilon_0 \varepsilon_{Si}} x_{dn}^2 + C. \tag{2.25b}$$

The built-in voltage is

$$V_b = V_n - V_p = \frac{q}{2\varepsilon_0 \varepsilon_{Si}}(N_d x_{dn}^2 + N_A x_{dp}^2). \tag{2.26}$$

The built-in voltage found in (2.26) must be the same as in (2.10).
The depletion widths are found by combining (2.21) and (2.26):

$$x_{dn} = \sqrt{\frac{2\varepsilon_0 \varepsilon_{Si} V_b}{q} \frac{N_A}{N_A N_D + N_D^2}} \quad \text{cm}, \tag{2.27}$$

$$x_{dp} = \sqrt{\frac{2\varepsilon_0 \varepsilon_{Si} V_b}{q} \frac{N_D}{N_A N_D + N_A^2}} \quad \text{cm}. \tag{2.28}$$

The total depletion width is:

$$x_d = x_{dn} + x_{dp} = \sqrt{\frac{2\varepsilon_0 \varepsilon_{Si} V_b}{q} \left[\frac{1}{N_A} + \frac{1}{N_D}\right]} \quad \text{cm}. \tag{2.29}$$

When $N_D \gg N_A$, the junction is described as one-sided abrupt and (2.29) simplifies to

$$x_d = x_{dn} + x_{dp} \cong x_{dp} \cong \sqrt{\frac{2\varepsilon_0 \varepsilon_{Si} V_b}{qN_A}} \cong \sqrt{\frac{1.3x10^7 V_b}{N_A}}. \tag{2.30}$$

Similarly, when $N_A \gg N_D$ (2.29) simplifies to

$$x_d = x_{dn} + x_{dp} \cong x_{dn} \cong \sqrt{\frac{2\varepsilon_0 \varepsilon_{Si} V_b}{qN_D}} \cong \sqrt{\frac{1.3x10^7 V_b}{N_D}}. \tag{2.31}$$

Therefore, when one side of the junction is heavily doped, the depletion region expands almost entirely into the lightly doped region.

The magnitude of electric field in the depletion region peaks at the metallurgical junction. The magnitude of peak field is

$$\left|E_{peak}\right| = \left|\frac{q}{\varepsilon_0 \varepsilon_{Si}} \int_0^{x_{dn}} N_D dx\right| = \left|\frac{q}{\varepsilon_0 \varepsilon_{Si}} \int_0^{x_{dn}} N_A dx\right| \quad V/\text{cm}. \tag{2.32}$$

The built-in voltage is found by integrating the electric field from x_{dp} to x_{dn}. For a step junction, it is the area of the triangle bounded by x_{dp}, E_{peak} and x_{dn} (Fig. 2.14). The relation between peak field and built-in voltage is then

$$E_{peak} = \frac{2V_b}{x_d}. \tag{2.33}$$

Small-Signal Junction Capacitance

Small-signal means that the ac voltage applied to measure capacitance is negligible compared to the built-in voltage, allowing measurement of capacitance with minimum disturbance of depletion conditions. A typical signal is a sine-wave of amplitude $\Delta V = 10$–$15\,\text{mV}$ and 1 kHz–1MHz frequency. Since the depletion region is practically void of free carriers, it can be considered as an insulator of dielectric constant ε_{Si} sandwiched between a p-type and n-type conductor. Under the assumption that the charge variation induced by the signal occurs at the depletion boundaries, as indicated in Fig. 2.15, the system can be treated as a parallel-plate capacitor. During one half of the signal, ΔV has the same polarity as V_b. The voltage across the depletion regions is then $V_b + \Delta V$ and x_{dp} and x_{dn} increase by Δx_{dp} and Δx_{dn}, exposing additional charge $-\Delta Q$ on the p-side and $+\Delta Q$ on the n-side (Fig. 2.15). Similarly, during the second half of the signal where ΔV is of opposite polarity to V_b, the depletion widths decrease by Δx_{dp} and Δx_{dn}, reducing the total charge by ΔQ at each depletion boundary.

The junction capacitance per unit area is[2]

[2] Unless otherwise stated, C denotes capacitance per unit area, cm^{-2}.

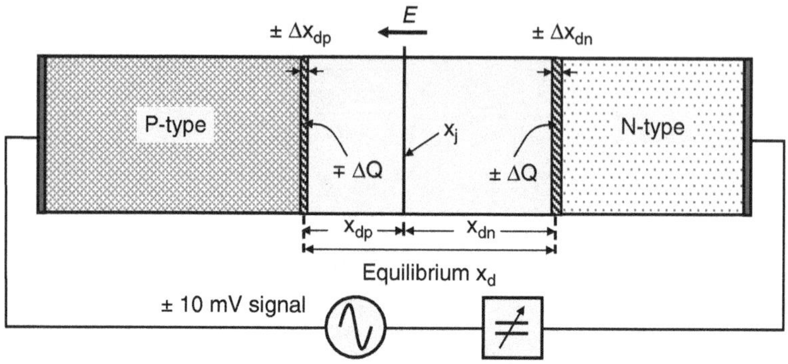

Fig. 2.15 Variation of space charge at the depletion boundaries in response to a small voltage signal

$$C_j = \frac{\varepsilon_0 \varepsilon_{Si}}{x_d} \quad \text{F/cm}^2. \tag{2.34}$$

It follows from (2.30) and (2.31) that the capacitance of a one-sided abrupt junction is

$$C_j = \sqrt{\frac{\varepsilon_0 \varepsilon_{Si} q N}{2V_b}}, \tag{2.35}$$

where N is the impurity concentration on the lightly-doped side.

Linearly Graded Junction

The linearly-graded junction is characterized by a concentration gradient a (2.1). Assuming the linear approximation to be valid throughout x_d, the concentrations at the depletion boundaries are found from the triangles in Fig. 2.2 as

$$N_A = N_D = \frac{1}{2} a x_d. \tag{2.36}$$

The Boltzmann approximation gives

$$V_b = \frac{kT}{q} \ln \frac{a^2 x_d^2}{4n_i^2}. \tag{2.37}$$

From the Poisson relation

$$V_b = \frac{q a x_d^3}{12 \varepsilon_0 \varepsilon_{Si}}. \tag{2.38}$$

Combining (2.37) and (2.38) results in

$$V_b = \frac{2kT}{3q} \ln \frac{3 \varepsilon_0 \varepsilon_{Si} a^2 V_b}{2 q n_i^3}. \tag{2.39}$$

Equation (2.39) can be solved iteratively to find V_b. The depletion width is then found from (2.38)

$$x_d = \left(\frac{12\varepsilon_{Si}V_b}{qa}\right)^{1/3} \quad \text{cm}. \tag{2.40}$$

The field distribution is found as (Fig. 2.16)

$$E(x) = \frac{qa}{2\varepsilon_0\varepsilon_{Si}}\left(x^2 - \frac{x_d^2}{4}\right). \tag{2.41}$$

The peak field is

$$E_{peak} = -\frac{qax_d^2}{8\varepsilon_0\varepsilon_{Si}} = \frac{3}{2}\frac{V_b}{x_d} \quad V/\text{cm}. \tag{2.42}$$

Comparing (2.42) to (2.33) shows that for equal depletion widths, the linearly-graded junction exhibits a lower peak field than the abrupt junction.

Integrating (2.41) with respect to x, subject to the arbitrary boundary condition that $V = 0$ at $x = -x_d/2$, the voltage as a function of distance is (Fig. 2.16)

$$V(x) = \frac{qa}{6\varepsilon_0\varepsilon_{Si}}\left[3\left(\frac{x_d}{2}\right)^2 x + 2\left(\frac{x_d}{2}\right)^3 - x^3\right]. \tag{2.43}$$

The small-signal capacitance per unit area is

$$C_j' = \frac{\varepsilon_0\varepsilon_{Si}}{x_d} = \left(\frac{\varepsilon_{Si}^2 qa}{12V_b}\right)^{1/3} \quad \text{F/cm}^2. \tag{2.44}$$

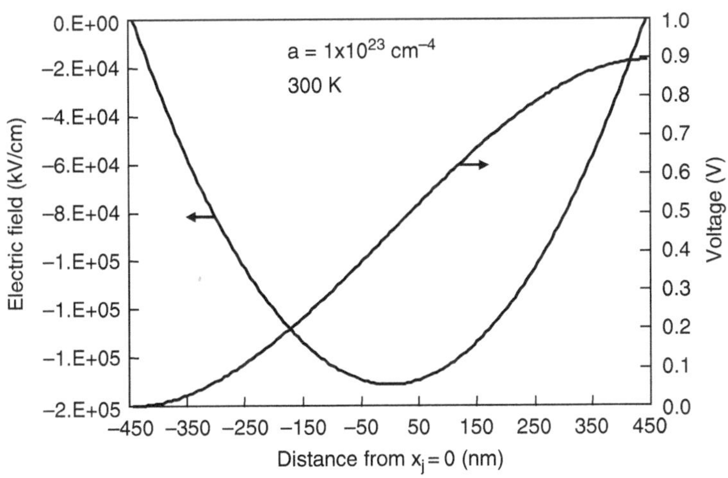

Fig. 2.16 Electric field and voltage distribution in a linearly-graded junction

Arbitrary Junction Profile

This section describes a numerical technique that can be applied to an arbitrary one-dimensional junction profile. Figure 2.17 shows a *PMOS* source to n-well profile for illustration. The second junction in the figure is formed between the n-well and a uniformly doped ~10 Ω-cm p-type substrate (*PMOS* and n-well is described in Chaps. 5 and 7).

The concentration versus depth is typically reported in the form of a graph or two columns in a table, taking small steps in depth, Δx. The junction, x_j, is found at the depth where N_A and N_D exactly cancel each other. To find the depletion boundaries:

1. Take the average of two consecutive net concentrations and multiply by $q.\Delta x$. This column yields the incremental charge in both sides of x_j.

$$\Delta Q_{p-side} = -q\Delta x\bar{N}_A; \Delta Q_{n-side} = q\Delta x\bar{N}_D \quad \mathrm{C/cm^2}, \tag{2.45a}$$

   where $\overline{N}_A, \overline{N}_D$ are the average net acceptor and donor concentrations in the interval Δx.
2. Multiply ΔQ by the average distance $\bar{x}$ of point x from x_j and divide by the dielectric constant. This column gives the incremental voltage ΔV in both sides of the junction:

$$\Delta V = \frac{\Delta Q}{\varepsilon_0 \varepsilon_{Si}} \bar{x} \quad V. \tag{2.45b}$$

3. Take the sum of all incremental voltages as a function of distance to x_j on both sides of the junction:

$$V(x)_{p-side} = \sum_{x}^{x_j} \Delta V(x); \; V(x)_{n-side} = \sum_{x_j}^{x} \Delta V(x). \tag{2.45c}$$

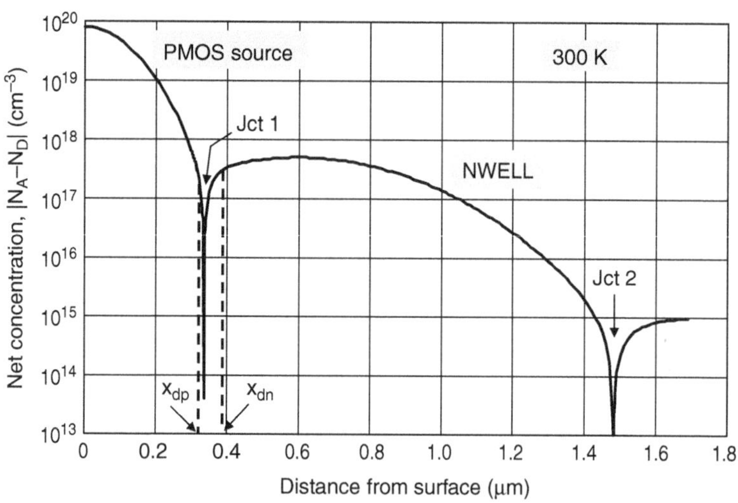

Fig. 2.17 A one-dimensional profile of a *PMOS* source or drain to n-well junction

4. Calculate the Fermi potential as a function of distance:

$$\phi_p(x) = -\frac{kT}{q}\ln\frac{\bar{N}_A(x)}{n_i};\ \phi_n(x) = \frac{kT}{q}\ln\frac{\bar{N}_D(x)}{n_i}. \tag{2.45d}$$

5. The points where $V(x)_{p-side} = \phi_p(x); V(x)_{n-side} = \phi_n(x)$ are the depletion boundaries, x_{dp}, and x_{dn}.
6. Verify neutrality by comparing the total charge on both sides of the junction:

$$\sum_{x_{dp}}^{x_j} \Delta Q(x) = -\sum_{x_j}^{x_{dn}} \Delta Q(x). \tag{2.45e}$$

Applied to Junction 1 in Fig. 2.17, the method defines x_{dp} and x_{dn}, respectively, at a distance of ~35 nm and ~50 nm from xj. The built-in voltage is found as $V_b = |V(x_{dn})| + |V(x_{dp})| = 0.894\,\text{V}$.

Effect of Curvature

A cylindrical junction-edge approximation is used for illustration (Fig. 2.18). The charge Q per unit area is higher at r_j than in the plane portion at x_j. This can be shown by comparing the conical section ABCD to the rectangular section A′B′C′D′ of the same metallurgical junction area. Since the electric field is proportional to Q, it will be higher on corners and edges than in the plane portion. The field increases as the edge or corner radius of curvature is reduced. This is sometimes referred to as the lightning-rod effect.

To first approximation, a one-sided abrupt n^+p junction of cylindrical edge of radius of curvature $r_j \cong x_j$ is assumed (Fig. 2.18).[3] Since the n-region is heavily

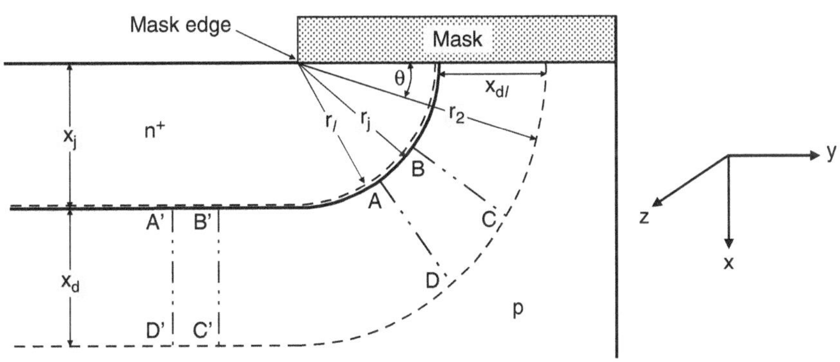

Fig. 2.18 Cylindrical junction approximation

[3] The superscript "+" indicates that the n-region is much higher doped than the p-region. The following qualifiers are used for the concentration ranges: n, p: 5×10^{15}–$10^{18}\,\text{cm}^{-3}$; n^+, p^+ : 10^{18}–$10^{20}\,\text{cm}^{-3}$; n^{++}, p^{++} :> $10^{20}\,\text{cm}^{-3}$; n^-, p^- : $5\times10^{14} - 5\times10^{15}\,\text{cm}^{-3}$; n^{--}, p^{--} :< $5\times10^{14}\,\text{cm}^{-3}$.

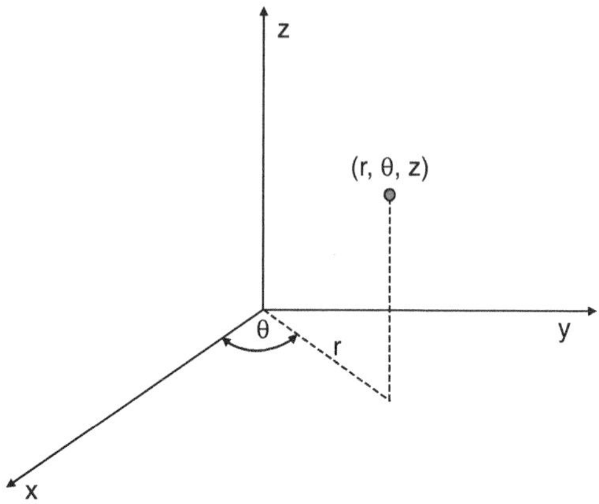

Fig. 2.19 Definition of cylindrical coordinates

doped, $x_{dn} \ll x_{dp}$ and $r_l \cong r_j$. Poisson's equation in Cartesian coordinates is

$$\nabla^2 V = \frac{\partial^2 V}{\partial x^2} + \frac{\partial^2 V}{\partial y^2} + \frac{\partial^2 V}{\partial z^2}, \tag{2.46}$$

and in cylindrical coordinates

$$\nabla^2 V = \frac{\partial^2 V}{\partial r^2} + \frac{1}{r}\frac{\partial V}{\partial r} + \frac{1}{r^2}\frac{\partial^2 V}{\partial \theta^2} + \frac{\partial^2 V}{\partial z^2}, \tag{2.47}$$

where x, y, z, θ and r are defined in Fig. 2.19.

Since the potential is symmetrical with respect to z and θ

$$\frac{\partial V}{\partial \theta} = 0; \frac{\partial V}{\partial z} = 0.$$

For $r_j \leq r \leq r_2 \rho \cong qN_A$ (2.47) simplifies to

$$\nabla^2 V = \frac{1}{V}\frac{d}{dr}\left(r\frac{dV}{dr}\right) = -\frac{qN_A}{\varepsilon_{Si}}. \tag{2.48}$$

Integrating (2.48) with the boundary conditions $\mathrm{E} = \mathrm{E}_{\mathrm{peak}}$ at $r = r_j$ and $\mathrm{E} = 0$ at $r = r_2$ gives

$$E_{peak} = \int_{r_j}^{r_2} \frac{qN_A}{\varepsilon_0 \varepsilon_{Si}} dr = \frac{qN_A}{2\varepsilon_{Si}} \frac{r_2^2 - r_j^2}{r_j}. \tag{2.49}$$

A second integration with the boundary conditions $V = 0$ at $r = r_j$, $V = V_b$ at $r = r_2$ gives [3]

$$V_b = \frac{qN_A}{2\varepsilon_0\varepsilon_{Si}}\left(r_2^2 \ln\frac{r_2}{r_j} - \frac{r_2^2 - r_j^2}{2}\right). \tag{2.50}$$

For an n$^+$ p junction, the built-in voltage V_b is

$$V_b \cong 0.55 + \frac{kT}{q}\ln\frac{N_A}{n_i}.$$

Equation (2.49) can be solved iteratively for r_2. The resulting depletion width is narrower at the edge than in the plane part of the junction.

A similar analysis can be made for spherical corners using spherical coordinates.

The edge of the junction constitutes one quarter of the assumed cylinder. The capacitance per unit edge-length is found by treating the cylinder as a coaxial transmission line, with the n$^+$ region as the core conductor and the p-region as the mantel, separated by the depletion layer as the "insulator" of dielectric constant ε_{Si}. The capacitance per unit edge-length is

$$C_{edge} = \frac{\pi\varepsilon_0\varepsilon_{Si}}{2\ln(r_2/r_j)} \quad F/\text{cm}. \tag{2.51}$$

From the above analysis it is concluded that the capacitance *per unit area* is larger at the junction edge than in its flat portion.

2.2.4 PN Junction in Forward Bias

At thermal equilibrium, the drift and diffusion current components are exactly balanced for each type of carrier and the net current crossing the junction is zero. When a bias voltage is applied to the junction, thermal equilibrium is disturbed. If the voltage remains constant, a non-equilibrium steady-state is established in a short time.

This section describes the junction characteristics under forward bias, that is, when the voltage on the n-side is negative with respect to the p-side or when the bias on the p-side is positive with respect to the n-side (Fig. 2.20). A forward voltage V_F reduces the barrier height, increasing the diffusion current component and reducing the drift component. The result is a considerable increase in the carrier concentrations because more carriers cross the junction by diffusion. Therefore, under forward bias $pn > n_i^2$.

The carriers diffuse from regions where they are majority carriers to regions where they become minority carriers (Fig. 2.21). This is called minority-carrier injection. Since at any time the region must remain electrically neutral, there will be a corresponding increase in majority carriers in each region to neutralize the injected excess minority carriers. The majority carriers are supplied by the contacts to neutralize and also recombine with minority carriers. The total current crossing any plane through the junction must be the same for all planes through the junction.

When the level of injected minority carriers is small compared to the majority-carrier concentration, one speaks of low-level injection. For example, the excess

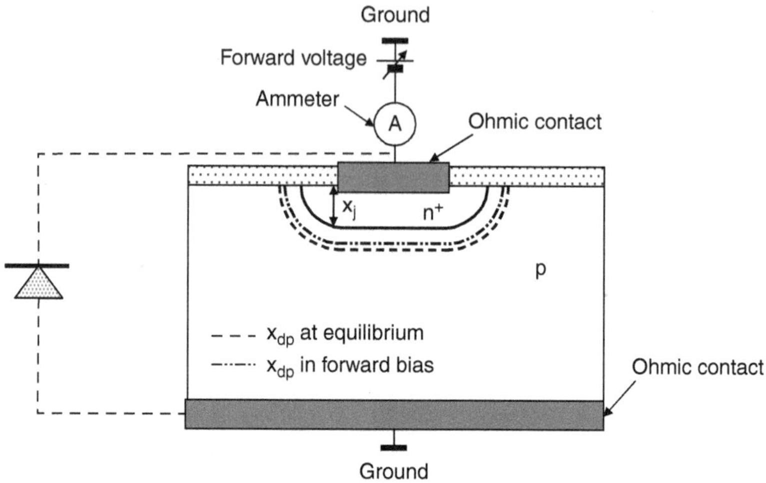

Fig. 2.20 PN junction under forward bias

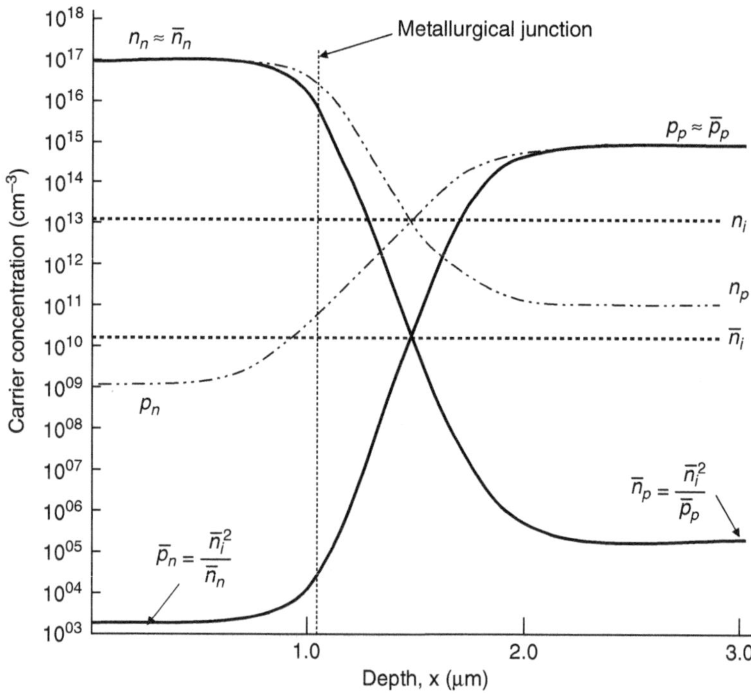

Fig. 2.21 Electron and hole distribution in an abrupt pn junction under forward bias

minority-carrier electron concentration in Fig. 2.21 is about 10^{11} cm^{-3}. This is about six orders of magnitude higher than the thermal-equilibrium electron concentration but about four orders of magnitude smaller than the majority-carrier hole concentration. A high-level injection condition occurs when the injected minority-carrier concentration becomes comparable to the thermal-equilibrium majority-carrier concentration.

2.2.4.1 Low-Level Injection

To simplify the discussion, it is assumed that the voltage seen at the depletion boundaries is the same as applied to the junction contacts, that is, there are no voltage drops in the regions outside the depletion. It is also assumed that there is no carrier loss due to recombination within the depletion region. The latter assumption is realistic since the carriers travel at high-speed (velocity saturation) through the depletion region.

At thermal equilibrium, the carrier concentrations at the depletion boundaries follow the Boltzmann distribution function as

$$\bar{n}_{p0} = \bar{n}_{n0} e^{-qV_b/kT} \quad \mathrm{cm}^{-3}, \tag{2.52}$$

$$\bar{p}_{n0} = \bar{p}_{p0} e^{-qV_b/kT} \quad \mathrm{cm}^{-3}. \tag{2.53}$$

In (2.52) and (2.53) the bar denotes thermal equilibrium and the subscript "0" depletion boundaries.

When a forward bias voltage V_F is applied, the barrier height is reduced to $V_b - V_F$. Since low-level injection is assumed, the majority-carrier concentrations are practically undisturbed and the new concentrations are

$$n_{p0} \cong \bar{n}_{n0} e^{-q(V_b - V_F)/kT}, \tag{2.54}$$

$$p_{n0} \cong \bar{p}_{p0} e^{-q(V_b - V_F)/kT}. \tag{2.55}$$

Combining (2.52), (3.53), (2.54) and (2.55) gives

$$n_{p0} \cong \bar{n}_{p0} e^{qV_F/kT}, \tag{2.56}$$

$$p_{n0} \cong \bar{p}_{p0} e^{qV_F/kT}. \tag{2.57}$$

Therefore, the excess minority carrier concentrations are

$$\Delta n_{p0} \cong \bar{n}_{p0}(e^{qV_F/kT} - 1) = \frac{n_i^2}{\bar{p}_{p0}}(e^{qV_F/kT} - 1), \tag{2.58}$$

$$\Delta p_{n0} \cong \bar{p}_{p0}(e^{qV_F/kT} - 1) = \frac{n_i^2}{\bar{n}_{n0}}(e^{qV_F/kT} - 1). \tag{2.59}$$

Assuming that all dopants are ionized, (2.58) and (2.59) simplify to

$$\Delta n_{p0} \cong \frac{n_i^2}{N_A}(e^{qV_F/kT} - 1), \tag{2.60}$$

$$\Delta p_{n0} \cong \frac{n_i^2}{N_D}(e^{qV_F/kT} - 1), \tag{2.61}$$

where N_A, N_D are, respectively, the acceptor and donor concentrations at the depletion boundaries. The ratio of excess minority carriers is

$$\frac{\Delta n_{p0}}{\Delta p_{n0}} \cong \frac{\bar{n}_{n0}}{\bar{p}_{p0}} \cong \frac{N_D}{N_A}. \tag{2.62}$$

The excess-carrier concentrations induced by a forward voltage V_F are in the same ratio as the majority equilibrium concentrations in the regions from which they are injected and are in the same ratio of minority-carrier concentrations in the regions into which they are injected. This result is of basic significance to the operation of bipolar transistors.

2.2.4.2 The Quasi-Fermi Level

In Chap. 1, the thermal equilibrium electron and hole concentrations were found as

$$\bar{n} = n_i e^{q\phi_b/kT}, \tag{2.63a}$$

$$\bar{p} = n_i e^{-q\phi_b/kT}, \tag{2.63b}$$

where ϕ_b is the Fermi potential defined as $(\mathrm{E_F} - \mathrm{Ei})/kT$. In the presence of injected carriers, the above relations do not apply and the Fermi level $\mathrm{E_F}$ loses its meaning. It is, however, convenient to define separate levels for electrons and holes that, when substituted in equations similar to (2.63a) and (2.63b) give the actual carrier concentration under biasing conditions. These levels are called the electron and hole quasi-Fermi levels, $\mathrm{E_{Fn}}$ and $\mathrm{E_{Fp}}$. The carrier concentrations are then expressed as

$$n = n_i e^{q(E_{Fn} - E_i)/kT}, \tag{2.64a}$$

$$p = n_i e^{q(E_i - E_{Fp})/kT}. \tag{2.64b}$$

The split in Fermi-levels in the presence of excess carriers is illustrated in Fig. 2.22.

Since the equilibrium concentrations are increased by the same amount, that is, $\Delta n = \Delta p$, the shift in the minority quasi-Fermi level is much larger than the majority quasi-Fermi level shift, as seen from (2.65).

$$|E_{Fn} - E_F| = kT\ln\left(1 + \frac{\Delta n}{\bar{n}}\right); |E_{Fp} - E_F| = kT\ln\left(1 + \frac{\Delta p}{\bar{p}}\right). \tag{2.65}$$

The quasi-Fermi levels show large gradients for minority carriers in the injection region. The higher the electron and hole current densities, j_n and j_p, the nearer the quasi-Fermi levels approach the respective band-edges. Within the space-charge

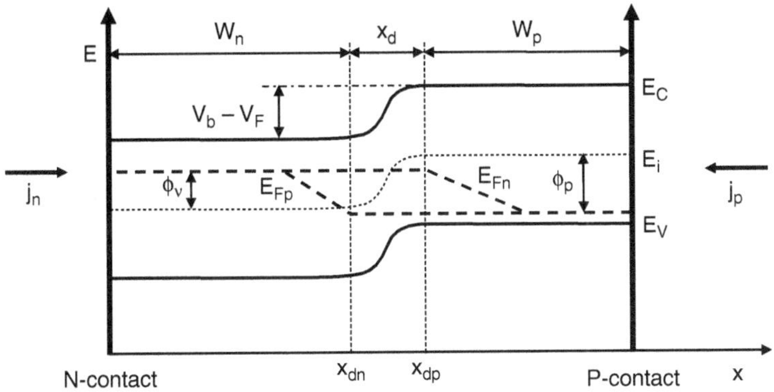

Fig. 2.22 Split of Fermi-level into quasi-Fermi levels E_{Fn} and E_{Fp} under forward bias

region, E_{Fn} and E_{Fp} are constant as a direct consequence of the assumption that no recombination occurs within this region. For wide regions W_n and W_p compared to the minority-carrier diffusion lengths L_p and L_n, that is, $W_n \gg L_p$ and $W_p \gg L_n$, all excess minority carriers recombine before they reach the contacts. The majority-carrier quasi-Fermi level practically coincides with its thermal equilibrium value, indicating that the field created by the injected carriers is negligible. The regions outside the depletion boundaries are described as quasi-neutral, indicating that the field is negligible but not zero.

2.2.4.3 Current–Voltage Relation

The one-dimensional continuity equation in steady-state is defined for minority-carrier electrons and holes as (Chap. 1, Sect. 1.6.2)

$$\frac{d^2 p_n}{dx^2} = \frac{\Delta p_n}{L_p^2}, \tag{2.66a}$$

$$\frac{d^2 n_p}{dx^2} = \frac{\Delta n_p}{L_n^2}. \tag{2.66b}$$

Consider, for example, the continuity relation for holes. The general solution is of the form

$$\Delta p_n(x) = C_1 e^{-x/L_p} + C_2 e^{x/L_p}, \tag{2.67}$$

where C_1 and C_2 are constants of integration to be determined. Since all excess carriers recombine at the ohmic contact placed at $x = W_n$, majority and minority carriers are maintained at their thermal equilibrium values at the contact, that is, $p_n = \overline{p}_n$. The excess hole concentration Δp_{n0} at $x = x_{dn}$ is given by (2.59). Defining $x = 0$ at x_{dn} and solving 2.66a subject to the above boundary conditions at $x = 0$ and $x = W_n$ gives

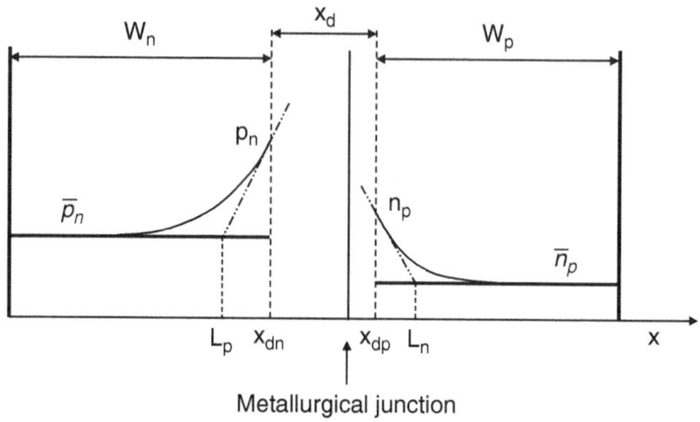

Fig. 2.23 Decay of excess minority concentration in wide regions. Slopes taken at $x = x_{\mathrm{dn}}$, $x = x_{\mathrm{dp}}$. Depletion greatly exaggerated

$$\Delta p_n = \bar{p}_{n0}(e^{qV_F/kT} - 1)\frac{\sinh(W_n - x/L_p)}{\sinh(W_n/L_p)} \quad \mathrm{cm}^{-3}. \tag{2.68}$$

The hole diffusion current density is given in one dimension by (Chap. 1, Sect. 1.5.4)

$$j_p = -qD_p\frac{d\Delta p}{dx} \quad A/\mathrm{cm}^2. \tag{2.69a}$$

The excess minority carriers decay with distance due to recombination with majority carriers (Fig. 2.23). One can assume that at $x = 0$ (the depletion boundary) the current consists of minority-carrier diffusion only, and as x increases, the majority-carrier current increases while the minority-carrier current decreases. The total current must be the same at any plane through the structure. As x approaches the contacts at W_n and W_p, the carrier concentrations must be at equilibrium values and the current consists of only majority carriers.

Let the boundary x_{dn} be at $x = 0$ for simplicity. Differentiating (2.68) at $x = 0$ and combining with (2.69a) gives the hole diffusion current density as

$$j_p = \frac{qD_p\bar{p}_n}{L_p\tanh(W_n/L_p)}(e^{qV_F/kT} - 1). \tag{2.69b}$$

A similar relation can be derived for minority carrier electron diffusion current density in the p-region

$$j_n = \frac{qD_n\bar{n}_p}{L_n\tanh(W_p/L_n)}(e^{qV_F/kT} - 1). \tag{2.69c}$$

Equations (2.68) and (2.69) apply to arbitrary n- and p-regions of width W_n and W_p. Two extreme cases can be distinguished, namely, the very wide structure in which $W_n \gg L_p(W_p \gg L_n)$ and the very narrow structure in which $W_n \ll L_p(W_p \ll L_n)$.

Wide Structure

For a wide n-region one can assume that at a certain distance from the depletion boundary, the excess minority carriers have decayed to zero and x can be set to infinity. Solving (2.67) subject to this boundary condition gives $C_2 = 0$. The excess minority concentration is then found to decay exponentially with distance beyond the depletion boundary as

$$\Delta p_n = \bar{p}_{n0}(e^{qV_F/kT} - 1)e^{-x/L_p}. \tag{2.70}$$

At a distance L_p from the depletion boundary, the excess carrier concentration drops to $1/e$ (37%) of its value at the depletion boundary. The gradient of excess minority-carrier concentration at $x = x_{dn}$ (set as $x = 0$) is

$$\left.\frac{d\Delta p_n}{dx}\right|_{x=0} = \frac{\bar{p}_n(e^{qV_F/kT} - 1)}{L_p}. \tag{2.71}$$

For $W_n \gg L_p$, $tanh\ (Wn/Lp) \cong 1$ and (2.69c) simplifies to

$$j_p = \frac{qD_p\bar{p}_n}{L_p}(e^{qV_F/kT} - 1), \tag{2.72}$$

$$\text{or } j_p \cong \frac{qD_p n_i^2}{L_p N_D}(e^{qV_F/kT} - 1). \tag{2.73}$$

A similar relation is found for the minority-carrier electron current density in a wide p-region

$$j_n \cong \frac{qD_n n_i^2}{L_n N_A}(e^{qV_F/kT} - 1). \tag{2.74}$$

The total current density is:

$$j = j_p + j_n = qn_i^2\left[\frac{D_p}{L_p N_D} + \frac{D_n}{L_n N_A}\right](e^{qV_F/kT} - 1). \tag{2.75}$$

Equation (2.75) can be written for the total current of junction area A as

$$I = I_0(e^{qV_F/kT} - 1) \quad A, \tag{2.76a}$$

where

$$I_0 = qAn_i^2\left[\frac{D_p}{L_p N_D} + \frac{D_n}{L_n N_A}\right] \quad A. \tag{2.76b}$$

The process-dependent term I_0 is sometimes called the saturation current. For $V_F > 3kT/q$, the 1 in (2.76a) can be neglected and the current increases exponentially with V_F (Fig. 2.24). For an ideal junction, the slope of the current–voltage plot is 60 mV/decade at 300 K.

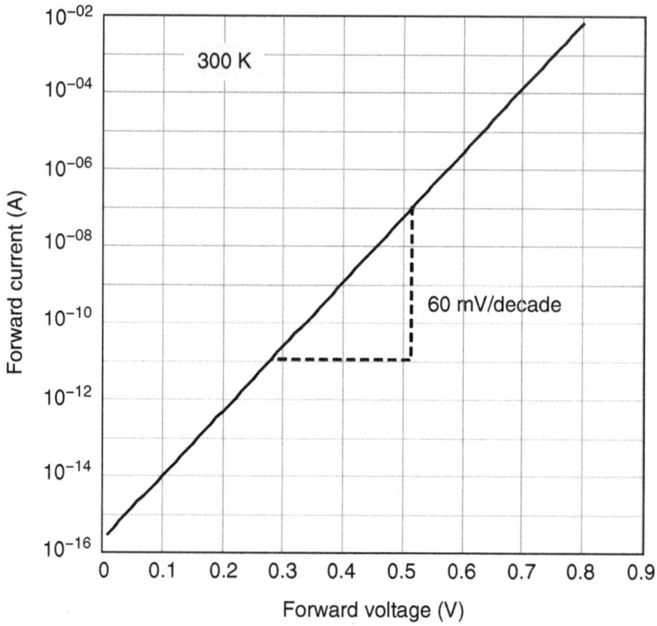

Fig. 2.24 Forward current–voltage characteristic of an ideal pn junction

Narrow Structure

This case is of particular importance because in most practical structures, the minority-carrier diffusion length is considerably larger than device dimensions and most recombination takes place at the contacts. For $Wn \ll Lp$, a series expansion of the hyperbolic functions in (2.68) reduces the relation to

$$\Delta p_n = \bar{p}_{n0}(e^{qV_F/kT} - 1)\left(1 - \frac{x}{W_n}\right), \tag{2.77}$$

which says that the excess minority-carrier hole concentration in a narrow n-region fall-off linearly with distance from the depletion boundary (Fig. 2.25).

The hole current is then

$$I_p = \frac{qAD_p n_i^2}{W_n N_D}(e^{qV_F/kT} - 1) \quad A. \tag{2.78}$$

Similarly, the minority-carrier electron current in a narrow p-region is

$$I_n = \frac{qAD_n n_i^2}{W_p N_A}(e^{qV_F/kT} - 1) \quad A. \tag{2.79}$$

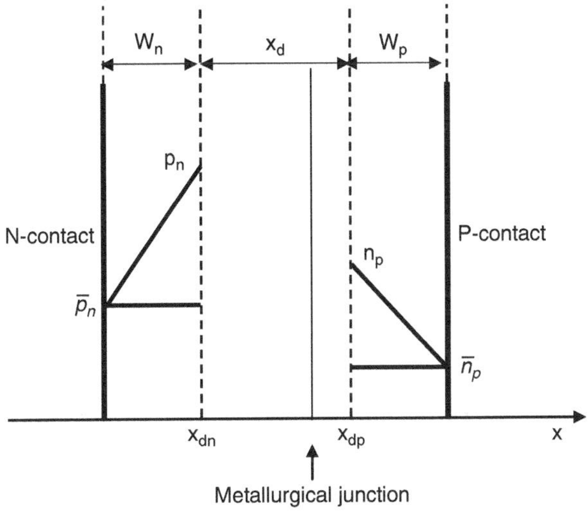

Fig. 2.25 Linear decay of excess minority concentration in narrow regions. Depletion region greatly exaggerated

The total current is given by (2.76a) with I_0 defined as

$$I_0 = qAn_i^2 \left[\frac{D_p}{W_n N_D} + \frac{D_n}{W_p N_A} \right] \quad A.$$

2.2.4.4 Injection Efficiency

In many applications, particularly bipolar transistors, it is important to increase the injection concentration of one type of minority carrier while reducing the injection concentration of the other. The injection efficiency for electrons is defined as

$$\gamma_n = \frac{I_n}{I_n + I_p} = \frac{I_n}{I}. \tag{2.80}$$

For $W_n \ll L_p$ and $W_p \ll L_n$, which is typically the case for modern bipolar transistors, (2.78–2.80) give:

$$\gamma_n = \frac{D_n / W_p N_A}{D_n / W_p N_A + D_p / W_n N_D} = \frac{1}{1 + D_p W_p N_A / D_n W_n N_D}. \tag{2.81}$$

For example, a higher efficiency for electrons is obtained by increasing N_D over N_A, or by increasing W_n over W_p. Similar relations are defined for the hole injection efficiency.

2.2.4.5 Non-Uniform Impurity Profile

Non-uniform impurity profiles outside the depletion boundaries create additional electric fields in the quasi-neutral n- and p-regions. The field induces a drift component of minority-carrier current in addition to the diffusion component discussed in the previous section. Depending on the profile, the drift component can be in the same direction as the diffusion current, hence enhancing the current, or opposite to it. It is now shown that both drift and diffusion components can be accounted for by substituting the integrals of N_D and N_A in the n- and p-regions for the products W_nN_D and W_pN_A in (2.69) and (2.78). Consider, for example, the injection of electrons into a narrow p-region with $W_p \ll L_n$. Let $N_A(x)$ and $\Delta n_p(x)$ be the position-dependent acceptor concentration and excess minority-carrier electrons in the p-region. The built-in field in the quasi-neutral region is (Chap. 1, Sect. 1.5.4.2)

$$E = -\frac{kT}{q}\frac{1}{N_A(x)}\frac{dN_A(x)}{dx} \tag{2.82}$$

Accounting for both diffusion and drift components and considering that $\Delta n_p \approx n_p$, the total electron current density is

$$j_n \approx q\mu_n(x)\Delta n_p(x)E(x) + qD_n(x)\frac{d\Delta n_p(x)}{dx}. \tag{2.83}$$

Substituting (2.82) into (2.83) and assuming an effective diffusion constant $\tilde{D}_n = \tilde{\mu}(kT/q)$ gives

$$j_n = \frac{q\tilde{D}_n}{N_A}\left[N_A\frac{d\Delta n_p}{dx} + \Delta n_p\frac{dN_A}{dx}\right]. \tag{2.84a}$$

or

$$j_n = \frac{q\tilde{D}_n}{N_A}\frac{d(N_A\Delta n_p)}{dx}. \tag{2.84b}$$

The solution to (2.84b) is [4]

$$N_A\Delta n_p = \frac{j_n}{qD_n}\int_C^x N_A dx, \tag{2.85}$$

where C is a constant of integration. For the narrow p-region, $\Delta n_p = 0$ at $x = W_n$ and (2.85) is written as

$$\Delta n_p = \frac{j_n}{qD_nN_A}\int_x^{W_p} N_A dx. \tag{2.86}$$

At the depletion boundary, the excess electron concentration is

$$\Delta n_{p0} = \frac{j_n}{qD_nN_{A0}}\int_0^{W_p} N_A dx, \tag{2.87}$$

where N_{A0} is the acceptor concentration at the depletion boundary. Combining the above result with (2.60) gives

$$j_n = \frac{qD_n n_i^2}{\int_0^{W_p} N_A dx}(e^{qV_F/kT} - 1). \tag{2.88a}$$

Similarly, for minority carrier holes injected into a narrow n-region:

$$j_p = \frac{qD_p n_i^2}{\int_0^{W_n} N_D dx}(e^{qV_F/kT} - 1). \tag{2.88b}$$

The integrals in the denominators are called the Gummel numbers. Equations (2.88a) and (2.88b) are applicable to low- and medium-doped regions. In the general case where n_i and μ vary with position, depending on dopant concentration and energy gap, (2.88a) and (2.88b) can be written in the form

$$j_n \cong \frac{q(kT/q)}{\int_0^{W_p} (N_A dx/\mu_n(x) n_i^2(x))}(e^{qV_F/kT} - 1), \tag{2.89a}$$

$$j_p \cong \frac{q(kT/q)}{\int_0^{W_n} (N_D dx/\mu_p(x) n_i^2(x))}(e^{qV_F/kT} - 1). \tag{2.89b}$$

The total current is then (2.76a):

$$I = I_0(e^{qV_F/kT} - 1),$$

with

$$I_0 = q(kT/q)\left[\frac{1}{\int_0^{W_n} (N_D dx/\mu_p(x) n_i^2(x))} + \frac{1}{\int_0^{W_p} (N_A dx/\mu_n(x) n_i^2(x))}\right]. \tag{2.90}$$

2.2.4.6 Small-Signal Impedance

For low-level injection and at low frequencies where capacitive effects are negligible, one can calculate the small-signal impedance r in forward bias, also called AC or dynamic resistance, in a simple way. Differentiating (2.76a) with respect to V_F at constant T gives

$$\frac{1}{r} = \left.\frac{\partial I_F}{\partial V_F}\right|_{T=const} = \frac{qI_0 e^{qV_F/kT}}{kT} \quad S, \tag{2.91}$$

or

$$r = \frac{kT}{qI_F} \quad \text{Ohm}. \tag{2.92}$$

r decreases with increasing forward current.

2.2.4.7 Charge Storage

There is a stored charge associated with injected minority carriers. Consider, for example, the injection of minority-carrier electrons into the wide p-region of a one-sided $n^+ p$ junction (Fig. 2.26). The excess minority electrons concentration decays with distance as

$$\Delta n_p = \Delta n_{p0} e^{-x/L_n}, \tag{2.93}$$

where x is the distance from the p-sided depletion boundary, and L_n is the electron diffusion length. The shaded area under the $q\Delta n_p(x)$ curve of Fig. 2.26 represents the stored charge Qs for a forward voltage $\mathbf{V_F}$. Its magnitude can be found by integrating (2.93) from 0 to infinity and multiplying by the junction area A and electron charge

$$Q_s = qA \int_0^{\infty} \Delta n_p dx = qA\Delta n_{p0} L_n \quad C. \tag{2.94}$$

To first approximation, the stored charge of injected minority-carrier holes into the heavily doped n$^+$-region can be neglected.

Transit time, τ

The diffusion electron current in a wide p-region I_n is found at $x = 0$ as

$$I = -qAD_n \frac{d\Delta n}{dx} = qAD_n \frac{\Delta n_{p0}}{L_n} \quad A. \tag{2.95}$$

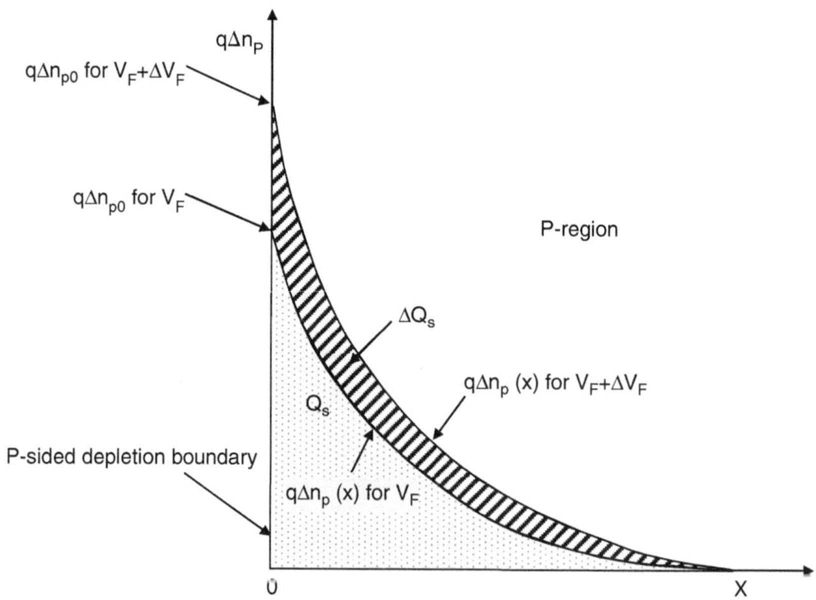

Fig. 2.26 Increase in minority-carrier stored charge associated with an increase in forward bias

Combining (2.94) and (2.95) gives the stored charge as

$$Q_s = \frac{IL_n^2}{D_n} = I\tau_n \quad C, \tag{2.96}$$

where τ_n is the transit time, that is, the average time for the charge Q_s to travel through a distance L_n. It can be seen from (2.96) that the transit time is proportional to the square for the diffusion length:

$$\tau_n = \frac{Q_s}{I} = \frac{L_n^2}{D_n} \quad s. \tag{2.97}$$

Diffusion Capacitance

At high frequencies, one must take two important capacitances into account, the depletion capacitance (Fig. 2.15, (2.35)), and a capacitance associated with the variation of charge storage of injected minority carriers. For a small increase in forward bias, ΔV_F, there is a corresponding amount of excess minority-electrons Δn and charge $\Delta Q_s = q\Delta n$ stored in the p-region (cross-hatched area in Fig. 2.26). The variation of stored charge with applied forward voltage represents a capacitance called the diffusion capacitance

$$C_D = \frac{dQ_s}{dV_F} \quad F. \tag{2.98}$$

Combining (2.91), (2.96) and (2.98) gives another definition of diffusion capacitance

$$C_D \cong \frac{\tau_n}{r_e} \quad F. \tag{2.99}$$

When the forward bias is dropped instantaneously to zero, the stored charge diffuses in both directions and disappears by recombination.

An equivalent circuit of the junction in forward bias is shown in Fig. 2.27. In parallel to the junction depletion capacitance C_j, there is a diffusion capacitance C_D that is typically orders of magnitude larger than the depletion capacitance.

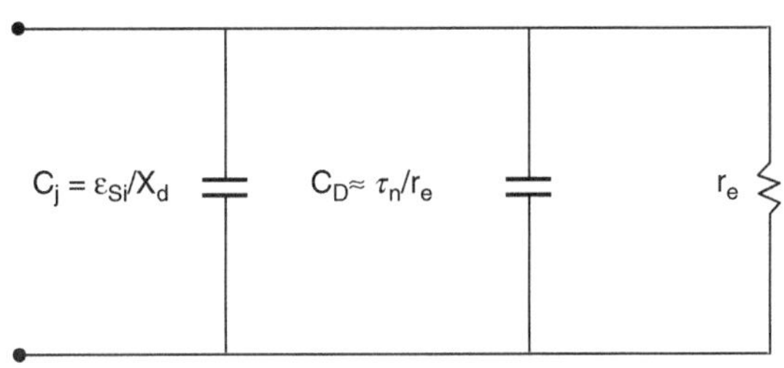

Fig. 2.27 Equivalent circuit of a forward biased junction

For a narrow p-region ($W_p \ll L_n$), such as the base of an npn transistor (Chap. 3), the stored charge is found as

$$Q_s = qA\Delta n_p \int_0^{W_p} \left(1 - \frac{x}{W_p}\right) dx = qA\Delta n_{p0} \frac{W_p}{2}. \tag{2.100}$$

The diffusion current is

$$I = -qAD_n \frac{\Delta n_{p0}}{W_p}. \tag{2.101}$$

Combining (2.100) and (2.101) defines the stored charge

$$Q_s = \frac{IW_p^2}{2D_n}. \tag{2.102}$$

The diffusion capacitance is

$$C_D = \frac{dQ_s}{dV_F} = \frac{W_p^2}{2r_e D_n}. \tag{2.103}$$

The transit time, that is, the time for electrons to travel through W_p, is

$$\tau_n = \frac{W_p^2}{2D_n}. \tag{2.104}$$

For a narrow p-region of width W_p, the transit time is proportional to the square of W_p.

Neutrality requires that an excess in majority-carrier holes $\Delta p_p = \Delta n_p$ be present to balance excess minority-carrier electrons. This means that, as long as excess minority carriers are present, there will be an equal amount of excess majority carriers. Without this constraint, excess majority carriers would disappear in a time, defined in Chap. 1 as the dielectric relaxation time, of the order of 10^{-11} s, while the minority carrier lifetime is six to ten orders of magnitude longer.

2.2.4.8 High-Level Injection

High-level injection effects are observed when the injected minority-carrier concentration becomes comparable to the thermal-equilibrium majority-carrier concentration. The effects are not all observed at the same injection level and the onset of high-level injection is not an abrupt event. The most important effects are:

1. Voltage drops outside the depletion region,
2. Decrease in injection efficiency,
3. Aided motion of minority carriers by the field created by the gradient in excess majority carriers.

There may also be some lowering of the energy-gap caused by the increase in majority carriers at high-level injection (Chap. 1).

The forward voltage, V_F, in (2.89a) and (2.89b) is assumed to appear at the depletion boundaries and not necessarily equal to the externally applied voltage, V_a. In practice V_F is always smaller than V_a because of the unavoidable voltage drops in the bulk of the n- and p-regions and at their contacts

$$V_F = V_a - I_F R, \tag{2.105}$$

where I_F is the forward current and R the sum of all resistances outside the depletion regions, including contact resistance. Taking this IR drop into account, (2.76) becomes

$$I = I_0(e^{q(V_A - I_F R)/kT} - 1). \tag{2.106}$$

There is a practical and fundamental reason why the forward voltage seen at the depletion boundaries cannot approach the junction built-in voltage. When the applied voltage is increased, I_F increases and so does $I_F R$. The difference between V_a and V_F increases (Fig. 2.28). The current becomes limited by R, or by Joule heating causing "catastrophic" failures in the junction leads, contacts, or the junction itself.

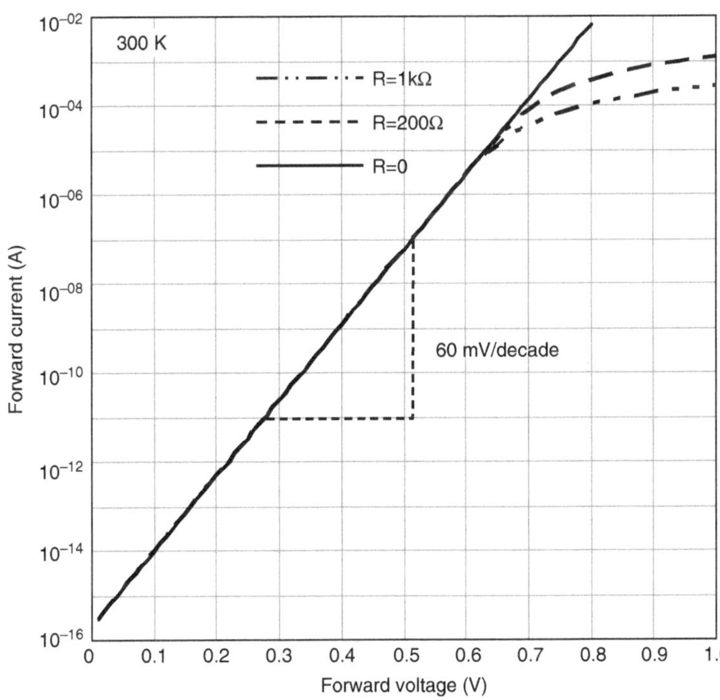

Fig. 2.28 Forward current–voltage characteristic of a pn junction with $R = 0$, 200 Ohm, and 1 kOhm

In most cases, however, other high-level effects come into play before a catastrophic failure occurs.

At high-level injection, the percentile increase in majority carriers becomes significant and must be taken into account. Charge neutrality requires that the gradients in minority and majority carriers be the same

$$\frac{d\Delta n_p}{dx} = \frac{d\Delta p_p}{dx}. \tag{2.107}$$

The gradient in excess majority carriers creates an electric field given by a relation similar to (2.82)

$$E_{grad} = \frac{kT}{q}\frac{1}{p}\frac{d\Delta p}{dx}. \tag{2.108}$$

The direction of the field is such as to aid the motion of minority carriers (electrons in this case). The electron current density then consists of a diffusion component and a field-aided drift component

$$j_n = q\mu_n n_p E_{grad} + qD_n\frac{dn_p}{dx} = qD_n\left[1+\frac{n_p}{p_n}\right]\frac{dn_p}{dx}. \tag{2.109}$$

For $W_p \ll L_n$ and an excess electron concentration Δn_{p0} at the depletion boundary, $dn_p/dx \cong \Delta n_{p0}/Wp$.

In the limiting case where $n_p \cong p_p$ the current density becomes

$$j_n \cong 2qD_n\frac{\Delta n_p}{W_p} \cong 2qD_n\frac{n_i^2}{W_p(\bar{p}_{p0}+(\Delta p_{p0}/2))}(e^{qV_F/kT}-1). \tag{2.110}$$

where $\Delta p_{p0} = \Delta n_{p0}$. Equation (2.110) shows that the "diffusion constant" appears to double. Also, the majority carrier concentration increases causing the injection efficiency to decrease.

The voltage externally applied to the junction, V_a, is the sum of forward voltage across the depletion region V_F, which is the voltage required to support the electric field created by the gradient in majority carriers in the quasi-neutral regions V_{grad}, and voltage losses due to the IR in the bulk and contact resistances

$$V_a = V_F + V_{grad} + IR. \tag{2.111}$$

The above relation can be written as

$$V_F = V_a - \int_0^{W_p} E_{grad}dx - IR. \tag{2.112}$$

Similar relations can be derived for minority-carrier holes injected into the n-region.

2.2.5 PN Junction in Reverse Bias

When a reverse bias is applied to a junction, carriers are pulled away from the junction, the depletion region widens, the barrier height increases and the peak electric field increases. For low peak fields, below $\sim 5 \times 10^4\,\mathrm{V/cm}$, the measured reverse current consists mainly of thermally-generated electron–hole pairs. There is some probability for carriers to gain sufficient kinetic energy from the electric field and generate electron–hole pairs by direct impact with silicon bonds. This latter process is referred to as impact ionization. For low peak fields, however, the probability for impact ionization is small and the measured impact-ionization current is negligible.

As the peak-field increases above $\cong 5 \times 10^4\,\mathrm{V/cm}$, the probability for impact ionization increases until a point is reached where the junction breaks down. At a peak field above $\cong 9 \times 10^5\,\mathrm{V/cm}$, another mechanism known as direct tunneling comes into play.

This section discusses the junction reverse-characteristics for low electric fields, where impact ionization and tunneling currents are negligible, and then for high electric fields.

2.2.5.1 Low-Peak Field

The current–voltage (IV) and capacitance–voltage (CV) characteristics of a reversed-biased pn junction are discussed for small fields where carrier generation by impact ionization can be neglected. There are two components to the measured reverse leakage current: electron–hole pair thermal generation and diffusion from outside the depletion region and thermal generation and drift from within the depletion region.

Depletion Width

An applied reverse voltage, V_a, increases the junction barrier from V_b to $V_R = V_b + |V_a|$, where V_R is the total reverse voltage. For a step- or linearly-graded junction, the depletion width is approximated by simply substituting $(V_b + |V_a|)$ for V_b in 2.29, 2.30, 2.31, and 2.40. For a junction of arbitrary profile, a similar procedure as outlined for thermal equilibrium can then be followed by satisfying the criterion in step 5 (2.45a–2.45e) that the points where

$$\left|V(x)_{p-side}\right| + \left|V(x)_{n-side}\right| = \left|\phi_p(x)\right| + \left|\phi_n(x)\right| + \left|V_a\right| \tag{2.113}$$

are the depletion boundaries, x_{dp}, and x_{dn} under reverse bias.

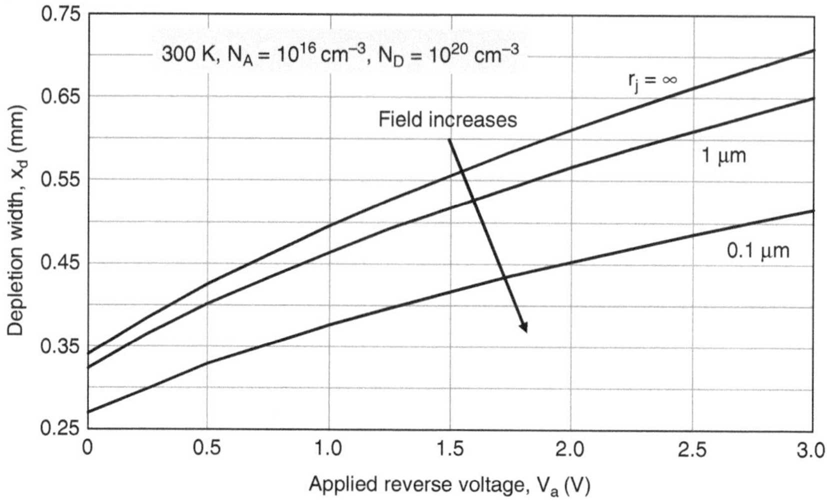

Fig. 2.29 Depletion width at junction edge as function of reverse voltage and junction radius of curvature, r_j

Effect of Curvature

The effect of radius of curvature on depletion width is shown in Fig. 2.29 for a one-sided step-junction and different junction curvatures r_j. Equation (2.49) is applied to generate the plots of depletion width versus reverse voltage by simply substituting V_b with $V_b + |V_a|$. For a given background concentration N_A and applied voltage V_a, the depletion width decreases as the radius of curvature in reduced. Note that this means that decreasing r_j leads to a higher electric field.

2.2.5.2 Reverse Leakage Current at Low Field

At low electric fields, the leakage current in a defect-free junction consists of thermal generations of electron–hole pairs outside and within the depletion region. Generated carriers outside the depletion region diffuse to the junction boundaries and are swept to the other side of the junction where they recombine. Carriers generated inside the depletion region drift to opposite sides of the junction and recombine. The two leakage components are discussed in this section.

Diffusion Leakage Component: Generation Outside Depletion

When a reverse bias is applied to the junction, the minority carrier concentrations drop to zero at the depletion boundaries. This is illustrated in Fig. 2.30 for wide and in Fig. 2.31 for narrow n- and p-regions. The drop in carrier concentration at

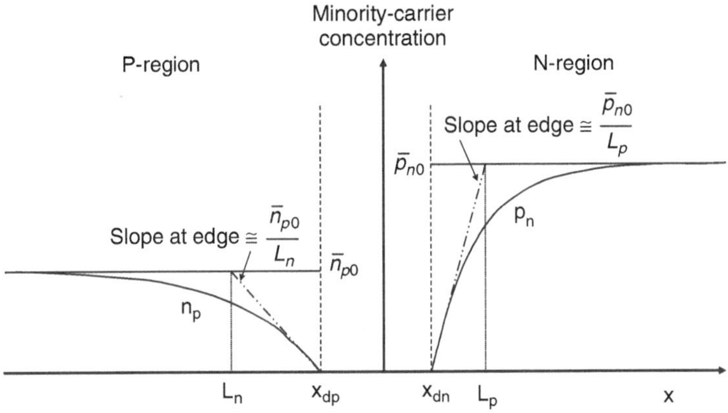

Fig. 2.30 Carrier flow in a reversed biased junction with wide p- and n-regions

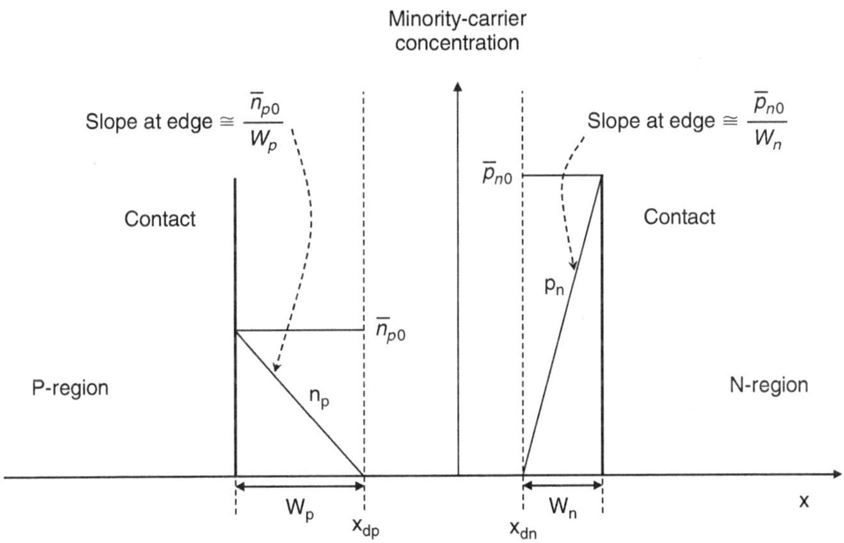

Fig. 2.31 Carrier flow in a reversed biased junction with narrow p- and n-regions

the depletion boundaries creates a concentration gradient whereby minority carriers diffuse from their corresponding quasi-neutral regions toward the depletion boundaries. Once at the boundaries, the carriers are under the influence of the large space-charge field and drift at approximately velocity saturation to the other side of the junction where they become majority carriers. The small electric field in the quasi-neutral regions causes excess majority carriers to drift to the corresponding contacts and recombine there. This current is limited by the rate of carriers diffusing toward the depletion boundaries. When the reverse voltage is larger than $\cong 3kT/q$ (negative in 2.76a), the exponential term becomes much smaller than the "1" and can be neglected.

The reverse current is therefore $-I_0$, the saturation current. For wide p- and n-regions (shown uniformly doped in Fig. 2.30 for simplicity), the reverse diffusion current is:

$$I_R = -I_0 = -qAn_i^2 \left[\frac{D_n}{L_n N_A} + \frac{D_p}{L_p N_D} \right] \quad A. \tag{2.114}$$

For a narrow junction, the most common case (Fig. 2.31), the reverse diffusion current is:

$$I_R = -I_0 = -qAn_i^2 \left[\frac{D_n}{W_p N_A} + \frac{D_p}{W_n N_D} \right]. \tag{2.115}$$

Equations (2.114) and (2.115) contain important information on the reverse diffusion current:

1. The current increases with temperature as does n_i^2, by a factor of about 4 every 10°C.
2. For narrow regions, the reverse current depends on the distance between contacts and depletion boundaries and not on the minority-carrier lifetime.
3. The reverse current does not directly depend on reverse voltage; this is why it is called "saturation current." There is, however, an indirect dependence on reverse voltage related to the widening of the depletion region and narrowing of already narrow n- and p-regions.
4. The current is limited by how fast the carriers diffuse from the quasi-neutral regions or contacts to the depletion boundaries rather than how fast they are generated.

Drift Leakage Current: Generation Within the Depletion Region

When a reverse bias voltage is applied to the junction, majority carriers are removed from the depletion region and its boundaries so that $pn \ll n_i^2$. The system tends to re-establish equilibrium by thermal generation of electron–hole pairs (Fig. 2.32).

In the presence of a strong field within the depletion region, carriers generated there drift toward the quasi-neutral regions at approximately saturation velocity. The probability for electron–hole pair recombination within the depletion region is practically zero. Electrons drift to the n-region and holes to the p-region where they become majority carriers. The excess majority carriers drift to the contacts where they recombine with minority carriers. Each pair contributes only one electronic charge to the external circuit.

For $p < n_i$, $n < n_i$, and equal electron and hole capture cross-sections, the generation rate is found from statistical analysis as [5, 6]

$$U = -\frac{\sigma v_{th} N_T n_i}{2\cosh(E_i - E_T/kT)} \quad cm^{-3}s^{-1}, \tag{2.116}$$

where

U – generation rate per unit volume (cm^{-3}s^{-1}),

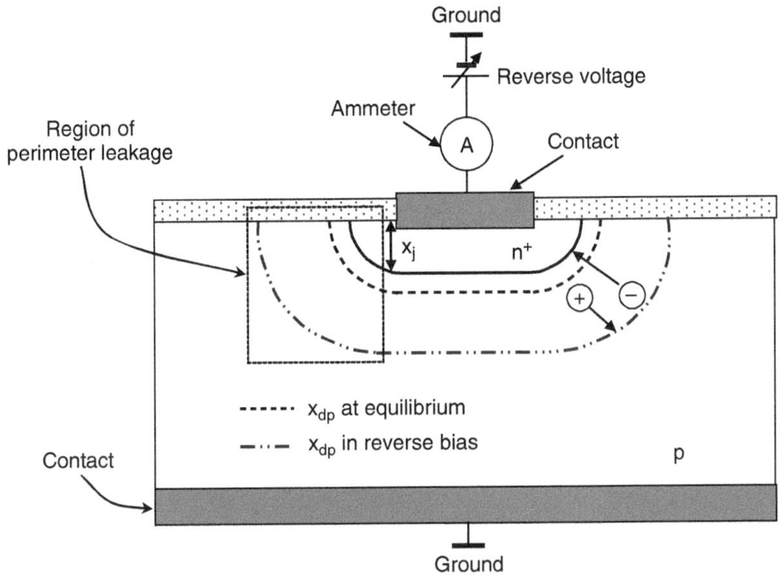

Fig. 2.32 Illustration of electron–hole pair generation and drift within the depletion region

v_{th} = thermal velocity, $\cong 10^7\,\text{cm/s}$,
σ = capture cross-section, $\cong 10^{-15}\,\text{cm}^2$,
N_T = density of generation sites or "traps" (cm^{-3}),
E_i = intrinsic energy level (eV),
E_T = energy level of generation site (eV).

The energy levels E_T are typically created by heavy metals, such as copper, gold, molybdenum, nickel, tungsten, titanium and zinc. They can also originate at point defects created by ion implantation or irradiation with other high-energy particles. The levels are located within the energy-gap, mostly in the vicinity of mid-gap E_i. The more E_T moves away from the mid-gap, the smaller the generation rate since the probability for transitions from and to one of the bands decreases rapidly. If one assumes an effective density of traps, N_{Teff}, located exactly at mid-gap and affecting the generation rate the same way as the distributed E_T, (2.116) simplifies to

$$U = \frac{1}{2}\sigma\, v_{th} N_{Teff}\, n_i. \tag{2.117}$$

The product $\sigma v_{th} N_{Teff}$ is the inverse of the lifetime τ. The generation rate can then be defined as

$$U = \frac{n_i}{2\tau}. \tag{2.118}$$

The generation current is proportional to the volume of the depletion region

$$I_{gen} = \frac{q}{2}\frac{n_i}{\tau}x_d A \quad A, \tag{2.119}$$

where A is the junction area in cm^2. The current increases with temperature as n_i, approximately by a factor of 2 every 10°C. It is essentially limited by the rate of generation within the depletion region. Since x_d widens as V_R increases, I_{gen} also increases. For a one-sided abrupt junction approximation, I_{gen} increases with the square-root of $(V_R + V_b)$. The junction perimeter in Fig. 2.32 includes the intercept of the junction with the surface. The generation at the surface is treated in the same way as the generation within the bulk of the depletion region [7]. The rate of generation is determined by the density of surface generation sites (interface "traps," N_{it}) and characterized by a parameter s defined as

$$s = \sigma v_{th} N_{it} \quad cm/s. \tag{2.120}$$

Since s has the dimension of cm/s, it is referred to as the surface recombination-generation velocity. Typical values for s are 10–100 cm/s. Interface states are distributed within the forbidden gap and also inside the bands. They are created by the discontinuity in the crystal lattice and strongly influenced by exposure of the surface to impurities and radiation during processing or device operation. In (2.120) N_{it} is the effective density of states, in cm^{-2}, at the interface between the silicon surface and the layer covering it, assumed to be located at mid-gap. The surface generation rate is

$$U_s = \frac{sn_i}{2} \quad cm^{-3}s^{-1}. \tag{2.121}$$

The surface generation current is proportional to the area of the intercepted depletion with the surface, A_s

$$I_s \cong \frac{qsn_iA_s}{2} \cong \frac{qsn_ix_{ds}P}{2} \quad A, \tag{2.122}$$

where x_{ds} is the depletion width at the surface and P the junction perimeter in cm.

2.2.5.3 Capacitance–Voltage Characteristics

The small-signal junction capacitance is measured in a similar way as described in Fig. 2.15. The small signal is superimposed on the reverse voltage and measured as the reverse voltage is swept from 0 to V_R.

The junction capacitance can be treated as a parallel-plate capacitor with the quasi-neutral p- and n-regions acting as the plates and the depletion region as the insulator of dielectric constant ε_{Si}. The capacitance per unit area is then

$$C_j = \frac{dQ}{dV} = \frac{\varepsilon_0 \varepsilon_{Si}}{x_d(V_R)} \quad F/cm^2, \tag{2.123}$$

where V_R is the total reverse voltage. For a one-sided step junction with uniform concentrations N in the lightly doped region

$$C_j = \sqrt{\frac{qN\varepsilon_0\varepsilon_{Si}}{2(V_a+V_b)}}, \tag{2.124}$$

where N is the dopant concentration in the lightly-doped region. In this case, the plot of $1/C_j^2$ versus V_a is a straight line and N can be directly extracted from the slope of the plot and V_b from the intercept with the V_a axis (Fig. 2.33). This relation between small-signal capacitance and applied reverse voltage can be extended to an n^+p or p^+n junction with an arbitrary profile in the lightly doped region. Consider, for example, an n^+p junction with an arbitrary distribution $N_A(x)$ in the p-region. A small increment dV_R in reverse voltage causes an increase dx_d in depletion width and a corresponding increase in electric field

$$dE = \frac{dQ'}{\varepsilon_0\varepsilon_{Si}} = -\frac{qN_A(x)dx_d}{\varepsilon_0\varepsilon_{Si}} \tag{2.125}$$

For small increments

$$dV_R \cong -x_d(V_R)dE = \frac{qN_A(x)d(x_d^2)}{2\varepsilon_0\varepsilon_{Si}}. \tag{2.126}$$

Substituting x_d^2 from (2.123) in (2.126) gives:

$$N_A(x) = -\frac{2}{q\varepsilon_0\varepsilon_{Si}}\frac{dV_R}{d(1/C_j^2)}. \tag{2.127}$$

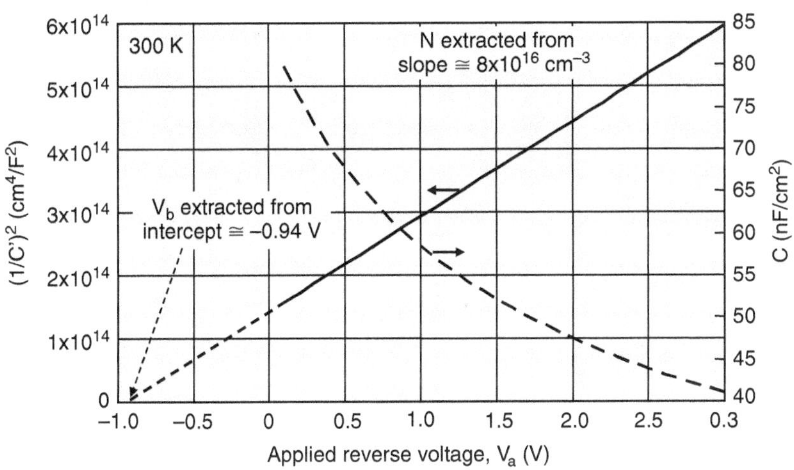

Fig. 2.33 Capacitance and inverse-square capacitance of a one-sided step junction and extraction of N and V_b

Equation (2.127) is a relation for the impurity concentration in the lightly doped p-region as a function of distance from the metallurgical junction. The profile is found experimentally by increasing the reverse voltage in increments ΔV_R and measuring the corresponding change in inverse-square capacitance $\Delta(1/C_j'^2)$ and then applying (2.127) to find N at each point. The corresponding distance from the metallurgical junction is approximated by (2.123).

2.2.5.4 High-Peak Field

A pn junction can sustain a limited reverse voltage before breaking down or conducting large reverse current. A typical reverse characteristic of a junction is shown in Fig. 2.34.

The breakdown voltage, BV, is measured at a specified reverse current, for example, 1 μA. For BV larger than ~8V, the mechanism for breakdown is avalanche by secondary impact ionization. For BV less than ~4V, the mechanism is by direct tunneling, or field ionization of electrons from the valence band to the conduction band. This section discusses the mechanisms of avalanche breakdown and tunneling.

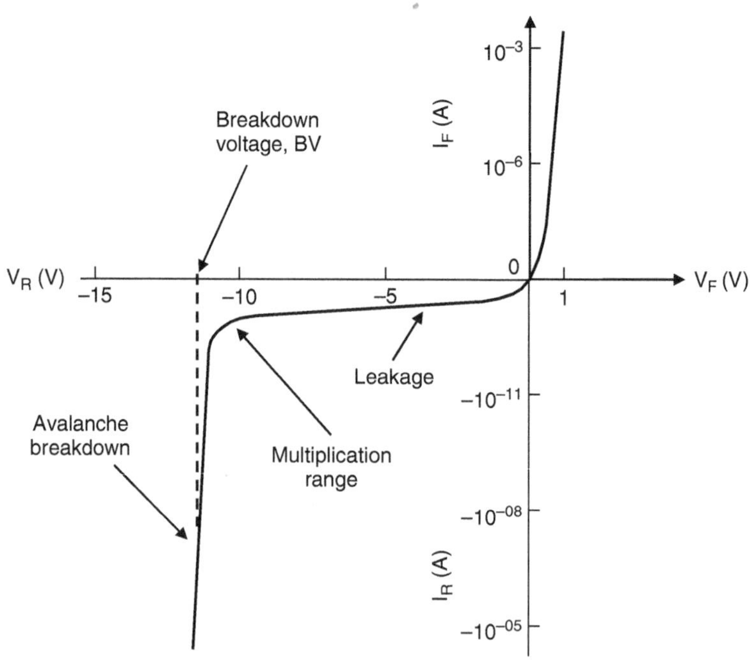

Fig. 2.34 Illustration of junction reverse and forward characteristics

Impact Ionization and Avalanche Breakdown

Initially, as the reverse voltage is increased, the reverse current consists of diffusion and generation reverse currents discussed above. Carriers entering the depletion region or generated within the region gain kinetic energy from the large electric field and drift at approximately saturation velocity through the depletion region, with electrons going to the n-region and holes to the p-region. The probability for carriers to recombine within the depletion region is practically zero. For electric fields above $\sim 5 \times 10^4\,\mathrm{V/cm}$, the carrier energy becomes comparable to the optical phonon energy (63 meV) so that the main mechanism for energy loss is by emission of optical phonons [8]. When the field increases above $\sim 10^5\,\mathrm{V/cm}$, a larger fraction of the carriers gains sufficient kinetic energy from the field to break covalent bonds and ionize silicon atoms. The generation of electron–hole pairs by this process is known as impact ionization. Figure 2.35 illustrates the mechanism of impact ionization. The bands are shown tilted because of the reverse voltage. The Fermi-level in the n- and p-regions (not shown) are separated by V_R.

Consider, for example, the path of one electron in the conduction band. In paths 1 and 2 the electron gains energy from the field and rises above the bottom edge of the conduction band, E_C. The electron loses its energy by optical phonon emission and the electron falls back to E_C. There is a finite probability that the electron free path is sufficiently large so that it gains more energy from the field. This is shown in path 3. If this energy reaches the threshold ionization energy, $E_I > E_g$, an electron–hole pair is generated by impact ionization. The original and the new electron are shown accelerated to the n-side while the generated hole is accelerated to the p-side. The incident electron is thus multiplied by the generation of secondary carriers in the junction and the reverse current increases.

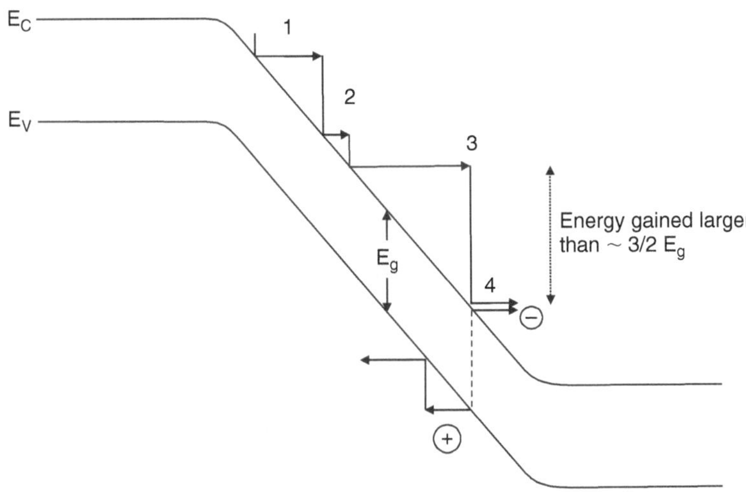

Fig. 2.35 Illustration of impact ionization in a pn junction

A simplified model is used to estimate the threshold ionization energy. For a collision to occur, energy and momentum must be conserved. The momentum of the incident carrier is redistributed between the two electrons and the hole:

$$m_n v_0 = m_n v_1 + m_n v_2 + m_p v_3. \tag{2.128}$$

v_O is the velocity of the incident electron before collision, and v_I, v_2, and v_3 the velocities of the three carriers after impact. The energy of the incident electron must be the sum of energy gap and kinetic energy of the three carriers after impact:

$$\frac{1}{2}m_n v_0^2 = E_g + \frac{1}{2}m_n v_1^2 + \frac{1}{2}m_n v_2^2 + \frac{1}{2}m_p v_3^2. \tag{2.129}$$

Assuming equal masses and $v_I = v_2 = v_3 = v_O/3$, the minimum energy for ionization, $E_I \cong 1.5E_g$. This shows that E_I increases with the semiconductor bandgap.

A hole can initiate a similar process. When electrons and holes have energies $\geq E_I$, an incident current is multiplied by the generation of secondary carriers in the junction.

The field-dependent ionization rate, $\alpha(E)$, is defined as the number of secondary carrier-pairs, N_S, generated by a single carrier along a path of 1 cm. For a depletion region of width $x_d = x_{dn} + x_{dp}$

$$N_S = \int_{x_{dn}}^{x_{dp}} \alpha(E)dx. \tag{2.130}$$

Under the simplifying assumption that electrons and holes have the same ionization rate α, an incident current I_0 creates $I_0 N_S$ secondaries, the current $I_0 N_S$ creates $(I_0 N_S)N_S$ secondaries, and so on. The total measured current is then [8]

$$I = MI_0 = I_0(1 + N_S + N_S^2 + \ldots\ldots) = \frac{I_0}{1 - N_S}. \tag{2.131}$$

M is called the multiplication factor found as

$$M = \frac{1}{1 - \int_{x_{dn}}^{x_{dp}} \alpha_i(E)dx}. \tag{2.132}$$

When the integral in the denominator of (2.132) approaches unity the multiplication factor increases rapidly. This is the onset of avalanche breakdown.

The ionization rate depends exponentially on the field and is approximated by a relation of the form [9–13]

$$\alpha_i(E) = ae^{-b/|E|} \quad cm^{-1}. \tag{2.133}$$

Table 2.1 Ionization parameters from [11]

	E-range (V/cm)	a (cm^{-1})	b (V/cm)
Electrons	$1.75 \times 10^5 \leq E \leq 6 \times 10^5$	7.03×10^5	1.231×10^6
Holes	$1.75 \times 10^5 \leq E \leq 4 \times 10^5$	1.582×10^5	2.036×10^6
Holes	$4.01 \times 10^5 \leq E \leq 6 \times 10^5$	6.71×10^5	1.693×10^6

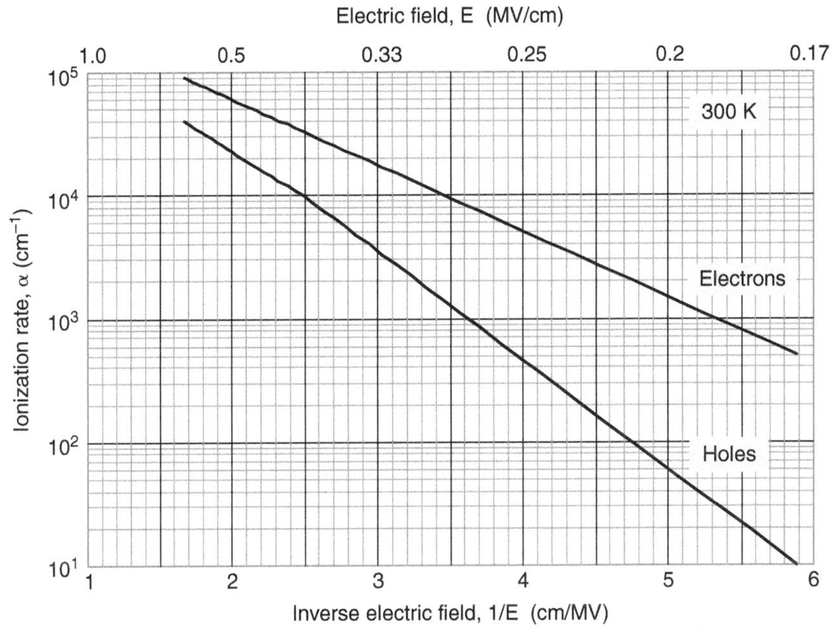

Fig. 2.36 Electron and hole impact ionization rates [11]

The parameters a and b are extracted from charge multiplication measurements and shown in Table 2.1 [11]. The ionization rates for electrons and holes are shown as a function of $1/E$ in Fig. 2.36.

Optical phonon generation and impact ionization are the two major mechanisms by which a hot carrier can lose its energy. For carrier energies smaller than the threshold ionization energy, E_I, the carrier loses its energy mainly to optical phonons. The reported mean-free path for optical-phonon scattering shows a wide spread in the range $5.0\,\text{nm} \leq l_r \leq 10\,\text{nm}$. Room-temperature values for l_r are given in [2, 8] as $\cong 7.6\,\text{nm}$ for electrons and $\cong 5.5\,\text{nm}$ for holes. As the carrier energy increases to a value $\geq E_I$, the probability for energy loss by impact ionization increases. The reported mean free path for impact ionization is in the range $10\,\text{nm} \leq l_I \leq 100\,\text{nm}$. The effective mean-free path is:

$$\frac{1}{l_{eff}} = \frac{1}{l_r} + \frac{1}{l_I}$$

The relative probability for ionization collision to optical-phonon collision is:

$$r = \frac{l_r}{l_I}$$

For simplicity, it is assumed that if the carrier energy is less than the threshold ionization energy, l_I is infinite, and for energies equal or larger than E_I, l_I is finite and constant, so that if l_r decreases, the ratio r decreases. The temperature dependence of ionization rate is therefore dominated by the carrier mean-free path for optical phonon generation and the temperature dependence of phonon energy [14, 15]. When the temperature increases, l_r decreases and hot carriers lose more energy to the crystal lattice by optical phonon scattering along their path within the depletion region. Consequently, for impact ionization to occur at the same rate, the carriers must pass through a greater potential difference, that is, larger reverse voltage, to acquire energies $\geq E_I$. The decrease in ionization rate with increasing temperature is illustrated for electrons in Fig. 2.37 [15].

Measurements of the multiplication near the breakdown voltage show that the multiplication factor in (2.132) can be empirically related to the reverse voltage V_R by

$$M \cong \frac{1}{1 - (V_R/BV)^n}, \tag{2.134}$$

where BV is the breakdown voltage and $n \cong 3$. The breakdown voltage decreases as the dopant concentration in the n- and p-regions increase.

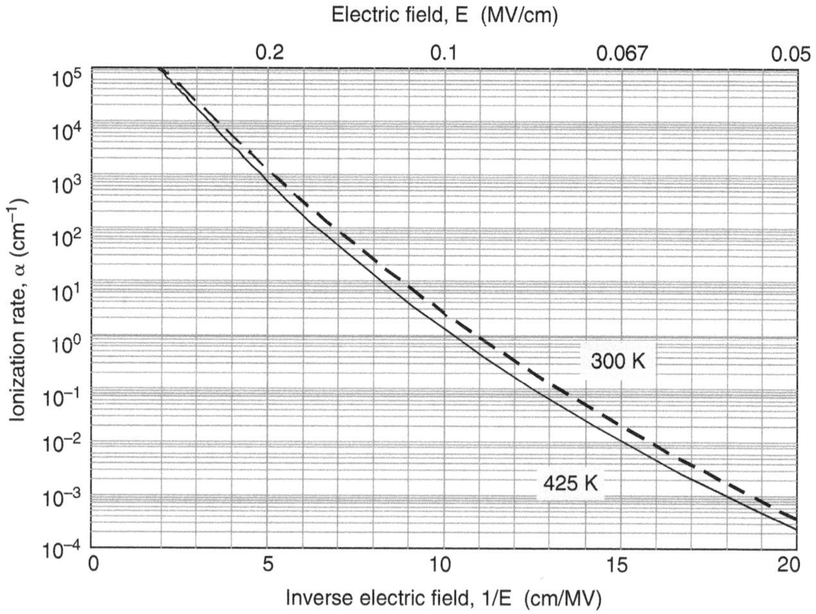

Fig. 2.37 Effect of temperature on electron ionization rate [15]

For a one-sided abrupt junction, BV depends mainly on the concentration in the lightly-doped region. This is because the depletion width in the heavily-doped side is negligible when compared to the total depletion width and a carrier traversing the depletion region has most of its path in the lightly-doped region. Consequently, the average rate at which it has ionizing collisions with the lattice depends not only on the field strength but also on the *path-length* in the depletion region. The breakdown voltage is independent of whether the region is n-type or p-type because in a "stand-alone" pn junction, BV is a property of both secondary electrons and secondary holes and not on the type of the primary carrier. The avalanche breakdown voltage is shown in Fig. 2.38 for a *planar* one-sided step junction as a function of background concentration and in Fig. 2.39 for a linearly-graded junction as a function of gradient [8, 16].

The dashed line in Fig. 2.38 follows an approximate empirical relation for a one-sided step junction adapted from [16]:

$$BV \cong 60\left(\frac{N}{10^{16}}\right)^{-0.7} \quad V. \tag{2.135}$$

The above relation is applicable to background concentrations smaller than $\sim 2 \times 10^{17}\,\text{cm}^{-3}$ where the mechanism of breakdown is by the process of impact ionization. The solid line is adapted from measured and calculated data from [4, 8, 16–18]. For a breakdown voltage less than 4V, which corresponds to a concentration above $\sim 10^{18}\,\text{cm}^{-3}$, the breakdown is dominated by tunneling. At intermediate breakdown

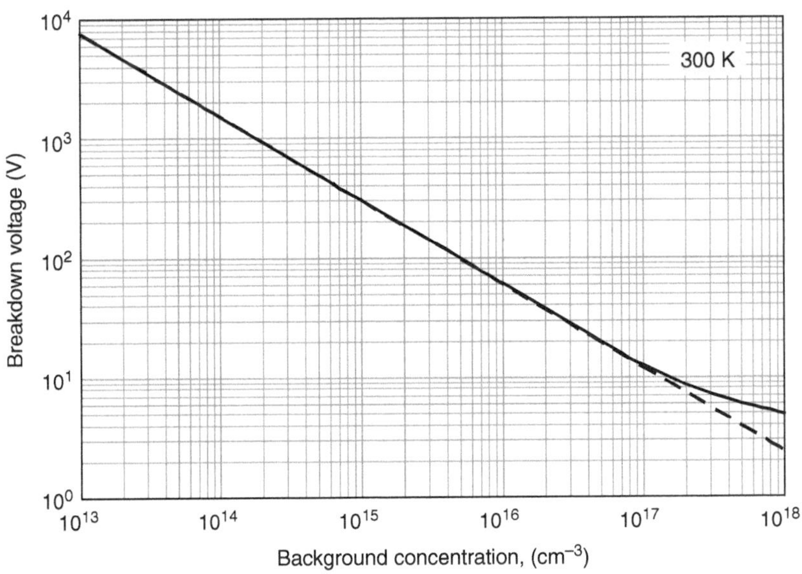

Fig. 2.38 Avalanche breakdown voltage for a one-sided step junction as a function of background concentration

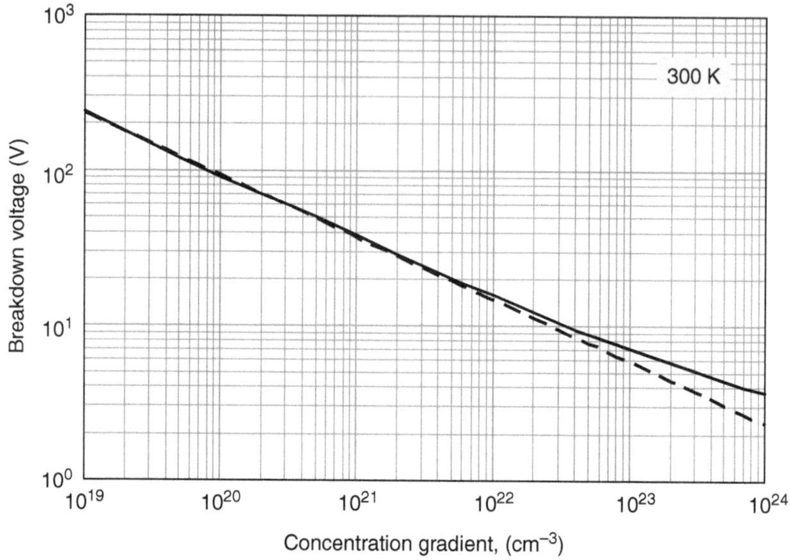

Fig. 2.39 Avalanche breakdown voltage for a linearly-graded junction as a function of gradient

voltages between 4V and 8V, there is a mixture of impact ionization and tunneling [8]. In this region, the breakdown voltage does not decrease at the same rate as in (2.135). This is because the depletion width at breakdown, and hence path-length, becomes comparable to the mean-free path for impact ionization, so the integrated ionization decreases. Thus, as N increases the "critical" field E_C for avalanche breakdown increases. The dependence of E_C on N can be approximated by

$$E_C \approx 3 \times 10^4 \ln \frac{N}{10^{10}} + 5.7 \times 10^{-13} N \quad V/cm, \tag{2.136}$$

where N is the background concentration.

For a linearly-graded junction of gradient a less than $\sim 2 \times 10^{22}\,\text{cm}^{-4}$, BV is approximated as [16]:

$$BV \cong 60 \left(\frac{a}{3 \times 10^{20}} \right)^{-0.4} \quad V. \tag{2.137}$$

This is shown by the dashed line in Fig. 2.39. For larger gradients tunneling comes into play. The solid line is adapted from measured and calculated data from [8, 16].

For a combination of linearly-graded region and uniform background, the breakdown voltage is intermediate between that of a one-sided step junction and linearly graded junction [19]. The semi-empirical relations derived in [19] apply to, for example, a junction formed between the PMOS n-well and the uniformly doped p-type substrate (Chap. 5). For an arbitrary junction profile, the breakdown voltage is found by combining (2.132), (2.133) and Poisson's equation.

Effect of Curvature on Breakdown

In planar diffused or implanted junctions with idealized spherical corners and cylindrical edges, as illustrated in Fig. 2.8, avalanche breakdown begins at corners and edges where the field is highest. For a one-sided step junction, the breakdown voltage is approximated by [3, 20, 21]

$$BV \cong 9.51 \times 10^{12} N^{-0.7} \{[(n+1-\gamma)\gamma^n]^{1/(n+1)} - \gamma\} \quad V, \tag{2.138}$$

where

$n = 1$ for cylindrical edges, $n = 2$ for spherical corners,
$\gamma = r_j / x_d$,
r_j = radius of curvature $\cong$ junction depth,
x_d = depletion width at breakdown, in the plane portion of the junction.

Figure 2.40 shows the breakdown voltage for a one-sided abrupt junction with cylindrical edge and spherical corner approximations as a function of background concentration. The breakdown voltage for an elliptical junction-edge approximation has also been analyzed, showing near 25% higher breakdown voltage than for a cylindrical shape [22]. Calculation of the breakdown voltage for arbitrary edge- and corner-shapes requires detailed numerical analysis.

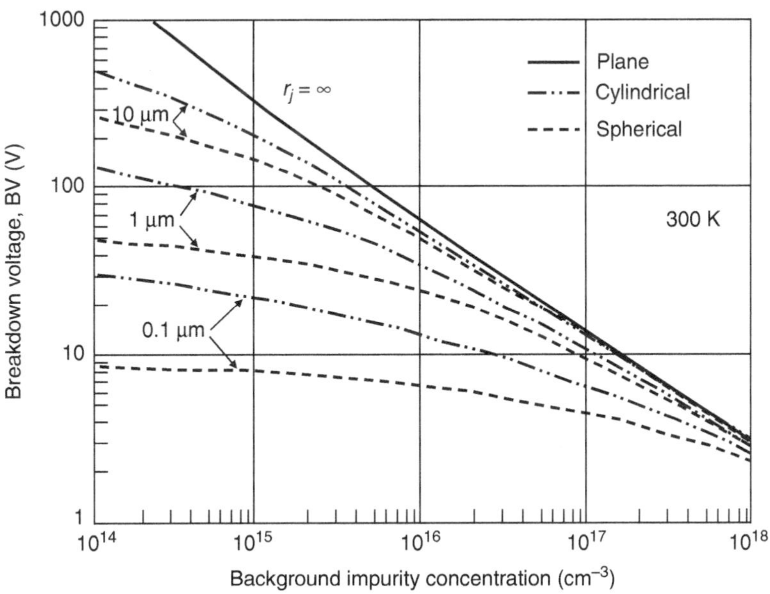

Fig. 2.40 Avalanche breakdown voltage as a function of background concentration for a one-sided abrupt junction with cylindrical edges and spherical corners; r_j is the junction radius of curvature as indicated in Fig. 2.18 (After Sze and Gibbons [20], © 1981, Bell Telephone Labs., Inc., reprinted with permission)

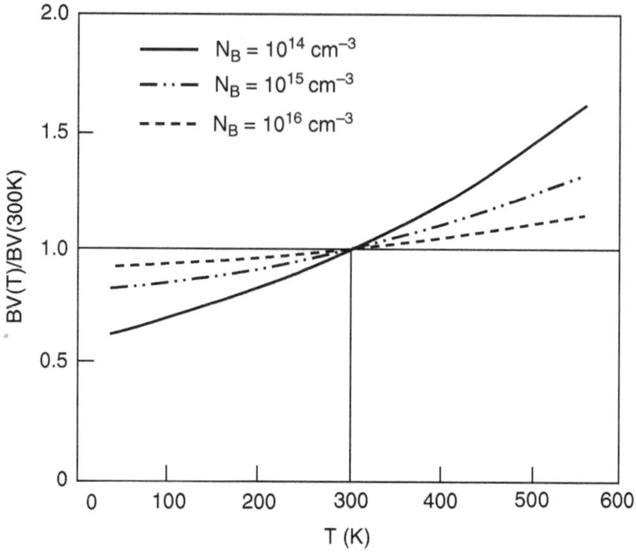

Fig. 2.41 Normalized avalanche breakdown voltage versus crystal temperature [14]

Effect of Temperature on Breakdown

As the lattice temperature increases, the ionization rate decreases due to an increase in the optical-phonon collision rate, as discussed in the previous section. The breakdown voltage therefore increases with temperature [14, 23]. The normalized breakdown voltage is shown as a function of temperature in Fig. 2.41. The ratio $BV(T)/BV(300\,\mathrm{K})$ increases at low concentrations. Although the probability for both impact ionization and optical-phonon scattering increases as the depletion region gets wider, the loss of energy to optical phonons is more significant.

Tunneling and Zener Breakdown

Tunneling is the penetration of electrons through a barrier rather than crossing over it. From quantum mechanics, it is known that electrons can penetrate into a potential barrier for a short distance. If the barrier in a pn junction, which is the distance between the valence band and conduction band edges, is thin enough, there is a finite probability for direct excitation of electrons from the valence band in the p-side into the conduction band in the n-side. This occurs when both sides of the junction are doped at high concentration, above $\sim 5 \times 10^{17}\,\mathrm{cm}^{-3}$, and both depletion regions are thin whereby a high peak field is obtained for a small reverse voltage. If the peak field approaches $\sim 10^6\,\mathrm{V/cm}$, the probability for electron tunneling increases rapidly and a large reverse tunneling current is measured in addition to the impact ionization current. The mechanism for this field-assisted band-to-band transitions

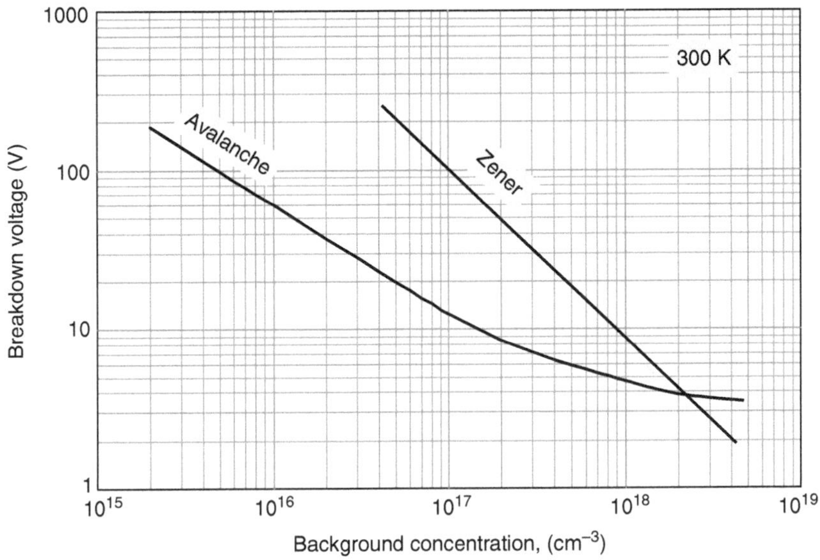

Fig. 2.42 Comparison of Zener and avalanche breakdown versus background concentration for a one-side step junction

is known as field ionization. As the concentrations increases above $\sim 10^{18}\,cm^{-3}$, the depletion width and hence the path-length traveled by hot carriers becomes so short that the probability for carriers to gain energy and create electron–hole pairs by impact ionization decreases to the point where tunneling becomes the dominant breakdown mechanism. This breakdown mechanism is also referred to as Zener breakdown after Clarence Zener [24].

Figure 2.42 compares Zener and avalanche breakdown as a function of background dopant concentration in a one-sided step junction.

To estimate the width of the barrier, consider the band bending in Fig. 2.43. The slope of the band may be approximated as qV_R/x_d where $V_R = V_a + V_b, V_a$ is the externally applied voltage, V_b the built-in voltage and x_d the depletion width. The slope is also $E_g/q\Delta x$ where E_g is the energy gap and Δx the barrier width. Therefore,

$$\Delta x \approx \frac{E_g}{q}\frac{x_d}{V_R}. \tag{2.139}$$

For a one-sided abrupt junction, the width of the depletion region is given by (2.30) and (2.31) as

$$x_d \cong \sqrt{\frac{2\varepsilon_0\varepsilon_{Si}V_R}{qN}}, \tag{2.140}$$

where N is the background concentration and V_R is the total reverse voltage ($V_R = V_b + |V_a|$). The barrier width is then

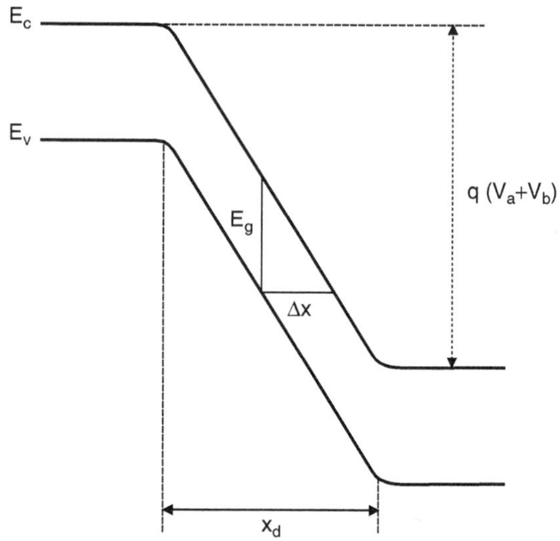

Fig. 2.43 Simplified model to extract the barrier width

$$\Delta x \approx \frac{E_q}{q}\sqrt{\frac{2\varepsilon_0\varepsilon_{Si}}{qNV_R}}. \tag{2.141}$$

Thus, the tunneling distance is inversely proportional to $\sqrt{NV_R}$.

A detailed discussion of tunneling in pn junctions can be found in Chap. 9 of [8]. The tunneling current density, j_T, is given as [8]

$$j_T = k_1 e^{-k_2},$$

with

$$k_1 = \frac{\sqrt{2m^*q^3}\,|E|\,V_R}{4\pi^3\hbar^2\sqrt{E_g}}, \tag{2.142}$$

$$k_2 = \frac{\pi\sqrt{m^*}E_g^{3/2}}{2\sqrt{2}q\,|E|\,\hbar}.$$

In (2.142) m^* is the effective electron mass, E_g the energy gap, V_R the reverse voltage, $\hbar = h/2\pi$ the reduced Planck constant, and E the electric field. For a one-sided step junction

$$|E| = \frac{qNx_d}{\varepsilon_{Si}}. \tag{2.143}$$

Substituting (2.140) and (2.143) in (2.142) simplifies the relation to:

$$J_T \approx 5.57\times10^{-2}\sqrt{NE_gV_R^3}\,e^{-(7.25\times10^{10}E_g^{1.5}/\sqrt{NV_R})}, \tag{2.144}$$

It is possible to distinguish between impact ionization and tunneling current by measuring their temperature dependence. The impact ionization current decreases with increasing temperature due to increased scattering, while the tunneling current increases because of the reduction in energy gap.

2.3 Contacts

When a metal is placed in contact with a semiconductor, an energy barrier is typically created between the two materials, resulting in a rectifying contact. To form an ohmic contact to the semiconductor, the contacted region must be doped to a high concentration to allow tunneling through the barrier, or have a very high recombination velocity such that carrier concentrations remain at thermal equilibrium [4]. This section discusses the properties of these two types of contact.

2.3.1 Rectifying Contacts, Schottky Barrier Diode

The rectifying contact may be the oldest semiconductor component used in electronics. The rectifying behavior of a metal-semiconductor contact was discovered by K. F. Braun in 1874 [25]. Figure 2.44 shows a current–voltage characteristic of a point contact between a metal wire and a lead-sulfide crystal taken in 1877 [25, 26].

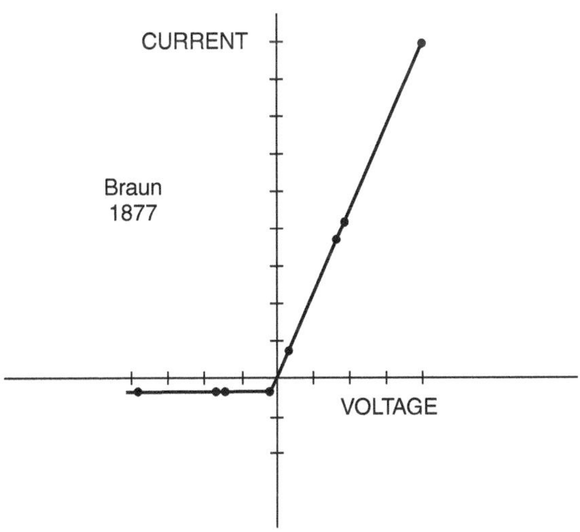

Fig. 2.44 Current–voltage characteristic of metal wire contact to lead sulfide crystal [27]. (Adapted from [2])

A large current is observed when the metal is positive and very little current when the metal is negative with respect to the semiconductor. This was puzzling since the metal has a high density of electrons and, with the metal negatively biased, electrons are crowded against the interface with the semiconductor. It is now known that a barrier exists between metal and semiconductor and the barrier increases with increasing reverse voltage. The rectifying contact is called a Schottky-Barrier Diode (*SBD*), after Walter Schottky who first developed a model for rectification [28].

The current–voltage characteristics of rectifying contacts are similar to those of pn junctions, but the underlying mechanisms for forward and reverse currents are fundamentally different. In particular, the forward current in a Schottky-barrier diode consists mainly of majority carriers while in pn junctions, minority-carrier injection is the primary mechanism. Schottky-barrier diodes are therefore advantageous where the switching speed is critical. The structure has also found applications in microwaves, *MESFET*s (Metal-Semiconductor Field-Effect Transistors), solar cells, photodetectors, and *MOSFET* source/drain contacts (Chap. 5). A review of rectifying metal-semiconductor contacts can be found in [29].

2.3.1.1 The Barrier

Starting with an isolated metal in vacuum, there is a barrier between the highest energy level occupied by electrons in the metal and the lowest energy level that can be occupied by an electron in vacuum. This barrier is called the workfunction ϕ_m of the metal (Fig. 2.45). The workfunction of a material is the minimum energy required to bring an electron from the Fermi level of the material to the vacuum level.

The main features of a model for a rectifying metal-semiconductor structure are illustrated for n-type silicon in Figs. 2.46 and 2.47 [28].

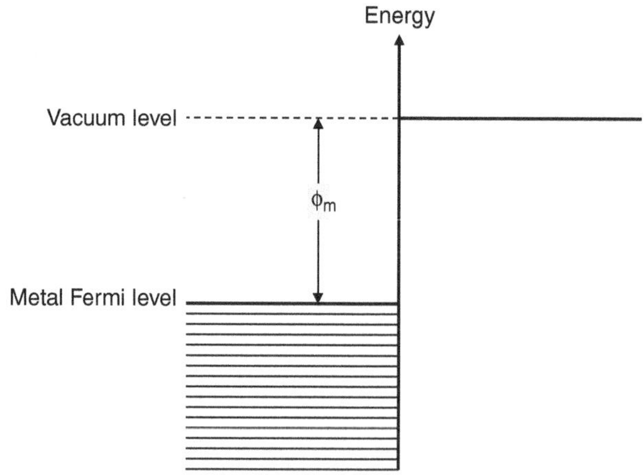

Fig. 2.45 Diagram of electron energy showing the metal workfunction, ϕ_m

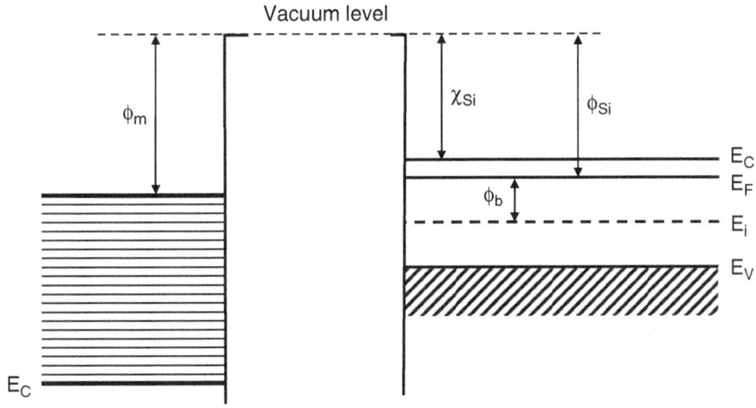

Fig. 2.46 Energy band diagram of separated metal and n-type *Si*

Figure 2.46 shows a simplified energy band diagram for separated metal and crystal. The minimum energy required to bring an electron from the bottom of the conduction band of the semiconductor to vacuum is called the electron affinity, χ_{Si}. ϕ_{Si} and ϕ_{Fn} are, respectively, the silicon workfunction and Fermi potential. For n-type silicon, the Fermi level is typically at a higher energy than the metal Fermi level. Because of this workfunction difference, the potential energy of an electron in vacuum next to the metal is larger than that of an electron next to silicon. When the two materials are brought in contact, electrons momentarily flow from silicon to the metal and thermal equilibrium is established (Figs. 2.47a and 2.47b).

A potential difference $(\phi_m - \phi_{Si})$ is established between metal and silicon. For a large spacing x between the two materials an electric field of approximately $(\phi_m - \phi_{Si})/x$ is created. Actually, part of the field must be within the metal and silicon. Since the electron concentration in the metal is very high, one can neglect the voltage drop and field penetration in the metal. In silicon, the potential drop and field penetration become significant as the spacing x decreases to a value of atomic dimension, δ (Fig. 2.47a). The gap is so thin that electrons can easily pass through it by tunneling. Eventually, as δ approaches zero, all potential drop occurs in silicon within a layer x_d, as shown in Fig. 2.47b. Within x_d, electrons are depleted leaving behind positively ionized fixed donors. Analysis of the barrier thus created is similar to that of a p^+n step junction.

In this idealized model, the barrier seen by electrons in the metal is the energy difference between the Fermi level and the conduction band edge at the surface

$$\begin{aligned} \phi_B &= \phi_m - \chi_{Si} \ (n\text{--}type), \\ \phi_B &= \frac{E_g}{q} - (\phi_m - \chi_{Si}) \ (p\text{--}type). \end{aligned} \tag{2.145}$$

The barrier seen by an electron at the bottom of the silicon conduction band is V_b in Fig. 2.47b [30].

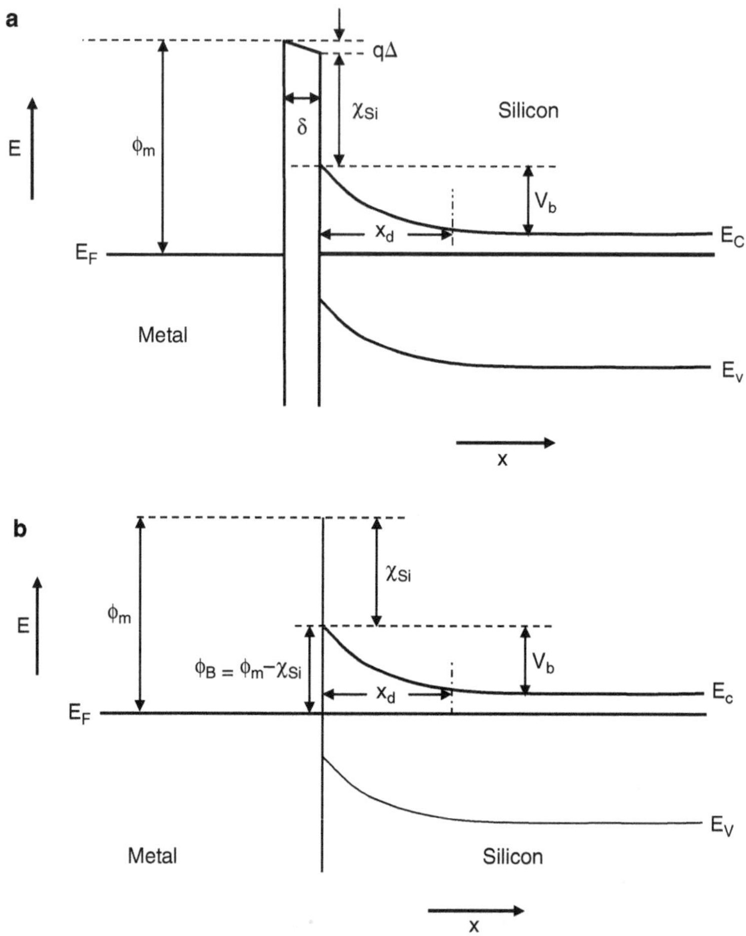

Fig. 2.47 **a** Energy band diagram of an idealized metal–silicon contact at thermal equilibrium, showing a thin gap σ. **b** Energy band diagram of an idealized metal to n-type silicon contact at thermal equilibrium for $\sigma = 0$

The barrier obtained from (2.145), however, typically differs from measured values. For example, the workfunction of aluminum is $\phi_{\mathrm{m}} \cong 4.25\,\mathrm{V}$ and the silicon affinity $\chi_{Si} \cong 4.05\,\mathrm{V}$. Equation (2.145) should then yield an idealized barrier height of 0.2 V while measured barriers range from 0.5 V to 0.9 V for aluminum. Bardeen explained the observed difference by including the effects of surface states in Schottky's model, noting that for high surface state concentrations, the barrier is almost independent of the metal workfunction [31].

Surface states are unavoidable because of the termination of the silicon lattice [32,33]. Surface states are of donor or acceptor type and assumed to be continuously distributed in energy within the bandgap.

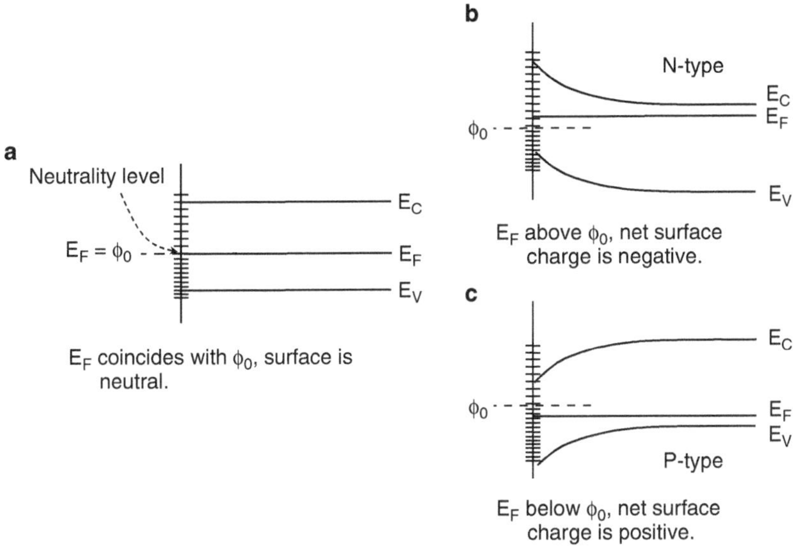

Fig. 2.48 Effect of Fermi-level position with respect to neutrality-level surface charge and band bending. **a** Neutral surface, **b** Net negative surface charge, **c** Net positive surface charge

Figure 2.48 illustrates schematically the distribution of surface states within the band gap and inside the conduction and valence band for a free silicon surface. Also shown is the charge neutrality level, ϕ_0. When the Fermi level E_F coincides with ϕ_0, surface states below E_F are filled and above are empty so that the net charge of all surface states is zero, that is, the surface is neutral. If E_F is below ϕ_0, the net surface charge is positive and if E_F is above ϕ_0, the net surface charge is negative.

When a metal is brought into contact with the silicon surface, several modifications to the surface will occur. In particular, the initial concentration of surface states will change because of the proximity of metal atoms and an interfacial layer of atomic dimension typically forms at the interface. The resulting band-diagram is shown in Fig. 2.49a, illustrating the presence of a gap of atomic thickness δ and an associated voltage drop Δ. Similar to the idealized case without surface states, all voltage drops within the space charge region in silicon when δ and Δ approach zero (Fig. 2.49b).

If the surface state density is high, a minor displacement in the Fermi-level from the neutrality level causes a large change in surface charge. If E_F drops slightly below ϕ_0, there is a large positive charge and part of the field lines terminating on the metal emanate from this charge rather than from the ionized impurities in silicon. Similarly, when E_F moves above ϕ_0, there is a large net negative charge at the surface and the barrier height increases. The Fermi level tends to be locked into (pinned to) the charge neutrality level so that the barrier becomes almost independent of the metal. Also, the observed metal workfunction becomes dependent on the surface state density. An apparent metal workfunction, ϕ_{m-app}, is then measured that is different from the vacuum workfunction ϕ_m. The barrier, therefore, depends

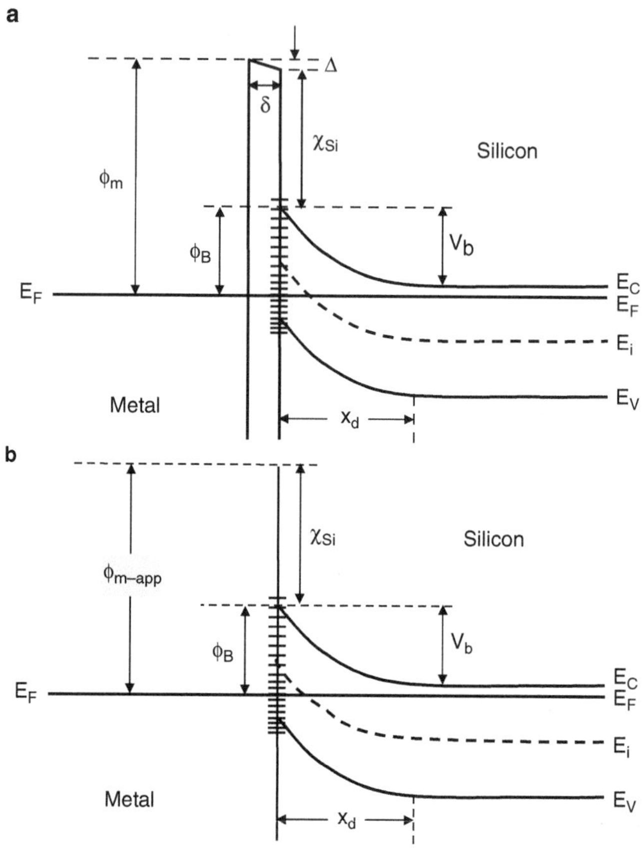

Fig. 2.49 Metal–silicon contact illustrating Fermi-level pinning by high surface-state density. **a** Gap present, **b** No gap

on the silicon surface properties. This is particularly the case when the surface is chemically treated.

Bardeen's theory on surface states is approximated as [34, 35]

$$\phi_B \approx \gamma(\phi_{m-app} - \chi_{Si}) + (1-\gamma)\left(\frac{E_g}{q} - \phi_0\right). \tag{2.146}$$

E_g is the energy gap, ϕ_0 the neutrality level, χ_{Si} the electron affinity in silicon, ϕ_{m-app} the apparent metal work function. γ is given by

$$\gamma = \frac{\varepsilon_{gap}}{\varepsilon_{gap} + q^2 \delta\, D_{it}}, \tag{2.147}$$

where ε_{gap} is the permittivity of the interfacial layer ($3.9\varepsilon_0$ in case of silicon-dioxide), δ the gap thickness, and D_{it} the density of surface states per eV and per unit area. For $D_{it} = 0$, $\gamma = 1$ and $\phi_B = \phi_m - \chi_{Si}$, as is the case for an idealized Schottky model. When D_{it} tends to infinity, γ tends to zero and $\phi_B = \frac{E_g}{q} - \phi_0$.

2.3.1.2 Image Force Barrier Lowering

As an electron approaches a conducting plane in vacuum, the conductor becomes polarized and exerts an attractive force on the electron. For large distances x from the conductor compared to interatomic dimensions, the conductor can be considered homogeneous [36]. An electrostatic field is produced as if there were a positive charge located at the mirror image of the electron in the plane of the metal-vacuum interface. Since the image charge is of opposite polarity, it lowers the potential energy of the approaching electron by $-q^2/4x$. Figure 2.50 illustrates this effect for a metal-semiconductor contact where ϕ_{B0} and V_{B0} indicate the barrier heights without consideration of image-force lowering.

The location of the saddle point x_m and the barrier lowering $\Delta\phi$ in Fig. 2.50 are given as [2,37]

$$x_m = \sqrt{\frac{q}{16\pi\varepsilon_0\varepsilon_{Si}E}}, \tag{2.148}$$

$$\Delta\phi = \sqrt{\frac{qE}{4\pi\varepsilon_0\varepsilon_{Si}}} = 2Ex_m, \tag{2.149}$$

where E is the externally applied field. $\Delta\phi$ depends on the maximum field that would exist without image-force lowering and is a function of applied bias. For an electric field of $10^5\,\mathrm{V/cm}$, $x_m \cong 1.75\,\mathrm{nm}$ and $\Delta\phi \cong 35\,\mathrm{mV}$. The barrier increases with increasing forward bias and decreases with increasing reverse bias.

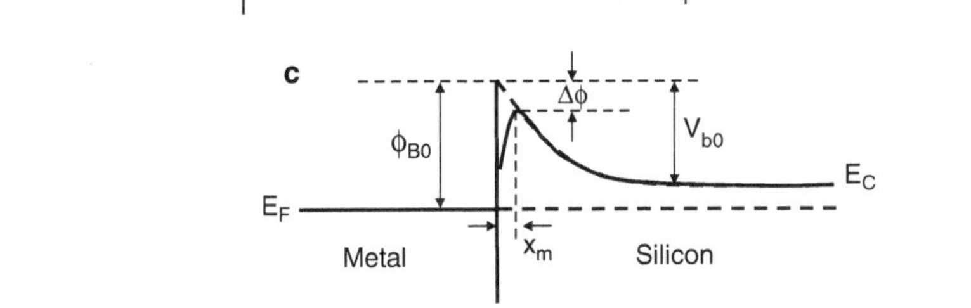

Fig. 2.50 Image-force barrier lowering. **a** Schottky barrier, **b** Image potential, **c** Resultant [35]

2.3.2 Current–Voltage Characteristics

The current–voltage characteristics are discussed here for n-type silicon. The discussion can be applied to p-type silicon with appropriate changes in polarities.

2.3.2.1 Forward Bias

When a forward voltage V_F is applied to the contact, the barrier seen by electrons in the silicon conduction band is reduced $V_b - V_F$ (Fig. 2.51).

The total current measured under forward bias is a combination of:

a) Thermionic emission of majority-carrier electrons from silicon conduction band into the metal,
b) Injection of minority-carrier holes from the metal interface into silicon,
c) Tunneling of electrons from the silicon conduction band into the metal.

Tunneling is only significant for dopant concentrations above $\sim 10^{18}\,\text{cm}^{-3}$, such as in Ohmic contacts.

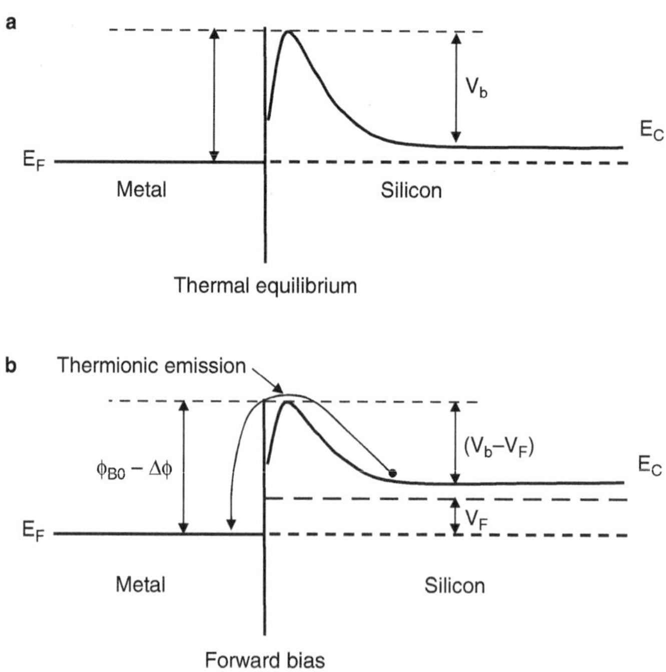

Fig. 2.51 Band diagram of Schottky barrier. **a** Thermal equilibrium, **b** Forward bias

Thermionic Emission

Thermionic emission from a metal can be described as the "evaporation" of electrons that possess energies larger than the metal workfunction, whereby the emission current density is given as [38]

$$j = AT^2 e^{-q\phi_m/kT} \quad A/cm^2, \tag{2.150}$$

where A is the Richardson constant defined as:

$$A = \frac{4\pi\, q\, m_0\, k^2}{h^3} \quad A/cm^2K^2. \tag{2.151}$$

In the above equations, T is the absolute temperature, ϕ_m the metal workfunction, q the electronic charge, m_0 the free-electron mass, k the Boltzmann constant, and h is Planck's constant. The theoretical value of A is then $\approx 120\,A.cm^{-2}K^{-2}$. Thermionic emission from the silicon conduction band into the metal can be described by an equation similar to (2.151) by substituting the effective mass m^* for m and the corresponding Richardson constant A^* for A [39]. For all practical purposes, it will be assumed that for electrons $A^* \approx 120$ and for holes, $A^* \approx 32\,A.cm^{-2}K^{-2}$ [39].

The Schottky barrier current–voltage characteristic under forward bias is similar to that of a pn junction. When the current is sufficiently low so that series resistances can be neglected, the current density is given by

$$j = j_s(e^{qV_F/nkT} - 1) \quad Acm^{-2}, \tag{2.152}$$

where

$$j_s = A^* T^2 e^{-\phi_b/kT} \quad Acm^{-2}. \tag{2.153}$$

J_s is the saturation current density and n is an ideality factor that ranges from 1.04 to 1.2 and is related to the increase in barrier height due to the reduction of image force lowering under forward bias.

It should be noted that electrons crossing the barrier from silicon into the metal possess energies that exceed the Fermi energy by about 1 eV. As they move in the metal, they lose their excess energy by collisions with optical phonons.

Barrier Height Measurement

One method to obtain the barrier height is to extract it from the forward characteristics defined by (2.152) and (2.153). The ideality factor is found from (2.152) by taking the ratio of two successive current measurements, thus eliminating j_s. The saturation current j_s is then extracted from (2.152). Finally, ϕ_B is found from (2.153) using an assumed value for the Richardson constant. Measurements are typically done at different current densities, keeping the current levels sufficiently low so that resistances in series with the diode can be neglected. An alternative method is to measure the diode AC resistance given by [40]

$$R = R_{Series} + \frac{nkT}{qI} \quad Ohm, \tag{2.154a}$$

where *nkT/q*I is the dynamic diode resistance obtained from (2.92), and R_s the series resistance given by

$$R_s = R_{Bulk} + \frac{\rho_{Bulk}}{4r} + R_C \quad Ohm. \tag{2.154b}$$

The first term in (2.154b) is the resistance of the quasi-neutral region from the depletion boundary to the ohmic silicon contact. The second term is the spreading resistance of the Schottky-barrier contact to silicon of resistivity ρ_{Bulk}, approximated by a circular shape of radius r. The third term is the resistance of the ohmic contact to silicon (Sect. 2.3.3). Ideally, a plot of R versus I should give a straight line with slope nkT/q and intercept R_s. In typical structures, however, R_s is found to be current-dependent because of changes in the effective current path within the semiconductor as the current is varied. In this case, R_s must be extracted numerically.

Minority-Carrier Injection and Charge Storage

The Schottky barrier is typically a majority-carrier, non-injecting diode, earning it the property of a high switching-speed device. There is, however, unavoidable minority-carrier injection from the metal-silicon interface into silicon. This injection current is only significant under high forward-bias conditions. If the barrier height exceeds half of the energy gap, the resulting band-bending indicates that the surface must be inverted, that is, for n-type silicon the concentration of minority-carrier holes at the surface must be larger than that of electrons. The downward band-bending away from the surface, however, represent a barrier to hole injection so that the injection ratio, that is, the ratio of minority-carrier current to total current, is extremely small, of the order of 10^{-4} for small forward bias. Above a critical current density where the voltage drop across the series resistance R_s becomes significant, the injection ratio begins to increase linearly with current density as [41]:

$$\frac{j_p}{j} = \frac{n_i^2 j}{b\, N_D^2\, j_s}, \tag{2.155}$$

where j_p is the minority-carrier hole current density, j the total current density, n_i the intrinsic concentration, b the ratio of electron to hole mobility, N_D the donor concentration in silicon, and j_s the saturation current density given by (2.153).

Minority-carrier injection in a Schottky barrier results in an undesirable increase in charge storage, which reduces the switching speed. Assuming a plane of surface recombination velocity σ beneath the contact, the charge storage can be approximated as [41]

$$Q \cong \frac{q n_i^2 D_p j}{N_D j_s \sigma} \quad C/cm^2. \tag{2.156}$$

2.3.2.2 Reverse Bias

The reverse characteristics of a Schottky barrier diode are treated in the same way as for a pn junction. The mechanisms responsible for the reverse current in Schottky-barrier diodes are, however, fundamentally different from those in pn junctions.

Reverse Current

The reverse current consists primarily of thermionic emission of electrons from the metal into silicon. Applying a reverse bias V_R increases the barrier from V_b to $V_b + V_R$ (Fig. 2.52). For V_R larger than about $-3kT/q$, the exponential in (2.152) can be neglected and the reverse current density $j_R = -j_s$. As for pn junctions, thermal generation within the depletion region and diffusion of minority carriers from the bulk toward the depletion region also contribute to the reverse current. Their magnitude is, however, negligible compared to j_s. Since the barrier height decreases with increasing reverse voltage, j_R does not saturate to the value j_s given by (2.152). Instead, j_R increases with reverse voltage as $j_R = j_s e^{\Delta\phi/kT}$ where $\Delta\phi$ is the image force barrier lowering defined by (2.149).

The electric field in (2.149) increases with increasing V_R and depends not only on the dopant concentration in silicon but also on the geometry of the structure and properties of the metal-silicon interface. The field is enhanced at contact corners, edges, and asperities caused by a non-uniform interface. One method to reduce

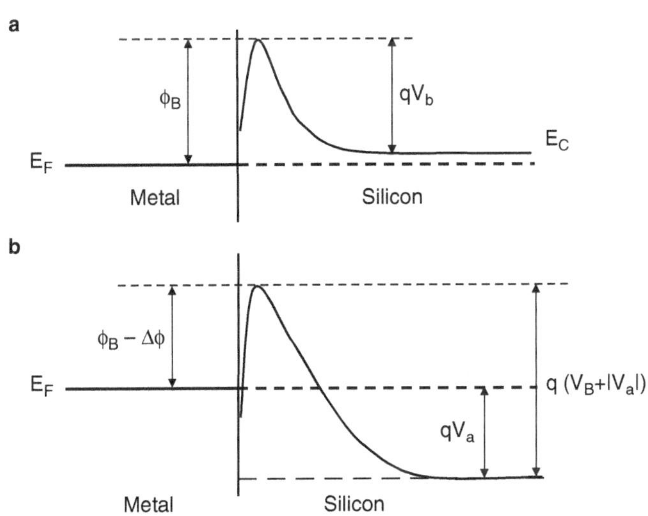

Fig. 2.52 Band diagram of Schottky barrier. **a** Thermal equilibrium, **b** Reverse bias

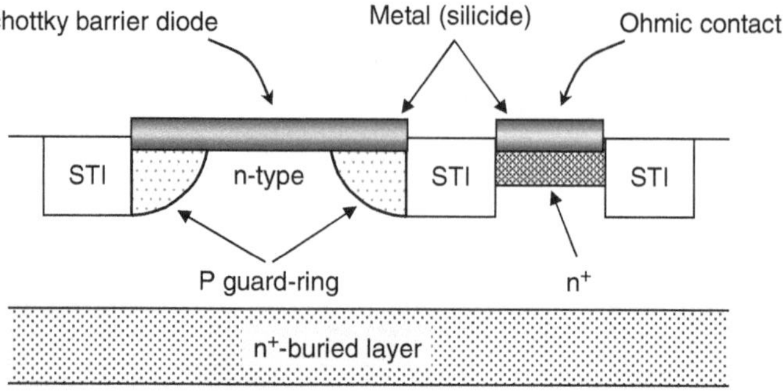

Fig. 2.53 Schottky-barrier diode with guard-ring

the corner and edge field is to place a guard ring around the contact, as shown in Fig. 2.53 for a silicide-silicon Schottky barrier bounded by oxide-filled shallow-trench isolation (*STI*, Chap. 7). The purpose of the buried layer in the figure is to reduce the series resistance. The reduction of field comes, however, at the cost of increased area and capacitance. The latter is exacerbated by an increase in minority carrier injection and storage.

Capacitance–Voltage Measurements

The barrier height can be obtained by plotting the inverse-square capacitance $1/C^2$ versus reverse voltage. The method is described in Sect. 2.1.4, (2.127), and illustrated for n-type silicon in Fig. 2.54.

For uniform N_D, the plot gives a straight line. The barrier height is extracted from the plot-intercept with the voltage axis as[4]

$$\phi_B \cong V_{\text{intercept}} + \phi_{Fn}, \tag{2.157}$$

where ϕ_B is the barrier height and ϕ_{Fn} the Fermi potential. The concentration in silicon is found from the slope of the plot.

An analogous treatment can be done for p-type silicon by appropriate changes in polarities.

[4] The method assumes that the depletion approximation is valid and the charge in silicon consists only of ionized impurities. A correction factor of kT/q can be added to ϕ_B, accounting for the tail in majority carriers at the depletion boundary.

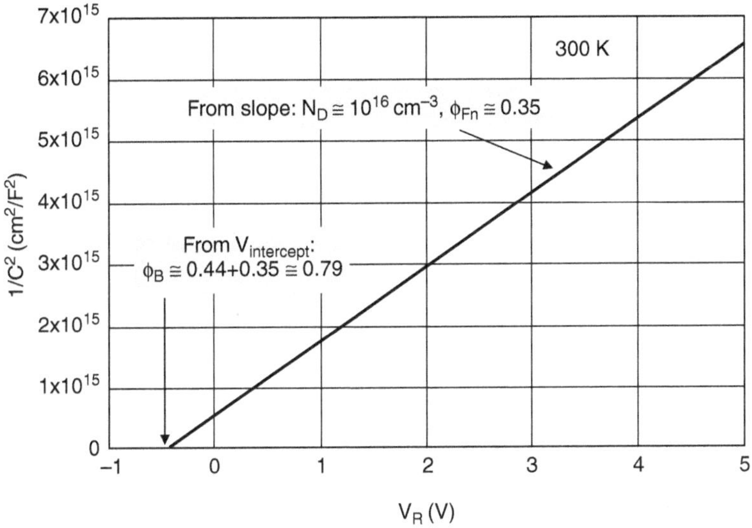

Fig. 2.54 Barrier-height extraction from $1/C^2$ versus V_R measurements

2.3.3 Ohmic Contacts

A contact is ohmic if its current–voltage characteristic is linear and symmetrical with respect to the origin (Fig. 2.55). The measured contact resistance is $\Delta V/\Delta I$. Understanding the physical nature of the contact and methods to reduce its resistance has become increasingly important as contact dimensions are reduced.

The contact resistance depends on the dopant concentration in silicon immediately beneath the contact, contact size, uniformity of current within the contact area, barrier height, interface properties, and measurement method.

In the absence of a barrier height or when the barrier height is very small, the contact would be ohmic. This situation is, however, seldom encountered. Therefore, relying on reducing the barrier height to make ohmic contacts is not realistic. The most practical way to reduce the contact resistance is to increase the dopant concentration beneath the contact. For concentrations less than about $10^{17}\,\mathrm{cm}^{-3}$, the forward and reverse currents are dominated by thermionic emission discussed in the preceding section. In this case, the depletion region is sufficiently wide so that tunneling of electrons through the barrier is negligible (Fig. 2.56a). As the concentration is increased, the depletion region narrows so that regions near the top of the barrier become sufficiently thin to allow tunneling (Fig. 2.56b). In the concentration range 10^{17} to about $10^{19}\,\mathrm{cm}^{-3}$, the conduction mechanism is a mixture of thermionic and tunneling current. At higher concentrations, the probability for electron to tunnel through the barrier in both directions becomes very high. The current is then dominated by tunneling through the barrier instead of thermionic emission over the barrier (Fig. 2.56c).

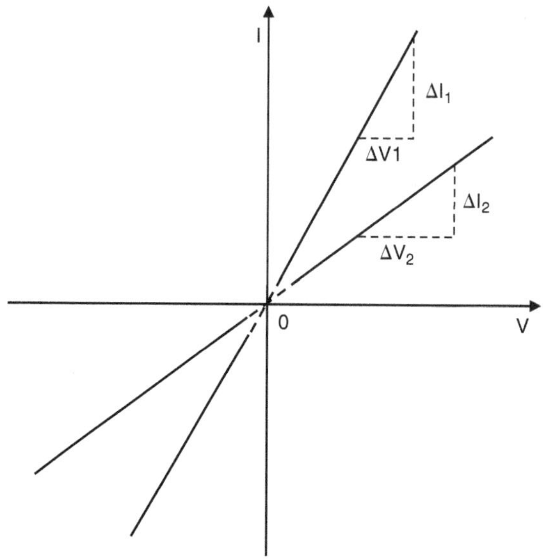

Fig. 2.55 Ohmic contacts of different resistances

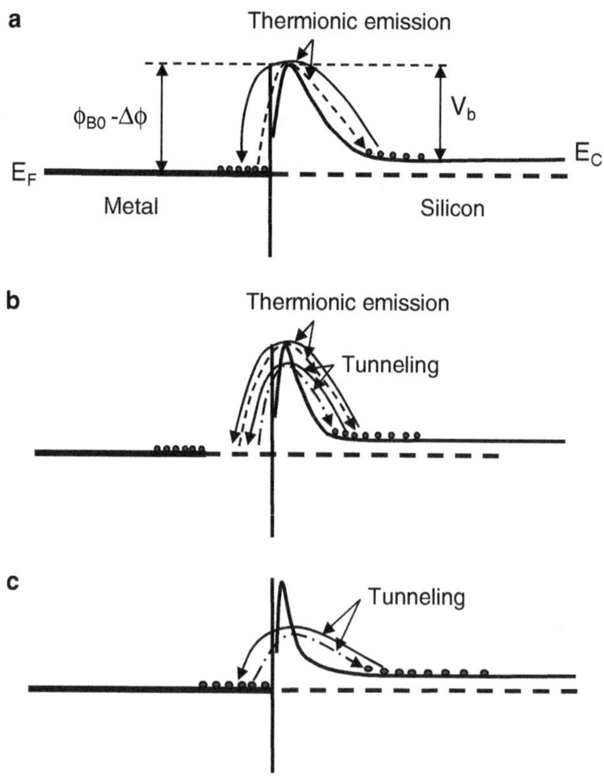

Fig. 2.56 Schematic to illustrate conduction mechanisms for different dopant concentration ranges. **a** Low, $<\approx 10^{17}\,cm^{-3}$, **b** Medium, $\approx 10^{17}$–$10^{19}\,cm^{-3}$, **c** High, $>\approx 10^{19}\,cm^{-3}$

2.3.3.1 Specific Interface Contact Resistance

The specific interface contact resistance (or contact resistivity) is defined as

$$\rho_c = \left.\frac{\partial V}{\partial j}\right|_{V=0} \quad \text{Ohm}-\text{cm}^2. \tag{2.158}$$

For thermionic emission, ρ_c is found from (2.152) and (2.153) as

$$\rho_c = \frac{k}{qAT} e^{q(\phi_B - \Delta\phi)/kT}. \tag{2.159}$$

For an effective barrier height of 0.57 eV, this yields a very high specific contact resistance of about 10 Ohm-cm^2.

When tunneling dominates, ρ_c is approximated by [42]

$$\rho_c = \frac{k}{qAT} e^{q(\phi_B - \Delta\phi)/E_{00}}. \tag{2.160}$$

where E_{00} is an energy that is characteristic of the tunneling probability, defined as [43]

$$E_{00} = \frac{q\hbar}{2}\sqrt{\frac{N_D}{m\varepsilon_{Si}}}. \tag{2.161}$$

E_{00} is shown as a function of dopant concentration in Fig. 2.57. Tunneling begins to dominate when $E_{00} > kT$, that is, at concentrations above $3 \times 10^{19}\,\text{cm}^{-3}$. [43].

The specific interface resistance is plotted in Fig. 2.58 as a function of inverse square-root of N_D. At concentrations below $10^{17}\,\text{cm}^{-3}$, ρ_c levels to

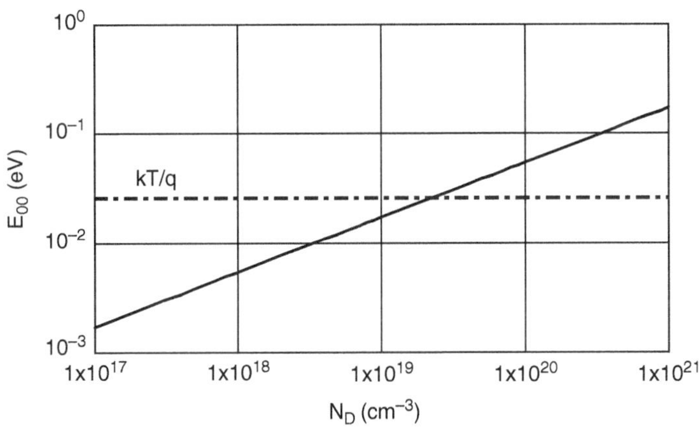

Fig. 2.57 Tunneling energy E_{00} versus dopant concentration. Tunneling becomes important when $E_{00} > kT$ [43,44]

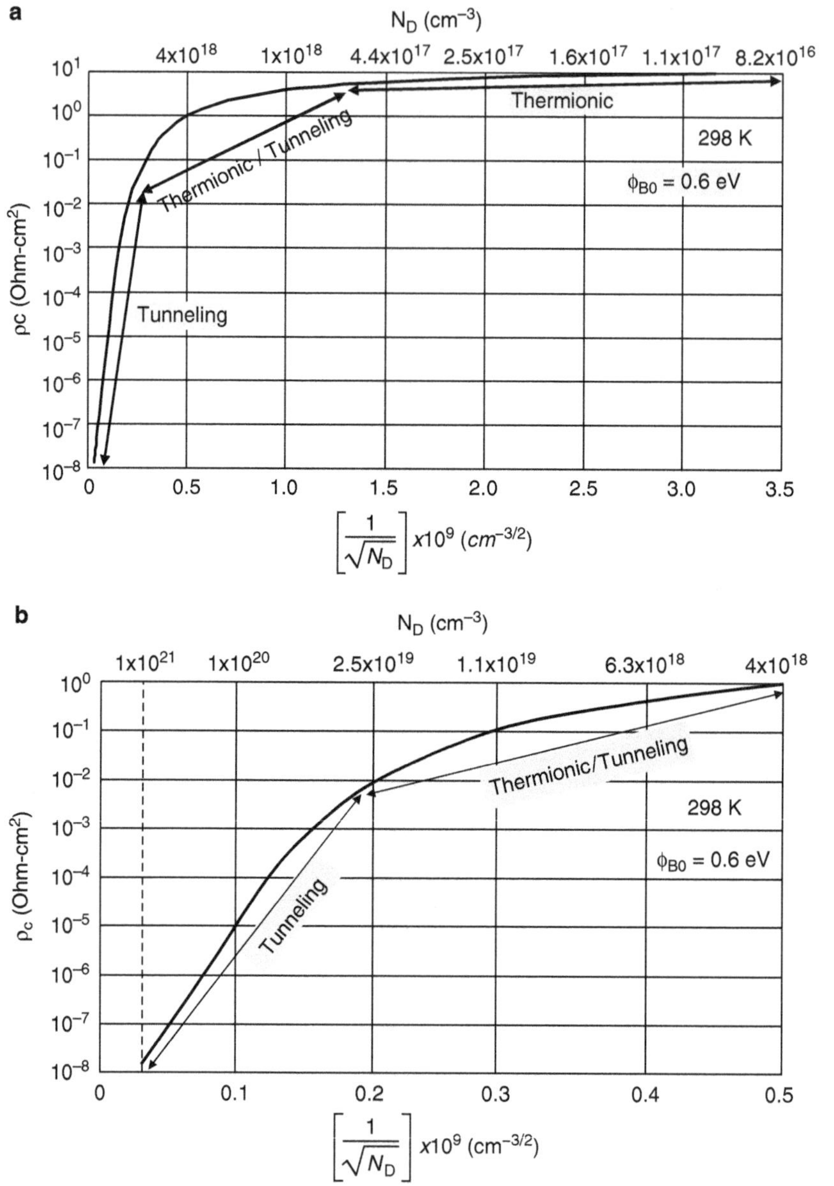

Fig. 2.58 a Dependence of ρ_c on dopant concentration showing the three different regions. **b** A blow-up of Fig. 2.58a showing details of the tunneling region

about 10 Ohm-cm^2 (Fig. 2.58a). There is a sharp decrease in ρ_c with in creasing concentrations above $\approx 3 \times 10^{19}\,\text{cm}^{-3}$ where tunneling begins to dominate (Fig. 2.58). The image-force barrier lowering in (2.149) is found by approximating the electric field as

$$E = \frac{Q_B}{\varepsilon_{Si}} = \sqrt{\frac{2qN_D(V_B - kT/q)}{\varepsilon_{Si}}}. \tag{2.162}$$

To achieve a specific interface resistance of $10^{-8}\,\text{Ohm}-\text{cm}^2$ the concentration must be $\geq 10^{21}\,\text{cm}^{-3}$. For this purpose, advanced doping and activation techniques must be developed to obtain a concentration above the normal solid-solubility limit (Chap. 7). A similar analysis for p-type silicon shows that the concentration must be $\geq 10^{20}\,\text{cm}^{-3}$ to achieve the same ρ_c.

2.3.3.2 Contact Resistance

For a uniform current density across the contact area and negligible resistance outside the metal-silicon interface, the contact resistance can be found by merely dividing ρ_c by the contact area. For $R_c = 10^{-8}\,\text{Ohm}-\text{cm}^2$ and a $0.25 \times 0.25\,\mu\text{m}^2$ contact, for example, the resistance would be 16 Ohms per contact. This is, however, an idealized condition. In most structures, the current density is not uniform across the contact area because of "current-crowding" effects near one edge of the contact due to the lateral flow of carriers as illustrated in Fig. 2.59 [45–49], and due to non-uniformities in the metal–silicon interface caused by, for example, residues or metal spikes.

The effective contact area is reduced and the local current density increases due to crowding near the contact edge (Fig. 2.59). A transmission-line model is used in [47,49] to estimate the resistance of a rectangular-shaped contact of width W and length L (Fig. 2.60). The model is valid for $W \gg L$, an infinitesimally thin junction and negligible metal resistance.

The contact resistance R_C is found as [47]

$$R_c = \frac{R_S L}{W}\left[\frac{\coth y}{y}\right], \tag{2.163}$$

where y is defined as

$$y = \sqrt{\frac{R_S L^2}{\rho_c}}. \tag{2.164}$$

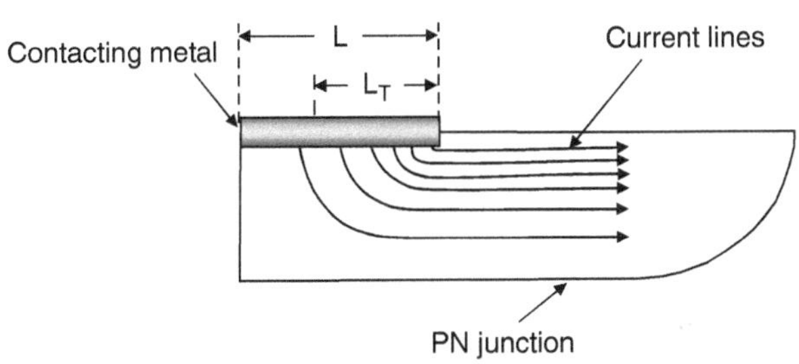

Fig. 2.59 Current crowding at contact edge. Width W is normal to paper. L is the contact length (parallel to direction of current), L_T is the "transfer length" [46]

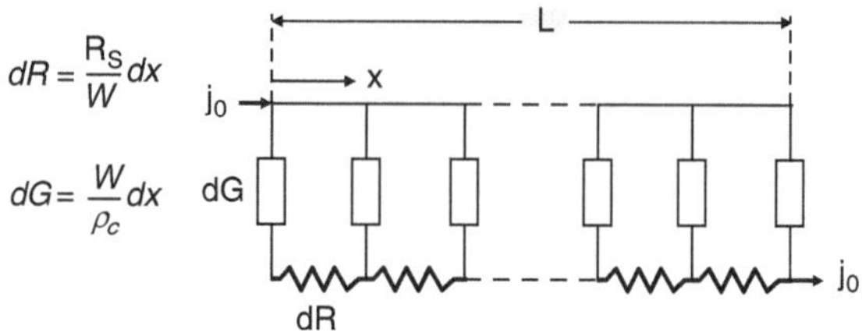

Fig. 2.60 Equivalent transmission-line for contact resistance [48]. *dR* and *dG* are, respectively, the elemental resistance and conductance. Width *W* is normal to paper

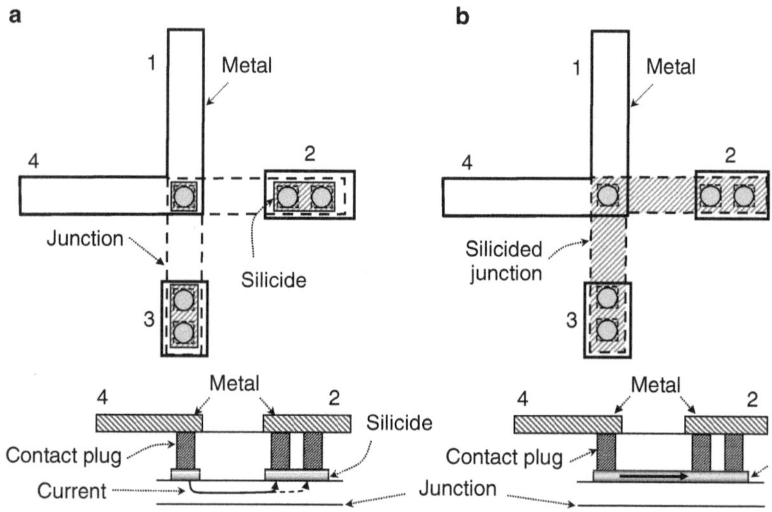

Fig. 2.61 Four-terminal configuration to measure contact resistance, illustrating multiple interfaces. **a** Current forced through the junction. **b** Most of current forced through silicide

R_S is the junction sheet resistance and ρ_c the specific interface resistance. The result is that the effective area $A = WL$ of the contact is reduced to

$$A_{eff} = \frac{A}{y \coth y}. \qquad (2.165)$$

The contact resistance is determined by contact size, interface specific resistance, and sheet resistance of the contacted film. Real contacts are more complex than the simple structure used to derive the above relations. Contacts are circular because minimum-size contacts are typically used for better control of size and uniformity and the contacts are rounded when patterned. The regions beneath and around the contact have finite dimensions leading to two- and three-dimensional effects not considered in the above derivations. Typical contacts consist of multiple interfaces, as illustrated in Fig. 2.61.

The interface between silicide and silicon can present problems caused by segregation of dopants into silicide, increasing the interface resistance. Also, the interface exhibits non-uniformities in electric field and current caused by spikes of silicide penetrating into silicon.

The contact plug consists of a core of tungsten or copper, separated from the silicide by a single or dual film of barrier metal. This interface of dissimilar materials contributes to the overall contact resistance. Finally, there is an interface between wiring metal and contact plug that needs to be considered.

The four-terminal configuration in Fig. 2.61 approximates the situation in a *MOSFET*, whereby the current in silicon is in the lateral direction.

The current is forced between two opposite metal lines, for example, from 4 to 2, and the voltage measured between the other two terminals. In the configuration on the left, all interfaces contribute to the measured contact resistance. On the right, the silicide to silicon interface does not contribute appreciably to the measured contact resistance.

The structure measures a contact resistance that is specific to the vertical and horizontal geometry of the contact. It is used to monitor the resistance and ensure that it remains within statistical process control limits. It is extremely difficult, however, to extract the specific contact resistance from this structure alone.

Other contact configurations include a layer of doped polysilicon between silicide and junction, such as in elevated source and drains or in the emitter of a bipolar transistor, as discussed in Chaps. 3 and 7. The polysilicon film typically serves as the dopant source to form the junction. In such cases, more sophisticated techniques must be introduced to extract the resistance of polysilicon and its interfaces.

2.4 Problems

1. In a step-junction, $N_D = 10^{18}\,cm^{-3}$ and $N_A = 10^{16}\,cm^{-3}$. Find for thermal equilibrium at 25 °C and 85 °C:

a. The built-in voltage,
b. The depletion widths, x_{dp} and x_{dn},
c. The total positive charge per unit area,
d. The peak field,
e. The capacitance per unit junction area.

2. A junction is formed by implanting and diffusing boron into a uniformly doped n-type region of concentration $10^{16}\,cm^{-3}$. The boron profile is Gaussian, with a peak of $5 \times 10^{18}\,cm^{-3}$ at a depth of 0.2 µm. The metallurgical junction depth is 0.8 µm.

a. Plot $|N_A - N_D|$ as a function of depth x in silicon.
b. Find the room-temperature thermal-equilibrium depletion width.
c. Approximate the forward bias voltage at which the excess minority-carrier concentration on the n-side is equal to the majority-carrier concentration.

d. For the forward bias found in c. what is the ratio of minority- to majority-carrier concentration at the depletion boundary in the p-side?

3. Consider a one-sided n^+p step junction having a junction depth of 0.3 μm and a uniform background concentration $N_A = 10^{17}\,cm^{-3}$. The effective density of generation-recombination sites in the p-region is $10^{10}\,cm^{-3}$. Assume that plane ohmic contacts are placed at the surface of the n^+-region and at a depth 0.8 μm below the surface in the p-region.

a. Calculate the electron current density for a forward bias voltage of 0.8V at 25 °C and 100 °C.
b. Punch-through occurs when the p-sided depletion region reaches the contact. The reverse voltage is increased until a current of $1\,\mu A/\mu m^2$ is measured. Would the main mechanism for this current be impact ionization, punch-through, or thermal-generation?
c. Repeat a., and b., assuming that the contact plane to the p-region is placed 0.5 μm beneath the surface.

4. Show that in the presence of injected carriers, the hole and electron current densities are given in one dimension by the following relations

$$j_n = \mu_n n \frac{dE_{Fn}}{dx},$$
$$j_p = \mu_p p \frac{dE_{Fp}}{dx},$$

where E_{Fn} and E_{Fp} are, respectively, the electron and hole quasi-Fermi levels.

5. A one-sided n^+p junction is formed by diffusing a heavily doped n-region to a depth of 0.5 μm into a 10 Ohm-cm p-substrate of thickness 725 μm. The density of recombination-generation sites in the p-region is $5 \times 10^{10}\,cm^{-3}$. A forward bias of 0.7 V is applied to the junction at 25 °C.

a. Will the minority-carrier electrons reach the backside of the substrate?
b. Estimate the time for the minority carriers to dissipate.

6. The drain of an NMOS is formed by implanting and diffusing heavily-doped arsenic through a mask opening into a p-type substrate. The metallurgical junction is 0.25 μm deep and the substrate is uniformly doped with boron at a concentration of $5 \times 10^{17}\,cm^{-3}$.

a. Assume a one-sided abrupt junction with cylindrical junction edges and neglect surface effects. Calculate the junction breakdown voltage at 25 °C.
b. Phosphorus is introduced through the same mask opening to extend the metallurgical junction by 0.15 μm on all sides Assume a linearly-graded junction of gradient $4 \times 10^{23}\,cm^{-4}$ is formed. Estimate the breakdown voltage of the extended junction.

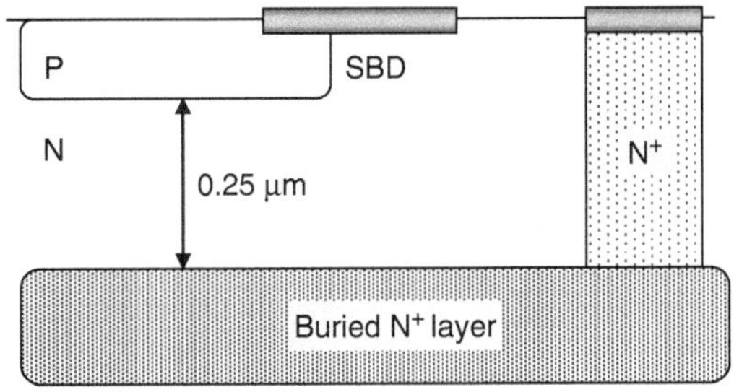

Fig. 2.62 PN junction clamped by SBD, problem 7

7. Consider the structure depicted in Fig. 2.62. The area of the p-region is $10\,\mu m^2$. The n-region is uniformly doped to a concentration of $5 \times 10^{16}\,cm^{-3}$. The silicide contacting the p-region extends into the n-region to form a Schottky-barrier diode (SBD) of barrier-height 0.8 eV and ideality factor $n = 1.1$. A forward bias voltage of 0.6V is applied to the junction at 85 °C. Assume that injected minority carriers recombine at the top-surface of the buried n^+ layer.

a. Find the SBD area that is necessary to ensure that only 10% of the forward current is due to minority carrier injection. Neglect series resistances.
b. For the SBD area found in a., estimate the leakage current at 85 °C for a reverse voltage of 2.5V applied to the junction. Neglect surface effects.

8. Only two points of capacitance versus reverse voltage measurements on a $100 \times 100\,\mu m^2$ Schottky-barrier are available: 5.09 pF at $V_R = 1V$ and 2.73 pF at $V_R = 5V$. Knowing that silicon is uniformly doped, find the concentration and barrier height.

9. Suggest test structures and define electrical test (E-test) plans to extract all pertinent parameters for

a. PN junctions,
b. Schottky-barrier diodes,
c. Ohmic contacts.

References

1. R. W. Dutton and Z. Yu, Technology CAD, Computer Simulation of IC Process and Devices, Kluwer Academic Publishers, 1993.
2. S. M. Sze, Physics of Semiconductors, John Wiley & Sons, 1981.
3. H. Armstrong, "A theory of voltage breakdown of cylindrical P–N junctions, with applications," IRE Trans. Electron Dev., ED-4, 15–16, 1957.
4. A. B. Phillips, Transistor Engineering, McGraw-Hill, New York, 1962.
5. W. Shockley, and W. T. Read, "Statistics of recombination of holes and electrons," Phys. Rev. 87, 835, 1952.
6. R. N. Hall, "Electron-hole recombination in Germanium," Phys. Rev., 87, 387, 1952.

7. A. K. Jonscher, Principles of Semiconductor Device Operation, John Wiley & Sons, New York, 1960.
8. J. L. Moll, Physics of Semiconductors, McGraw-Hill, New York, 1964.
9. A. G. Chynoweth, "Ionization rates for electrons and holes in silicon," Phys. Rev., 109 (5), 1537–1540, 1958.
10. G. A. Baraff, "Distribution functions and ionization rates for hot electrons in silicon," Phys. Rev., 128 (6), 2507–2517, 1962.
11. R. Van Overstraeten and H. DeMan, "Measurement of the ionization rates in diffused silicon p-n junctions," Solid State Electron., 13 (5), 583–608, 1970.
12. W. N. Grant, "Electron and hole ionization rates in epitaxial silicon at high electric fields," Solid State Electron., 16, 1189–1203, New York 1973.
13. A. D. Sutherland, "An improved empirical fit to Baraff's universal curves for the ionization coefficients of electron and hole multiplication in semiconductors," IEEE Trans. Electron. Dev., ED-27 (7), 1299–1300, 1980.
14. C. R. Crowell and S. M. Sze, "Temperature dependence of avalanche multiplication in semiconductors," Appl. Phys. Lett., 9 (6), 242–244, 1966.
15. S. Reggiani, E, Gnani, M. Rudan, G. Baccarani, C. Corvasce, D. Barlini, M. Ciappa, W. Fichtner, M. Denison, N. Jensen, G. Groos, and M. Stecher, "Measurement and modeling of the electron impact-ionization coefficient in silicon up to very high temperature," IEEE Trans. Electron. Dev., 52 (10), 2290–2299, 2005.
16. S. M. Sze and G. Gibbons, "Avalanche breakdown voltage of abrupt and linearly graded p-n junctions in Ge, Si, GaAs and GaP," Appl. Phys. Lett., 8, 111, 1966.
17. K. G. McKay, "Avalanche breakdown in silicon," Phys. Rev., 94, 877–884, 1954.
18. R. M. Warner, "Avalanche breakdown in silicon diffused junctions," Solid State Electron., 15, 1303, 1972.
19. J. H. He, X. Zhang, and Y. Wang, "Equivalent doping profile transformation: a semi-empirical analytical method for predicting breakdown characteristics of an approximate single-diffused parallel-plane junction," IEEE Trans. Electron. Dev., 48 (12), 2763–2768, 2001.
20. S. M. Sze and G. Gibbons, "Effect of junction curvature on breakdown voltages in semiconductors," Solid State Electron., 9, 831–840, 1966.
21. S. K. Ghandhi, Semiconductor Power Devices, John Wiley, New York, 1977.
22. D. Krizaj and S. Amon, "Breakdown voltage of elliptic pn junctions," Fifth European Conference on Power Electronics and Applications, Vol. 2, pp. 293–296, Sept. 13–16, 1993.
23. R. B. Fair and W. W. Hayden, "Zener and Avalanche Breakdown in As-implanted low-voltage Si n-p junctions," IEEE Trans. Electron. Dev., ED-23(5), 512–518, 1976.
24. C. Zener, "A theory of electrical breakdown voltages of solid dielectrics," Proc. Roy. Soc. London, A145, 523–529, 1934.
25. F. Braun, "Ueber die Stromleitung durch Schwefelmetalle," Ann. Phys. J. C. Poggendorff. Phys. Chem., 153, 556–563, 1874.
26. F. Braun, "Ueber Abweichungen vom Ohm'schen Gesetz in metallisch leitenden Koerpern," Ann. Phys. G. Wiedemann, 1, 95–110, 1877.
27. C. A. Mead, "Physics of interfaces," in *Ohmic Contacts to Semiconductors* (B. Schwartz, ed.), Electrochem. Soc., New York, 3–16, 1969.
28. W. Schottky, "Halbleitertheorie der Sperrschicht," Naturwissenschaften, 26, 843, 1938; Z. Phys. 113, 367, 1939; 118, 539, 1942.
29. H. K. Henisch, "Rectifying Semiconductor Contacts," Clarendon, Oxford, 1957.
30. M. M. Atalla, "Metal-semiconductor Schottky barriers, devices and applications," Proc. Munich Symp. Microelectronics, pp. 123–157, October 1966.
31. J. Bardeen, "Surface states and rectification at a metal semi-conductor contact," Phys. Rev., 71 (10), 717–727, 1947.
32. W. Shockley, "On the surface states associated with a periodic potential," Phys. Rev., 56 (4), 317–323, 1939.
33. W. H. Brattain and W. Shockley, "Density of surface states on silicon deduced from contact potential measurements," Phys. Rev., 72, 345, 1947.

34. A. M. Cowley and S. M. Sze, "Surface sates and barrier height of metal-semiconductor systems," J. Appl. Phys., 36, 3212, 1965.
35. E. H. Rhoderick, "The physics of Schottky barriers," Third Solid-State Device Conf., pp. 1153–1168, Exeter, 1969.
36. A. J. Dekker, Solid State Physics, Prentice-Hall, New Jersey, USA, 1965.
37. V. L. Rideout and C. R. Crowell, "Effects of image force and tunneling on current transport in metal-semiconductor (Schottky barrier) contacts," Solid-State Electron., 13, 993–1009, 1970.
38. H. A. Bethe, "Theory of boundary layer of crystal rectifiers," MIT Radiation Lab. Rept., 43/12, 1942.
39. C. R. Crowell, "The Richardson constant for thermionic emission in Schottky barrier diodes," Solid-State Electron., 8, 395–399, 1965.
40. A. M. Cowley, "Titanium-Silicon Schottky barrier diodes," Solid-State Electron., 12, 403–414, 1970.
41. D. L. Scharfetter, "Minority carrier injection and charge storage in epitaxial Schottky barrier diodes," Solid-State Electron., 8, 299–211, 1965.
42. A. Y. C. Yu, "Electron tunneling and contact resistance of metal-silicon contact barriers," Solid-State Electron., 13, 239–247, 1970.
43. F. A. Padovani and R. Stratton, "Field and thermionic-field emission in Schottky barriers," Solid-State Electron., 9, 695–707, 1966.
44. D. K. Schroeder and D. L. Meier, "Solar cell contact resistance: A review," IEEE Trans. Electron. Dev., ED-31 (5), 637–647, 1984.
45. D. P. Kennedy and P. C. Murley, "A two-dimensional mathematical analysis of the diffused semiconductor resistor," IBM J. Res. Dev., 12, 242–250, 1968.
46. H. Murrmann and D. Widmann, "Current crowding on metal contacts to planar devices," IEEE Trans. Electron. Dev., ED-16, 1022–1024, 1969.
47. H. Murrmann and D. Widmann, "Messung des Uebergangswiderstandes zwischen Metall und Diffusionsschichet in Si Planarelementen," Solid-State Electron., 12, 879–886, 1969.
48. H. Murrmann and D. Widmann, "Current crowding on metal contacts to planar devices," IEEE Trans. Electron. Dev., ED-16, 1022–1024, 1969.
49. H. H. Berger, "Models for contacts to planar devices," Solid-State Electron., 15, 145–158, 1972.

Chapter 3
The Bipolar Transistor

3.1 Introduction

The concepts and derivations developed for the pn junction in Chap. 2 are directly applicable to the bipolar transistor since the transistor is formed by placing two pn junctions back-to-back, arranged vertically or laterally. The resulting structure can be described as a three-layer sandwich of p-type and n-type material (Fig. 3.1). Since the merged center can be p-type or n-type, there are two kinds of bipolar junction transistors (*BJT*): *NPN* and *PNP*. The center layer is called the base of width W_b. When the transistor is operated as an amplifier, one of the junctions is forward-biased while the other is reverse-biased. The outer layer of the junction that is forward-biased is called the emitter, because it emits (injects) minority carriers into the base. If the base region is narrow enough, the injected minority carriers traverse the base at a certain speed and reach the reverse-biased junction where they are collected. Therefore, the outer layer of the junction that is reverse biased is called the collector.

The currents into or out of the three layers of the transistor are called the base current, I_B, the emitter current, I_E, and the collector current, I_C. Both carrier polarities take part in transistor action.

When the distance between the two junctions is much larger than the minority-carrier diffusion length in the base, transistor action is not possible because injected minority carriers recombine before reaching the collector. Whenever two or more pn junctions share the same piece of silicon having a size comparable to the minority-carrier diffusion length, there is the possibility of bipolar action. In some cases, such as when suppressing latch-up in *CMOS* (Chap. 8), it is necessary to reduce the bipolar gain. High-performance transistors are, however, optimized for gain and speed by maximizing the emitter-base injection ratio, minimizing the minority-carrier transit time through the base, and minimizing parasitic capacitances and resistances outside the active regions.

B. El-Kareh, *Silicon Devices and Process Integration: Deep Submicron and Nano-Scale Technologies,*
© Springer Science+Business Media, LLC 2009

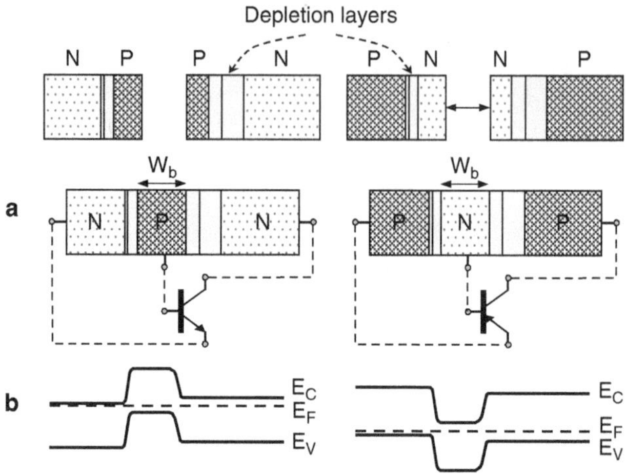

Fig. 3.1 **a** Formation of a *NPN* and *PNP* bipolar transistors by placing two pn junctions back to back. **b** Band diagrams shown schematically at thermal-equilibrium condition

3.2 Transistor Action, a Qualitative Description

The basic bipolar transistor characteristics are described for an idealized NPN structure under several simplifying assumptions. Real structures are then gradually introduced, leading to state-of-the-art technologies and designs. By interchanging n for p and p for n and reversing voltage polarities and current directions, the discussion and derivations are equally applicable to a PNP structure. In this sense, NPN and PNP transistors are said to be complementary.

3.2.1 Nomenclature and Regions of Operation

The bipolar structure is a 3-terminal device. If the voltages or currents are known for two terminals, then, as a consequence of Kirchhoff's laws, they are also known for the third. The transistor has therefore only two independent voltages and two independent currents. The substrate on which the transistor is constructed constitutes a fourth terminal that is typically biased at ground and only considered in the analysis when parasitic capacitances and currents come into play.

Figure 3.2a defines the conventional symbols, voltage polarities and current directions in *NPN* and *PNP* transistors. The conventional current points in the direction of positive charge flow. To avoid confusion, the electron and hole currents will also be indicated. The voltage symbols define the voltage polarity with the first letter in the subscript indicating the positive pole. When one of the terminals floats, that is, is open, while the other two are biased, the "open" is indicated by an "*O*" in the subscript (Fig. 3.2b). For example, V_{CEO} indicates that the collector is positive

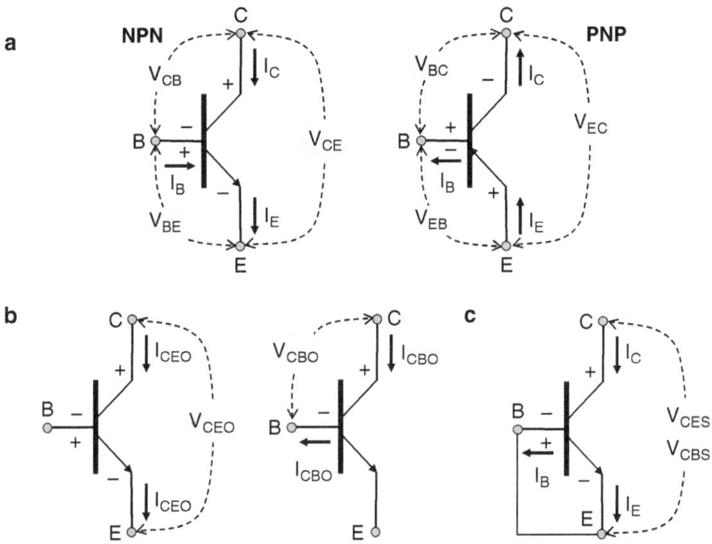

Fig. 3.2 *NPN* and *PNP* transistor conventions and symbols. **a** Current direction and voltage polarities; current points in direction of positive charge flow, first letter in voltage subscript indicates positive pole. **b** Subscript "*O*" indicates open terminal. **c** Subscript "*S*" indicates second and third terminals shorted

Table 3.1 Four modes of transistor operation [1]

Mode	Emitter-base bias	Collector-base bias
Off	Reverse	Reverse
Forward active	Forward	Reverse
On, or saturated	Forward	Forward
Reverse active	Reverse	Forward

with respect to the emitter and the base is open. Similarly, V_{CBO} says that the collector is positive with respect to the base and the emitter is open. If two terminals are shorted while the third is biased, this is shown as an "*S*" in the subscript (Fig. 3.2c). For example, V_{CES} or V_{CBS} mean that the collector is biased positive with the emitter shorted to base or base shorted to emitter, depending on whether the emitter or base is contacted. In most cases, V_{CES} and V_{CBS} result in the same current-voltage characteristics.

The four modes of transistor operation are summarized in Table 3.1. When both the emitter and collectors are reverse-biased, the transistor is off (cut-off, or logical off). The current measured consists of the junction reverse currents discussed in Chap. 2.

The transistor is in the forward active region when the emitter is forward biased and the collector reverse biased. When both the emitter and collector are forward-biased, the transistor is said to be in the "on," or saturation mode. This state approximates a closed switch and corresponds to a logical "on." Finally, when the collector is forward biased and the emitter reverse biased, the transistor is in the reverse active mode. The four modes of operation are discussed in this chapter.

3.2.2 Idealized Structure

Consider the *NPN* transistor in Fig. 3.3, biased in the active mode, as in an amplifier. For a qualitative description of transistor action and relations between terminal currents, an idealized structure is first described with the following simplifying assumptions:

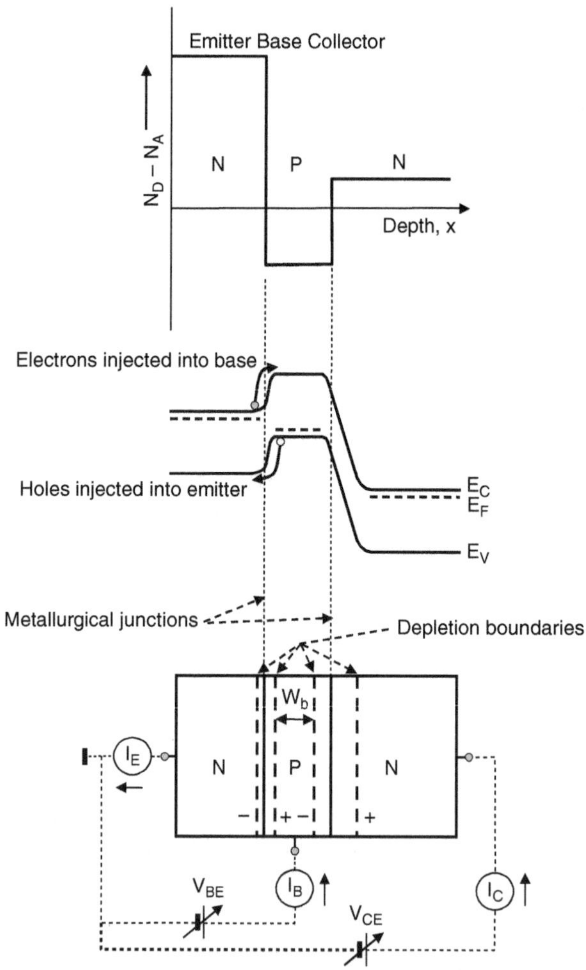

Fig. 3.3 Ideal *NPN* transistor biased in the active mode. The bias polarities are shown at the depletion boundaries. W_b is the neutral base-width. The emitter-base barrier height is reduced by V_{BE}; the collector-base barrier height is increased by V_{CE}

(a) The structure has a uniform cross-sectional area and all variables depend only on the direction normal to the emitter.
(b) The dopant concentrations in all regions are uniform and change abruptly from one region to the other.
(c) The width of the base region is very small compared to the minority-carrier diffusion length, so that recombination of minority carriers in the base is negligible. Thus, all injected minority carriers from the emitter into the base reach the collector.
(d) Resistances outside the active transistor region are negligible.
(e) All minority-carrier holes injected from the base into the emitter reach the emitter contact.
(f) The physical base-width is larger than the sum of both depletion regions in the base (at the emitter-base and collector-base junctions).
(g) The electric fields are moderate so that impact ionization is negligible.
(h) Leakage currents are negligible.

Electrons are injected from the emitter into the base. The minority-electron concentration at the boundary of the emitter-base depletion region (in the base) increases above the equilibrium concentration (Fig. 3.4). The excess electrons are transported by diffusion through the field-free (neutral) base region.[1] Once the electrons reach the collector-base depletion boundary, they are swept quickly into the n-type collector.

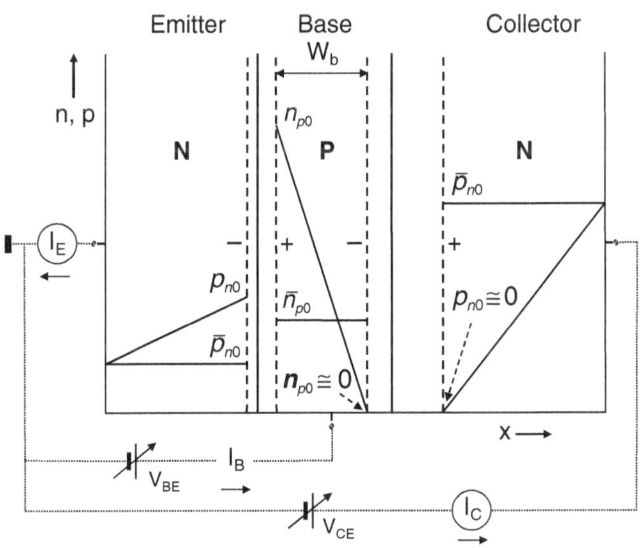

Fig. 3.4 Distribution of minority carriers in the active bias condition. p_{n0}, n_{p0} are the minority-carrier concentrations at the depletion boundaries under applied bias conditions and $\overline{p}_{n0}, \overline{n}_{p0}$ are the concentrations at thermal equilibrium

[1] In a non-uniform base, the region is more correctly referred to as quasi-neutral because of the existence of a built-in field and hence charge separation.

The minority-electron concentration at the collector-base depletion boundary (in the base) decreases far below the thermal-equilibrium concentration. Consequently, a gradient in minority-carrier concentration is created in the base, giving rise to electron diffusion current, I_n. Under the above assumptions, the collector-base leakage current can be neglected and $I_n \approx I_C$. Thus, the collector current is under direct control of the emitter-base forward voltage because that voltage determines the concentration of excess minority carriers and hence the electron diffusion current. The transistor is therefore similar to a valve in which the collector current is controlled by the emitter-base voltage. A small change in emitter-base voltage results in a large change in collector current.

Holes are also injected from the base into the emitter. Since the physical width of the emitter is assumed to be smaller than the diffusion length of holes in the emitter, the emitter contact acts as a "sink" for the injected holes, that is, electrons recombine instantaneously with holes reaching the emitter contact. The concentration gradient of holes in the emitter gives rise to a hole diffusion current, I_p. Under the above simplifying assumptions, the hole current constitutes the base current, that is, $I_B \approx I_p$. From Kirchhoff's current law, the emitter current is the sum of base and collector current:

$$I_E = I_B + I_C \approx I_p + I_n. \tag{3.1}$$

The electron injection efficiency γ is defined as the ratio of injected electron current to the total emitter current:

$$\gamma = \frac{I_n}{I_n + I_p}. \tag{3.2}$$

The base transport factor α_T is the ratio of injected electron current to total collector current:

$$\alpha_T = \frac{I_C}{I_n}. \tag{3.3}$$

Under the assumption of no recombination in the base, $\alpha_T \approx 1$. The ratio of collector to emitter current is the current gain α:

$$\alpha = \frac{I_C}{I_E} = \gamma\alpha_T \approx \gamma. \tag{3.4}$$

α is determined directly by measuring the emitter and collector currents in the grounded-base configuration (emitter-base forward-biased, collector-base reverse-biased). The ratio of collector to base current is referred to as the current gain β:

$$\beta = \frac{I_C}{I_B}. \tag{3.5}$$

The relation between current gains is

$$\alpha = \frac{\beta}{\beta + 1}, \tag{3.6}$$

$$\beta = \frac{\alpha}{1-\alpha}. \tag{3.7}$$

3.2.3 Ebers-Moll Equations

The first and simplest large-signal circuit model that describes the bipolar modes of operation in Table 3.1 was developed by J. J. Ebers and J. L Moll [2]. Under the simplifying assumptions made in 3.1.2, the general form of transistor operation can be treated as a superposition of forward and reverse modes of operation (Fig. 3.5). Treated separately, the current–voltage characteristics of the base-emitter and base-collector junctions are described as (Chap. 2)

$$I_{BE} = I_{Esat}\left(e^{qV_{BE}/kT} - 1\right), \tag{3.8}$$

$$I_{BC} = I_{Csat}\left(e^{qV_{BC}/kT} - 1\right), \tag{3.9}$$

where I_{Esat} and I_{Csat} are, respectively, the emitter and collector junction saturation currents.

Assume, for example, that both junctions are forward-biased. The emitter current is the algebraic sum of I_{BE} in (3.8) and electron fraction injected from the collector into the base that reaches the emitter. Following the conventional sign, this current is

$$I_E = -I_{BE} + \alpha_R I_{BC}, \tag{3.10}$$

where α_R is the grounded-base current gain in the reverse-active mode. Similarly, the collector current is the algebraic sum of I_{BC} in (3.9) and the electron fraction of I_{BE} that reaches the collector

$$I_C = -I_{BC} + \alpha_F I_{BE}, \tag{3.11}$$

where α_F is the grounded-base current gain in the forward-active mode. The base current is

$$I_B = (1-\alpha_F)I_{BE} + (1-\alpha_R)I_{BC}. \tag{3.12}$$

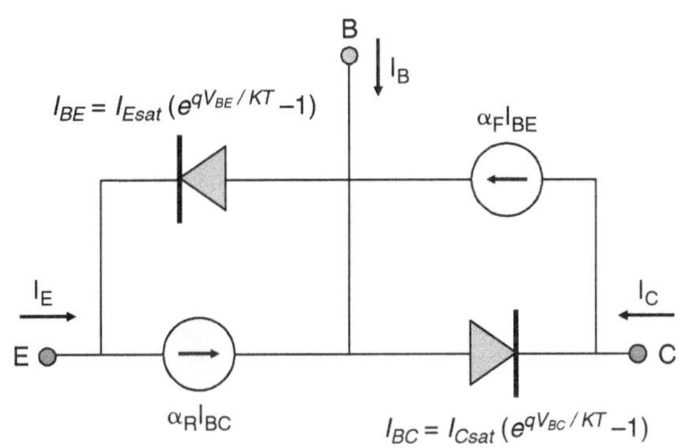

Fig. 3.5 Basic Ebers-Moll *NPN* transistor model

By combining (3.9)–(3.11), the basic equations for the Ebers-Moll model are derived for an NPN transistor as

$$I_E = -I_{Esat}(e^{qV_{BE}/kT} - 1) + \alpha_R I_{Csat}(e^{qV_{BC}/kT} - 1), \tag{3.13}$$

$$I_C = -I_{Csat}(e^{qV_{BC}/kT} - 1) + \alpha_F I_{Esat}(e^{qV_{BE}/kT} - 1). \tag{3.14}$$

Similar equations apply to a *PNP* transistor by appropriate changes in polarities. The currents are described by four parameters, I_{Esat}, I_{Csat}, α_F, and α_F. The number of parameters is reduced by one through reciprocity principle [3]

$$\alpha_F I_{Esat} = \alpha_R I_{Bsat}. \tag{3.15}$$

The reciprocity property becomes evident by considering that the base impurity profile in the *idealized* transistor is the same for forward and reverse injection. It can be shown, however, that the base impurity profile and area need not be uniform for (3.15) to be applicable [1]. When the transistor is operated in the forward active mode, the collector is reverse-biased and the exponential term in (3.9) is negative. This term goes rapidly to zero and $I_{BC} = -I_{Csat}$. The emitter current is then

$$I_E = -I_{BE} - \alpha_R I_{Csat}. \tag{3.16}$$

The second term in (3.16) is typically very small.

3.2.4 Collector Saturation Voltage, V_{CEsat}

When both junctions are forward-biased, the collector to emitter voltage is very small since it represents the sum of two opposing potentials. This voltage is referred to as the collector saturation voltage, V_{CEsat}. It is an important parameter in grounded-emitter switching applications.

As the forward voltage, V_{BE} or V_{BC} increases above $\sim 3\text{kT/q}$, the exponential term in (3.8) and (3.9) becomes very large and the 1 in the parentheses can be neglected. The emitter forward voltage can then be extracted from (3.8) and (3.10) as

$$V_{BE} = \frac{kT}{q} \ln \frac{I_{BE} + \alpha_R I_{BC}}{I_{Esat}}. \tag{3.17}$$

Similarly, (3.9) and (3.11) give the collector forward voltage as

$$V_{BC} = \frac{kT}{q} \ln \frac{I_{BC} + \alpha_F I_{BE}}{I_{Csat}}. \tag{3.18}$$

The collector to emitter voltage can be expressed in terms of I_C and I_B in the grounded-emitter configuration by combining (3.15), (3.17), and (3.18), and employing $I_E = I_B + I_C$ [1,4]

$$V_{CEsat} = \pm \frac{kT}{q} \ln \frac{\alpha_R[1 - I_C(1 - \alpha_F)/\alpha_F I_B]}{1 + I_C(1 - \alpha_R)/I_B}, \tag{3.19}$$

where the plus sign applies to a PNP transistor and the minus sign to an NPN transistor.

3.3 Planar Transistor, Low-Level Injection

The main features of an earlier version of a planar transistor are shown in Fig. 3.6.

The starting material is a p-type wafer that is lightly doped to ensure a low collector to substrate parasitic capacitance. A heavily-doped buried n-layer is implanted into the substrate.[2] The purpose of this layer is to reduce the overall collector

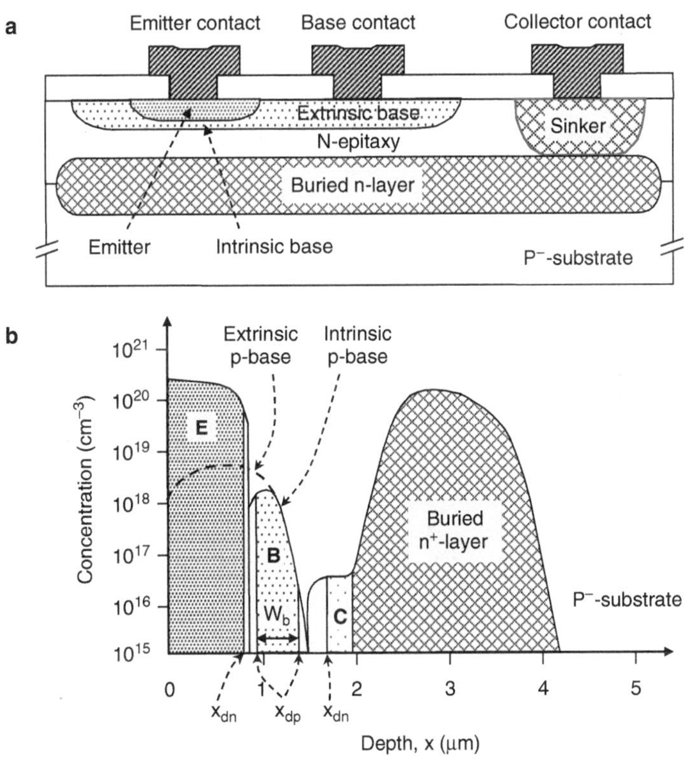

Fig. 3.6 Main features of a planar *NPN* transistor. **a** Schematic cross-section, **b** Approximate profile: cut-line through intrinsic base. W_b is the neutral base-width; x_{dn}, x_{dp} are, respectively, the depletion boundaries in the n-region and p-region

[2] Advanced patterning and other enabling processes are described in Chap. 7. Conventional unit processes, such as implantation, are detailed in [5].

resistance. A lightly n-type doped epitaxial layer is then grown. The epitaxial thickness and concentration are optimized primarily for transistor breakdown, capacitance and high-current density effects, as will be discussed in Sect. 3.4. To reduce the vertical resistance between the top collector contact and buried n-layer, a heavily n-type doped sinker is patterned, implanted and driven deep to merge with the buried layer. This is done before base and emitter formation to avoid excessive dopant diffusion during the sinker thermal cycle. The p-type base is then patterned, implanted and diffused, followed by a similar process for the heavily n-type doped emitter formation. Finally, metal contacts are formed directly on the emitter, base and collector (Fig. 3.6a).

The intrinsic base is the "active region" immediately under the emitter. This region is "pinched" by the emitter because the remainder of the base above it is compensated by the emitter. The base region outside the active area is called the extrinsic base.

The impurity profiles of emitter, base, collector and substrate are shown schematically in Fig. 3.6b. The emitter is heavily doped to increase its Gummel number and reduce the injection of holes from the base into the emitter ((2.88), Chap. 2). In earlier transistor versions, the junction depth of the emitter, x_{jE}, was larger than the diffusion length of minority-carrier holes in the emitter, L_{pE}. In this case, the injection of holes into the emitter depended on L_{pE} rather than on x_{jE}.

The intrinsic base is optimized for gain and speed. Thus, the base is narrow and formed with a concentration gradient that creates an electric field in a direction to accelerate minority carriers toward the collector. The extrinsic base (outside the emitter region, shown dotted in Fig. 3.6b) is optimized to reduce the base resistance while limiting the emitter-base tunneling leakage at the boundary near the surface.

The collector region immediately under the active base is lightly doped. The thickness and concentration of that region are a trade-off between base-push effect at high-current densities (Sect. 3.4.2), base-collector breakdown voltage, collector resistance and base-collector capacitance.

3.3.1 Low-Level Injection Parameters

Low-level injection conditions apply when the concentration of injected minority carriers into the base is sufficiently low that the increase in majority carrier concentrations remains negligible when compared to the thermal-equilibrium majority-carrier concentration (Chap. 2, Fig. 2.21).

3.3.1.1 Electron Injection into the Base

Consider an *NPN* transistor and assume initially that the base and emitter are uniformly doped. The thermal-equilibrium minority-carrier concentrations in the base $\overline{n}_p$ and emitter $\overline{p}_n$ are then flat, as shown in Fig. 3.7. When the transistor is biased in the forward-active mode, the emitter injects electrons into the base and the base injects holes into the emitter, raising the minority-carrier concentrations at the boundaries of the base-emitter depletion regions, x_{dp} and x_{dn} (Chap. 2). The excess concentrations are shown in Fig. 3.7 as Δn_{p0} and Δp_{n0}. The resulting concentration gradients cause the excess carriers to diffuse away from the depletion boundaries. Since the emitter is higher doped than the base, the concentration of excess holes in the emitter is smaller than that of excess electrons in the base (Chap. 2). Also, since the emitter is assumed to be wider than the hole diffusion length L_{pE}, the excess holes recombine before reaching the emitter contact.

In optimized NPN transistors, the base width W_b is considerably smaller than the electron diffusion length in the base, L_n. Thus, there is negligible recombination of minority-carrier electrons in the base. With the collector-base junction reversed-biased and at low-level injection, it can be assumed that the electron concentration drops to zero at the collector-base depletion boundary, that is, the electrons are "instantaneously" swept across the base-collector depletion region and $n = 0$ at $x = W_b$. This assumption does not hold for high-level injection, as will be discussed in (Sect. 3.4.2). Thus, the distribution of electrons in the base is linear, as shown in Fig. 3.7. The electron current can now be found in the same manner as for a pn junction with uniform dopant concentration described by (2.79) in Chap. 2

$$I_n = \frac{qAD_n n_i^2}{N_A W_b}(e^{qV_F/kT} - 1), \tag{3.20}$$

where A is the area of the emitter-base active region and the product $N_A W_b$ in the denominator of (3.20) is referred to as the Gummel number in the base. Similarly, the injection of holes into the emitter is described by

$$I_p = \frac{qAD_p n_i^2}{N_D L_{pE}}(e^{qV_F/kT} - 1), \tag{3.21}$$

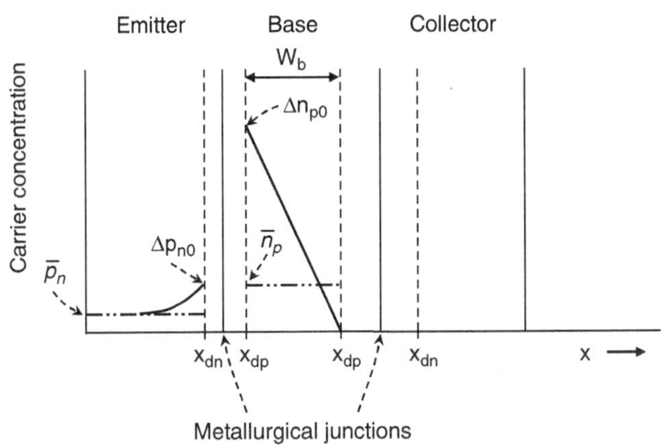

Fig. 3.7 Schematic representation of minority-carrier distribution in the base and emitter of an *NPN* transistor biased in the forward-active mode

and the product $N_D L_{pE}$ is related to the Gummel number in the emitter. Recall that diffusivity D is related to mobility μ by the Einstein relation (Chap. 1)

$$D_n = \mu_n \frac{kT}{q}; \quad D_p = \mu_p \frac{kT}{q}. \tag{3.22}$$

For uniform profiles, diffusivity and hence mobility is constant. The intrinsic carrier concentration, n_i, depends on temperature and energy gap (Chap. 1). Since for silicon transistors (as opposed to silicon-germanium, discussed later), the base is typically doped in the range 10^{18}–$10^{19}\,\mathrm{cm}^{-3}$ and the emitter in the range 10^{20}–$10^{21}\,\mathrm{cm}^{-3}$, there will be substantial energy-gap lowering in the base and emitter and n_i will depend on dopant concentration (Figs. 1.22 and 1.23, Chap. 1).

In real structures, the base and emitter are not uniformly doped and μ_n, μ_p and n_i are position dependent. The base is graded in a direction to create a built-in field that accelerates minority carriers toward the collector, thus enhancing gain and speed. For low-level injection, the built-in field is expressed for a p-type base as (Chap. 1)

$$E = \frac{kT}{q}\frac{1}{p(x)}\frac{dp}{dx} \cong \frac{kT}{q}\frac{1}{N_A(x)}\frac{dN_A}{dx}. \tag{3.23}$$

The built-in field induces a drift current component in addition to the diffusion current component. Both drift and diffusion currents contribute to current density

$$j_n = q\tilde{\mu}_n nE + q\tilde{D}_n \frac{dn}{dx}, \tag{3.24}$$

where $\tilde{\mu}_n$ and $\tilde{D}_n$ denote averaged values for electron mobility and diffusivity and have a weak dependence on dopant concentration. Combining the above equations gives

$$j_n = q\tilde{D}_n\left(\frac{n}{p}\frac{dp}{dx} + \frac{dn}{dx}\right) = \frac{q\tilde{D}_n}{p(x)}\frac{d(pn)}{dx}. \tag{3.25}$$

For negligible recombination in the base, j_n is practically constant and

$$n(x) = \frac{j_n}{q\tilde{D}_n}\frac{1}{n(x)}\int_x^{W_b} p(x)dx. \tag{3.26}$$

By setting $x = 0$ at the emitter-base depletion boundary in the base and denoting $n = n_0, p = p_0$ at this boundary, the current density can be expressed as

$$j_n = \frac{q\tilde{D}_n p_{p0} n_{p0}}{\int_0^{W_b} p_p(x)dx} \cong \frac{q\tilde{D}_n p_{p0} n_{p0}}{\int_0^{W_b} N_A(x)dx}. \tag{3.27}$$

The integrated dopant concentration in (3.27) is called the Gummel number in the base. The product $p_{p0} \cdot n_{p0}$ is found from the Boltzmann approximations (2.56 and 2.57, Chap. 2) as

$$p_{p0}n_{p0} = n_i^2 e^{qV_F/kT}. \tag{3.28}$$

The thermal-equilibrium intrinsic-carrier concentration n_i depends on energy-gap lowering ΔE_g as (1.52, Chap. 1)

$$n_i^2 = n_{i0}^2 e^{\Delta E_g}, \tag{3.29}$$

where n_{i0} is the intrinsic carrier concentration at low to moderate dopant concentrations, and ΔE_g the position-dependent energy gap lowering.

For a position-dependent mobility, the electron current density is expressed as [6]

$$j_n \cong \frac{q}{(kT/q)\int\limits_0^{W_b} [N_A(x)dx] / [\mu_n(x)\, n_i^2(x)]} \left(e^{qV_F/kT} - 1\right). \tag{3.30}$$

For the special case of a uniformly doped base, (3.30) reduces to the form of (3.20).

It should be noted that as V_F increases, the emitter-base barrier height, and hence the depletion width decreases, resulting in an increase in the base Gummel number and a reduced electron injection. This effect can be more pronounced in *SiGe* transistors (Sect. 3.7).

3.3.1.2 Hole Injection into the Emitter

When the emitter width is larger than the diffusion length of holes in the emitter, L_{pE}, the injection of holes into the emitter is

$$j_n \cong \frac{q}{(kT/q)\int\limits_0^{L_{pE}} [N_D(x)dx] / [\mu_p(x)\, n_i^2(x)]} \left(e^{qV_F/kT} - 1\right). \tag{3.31}$$

$x = 0$ is chosen here at the emitter-base depletion boundary on the emitter-side. For an emitter area A, the total emitter current is

$$I_E = (j_n + j_p) \,.\, A = I_0(e^{qV_F/kT} - 1). \tag{3.32}$$

I_0 is the saturation current of the emitter-base junction. It is equal to the sum of the pre-exponential factors in (3.30) and (3.31).

Since the concentration in the emitter is considerably higher than in the base, a larger energy-gap lowering occurs in the emitter than in the base, which increases n_i and hence j_p. The diffusion length L_{pE} depends on the life-time of holes in the emitter, τ_{pE}, which is a function of the recombination rate of minority carriers in the emitter. In other words, a higher recombination rate means a shorter life-time and a smaller diffusion length, hence larger j_p. The two main mechanisms that determine the hole lifetime in the emitter are the Shockley-Read-Hall recombination and Auger recombination, both discussed in Chap. 1. Auger recombination becomes significant in the highly doped emitter.

3.3.1.3 Injection Ratio

The ratio of electron to hole injection is called the injection ratio, expresses as

$$\frac{J_n}{j_p} = \frac{\int_0^{L_p} [n(x)dx] / \left[\mu_p(x) n_i^2(x)\right]}{\int_0^{W_b} [p(x)dx] / \left[\mu_n(x) n_i^2(x)\right]} \cong \frac{\int_0^{L_p} [N_D(x)dx] / \left[\mu_p(x) n_i^2(x)\right]}{\int_0^{W_b} [N_A(x)dx] / \left[\mu_n(x) n_i^2(x)\right]}. \quad (3.33)$$

For a uniform injection across the junction, negligible leakage current, and negligible multiplication at the collector-base junction, this ratio also defines the current gain $\beta = I_C/I_B$. The gain increases as the Gummel number in the emitter is increased and the Gummel number in the base is reduced. An increase in emitter concentration, however, tends to reduce the injection ratio by

1. Lowering the energy gap and hence increasing n_i,
2. Reducing the minority-carrier lifetime in the emitter because of an increase in Auger recombination, and
3. Reducing the minority-carrier mobility, and hence L_{pE}.

A plot of j_C, j_B versus V_{BE} is shown for a real structure in Fig. 3.8 together with the ratio $j_C/j_B = \beta$. Such a plot is referred to as a Gummel plot. In the absence of bandgap narrowing and Auger recombination, the gain would be considerably higher than shown in the figure [7].

It can be seen that the current gain decreases at low and high V_{BE}. The decrease at low V_{BE} is because the recombination currents within the neutral base region and recombination-generation currents at the emitter-base and collector-base junctions become an appreciable fraction of the total base current. They typically have negligible impact on emitter and collector currents. The net result is a decrease in current gain. The drop in gain at higher current densities is attributed to high-level injection effects, including voltage drops across extrinsic resistances, base conductivity modulation, and base push effect (Sect. 3.4.2). The extrapolated dashed lines in Fig. 3.8 show the ideal current-voltage characteristics in the absence of leakage currents and high-level injection effects. Ideally, at 300 K, V_{BE} increases by 60 mV per decade of current (Chap. 2).

3.3.1.4 Base Current Components

All current components are needed to determine the actual transistor gain. In the active mode, the most important base current components are

1. Hole injection from base into emitter as discussed above,
2. Electron-hole pair generation at the collector-base junction,
3. Excess electron–hole pair recombination within the neutral base region,
4. Recombination at the surface intercept of the emitter-base junction.

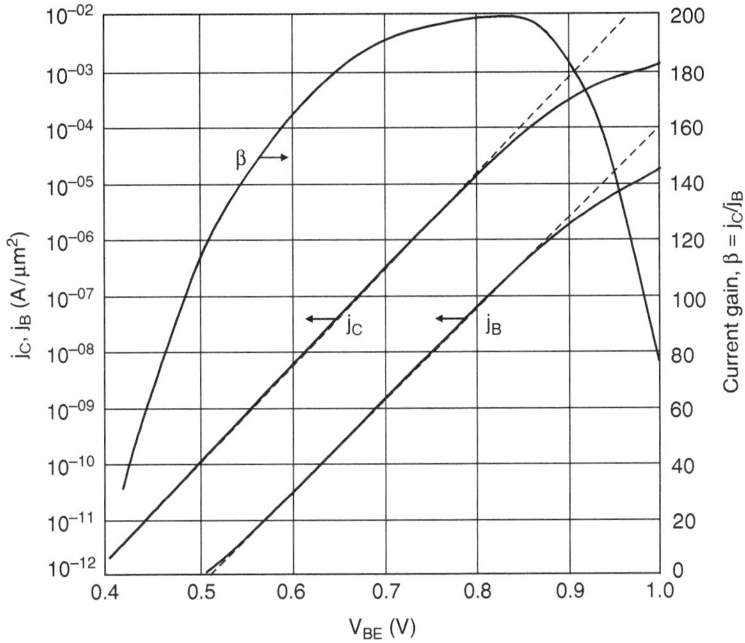

Fig. 3.8 Gummel plot measured on a real structure

Electron-Hole Pair Generation at the Collector-Base Junction

For small collector-base reverse voltages, where impact ionization is negligible, the reverse base-collector current consists mainly of thermal generation of electron–hole pairs [8]. Holes drift to the base, in a direction opposite to the injected base current, and electrons drift to the collector. There are two parts to the generation within the collector-base depletion region, one in the silicon bulk, away from the junction-dielectric interface at the surface, and the other at or near the interface. Both components are associated with "trap-levels" located within the forbidden gap. Generation within the bulk depletion, $I_{gen\text{-}bulk}$, is approximated by (Chap. 2)

$$I_{gen-bulk} \cong \frac{1}{2} q \frac{n_i}{\tau} x_{d-bulk} A_j \quad \text{A}, \tag{3.34}$$

where A_j is the area of the collector base junction in the bulk, $x_{d\text{-}bulk}$ the bulk depletion width, and τ_0 the effective lifetime within the depletion region, assumed the same for electrons and hole, which is approximated by

$$\tau_0 \cong \frac{1}{\sigma v_{th} N_T} \quad \text{s}, \tag{3.35}$$

where σ is the carrier capture cross-section ($\sim 10^{-15}\,\text{cm}^2$), v_{th} the thermal velocity ($10^7\,\text{cm/s}$ at 300 K), and N_T the effective density of "traps" (cm^{-3}). Generation at or just under the surface intercept of the depletion region can be described in terms of a surface recombination velocity [9, 10] as

$$I_{gen-surf} \cong \frac{1}{2} q n_i s_0 x_{d-surf} P_j \quad \text{A}. \tag{3.36}$$

$I_{gen\text{-}surf}$ is the surface generation current, P_j is the perimeter of the collector-base junction at the surface, $x_{d\text{-}surf}$ the collector-base depletion width at the junction-dielectric interface, and s_o the surface recombination velocity, defined as

$$s_0 \cong \sigma v_{th} N_{it} \quad \text{cm/s}, \tag{3.37}$$

where N_{it} is the effective interface trap density per unit area. Because the thermally generated base current is in a direction opposite to the injected hole current, the net base current decreases or can even reverse polarity at very low-level injection and high generation of electron–hole pairs. This is observed as an apparent increase in transistor gain.

Recombination within the Neutral Base Region

Since the base width is typically much smaller than the minority-carrier diffusion length in the base, recombination in the intrinsic neutral base is negligible in silicon bipolar transistors. Also, in well-designed transistors, emitter sidewall injection and recombination in the bulk of the extrinsic base can be neglected. Heterojunction bipolar transistors (*HBT*), however, are found to exhibit recombination currents in the neutral base that can be sufficiently high to reduce the injected emitter current that reaches the collector, which degrades transistor gain.

Surface Recombination at the Emitter-Base Junction

Interface traps and traps just beneath the surface intercept of the emitter-base depletion region can give rise to an appreciable surface recombination current approximated for a forward voltage V_F by [9–11]

$$I_{rec-surf} \cong I_{gen-surf}\, e^{V_F/2kT} \quad \text{A}. \tag{3.38}$$

The factor of 2 in the denominator results from the condition for maximum recombination current when $p = n$ in the depletion region [9–11]. The recombination current can seriously impact the gain if the surface state density is high. The impact on current gain is more serious at low collector current and for small emitter sizes.

High interface trap densities are observed when the surface is not properly passivated with oxide (Chap. 4) or, in some applications, the emitter-base junction becomes reverse-biased and a high two-dimensional field is created near the surface intercept where surface traps can be generated. In the latter case, the emitter-base profile must be optimized to reduce the field. Also, appropriate limits on magnitude and duty-cycle[3] of emitter-base reverse bias are typically imposed.

3.3.2 Collector-Base Reverse Characteristics

The output characteristics in the forward-active and reverse-active modes of operation described in Table 3.1 are shown for an *NPN* in Fig. 3.9. The collector current is plotted as a function of V_{CE}, with I_B as a parameter that is typically increased in regular steps. The saturated region discussed in Sect. 3.2.4 is shown cross-hatched on the left of the dotted line in forward active mode. The cut-off region is below the line for $I_B = 0$, equivalent to an open base. In this case,

$$I_{CEO} = \beta I_{CBO}, \tag{3.39}$$

where I_{CBO} is the reverse current caused by electron–hole pair generation at the collector-base junction with the emitter open. I_{CBO} is the sum of thermal generation within the collector-base depletion region, thermal generation in the collector outside the depletion region within a diffusion length from the junction (for a wide collector), and generation by impact ionization at high reverse fields (Chap. 2). I_{CEO} in (3.39) is the collector-emitter reverse current with the base open. It is an

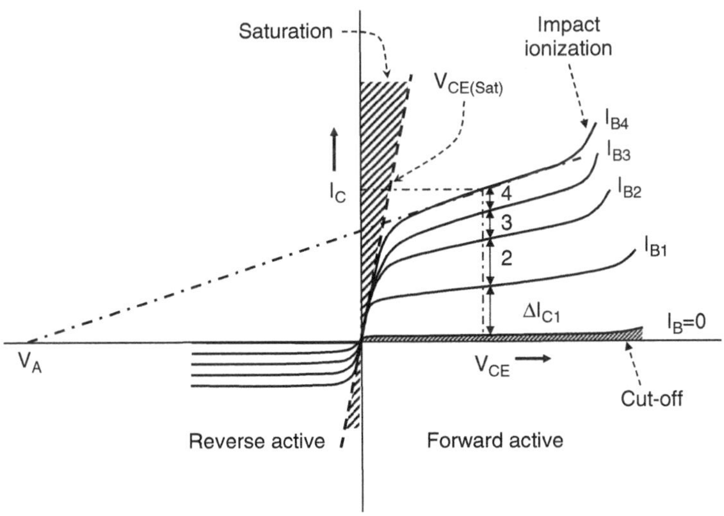

Fig. 3.9 Output characteristics for *NPN* in the four modes of operation. Base-current is increased in constant increments ΔI_B. For a given V_{CE}, the current gain $\beta = \Delta I_C/\Delta I_B$. V_A is the Early voltage

[3] The duty cycle is the fraction of time that the emitter-base junction is reverse biased. In periodic reverse-bias cycles, the duty cycle is the ratio of the duration of the reverse-bias pulse τ to the cycle duration T.

amplification of I_{CBO} by the current gain β. This can be visualized by considering that generated holes are driven into the base and, since the base is open, they are injected into the emitter as a small current increment $\Delta I_B = I_{CBO}$, increasing the base current that is amplified by β.

Modern transistors are asymmetrical. They are optimized for high gain and speed when operated in one direction of current, typically with the emitter on top and the collector at the bottom of the structure, while the reverse-active mode of operation exhibits much smaller gain and speed and is of little interest to circuit applications.

When measuring the current gain at a fixed collector voltage with increments ΔI_B of base current, transistor gain is found as $\beta = \Delta I_C / \Delta I_B$. The decrease in ΔI_C at higher base current steps is caused by high-level injection effects discussed in Sect. 3.4. The rapid increase in I_C on the far right of the plots is due to multiplication by impact ionization at high fields.

In the region between saturation and impact ionization, the plots exhibit a slope showing, for the same base current, an increase in collector current with increasing collector voltage. This is the result of base-width modulation and is best described by the slope intercept V_A on the voltage axis, as illustrated in Fig. 3.9 and discussed in the next section.

3.3.2.1 Base-Width Modulation

Ideally, the transistor output characteristics in the range between saturation and impact ionization should be flat with zero conductance, that is, $\partial I_C / \partial V_{CE} = 0$ for constant V_{BE}. In reality, the collector current increases monotonically as V_{CE} increases in this range because the collector-base depletion width increases, reducing the active neutral base-width W_b. The characteristic exhibits a slope that is best described by drawing a tangent at a certain point of the output and extending it to intercept the voltage axis at a point V_A, as shown in Fig. 3.9. V_A is referred to as the Early voltage, after J. M. Early [12], and is negative for *NPN*, positive for *PNP* transistors. The effect is explained as follows: W_b is the region of non-depleted base (Figs. 3.1–3.3). The base Gummel number is the integral of dopant concentration between the two depletion boundaries defining W_b. When the reverse base-collector voltage is increased, the depletion region expands into both the collector and base. For a given voltage, the depletion width depends on the impurity profile in both regions. As the depletion width expands into the base, W_b decreases, resulting in:

1. An increase in slope of excess injected minority carriers (Fig. 3.10),
2. A reduction in the Gummel number,
3. A reduction in the base transit time, that is, the time for minority carriers to traverse the base decreases.

There is also an associated decrease in collector-base capacitance and a small increase in collector-base reverse current.

The collector voltage modulates the base-width and hence transistor gain and speed. For a reverse collector voltage V_{CE}, the slope at any point of the output is [12, 13]

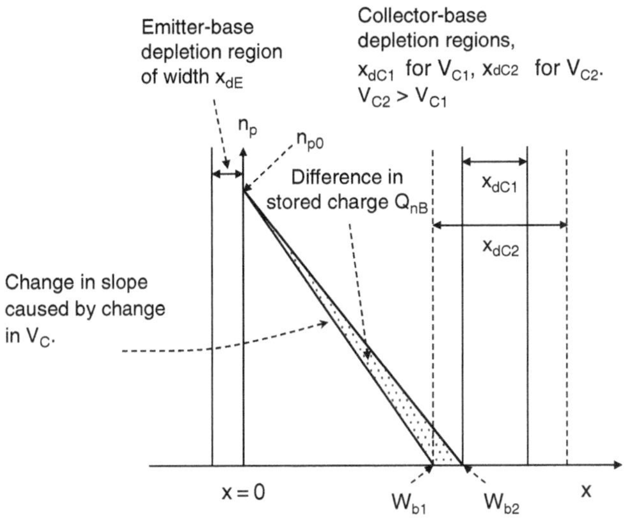

Fig. 3.10 Base-width modulation by collector voltage

$$\frac{\partial I_C}{\partial V_{CE}} = \frac{I_C}{|V_A| + |V_{CE}|}. \tag{3.40}$$

Assuming a constant emitter-base junction area, the forward Early voltage is then defined as

$$|V_A| = \frac{j_C}{\partial j_C/\partial V_C} - |V_C|. \tag{3.41}$$

In the vicinity of saturation or impact ionization, $\partial I_C/\partial V_C$ increases and hence the magnitude of V_A decreases. As the region of impact ionization is approached, V_A can even become positive (NPN) or negative (PNP). For a sufficiently wide range between saturation and impact ionization, the slope for a given base current is assumed to be constant. Outside the two regions, V_A will depend on V_{CE}. Also, the slopes taken for different base currents do not necessarily intersect the voltage axis at the same point. It is therefore recommended to define both the collector voltage range and base current when V_A is extracted.

As can be seen from (3.33), β increases as the Gummel number in the base $\int_0^{W_b} N_A(x)dx$ decreases. For a fixed emitter-base forward voltage $V_F = V_{BE}$, the change in β is related to the change in collector current density. From (3.30) the slope is then found as [12–15]:

$$\frac{\partial j_C}{\partial V_C} = \frac{q\tilde{D}_n\tilde{n}_i^2 N_A(W_b)(e^{qV_{BE}/kT} - 1)}{\left[\int_0^{W_b} N_A(x)dx\right]^2}\frac{\partial W_b}{\partial V_C} = \frac{j_C N_A(W_b)}{\int_0^{W_b} N_A(x)dx}\frac{\partial W_b}{\partial V_C}. \tag{3.42}$$

For simplicity, the position-dependent diffusivity and intrinsic-carrier concentration are replaced by their averaged values in (3.42). At low-level injection, $p(x) \cong N_A(x)$. Substituting in 3.41 gives

$$|V_A| = \frac{q\int_0^{W_b} N_A(x)dx}{qN_A(W_b)(\partial W_b/\partial V_C)} - |V_C|. \tag{3.43}$$

In (3.43), the numerator is the integrated majority-carrier hole-charge in the neutral base, Q_B. The denominator defines the change in hole-charge at the base boundary, W_b. This change is expressed as $\partial Q_b/\partial V_C = C_{jC}$, where C_{jC} is the collector-base depletion capacitance. The Early voltage is thus found as

$$|V_A| = \frac{Q_B}{C_{jC}} - |V_C|. \tag{3.44}$$

The dependence of V_A on majority-carrier charge Q_B in the intrinsic base suggests a direct relation between V_A and the sheet resistance of the intrinsic base. Since the intrinsic active base is "pinched" between the depletion boundaries at the emitter-base and collector-base, its sheet resistance is sometimes referred to as the base "pinch" resistance, defined as

$$R_{pinch} = \frac{1}{\tilde{\mu}_p Q_B}, \tag{3.45}$$

where $\tilde{\mu}_p$ is the effective hole mobility in the base. A higher pinch-resistance results in higher gain and speed but lower magnitude of V_A.

One important integration trade-off can be already seen from the analysis of transistor gain and base-width modulation. To increase the gain and speed, the Gummel number is decreased by reducing the dopant concentration or the base width, or both. This, however, degrades the Early voltage. A good figure of merit for analog designs is the βV_A product that should be as high as possible to increase linearity.

When the base Gummel number is reduced, a critical point can be reached where for a fixed collector voltage, the collector-base depletion region spreads through the base and merges with the emitter-base depletion region. The voltage where the merger occurs is referred to as *voltage punch-through*, V_{PT}. At this point, the full base becomes depleted and further increase in collector voltage induces direct injection from emitter to collector, resulting in a large, uncontrolled increase in collector current that is only limited by series resistances. For uniformly doped base, voltage punch-through is

$$V_{PT} = \frac{qN_A W_b^2}{2\varepsilon_{Si}}. \tag{3.46}$$

V_{PT} increases as the base concentration or base-width increases.

3.3.2.2 Multiplication and Transistor Breakdown

Multiplication at the collector is the increase in collector current by impact ionization. Transistor breakdown can occur either by impact ionization or voltage punch-through. When the breakdown voltage of the collector-base junction is smaller than the punch-through voltage, the mechanism for collector-base breakdown voltage is the same as for an isolated *pn* junction discussed in Chap. 2. In this case, the collector-base breakdown voltage with the emitter open is the same as the breakdown voltage with shorted emitter-base, $BV_{CBO} = BV_{CES}$. If punch-through occurs at a lower voltage than avalanche breakdown, then at punch-through the emitter begins to follow the voltage on the collector and, as the collector voltage is further increased, the emitter-base junction breaks down. At this point

$$BV_{CBO} = V_{PT} + BV_{EBO}. \tag{3.47}$$

Of particular importance is the grounded-emitter transistor breakdown, BV_{CEO} with $I_B = 0$. Under this bias condition, the collector-emitter voltage is divided between the collector-base and emitter-base capacitances. The collector-base junction is reverse-biased and the emitter-base junction slightly forward-biased (Fig. 3.11). The transistor can be treated as two capacitors in series, C_{jE} and C_{jC}, where

$$V_{BE} = \frac{C_{jC}}{C_{jE} + C_{jC}} V_{CE}. \tag{3.48}$$

The total collector current is

$$I_C = I_{CBO} + \alpha I_E. \tag{3.49}$$

Since $I_C = I_E = I_{CEO}$, the above equation can be written as

$$I_{CEO} = \frac{I_{CBO}}{1 - \alpha} \cong \beta I_{CBO}. \tag{3.50}$$

As the collector-base reverse voltage is increased, impact ionization can become significant and carriers entering the region or generated within the region are multiplied by the factor M, approximated in (2.132) in Chap. 2 and repeated here for convenience

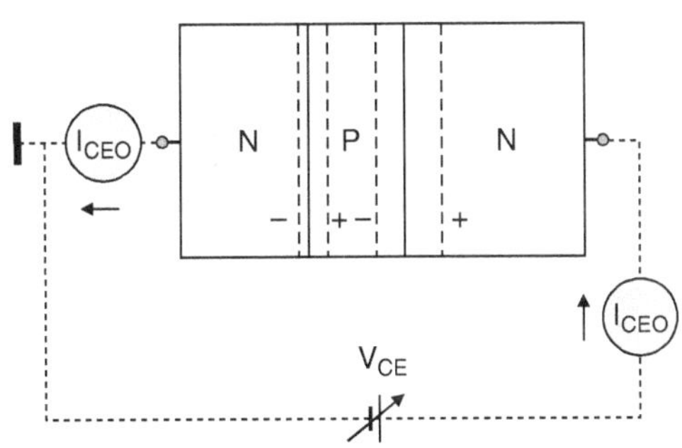

Fig. 3.11 Voltage polarities under BV_{CEO} condition

$$M \cong \frac{1}{1-(V_R/BV)^n}, \tag{3.51}$$

where V_R is the reverse voltage, BV the avalanche breakdown voltage of the collector-base junction and n typically ranges from 2–6, depending on the transistor profile and geometry. The collector current in (3.49) is multiplied by M and (3.50) can be written as

$$I_{CEO} = \frac{MI_{CBO}}{1-M\alpha}. \tag{3.52}$$

When $M\alpha$ approaches 1, I_{CEO} tends to infinity and becomes only limited by the external resistance. This is the onset of transistor breakdown. By substituting BV_{CEO} for V_R and BV_{CBO} for BV in (3.51) and solving for BV_{CEO}, a good approximation for the transistor breakdown is found as

$$BV_{CEO} = (1-\alpha)^{1/n} BV_{CBO} \approx \frac{BV_{CBO}}{\sqrt[n]{\beta}}. \tag{3.53}$$

In all high-performance transistors, BV_{CEO} is considerably smaller than BV_{CBO}. It should be emphasized that BV_{CBO} in (3.53) is the collector-base breakdown voltage immediately under the intrinsic base. The measured BV_{CBO} does not always represent this "intrinsic" breakdown and can sometimes be lower, such as at edges or corners of the base (Chap. 2).

For a given BV_{CBO}, increasing the gain reduces transistor breakdown. Thus, there is a trade-off between β and BV_{CEO}. There is another important trade-off between BV_{CEO} and transistor speed. As will be shown in Sect. 3.4, a local increase in collector dopant concentration immediately beneath the intrinsic base is necessary to improve transistor speed. This reduces transistor breakdown in two ways: (a) BV_{CBO} decreases and, from (3.53), BV_{CEO} decreases accordingly. (b) For a given base profile, an increase in collector concentration reduces the base width, further increasing transistor speed and gain. An increase in gain reduces BV_{CEO}. The Early voltage is also reduced.

3.3.3 Emitter-Base Reverse Characteristics

In high-performance bipolar transistors, the emitter is degenerately doped and the base profile is graded with a peak concentration near the surface ranging from 3×10^{18}–$10^{19}\,\text{cm}^{-3}$ (Fig. 3.6). At these concentrations, the emitter-base reverse current becomes a mixture of thermal generation, tunneling and impact ionization (Chap. 2). For small reverse voltages, tunneling and thermal generation are predominant and impact ionization is negligible. As the reverse voltage is increased, the probability for tunneling and impact ionization increases until breakdown is reached where the individual contributions of tunneling and impact ionization currents depend on the impurity profiles. The emitter-base reverse characteristics are illustrated

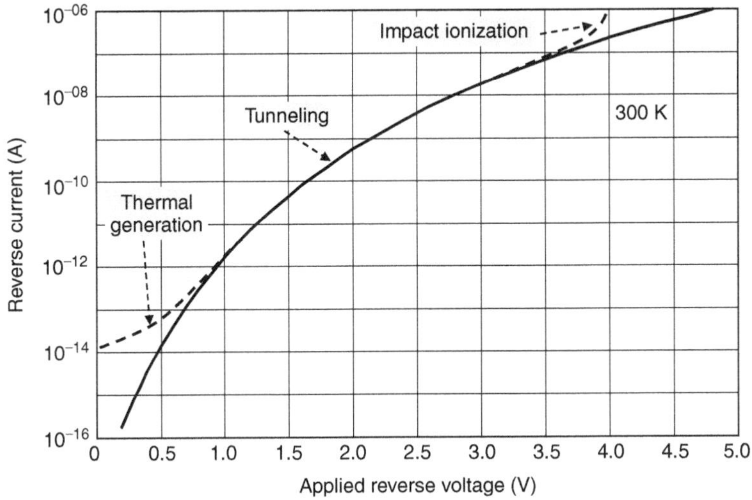

Fig. 3.12 NPN emitter-base reverse characteristics at 300 K. $A_E = 1 \times 1\,\mu m^2$; $X_{jE} = 50\,nm$; $N_D = 10^{20}\,cm^{-3}$; $N_A = 5 \times 10^{18}\,cm^{-3}$

for an NPN transistor in Fig. 3.12. Since the base and emitter concentrations peak near the silicon surface, this is a region of high electric field where carriers can become hot (Chap. 1). Hot-carriers incident on the surface are known to create and populate interface states that increase the emitter-base recombination-generation current and hence increase the base current which reduces the gain (Sect. 3.3.1.4).

3.3.3.1 Reverse Early Voltage

The derivation of the reverse Early voltage is similar to that of the forward Early voltage. An expression similar to (3.43) is found as

$$|V_{A(reverse)}| = \frac{q\int_0^{W_b} N_A(x)dx}{qN_{A(0)}(\partial W_b/\partial V_E)}, \tag{3.54}$$

where $N_{A(0)}$ is the peak base-concentration near the emitter-base junction. Since $N_{A(0)}$ is typically much larger than $N_{A(Wb)}$ in (3.42), the reverse Early voltage must be smaller than the forward V_A.

3.3.3.2 Reverse Punch-Through Voltage

The reverse punch-through voltage, $V_{PT(reverse)}$ is defined as the emitter-base voltage at which the base becomes totally depleted. It is found by integrating Poisson's equation from 0 at the emitter-base junction to W_b at the collector-base junction:

$$\left|V_{PT(reverse)}\right| = \int_0^{W_b} \frac{qN_A(x)x}{\varepsilon_{Si}}\,\mathrm{d}x. \tag{3.55}$$

For example, for $W_b = 0.1\,\mu\text{m}$ and a Gaussian base profile with a peak concentration $N_{A(0)} = 2.5 \times 10^{18}\,\text{cm}^{-3}$, and $N_{A(Wb)} = 5 \times 10^{16}\,\text{cm}^{-3}$, the reverse $V_{PT} \approx 3.3\,\text{V}$. In this case, reverse punch-through occurs before the emitter-base breaks-down while the forward punch-through voltage remains considerably larger than the collector-base breakdown voltage (Problem 3).

3.3.4 Polysilicon Emitter and Interface Oxide

To improve performance and density, the vertical and horizontal geometry of the transistor must be scaled to smaller dimensions. This requires, among others, ultra-shallow emitters to be formed in the range below 50 nm. Ultra-shallow junctions are not easy to achieve by direct implantation into single crystal silicon because of the generation of point defects and resulting transient-enhanced diffusion, *TED* [5]. While new techniques, such as pulsed plasma doping (*PLAD*), have been suggested to form ultra-shallow junctions (Chap. 7, [16]), the close proximity of contact metal or silicide to the metallurgical junction can induce paths of high leakage current or even shorts.

Issues with implant damage and metal penetration are resolved by introducing a polysilicon emitter. After forming the base, the single-crystal emitter regions are exposed and a 100–250 nm undoped polysilicon film is deposited that comes in direct contact with the single-crystal emitter surface in the exposed region. The film is implanted, typically with arsenic (As) for *NPN* and boron for *PNP*, with a precise dose and an energy that places most of the implant damage inside the polysilicon film. The structure is then subjected to a high-temperature drive-in cycle that diffuses the impurities from polysilicon into monosilicon to form the ultra-shallow emitter in the single-crystal silicon. This two-step emitter process restricts the implant damage to polysilicon and reduces the likelihood of metal penetration into the single-crystal emitter. Polysilicon emitters were first introduced in the 1970s [17–19]. They were initially doped as deposited and utilized as a diffusion source to form shallow emitter junctions [19]. After patterning the doped polysilicon, aluminum contacts could be made without metal penetration into the monocrystal emitter. The precision of implantation was then introduced in [20], resulting in a surprisingly large increase in injection ratio and current gain when compared to earlier versions. Polysilicon emitter contacts can also be made to an already formed emitter in the monocrystal, resulting again in an increase in current gain [21]. In all structures, the emitter-base metallurgical junction resides in the single-crystal silicon and the emitter is typically less than 50-nm deep.

State of the art bipolar transistors are typically of the "double-poly" type, where polysilicon is utilized as a contacting pad and diffusion source for both the

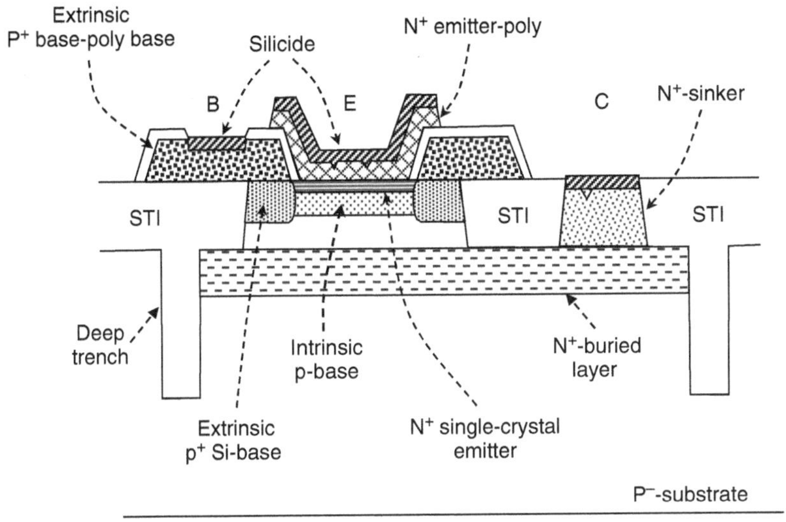

Fig. 3.13 Schematic cross-section of a double-poly *NPN* [22]

shallow emitter and extrinsic base (Fig. 3.13 [22]). In this structure, the arsenic- or phosphorus-doped emitter is self-aligned to the base and separated from it by an insulating film. The boron-doped polysilicon base is contacted over the field oxide, and the extrinsic base abuts the dielectric isolation, further reducing parasitic resistances and capacitances. Self-alignment and dielectric isolation significantly reduce transistor size and increase speed.

3.3.4.1 Interface Oxide, IFO

When the emitter junction depth, x_{jE}, is larger than the minority-carrier diffusion length, the injected minority carriers recombine fully within the emitter bulk and the base current can then be approximated by (3.21). Bipolar scaling to smaller vertical and horizontal geometry requires that the emitter junction depth be reduced. In ultra-shallow emitters, however, the minority-carrier diffusion length is typically larger than x_{jE} so that a substantial fraction of injected minority carriers reach the emitter surface and the injection becomes dominated by the properties of the polysilicon–monosilicon interface rather than those of the emitter "bulk." In this case, the calculation of minority-carrier injection into the emitter becomes more complicated than in (3.21) because of mechanisms related to polysilicon morphology and the discontinuity at the interface. The surface mechanisms that control the gain have therefore received considerable attention.

Several models have been suggested to explain surface mechanisms [23–28]. The presence of a thin interfacial oxide (*IFO*) layer between polysilicon and monosilicon was first recognized and analyzed in [23]. The *IFO* was grown as an unavoidable

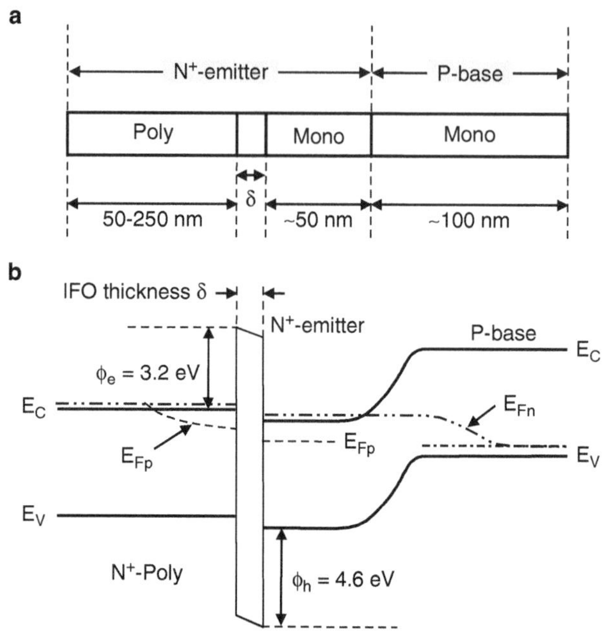

Fig. 3.14 Schematic of a one-dimensional *NPN* structure and band diagram under forward biased emitter-base junction

native oxide film during exposure of the wafer surface to air. An associated increase in the injection efficiency was observed and explained by treating the *IFO* as a tunneling barrier to carriers injected from the base. The *IFO* of thickness δ is shown schematically for an *NPN* emitter-base region in Fig. 3.14a and the corresponding band-diagram in Fig. 3.14b.

Since the polysilicon and monosilicon emitters are degenerately doped, their Fermi levels lie close to the conduction band. The quasi-Fermi levels E_{Fp} and E_{Fn} reflect, respectively, the injection of holes into the emitter and electrons into the base (Chap. 1). The total base current consists of injection of minority carriers into the monosilicon emitter, recombination at the emitter oxide interface, tunneling through the oxide, and recombination within the polysilicon emitter.

For ultra-shallow emitters (20–50 nm), it can be assumed that minority-carrier recombination in the emitter bulk is negligible. For a uniform *IFO* of thickness δ less than about 1 nm grown on an ultra-shallow junction, the base current is found to be limited by tunneling [29, 30]. A simplified relation for the tunneling probability through the *IFO* is derived in [29, 30]. It is approximated as

$$P \approx e^{-\alpha_T \delta \sqrt{\phi}}, \tag{3.56}$$

where a rectangular barrier of effective barrier height ϕ and thickness δ is assumed, and α_T is defined as

$$\alpha_T = 2\sqrt{\frac{2m^*}{\hbar^2}}. \tag{3.57}$$

Assuming that the effective mass in the oxide m^* is equal to the free electron mass m_0 and defining the *IFO* thickness δ in Å, ϕ in V, one finds $\alpha_T \approx 1$ [27].

The effective barrier height for electrons, $\phi_{e\text{-}eff}$, and for holes, $\phi_{h\text{-}eff}$, are slightly smaller than ϕ_e and ϕ_h in Fig. 3.14. This is due to the triangular barrier shape in the *IFO* caused by the applied voltage. For $\phi_{e\text{-}eff} \approx 3.0\,\text{V}$ and $\phi_{h\text{-}eff} \approx 4.4\,\text{V}$ and an *IFO* thickness $\delta = 4$ Å, the tunneling probability is $\sim 10^{-3}$ for electrons and 2×10^{-4} for holes, which is sufficiently high to cause substantial tunneling current through the *IFO*.

Typical polysilicon-emitter *NPN* and *PNP* profiles are shown in Fig. 3.15 [31,32]. The increase in concentration at the polysilicon–monosilicon interface is attributed to segregation of impurities within or near the *IFO*. It also serves as a "marker" to locate the interface. The role of the selectively implanted collector, *SIC*, is discussed in Sect. 3.4.2.

The *IFO* must be sufficiently thin to allow controlled diffusion of dopants from the polysilicon through the thin oxide, but sufficiently thick to increase the emitter efficiency by retarding injection from the base into the emitter without appreciably reducing the transport of majority carriers from polysilicon to the monosilicon, that is, without appreciably increasing the emitter resistance [33, 34]. An optimized *IFO* thickness satisfying the above requirements is below 1 nm. For such a thickness, both I_B and I_C exhibit near-ideal exponential characteristics. It follows that the *IFO* thickness must be precisely controlled if grown intentionally. The thickness control has been demonstrated by special treatment of the surface and low-temperature Chemical-Vapor-Deposition (*CVD*) [35], by low-temperature Rapid-Thermal-Oxidation (*RTO*) [32], and by Atomic-Layer-Deposition (*ALD*) [36]. *CVD* is described in [5], *RTO* and *ALD* are discussed in Chap. 7.

In the absence of *IFO*, there is the possibility of epitaxial realignment of the deposited polysilicon during thermal cycles following deposition [37, 38]. In regions of epitaxial realignment, the deposited film becomes a single-crystal extension of the monosilicon emitter and the local injection parameters follow those of a single-crystal emitter. Also, the emitter may be shallower in epitaxially realigned regions for the following reason: As dopants diffuse locally into the original silicon crystal, they must be replenished from regions around and above that location. Replenishing is faster with polysilicon than with epitaxial realignment because dopants diffuse more rapidly along grain boundaries in polysilicon than in single-crystal silicon [39].

Epitaxial realignment also occurs when a native oxide film, that was present during polysilicon deposition, "breaks-up" into islands after subjecting the structure to high-temperature rapid-thermal annealing [37]. In this case, the deposited film is a mixture of monosilicon extensions and polysilicon on *IFO* islands, resulting in a nonuniform emitter depth and injection parameters. The average current gain in such structures is typically lower than with a uniform *IFO*.

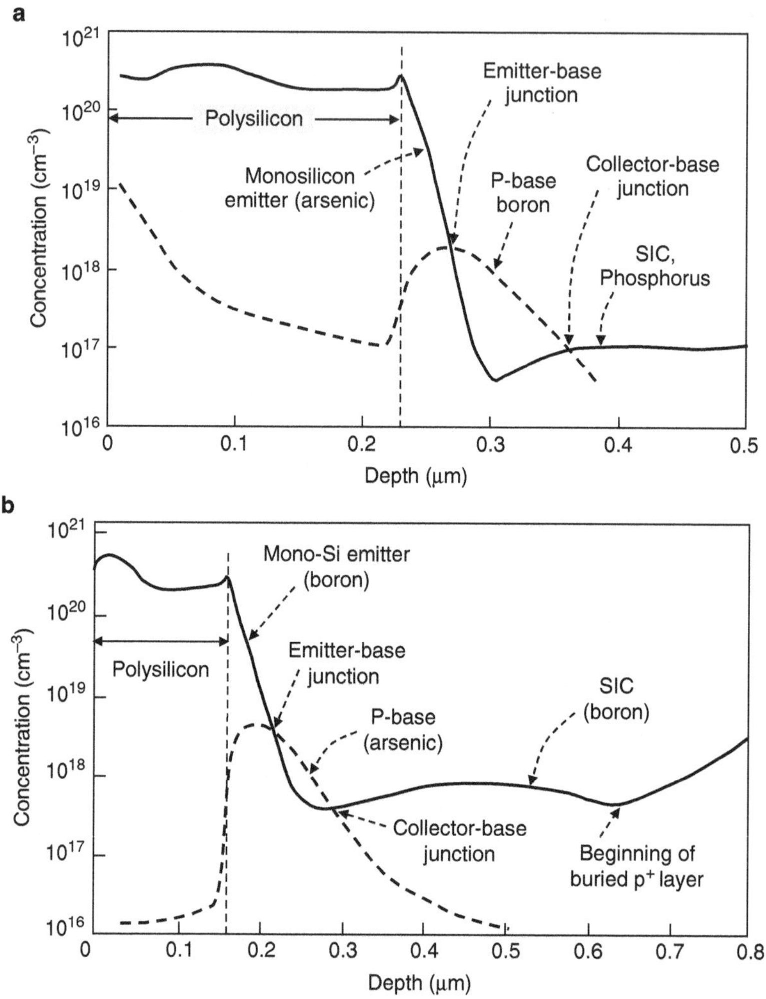

Fig. 3.15 Examples of polysilicon-emitter transistors. **a** *NPN*, **b** *PNP*. Impurity profiles reconstructed from *SIMS* data [31,32]. *SIC*: Selectively implanted collector

3.3.4.2 Narrow-Emitter Effects

The emitter area and perimeter behave differently in the forward active mode. While the injection parameters in the center of the emitter area depend essentially on the one-dimensional vertical profile, two dimensional effects at the emitter edge begin to play a dominant role as the emitter width is reduced. For small-size emitters, even three-dimensional effects at emitter corners become significant. These effects contribute to increase the injection from the base into the emitter and reduce transistor gain as the emitter is scaled to smaller dimensions. Among the mechanisms that are responsible for the degradation in gain are:

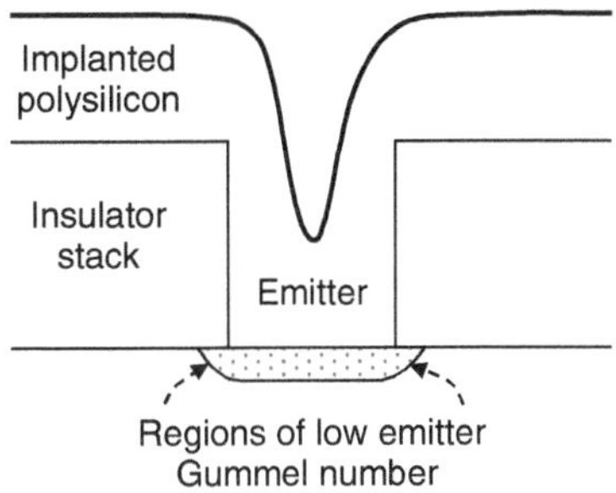

Fig. 3.16 Reduced edge concentration due to aspect-ratio and shadowing effects [40]

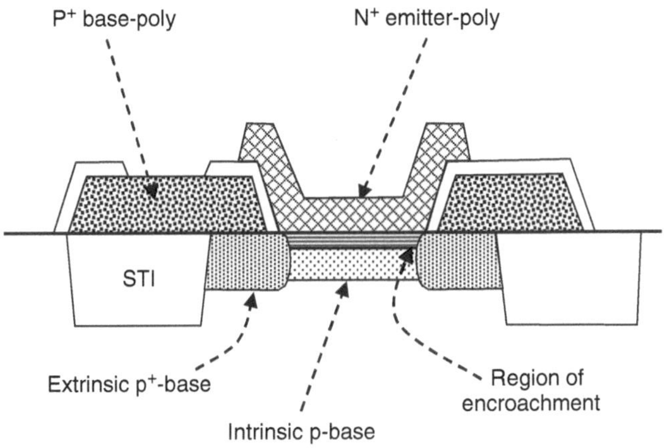

Fig. 3.17 Encroachment of the extrinsic base into the emitter and intrinsic base [44]

1. At the perimeter, the emitter concentration decreases vertically and laterally, reducing the effective Gummel number and increasing the lateral base injection into the emitter (Fig. 3.16). This effect alone is, however, not significant in the presence of an *IFO* where tunneling through the oxide is the limiting factor.
2. For an implanted polysilicon emitter, shadowing and aspect-ratio effects reduce the implanted dose in the polysilicon sidewalls, resulting in less dopant diffusion into the emitter perimeter (Fig. 3.16) This further reduces the emitter Gummel number and increases the base-injection [40, 41]. The grains in polysilicon are known to grow in a columnar arrangement, that is, with boundaries normal to the emitter surface. While this arrangement enhances dopant diffusion normal to the surface, the lateral diffusion to the edges must occur through the grain itself, a slower process, resulting in a lower Gummel number at the edge [38]. This "edge effect" is not observed with in-situ doped (doped while grown) emitters.
3. For a self-aligned emitter-base structure as shown in Fig. 3.17, the distance between intrinsic base and out-diffused extrinsic base is a trade-off between transistor gain and base resistance. A large distance causes the base resistance to increase. If sufficiently close to the emitter boundary, the extrinsic base can encroach into the emitter perimeter. This can result in an increase in the

intrinsic-base Gummel number and reduction in the emitter Gummel number at the boundary by dopant compensation [42–44].

4. As the emitter width is reduced, such as in narrow stripes, the above mechanisms become more pronounced.

3.3.4.3 Temperature Dependence of Current Gain

The current gain β is typically equal to the injection ratio. In the absence of an *IFO*, β is related to the ratio of n_i^2 in the emitter to n_i^2 in the base (3.33). Since there is more bandgap lowering in the degenerately doped emitter (Figs. 1.22 and 1.23), n_i^2 is larger in the emitter than in the base (Fig. 3.18). The rate at which n_i^2 increases with temperature is, however, higher in the base than in the emitter and the ratio of n_i^2 in the emitter to n_i^2 in the base decreases as temperature increases, as shown in Fig. 3.18. This results in an increase of β with temperature [45].

The situation is different with ultra-shallow emitters, where the surface plays a dominant role in the presence of an interfacial oxide layer. The complex combination of tunneling-limited base current and dopant segregation causing band-bending at the *IFO* boundaries can result in less increase or even decrease in gain as the temperature is increased [23]. The gain does not only depend on the difference in bandgap lowering but also on band bending and a temperature term in the tunneling current that will not be discussed further in this chapter.

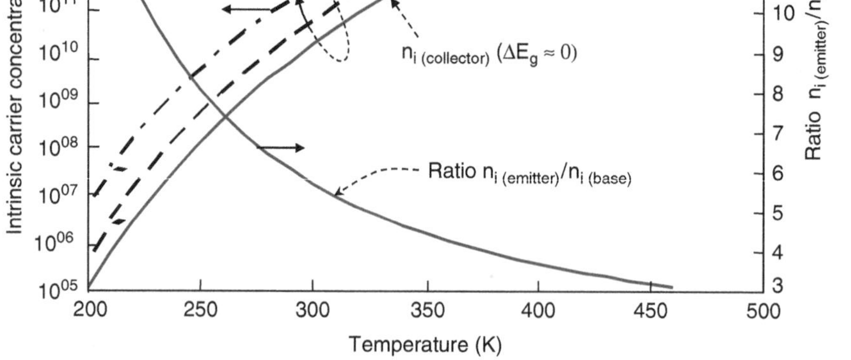

Fig. 3.18 Temperature dependence of n_i for estimated bandgap lowering in the emitter and base. The ratio of $n_{i(emitter)}$ to $n_{i(base)}$ decreases with increasing temperature causing an increase in β

3.3.5 Transistor Resistances

The transistor injection parameters are related to the local internal potentials at the emitter-base and collector-base depletion boundaries. Series resistances outside the depletion boundaries can strongly influence the transistor static and dynamic characteristics. These resistances become increasing important as the horizontal and vertical transistor geometry is scaled down. All series resistances impact transistor conductance, frequency response, and switching speed. Voltage drops across the emitter and base series resistances R_E and R_B are additive to the intrinsic base input voltage V_{BE} in (3.17) as

$$V_{BE} = \frac{kT}{q}\ln\frac{|I_E| + |\alpha_R I_C|}{|I_{Esat}|} + |I_E|R_E + |I_B|R_B. \tag{3.58}$$

Equation (3.58) applies to both *NPN* and *PNP* transistors. Also, voltage drops across the emitter and collector series resistances increase the collector saturation voltage in (3.19) as:

$$V_{CEsat} = \pm\frac{kT}{q}\ln\frac{\alpha_R[1 - I_C(1-\alpha_F)/\alpha_F I_B]}{1 + (I_C(1-\alpha_R)/I_B)} + I_E R_E + I_C R_C, \tag{3.59}$$

where the plus sign is for *PNP* and the minus sign for *NPN*. The voltage drops are additive for both *NPN* and *PNP*.

3.3.5.1 Emitter Resistance

As the emitter area A_E is reduced, the emitter series resistance increases approximately as $1/A_E$. In polysilicon-emitter transistors, the emitter resistance R_E consists of four components (wiring resistances not included):

1. Contact resistance between metal/silicide and polysilicon,
2. Polysilicon–monosilicon interface resistance,
3. Vertical resistance in polysilicon,
4. Vertical resistance of the single-crystal emitter.

The contact resistance between metal/silicide to polysilicon depends on the barrier between the two materials. It is minimized by increasing the dopant concentration in polysilicon, particularly at the interface between silicide and polysilicon. For a typical contact resistivity of 5×10^{-7} Ω-cm^2 and a contact size of $0.25\,\mu\text{m}^2$, the contact resistance is as high as 800 Ω.

The emitter resistance depends strongly on the polysilicon–monosilicon interface properties. If the *IFO* thickness increases above $\sim$1 nm, the emitter resistance increases because of the rapid decrease in the tunneling probability of majority carrier across the *IFO*.

The vertical polysilicon resistance depends on dopant concentration and morphology of grains. For columnar grain-growth, the resistance depends mostly on the conducting properties of individual grains. It can be reduced by decreasing the polysilicon thickness. For cobalt and titanium silicides (Chap. 7), however, the minimum polysilicon thickness is limited by metal penetration along grain boundaries.

Finally, the vertical resistance in the monocrystal emitter depends on emitter doping concentration and depth. For ultra-shallow emitters (20–50 nm), this resistance becomes negligible.

Measurement of Emitter Resistance

Measurement of transistor series resistances is not a trivial task because the techniques require high current densities where several effects, such as current crowding, conductivity modulation, and base push-out (Sect. 3.3.2), occur simultaneously.

Several static and dynamic methods to measure emitter resistance have been reported [4, 46–50]. AC measurements can be slow, tedious and complex, especially because of the sensitivity of measurement to parasitic capacitances. The dc techniques are preferred because they are fast, require simple instrumentation, and are easy to apply. One widely used dc measurement method is the floating-collector technique shown in the inset of Fig. 3.19 [4]. An emitter current is forced by forward-biasing the base-emitter junction with the collector connected to a high-impedance voltmeter, whereby the collector appears as floating. Thus, electrons injected from the emitter into the base reach the collector and are then injected back into the base. This is the condition for voltage saturation since both the emitter and collector junctions are forward biased. The collector saturation voltage $V_{CE(sat)}$ in (3.59) is comprised of emitter and collector junction voltages and voltage drops across resistors in series with the collector and emitter. The emitter resistance is obtained from measuring the base current as a function of V_{CEsat} with the collector "open." Setting $I_C = 0$ in (3.59) and changing the logarithm to a positive value gives

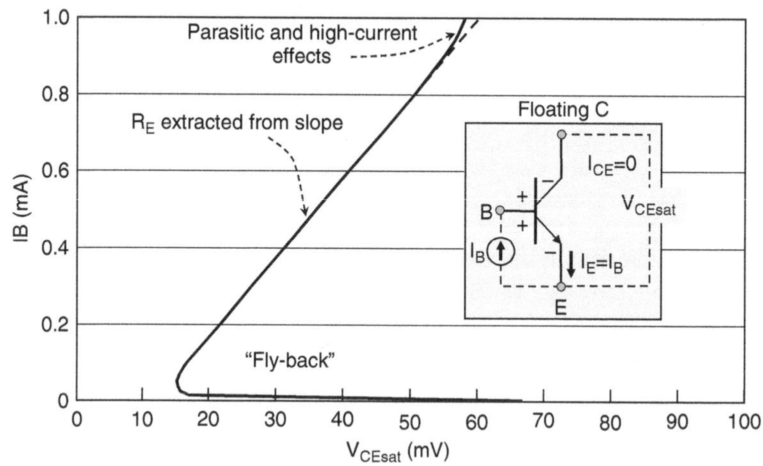

Fig. 3.19 Measurement of emitter-resistance by the floating collector method

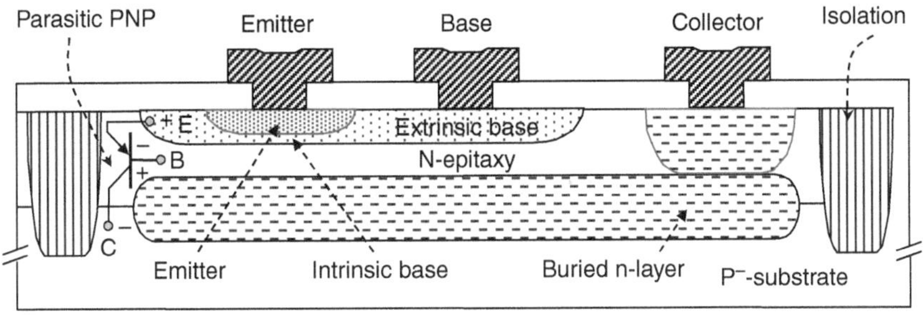

Fig. 3.20 Illustration of parasitic substrate *PNP* transistor in a junction-isolated *NPN*

$$V_{CEsat} = \frac{kT}{q} \ln\left(\frac{1}{\alpha_R}\right) + I_E R_E. \tag{3.60}$$

A plot of V_{CEsat} versus I_B with $I_C = 0$ is shown schematically in Fig. 3.19. The "fly-back" at low-current is caused by the decrease in α_R as the current is reduced and recombination effects become more pronounced. At higher currents, high-level injection effects come into play. Also, when both the emitter-base and collector-base junctions are forward biased and the transistor is not fully isolated, there is the possibility of a parasitic transistor formed between base and substrate, as illustrated for an NPN structure in Fig. 3.20. In this case, a parasitic *PNP* transistor is formed with the *NPN* base acting as the parasitic emitter, the NPN collector as the parasitic base, and the substrate as the parasitic collector. The parasitic *PNP* transistor must be considered when extracting the emitter resistance.

3.3.5.2 Base Resistance

The base resistance can be divided into two regions, intrinsic and extrinsic. The intrinsic base is the region immediately under the emitter. It is also called active base because most of the injection occurs in this region, or pinched-base to indicate that the emitter pinches the base to a very thin region. The extrinsic base is the remaining base region extending to and including the base contacts.

A simple method to extract the base resistance is to measure the deviation ΔV_{BE} of the forward characteristic from the ideal exponential behavior, as shown in Fig. 3.21 [1, 48]. ΔV_{BE} is caused by voltage drops across the extrinsic base resistance ($I_B R_{Bext}$), across the intrinsic base resistance ($I_B R_{Bint}$), and across the emitter resistance ($I_E R_E$).

At intermediate currents, both I_B and I_C exhibit ideal characteristics, with V_{BE} increasing at 300 K by about 60 mV per decade of current. The deviation ΔV_{BE} at high currents is caused by *IR* drops in the base and emitter and by high-level injection effects such as conductivity modulation and base-push effect discussed in Sect. 3.3. It can be shown that at the emitter current level where the deviation begins, ΔV_{BE} depends predominantly on the *IR* drops in the base and emitter. In this case, high-level injection effects can be neglected and

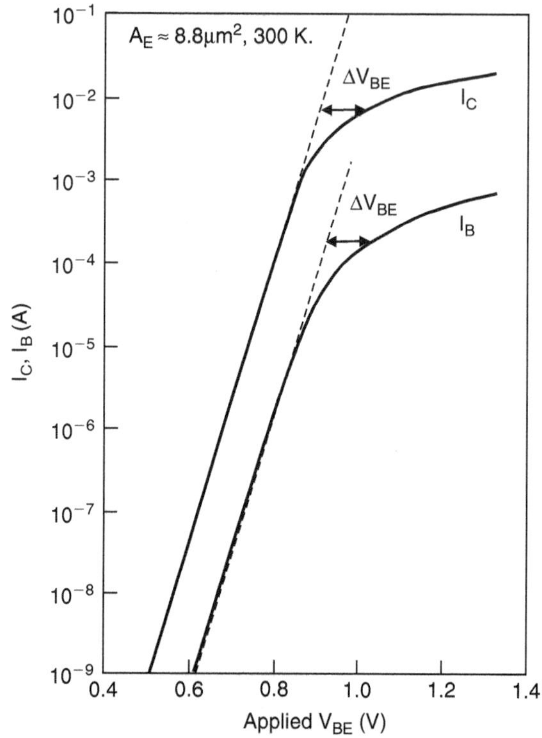

Fig. 3.21 Gummel plots of a typical *NPN* transistor. ΔV_{BE} is caused by voltage drops across the base and emitter resistances

$$\Delta V_{BE} = I_B R_{B(ext)} + I_B R_{B(\text{int})} + I_E R_E. \tag{3.61}$$

The emitter resistance R_E can be extracted by the method described in Fig. 3.19 or by, e.g., a method described in [48].

Given the transistor impurity profiles, the sheet resistance of both the intrinsic and extrinsic base can be numerically or analytically calculated. They can also be directly measured on four-point probe or specially designed structures as discussed in [51]. Calculation of the extrinsic base sheet resistance is straight-forward. The base profile is obtained from simulated, *SIMS*, or *SRP*[4] impurity profiles, or approximated by an appropriate distribution function such as a Gaussian or exponential. The sheet resistance is then found as:

$$\frac{1}{R_{BSext}} = \sum_{0}^{x_{dB}} \sigma(x)\Delta x \quad \text{Ohm/square}, \tag{3.62a}$$

[4] Secondary Ion Mass Spectroscopy (SIMS) and Spreading Resistance Profiling (SRP) techniques are used to measure the concentration versus depth.

or

$$\frac{1}{R_{BSext}} = \int_{0}^{x_{dB}} \sigma(x)\, dx \quad \text{Ohm/square.} \tag{3.62b}$$

In the above equations, 0 is taken at the surface of the base, x_{dB} is the depth of the base-sided depletion boundary at the collector-base junction, Δx the increment in depth obtained from the *SIMS* or *SRP* profile, and $\sigma(x)$ the position dependent conductivity defined as

$$\sigma(x) = q\mu(x)N(x). \tag{3.63}$$

$N(x)$ and $\mu(x)$ are, respectively, the position-dependent dopant concentration and majority-carrier mobility.

The extrinsic base resistance R_{Bext} depends on transistor geometry. For a symmetrically arranged double-base-contact as in Fig. 3.22, R_{Bext} has half the value of either extrinsic side.

Without applied bias, the intrinsic base sheet resistance R_{BSint} can be found in the same manner as for the extrinsic base. The summation in (3.62a) or integral in (3.62b) is taken between the emitter-base and collector-base depletion boundaries in the base. From the impurity profiles in Fig. 3.15, it can be seen that $R_{BSint} \gg R_{BSext}$. When the transistor is biased, for example, in the forward-active mode, the base current causes a lateral voltage drop in the intrinsic base that changes the emitter current profile (Fig. 3.22). Let y be the direction parallel to the floor of the emitter. The base current develops a transverse voltage drop $\Delta V_{BE}(y)$ along the intrinsic base of polarity opposite to that of the emitter-base forward voltage, V_{BE}, gradually reducing the magnitude of V_{BE} from emitter center to emitter edge. Thus, the forward voltage becomes position-dependent, being smallest at the point furthest away from the base contact relative to the emitter center. In the case of a double-base contact, this point would lie in the center-line of the emitter so that the injected current density decreases from edge to center, eventually crowding at the emitter edges facing the base contacts, as shown by the arrows in Fig. 3.22. This is why power transistors that operate at high currents densities are designed with a narrow emitter and a large perimeter to area ratio.

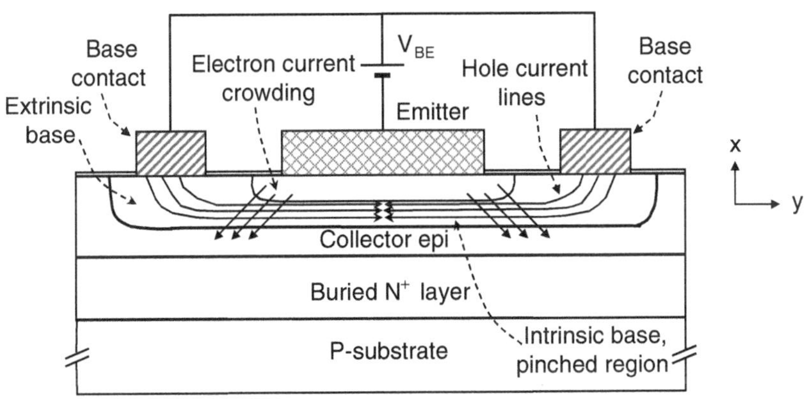

Fig. 3.22 Illustration of current crowding in an NPN transistor

Since V_{BE} is not constant, the effective intrinsic base resistance $R_{B(int)}$ is found from the average ratio of ΔV_{BE} divided by the base current that produces it [1]. The resistance thus found is also called the "base-spreading resistance" because of the distributed effect throughout the intrinsic base. The effective base resistance decreases as I_B increases because the "injection front" moves gradually from center to edge, reducing the effective I_B path-length.

The effective base resistance can be estimated using a simple model shown in Fig. 3.23 [52]. The intrinsic base is assumed to be a square of sheet resistance R_{BSint}. For a symmetrically arranged double-base contact *NPN*, the analysis is simplified by considering only one half of the transistor. Assume that $R_{BSint} = 24\,\mathrm{k}\Omega$/square. Without applied bias, the intrinsic base in half of the transistor would have a resistance $R_{BO} = 24/2 = 12\,\mathrm{k}\Omega$ (shunting effects by base polysilicon and monosilicon material around the active emitter are neglected).

Assume now that the transistor is biased in the forward-active mode. Since the current and voltage are distributed along the intrinsic base, the effective base resistance can only be expressed as an average value $R_{Bint} < R_{BO}$.

An estimate of R_{Bint} as a function of I_C can be made by slicing the emitter into a number of identical elemental transistors. For illustration, only three slices are delineated in Fig. 3.23. Each transistor is assumed to have zero internal base resistance,

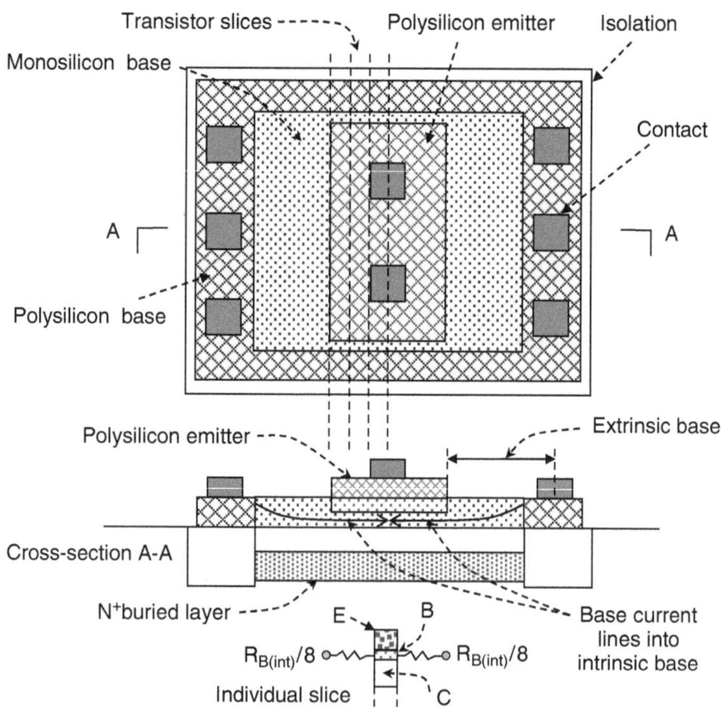

Fig. 3.23 Symmetrically-arranged *NPN* with double-base contact sliced into elemental transistors to calculate $R_{B(int)}$

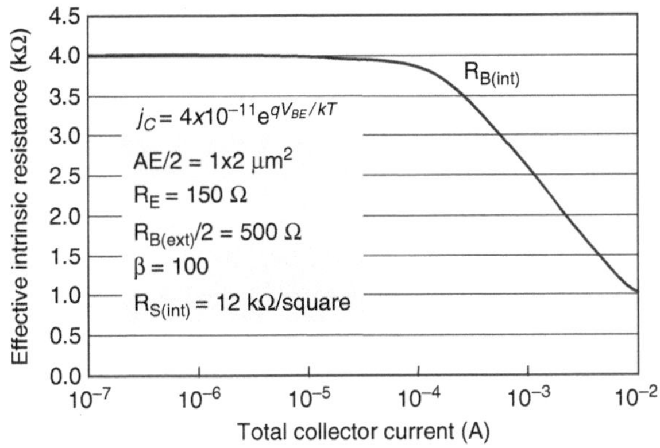

Fig. 3.24 Decrease in $R_{B(int)}$ resistance with increasing I_C

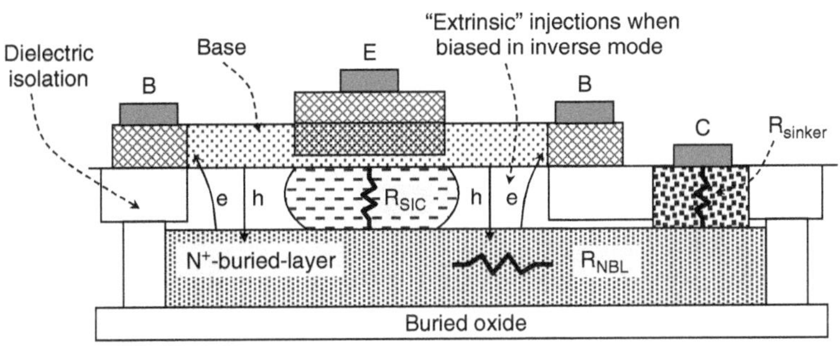

Fig. 3.25 Three main components of collector resistance

but to be connected to an external resistance of $R_{B(int)}/6$ on each side of the slice. Calculations are made by assuming a collector current in the center transistor and working backward to the edge transistor [52].

Figure 3.24 shows results obtained for an *NPN* transistor under the assumptions given in the inset. It can be seen that $R_{B(int)} \approx R_{B0}/3$ at low I_C, as derived in [1]. At high I_C, $R_{B(int)}$ decreases rapidly, and the base resistance approaches $R_{B(ext)}$ because the current crowds at the intrinsic-base boundary.

3.3.5.3 Collector Resistance

The collector resistance essentially consists of three parts: The sinker resistance, R_{sinker} that also includes the contact resistance, the N^+-buried-layer resistance, R_{NBL}, and the resistance of the selectively-implanted collector, R_{SIC} (Figs. 3.15 and 3.25; Sect. 3.4.2).

The resistances can be directly extracted from measurements on specially designed structures. They can also be calculated from measured *SIMS* or *SRP* concentrations versus depth.

Given the impurity profiles, R_{sinker} and R_{SIC} can be obtained from the conductivity and the geometry of the sinker and the selectively implanted collector as

$$R = \frac{1}{L \times W} \int_0^{x_{REF}} \frac{dx}{\sigma(x)}. \tag{3.64}$$

The origin $x = 0$ is at the surface. L and W are the length and width of the region and x_{REF} the depth from the surface where the net concentration reaches a reference level, typically chosen in the range 1×10^{17}–$5 \times 10^{17}\,\mathrm{cm}^{-3}$. The conductivity σ is found from (3.63). R_{NBL} is found from the N^+-buried-layer sheet resistance and geometry.

The "fly-back" method in Fig. 3.19 used for measuring the emitter resistance does not give accurate results when applied to the collector, even with a fully-isolated transistor. It is complicated by the presence of process-sensitive "extrinsic" electron and hole injection currents that must be taken into account when the transistor is biased in the inverse mode, as shown by the "*e*" and "*h*" arrows in Fig. 3.25.

3.4 High-Level Injection Effects

There are two high-level injection effects in a bipolar transistor biased in the forward-active mode, one at the emitter-base junction, referred to as base conductivity modulation, and the other at the collector-base junction, referred to as base push-effect or Kirk-effect.

3.4.1 Base Conductivity Modulation

For a large emitter-base forward bias voltage, the injected excess minority carrier concentration can become comparable to or even exceed the dopant concentration in the base. To maintain charge neutrality within the base, the majority carrier concentration must increase by the same amount of increase in minority carrier concentration at every point in the base. The excess majority carriers increase the base conductivity and decrease the effective intrinsic-base resistance, $R_{B(int)}$. As a consequence, the injection efficiency decreases, contributing to the drop in β at high-current densities (Fig. 3.8) [53]. In addition to conductivity modulation, the gradient in majority carriers creates a field that enhances the drift current component of minority carriers. In the limit, this field doubles the effective minority-carrier diffusion constant (Chap. 2).

Consider, for example, an *NPN* structure having a base concentration N_{AO} at the depletion boundary near the emitter. Neutrality requires that $\Delta n_{pO} = \Delta p_{pO}$. Also, because $\Delta n_{p0} \gg \bar{n}_{p0}$, the approximation $\Delta n_{pO} \approx n_{pO}$ can be made. Equation (3.28)

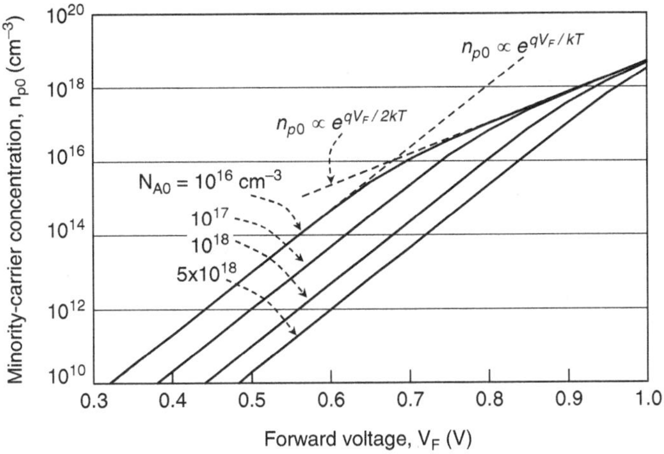

Fig. 3.26 Effect of high-level injection on slope of excess minority-carrier concentration versus forward voltage

then yields

$$n_{p0} = \frac{n_i^2}{n_{p0} + N_{A0}} e^{qV_F/kT}, \tag{3.65}$$

or

$$n_{p0} = \frac{1}{2}\sqrt{N_{A0}^2 + 4n_i^2 e^{qV_F/kT}} - \frac{N_{A0}}{2}. \tag{3.66}$$

At low-level injection, $n_{po} \ll N_{A0}$ and (3.65) simplifies to (2.60) where n_{p0} is proportional to $e^{qV_F/kT}$. At high-level injection, n_p approaches p_p and $n_{p0} \propto n_i e^{qV_F/2kT}$ (Chap. 2). The $e^{qV_F/kT}$ and $e^{qV_F/2kT}$ limits are illustrated in Fig. 3.26 for different base dopant concentrations.

Crowding at the emitter edges increases the current density and reinforces conductivity modulation, resulting in a faster decrease in current gain in those regions. It can be seen from the plots that the effect becomes less pronounced as the dopant concentration increases above $\sim 10^{18}\,\mathrm{cm}^{-3}$. For example, conductivity modulation is negligible in the heavily-doped $(10^{20}\text{–}10^{21}\,\mathrm{cm}^{-3})$ emitter and more pronounced in the collector where the *SIC* concentration is low, typically in the range $10^{16}\text{–}10^{17}\,\mathrm{cm}^{-3}$.

3.4.2 Base-Push Effect (Kirk Effect)

So far, it was assumed that injected minority carriers that reach the collector depletion boundary are immediately swept through the depletion region. This assumption is not strictly valid since carriers do not traverse the collector-base depletion region at infinite velocity. The maximum carrier velocity in the collector depletion region is the saturation velocity $v_{sat} \approx 10^7$ cm/s (Chap. 1). For an *NPN* structure, the current density across the collector-base depletion is

$$j_n = q\Delta n v_{sat}. \tag{3.67}$$

A finite electron concentration must therefore exist within the collector-base depletion to sustain the current.

In transistors where the injected current in the base is diffusion-limited, the carrier velocity in the base is 100 to 1000 times smaller than v_{sat}. In this case, one can assume that at low-level injection, Δn at the base-collector depletion boundary is ~0. In a transistor having a graded-base with a built-in field in a direction that accelerates the transport of minority carriers from emitter to collector, this approximation may not be valid. Consider, for example, an *NPN* transistor of microwave performance (high-speed) having a base profile of the form

$$N_A = N_{A(0)} e^{-Cx}, \tag{3.68}$$

where $N_{A(0)}$ is the base concentration at the emitter-base depletion boundary. For such a profile, the built-in field is constant (Chap. 1) and

$$E = -\frac{kT}{q}\frac{1}{N_A}\frac{dN_A}{dx} = C\frac{kT}{q}. \tag{3.69}$$

For example, for $N_{A0} = 10^{19}\,\text{cm}^{-3}$, $W_b = 60\,\text{nm}$ and $C \approx 7.7 \times 10^5\,\text{cm}^{-1}$, $N_{A(Wb)} = 10^{17}\,\text{cm}^{-3}$ and $E \approx 2 \times 10^4\,\text{V/cm}$. At such a high field, minority carriers drift at saturation velocity, the same velocity as in the collector-base depletion layer. The flow of minority carriers effectively increases the negative charge density in the base-side of the collector-base depletion layer and decreases the positive charge density in the collector-side of the depletion layer. To maintain space-charge neutrality, the depletion width must expand in the collector and shrink in the base. Therefore, as the current is increased, the base-width widens. In the limit, the depletion boundary in the collector moves toward the N^+-buried-layer where it becomes "pinned" [54,55]. The low-level and high-level injection boundaries are shown schematically in Fig. 3.27.

The increase in base-width at high-level injection causes the current gain and transistor speed to decrease. It is referred to as the base-push or Kirk-effect.

The onset of high-level injection at the collector can be arbitrarily defined as the current density at which Δn becomes equal to the collector dopant concentration at the collector depletion boundary, N_{DO}. The current density at the onset can then be estimated from (3.67). To prevent excessive base-widening at a specified current density j_C,N_{DO} must be kept greater than j_C/qv. Typical epitaxial collectors are grown with uniform N_D in the range $10^{16} - 5 \times 10^{16}\,\text{cm}^{-3}$. The concentration must be sufficiently low to ensure a minimum collector-base breakdown voltage and to reduce the collector base capacitance. A low concentration also reduces the spread of the depletion region into the base as the reverse voltage increases

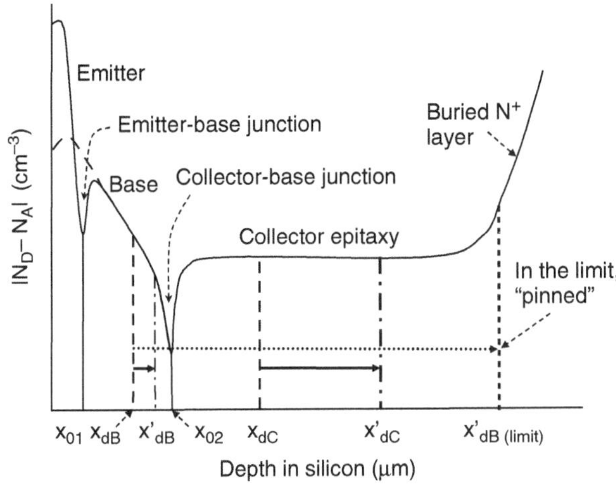

Fig. 3.27 Schematic illustration of base-push effect. x_{01}, x_{02} are the metallurgical junctions, x_{dB}, x_{dC} the depletion boundaries at low-level injection, and x'_{dB}, x'_{dC} the displaced boundaries as I_C increases

(Chap. 2), thus minimizing the impact on Early voltage. At such a low concentration, however, the onset of base push-effect occurs at a current density of only about $0.16 - 0.8\,\mathrm{mA}/\mu\mathrm{m}^2$ (3.67). To increase the current density while maintaining a low collector-base capacitance, N_{D0} is locally increased by selectively implanting the collector (*SIC*) immediately under the intrinsic base region. Thus, the *SIC* concentration is a trade-off between collector-base breakdown voltage, Early voltage, and current density at onset of the Kirk-effect. In some designs, the collector epitaxy is grown near intrinsic and the *SIC* concentration chosen in the range of about $10^{17} - 5 \times 10^{17}\,\mathrm{cm}^{-3}$.

The plot in Fig. 3.8 shows that the gain falls-off at high collector current densities. The base-push effect, base-conductivity modulation and emitter crowding are among the important mechanisms that contribute to the fall-off. Other mechanisms are the increase in base Gummel number as the emitter-base junction is forward biased and the increased injection of holes from the base into the emitter. A thorough analysis of all the mechanisms is only possible with numerical simulations because they typically occur simultaneously and, in many cases, interact with each other.

3.5 Frequency Response of Current Gain

The time required for carriers to travel from emitter to collector is called the transit time, τ. The delay terms that contribute to the overall transit time τ are shown in Fig. 3.28. The critical delay terms that affect the transistor frequency response and their trade-offs with other parameters are discussed in this section.

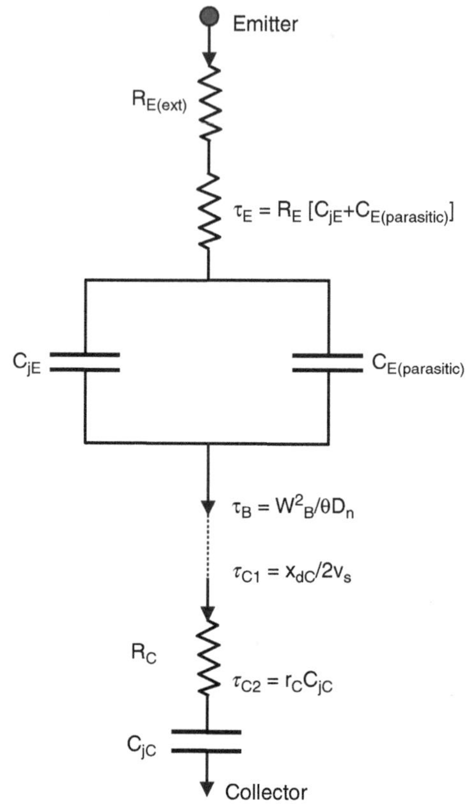

Fig. 3.28 Critical transistor time-delay terms

3.5.1 Emitter Delay, τ_E

The emitter time-delay is the product of emitter resistance and emitter capacitance

$$\tau_E = [R_{Eext} + r_{Eint}][C_{jE} + C_{E(parasitic)}]. \tag{3.70}$$

R_{Eext} includes the emitter contact resistance, the vertical resistance in the polysilicon and monosilicon emitter, and an additional resistance that may be caused by the presence of an interface oxide. The dynamic emitter resistance r_{Eint} represents the slope of the forward-biased emitter-base junction. It is obtained by partial differentiation of (3.20) as

$$\frac{1}{r_{Eint}} \cong \left.\frac{\partial I_E}{\partial V_{BE}}\right|_{V_{CE}=const} = \frac{qI_E}{kT}. \tag{3.71a}$$

As can be seen, r_{Eint} decreases with increasing emitter current as

$$r_{Eint} = \frac{kT}{qI_E} \cong \frac{kT}{qI_C} = \frac{1}{g_m}, \tag{3.71b}$$

where g_m is the transconductance in Siemens.

C_{jE} is the emitter-base depletion capacitance consisting of area and perimeter components. It can be measured on specially designed test structures or calculated given the base and emitter impurity profiles. For low to moderate forward bias, an estimate of this capacitance is

$$C_{jE} = \sqrt{\frac{q\varepsilon_0\varepsilon_{Si}N_B}{1 - V_{BE}}} \quad \text{F/cm}^2. \tag{3.72}$$

The "1" in the denominator of (3.72) is an approximation for the emitter-base built-in voltage; N_B is the base concentration at the emitter boundary; and V_{BE} is the small forward voltage. The above approximation for C_{jE} ceases to be valid when the injected minority carrier concentration approaches N_B.

$C_{E(parasitic)}$ in Fig. 3.28 is the emitter-base capacitance outside the active region, for example, the overlap capacitance between the polysilicon emitter and the extrinsic base. This capacitance can be minimized by controlling the overlap and increasing the insulator thickness between the overlapping polysilicon and base.

Equations (3.69)–(3.71) show that the emitter time delay τ_E decreases with increasing emitter current. To minimize τ_E, the emitter series resistance must be reduced, and the emitter capacitance be made as small as possible by reducing the emitter size and base dopant concentration at the emitter boundary. When optimizing τ_E, however, there is a trade-off between speed, current-carrying capability, Early voltage, and punch-through voltage.

3.5.2 Base Transit Time, τ_B

The time required for injected minority carriers to travel through the base is the base transit time, τ_B. The base transit time is also found as the ratio of injected (stored) base charge Q_{nB} (Fig. 3.10) to emitter current

$$\tau_B = \frac{Q_{nB}}{I_E} \quad \text{s}. \tag{3.73}$$

The injected electron concentration at the emitter-base depletion boundary in the base is defined as (Chap. 2)

$$n_{p0} \approx \bar{n}_{p0} e^{qV_F/kT}. \tag{3.74}$$

Assuming a uniformly-doped base and low-level injection, the mobility in the base is constant and $n_p \approx 0$ at W_b. Also, since the base width is much smaller than the minority diffusion length, the assumption of zero recombination in the base is valid. Thus, the excess minority carrier concentration drops linearly from emitter to collector and the carriers are transported by diffusion. The charge stored in the base is therefore

$$Q_{nB} \approx \frac{1}{2} q n_{p0} W_b. \tag{3.75}$$

The electron current is then

$$j_E = qD_n \frac{n_{p0}}{W_b}. \quad (3.76)$$

The base transit time is then found as

$$\tau_B = \frac{qn_{p0}W_b}{2(qD_n n_{p0}/W_b)} = \frac{W_b^2}{2D_n} \quad s. \quad (3.77)$$

Thus, the base transit time is proportional to the square of the base-width and hence very sensitive to any changes in base-width.

For a nonuniform base profile, the situation is more complex because several factors come into play:

1. There is a built-in field in a graded base that induces a drift current component that can accelerate or decelerate minority carriers, depending on the gradient sign (3.69).
2. The carrier mobility and hence diffusivity is no longer constant but varies with base concentration and electric field.
3. The energy-gap lowering varies across the base because it depends on base concentration. This induces an electric field whose sign is opposite to the sign of the built-in field and can retard or accelerate carriers, depending on the gradient sign.

For an arbitrary base profile, the electron current is expressed by the drift-diffusion relation [56, 57]

$$j_E = qD_n(x)\frac{dn_p(x)}{dx} + q\mu_n(x)E(x)n_p(x), \quad (3.78)$$

where

$$E(x) = \frac{kT}{q}\left(\frac{1}{N_A(x)}\frac{dN_A(x)}{dx} - \frac{1}{n_i^2(x)}\frac{dn_i^2(x)}{dx}\right). \quad (3.79)$$

The first term in the above equation is the built-in field caused by the gradient in dopant concentration and the second term the field due to the position-dependent bandgap narrowing approximated by (1.53) and the resulting increase in intrinsic-carrier concentration. It can be shown that [57]

$$n(x) = \frac{j_{nE}}{q}\frac{n_i^2(x)}{N_A(x)}\int_x^{W_b}\frac{1}{D_n(x)}\frac{N_A(x)}{n_i^2(x)}dx. \quad (3.80)$$

Integrating $n(x)$ and dividing by j_{nE}/q gives τ_B as

$$\tau_B = \frac{Q_B}{j_{nE}} = \int_0^{W_b}\frac{n_i^2(x)}{N_A(x)}\int_x^{W_b}\frac{1}{D_n(v)}\frac{N_A(v)}{n_i^2(v)}dv.dx. \quad (3.81)$$

For a uniform base, the above equation simplifies to (3.77). Typical base profiles are Gaussian or exponential. In both cases, the peak base concentration under the emitter should be close to the emitter-base junction and not deeper to avoid increasing τ_B by a retarding field.

In high-performance transistors, the base-width is scaled down to aggressive dimensions in the range 20–50 nm, as in *SiGe* Heterojunction Bipolar Transistors (*HBT*) [14, 58, 59]. The power-supply voltage, however, does not decrease at the same rate. This results in an increase in electric field in the base and its gradient at the base-collector junction [60]. Simulations show that for such high electric fields and ultra-thin base of width comparable to the carrier mean-free path, the drift-diffusion approximation is no longer valid and two nonequilibrium transport effects become important (Chap. 5), namely velocity overshoot at the collector-base junction and quasi-ballistic transport in the base. Under these conditions, the base transit time depends linearly on W_b rather than on the square of W_b [60, 61].

3.5.3 Collector Delay, τ_C

The collector delay τ_C consists of two parts, one associated with charging and discharging the collector-base depletion capacitance, τ_{CjC} and the other with the transit time through the collector depletion layer, τ_{xdC}.

3.5.3.1 Collector-Base Capacitance Delay, τ_{CjC}

The collector-base junction capacitance C_{jC} is charged through both the emitter and collector series resistances, including the emitter dynamic resistance r_{Eint}. The associated delay time is

$$\tau_{CjC} = \left[R_{Eext} + r_{E\mathrm{int}} + R_C\right] C_{jC}. \tag{3.82}$$

C_{jC} encompasses the full collector base junction. In a nonuniformly doped collector, such as with a selectively-implanted collector (*SIC*), the capacitances of both *SIC* and non-*SIC* regions must be included in C_{jC}. For a near-intrinsic epitaxy, the capacitance per unit area outside the *SIC* region can be estimated as $C_{epi} = \varepsilon_{Si}/t_{epi}$, where t_{epi} is the distance between base and the point in the up-diffused buried layer where the concentration reaches about $10^{17}\,\mathrm{cm}^{-3}$.

Note that parasitic vertical and lateral capacitances between collector and substrate and collector to adjacent structures exist. They do not, however, contribute to the emitter-collector delay time.

3.5.3.2 Transit Time through the Collector-Base Depletion Layer, τ_{xdC}

Injected carriers crossing the collector-base depletion boundary are subjected to a high electric field, typically larger than $10^4\,\mathrm{V/cm}$. Thus, the carriers travel through the collector-base depletion region at saturation velocity, $v_{sat} \approx 10^7\,\text{cm/s}$. For a depletion width x_{dC}, one would expect the carrier transit time to be x_{dC}/v_{sat}. The actual delay, however, is found to be half this value, that is, $x_{dC}/2v_{sat}$ [62–66]. This can be shown by a simple charge-control analysis [62]. Consider, for example, an *NPN* structure with uniformly-doped collector and fixed collector-base reverse voltage. For a constant velocity v_{sat}, the collector current is $j_C = qv_{sat}n$. Current continuity requires that n and hence the mobile charge density ρ_C also be constant, where

$$\rho_C = qn = \frac{j_C}{v_{sat}}. \tag{3.83}$$

As ρ_C travels through the depletion layer, it adds to the fixed negative charge in the base, shortening the depletion layer in the base, and subtracts from the fixed positive charge in the collector, widening the depletion layer in the collector (Sect. 3.4.2). Under the assumption that the depletion layer expands mostly into the collector, the mobile charge per unit area can be approximated as $Q_C = \rho_C.x_{dc}$. Half of this charge is neutralized by shortening the depletion layer in the base. The actual delay time is found by dividing this charge by j_C [62]

$$\tau_{xdC} = \frac{\rho_C x_{dC}}{J_C} = \frac{\rho_C x_{dC}}{2\rho_C v_{sat}} = \frac{x_{dC}}{2v_{sat}}, \tag{3.84}$$

which is in agreement with the results in [63–66].

3.5.3.3 Gain-Bandwidth Product, f_T

For a given current density, the current gain remains almost constant up to a certain frequency where the carriers can no longer respond to the varying signal, at which point the gain β begins to fall-off. A widely accepted figure of merit of transistor speed is the grounded-emitter gain-bandwidth product

$$f_T = \beta f, \tag{3.85}$$

where f is the operating frequency and f_T is the frequency at which the grounded-emitter current gain is unity. There is a wide frequency range where the rate of decrease in β follows 6 dB per octave of frequency. It can be shown that in this range, extraction of f_T can be simplified by measuring β at one frequency and extrapolating to the frequency where $\beta = 1$ [1]. For example, if β falls to 20 at 5 GHz, then $f_T = 100\,\text{GHz}$. The relation between f_T and delay time is

$$f_T = \frac{1}{2\pi\tau} \quad \text{Hz}, \tag{3.86}$$

where τ is the sum of all time delays discussed above

$$\tau = \tau_E + \tau_B + \tau_{CjC} + \tau_{xdC}. \tag{3.87}$$

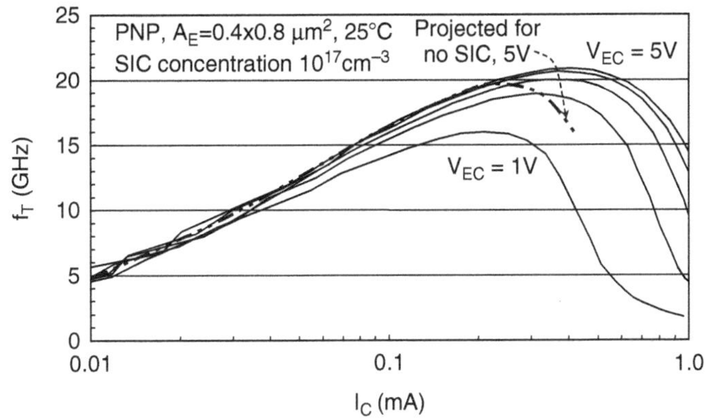

Fig. 3.29 Measured f_T versus I_C for *PNP* at $V_{EC} = 1$–$5\,\text{V}$

Neglecting the parasitic emitter capacitance in (3.70), the total delay time can be approximated as

$$\tau = \left(R_E + \frac{kT}{qI_C}\right)(C_{jE} + C_{jC}) + \frac{W_b^2}{\theta D_n} + R_C C_{jC} + \frac{x_{dc}}{2v_{sat}}, \tag{3.88}$$

where θ is an adjustment factor of 2–6 for the built-in field in a graded-base structure. A measured plot of f_T versus collector current in a *PNP* is shown in Fig. 3.29. f_T is low at small collector currents because the dynamic emitter resistance kT/qI_C in the first term of (3.88) dominates. As the collector current increases, f_T increases monotonically and reaches a peak at high current density above which it begins to decrease, mainly because of the Kirk-effect discussed in Sect. 3.4.2.

The dotted plot in Fig. 3.29 is a projection of f_T versus I_C without the selectively implanted collector (*SIC*). Increasing the collector dopant concentration by *SIC* immediately beneath the active base delays the Kirk-effect and shifts the current at which f_T begins to fall-off to a higher value. The increase in f_T at higher collector reverse voltage is the result of W_b shortening as the depletion region moves into the base. Also, an increase in the total collector-base depletion width reduces C_{jC}. It should be noted that, as the base-width is reduced below ~100 nm, the base transit time can drop below the collector transit time and the last term in (3.88) becomes more important.

Similar plots measured on an *NPN* structure of comparable geometry to that in Fig. 3.29 show higher values for peak f_T than for *PNP*. This is mainly due to the higher electron mobility.

3.5.3.4 Maximum Frequency of Oscillation, f_{max}

The power gain G of an amplifier is the ratio of output to input power. It is the product of voltage gain and current gain. For a grounded-emitter transistor, the frequency dependence of G is derived as [67]

$$G = \frac{f_T}{8\pi f^2 r_B C_{jC}}. \tag{3.89}$$

where r_B is the base input resistance, f the frequency of operation, and C_{jC} the collector-base capacitance. The output resistance is

$$\frac{1}{\omega_T C_{jC}} = \frac{1}{2\pi f_T C_{jC}},$$

where ω is the angular frequency. A transistor figure of merit relevant to *RF* and microwave applications is the frequency at which the amplifier power-gain drops to unity. This is referred to as the maximum frequency of operation, f_{max}. By setting $G = 1$ and $f = f_{max}$ in (3.89), the relation between f_{max} and f_T is obtained as [67]

$$f_{\max} = \sqrt{\frac{f_T}{8\pi r_B C_{jc}}}. \tag{3.90}$$

Thus, to increase f_{max}, f_T should be increased and r_B, and C_{jC} should be reduced. For fixed intrinsic and extrinsic base sheet resistances, reducing the base resistance can only be achieved by optimizing the transistor layout, that is, by reducing the width of emitter stripes to reduce the contribution of intrinsic base resistance, by implementing multiple base leads and contacts, and by minimizing the distance between the extrinsic and intrinsic base.

3.6 The Transistor as a Switch

When operating as a switch, the transistor has two states of conduction: it is off or on. An ideal switch has a very high resistance in the off state, approaching that of an open-circuit, and a very low resistance in the on-state, approaching that of a short-circuit. Also, the response to a switching signal should be nearly instantaneous to achieve a rapid transition from one state to the other. The transistor is, however, far from being an ideal switch. Consider, for example, the switching circuit for a grounded-emitter PNP transistor in Fig. 3.30a.

The collector response to an ideal base current pulse of sufficient magnitude to turn-on the transistor is shown in Fig. 3.30b. It can be seen that the collector does not immediately respond to the signal. Also, there is a delay in collector response to a base current pulse of magnitude sufficient to turn-off the transistor. The switching cycle can be divided into four periods, the delay time t_d, the rise time t_r, the storage time t_s, and the fall time t_f. The turn-on and turn-off times are defined as:

$$t_{on} = t_d + t_r, \tag{3.91a}$$

$$t_{off} = t_s + t_f. \tag{3.91b}$$

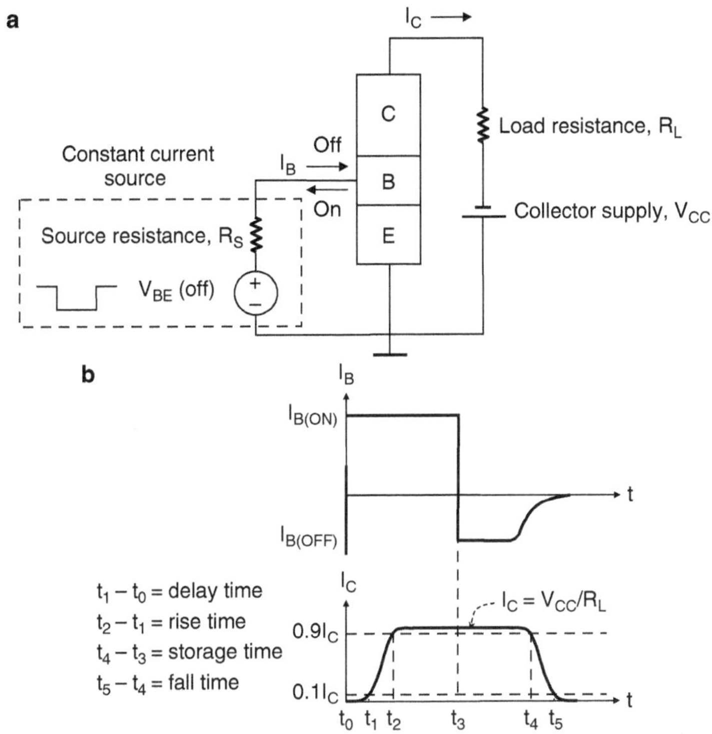

Fig. 3.30 Switching cycle from grounded-emitter *PNP* transistor: **a** Switching diagram; **b** Base and collector currents during turn-on and turn-off transients

Since switching involves large signals, a current analysis of turn-on and turn-off times becomes very complex because of the high degree of nonlinearity during the transitions. The charge-controlled method is preferred because of its simplicity [1, 11, 68].

3.6.1 Delay Time, t_d

An example of grounded-emitter *PNP* output characteristics is shown in Fig. 3.31. In the active mode, the collector current is controlled by the base currents. The load line is the locus of allowable points that satisfy $V_{EC} = I_C R_{load}$. Point *A* is the boundary between the off-state and the active region. At this point, $V_{EB} = 0$, $V_{EC} \approx V_{BC}$, and the base current consists only of reverse base-collector leakage current.

The delay time t_d is the time required to bring the transistor from the off state to the onset of the active region at point *A* in Fig. 3.31. For $V_{BE} = 0$ at the beginning of switching, the delay time is zero. If, however, the initial condition is $V_{BE} > 0$, that is reverse-biased base-emitter junction, there will be a delay time associated

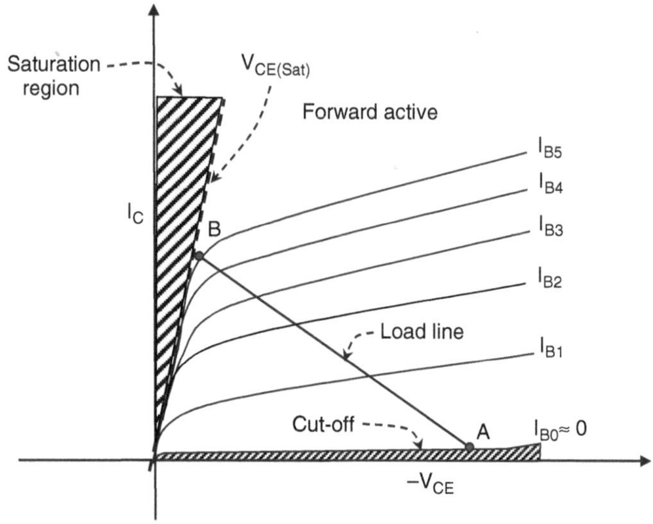

Fig. 3.31 Output characteristics and operating regions in grounded-emitter *PNP* transistor. Magnitude of base current is increased by increments ΔI_B

with charging the emitter-base depletion region when the base voltage goes from reverse to zero. This is because reducing the forward voltage by ΔV_{BE} causes a corresponding change in the emitter-base depletion width Δx_{dE} and charge must be provided by the base contact to neutralize the excess ionized impurities in the depletion layer. This charge is $\Delta Qx_{dE} = qN_D\Delta x_{dE}$. The ratio $\Delta Q_{xdE}/\Delta V_{BE}$ constitutes an emitter capacitance C_{jE} that contributes to the overall *RC* delay. Similarly, a change in collector-base voltage is associated with a charge ΔQx_{dC} and a collector capacitance C_{jC}. The base current necessary to charge the depletion capacitances is

$$I_B = \frac{dQ_{xdE}}{dt} + \frac{dQ_{xdC}}{dt}. \tag{3.92}$$

If the voltage dependence of the emitter-base and collector-base capacitances is known, the delay-time t_d can be found by integrating (3.92) and making appropriate changes in limits

$$\int_0^{t_d} I_B dt = \int_{-V_{BE1}}^{0} C_{jE} dV_{BE} + \int_{-V_{BC1}}^{-V_{CC}} C_{jC} dV_{CE}, \tag{3.93}$$

where V_{BEI} and V_{BCI} are, respectively, the initial emitter-base and collector-base voltages, V_{CC} the steady-state collector-base voltage, and $V_{BCI} = (V_{CC} + V_{BEI})$. The delay time can be reduced by minimizing the emitter-base and collector-base junction capacitances. For fixed impurity profiles, the emitter and collector should be designed as small as possible to achieve high-switching speed.

3.6.2 Rise Time, t_r

The rise time t_r is the time required to bring the collector current from 10% to 90% of its final value. This is approximately the time required for a transition from the onset of the active region at point A to the onset of saturation at point B in Fig. 3.31. During this transition, the minority charge in the base rises to its final value shown in Fig. 3.32, and the emitter-base and collector-base depletion regions shrink further.

The linear decay of stored charge, shaded triangle in Fig. 3.32, applies to a typical case where W_b is much smaller than the diffusion length of holes in the base. Since at every point in the base, the minority carriers must be neutralized by majority carriers, the base contact must supply majority carriers of total charge $Q_{nB} = -Q_{pB}$.

During the transition, the total base current is

$$I_B = \frac{dQ_{xdE}}{dt} + \frac{dQ_{xdC}}{dt} + \frac{dQ_B}{dt} + \frac{Q_B}{\tau_B}. \tag{3.94}$$

The first and second terms on the right are associated with charging the emitter-base and collector base capacitances as the depletion regions shrink. The third term is the base current necessary to replenish recombination of holes in the base. The fourth term is the base current needed to charge the base to the level where the collector current reaches its value at point B. τ_B is the base transit time, defined for a uniform base in (3.73) and (3.81). The time at which I_C reaches the collector current at point B in Fig. 3.31 can be approximated as [1, 11]

$$t_r \approx \beta \tau_{eff} \ln \frac{1}{1-(0.9I_C/\beta I_B)}, \tag{3.95}$$

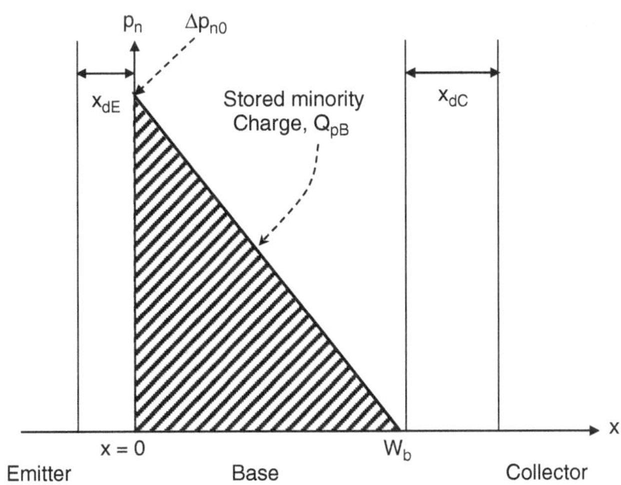

Fig. 3.32 Stored minority-carrier charge in *PNP* base

where τ_{eff} is an effective transistor time constant that also includes the terms $1/f_T$ and the time constant $R_L C_{jC}$ related to charging through the load resistance R_L in Fig. 3.30. As f_T increases, τ_{eff} and hence the rise-time decreases.

The stored minority-carrier charge in the emitter, Q_E, was neglected so far on the grounds that the emitter is doped much heavier than the base and the concentration of minority-carriers injected into the emitter is considerably smaller than that injected into the base. In this case, the stored charge in the emitter, Q_E, constitutes only a small fraction of the stored charge in the base, Q_B. For an ultra-thin base, however, Q_B becomes very small and Q_E can no longer be neglected when compared to Q_B and the contribution of Q_E must be included in (3.94):

$$I_B = \frac{dQ_{xdE}}{dt} + \frac{dQ_{xdC}}{dt} + \frac{dQ_B}{dt} + \frac{Q_B}{\tau_B} + \frac{dQ_E}{dt} + \frac{Q_E}{\tau_E}. \tag{3.96}$$

3.6.3 Storage Time, t_s

When both the emitter-base and collector-base junctions become forward biased, the transistor is in saturation (Fig. 3.31). The charge in the base can then be represented by two stored-charge triangles Q_{pB1} and Q_{pB2}, as shown in Fig. 3.33.

Qp_{B1} is the charge of excess holes injected from the emitter into the base (Fig. 3.32). Qp_{B2} is the charge of excess holes injected from the collector into the base when the transistor is in saturation. The total charge of injected holes is $Q_{pB} = Q_{pB1} + Q_{pB2}$. Also shown in the figure is the minority-electron charge injected from the base into the emitter and collector. Since the dopant concentration in the collector is typically low, there is considerable electron charge storage in the collector. The concentration of excess electrons in the collector is assumed to drop to zero near the P$^+$-buried-layer (*PBL*) where the recombination-rate is high. While

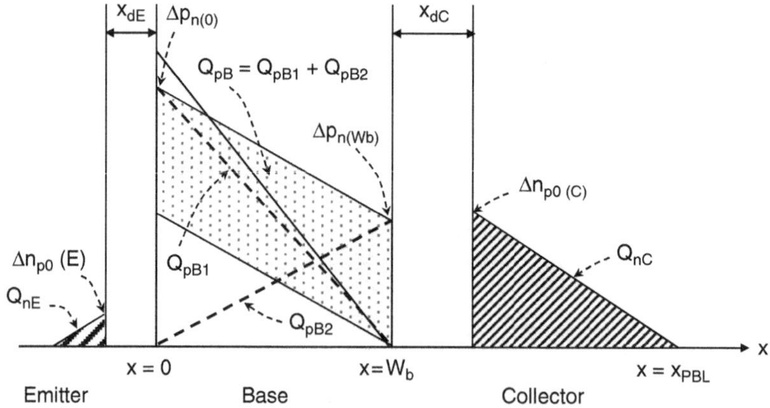

Fig. 3.33 Excess hole charge in base of *PNP* in saturation. Also shown is the excess electron charge in the emitter and collector

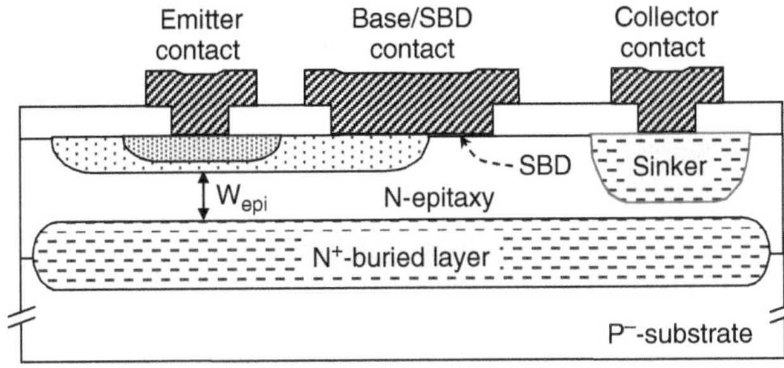

Fig. 3.34 Illustration of an *NPN* structure with a Schottky-clamp

minority-carrier storage in the emitter is small, it may not be negligible when compared to the total stored charge when the base is ultra-thin.

At time t_3 in Fig. 3.30b, the base is "instantaneously" switched from the "on" to "off" level. The collector current remains essentially constant until time t_4 where V_{CB} drops to zero. This is the boundary between saturation and active regions that corresponds to point *B* in Fig. 3.31. The storage time can be approximated as [1, 11]

$$t_s \approx \tau_s \ln \frac{I_{B(on)} - I_{B(off)}}{(I_C/\beta) - I_{B(off)}}, \tag{3.97}$$

where τ_s is approximately the sum of forward base-transit time, reverse base-transit time, and minority-carrier lifetime in the collector, and $I_{B(on)}$, $I_{B(off)}$ are defined in Fig. 3.30b.

To minimize the storage time, it is important to reduce the minority-carrier lifetime in the collector and the distance W_{epi} between the collector and the heavily-doped buried layer (Fig. 3.34). In earlier transistor versions, the minority-carrier lifetime was reduced by introducing heavy metals, such as gold in the collector to create recombination sites (Chap. 1). This method is, however, not practical in scaled-down modern transistors. The stored charge in the collector can be considerably reduced by clamping the base-collector junction with a Schottky-barrier diode (*SBD*) of appropriate barrier height and area, as shown for an NPN structure in Fig. 3.34. When the collector-base junction is reverse-biased, the *SBD* is also reverse biased. The current measured is then the reverse-bias leakage current of both the base-collector junction and *SBD* (Chap. 2, Sect. 2.3.1). When the transistor enters saturation, both the collector-base junction and *SBD* are in forward bias. Since the *SBD* is essentially a majority-carrier device, there is negligible charge storage associated with it. The barrier height and area of the *SBD* are typically chosen such that the fraction of current that the *SBD* carries is at least 0.9 (Problem 10).

3.6.4 Fall Time, t_f

The fall time is the time required for the collector current to fall from $0.9I_C$ to $0.1I_C$, that is, $t_f = t_5 - t_4$ (Fig. 3.30b). This is the time required for the operating point to traverse the active region from point B to point A in Fig. 3.31. During this transition, the excess minority-carrier charge is removed from the base (Fig. 3.32). The approximation of the fall-time is therefore similar to that of the rise time, resulting in an effective time-constant similar to that in (3.95) [1,69]

$$t_f \approx \beta \tau_{eff} \ln \frac{I_C/\beta I_{B(ON)}}{(0.1 I_C/\beta I_{B(on)}) - 1}. \tag{3.98}$$

3.7 Silicon-Germanium Transistor

A typical silicon-germanium (*SiGe*) transistor consists of a *SiGe* base sandwiched between a silicon-emitter and a silicon-collector. Since the emitter-base and collector-base junctions are formed between dissimilar materials, *SiGe* transistors are said to belong to the family of heterojunction bipolar transistors (HBT). The main objective of adding germanium is to tailor the bandgap in the base, increasing the flexibility in optimizing current-gain, speed, Early-voltage, and noise. The properties and advantages of *SiGe* transistors are discussed in this section.

3.7.1 SiGe Film Deposition and Properties

The *SiGe* film is typically deposited by low temperature epitaxial growth through the reduction of dichlorosilane (SiH_2Cl_2) by hydrogen. The low-temperature is required to avoid excessive dopant diffusion and maintain a narrow-base. Germanium is incorporated in-situ by adding germane (GeH_4) to the gas stream (Chap. 7). The film can be deposited by ultra-high vacuum chemical-vapor deposition (*UHV-CVD*) at a temperature as low as 450 °C [69,70], or by conventional low-pressure or atmospheric-pressure CVD at a temperature near 600 °C [71–73]. The film grows as single-crystal silicon over monosilicon, and as a polysilicon film over insulators and polysilicon. Low-temperature epitaxy requires high-purity hydrogen to avoid traces of oxygen and water. The film can also be deposited self-aligned to the base by selective epitaxial growth (*SEG*) [74–76]. In this case, the film nucleates only over exposed silicon. In-situ doping is performed by adding, for example, B_2H_6, AsH_3, or PH_3 in the gas-stream.

A typical *NPN SiGe*-base profile is shown in Fig. 3.35. The germanium concentration is gradually increased and then decreased to obtain a triangular shape. As will be discussed later in this section, the triangular *Ge* profile enhances the field that aids the transport of carriers through the base, reducing the base transit-time. Other *Ge* shapes, such as box or trapezoidal, are also achievable. The choice of *Ge* shape depends on which parameter should be optimized.

The optimum growth temperature is ~550 °C for a *Ge* mole fraction up to 0.15. The Ge transition at the SiGe/Si interface is very steep. The *SiGe* layer is typically

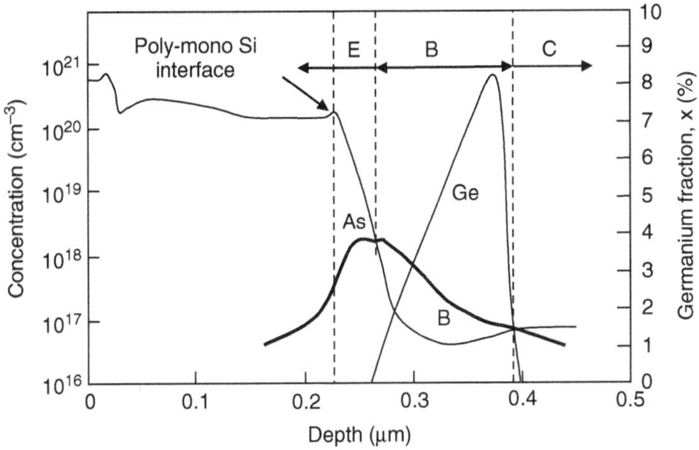

Fig. 3.35 *NPN SiGe*-base profile reconstructed from *SIMS* data

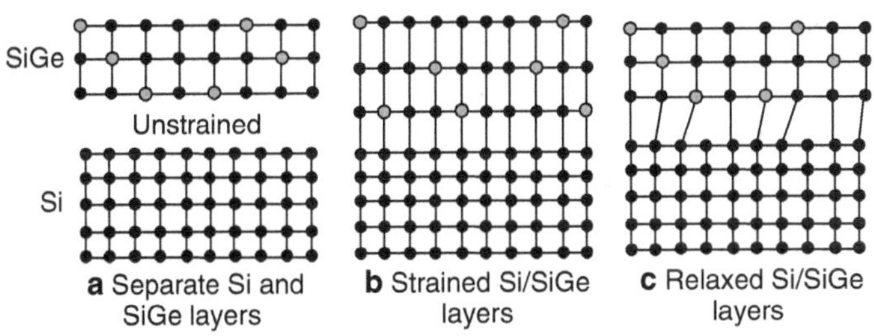

Fig. 3.36 Illustration of lattice mismatch, pseudomorphic layer and stress-relief relaxation. **a** Separate *Si* and *SiGe* crystals; **b** Pseudomorphic layers, bi-axial strain, tetragonal elongation; **c** Relaxed layers

separated from the *Si*-emitter and *Si*-collector regions by very thin intrinsic *Si*-films called vertical spacers. The primary purpose of the spacers is to make sure that emitter-base and collector-base depletion boundaries remain within the main Ge distribution, as discussed below. Also, an intrinsic buffer layer between *SiGe* and the Si-collector reduces the collector-base capacitance outside the selectively implanted collector (*SIC*) region. The *SiGe* film is also capped with an undoped silicon layer that serves three main objectives [77]: (a) terminating the crystal with Si rather than *SiGe* to achieve better compatibility with subsequent processing, such as oxidation; (b) providing some flexibility in tailoring the impurity profiles and adjusting the peak-field and capacitance at the emitter-base junction; (c) improving thermal stability by preventing misfit dislocations at higher processing temperatures [77–79].

A key deposition condition is to have the *SiGe* lattice adapt to the silicon substrate. This condition is called pseudomorphic, indicating that the deposited SiGe crystalline layer adopts the lattice of the Si layer upon which it grows (Fig. 3.36).

The lattice constant of a *Ge* crystal is $a_{Ge} = 0.564613$ nm and that of silicon is $a_{Si} = 0.543095$ nm. The lattice mismatch between *Si* and *Ge* is ~4.2% at 25 °C and increases slightly with increasing temperature. A bulk $Si_{1-x}Ge_x$ crystal containing a *Ge* mole fraction x will have a latticc constant that lies between that of a purely *Ge* and a purely *Si* crystal. The lattice constant can be approximated from an empirical rule, called Vegard's law, stating that for a given temperature, a linear relation exists between the crystal lattice constant of an alloy and the concentrations of the constituent elements. For low *Ge* concentration ($x < 20\%$), the *SiGe* lattice constant is [80]

$$a_{Si_{1-x}Ge_x} \cong a_{Si} + x(a_{Ge} - a_{Si}). \tag{3.99}$$

The pseudomorphic *SiGe* film in Fig. 3.36b is under compressive bi-axial in-plane stress and tensile stress normal to the surface, resulting in the tetragonal elongation as indicated, in accordance with Poisson's ratio. Since the silicon wafer is very thick, its lattice remains essentially unchanged.

Strain compensation by stress-relaxation (broken bonds) results in poor crystalline quality. It is accompanied by crystal defects such as dislocations and stacking faults, as schematically shown in Fig. 3.36c. It is therefore important to avoid strain-relaxation during film depositions or during subsequent thermal cycles. $Si_{1-x}Ge_x$ is miscible in all proportions, but stability of the film depends on the film thickness and the amount of strain created by adding *Ge*. Because strain is proportional to *Ge* concentration, stability depends essentially on the integrated *Ge* content rather than the detailed *Ge* distribution. The greater the strain, the thinner the film must be to remain stable [81–83]. Figure 3.37 shows the critical thickness above which the film becomes unstable [81, 82].

A film that has a thickness and *Ge* concentration below the curve is stable. As the film thickness increases above the curve, its strain-energy increases above the

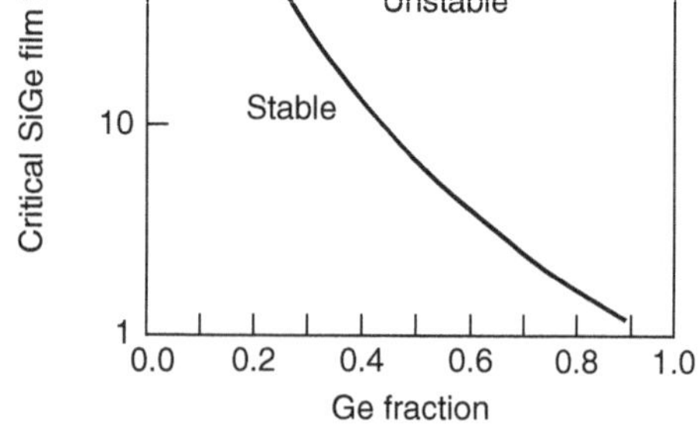

Fig. 3.37 Critical *SiGe* film thickness as a function of *Ge* fraction

"critical" limit. The film may still be pseudomorphic. It is said to be metastable since, when subjected to high-temperature processing, it will relax to the smaller *Si* lattice constant, generating defects.

3.7.2 Bandgap Lowering

The dependence of silicon bandgap on dopant concentration was discussed in Chap. 1. The bandgap was found to decrease with increasing dopant concentration above $\sim 5 \times 10^{17}\,\text{cm}^{-3}$. Since the base concentration typically ranges from 10^{18}–$1 \times 10^{19}\,\text{cm}^{-3}$, one would expect to see an appreciable decrease in bandgap, particularly near the emitter-base junction where the concentration is highest [84].

In a $Si_{1-x}Ge_x$ film grown pseudomorphic on Si, there is an additional reduction in bandgap caused by the strain created in the *SiGe* film [85–88]. In the range $x \leq 0.3$, the band offset ΔE_g is found to increase almost linearly with increasing *Ge* mole fraction (Fig. 3.38).

Most of the band-offset occurs in the valence band. It remains below 0.02 eV in the conduction band, as shown schematically in Fig. 3.39 [86]. Several methods have been suggested to measure the band offset in *SiGe* crystals [89–94]. Among them are the capacitance-voltage technique [90], and the temperature dependence of collector current method [91, 94]. While the methods and results vary, there is good agreement in the general trend of band offset versus *Ge* content shown in Fig. 3.38.

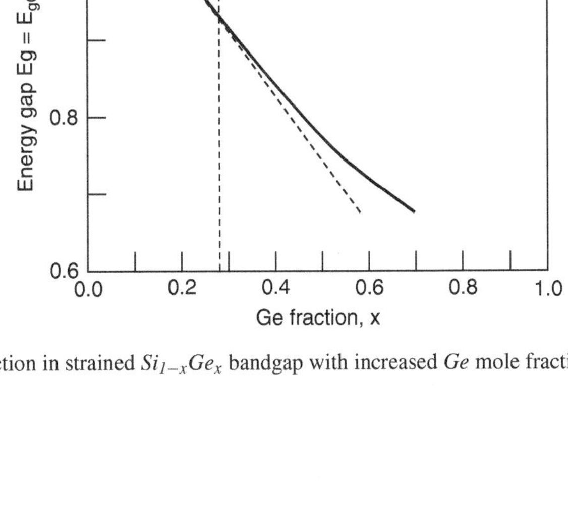

Fig. 3.38 Reduction in strained $Si_{1-x}Ge_x$ bandgap with increased *Ge* mole fraction [85, 86]

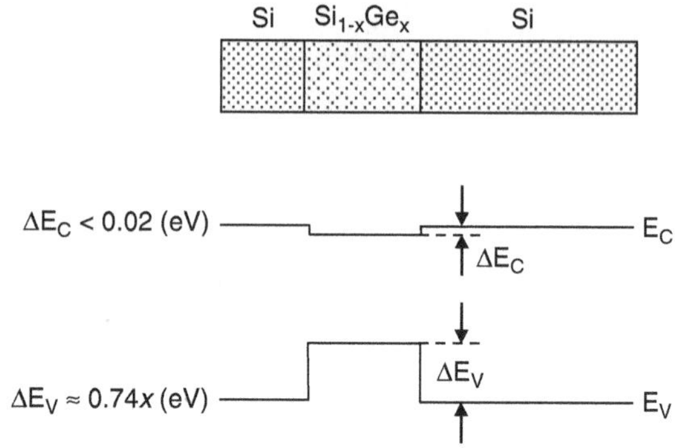

Fig. 3.39 Schematic band offset for $Si_{1-x}Ge_x/Si$ film [86]

3.7.3 Density of States

In Chap. 1, the density of states was found to depend on effective mass as (1.24)

$$N_C.N_V = 4\left(\frac{2\pi kT}{h^2}\right)^3 (m_n^* m_p^*)^{3/2} (\text{cm}^{-6}), \tag{3.100}$$

where N_C, N_V are, respectively, the density of states in the conduction and valence bands, and m_n^*, m_p^* the electron and hole effective mass. Since the effective mass depends on the curvature of the E-k plot and hence on crystal orientation, one would expect its magnitude and hence the density of states in a specific crystallographic direction to change with strain-induced band perturbation. The change in density of states can be inferred from measurements of collector current versus Ge content and comparison with the values obtained from energy-gap lowering alone [77, 95, 96]. The collector current is found to be two to three times smaller than predicted by bandgap lowering alone. The drop is attributed to a reduction in effective mass and hence density of states as the *Ge* mole fraction increases [77]. Simulated density of states for strained and unstrained *SiGe* as a function of *Ge* mole fraction are shown in Fig. 3.40 [99]. The intrinsic carrier concentration is found in Chap. 1, (1.36), as

$$n_i^2 = N_C N_V e^{-E_g/kT}. \tag{3.101}$$

Thus, reducing the product $N_C.N_V$ decreases n_i^2 while reducing E_g increases n_i^2. Both effects must be taken into account when calculating carrier injection and gain, as discussed below.

3.7.4 Mobility

Alloying germanium with silicon and inducing strain in *SiGe* introduces new scattering mechanisms that affect both minority and majority carrier mobility. Because

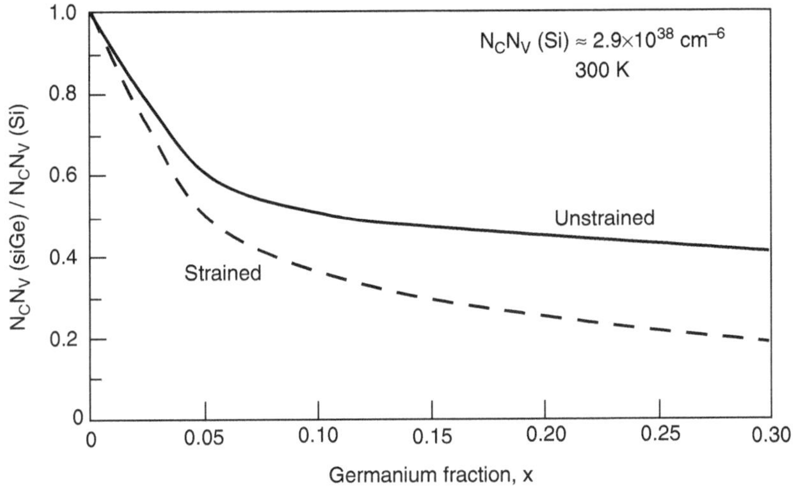

Fig. 3.40 Simulated density of states in *SiGe* [99]

of the small base dimensions and the multiple effects involved, most of the mobility studies are made by the Monte-Carlo (MC) method rather than by other numerical or analytical techniques. The results are summarized here for near-intrinsic SiGe where ionized-impurity scattering can be neglected.

3.7.4.1 Low-Field Mobility in Unstrained SiGe

The low-field mobility in a particular crystallographic direction is inversely proportional to the effective mass. This is given in Chap. 1, (1.69), as

$$v_{dx} = \frac{q\tau_x}{m_x^*} E_x = \mu E_x, \tag{3.102}$$

where subscript x means that the values are taken in the x-direction. v_x is the drift velocity, τ_x the mean-time between collisions, E_x the electric field, and m_x^* the effective mass. From (3.100) and Fig. 3.40, one would then expect the low-field mobility to increase with increased *Ge* content. The deformation of the lattice by *Ge*, however, introduces an additional phonon-scattering mechanism, referred to as alloy-scattering that degrades mobility. The electron mobility in unstrained, lightly-doped bulk *SiGe* is indeed found to decrease with increasing *Ge* mole fraction as shown in Fig. 3.41 [97, 98]. In contrast, simulations of hole mobility show little degradation by alloy scattering [99, 100]. This is not in agreement with earlier results that predict a substantial drop in hole mobility in the same range of *Ge* concentration [101]. The discrepancy is attributed to the different assumptions and approximations made in these studies.

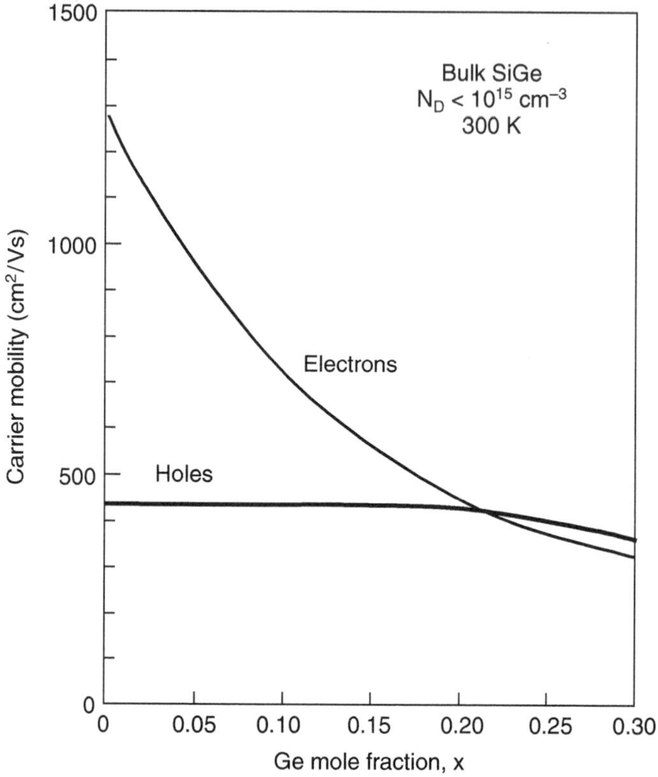

Fig. 3.41 Lattice mobility of electrons and holes in unstrained *SiGe* with increasing *Ge* mole fraction [97,99]

3.7.4.2 Low-Field Mobility in Strained SiGe

Applying a strain to the crystal distorts the bands and causes a change in the carrier effective mass of carriers and the relative distribution of carriers within the bands, a piezoresistance effect described in [102, 103].

Let the $x-y$ plane coincide with the (100) plane, parallel to the plane of the transistor base. The z-axis is normal to that plane, in the direction of injected carriers. The axes are shown on the left of Fig. 3.42.

In silicon there are six conduction band valleys, two along each of the $\langle 100 \rangle$ crystallographic directions (Chap. 1, Fig. 1.14). Without strain, the probability of finding an electron at the bottom of the conduction band is the same for the six valleys. The six valleys are denoted by $\Delta 6$ and said to have a six-fold degeneracy. Also shown in the figure are three valence bands, the light-hole band, the heavy-hole band and the split-off valence band that have maxima at momentum $k=0$ (Chap. 1). The maxima of the light-hole and heavy-hole valence bands coincide at $k=0$ (Fig. 3.43).

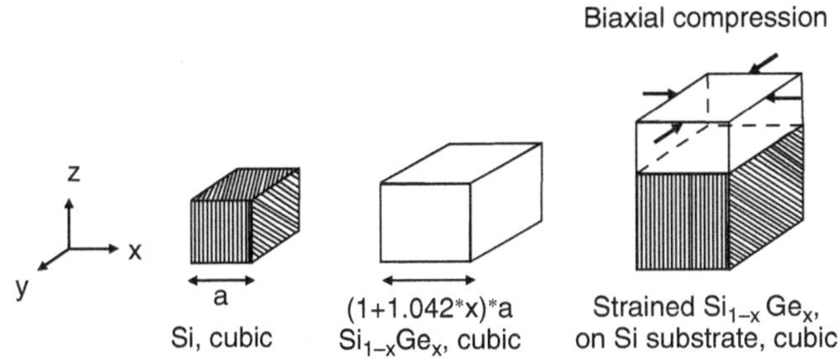

Fig. 3.42 Illustration of bi-axial strain in *SiGe*

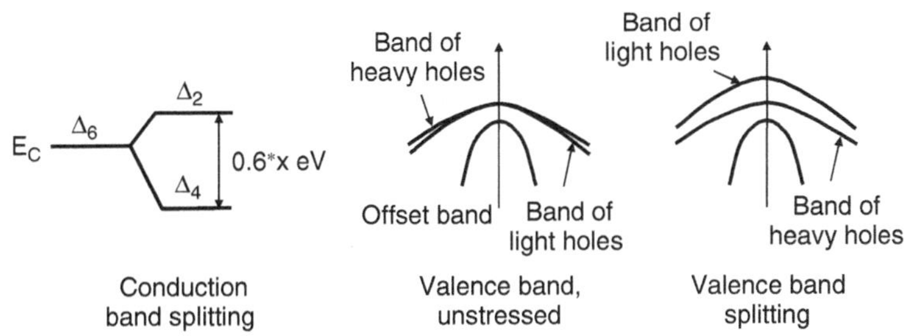

Fig. 3.43 Illustration of band-splitting under strain [104]

A silicon-germanium film grown on a silicon substrate is under bi-axial compressive strain, that is, compressive in a direction parallel to the silicon surface (Fig. 3.42). The strain is tensile in a direction normal to the surface. Elastic strain destroys the cubic symmetry in silicon and causes a split in the band energies. The four in-plane conduction band valleys ($\Delta 4$) shift down in energy and the two out-of-plane valleys ($\Delta 2$) shift up (Fig. 3.43). The total shift is approximated by $\Delta E \approx 0.6x$, where x is the *Ge* mole fraction [104]. Therefore, practically all electrons will occupy the $\Delta 4$ valleys. Since the electron effective mass is small $(0.19\, m_o)$ in the direction normal to an axis and high $(0.98\, m_o)$ along the axis (Chap. 1), almost all injected electrons travel in the direction of small effective mass and hence high mobility. Mobility enhancement is, however, mostly offset by the degradation of electron mobility due to alloy scattering [98]. Also, as the base dopant concentration increases to the range of 10^{18}–$5 \times 10^{19}\,\text{cm}^{-3}$ in high-performance transistors, ionized-impurity scattering becomes important [105, 106]. The net, however, is an increase in the low-field electron mobility μ_z in a direction normal to the base of an *NPN* structure, as shown in Fig. 3.44a [97]. The electron mobility parallel to the base is found to degrade under compressive strain.

The light-hole and heavy-hole bands are degenerate in unstrained silicon. Under strain, the light-hole band moves up and the heavy-hole band down in energy, as illustrated in Fig. 3.43. The split in energy is approximated by $0.166x$ [107]. This split is sufficiently large to cause practically all holes to occupy the light-hole bands where the mobility is high.

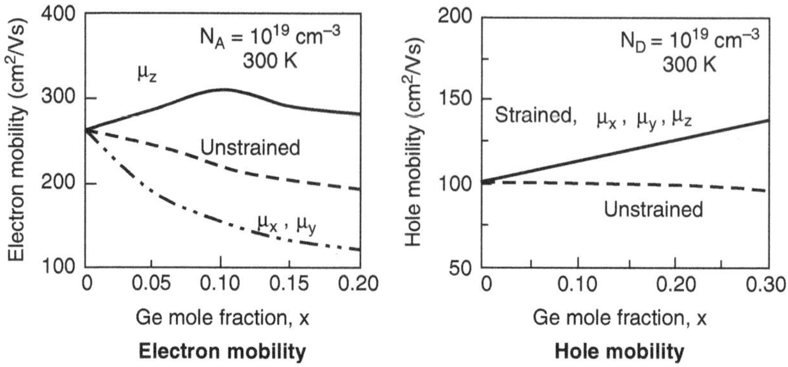

Fig. 3.44 Low-field electron and hole mobilities in *SiGe* under strain for $N = 10^{19}\,\text{cm}^{-3}$ [97, 99]

Fig. 3.45 Equilibrium energy band-diagram of an *NPN* structure with uniform *Ge* and doping concentration

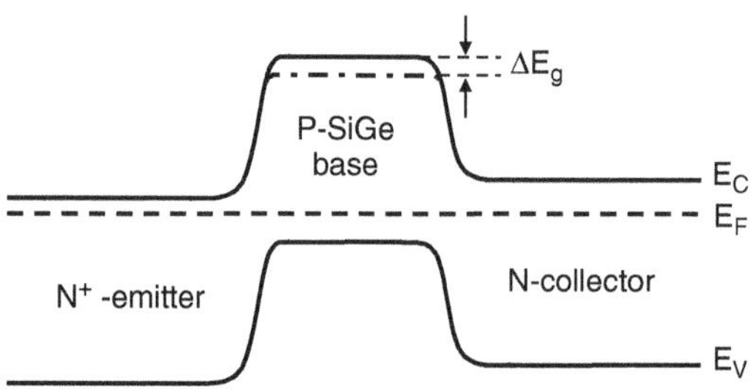

Both the in-plane and out-of-plane hole mobilities are found to increase under stress (Fig. 3.44). This improves the *PNP* performance and reduces the *NPN* base resistance.

Strain-induced mobility enhancement is further discussed in Chap. 5.

3.7.5 Transistor Parameters

The combination of *SiGe* and polysilicon emitters with controlled interface oxide provides designers with the flexibility of independently optimizing transistor parameters, such as current gain, Early-voltage, and speed.

3.7.5.1 Current Gain

The total *Ge* dose in the neutral-base region controls the injected emitter current and hence the current gain through bandgap reduction.

Consider, for example, a simplified *NPN* structure having uniform doping and *Ge* concentration across the base. The equilibrium band diagram of this structure is shown schematically in Fig. 3.45.

The Fermi level must align throughout the structure. Since, for a given base dopant concentration the valence band must remain at a fixed energy with respect to the Fermi level, bandgap lowering forces the conduction band in the base to be lower by ΔE_g, although the band-offset occurs mostly in the valence band.

It is assumed (without proof) that the mechanisms of dopant-induced and *Ge*-induced barrier lowering are additive. Under forward active bias, the collector current can then be expressed as:

$$j_{C(Ge)} = \frac{qD_{n(Ge)}n_i^2}{N_A W_b}(e^{qV_F/kT} - 1), \tag{3.103}$$

where

$$n_i^2 = n_{i0}^2 e^{\Delta E_g}. \tag{3.104}$$

n_{i0} is the intrinsic concentration in intrinsic, pure silicon, $D_{n(Ge)}$ is the electron diffusion constant in the doped SiGe base, and ΔE_g is the sum of dopant-induced and *Ge*-induced bandgap lowering

$$\Delta E_g = \Delta E_{g(dope)} + \Delta E_{g(Ge)}. \tag{3.105}$$

$\Delta E_{g(dope)}$ and $\Delta E_{g(Ge)}$ are, respectively, the dopant-induced and *Ge*-induced bandgap lowering. Compared to a similar silicon structure (without *Ge*), the improvement in gain is

$$\frac{j_{C(Ge)}}{J_{C(Si)}} = \frac{\mu_{nGe)}}{\mu_{n(Si)}}\frac{(N_C N_V)_{Ge}}{(N_C N_V)_{Si}} e^{\Delta E_g}, \tag{3.106}$$

where $\mu_{n(Ge)}, \mu_{n(Si)}$ are, respectively, the electron mobility in the *SiGe* and the *Si* base, and $(N_C N_V)_{Ge}, (N_C N_V)_{Si}$ the density of states in *SiGe* and *Si*. Since the density of states decreases with the increasing *Ge* mole concentration, the electron mobility increases in the direction of electron flow. The main improvement, however, comes from the increase in intrinsic carrier concentration associated with bandgap lowering in the base.

For nonuniform boron and *Ge* concentrations in the base, (3.103) must be rearranged to take into account the position-dependence of $n_i, \mu_{n(SiGe)}$, and N_A, and the field induced by the nonuniform base profile

$$j_{C(Ge)} = \frac{q}{kT/q\int_0^{W_b} N_A(x)dx/\mu_{n(Ge)}(x).n_{i(Ge)}^2(x)}.(e^{qV_F/kT} - 1). \tag{3.107}$$

For a linearly graded *Ge* profile, the bandgap decreases linearly from emitter to collector as

$$E_g(x) = E_{g(0)} - (E_{g(0)} - E_{g(Wb)})\,\frac{x}{W_b}. \tag{3.108}$$

The band-diagram for a linearly graded Ge profile and uniform base doping is shown schematically in Fig. 3.46 [108, 109]. Defining the ratios of mobility and density of states as

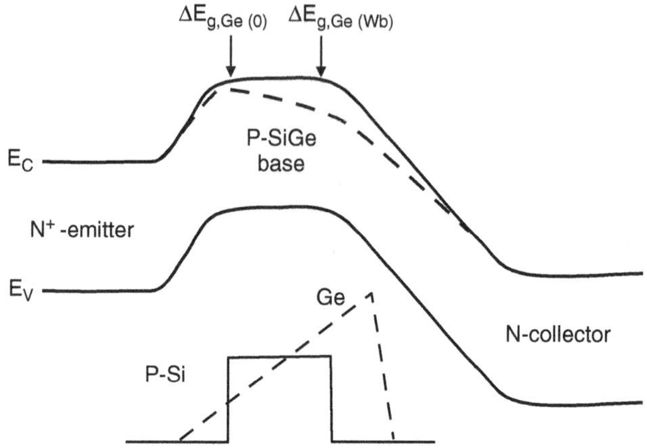

Fig. 3.46 Energy band diagram for silicon and graded-base *SiGe NPN* transistor in the forward active mode [108, 109]

$$\eta = \frac{\mu_{niGe)}}{\mu_{n(Si)}}; \gamma = \frac{(N_C N_V)_{Ge}}{(N_C N_V)_{Si}},$$

and taking the position-averaged values $\tilde{\eta}$, $\tilde{\gamma}$ of these ratios, the improvement in collector current in the forward active mode and low-level injection can be expressed as [108]

$$\frac{j_{C(SiGe)}}{J_{C(Si)}} = \tilde{\gamma}\tilde{\eta} \frac{\Delta E_{g,Ge(Wb)} - \Delta E_{g,Ge(0)}}{kT(1 - e^{-(\Delta E_{g,Ge(Wb)} - \Delta E_{g,Ge(0)})/kT})} e^{\Delta E_{g,Ge(0)}/kT}, \tag{3.109}$$

The above relation shows the strong dependence of collector current and hence current gain β on the *Ge* concentration at the emitter-base junction. The relation also shows that at a fixed collector current, the forward voltage V_{BE} is lower for *SiGe* than for *Si* (Problem 15).

3.7.5.2 Early Voltage and βV_A Product

The Early voltage V_A and βV_A product are important figures of merit for analog designs. The Early voltage is defined by (3.41)–(3.44). It is a measure of conductance, that is, the rate of increase in collector current with increasing collector voltage

$$V_A = \frac{J_C}{dJ_C/dV_{CB}|_{V_{BE}}} - V_{CE} \cong J_C \frac{1}{dj_C/dV_{CB}|_{V_{BE}}} = \frac{J_C}{(dJ_C/dW_b)(dW_b/dV_{CB})|_{V_{BE}}}. \tag{3.110}$$

For a uniform base, there is a trade-off between V_A and β, but the βV_A product remains essentially constant, as can be seen from (3.41)–(3.44). For a varying bandgap across the base and a nonuniform base profile, the collector current is defined by the general relation (3.107). The low-level injection current gain is then

$$\beta = \frac{j_C}{J_B} = \frac{q}{J_B\left(\int_0^{W_b}(N_A(x))/(n_i^2(x)D_n(x))\,\mathrm{d}x\right)}. \tag{3.111}$$

The Early voltage is found as [110]

$$V_A = \frac{qn_i^2(W_b)D_n(W_b)}{C_{jc}}\left(\int_0^{W_b}(N_A(x))/(n_i^2(x)D_n(x))\,\mathrm{d}x\right). \tag{3.112}$$

The integral in the above equations is dominated by the region with smallest n_i^2 in the base. Thus, for a linear *Ge* profile shown in Fig. 3.35, β will depend mainly on the region where the *Ge* fraction x is lowest, that is, near the emitter. The Early voltage, V_A, depends on the ratio $n_i^2(W_b)/n_i^2(0)$, where $n_i(Wb)$ is the intrinsic carrier concentration at W_b (near the collector) and $n_i(0)$ the intrinsic carrier concentration at $x = 0$, the depletion boundary in the base near the emitter. The βV_A product, however, depends mainly on $n_i^2(W_b)$ [110]

$$\beta V_A = \frac{q^2 n_i^2(W_b)D_n(W_b)}{J_B C_{jc}}. \tag{3.113}$$

The ratio $(\beta V_A)_{SiGe}/(\beta V_A)_{Si}$ for constant V_{BE} is [108]

$$\frac{(\beta V_A)_{SiGe}}{(\beta V_A)_{Si}} = \tilde{\gamma}\tilde{\eta}e^{\Delta E_g(W_b)/kT}. \tag{3.114}$$

Thus, the Early voltage can be independently adjusted by "tuning" the *Ge* concentration near the collector, an important flexibility in analog designs.

3.7.5.3 Base Transit Time

The base transit time τ_B is one of the delay components that limit device speed. Bandgap lowering by *SiGe* can reduce the base transit time in two ways: (a) by increasing the base dopant concentration to high levels and hence allowing the base-width to be reduced without compromising the intrinsic base sheet resistance or current gain, (b) by properly grading the *Ge* profile to create a built-in field that accelerates minority carriers in the base. In the latter case, *Ge* is graded so that the bandgap is smaller near the collector than near the emitter. For a linearly graded *Ge* profile in the base, the energy gap varies as in (3.108), implying that grading the *Ge* concentration induces a constant field of magnitude

$$E_{Ge} = \frac{E_{g(0)} - E_g(Wb)}{q\,W_b}. \tag{3.115}$$

The induced field is assumed to add algebraically to the field created by grading the impurity profile in the base.

As the base width is reduced to ultra-thin dimensions, however, τ_B becomes less significant when compared to other delay components. As the field increases in the base, the minority-carrier mobility degrades and, eventually velocity saturation is reached. Also, for an ultra-thin base, the probability for quasi-ballistic transport and velocity overshoot increases (Chap. 5). The transit time through the base becomes very small and the transit time through the collector space-charge region becomes more important [60, 64].

A detailed analysis of base transit time for a nonuniformly doped silicon base with uniform bandgap is given in [111] as

$$\tau_B \approx \frac{1}{\tilde{D}_n} \int_0^{W_b} \frac{1}{N_A(x)} \left[\int_x^{W_b} N_A(y)\, dy \right] dz, \tag{3.116}$$

where $\tilde{D}$ is the effective minority-carrier diffusion constant in the base. For a varying energy gap in the base, the above relation is extended to [6]:

$$\tau_B \approx \int_0^{W_b} \frac{n_i^2(z)}{N_A(z)} \left[\int_x^{W_b} \frac{N_A(y)}{\tilde{D}_n(y) n_i^2(y)} dy \right] dz. \tag{3.117}$$

For a uniformly doped base with a uniform bandgap, the base transit time simplifies to $\tau_B \cong W_b^2/2D_n$.

3.7.6 Transistor Optimization

A transistor is optimized for specific applications. Digital designs typically require high speed and gain and small transistor-size for high packing density, while high Early voltage, high βV_A product, and low noise are more important to analog designs. Transistor size is not very important in most analog designs because capacitors, resistors and inductors occupy most of the chip area and the fraction occupied by transistors is small. Mixed analog-digital designs require process compatibility with *CMOS*, limiting the flexibility in optimization. Power transistors require high voltage and current capabilities and power management. In all cases, the transistor should be designed for high reliability and yield while minimizing leakage and power.

3.7.6.1 Base Profile Optimization

In a *SiGe* transistor, the base transit time is minimized by thinning the base to the limit allowable by process and supply voltage. The base dopant concentration is simultaneously increased to maintain an acceptable intrinsic base sheet resistance, and to limit the spread of collector-base and emitter-base depletions into the base.

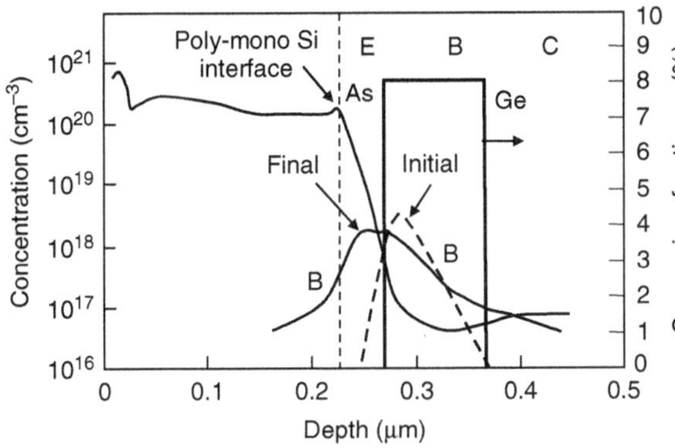

Fig. 3.47 Displacement of emitter-base and collector-base junctions with respect to *Ge* boundaries

The dopant profile is optimized by grading the base to induce a built-in field that accelerates the injected minority carriers. Typical base profiles are Gaussian with a peak near the emitter boundary and a main section in the quasi-neutral base that can be approximated by an exponential distribution. Since the retrograde region (decreasing toward the surface) of the base profile induces a decelerating field, it must be minimized.

Placing the emitter-base junction at the peak base concentration to eliminate the retrograde region, however, brings with it a problem with tunneling-leakage and reliability since it increases the field at the emitter-base junction. The high field can degrade the surface by creating traps at the interface or in the oxide covering the surface.

Another important consideration is the placement of emitter-base and collector-base junctions with respect to the *Ge* profile (Fig. 3.47).

As grown, the initial base boron profile is shown mostly confined within the abrupt *Ge* boundaries. During additional thermal cycles, however, boron diffuses faster than *Ge* into the emitter and collector, which causes a displacement of the junction with respect to the *SiGe-Si* heterojunctions. Because the displacement is beyond the region of energy-gap lowering by *Ge*, it creates parasitic barriers at both junctions. At the emitter-base junction, minority carriers are injected from the emitter into silicon rather than *SiGe* of a lower bandgap. At the collector, minority carriers must be transported over the *SiGe-Si* band offset. Both parasitic barriers degrade transistor gain, speed, and Early voltage [112–116]. For optimum transistor performance, parasitic barriers are minimized by forming vertical *Si* spacers between the *SiGe* film and emitter and collector and the *SiGe-Si* interfaces are kept within the junction depletion regions.

Suppression of Boron Diffusion by Carbon

The undesired boron out-diffusion can be suppressed by incorporating substitutional carbon in thin spacers between *SiGe* and the emitter and the collector. The

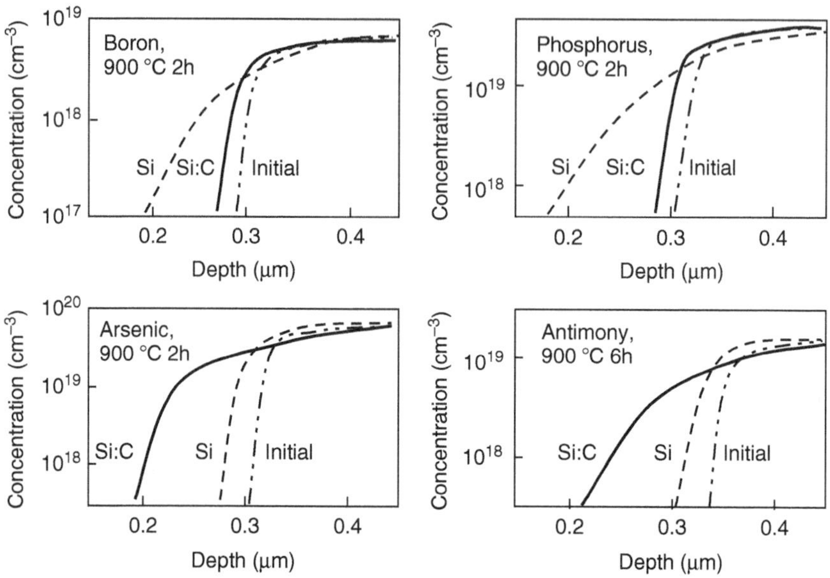

Fig. 3.48 Diffusion of *B, P, As*, and *Sb* from highly doped substrate into *Si* and *SiGe* layers (adapted from [117])

spacer films can be grown as undoped $Si_{1-y}C_y$ while minimizing the spacer thickness to avoid strain-induced defects [117, 118]. Substitutionally incorporated carbon (0.5–1%) effectively retards boron (and phosphorus) out-diffusion. This is shown in Fig. 3.48 [119]. In contrast, the presence of carbon enhances the diffusion of arsenic and antimony. The effects are explained by interactions of carbon and point defects. Boron and phosphorus diffuse by the interstitial mechanism while arsenic and antimony diffuse by the vacancy mechanisms. The presence of carbon causes a deficiency in silicon interstitials, retarding boron and phosphorus diffusion. The simultaneous generation of vacancies enhances the diffusion of arsenic and antimony [119].

The capability of reducing the base width to ultra-thin dimensions below 50 nm with *SiGe* and controlling the base profile by incorporating carbon to suppress out-diffusion from the base allows significant performance enhancement over Si-transistors.

3.7.6.2 Collector Profile Optimization

The collector profile affects the transistor delay and breakdown voltage. An increase in collector concentration beneath the intrinsic base increases the current density at which the onset of the Kirk-effect occurs. It also reduces the collector series resistance and transit time through the collector depletion layer, and reduces the base-width by compensating ("pinching") part of the base out diffusion. This results

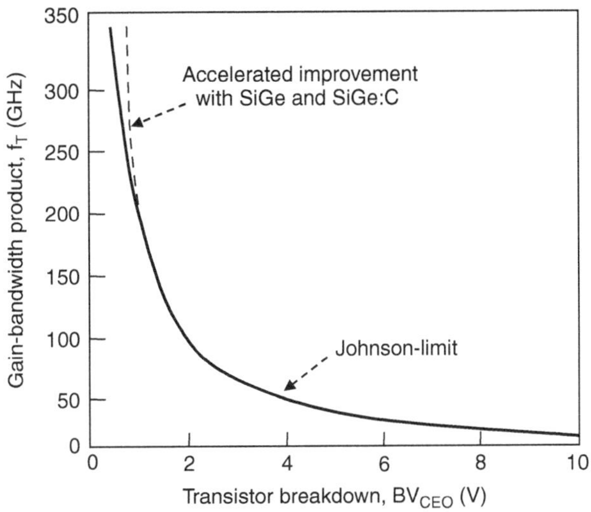

Fig. 3.49 The Johnson limit and its extension to higher values with *SiGe* and *SiGe:C* transistors

in a net improvement in cutoff frequency f_T. Improving f_T, however, comes with the penalty of increased collector-base junction capacitance and decrease in transistor breakdown voltage. Thus, when optimizing the collector profile, there is a trade-off between f_T and BV_{CEO}. The product of peak f_T and BV_{CEO} is commonly referred to as the Johnson limit. This limit predicts that the $f_T BV_{CEO}$ product for silicon bipolar transistors cannot exceed ~200 GHz-V [120]. The limit should, however, be considered as a guide-line. Recent data on *SiGe* transistors demonstrate that the limit can be considerably higher, as shown in Fig. 3.49 [121]. A detailed analysis of transistor delay components shows that the $f_T BV_{CEO}$ product is not constant and can even reach values near 500 GHz-V [122].

Interactions between process and device parameters and their trade-offs discussed so far are summarized in Fig. 3.50.

Increasing the *IFO* thickness reduces the base current and hence increases the current gain β. A thicker *IFO*, however, increases the surface state density and hence the low-frequency noise and, as the *IFO* thickness increases above ~1 nm, the emitter resistance increases appreciably. BV_{CEO} decreases because it is dependent on the inverse root of β (3.53).

As the concentration Ge_E increases at the base depletion boundary of the emitter-base junction, the bandgap decreases, resulting in higher minority-carrier injection into the base and higher β. There may be an associated decrease in Early voltage if the Ge_C/Ge_E ratio gets appreciably lower.

The base transit time is shorter for smaller W_b, leading to higher f_T. If by reducing W_b the integrated dopant concentration in the quasi-neutral decreases, β will increase. In an optimized structure, however, the base is typically doped at a higher level as W_b is reduced, keeping the Gummel number almost constant. The Early

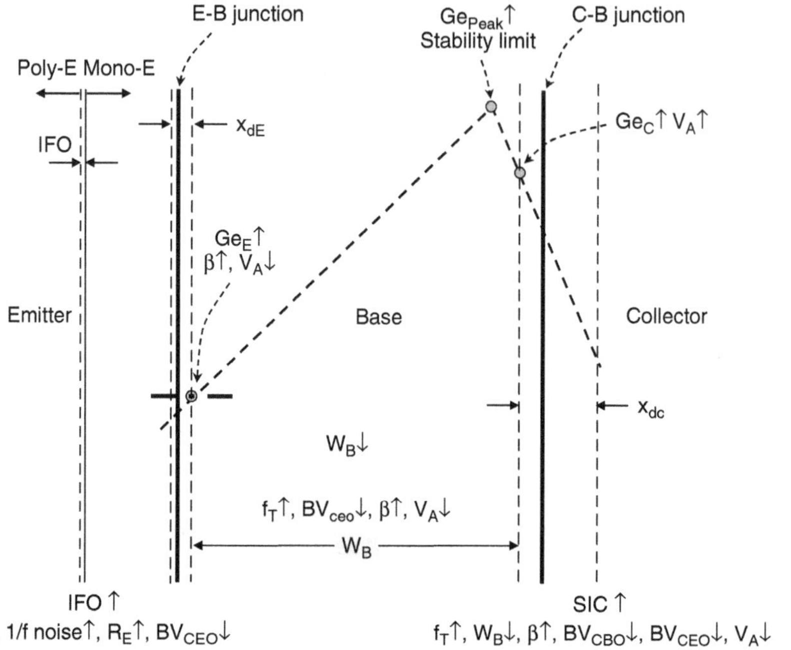

Fig. 3.50 Interactions between *SiGe* process and device parameters

voltage V_A tends to decrease as the base becomes narrower, but the change can be offset by increasing the *Ge* concentration at the collector boundary. The offset can be qualitatively described as a result of two competing effects: as the collector voltage increases, the depletion spreads into the base, reducing the base Gummel number, therefore increasing the collector current and reducing V_A. At the same time the integrated *Ge* concentration decreases, reducing the effect of bandgap lowering on collector current which increases V_A. As can be seen in Fig. 3.50, this also results in a higher Ge_C/Ge_E ratio as the depletion spreads into the base.

Selectively implanting the collector reduces the local collector resistance and "pushes" the onset of Kirk effect to higher current densities. The retrograde *SIC* profile displaces the collector-base junction slightly inside the base, reducing the base-width. The net is an increase in β and f_T but a decrease in BV_{CBO}, BV_{CEO}, and V_A.

3.8 Problems

The temperature is 300 K unless otherwise stated.

1. The applied voltage to a grounded-emitter NPN is fixed at 1.5 V and the base-emitter junction is gradually forward-biased. At $V_{BE} = 0.6$V, the collector current is 100 nA. For a total collector resistance of 1 KOhm, find the collector current at onset of saturation. Assume low-level injection and negligible impact of emitter and base resistances.

2. The base and collector of an NPN transistor are uniformly doped at a concentration $N_A = N_D = 2.5 \times 10^{17}\,cm^{-3}$, and the emitter is degenerately doped. The distance between collector-base and emitter-base metallurgical junctions is $0.1\,\mu m$. Will punch-through or avalanche breakdown occur first?

3. Consider an NPN transistor of a base-width $W_b = 0.1\,\mu m$. Assume a Gaussian base profile with a peak concentration $N_A(0) = 2.5 \times 10^{18}\,cm^{-3}$ at the emitter-base junction and $N_A(W_b) = 5 \times 10^{16}\,cm^{-3}$ at the collector-base junction. The emitter is degenerately doped. Will reverse punch-through or breakdown voltage occur first?

4. Calculate the base sheet resistance in problem 2 for a collector-base reverse voltage of 2.5 V and $V_{BE} = 0.4\,V$.

5. Show that for a transistor in the forward active mode, β is approximately proportional to the base sheet resistance.

6. Show that for a graded base profile as in Fig. 3.6, the Early-voltage is lower in the reverse than in the forward mode.

7. Assume that the onset of current crowding occurs when the voltage drop along the intrinsic base is kT/q. For a $2 \times 2\,\mu m^2$ emitter and two symmetrically arranged base contacts, estimate the base current at onset of crowding in problem 2. Neglect voltage drops in the extrinsic base.

8. Assume that the base, emitter, and collector of an NPN structure are uniformly doped, respectively, at a concentration of 10^{18}, 10^{20}, and $10^{17}\,cm^{-3}$. The depth of the emitter junction and the width of the metallurgical base are both $0.1\,\mu m$. The effective density of generation-recombination centers in the base, emitter and collector are, respectively, 5×10^{12}, 10^{14} and $10^{12}\,cm^{-3}$. Find:

(a) The minority-carrier lifetime in the base and emitter.
(b) The energy gap and n_i^2 in the base and emitter.
(c) The low-level injection base sheet resistance.
(d) The collector and base current densities, β and the ac (dynamic) emitter resistance for a forward base-emitter voltage of 0.7 V and a reverse collector-base voltage of 2.5 V.
(e) The average minority-carrier velocity and base transit-time for the condition in c).
(f) The transit time through the collector depletion layer.

9. Assume the emitter-base and collector-base junction areas in problem 3.3 to be, respectively $1\,\mu m^2$ and $10\,\mu m^2$. The emitter, base, and collector resistances at $V_{BE} = 0.7\,V$ are, respectively, 50, 1000, 500 Ohm.

(a) For $V_{BE} = 0.8\,V$, and $V_{CE} = 2.5\,V$, find the emitter-base and collector-base junction capacitances. Neglect the effect of injected carriers.
(b) Estimate f_T and f_{max} at $V_{BE} = 0.8\,V,\ V_{CE} = 2.5\,V$.
(c) Estimate the forward voltage at onset of base-conductivity modulation.
(d) Estimate the current at the onset of the Kirk-effect.

10. The base and collector regions of an NPN transistor are doped uniformly with $N_A = 10^{18}\,cm^{-3}$ and $N_D = 5 \times 10^{16}\,cm^{-3}$. A degenerately doped buried N^+ layer is placed $0.4\,\mu m$ beneath the base. The base-collector junction is shunted with a Schottky-barrier of barrier height 0.7 eV. Find the Schottky barrier area required to ensure that a maximum of 10% of the forward current is due to minority-carrier injection when the collector-base junction becomes forward-biased.

11. A transistor is fabricated with an exponential base profile having a concentration of $5 \times 10^{18}\,cm^{-3}$ at the emitter-base depletion boundary and $10^{17}\,cm^{-3}$ at the collector-base depletion boundary in the base. The base width W_b is $0.1\,\mu m$. Calculate for $V_{CE} = 1\,V,\ V_{BE} = 0.7\,V$:

(a) The base sheet resistance.
(b) The built-in field in the base.
(c) The minority-carrier drift-velocity in the base.
(d) The base transit time.

12. In a PNP transistor, the collector current is kept constant while the temperature is increased. I_B decreases in magnitude, passes through zero and changes polarity. What physical effects account for this behavior?

13. Self-heating is the increase in transistor temperature during operation as a result of insufficient dissipation of power generated by the transistor. A transistor operates in the forward active mode and its temperature rises gradually as the collector voltage is increased.

(a) For a constant V_{BE} how does the increase in temperature affect I_C?
(b) How would this effect change if I_B is kept constant while V_{CE} is increased?

14. For a constant V_{CE}, how would V_{BE} vary as I_B increases?

15. The emitter, base and collector of a SiGe NPN transistor are uniformly doped with, respectively, $N_D = 10^{19}$, $N_A = 10^{19}$ and $N_D = 10^{18}\,cm^{-3}$. The Ge concentration in the base is also uniform at 8%. The metallurgical base-width is 50 nm.

(a) Estimate the magnitude of n_i^2 in the emitter and base.
(b) Estimate the required base-emitter forward voltage V_{BE} for a collector current density $J_C = 10\,\mu A/\mu m^2$ at $V_{CE} = 2\,V$. How would the result change if Ge were not present in the base?
(c) Find the base transit time under the above conditions.

16. Assume the Ge profile in problem 3.10 to be linearly graded with a concentration of 1% and 10%, respectively, at the emitter-base and collector-base depletion boundaries.

(a) Calculate the field induced by grading Ge
(b) Estimate the minority-carrier drift velocity and base transit time.

17. The boron concentration in problem 3.11 is $2 \times 10^{19}\,\mathrm{cm}^{-3}$ at the emitter-base depletion boundary and $10^{18}\,\mathrm{cm}^{-3}$ at the collector-base depletion boundary.

(a) Calculate the field induced by grading both N_A and Ge.
(b) Estimate the minority-carrier drift velocity and base transit time.

18. Derive (3.66).

References

1. A. B. Phillips, Transistor Engineering, McGraw-Hill, New York, 1962.
2. J. J. Ebers and J. L. Moll, "Large-signal behavior of junction transistors," Proc. IRE, 42, 1761–1772, 1954.
3. P. Gray, D. DeWitt, A. R. Boothroyd, and J. F. Gibbons, Physical Electronics and Circuit Models of Transistors, SEEC, Vol. 2, p. 181, John Wiley, New York, 1964.
4. I. Getreu, Modeling the Bipolar Transistor, Tektronix, Inc., Beaverton, Oregon, 1976.
5. B. El-Kareh, Fundamentals of Semiconductor Processing Technologies, Kluwer Academic Publishers, Boston, 1997.
6. H. Kroemer, "Two integral relations pertaining to the electron transport through a bipolar transistor with a nonuniform energy gap in the base region," Solid-State Electronics, 28 (11), 1101–1103, 1985.
7. E. J. McGrath and D. H. Navon, "Factors limiting current gain in power transistors," IEEE Trans. Electron Dev., ED-24 (10), 1255–1259, 1977.
8. C. T. Sah, R. N. Noyce, and W. Shockley, "Carrier generation and recombination in p-n junctions and p-n junction characteristics," IEEE Trans. Electron Dev., ED-45 (9), 1228–1238, 1957.
9. C. T. Sah, "Effect of surface recombination and channel on p-n junction and transistor characteristics," IEEE Trans. Electron Dev., ED-9 (1), 94–108, 1962.
10. P. J. Coppen and W. T. Matzen, "Distribution of recombination current in emitter-base junctions of silicon transistors," IEEE Trans. Electron Dev., ED-52 (1), 75–81, 1962.
11. J. L. Moll, Physics of Semiconductors, McGraw-Hill Physical and Quantum Electronics Series, New York, 1964.
12. J. M. Early, "Effects of space-charge layer widening in junction transistors," Proc. IRE, 40, 1401–1406, 1952.
13. R. C. Jaeger and A. J. Brodersen, "Self consistent bipolar transistor models for computer simulations," Solid-State Electronics, 21 (10), 1269–1272, 1978.
14. J. D. Cressler and G. Niu, Silicon-Germanium Heterojunction Bipolar Transistors, Artech House, Boston, 2003.
15. R. S. Muller, T. I. Kamins, and M. Chan, Device Electronics for Integrated Circuits, John Wiley & Sons, New York, 2003.
16. M. Takase, K. Yamashita, A. Hori, and B. Mizuno, "Shallow source/drain extensions for pMOSFETs with high activation and low process damage fabricated by plasma doping," IEEE IEDM Tech. Dig., 475–478, 1997.
17. M. Takagi, K. Nakayama, C. Tevada, and H. Kamioko, "Improvement of shallow base transistor technology by using a doped polysilicon diffusion source," J. Japan. Soc. Appl. Phys. (Suppl.), 42, 101–109, 1972.
18. K. Tsukamoto, Y. Akasaka, and K. Horie, "Arsenic implantation into polycrystalline silicon and diffusion to silicon substrate," J. Appl. Phys. 48, 1815, 1977.

19. J. Graul, A. Glasl, and H. Murrmann, "Ion implanted bipolar high-performance transistors with POLYSIL emitter," IEEE IEDM Tech. Dig., 450–454, 1975.
20. J. Graul, A. Glasl, and H. Murrmann, "High-performance transistors with arsenic-implanted poly emitters," IEEE J. Solid-State Circuits, SC-11 (4), 491–495, 1976.
21. T. H. Ning and R. D. Isaac, "Effect of emitter contact on current gain of silicon bipolar devices," IEEE Trans. Electron. Dev., ED-27 (11), 2051–2055, 1980.
22. T. H. Ning and D. D. Tang, "Bipolar trends," Proc. IEEE, 74 (12), 1669–1677, 1986.
23. H. C. De Graaff and J. G. De Groot, "The SIS tunnel emitter: A theory for emitters with thin interface layers," IEEE Trans. Electron. Dev., ED-26, 1771–1776, 1979.
24. A. A. Eltoukhy and D. J. Roulston, "Minority-carrier injection into polysilicon emitters," IEEE Trans. Electron. Dev., ED-29 (6), 961–964, 1982.
25. A. A. Eltoukhy and D. J. Roulston, "The role of interfacial layer in polysilicon emitter bipolar transistors," IEEE Trans. Electron. Dev., ED-29 (12), 1862–1869, 1982.
26. T. H. Ning and R. D. Isaac, "Effect of emitter contact on current gain of silicon bipolar devices," IEEE Trans. Electron. Dev., ED-27 (11), 2051–2055, 1980.
27. C. C. Ng and E. S. Yang, "A thermionic diffusion model for polysilicon emitter," IEEE IEDM Tech. Digest, 32–35, 1986.
28. H. Schaber and T. F. Meister, "Technology and physics of polysilicon emitters," Proc. IEEE BCTM, 75–81, 1989.
29. V. Kumar and W. E. Dahlke, "Characteristics of Cr-SiO2-nSi tunnel diodes," Solid-State Electron., 20 (2), 143–152, 1977.
30. S. M. Sze, Physics of Semiconductor Devices, John Wiley & Sons, New York, 1981.
31. S. Ratanaphanyarat, W. Rausch, M. Smadi, Mary Jo Saccamango, S. N. Mei, Shao-Fu Chu, P. A. Ronsheim, and J. O. Chu, "Effect of emitter contact materials on high-performance vertical p-n-p transistors," IEEE Electron. Dev. Lett., 12 (6), 261–263, 1991.
32. B. El-Kareh, S. Balster, W. Leitz, P. Steinmann, H. Yasuda, M. Corsi, K. Dawoodi, C. Dirnecker, P. Foglietti, A. Haeusler, P. Menz, M. Ramin, T. Scharnagl, M. Schiekofer, M. Schober, U. Schulz, L. Swanson, D. Tatman, M. Waitschull, J. W. Weijtmans, and C. Willis, "A 5V complementary-SiGe BiCMOS technology for high-speed precision analog circuits," IEEE BCTM, 211–214, 2003.
33. E. Crabbé, S. Swirhun, J. del Alamo, R. F. Pease, and R. M. Swanson, "Majority and minority carrier transport in polysilicon emitter contacts," IEEE IEDM Tech. Dig., 28–31, 1986.
34. E. F. Chor, P. Ashburn, and A. Brunnschweiler, "Emitter resistance of arsenic- and phosphorus-doped polysilicon emitter transistors," IEEE Electron. Dev. Lett. EDL-6 (10), 516–518, 1985.
35. C. M. Camalleri, S. Lorenti, D. Cali', P. Vasquez, and G. Ferla, "Control of amount and uniformity at the interface of an emitter region of a monocrystalline silicon wafer and a polycrystalline layer formed by chemical vapor deposition," United States patent 6 642 121, November 4, 2003.
36. T. Suntola and J. Antson, "Method for producing compound thin films," US Patent No. 4 058 430, Nov. 25, 1975.
37. B. Y. Tsaur and L. S. Hung, "Epitaxial alignment of polycrystalline Si films on (100) Si," Appl. Phys. Lett. 37 (10), 648–651, 1980.
38. M. Y. Ghannam and R. W. Dutton, "Solid phase epitaxial regrowth of boron-doped polycrystalline silicon deposited by low-pressure chemical vapor deposition," Appl. Phys. Lett. 51 (8), 611–613, 1987.
39. T. Kamins, Polycrystalline Silicon for Integrated Circuits and Displays, Kluwer Academic Publishers, Boston, 1998.
40. J. N. Burghartz, J. Y.-C. Sun, C. L. Stanis, S. R. Mader, and J. D. Warnock, "Identification of perimeter depletion and emitter plug effects in deep-submicrometer shallow-junction polysilicon emitter bipolar transistors," IEEE Trans. Electron. Dev., 39 (6), 1477–1489, 1992.
41. Y. Tamaki, F. Murai, K. Sagara, and A. Anzai, "A 100 nm emitter transistor fabricated with direct EB writing for high-speed bipolar LSIs," Symp. VLSI Technol., 31–32, 1987.
42. E. H. Stevens, "Saturation currents in smaller geometry bipolar transistors," IEEE Trans. Electron. Dev., ED-31 (1), 80–82, 1984.

43. D. D. Tang, T. C. Chen, C. T. Chuang, G. P. Li, J. M. C. Stork, M. B. Ketchen, E. Hackbarth, and T. H. Ning, "Design consideration for high-performance narrow-emitter bipolar transistor," IEEE Electron. Dev. Lett., EDL-8 (4), 174–175, 1987.
44. G. P. Li, C. T. Chuang, T. C. Chen, and T. H. Ning, "On the narrow-emitter effect of advanced shallow profile bipolar transistors," IEEE IEDM Tech. Dig., 174–177, 1987.
45. W. L. Kaufmann and A. A. Bergh, "The temperature dependence of ideal gain in double diffused silicon transistors," IEEE Trans. Electron. Dev., ED-15 (10), 732–735, 1968.
46. W. M. C. Sansen and R. G. Meyer, "Characterization and measurement of the base and emitter resistance of bipolar transistors," IEEE J. Solid-State Circuits, SC-7 (6), 492–498, 1972.
47. W. F. Filenski and H. Beneking, "New technique for determination of static emitter and collector series resistances in bipolar transistors," Electronics letters, 17 (14), 503–504, 1981.
48. T. H. Ning and D. D. Tang, "Method for determining the emitter and base series resistances of bipolar transistors," IEEE Trans. Electron Dev., ED-31 (4), 409–412, 1984.
49. A. Neugroschel, "Measurement of the low-current base and emitter resistances of bipolar transistors," IEEE Trans Electron Dev., ED-34 (4), 817–822, 1987.
50. J.-S. Park, A. Neugroschel, V. dela Torre, and P. J. Zdebel, "Measurement of collector and emitter resistances in bipolar transistors," IEEE Trans. Electron. Dev., ED-38 (2), 365–371, 1991.
51. B. El-Kareh and R. J. Bombard, Introduction to VLSI Silicon Devices, Kluwer Academic Publishers, Boston, 1985.
52. P. E. Gray, D. DeWitt, A. R. Boothroyd, and J. F. Gibbons, Physical Electronics and Circuit Models of Transistors, Semiconductor Electronics Education Committee, Vol. 2, John Wiley and Sons, New York, 1964.
53. W. M. Webster, "On the variation of junction transistor current amplification factor with emitter current," Proc. IRE, 42 (6), 914–920, New York, 1954.
54. S. K. Ghandhi, The Theory and Practice of Microelectronics, John Wiley and Sons, New York, 1968.
55. C. T. Kirk, Jr., "A theory of transistor cut-off frequency fall-off at high current densities," IEEE Trans. Electron. Dev., ED-9 (2), 164–174, 1962.
56. R. J. Van Overstraeten, H. J. DeMan, and R. P. Mertens, "Transport equation in heavy doped silicon," IEEE Trans. Electron Dev., ED-20 (3), 290–298, 1973.
57. K. Suzuki, "Optimum base doping profile for minimum base transit time," IEEE Trans. Electron. Dev., ED-38 (9), 2128–2133, 1991.
58. K. Suzuki, T. Fukano, H. Ishiwari, T. Yamazaki, M. Taguchi, T. Ito, and H. Ishikawa, "50-nm ultra-thin base silicon bipolar device fabrication based on photo epitaxial growth," Tech. Digest of 1989 Symposium on VLSI Technology, pp. 91–92.
59. C. A. King, J. L. Hoyt, C. M. Gronet, J. F. Gibbons, M. P. Scott, and J. Turner, "$Si/Si_{1-x}Ge_x$ heterojunction bipolar transistors produced by limited reaction processes," IEEE Electron. Device Lett., EDL-10 (2), 52–54, 1989.
60. W. Lee, S. E. Laux, M. V. Fischetti, and D. D. Tang, "Monte Carlo simulation of non-equilibrium transport in ultra-thin base Si bipolar transistors," IEEE IEDM Tech. Dig., 473–476, 1989.
61. P. Rohr, F. A. Lindholm, and K. R. Allen, "Questionability of drift-diffusion transport in the analysis of small semiconductor devices," Solid-State Electron., 17 (7), 729–734, 1974.
62. R. G. Meyer and R. S. Muller, "Charge-control analysis of the collector-base space-charge-region contributions to bipolar-transistor time constant, τ_T," IEEE Trans. Electron. Dev., ED-34 (2), 450–452, 1987.
63. R. D. Thornton, D. DeWitt, P. E. Gray, and E. R. Chenette, Characteristics and Limitations of Transistors, Semiconductor Electronics Education Committee, Vol. 4, John Wiley and Sons, 1966.
64. S. E. Laux and W. Lee, "Collector signal delay in the presence of velocity overshoot," IEEE Electron. Device Lett., EDL-11 (4), 174–176, 1990.
65. J. M. Early, "P-N-I-P and N-P-I-N junction transistor triodes," Bell. Syst. Tech. J., 33, 517–533, 1954.

66. F. N. Trofimenkoff, "Collector depletion region transit time," Proc. IEEE, 52 (1), 86–87, 1964.
67. L. J. Giacoletto, "Study of p-n-p alloy junction transistor from d-c. through medium frequencies," RCA Rev., 15 (4), 506–562, 1954.
68. R. Beaufoy and J. J. Sparkes, "The junction transistor as a charge-controlled device," ATE J., 13, 310–327, 1957.
69. B. S. Meyerson, "Low-temperature silicon epitaxy by ultrahigh vacuum/chemical vapor deposition," Appl. Phys. Lett. 48 (12), 797–799, 1986.
70. G. L. Patton, E. James, H. Comfort, B. Ernard, B. S. Meyerson, E. F. Crabbe, G. Erald, J. Scilla, E. De Fresart, J. M. Stork, J. Y.-C. Sun, D. Harame, and J. Burghartz, "75-GHz f_T SiGe-base heterojunction bipolar transistors," IEEE Electron Dev. Lett., 11 (4), 171–173, 1990.
71. J. F. Gibbons, C. M. Gronet, and K. E. Williams, "Limited reaction processing: silicon epitaxy," Appl. Phys. Lett., 47 (7), 721–723, 1985.
72. J. L. Hoyt, C. A. King, D. B. Noble, C. M. Gronet, and J. F. Gibbons, "Limited reaction processing: growth of $Si_{1-x}Ge_x$/Si for heterojunction bipolar transistor applications," Thin Solid Films, 184 (1–2), 93–106, 1990.
73. T. O. Sedgwick, M. Berkenblit, and T. S. Kuan, "Low-temperature selective epitaxial growth of silicon at atmospheric pressure," Appl. Phys. Lett., 54 (26), 2689–2691, 1989.
74. W. B. de Boer and D. J. Meyer, "Low-temperature chemical vapor deposition of epitaxial Si and SiGe layers at atmospheric pressure," Appl. Phys. Lett., 58 (12), 1286–1288, 1990.
75. A. Pruijmboom, D. Terpstra, C. E. Timmering, W. B. de Boer, M. J. J. Theunissen, J. W. Slotboom, R. J. E. Hueting, and J. J. E. M. Hageraats, "Selective-epitaxial base technology with 14 pcs ECL-gate delay," IEEE IEDM Tech. Dig., 747–750, 1995.
76. K. Washio, E. Ohue, K. Oda, M. Tanabe, H. Shimamato, and T. Onai, "A selective epitaxial SiGe HBT with SMI electrodes featuring 9.3-ps ECL-gate delay," IEEE IEDM Tech. Dig., 795–798, 1997.
77. J. D. Cressler "SiGe HBT technology,: A new contender for Si-based RF and microwave circuit applications," IEEE Trans. Microw. Theory Tech., 46 (5), 572–589, 1998.
78. D. C. Houghton, C. J. Gibbings, C. G. Tuppen, M. H. Lyons, and M. A. G. Halliwell, "The structural stability of uncapped versus buried $Si_{1-x}Ge_x$ strained layers through high temperature processing," Thin Solid Films, 183 (1–2), 171–182, 1989.
79. D. B. Noble, J. L. Hoyt, and J. F. Gibbons, "Thermal stability of $Si/Si_{1-x}Ge_x/Si$ heterojunction bipolar transistor structures grown by limited reaction processing," Appl. Phys. Lett., 55 (19), 1978–1980, 1989.
80. A. R. Denton and N W. Ashcroft, "Vegard's law," Phys. Rev. A 43, 003161, 1991.
81. J. W. Matthews and A. E. Blakeslee, "Defects in epitaxial multilayers. I. Misfit dislocations," J. Crystal Growth, 27, 118–125, 1974.
82. J. W. Matthews and A. E. Blakeslee, "Defects in epitaxial multilayers. III. Preparation of almost perfect multilayers," J. Cryst. Growth, 32 (2), 265–273, 1976.
83. R. People and J. C. Bean, "Calculation of critical layer thickness versus lattice mismatch for Ge_xSi_{1-x}/Si strained-layer heterostructures," Appl. Phys. Lett., 47 (3), 322–324, 1985.
84. J. W. Slotboom and H. C. DeGraaff, "Measurement of bandgap narrowing in Si bipolar transistors," Solid-State Electron., 19 (10), 857–862, 1976.
85. R. People, "Indirect band gap of coherently strained Ge_xSi_{1-x} bulk alloys on $\langle 001 \rangle$ silicon substrates," Phys. Rev. B, 32, 1405–1408, 1985.
86. D. V. Lang, R. People, J. C. Bean, and A. M. Sergent, "Measurement of the band gap of Ge_xSi_{1-x} strained-layer heterostrcutures," Appl. Phys. Lett. 47 (12), 1333–1335, 1985.
87. R. People, "Physics and applications of Ge_xSi_{1-x} strained-layer heterostructures," IEEE J. Quantum Electron., WE-22 (9), 1696–1710, 1986.
88. S. S. Iyer, G. L. Patton, J. M. C. Stork, B. S. Meyerson, and D. L. Harame, "Heterojucntion bipolar transistors using Si-Ge alloys," IEEE Trans. Electron. Dev., 36 (10), 2043–2064, 1989.

89. J. C. Brighten, I. D. Hawkins, A. R. Peaker, E. H. C. Parker, and T. E. Whall, "The determination of valence band discontinuities in 91. $Si/Si_{1-x}Ge_x/Si$ heterojunctions by capacitance-voltage techniques," J. Appl. Phys., 74 (3), 1894–1899, 1993.
90. Y. T. Tang and J. S. Hamel, "An electrical method for measuring the difference in bandgap across the neutral base in SiGe HBT's," IEEE Trans. Electron. Dev., 47 (4), 797–804, 2000.
91. B. Le Tron, M. D. R. Hashim, P. Ashburn, M. Mouis, A. Chantre, and G. Vincent, "Determination of bandgap narrowing and parasitic energy barrier in SiGe HBTs integrated in a bipolar technology," IEEE Trans. Electron. Dev., 44 (5), 715–722, 1997.
92. C. H. Gan, J. A. Del Alamo, B. R. Bennett, B. S. Meyerson, E. F. Crabbe, C. G. Sodini, and L. R. Reif, "$Si/Si_{1-x}Ge_x$ valence band discontinuity measurements using semiconductor-insulator-semiconductor (SIS) heterostructures." IEEE Trans. Electron. Dev., 41 (12), 2430–2439, 1994.
93. K. Nauka, T. I. Kamins, J. E. Turner, C. A. King, J. L. Hoyt, and J. F. Gibbons, "Admittance spectroscopy measurements of band offsets in $Si/Si_{1-x}Ge_x/Si$ heterostructures," Appl. Phys. Lett., 60 (2), 195–197, 1992.
94. C. King, J. Hoyt, and J. Gibbons, "Bandgap and transport properties of $Si_{1-x}Ge_x$ by analysis of nearly ideal $Si/Si_{1-x}Ge_x$ heterojunction bipolar transistors," IEEE Trans. Electron. Dev., 36 (10), 2093–2104, 1989.
95. E. Prinz, P. M. Garone, P. V. Schwartz, X. Xiao, and J. C. Sturm, "The effect of base-emitter spacers and strain-dependent densities of states in $Si/Si_{1-x}Ge_x$ heterojunction bipolar transistors," IEEE IEDM Tech. Dig., 639–642, 1989.
96. D. M. Richey, J. D. Cressler, and A. J. Joseph, "Scaling issues and profile optimization in advanced UHV/CVD SiGe HBTs," IEEE Trans. Electron. Dev., 44 (3), 431–440, 1997.
97. L. E. Kay and T. W. Tang, "Monte Carlo calculation of strained and unstrained electron mobilities in $Si_{1-x}Ge_x$ using improved ionized-impurity model," J. Appl. Phys. Lett., 70 (3), 1483–1488, 1991.
98. M. V. Fischetti and S. E. Laux, "Band structure, deformation potentials, and carrier mobility in strained Si, Ge, and SiGe alloys," J. Appl. Physics, 80 (4), 2234–2252, 1996.
99. T. Manku, J. M. Gregor, A. Nathan, D. J. Roulston, J.-P. Noel, and D. C. Houghton, "Drift hole mobility in strained and unstrained doped $Si_{1-x}Ge_x$ alloys," IEEE Trans. Electron. Dev., 40 (11), 1990–1996, 1993.
100. T. Manku and A. Nathan, "Lattice mobility of holes in strained and unstrained $Si_{1-x}Ge_x$ alloys," IEEE Electron. Dev. Lett., 12 (12), 704–706, 1991.
101. G. Busch and O. Vogt, "Elektrische Leitfaehigkeit und Halleffekt von GeSi-Legierungen," Helv. Phys. Acta., 33, 437–458, 1960.
102. R. W. Keyes, "High-mobility FET in strained silicon," IEEE Trans. Electron. Dev., ED-33 (6), 863, 1986.
103. C. S. Smith, "Piezoresistance effects in germanium and silicon," Phys. Rev., 94, 42–49, 1954.
104. R. People and J. C. Bean, "band alignment of coherently strained Ge_xSi_{1-x}/Si heterostructures on $\langle 100 \rangle$ Ge_ySi_{1-y} substrates," Appl. Phys. Lett. 48 (8), 538–540, 1986.
105. D. B. M. Klaassen, "A unified mobility model for device simulation – I. Model equations and concentration dependence," Solid-State Electron., 35 (7), 953–959, 1992.
106. D. B. M. Klaassen, "A unified mobility model for device simulation – II. Temperature dependence of carrier mobility and lifetime," Solid-State Electron., 35 (7), 961–967, 1992.
107. T. Manku and A. Nathan, "Effective mass for strained p-type $Si_{1-x}Ge_x$," J. Appl. Phys., 69 (12), 8414–8416, 1991.
108. D. L. Harame, J. H. Comfort, J. D. Cressler, E. F. Crabbé, B. S. Meyerson, and T. Tice, "Si/SiGe epitaxial base transistors – Part I: materials, physics, and circuits," IEEE Trans. Electron. Dev., 42 (3), 455–468, 1995.
109. S. L. Salmon, J. D. Cressler, R. C. Jaeger, and D. L. Harame, "The influence of Ge grading on the bias and temperature characteristics of SiGe HBTs for precision analog circuits," IEEE Trans. Electron. Dev., 47 (2), 292–298, 2000.
110. E. J. Prinz and J. C. Sturm, "Current gain–Early voltage products in heterojunction bipolar transistors with nonuniform base bandgaps," IEEE Trans. Electron. Device Lett., 12 (12), 661–663, 1991.

111. J. L. Moll and I. M. Ross, "The dependence of transistor parameters on the distribution of base layer resistivity," Proc. IRE, 44 (1), 72–78, 1956.
112. J. W. Slotboom, G. Streutker, A. Pruijmboom, and D. Gravensteijn, "Parasitic energy barriers in SiGe HBTs," IEEE Electron. Dev. Lett., 12 (9), 486–488, 1991.
113. E. Prinz, P. Garone, P. Schwartz, X. Xiao, and J. Sturm, "The effects of base dopant out-diffusion and undoped $Si/Si_{1-x}Ge_x$ junction spacer layers in $Si/Si_{1-x}Ge_x$ heterojunction bipolar transistors," IEEE Electron. Dev. Lett., 12 (2), 42–44, 1991.
114. A. Gruhle, "The influence of emitter-base junction design on collector saturation current, ideality factor, Early voltage, and device switching speed," IEEE Trans. Electron. Dev., 41 (2), 198–203, 1994.
115. R. J. E. Hueting, J. W.Slotboom, A. Pruijmboom, W. B. de Boer, C. E, Timmering, and N. E. B. Cowern, "On the optimization of SiGe-base bipolar transistors," IEEE Trans. Electron. Dev., 43 (9), 1518–1524, 1996.
116. G. Niu and J. D. Cressler, "The impact of bandgap offset distribution between conduction and valence bands in Si-based graded bandgap HBTs," Solid-State Electron., 43 (12), 1999.
117. H.J. Osten, R. Barth, G. Fischer, B. Heinemann, D. Konoll. G. Lippert, H. Rücker, P. Schley, and W. Röpke, "Carbon-containing group IV heterosctructures on Si: properties and device applications," Thin Solid Films, 321 (1–2), 11–14, 1998.
118. J. L. Hoyt, T. O. Mitchel, K. Rim, D. V. Singh, and J. F. Gibbons, "Comparison of $Si/Si_{1-x-y}Ge_xC_y$ and $Si/Si_{1-y}C_y$ heterojunctions grown by rapid thermal chemical vapor deposition," Thin Solid Films, 321 (1–2), 41–46, 1998.
119. H. Rücker, B. Heinemann, D. Bolze, D. Knoll, D. Krüger, R. Kurps, H. J. Osten, P. Schley, B. Tillack, and P. Zaumseil, "Dopant diffusion in C-doped Si and SWiGe: physical model and experimental verification," IEEE IEDM, 345–348, 1999.
120. E. O. Johnson, "Physical limitations on frequency and power parameters of transistors," RCA Rev., 163–177, 1975.
121. L. Lanzerotti, et al., "A low-complexity 0.13 mm SiGe BiCMOS technology for wireless and mixed signal applications," IEEE BCTM, 237–240, 2004.
122. K. K. Ng, M. R. Frei, and C. A. King, "Reevaluation of the $f_T BV_{ceo}$ limit on Si bipolar transistors," IEEE Trans. Electron. Dev., 43 (8), 1854–1855, 1998.

Chapter 4
The MOS Structure

4.1 Introduction

The *MOS* structure consists of a semiconductor covered by an insulator upon which a conductive electrode is deposited (Fig. 4.1). The term *MOS* stands for Metal-Oxide-Silicon and stems from earlier technologies that utilized aluminum, silicon-dioxide (or simply oxide), and silicon to form the capacitor between source and drain of an *MOS* Field-Effect Transistor, *MOSFET* (Chap. 5). The need for a gate-conductor that can withstand high-temperature annealing and allow self-alignment of gate to source-drain led to the development of heavily doped n-type or p-type polysilicon gate-conductors to replace aluminum. While doped polysilicon is the gate-conductor of choice for oxide thickness above $\sim$3 nm, its low conductivity compared to metals begins to seriously impact *MOSFET* performance as device dimensions are reduced to the nanoscale range. Metal-gates, such as tungsten, molybdenum, and fully-silicided polysilicon have therefore become necessary to overcome this limitation. Also, as the silicon-dioxide thickness is reduced below $\sim$2 nm, the level of power-consumption caused by tunneling current through the oxide becomes prohibitive. For such dimensions, it is necessary to replace silicon-dioxide with alternate dielectrics of higher dielectric constant.

As will be discussed in Chap. 5, there are several advantages of replacing silicon with semiconductor variants such as silicon-germanium (*SiGe*) and silicon-carbon ($Si:C$) alloys, or germanium. In this chapter, however, the term *MOS* will be used to describe all of the above combinations of gate-conductor, insulator, and semiconductor.

The *MOS* structure is a fast and effective two-terminal device to study properties of the semiconductor, insulator, and their interfaces [1–3]. It is widely utilized to measure parameters directly related to the *MOSFET*, and to monitor individual processing steps.

The chapter begins with a detailed discussion of the physics of *MOS* structures since this lays the groundwork for an understanding of *MOSFET* operation and process-device interactions. Experimental techniques are described to extract important process and device parameters.

B. El-Kareh, *Silicon Devices and Process Integration: Deep Submicron and Nano-Scale Technologies*,
© Springer Science+Business Media, LLC 2009

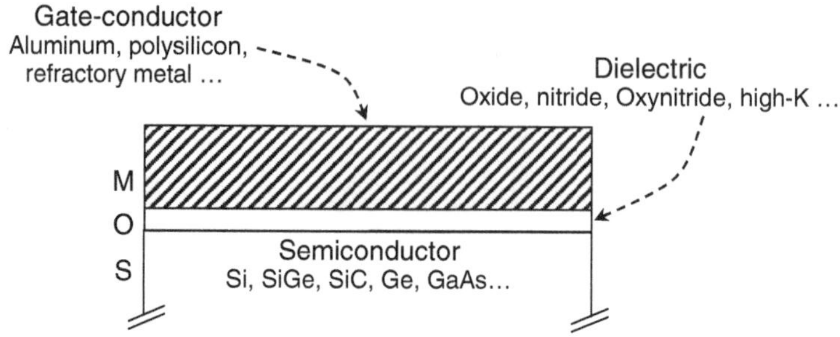

Fig. 4.1 The *MOS* structure

4.2 Physics of an Ideal *MOS* Structure

The theory of an *MOS* structure is best developed by starting with an idealized structure that exhibits zero current through the dielectric under all static conditions, zero charge within the dielectric bulk and interfaces, and zero contact potential between gate and semiconductor. This implies that in the absence of an external stimulus, such as applied voltage, light or temperature gradient, the semiconductor surface is undisturbed. A uniformly doped semiconductor is assumed for simplicity. The energy-band diagram for this case is shown in Fig. 4.2 for p-type and n-type silicon.

In both diagrams, majority carriers are exactly balanced by ionized impurities and the silicon energy bands are shown flat from bulk to surface. This is referred to as the flatband condition. $q\phi_m$ is the workfunction of the gate conductor. The workfunction of a material is the minimum energy required to lift an electron from E_F to the vacuum level. Obviously, a single gate material cannot simultaneously satisfy the flatband condition for both p-type and n-type silicon, so the workfunctions are shown different in Fig. 4.2. $q\chi_{Si}$ and $q\chi_{ox}$ are, respectively, the electron affinity of silicon and oxide. $q\chi$ is the minimum energy required to lift an electron from E_C to the vacuum level At any point in the semiconductor, the Fermi-potential is[1]

$$\phi_b = \frac{E_F - E_i}{q} \quad \text{V}. \tag{4.1}$$

The semiconductor workfunction is then (Fig. 4.2)

$$\phi_{Si} = \chi_{Si} + \left(\frac{E_g}{2q} - \phi_b \right) \quad \text{V}. \tag{4.2}$$

The workfunction difference (or contact potential) is

$$\phi_{ms} = \phi_m - \phi_{Si}. \tag{4.3}$$

[1] Unless otherwise stated, E, E_F, E_i, E_C, E_V, and Eg, are energies expressed in eV; ϕ, ψ, χ, are expressed as potentials in V. Energies and potentials have the same numerical value but different units. Example: kT is expressed in eV and kT/q in V.

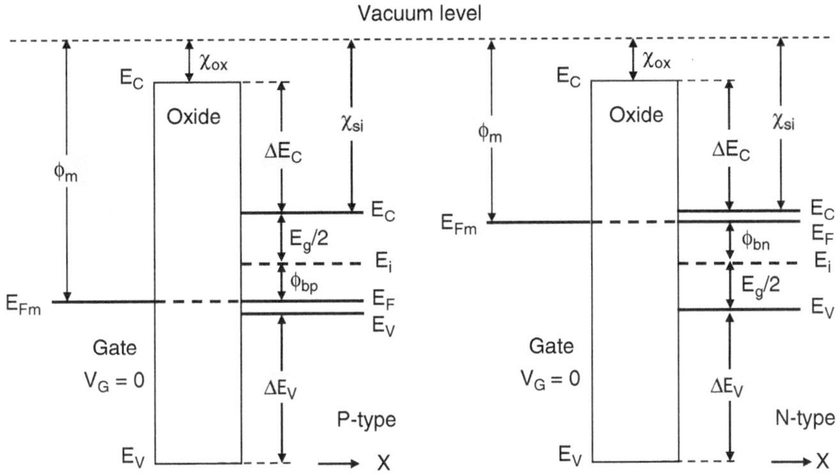

Fig. 4.2 Energy-band diagram for an idealized *MOS* structure

For zero contact potential, as idealized in Fig. 4.2, $\phi_m = \phi_{Si}$ and ϕ_{ms}=0. ΔE_C, ΔE_V are, respectively, the conduction-band and valence-band offsets between semiconductor and dielectric.

4.2.1 Description of Semiconductor Surface Conditions

In more practical *MOS* structures, $\phi_{ms} \neq 0$ and there will be a difference in Fermi potential between bulk and surface of the semiconductor. The semiconductor potentials are then defined in Fig. 4.3. Surface potentials are denoted with subscript "*s*" and bulk potential with subscript "*b*." Within the space-charge region the potential is

$$\psi(x) = \phi(x) - \phi_b \quad \text{V}. \tag{4.4}$$

The Fermi-potential is positive when the Fermi-level is above the intrinsic level and negative when it is below the intrinsic level. Deep in the bulk, the Fermi-potential for p-type silicon is $\phi_{bp} < 0$, and for n-type silicon $\phi_{bn} > 0$ (Fig. 4.2). At the surface, the Fermi-potential is ϕ_s. The surface potential $\psi_s = \phi_s - \phi_b$ is negative when the bands bend upward, positive when the bands bend downward, and zero at flatband. The surface electron and hole concentrations are defined as n_s and p_s. x is positive when pointing into silicon.

When a voltage is applied to the gate with respect to silicon, charge is induced at the gate-insulator interface. For a metal, the charge consists of accumulation or depletion of electrons, depending on the voltage polarity. Since the electron concentration in metals is very high ($\sim 10^{23}$ electrons/cm^3), the depleted depth is infinitesimally small and the charge is assumed to totally reside at the

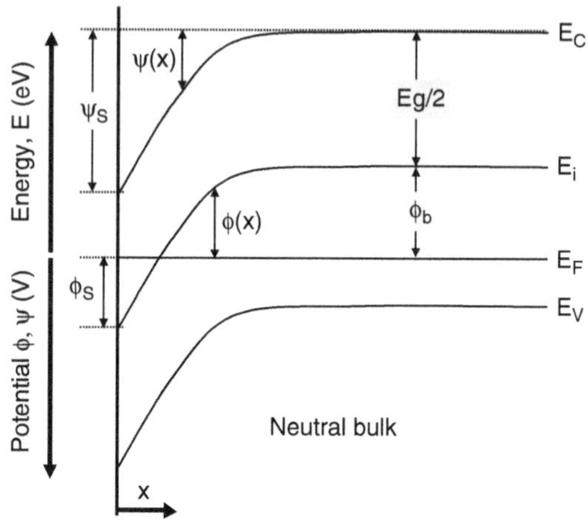

Fig. 4.3 Definition of potentials

metal-insulator interface. This is not exactly the case for a polysilicon gate because, even when degenerately doped, the polysilicon carrier concentration does not exceed 10^{20}–10^{21} cm^{-3}. While still much smaller than for moderately-doped silicon, the depletion depth in polysilicon can be significant. Neutrality requires that the induced gate-charge per unit area, Q_m, be balanced by a charge of opposite polarity Q_s in the semiconductor[2]

$$Q_s = -Q_m \quad \mathrm{C/cm^2}. \tag{4.5}$$

The band diagrams for p-type and n-type silicon are shown in Fig. 4.4 for the different bias conditions. The flatband diagram is shown again in Fig. 4.4a for reference. An undisturbed surface means $p_s \approx N_A$ for p-type, $n_s \approx N_D$ for n-type silicon, and $Q_s = 0$.

A difference of n_s and p_s from the bulk values n_b and p_b indicates that there is band bending near the silicon surface, in a direction that depends on the polarity of gate voltage. The silicon-Fermi level remains flat under all conditions because there is no significant current normal or parallel to the silicon surface. Under all steady-state bias conditions $p_s n_s = n_i^2$. Steady-state means that surface potentials and charges are time-independent.

Accumulation describes an increase in the majority carrier concentration at the surface. An accumulating voltage has the polarity to attract majority carriers to the surface. It results in an upward band-bending for p-type and downward band-bending for n-type silicon to reflect an increase in majority carriers, that is, $p_s > N_A$, $n_s > N_D$ (Fig. 4.4b). The amount of band-bending represents the surface potential ψ_s. The gate voltage is divided between the voltage across the oxide, V_{ox}, and the voltage between surface and bulk of silicon

$$V_G = V_{ox} + \psi_s \quad \mathrm{V}. \tag{4.6}$$

[2] Unless otherwise stated, Q denotes charge per unit area in C/cm^2.

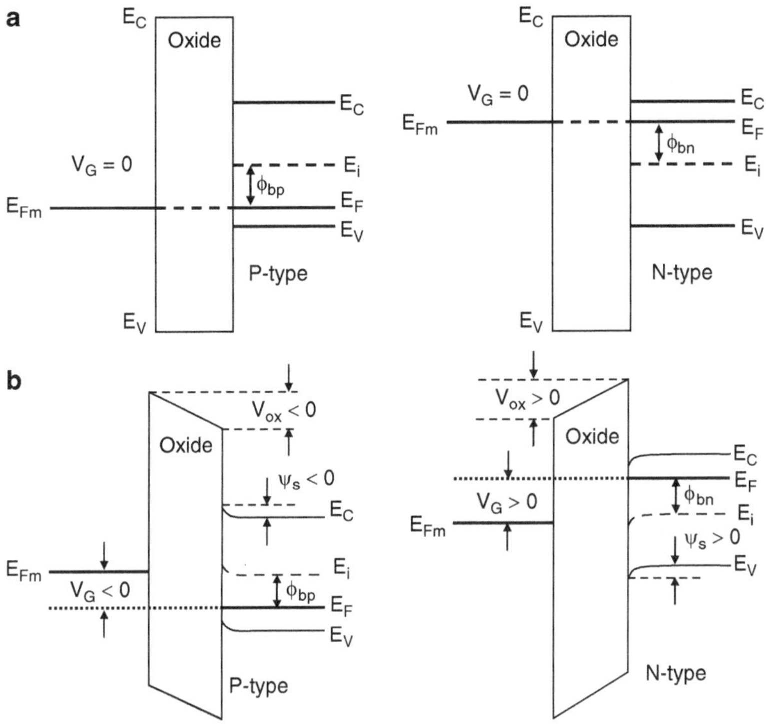

Fig. 4.4 **a** Flatband, **b** Accumulation

If the polarity of gate-voltage is such as to repel majority carriers from the surface, a depleted (space-charge) region of depth x_d forms at the surface. This is similar to the depletion region in a one-sided step junction. The corresponding band-diagram is shown in Fig. 4.4c.

For low gate voltages of this polarity, the space charge consists predominantly of fixed, uncompensated ionized impurities, negative for p-type and positive for n-type silicon. Minority carriers are also attracted to the surface but their concentration is negligible when compared to that of the uncompensated ionized impurities. For p-type, this condition can be expressed as $p_s < N_A$, $n_s \ll N_A$, $n_s \ll p_s$.

As the depleting gate voltage is increased, the majority carrier concentration at the surface decreases further, and the depletion region extends deeper into the bulk, while the minority concentration increases at the surface. A point is reached where the electron and hole concentrations exactly balance at the surface and $p_s = n_s$. This is the intrinsic condition. The corresponding band-diagram is illustrated in Fig. 4.4d with the Fermi-level shown coincident with the intrinsic level at the surface. This condition is reached when $\psi_s = \phi_b$.

With further increase in band-bending, the space-charge region continues to expand deeper into silicon and the Fermi level crosses the intrinsic level at the surface, indicating a reversal in polarity of carriers at the surface. The bulk minority carriers

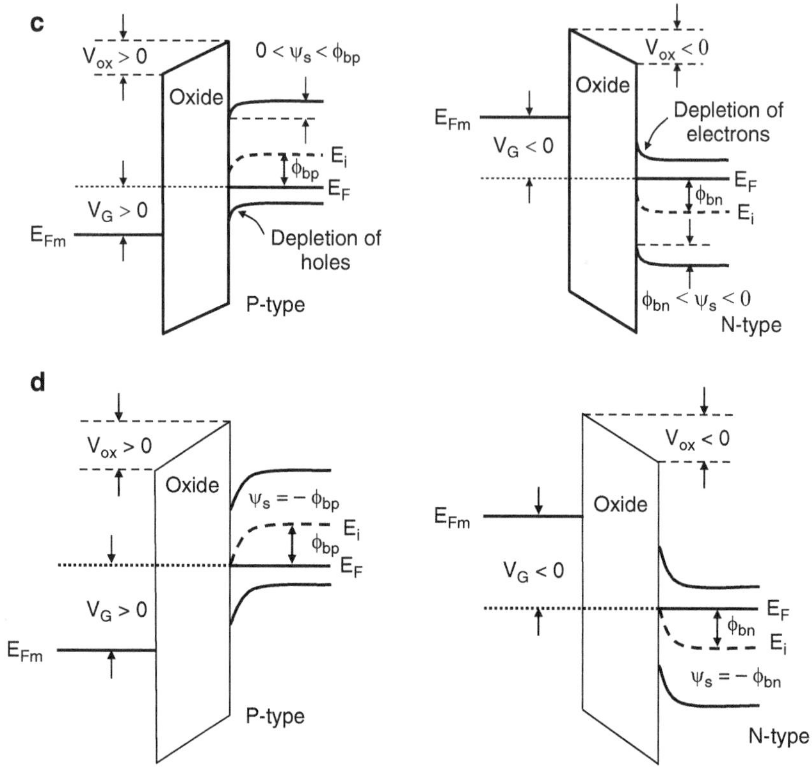

Fig. 4.4 (continued) **c** Depletion, **d** Intrinsic condition

become majority carriers at the surface, that is, $n_s > p_s$ for p-type, $p_s > n_s$ for n-type silicon (Fig. 4.4e), and the surface is said to be inverted. Initially, as the Fermi-level just crosses the intrinsic level, the inversion is weak since the inversion-carrier concentration is still small compared to the ionized impurity concentration.

As the inverting gate voltage is further increased, the inversion-carrier concentration increases to the point where it becomes approximately equal to the bulk majority-carrier ionized impurity concentration. This condition is reached when $\psi_s \approx 2\phi_b$. It is defined as the onset of strong inversion. The corresponding band-diagram is shown in Fig. 4.4f. An increase in the inverting gate voltage beyond this point results in a rapid increase in the inversion layer concentration and negligible change in depletion depth. The inversion layer that is formed is very thin, of the order of 10 nm, and acts as a conductive sheet which shields the sub-surface depletion region from changes in gate electric field. The depleted region therefore reaches its maximum depth and does not expand appreciably upon further increase in gate voltage beyond strong inversion.

At this point, it is appropriate to discuss the mechanisms that are responsible for the supply and transport of carriers from and to the surface. When a depleting voltage is applied, majority carriers are repelled from the surface into the bulk where

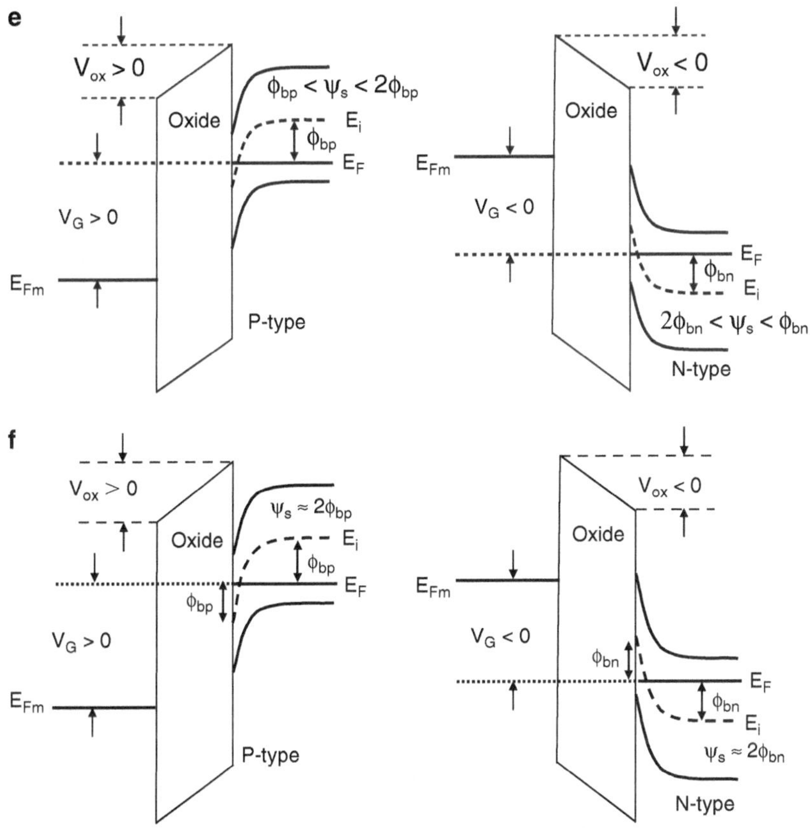

Fig. 4.4 (continued) **e** Weak inversion, **f** Onset of strong inversion

they are annihilated by recombination. Recombination with minority carriers occurs at the contact to the substrate. The time for majority carriers to flow from or to the surface is comparable to the "dielectric relaxation time," in the order of picoseconds for typical substrates. Therefore, one can assume that under normal operation the transport of majority carriers from or to the surface is "instantaneous" when compared to the speed at which the voltage is varied. This is not the case for minority carriers. As will be shown in the following sections, the semiconductor surface-bulk system behaves like a reverse-biased step pn junction when the surface is being depleted. Therefore, in the absence of light or impact ionization, minority carriers can be supplied to the surface only by thermal generation within the depletion region or within a minority-carrier diffusion length from the depletion boundary. The rate at which minority carriers are thermally generated depends on the minority-carrier lifetime which is typically in the microsecond to millisecond range. This is a relatively slow process. Therefore, when an inverting voltage is applied, minority carriers are not "instantaneously" supplied. There is typically a time-lag between demand (applied voltage) and supply (thermal generation) of minority carriers. This is particularly the case when the onset of strong inversion is approached where a large supply of bulk minority-carriers is required. Depending on the minority-carrier lifetime, the time-lag between applied voltage and supply of minority carriers may reach several seconds. With this in mind, one should consider the energy-band diagrams in Fig. 4.4 as being established after the required minority carriers have been supplied, by thermal generation or other means such as light.

4.2.2 Surface Charge and Electric Field [4–7]

The one-dimensional Poisson equation relates the potential as a function of depth x to the space-charge density as (Chap. 2)

$$\frac{d^2\phi}{dx^2} = -\frac{\rho(x)}{\varepsilon_0 \varepsilon_s}, \tag{4.7}$$

where ε_0 is the permittivity of free space ($8.86 \times 10^{-14}\,\mathrm{F/cm}$) and ε_s the relative dielectric constant of the material,[3] and $\rho(x)$ is defined as

$$\rho(x) = q\left[N_D^+ - N_A^- + p(x) - n(x)\right] \quad \mathrm{C/cm^3}. \tag{4.8}$$

Deep in the semiconductor bulk, charge neutrality requires that

$$N_D^+ - N_A^- = p_b - n_b \quad \mathrm{cm^{-3}}, \tag{4.9}$$

where p_b and n_b are, respectively, the bulk equilibrium hole and electron concentrations. For simplification, the following dimensionless "potentials" are introduced:

$$u = \frac{q\phi}{kT}; v = \frac{q\psi}{kT}. \tag{4.10}$$

With the above simplifications, the surface conditions are summarized for p-type and n-type silicon in Table 4.1.

Assuming a uniformly-doped semiconductor, the Boltzmann approximation gives:

$$\begin{aligned} \bar{n}_b &= n_i e^{u_b}\ ; \bar{p}_b = n_i e^{-u_b}, \\ n(x) &= n_i e^{u(x)}\ ; p(\mathrm{x}) = n_i e^{-u(x)}, \\ n(x) &= n_b e^{v(x)}\ ; p(\mathrm{x}) = n_b e^{-v(x)}, \end{aligned} \tag{4.11}$$

[3] When an atom is placed in an electric field, its electron cloud is shifted slightly, giving rise to an electric dipole. The ease with which dipoles can be formed in a dielectric (its polarization) is expressed by its relative dielectric constant. The polarization adds to the surface charge and hence increases the capacitance of the structure.

Table 4.1 Dimensionless potential for surface conditions

Condition	P-type	N-type
Flatband	$v_s = 0$	$v_s = 0$
Accumulation	$v_s < 0$	$v_s > 0$
Depletion	$0 < v_s < -u_b$	$0 > v_s > -u_b$
Intrinsic	$v_s = -u_b$	$v_s = -u_b$
Weak inversion	$-u_b < v_s < -2u_b$	$-u_b > v_s > -2u_b$
Strong inversion	$v_s = -2u_b$	$v_s = -2u_b$

$$n_s = n_i e^{u_s}\ ; p_s = n_i e^{-u_s},$$
$$n_s = n_b e^{u_s}\ ; p_s = n_b e^{-u_s}.$$

Equation (4.8) can now be written as

$$\rho(x) = qn_i\left[e^{u_b} - e^{-u_b} - e^{u(x)} + e^{-u(x)}\right], \tag{4.12}$$

or

$$\rho(x) = 2qn_i\left[\sinh u_b - \sinh u(x)\right]. \tag{4.13}$$

Poisson's equation takes the form

$$\frac{d^2u}{dx^2} = \frac{2q^2n_i}{\varepsilon_0\varepsilon_s kT}\left[\sinh u(x) - \sinh u_b\right]. \tag{4.14}$$

The factor in front of the bracket has the dimension cm^{-2} and is defined as $1/L_i^2$, where

$$L_i = \sqrt{\frac{\varepsilon_0\varepsilon_s kT}{2q^2n_i}} \quad \text{cm} \tag{4.15}$$

is the intrinsic Debye length. For Si at 25°C, $L_i \approx 2.5\times10^{-3}$ cm. The general solution of (4.13) is [5]

$$\frac{du}{dx} = Sgn(u_b - u_s)\frac{F\left[u_b, u(x)\right]}{L_i}, \tag{4.16}$$

where $Sgn(u_b - u_s) = +1$ for $u_s < u_b$, $Sgn(u_b - u_s) = -1$ for $u_s > u_b$, and

$$F\left[u(x), u_b\right] = \sqrt{2}\left\{\cosh u(x) - \cosh u_b + \left[u_b - u(x)\right]\sinh u_b\right\}^{1/2}. \tag{4.17}$$

At the surface, $u(x) = u_s$. $F(u_s,\ u_b)$ is plotted in Fig. 4.5 as a function of u_s for various values of u_b. $F(u_s,\ u_b)$ can be considered as the dimensionless field that is always positive (Fig. 4.5); a polarity is assigned by $Sgn(u_b - u_s)$. As u_s approaches u_b at flatband $F(u_s,\ u_b)$ goes to zero. The electric field at the surface is

$$E_s = -\frac{d\phi_s}{dx} = \frac{kT}{q}\frac{du_s}{dx} = Sgn(u_s - u_b)\frac{kT}{qL_i}F(u_b,\ u_s). \tag{4.18}$$

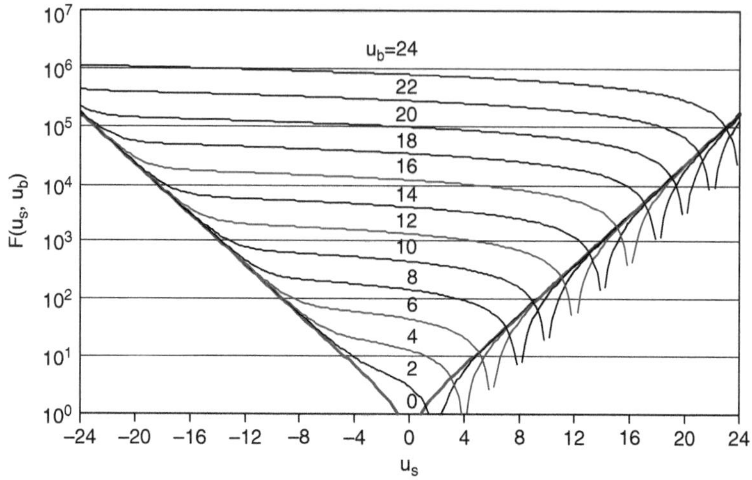

Fig. 4.5 Plots of *F(us, ub)* as a function u_s for various values of $\boldsymbol{u_b}$ (Adapted from [6])

For Si at 300 K, $E_s \approx 10F(u_b, u_s)$ V/cm. The space-charge density is found as

$$Q_s = \varepsilon_0 \varepsilon_{Si} E_s = Sign(u_s - u_b)\frac{\varepsilon_0 \varepsilon_{Si} kT}{qL_i} F(u_b, u_s) \quad \mathrm{C/cm^2}. \tag{4.19}$$

4.2.3 Approximations [7]

Approximations of the F-function are made here for a p-type substrate. Similar relations are found for an n-type substrate by appropriate changes in polarities.

4.2.3.1 Accumulation

For a p-type substrate, $u_b \ll -1$ and

$$\cosh u_b = -\sinh u_b \approx \frac{e^{-u_b}}{2}. \tag{4.20a}$$

In accumulation, $v_s < 0$ and $u_s < u_b$. Since $v_s = u_s - u_b$

$$\cosh u_s \approx \frac{e^{-u_s}}{2} = \frac{e^{-u_b} \cdot e^{-v_s}}{2}. \tag{4.20b}$$

$$F(u_s, u_b) \approx \sqrt{2}\left[\frac{e^{-u_b}.e^{-v_s}}{2} - \frac{e^{-u_b}}{2} - v_s\frac{e^{-u_b}}{2}\right]^{1/2}, \tag{4.20c}$$

or

$$F(u_s, u_b) \approx \frac{\sqrt{2}e^{-u_b}}{2}\left[e^{-v_s} - 1 - v_s\right]^{1/2}. \tag{4.20d}$$

In strong accumulation where v_s becomes more negative, the F-function simplifies to

$$F(u_s, u_b) \approx e^{-u_b}\, e^{-u_b}. \tag{4.20e}$$

E_s and Q_s become proportional to e^{-v_s}.

4.2.3.2 Depletion

This condition is best approximated for a near-intrinsic surface where $u_s \approx 0$, $v_s \approx -u_b$ $(u_b \ll -1)$. In this case,

$$\cosh u_s \approx 1\ ;\ \sinh u_b \approx -\frac{e^{-u_b}}{2}\ ;\ \cosh u_b \approx \frac{e^{-u_b}}{2}, \tag{4.21a}$$

$$F(u_s, u_b) = -\sqrt{2}\left[1 + \frac{e^{-u_b}}{2}(v_s - 1)\right]^{1/2} \text{ or} \tag{4.21b}$$

$$F(u_s, u_b) = -\left[2 + e^{-u_b}(v_s - 1)\right]^{1/2}. \tag{4.21c}$$

For $v_s \gg 2$, $e^{-u_b}.v_s \gg 2$, and

$$F(u_s, u_b) = -e^{-u_b/2}.\sqrt{v_s}. \tag{4.21d}$$

The surface space charge density Q_s is then

$$Q_s = -\frac{\varepsilon_0 \varepsilon_{Si} kT}{qL_i} e^{-u_b/2}.\sqrt{v_s} = -\sqrt{2\varepsilon_0 \varepsilon_{Si} kT n_i e^{-u_b}.v_s}. \tag{4.22}$$

For $n_i e^{-u_b} \approx N_A$ and $v_s = q\psi_s/kT$, the above relation simplifies to

$$Q_s = -\sqrt{2\varepsilon_0 \varepsilon_{Si} q N_A.\psi_s} \quad \mathrm{C/cm^2}. \tag{4.23}$$

Equation (4.23) is similar to the relation obtained with the depletion approximation for a one-sided step junction.

4.2.3.3 Inversion

For a p-type substrate $u_b \ll -1$, $u_s > -u_b$, $v_s \gg 0$. Therefore,

$$\cosh u_b \approx \frac{e^{-u_b}}{2};\ \sinh u_b \approx -\frac{e^{-u_b}}{2} \text{ and} \tag{4.24a}$$

$$\cosh u_s \approx \frac{e^{u_s}}{2} = \frac{e^{v_s}.e^{u_b}}{2}. \tag{4.24b}$$

The F-function is now

$$F(u_s, u_b) \approx -\sqrt{2}\left[\frac{e^{v_s}.e^{u_b}}{2} - \frac{e^{-u_b}}{2} + \frac{v_s e^{-u_b}}{2}\right]^{1/2}. \quad (4.24c)$$

This can be simplified to

$$F(u_s, u_b) \approx -\left[e^{v_s}e^{u_b} + e^{-u_b}(v_s - 1)\right]^{1/2}. \quad (4.24d)$$

In the range $0 << v_s < -2u_b$, the product $e^{v_s}e^{u_b}$ is small compared to the second term in (4.24d) and the function further simplifies to

$$F(u_s, u_b) \approx -\sqrt{e^{u_b}(v_s - 1)}. \quad (4.24e)$$

For $v_s \approx -2u_b$, the magnitude of the F-function is roughly proportional to the square-root of v_s:

$$F(u_s, u_b) \cong -\sqrt{e^{u_b}v_s}. \quad (4.24f)$$

When v_s increases above $-2u_b$, the first term in the bracket of (4.24d) dominates and

$$F(u_s, u_b) \cong -\sqrt{e^{u_b}e^{v_s}}. \quad (4.24g)$$

In summary, in the weak inversion regime, when v_s increases from $-u_b$ to the onset of strong inversion ($v_s = -2u_b$), $|F(u_s, u_b)|$, $|E_s|$, and $|Q_s|$ increases approximately as the square-root of v_s. As v_s increases above the onset of strong inversion, the values increase roughly as $e^{v_s/2}$.

4.2.4 Excess Surface Carrier Concentrations [4]

When the surface potential is varied from zero, there is a change in the hole and free-electron surface concentration with respect to flatband. The change is positive if the carrier concentration increases above its value at flatband, and negative if it decreases. Let Δp and Δn, respectively, denote the change in hole and electron concentration per unit area. Then

$$\Delta p = \int_0^\infty (p - \bar{p})dx = n_i \int_0^\infty (e^{-u(x)} - e^{-u_b})\, dx. \quad (4.25a)$$

Similarly,

$$\Delta n = \int_0^\infty (n - \bar{n})dx = n_i \int_0^\infty (e^{u(x)} - e^{u_b})\, dx. \quad (4.25b)$$

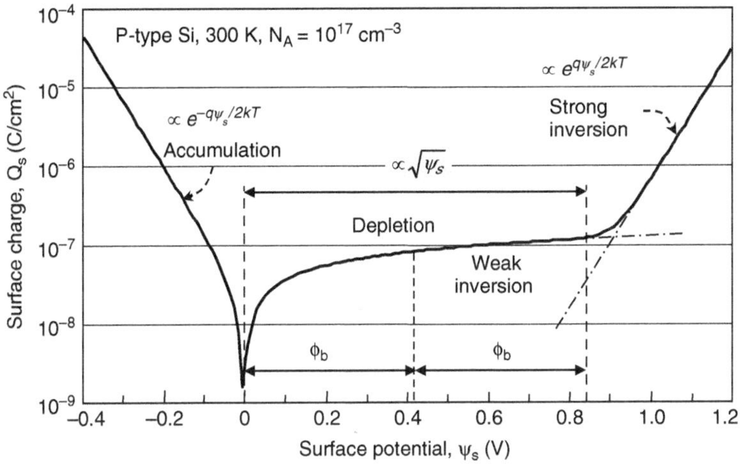

Fig. 4.6 Variation of surface charge as a function of surface potential

Combining (4.16) and (4.25) and noting that $u = u_s$ at $x = 0$ and $u = u_b$ at $x = \infty$ yields

$$\Delta p = n_i L_i \int_{u_s}^{u_b} \frac{e^{-u(x)} - e^{-u_b}}{F[-u_b, -u(x)]} du \quad \text{holes/cm}^2, \tag{4.26a}$$

$$\Delta n = n_i L_i \int_{u_s}^{u_b} \frac{e^{u(x)} - e^{u_b}}{F[u_b, u(x)]} du \quad \text{electrons/cm}^2. \tag{4.26b}$$

A negative integral means a decrease in carrier concentration. A plot of Q_s versus ψ_s is shown in Fig. 4.6 for a p-type substrate of uniform concentration $N_A = 10^{17}\,\text{cm}^{-3}$.

For negative ψ_s, the surface is accumulated (Δp positive). In depletion, the surface charge consists of fixed, uncompensated ionized acceptors (Δp negative). In weak inversion (Δn positive), the electron concentration is larger than the hole concentration but negligible compared to the ionized impurity concentration. In strong inversion, the electron concentration increases above the ionized impurity concentration. Note that the same plot applies to n-type silicon with $N_D = 10^{17}\,\text{cm}^{-3}$ by simply changing the sign of surface potential and the role of electrons and holes.

4.2.5 *MOS* Capacitance

The *MOS* capacitance can be treated as two capacitances in series, the oxide capacitance C_{ox} and the silicon capacitance C_{Si} (Fig. 4.7).

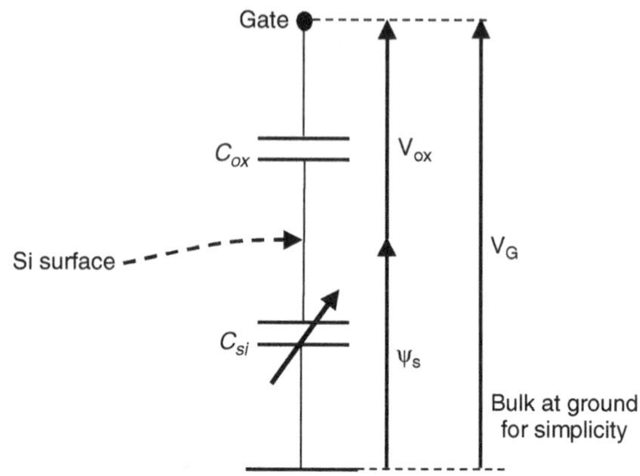

Fig. 4.7 Equivalent circuit of ideal MOS capacitor

C_{ox} is the dielectric capacitance. For oxide, it is defined as

$$C_{ox} = \frac{\varepsilon_0 \varepsilon_{ox}}{t_{ox}} \quad \mathrm{F/cm^2}, \tag{4.27}$$

where ε_{ox} and t_{ox} are, respectively, the oxide dielectric constant and thickness.

For an ideal structure, C_{ox} is independent of voltage and frequency. It is treated as a fixed capacitance. C_{Si} is the silicon capacitance between surface and bulk and depends on surface potential and hence on gate voltage.

4.2.5.1 Equivalent Capacitance

The equivalent *MOS* capacitance is found from

$$\frac{1}{C} = \frac{1}{C_{ox}} + \frac{1}{C_{Si}}, \text{ or} \tag{4.28a}$$

$$C = \frac{C_{ox}}{1 + C_{ox}/C_{Si}}. \tag{4.28b}$$

As C_{Si} increases, C increases. In the limit, $C \approx C_{ox}$.

4.2.5.2 Dielectric Field

For a given gate voltage, the field in the oxide is defined as

$$E_{ox} = \frac{V_{ox}}{t_{ox}} \quad \mathrm{V/cm}, \tag{4.29}$$

where V_{ox} is the voltage across the oxide. In an ideal structure, the field is uniform throughout the oxide. For dual or multiple insulators, however, the field changes from one dielectric to the other because of the difference in dielectric constant. For example, in a dual-insulator composed of oxide and silicon-nitride

$$V_G - \psi_s = E_{ox} t_{ox} + E_n t_n, \quad (4.30)$$

where E_n and t_n are, respectively, the constant field in the nitride and the nitride thickness. Continuity of the displacement vector yields

$$\varepsilon_0 \varepsilon_{ox} E_{ox} = \varepsilon_0 \varepsilon_n E_n. \quad (4.31)$$

4.2.5.3 Equivalent Oxide Thickness

For dielectrics other than silicon-dioxide, it is practical to define an equivalent oxide thickness t_{eq}, also known as *EOT*, which yields the same dielectric capacitance as that of the actual materials used. This simplifies the characterization of composite dielectrics and allows an easy and consistent way to compare dielectric materials, such as high-K (high dielectric constant) materials introduced in Chap. 5. Equation (4.30) can be written as

$$V_G - \psi_s = E_{ox} t_{eq} = E_{ox} t_{ox} + E_n t_n. \quad (4.32)$$

Combining with (4.31) gives

$$t_{eq} = t_{ox} + \frac{\varepsilon_{ox}}{\varepsilon_n} t_n \cong t_{ox} + \frac{3.9}{7} t_n = t_{ox} + 0.56 t_n. \quad (4.33)$$

In general, t_{eq} of a material of dielectric constant ε_x is

$$t_{eq} = \frac{\varepsilon_{ox}}{\varepsilon_x} t_x, \quad (4.34)$$

where t_x is the thickness of the material.

4.2.5.4 Carrier Response to Varying Gate Voltage

The gate voltage is divided between C_{ox} and C_{Si} as

$$V_G = V_{ox} + \psi_s = -\frac{Q_s}{C_{ox}} + \psi_s. \quad (4.35)$$

A change in gate voltage causes a change in V_{ox}, ψ_s, and in the total surface charge Q_s. For an accumulating gate voltage, Q_s consists of only excess majority carriers (Fig. 4.6). In depletion, the charge consists of mainly fixed, ionized impurities that are uncompensated by majority carriers. In both cases, the contribution of minority carriers to the total charge is negligible. Since majority carriers respond almost instantaneously to voltage-changes, steady-state is reached at all practical rates of change $\Delta V_G/\Delta t$. As the onset of strong inversion is approached, however, minority carriers become an increasing fraction of the total charge. Since, in the absence of light or other source of energy, minority carriers can only be supplied by thermal generation; their response-time is orders of magnitude longer than that of majority carriers. Initially, as the inverting gate voltage is increased, minority carriers do not immediately respond to the change unless $\Delta V_G/\Delta t$ is very small. Instead, the depleted region expands to satisfy neutrality by "exposing" more ionized impurities. Steady-state is then slowly achieved when the gate voltage is kept constant for a certain time, allowing minority carriers to be generated and the depletion region to relax to its steady state value.

Since the capacitance is voltage-dependent, there are two capacitance values associated with an *MOS* structure, the static and the differential capacitance. The static capacitance is the ratio of total charge to gate voltage

$$C_{static} = \frac{Q_s}{V_G} \quad \text{F/cm}^2. \tag{4.36}$$

The differential capacitance is defined as

$$C = \frac{\Delta Q_s}{\Delta V_G} \quad \text{F/cm}^2. \tag{4.37}$$

Because the capacitance is a nonlinear function of voltage, the two capacitances will be different. The differential capacitance is the more important of both values. Throughout this chapter, C will denote the differential capacitance per unit area.

4.3 Calculation of Capacitance

In the preceding sections, first-order quantitative relations were derived for accumulation, flatband, depletion and inversion. An exact analysis of charge and field is, however, needed for transitions between those regions. This can be provided for the low-frequency limit in which both majority and minority carriers follow the signal. The situation becomes more complex at high frequencies where some approximations must be made to ensure a smooth transition from depletion to inversion.

4.3.1 Calculation of Low-Frequency Capacitance

The voltage across the dielectric is $V_G - \psi_s$ and the charge induced by the gate is

$$Q_m = -Q_s = (V_G - \psi_s)\, C_{ox} \quad \text{C/cm}^2, \tag{4.38}$$

where $V_G - \psi_s = V_{ox}$, and C_{ox} is the equivalent-oxide capacitance.

The differential silicon capacitance is a function of ψ_s and defined as

$$C_s(\psi_s) = -\frac{dQ_s}{d\psi_s} = \frac{dQ_s}{d\phi_s} = \frac{q}{kT}\frac{dQ_s}{du_s} \quad \text{F/cm}^2, \tag{4.39}$$

and the total capacitance is then

$$C(\psi_s) = \frac{C_{ox}}{1 + C_{ox}/C_s(\psi_s)} \quad \text{F/cm}^2. \tag{4.40}$$

Combining (4.39) with (4.19) gives

$$C_s(u_s, u_b) = \frac{\varepsilon_0 \varepsilon_{Si}}{L_i}\left|\frac{d}{du_s}F(u_s, u_b)\right|, \tag{4.41}$$

$$C_s(u_s, u_b) = \frac{\varepsilon_0 \varepsilon_{Si}}{L_i}\left|\frac{\sinh u_s - \sinh u_b}{F(u_s, u_b)}\right|, \tag{4.42a}$$

$$C_s(u_s, u_b) = \frac{\varepsilon_0 \varepsilon_{Si}}{L_i}\left|\frac{(n_s - p_s/2n_i) - (\bar{n} - \bar{p}/2n_i)}{F(u_s, u_b)}\right|. \tag{4.42b}$$

A calculated low-frequency *MOS* capacitance as a function of surface potential ψ_s is shown in Fig. 4.8a for p-type silicon with $N_A = 10^{17}\,\text{cm}^{-3}$ ($\phi_b \approx 0.41\,\text{V}$ at 300 K) and $t_{eq} = 10\,\text{nm}$, indicating important regions of the curve.

The plot is first calculated with (4.42) as a function of u_s. It is then translated to ψ_s using the relation $\psi_s = (kT/q)(u_s - u_b)$ in (4.4). Care must be taken not to let u_s be exactly equal to u_b at flatband since both numerator and denominator of 4.42a will go to zero. Instead, u_s is varied infinitesimally by $u_s \pm \delta u_s$ around that point. The capacitance of the structure in Fig. 4.8a is plotted as a function of gate voltage in Fig. 4.8b, using (4.18) to calculate the surface charge density and (4.35) to relate the gate voltage V_G to surface potential.

4.3.2 Description of the Low-Frequency CV-Plot

Simple expressions for the silicon capacitance are derived here for the conditions of flatband, strong accumulation, depletion, and strong inversion. These expressions will become very useful when extracting process and device parameters from the *CV*-plot, as discussed in the following sections.

4.3.2.1 Flatband

At flatband, $u_s = u_b$, $v_s = 0$, and $F(u_s,\ u_b) = 0$. Substituting in (4.42a) results in $C_s = 0/0$, an indeterminate relation. It is therefore necessary to express (4.42)

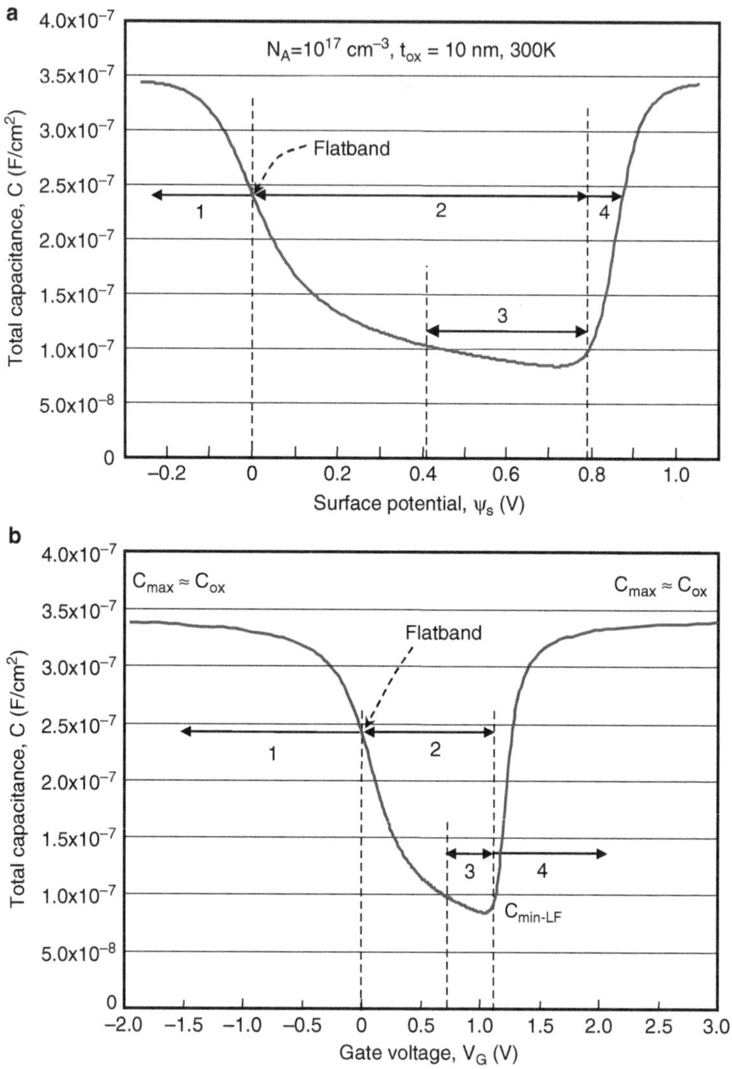

Fig. 4.8 Calculated low-frequency *CV*-plot for $N_A = 10^{17}\,\text{cm}^{-3}$ and $t_{ox} = 10\,\text{nm}$, **a** versus surface potential, **b** versus gate voltage. Important regions of the curve are: 1. accumulation, 2. depletion, 3. weak inversion, 4. strong inversion

in terms of the dimensional surface potential v_s and expand the exponential of v_s around $v_s = 0$ [5]. The silicon capacitance at flatband, C_{sFB}, is then obtained as the limit of $C_s(u_s, u_b)$ as $u_s \to 0$:

$$C_{sFB} = \frac{\varepsilon_0 \varepsilon_{Si}}{L_e} \quad \text{F/cm}^2, \tag{4.43}$$

where L_e is the extrinsic Debye length. For p-type silicon at flatband,

$$L_e = \sqrt{\frac{\varepsilon_0 \varepsilon_{Si} kT}{q^2(n+p)}} \approx \sqrt{\frac{\varepsilon_0 \varepsilon_{Si} kT}{q^2 N_A}} \quad \text{cm}, \tag{4.44a}$$

and for n-type silicon

$$L_e = \sqrt{\frac{\varepsilon_0 \varepsilon_{Si} kT}{q^2(n+p)}} \approx \sqrt{\frac{\varepsilon_0 \varepsilon_{Si} kT}{q^2 N_D}} \quad \text{cm}. \tag{4.44b}$$

The Debye length can be visualized as the average depth at which carriers follow the varying voltage signal. One can imagine a "plate" placed at a depth L_e and treat the capacitor between surface and plate as a parallel-plate capacitor. The total capacitance at flatband is

$$C_{FB} = \frac{C_{sFB} C_{ox}}{C_{sFB} + C_{ox}} \quad \text{F/cm}^2. \tag{4.45}$$

For an ideal structure, that is, zero charge within the dielectric bulk and interfaces, and zero contact potential between gate and semiconductor, C_{FB} is found at $V_G = 0$.

4.3.2.2 Strong Accumulation

The F-function is approximated for strong accumulation in (4.20e) and the silicon capacitance is found by substituting (4.20e) in (4.42b) as

$$C_{sAcc} = \frac{\varepsilon_0 \varepsilon_{Si}}{L_i} \frac{(n_s - p_s/2n_i) - (n_b - p_b/2n_i)}{e^{-u_b/2} e^{-v_s/2}} \quad \text{F/cm}^2. \tag{4.46}$$

For p-type silicon in strong accumulation, $p_b = N_A >> n_b$, $p_s/N_A >> n_s/N_A$, and

$$e^{-u_b/2} = \sqrt{\frac{N_A}{n_i}}. \tag{4.47}$$

Combining with (4.43), (4.46) simplifies to

$$C_{sAcc.} = \frac{C_{sFB}}{\sqrt{2}} e^{v_s/2} \quad \text{F/cm}^2. \tag{4.48a}$$

For n-type silicon,

$$C_{sAcc.} = \frac{C_{sFB}}{\sqrt{2}} e^{-v_s/2} \quad \text{F/cm}^2. \tag{4.48b}$$

If one defines an effective Debye length associated with the free carrier density at the surface, the result can be interpreted as equivalent to a reduction of the effective Debye length and hence an increase in silicon capacitance as the surface majority-carrier concentration increases. From (4.28b), it can be seen that as C_s increases in accumulation, C approaches the maximum value of C_{ox}.

4.3.2.3 Depletion

In depletion, the F-function is approximated as (4.21d)

$$F(u_s, u_b) = -e^{-u_b/2}\sqrt{v_s} = \sqrt{\frac{N_A}{n_i}}\sqrt{v_s}. \tag{4.49}$$

Substituting this into 4.42b and considering that for p-type $p_b = N_A >> n_b$ and in depletion $n_s << N_A, p_s << N_A$ gives

$$C_{sDep.} = \frac{\varepsilon_0 \varepsilon_{Si}}{x_d}. \tag{4.50}$$

x_d is the depletion width defined as

$$x_d = \sqrt{\frac{2\varepsilon_0 \varepsilon_{Si} \psi_s}{qN_A}} = \sqrt{\left|\frac{4\varepsilon_0 \varepsilon_{Si} kT}{q^2 N_A} \ln \frac{N_A}{n_i}\right|} \quad \text{cm}. \tag{4.51a}$$

Similarly, for n-type silicon

$$x_d = \sqrt{\frac{2\varepsilon_0 \varepsilon_{Si} \psi_s}{qN_D}} = \sqrt{\left|\frac{4\varepsilon_0 \varepsilon_{Si} kT}{q^2 N_D} \ln \frac{N_D}{n_i}\right|}. \tag{4.51b}$$

Combining (4.50) with (4.28b) yields the total capacitance as

$$C = \frac{C_{ox}}{1 + C_{ox}/C_{sDep}}. \tag{4.52}$$

In this range, the total capacitance decreases as the depleting gate voltage and hence surface potential increases.

4.3.2.4 Strong Inversion

In strong inversion, $|v_s| > 2|u_b|$ and the excess surface charge is dominated by the inversion charge density. The F-function for this mode is approximated by (4.24g). Substituting in (4.42b) for the silicon capacitance, and considering that $n_s >> p_s$, $n_s >> N_A$ yields

$$C_s(u_s, u_b) \cong \frac{\varepsilon_0 \varepsilon_{Si}}{L_i} \left| \frac{n_s/2n_i}{e^{u_b/2} e^{u_s/2}} \right| \quad \text{F/cm}^2. \tag{4.53}$$

From (4.12),

$$\begin{aligned} n_s &= \bar{n} e^{v_s} = \frac{n_i^2}{N_A} e^{v_s} \\ e^{u_b/2} &= \sqrt{\frac{n_i}{N_A}} \end{aligned}. \tag{4.54}$$

Substituting in (4.53) and manipulating gives the silicon capacitance in inversion, C_{inv}, as a function of flatband capacitance as

$$C_{inv} = \frac{C_{sFB}}{\sqrt{2}} \frac{n_i}{N_A} e^{q\psi_s/2kT} \quad \text{F/cm}^2. \tag{4.55a}$$

Similarly, for n-type silicon

$$C_{inv} = \frac{C_{sFB}}{\sqrt{2}} \frac{n_i}{N_D} e^{-q\psi_s/2kT}. \tag{4.55b}$$

Since the rate of change in surface potential is assumed to be sufficiently small to allow minority carriers to follow the signal, the response to the varying signal occurs at the surface where the inversion layer is formed. The silicon capacitance is proportional to $exp(q|\psi_s|/2kT)$. The total capacitance is

$$C = \frac{C_{ox}}{1 + C_{ox}/C_{inv}}. \tag{4.56}$$

As the inversion layer forms and C_{inv} increases, the total capacitance approaches its maximum value of approximately C_{ox}. For polysilicon gates, the measured maximum capacitance in inversion is, however, frequently found to be smaller than C_{ox}. One component of the difference is attributed to the depletion of the polysilicon gate when silicon is inverted. A positive V_G with respect to silicon is equivalent to a negative voltage on silicon with respect to the gate. For an n-type polysilicon gate on p-type silicon, this results in depletion or even inversion of the gate when silicon is inverted. If the polysilicon is heavily doped, the depletion depth in the gate is very small. As the dielectric is thinned, however, the contribution of depletion depth to the total equivalent oxide thickness becomes appreciable. This results in a thicker electrically-measured equivalent oxide thickness than physically deposited. A similar reasoning applies to p^+-polysilicon on n-type silicon.

4.3.2.5 Minimum Capacitance

The *CV*-plot in Fig. 4.8a goes through a minimum near $\psi_s = -2\phi_b$ before rising again. This can be qualitatively explained by considering that during depletion, it is the majority carriers that respond to a positive voltage increment δV by "exposing" more ionized acceptors, thus increasing the depletion width and reducing the silicon capacitance (4.50). As more inversion electrons become available to respond to the signal, however, the effective Debye-length decreases, increasing the silicon capacitance. The two opposite effects result in an inverse saddle-point. As the surface potential further increases, the strong inversion layer that forms at the surface constitutes a thin conductive film that shields the sub-surface from the field induced by the gate, so that the depletion region reaches a maximum depth and does not further expand. The minimum normalized surface potential u_{sMin} is found from the condition

$$\frac{dC_s}{d\psi_s} = \frac{q}{kT}\frac{dC_s(u_{sMin})}{du_s} = 0.$$

Differentiating (4.42a) *at* $u_s = u_{sMin}$ and using the relation

$$\frac{\partial F(u_s, u_b)}{\partial u_s} = \frac{\sinh u_s - \sinh u_b}{F(u_s, u_b)},$$

gives [5]

$$(\cosh u_{sMin})^{1/2} = \frac{\sinh u_{sMin} - \sinh u_b}{F(u_{sMin}, u_b)}. \tag{4.57}$$

In (4.57) u_{sMin} is a transcendental function of u_b. It can be solved iteratively. Substituting in (4.42a) gives the minimum capacitance

$$C_{sMin} = \frac{\varepsilon_0 \varepsilon_{Si}}{L_i}(\cosh u_{sMin})^{1/2}. \tag{4.58}$$

The minimum silicon capacitance depends implicitly on dopant concentration through (4.57).

4.3.2.6 Inversion Carrier Distribution

There are three ways used to represent the inversion carrier distribution: sheet of infinitesimal thickness, classical Fermi distribution with Boltzmann approximation, or quantum-mechanical (Fig. 4.9). In the analysis so far, it was assumed that all excess minority-carrier charge in an inversion layer constitutes a conductive sheet of infinitesimal thickness located directly below the surface, and only the area density of carriers was considered without accounting for carrier distribution within the layer.

In the classical model, the carrier density follows the Boltzmann approximation of the Fermi-Dirac distribution function (Chap. 1). For a uniformly doped bulk, the electron and hole concentrations as a function of depth are

$$n(x) \cong n_i e^{q\phi(x)/kT}, \quad p(x) \cong n_i e^{-q\phi(x)/kT},$$

where n_i is the intrinsic carrier concentration and $\phi(x)$ the Fermi potential as a function of depth (Fig. 4.3). In this model, the carrier concentration decays exponentially from surface to bulk (Fig. 4.9a).

Quantum effects in inversion layers involve solving Schrödinger's and Poisson's equations self-consistently [8–16]. Details of quantum-mechanical (*QM*) calculations are complex and beyond the scope of this book. The results are, however, of increasing significance in deep submicron and nanoscale technologies as the effective oxide thickness is reduced and the semiconductor dopant concentration increased, creating large electric fields at the silicon surface. Compared to the classical model, this results in broadening of the inversion layer, reduction in inversion-layer charge density at a given gate voltage, and decrease in the oxide

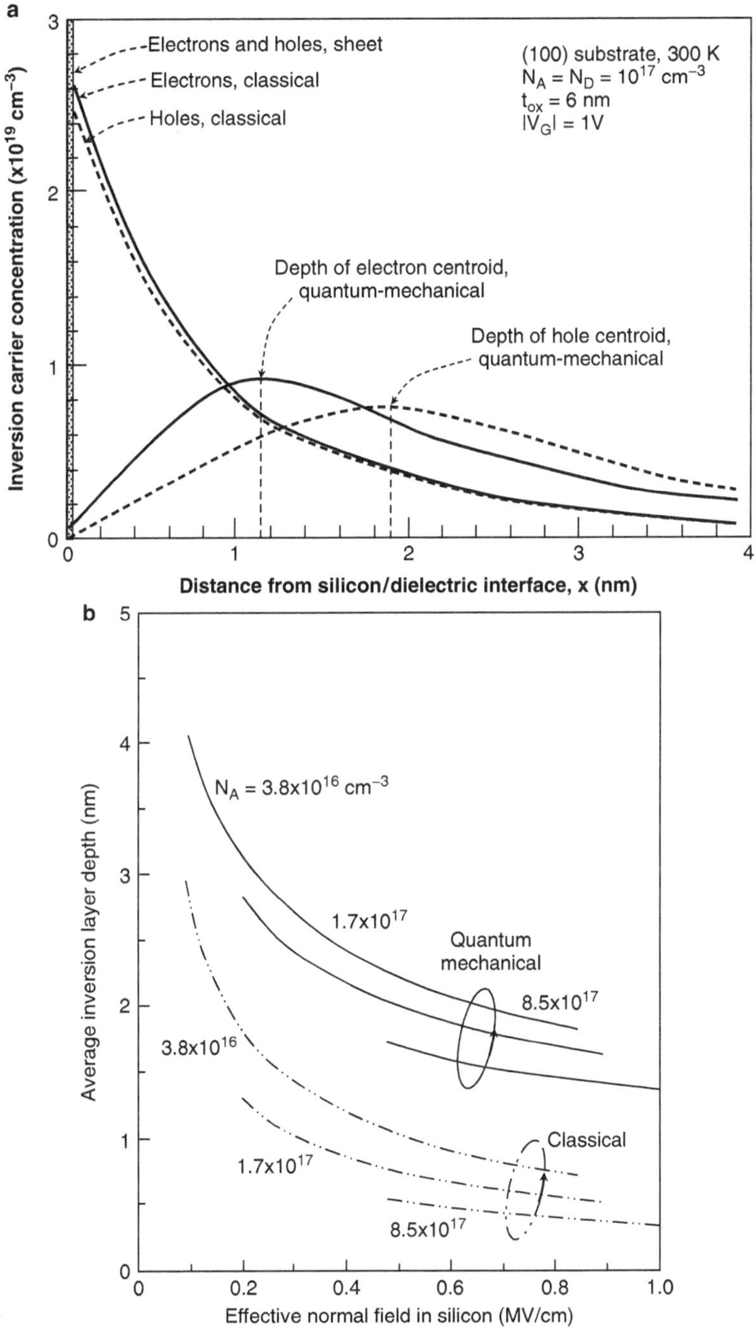

Fig. 4.9 **a** Predicted classical and quantum-mechanical inversion carrier distribution for electrons and holes (Adapted from [10]). **b** Predicted classical and quantum-mechanical average inversion layer depth as a function of effective surface field for three channel dopant concentrations (Adapted from [12])

capacitance due to the shift of the inversion layer charge away from the silicon surface [10–15]. There is also a decrease in surface mobility and increase in the magnitude of the threshold voltage, as will be discussed in the next chapter. The average inversion depth can be defined as

$$\bar{x}_{inv} = \frac{\int_0^{x_b} x\, n_{inv} dx}{\int_0^{x_b} n_{inv} dx} \quad \text{cm}, \tag{4.59}$$

where x_b is the depth of the neutral region under the depletion region. The main difference between the classical and quantum-mechanical models lies in the inversion-carrier distribution and average inversion-layer depth, as shown in Fig. 4.9 a and b.

The projected difference in average depth between *QM* and classical distributions is $\sim$1.2 nm [12]. At a finite depth of 1.2 nm, the inversion capacitance contributes an additional thickness of

$$\Delta t_{eq} = \frac{\varepsilon_{ox}}{\varepsilon_{Si}} \Delta x \approx \frac{\Delta x}{3} \approx 0.4 \text{ nm} \tag{4.60}$$

to the total equivalent oxide thickness, t_{eq}. The ratio of additional thickness to the equivalent oxide thickness t_{eq} increases as the oxide thickness is reduced below 5 nm. A similar analysis applies to the accumulation layer. The difference in centroid depth between electrons and holes result in a larger capacitance when the quantum-confined surface carriers are electrons than when they are holes [14].

Both the quantum effect shown in Fig. 4.9a and the depletion in a polysilicon gate described in Sect. 4.3.2.4 contribute to an increase in the effective equivalent-oxide thickness obtained from *CV* measurements and referred to as Capacitance Equivalent Thickness, *CET*.

The large perpendicular field leads to significant bending of the bands at the silicon surface. With enough band-bending, carriers can be confined within a sufficiently narrow potential well that their motion perpendicular to the interface becomes quantized. This causes the conduction band to split into discrete energy levels near the surface. The first allowed energy level for electrons in the well no longer coincides with the bottom of the conduction band but is located above the conduction band in a p-substrate. Similarly, the first energy level for holes will be located below the valence band in an n-substrate. This shifts the average location of the carriers beneath the surface. In addition to the shift in the inversion-carrier centroid, quantum-effects cause the bandgap to widen, as illustrated in Fig. 4.10 [8, 11]. An effect of widening of the effective bandgap is the increase in ϕ_b. Therefore, $\psi_s \approx 2\phi_b$ (onset of strong inversion) increases when the band-gap widens and the gate voltage at onset of strong inversion, defined as the threshold voltage V_T, increases. In addition, band-splitting has an impact on the inversion charge density and carrier mobility, which will be discussed in the next chapter.

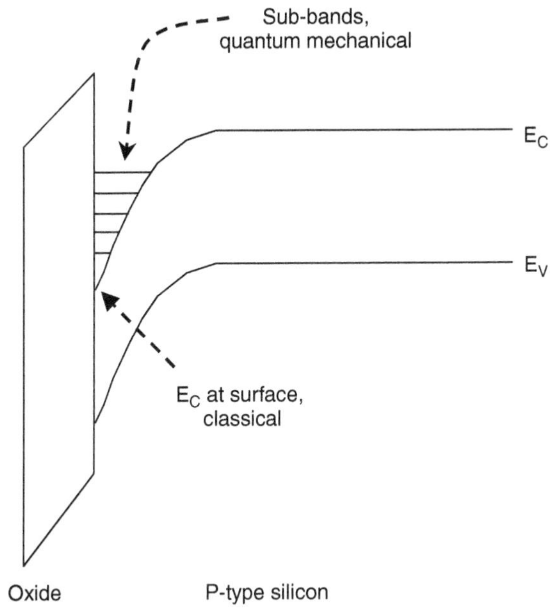

Fig. 4.10 Splitting of the conduction band into sub-bands at high electric field. Inversion electrons are distributed above the bottom of the conduction band, increasing the effective energy gap [8]

4.3.3 Calculation of High-Frequency Capacitance

Since excess minority carriers must be thermally generated (or recombine), they cannot respond to a high-frequency change in voltage. Therefore, at high frequency (HF), only majority carriers are able to follow the signal. For an ideal structure, the calculation of high- and low-frequency *CV* plots are identical in the range from accumulation to just before the onset of strong inversion. Since the concentration of minority carriers is negligible in this range, their inability to follow the high-frequency signal has little impact on the *CV* plot. When $|\psi_s| \geq 2|\phi_b|$, a large excess in minority carriers is required but the carriers cannot relax rapidly enough to follow the signal variation and the equilibrium theory is no longer valid. Consider, for example, a p-type substrate. If the gate voltage is suddenly increased from $V_G = 0$ to $V_G > V_T$ (strong inversion), a positive charge Q_m is induced at the gate-insulator interface. This charge must be neutralized by a negative charge $-Q_m$ (per unit area) in silicon. Since electrons are initially unavailable to establish equilibrium, the depletion region must expand deeper than at equilibrium to "expose" more negatively-charged bulk acceptor ions, Q_b, to satisfy neutrality. Initially,

$$Q_m = -Q_{b-deep}, \tag{4.61}$$

where $Q_{b\text{-}deep}$ is the bulk charge per unit area in deep depletion. If the gate voltage is kept fixed at V_{G-1}, long enough for electrons to be generated, the electron charge Q_n increases and the bulk charge decreases from $Q_{b\text{-}deep}$ to its steady-state value Q_b, as predicted by equilibrium theory. The negative charge is now the sum of bulk and inversion electron charge

$$Q_m = -(Q_b + Q_n). \tag{4.62}$$

The steady-state of (4.62) could have been established more rapidly by shedding a light-pulse on the structure to accelerate the generation of minority carriers.

If now a small high-frequency ac-signal δV_G of typically ± 10–$15\,\text{mV}$ amplitude is superimposed on V_G, electrons are unable to follow the varying signal. The total inversion charge Q_n is only a function of the applied *DC* bias on the gate. In response to the signal, the depletion region expands infinitesimally during the positive half and contracts infinitesimally during the negative half of the cycle so that

$$dQ_m = -dQ_b = qN_A dx_d. \tag{4.63}$$

Noting that

$$C_{Si} = qN_A \frac{dx_d}{d\psi_s}, \tag{4.64}$$

and combining (4.64) with (4.51a) gives the silicon capacitance in (4.50). The total capacitance is then found with (4.28b).

As the gate voltage is increased by another increment and equilibrium is reached again, Q_n increases to a higher steady-state value, but the depletion region does not expand further above its maximum value and C_{Si} remains at a constant, voltage-independent minimum level.

Note that there are two voltages in the high-frequency analysis, a very slow varying *DC* bias voltage to ensure that equilibrium is established at every bias-point in strong inversion, and an ac-signal small enough to ensure that the static condition is not significantly disturbed, and fast enough so that only majority carriers follow the signal.

High- and low-frequency *CV*-plots are shown in Fig. 4.11 for p-type silicon with $N_A = 10^{17}\,\text{cm}^{-3}$ and $t_{ox} = 10\,\text{nm}$. It can be seen that in accumulation, depletion and a large part of weak inversion, the two plots are essentially identical. The minority-carrier response-time determines the difference between the two plots in strong inversion. The *HF*-plot is constructed in two steps. First, the gate-voltage dependent capacitance in accumulation, depletion and weak inversion is calculated in the same procedure as for low-frequency, and then the gate-voltage independent capacitance is calculated in strong inversion.

The two parts are made to join smoothly at a gate voltage that corresponds to $\psi_s \approx 2\phi_b$ where the electron concentration is equal to N_A.

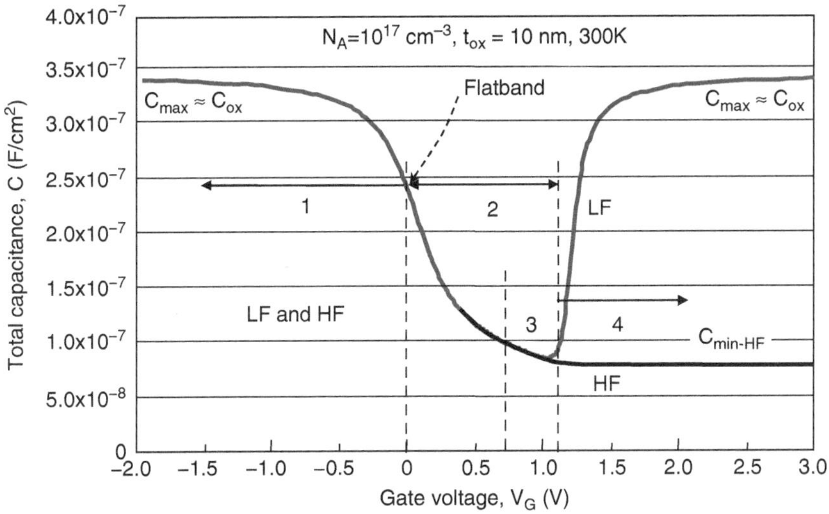

Fig. 4.11 Low-frequency (LF) and high-frequency (HF) plots

4.4 Measurement of *MOS* Capacitance

The *MOS* capacitance versus voltage (CV) plot is traced by sweeping the gate voltage and extracting the instantaneous differential capacitance for each gate voltage. The low- and high-frequency measurements complement each other for the extraction of important process and device parameters.

4.4.1 Low-Frequency, or Quasi-Static CV Measurement

When $\Delta V_G/\Delta t$ is sufficiently small that the rate of minority-carrier generation-recombination can keep up with the varying surface potential, charge-exchange with the inversion layer is in step with the measuring signal. The variation $\Delta Q_s/\Delta t$ in response to $\Delta V_G/\Delta t$ then occurs at the semiconductor surface. Frequencies below ~10 Hz are required to satisfy the low-frequency condition. Low-frequency *CV* plots are most commonly measured with the quasi-static (or triangular) voltage-ramp technique. A voltage-ramp generator applies a time-varying gate voltage of linear rate ***a***, typically 50 mV/s or below. The resulting displacement current through the *MOS* structure is measured with an electrometer. The displacement current is directly proportional to the differential capacitance. The displacement gate current density, j_G, is defined as

$$j_G = \frac{dQ_G}{dt} = \frac{dQ_G}{dV}\frac{dV}{dt} = C\frac{dV}{dt} \quad \mathrm{A/cm^2}. \tag{4.65}$$

For a linear voltage-ramp rate $a = \Delta V_G/\Delta t$, the capacitance can be extracted as

$$C = \frac{j_G}{a} \quad \mathrm{F/cm^2}. \tag{4.66}$$

The *MOS* area must be large enough to ensure that the displacement current can be accurately measured with the ammeter being used.

4.4.2 High-Frequency CV Measurement

To measure the high-frequency *CV*-plot, a small sinusoidal voltage signal of constant amplitude is superimposed on a slow-varying gate dc-voltage. The small-signal amplitude is typically $\pm 15\,\mathrm{mV}$ and the frequency typically $100\,\mathrm{kHz}-1\,\mathrm{MHz}$. The response to the signal gives the differential capacitance. The *MOS* structure is placed in a shielded-light enclosure to avoid distortion of the plot by photon-generation of electron-hole pairs. The voltage sweep can be started at any point, but typically from strong accumulation to strong inversion and in the reverse direction. To ensure equilibrium at every point in strong inversion while varying the gate voltage at a time-efficient sweep-rate, a light source can be briefly turned on to generate minority carriers and then off so that minority carriers in excess of equilibrium can dissipate. A measured high-frequency *CV*-plot is shown in Fig. 4.12 for illustration. The plot is shown for an excursion from accumulation to strong inversion.

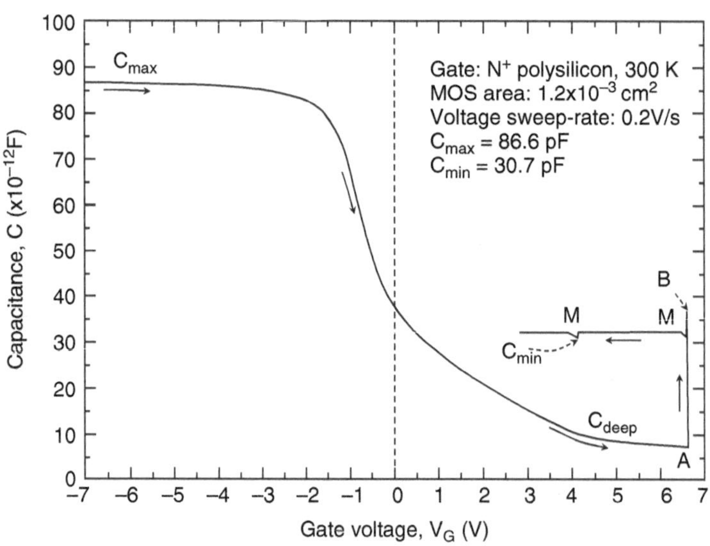

Fig. 4.12 Measured high-frequency *CV*-plot. Light pulse turned-on at point *A*. Capacitance relaxes and overshoots to point *B*. V_G stopped for measurement of C_{min} at points ***M***

At a voltage sweep-rate of 0.2 V/s, minority carriers are not able to follow the signal. The capacitance therefore goes to deep depletion (C_{deep}). After stopping the sweep at point A, the capacitance would have relaxed with time to its equilibrium value at C_{min}. Relaxation is, however, accelerated with a light pulse. Because of the large excess of carriers, the capacitance overshoots to a value higher than C_{min} at point B. It drops to the equilibrium value after turning off the light. As the voltage is swept in reverse direction, the capacitance increases slightly due to the displacement current. When the voltage sweep is stopped again, the capacitance settles at the equilibrium value C_{min}.

4.5 Non-Uniform Impurity Profile

So far, the discussion focused on uniformly-doped substrates. Typical structures are, however, doped by implantation at multiple energies and doses to adjust the threshold voltage at onset of strong inversion plus other sub-surface parameters that will be discussed in the next chapter. The result is a non-uniform profile as illustrated in Fig. 4.13 for a boron doped device.

The properties of *MOS* structures with known non-uniform profiles can be accurately predicted by one-dimensional computer simulations. In many cases, however, approximations can be made to greatly simplify calculations.

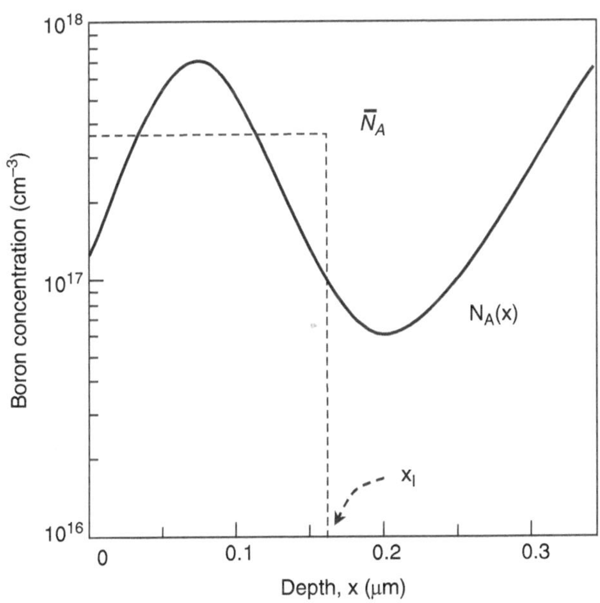

Fig. 4.13 Implanted boron profile on p-substrate, shown after activation anneal, and "box" approximation

4.5.1 Profile Approximations

In the box approximation, the actual profile of Fig. 4.13 would be treated as an equivalent uniform concentration $\bar{N}_A$ to a depth x_I so that $\bar{N}_A.x_I = \phi_I$, where ϕ_I is the threshold-adjust implanted dose. For $\bar{N}_A \approx 3.5 \times 10^{17}\,\text{cm}^{-3}$, the equilibrium depletion depth in inversion is $\approx$57 nm, well inside the box. The relations for accumulation, depletion and inversion derived earlier for a uniform substrate therefore apply.

In the extreme case where the implanted dose is distributed within a very narrow region near the surface of a uniformly-doped substrate of concentration N_A, the bulk charge Q_b in (4.62) is simply replaced by $(Q_b \pm q\phi_I)$, where ϕ_I is the implanted dose. $q\phi_I$ is positive if it is of the same polarity as Q_b, and negative if it is of opposite polarity. The latter is referred to as counter-doping. The analysis is then performed for the uniform concentration N_A, by just adding $q\phi_I$ to the bulk charge and $q\phi_I/\varepsilon_0\varepsilon_{Si}$ to the surface field.

4.5.2 Surface Conditions

Simplified analytical techniques are introduced in this section to more accurately describe surface conditions for a slowly-varying non-uniform profile.

4.5.2.1 Flatband and Accumulation

For a non-uniform profile, the concept of flatband is not strictly valid because the concentration gradient creates a built-in field and hence a non-zero surface potential when $V_G = 0$. For a slowly-varying distribution where the gradient is not too large, "flatband" can be defined as the condition where the overall space-charge in silicon is zero and majority carriers and ionized-impurities exactly balance each other, although they are slightly separated. The impurity concentration at the surface is then used as a reference for accumulation and to calculate the Fermi potential and extrinsic Debye length at flatband.

4.5.2.2 Depletion

In depletion, Poisson's equation must be solved for the non-uniform impurity distribution. Integrating the one-dimensional Poisson equation gives the peak field as [17]

$$E_{peak} = \frac{q}{\varepsilon_0 \varepsilon_{S_i}} \int_0^{x_d} N_A(x)dx + \left.\frac{d\psi}{dx}\right|_{x=x_d} \quad \text{V/cm}, \tag{4.67}$$

where $x = 0$ at the silicon surface, x_d is the depletion depth, $N_A(x)$ the net boron concentration as a function of depth, and $d\psi/dx$ at $x = x_d$ is the slope of the intrinsic level at the depletion boundary. A second integration gives the surface potential as

$$\psi_s = \frac{q}{\varepsilon_0 \varepsilon_{S_i}} \int_0^{x_d} x N_A(x) dx + x_d \left.\frac{d\psi}{dx}\right|_{x=x_d} + \psi(x_d) \quad \text{V}, \tag{4.68a}$$

where $\psi(x_d)$ is the difference in potential between the intrinsic level at the depletion boundary and the bulk intrinsic level. Assuming a uniform concentration in the bulk at $x \geq x_d$,

$$\psi(x_d) = \frac{kT}{q} \ln \frac{N_{Bulk}}{N_A(x_d)}; \left.\frac{d\psi}{dx}\right|_{x=x_d} = -\frac{kT}{q} \frac{1}{N_A(x_d)} \left.\frac{dN_A}{dx}\right|_{x=x_d}. \tag{4.68b}$$

For an implanted profile fully located within the depletion region, the two last terms in (4.68a) become zero.

4.5.2.3 Strong Inversion

For a non-uniform impurity profile, the onset of strong inversion is defined as the condition where the excess minority-carrier concentration at the surface is equal to the majority-carrier concentration at the depletion boundary [17]. Assuming a uniform background concentration N_B, the potential $\psi(x)$ is referred to the bulk intrinsic potential as (Fig. 4.14)

$$\psi(x) = \frac{kT}{q} [E_i(x) - E_{i-bulk}] \quad \text{V}. \tag{4.69}$$

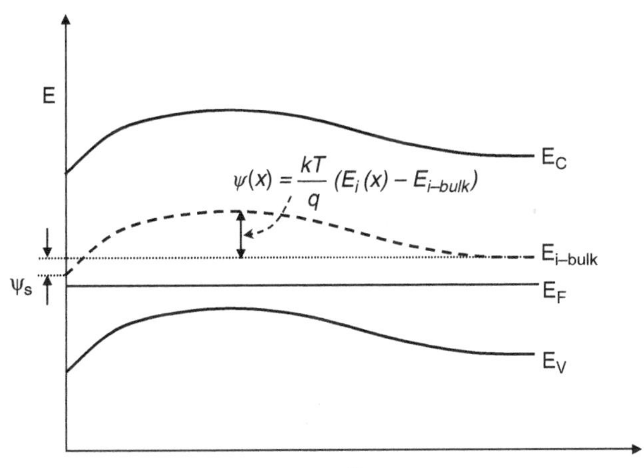

Fig. 4.14 One-dimensional energy band diagram for a non-uniformly doped p-type substrate

With the above definition, the surface potential at inversion is

$$\psi_{inv} = \frac{kT}{q} \ln \frac{N_A(x_{dmax})\, N_B}{n_i^2}. \tag{4.70}$$

The value of x_{dmax} can be obtained from the implicit relation

$$\psi_{inv} = \psi_{s-1} + \psi_{s-2} + \psi_{s-3} \tag{4.71a}$$

where

$$\psi_{s-1} = \frac{q}{\varepsilon_0 \varepsilon_{Si}} \int_0^{x_{d\max}} x N_A(x) dx, \tag{4.71b}$$

$$\psi_{s-2} = \frac{kT}{q} \frac{x_{d\max}}{N_A(x_{d\max})} \left.\frac{dN_A}{dx}\right|_{x=x_{d\max}}, \tag{4.71c}$$

$$\psi_{s-3} = \frac{kT}{q} \ln \frac{N_B}{N_A(x_{d\max})}. \tag{4.71d}$$

Equation (4.71b) takes into account the integrated ionized impurity concentration within the depletion layer. Equations (4.71c) and (4.71d) account for the non-zero electric field and potential at the depletion boundary. Note that (4.71b) can be solved in a closed-form for an exponential, Gaussian, or erfc profile.

4.6 Non-Ideal *MOS* Structure

The discussion so far focused on an ideal structure with zero contact potential between gate and semiconductor, and a perfect non-conducting insulator void of electric charge and traps. In real structures, however, the workfunction of the gate is typically different than that of the semiconductor, resulting in a non-zero contact potential between the two materials. Also, the insulator exhibits imperfections, such as charges and traps of different origins within its bulk and at its interfaces, and current conduction of varying degrees. These non-idealities strongly affect the characteristics of *MOS* and *MOSFET* structures.

4.6.1 Workfunction Difference

The workfunction is defined for n-type and p-type silicon by (4.2)

$$\phi_{Si} = \chi_{Si} + \left(\frac{E_g}{2q} - \phi_b\right) \quad \text{V}.$$

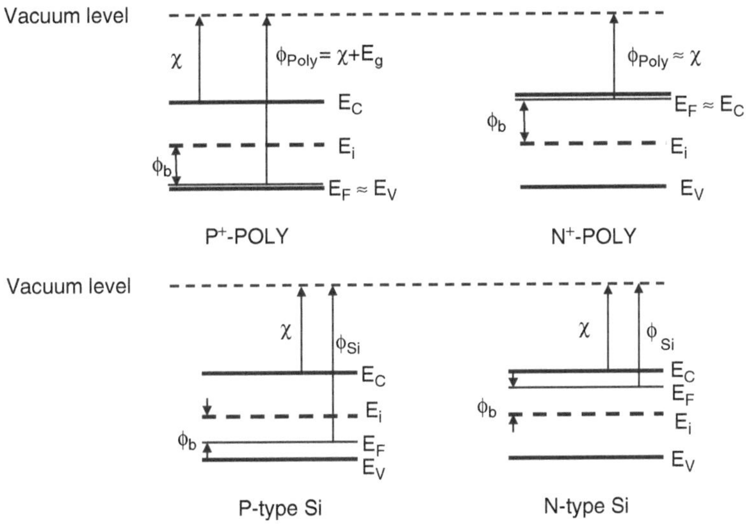

Fig. 4.15 Illustration of polysilicon and silicon workfunction

For a degenerately-doped n^+-polysilicon gate, the Fermi-level approximately coincides with the bottom of the conduction band and the workfunction is essentially the electron affinity of silicon (Fig. 4.15). For a p^+-polysilicon gate, the Fermi-level coincides with the top of the valence band and the workfunction is the sum of electron affinity and E_g/q. The workfunction difference for a degenerately-doped n^+-polysilicon gate and p-type silicon is then (4.3)

$$\phi_{ms} = \frac{E_g}{2q} + |\phi_b|, \tag{4.72a}$$

and for p^+-polysilicon gate on n-type silicon

$$\phi_{ms} = \frac{E_g}{2q} - |\phi_b|. \tag{4.72b}$$

The energy-band diagram is shown in Fig. 4.16 for an *MOS* structure with separated n^+-polysilicon gate, SiO_2 and p-type silicon.

Assuming, for example, $N_A = 10^{17}\,\mathrm{cm}^{-3}$, the silicon workfunction is 5.02 V and the workfunction difference $\phi_{ms} = -0.97\,\mathrm{V}$. When the regions are merged to form an *MOS* structure, thermal equilibrium is established in the three materials by transfer of electrons from polysilicon to silicon until the Fermi-levels align. The silicon sustains a voltage drop due to the charge stored on each side of it. This leaves the polysilicon gate 0.97 V more positive than the silicon bulk and causes the silicon surface to be depleted and the bands to bend as shown in Fig. 4.17.

If the oxide is void of charge, the voltage that must be applied to the gate to establish flatband is the workfunction difference. This voltage is referred to as the flatband

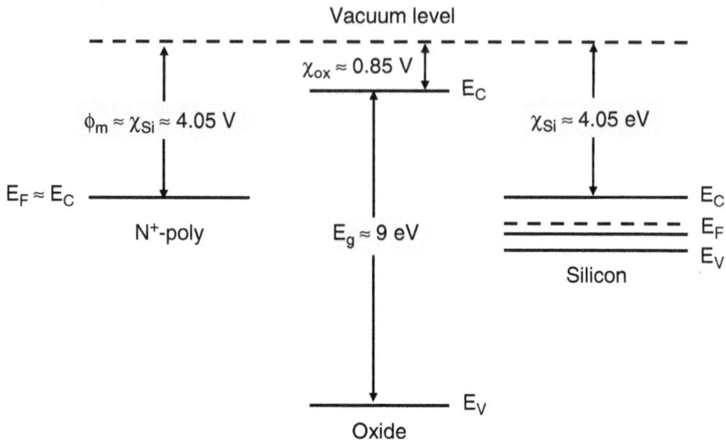

Fig. 4.16 Band diagram for separated n$^+$-poly, SiO_2 and Si

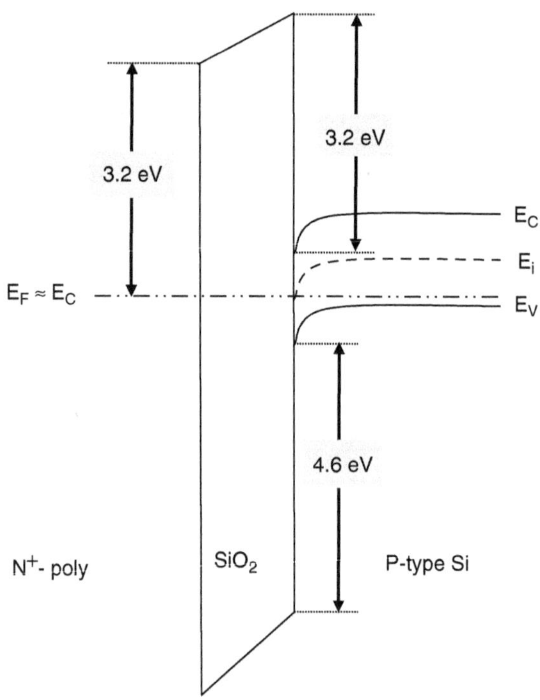

Fig. 4.17 *MOS* band-diagram for n$^+$-poly, oxide and p-type *Si* in equilibrium. Workfunction difference causes band-bending

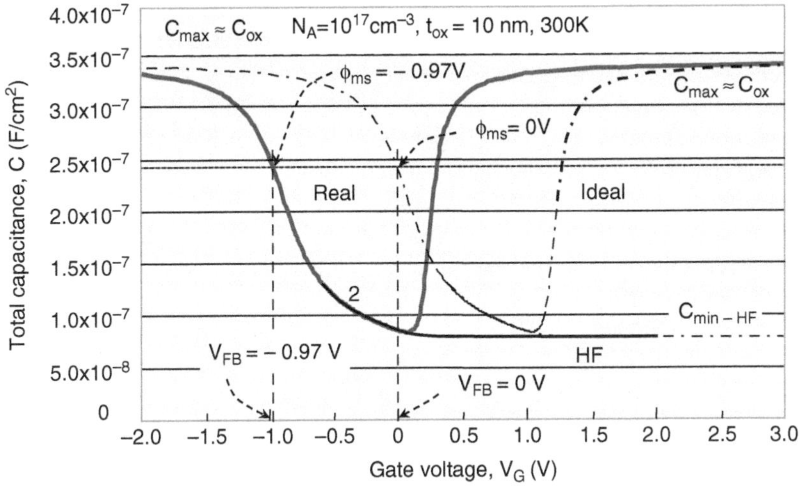

Fig. 4.18 Parallel shift of *MOS CV*-plot by workfunction difference

voltage, V_{FB}. In this case, $V_{FB} = -0.97\,\text{V}$. The workfunction difference alone causes a parallel-shift of $-0.97\,\text{V}$ in the *CV*-plot and hence in threshold voltage, as shown in Fig. 4.18. Note that electrodes other than polysilicon are being developed in conjunction with high-*K* insulators for deep submicron and nanoscale technologies.

4.6.2 Dielectric Charge

Dielectric charges of different origins may be created during processing or as a result of electrical stress. They were first analyzed for silicon-dioxide, the best understood dielectric, and classified as [18]

Q_{it}, interface trapped charge, located at the oxide-silicon interface,
Q_f, fixed charge in the oxide near the silicon surface,
Q_{ot}, oxide trapped charge, located in the bulk of the oxide,
Q_m, mobile charge.

The above nomenclature also applies today, whereby the word "oxide" encompasses other insulators, such as multiple or high-*K* dielectrics.

The relative importance of the charge components depends on process and stress-history. Their presence can cause a shift and, in some cases, distortion of the *CV*-plot and influence device characteristics.

A fifth component Q_{itm} is introduced here to describe the dielectric charge density at its interface with the gate. This component has become important with the introduction of new gate and dielectric materials in new technologies.

4.6.2.1 Interface Trapped Charge, Q_{it}

Interface traps (or surface states) were introduced in Chap. 2 in conjunction with Bardeen's theory on a barrier formed at a metal-semiconductor contact (Sect. 2.3.1). This section discusses traps that are generated and charged at the silicon-oxide interface of an *MOS* structure during processing, or when subjected to electrical stress or radiation. The silicon-oxide interface is a 0.5- to 1-nm thick transition region between single-crystal silicon and the stoichiometric amorphous silicon-dioxide network. The composition of the transition region depends on process conditions and dielectric composition. It is typically silicon-rich but can also be an oxygen-rich silicon-oxide compound, particularly in thermally nitrided SiO_2 [19,20].

Interface states result from the termination of the periodic arrangement of silicon atoms and the discontinuity in the periodic potential. In a simplified model, each surface silicon atom shares only three bonds with adjacent atoms and, in a freshly-cleaved crystal, the fourth bond is left "dangling" creating an allowed electronic state within the forbidden gap. Since the density of silicon atoms at the surface is $\sim 10^{15}\,\mathrm{cm}^{-2}$, one would expect to find the density of interface states in freshly-cleaved silicon to be of the same magnitude. One would also expect that $\{100\}$-oriented surfaces exhibit approximately one-third the density of states of that obtained on $\{111\}$ surfaces since the density of surface atoms is approximately in the same ratio. At such high densities of interface states, typical devices would not operate properly. The unsaturated bonds exhibit strong attraction to foreign impurities, such as oxygen, heavy metals or mobile ions. For example, when exposed to air at room temperature, dangling bonds adsorb oxygen, water, or carbon dioxide, and a layer of silicon dioxide, called "native oxide" is rapidly formed. Interface states created by impurities are sometimes referred to as "slow states" because of their very slow response to changes in potential when compared to electronic states that can respond in μs.

When the surface is oxidized at elevated temperature, most of the dangling bonds are saturated and the interface-state density is reduced considerably. In a defect-free *Si*-SiO_2 interface, the tetrahedral bonding of crystalline silicon atoms is accommodated by bonding with oxygen atoms of the oxide. A small fraction of surface silicon atoms remains, however, not bonded to oxygen atoms. Instead, unpaired electrons are localized on the defect silicon atom forming a hybrid orbital in a direction normal to the (111) plane [21,22]. This "dangling bond," referred to as a P_b center, can be detected by electron-spin resonance (*ESR*). A model for P_b centers is illustrated for (111), (110) and (100) planes in Fig. 4.19 [21,22] There is good agreement between the density of P_b centers measured by *ESR* and the electrically measured trap-density at the *Si*-SiO_2 interface. Other important interfacial defects, such as stretched *Si-O* bonds and stretched *Si-Si* bonds have also been proposed as origins of interface states [23].

Immediately after oxidation, the interface-state density is in the order of $10^{12}\,\mathrm{cm}^{-2}$. It is reduced by over two orders of magnitude after post-metallization anneal at about 450°C in a hydrogen-containing atmosphere. The interface-state density increases when the surface is subjected to any bond breaking processes,

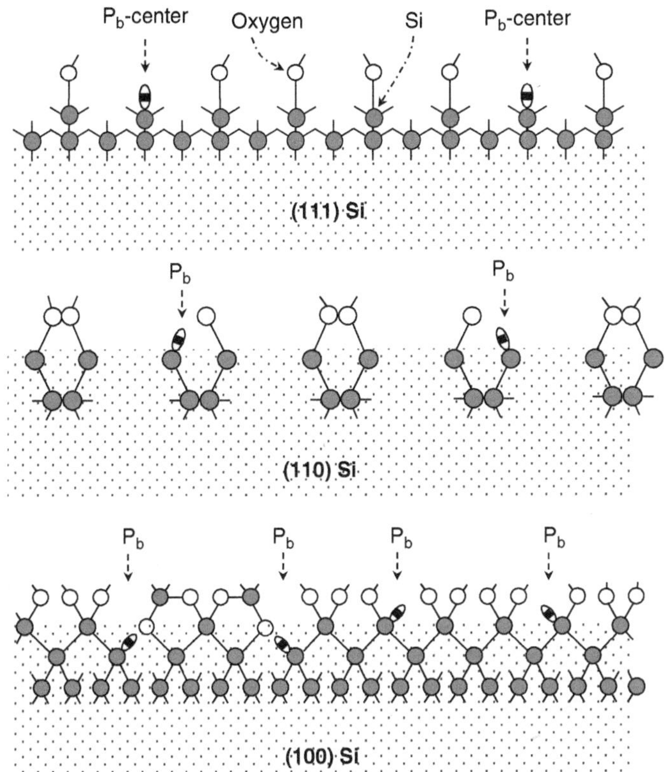

Fig. 4.19 Illustration of dangling bond model for P_b center on (111), (110), and (100) silicon surface [21, 22]

such as plasma processing (Chap. 7), hot-carrier injection or Fowler-Nordheim (*FN*) tunneling write/erase in EEPROMs (Chap. 8), electrical stress, or high-energy radiation. *FN* tunneling is discussed in Sect. 4.8.1.

Interface states can be of donor or acceptor type and are typically distributed continuously in energy within the bandgap, as schematically illustrated in Fig. 2.48. Above the Fermi-level, a donor level is positively charged while an acceptor level is neutral. Similarly, below the Fermi-level a donor level is neutral while an acceptor level is negatively charged. The net interface charge Q_{it} depends therefore on the position of the Fermi-level in the bandgap. A large fraction of Q_{it} can be neutralized by low-temperature (400–450 °C) annealing in hydrogen or forming gas, for example, 10% hydrogen and 90% nitrogen. Processes that break the hydrogen bonds and cause migration of hydrogen away from the interface tend to increase *Qit* [24].

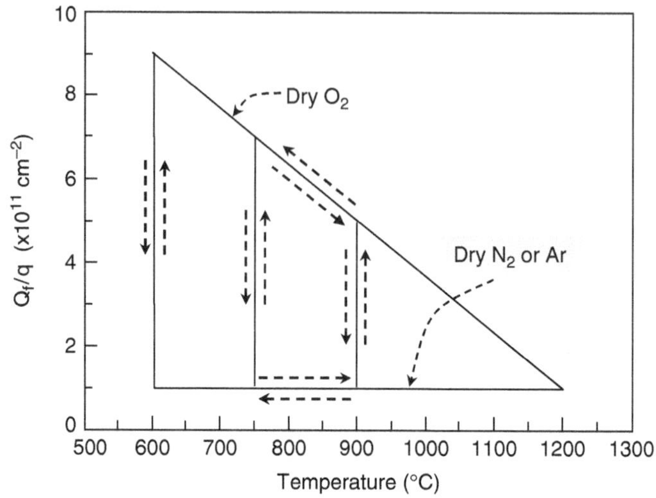

Fig. 4.20 The "Deal triangle." Annealing behavior of oxide fixed charge as a function of temperature and ambient atmosphere. Arrow show reversible paths [25]

4.6.2.2 Fixed Oxide Charge, Q_f

Fixed oxide charge is a predominantly positive charge located in the oxide less than 1 nm from the Si-SiO_2 interface. Unlike interface states, fixed oxide charge centers do not change their state by communicating with the underlying silicon. Their density depends on oxidation temperature and ambient atmosphere and decreases with increasing final oxidation temperature. Q_f can also be lowered by annealing in a nitrogen or argon atmosphere after oxidation [25]. The well-known "Deal-triangle" in Fig. 4.20 shows the annealing behavior of fixed oxide charge [25].

Oxidation at 1200°C yields the lowest Q_f. If, however, the processing requirements do not allow oxidation at such a high temperature, a post-oxidation anneal at a low-temperature of ~450°C in dry nitrogen or argon reduces the density of fixed charged $N_f = Q_f/q$ to its lowest value of $5 \times 10^{10}\,\text{cm}^{-2}$. The processes are reversible. For example, a first oxidation can be performed at 600°C, resulting in $N_f = 9 \times 10^{11}\,\text{cm}^{-2}$, then followed by oxidation at 900°C where N_f drops to $\sim 5 \times 10^{11}\,\text{cm}^{-2}$. Subsequently annealing at ~450°C in dry nitrogen or argon further lowers N_f to $5 \times 10^{10}\,\text{cm}^{-2}$.

4.6.2.3 Oxide Trapped Charge, Q_{ot}

This charge is associated with oxide defects creating traps within the insulator. The traps are initially neutral and can be charged by ionizing radiation or by injecting electrons or holes into the insulator. Q_{ot} can therefore be positive or negative. Injection can occur by Fowler-Nordheim tunneling from either the substrate or the

Table 4.2 Average E_A and μ_0 for Na^+, K^+, and Li^+ in SiO_2 [30]

Metal ion	μ_0 (cm^2/s)	E_A(eV)
Na^+	3.52×10^{-4}	0.44
K^+	4.5×10^{-4}	0.47
Li^+	2.5×10^{-3}	1.04

gate or by avalanche injection of hot-carriers from the substrate (Sect. 4.8.2). Oxide charge can be partially annealed at temperatures as low as 500°C, but neutral traps may not be eliminated.

4.6.2.4 Mobile Charge, Q_m

Mobile ions in the oxide cause instabilities in the *MOS* structure [25–29]. The dominant mobile ion in silicon-dioxide has been found to be sodium. Potassium and lithium may also present, but to a smaller extent [30, 31]. Because they are metallic, these ions are positive in SiO_2. At low concentrations, the drift mobilities of sodium, potassium and lithium ions follow the Arrhenius relation

$$\mu = \mu_0 e^{-E_A/kT}. \tag{4.73}$$

Table 4.2 summarizes the values for E_A and μ_0 for sodium, potassium, and lithium [30].

For an aluminum gate, the drift of sodium is asymmetrical. Under bias-temperature stress, the ions move faster from the silicon-oxide interface to the metal-oxide interface than in the opposite direction. This is attributed to the higher energy-barrier to sodium migration from the metal-oxide interface. The asymmetry is not observed with polysilicon gates.

4.6.2.5 Effective Oxide Charge, Q_{eff}

Assume initially that all insulator charges can be lumped into one charge-sheet of density $\sigma(\mathrm{C/cm^2})$, located at the silicon-oxide interface. A space-charge Q_s of opposite polarity is formed in silicon with $Q_s = -\sigma$. The combined effect of all charges is to shift the *CV*-plot and hence the flatband voltage by

$$\Delta V_{FB} = -\frac{Q_{it} + Q_{ot} + Q_f + Q_m}{C_{ox}} = -\frac{\sigma}{C_{ox}} \quad \mathrm{V}, \tag{4.74}$$

where σ is the algebraic sum of all charges. ΔV_{FB} is negative for a net positive charge density and positive for a net negative charge density. The flatband voltage is then

$$V_{FB} = \phi_{ms} - \frac{\sigma}{C_{ox}} \quad \mathrm{V}. \tag{4.75}$$

The threshold voltage (gate voltage where $\psi_s = -2\phi_b$) is defined as

$$V_T = -\frac{Q_b}{C_{ox}} \pm \psi_s + V_{FB} \quad \text{V}. \tag{4.76}$$

Q_b is the concentration of ionized charge per unit area within the depletion region, defined for p-type silicon as

$$Q_b = \sqrt{2\varepsilon_0 \varepsilon_{Si} q N_A \psi_S} \quad \text{C/cm}^2.$$

The surface potential ψ_s is positive for p-type and negative for n-type silicon at onset of strong inversion. A positive charge at the silicon-oxide interface therefore causes a negative shift in V_T (Fig. 4.21).

For a charge-sheet σ located at a distance x_1 from the silicon surface and x_2 from the gate-insulator interface, the image charge of σ is shared between gate and silicon (Fig. 4.22). Let Q_s and Q_m be, respectively, the charge imaged on the silicon and gate, then

$$Q_s + Q_m = -\sigma \quad \text{C/cm}^2, \tag{4.77}$$

and the lever rule gives

$$Q_s x_1 + Q_m x_2 = (Q_s - \sigma)x_2 \text{ or} \tag{4.78}$$

$$Q_s = -\frac{x_2}{x_1 + x_2}\sigma = -\frac{x_2}{t_{ox}}\sigma \quad \text{C/cm}^2, \tag{4.79}$$

where t_{ox} is the oxide thickness.

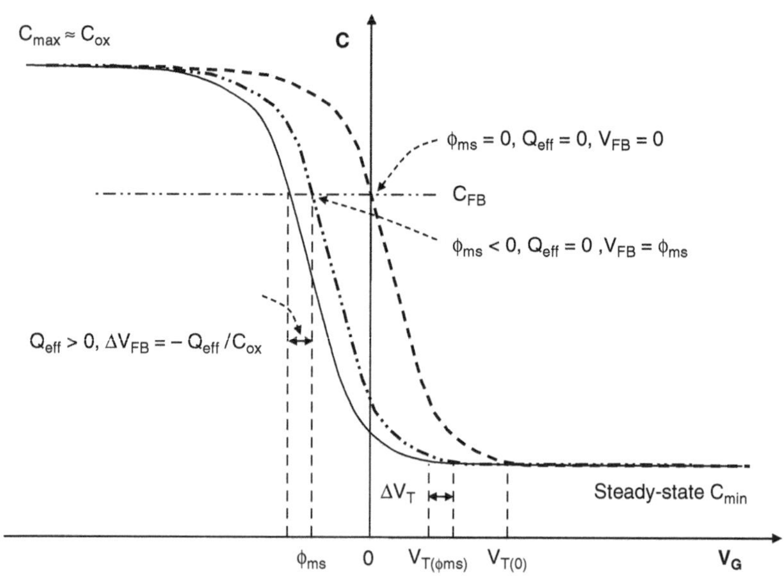

Fig. 4.21 Shift in *CV*-plot caused by oxide charge. For parallel shift, $\Delta V_T = \Delta V_{FB}$

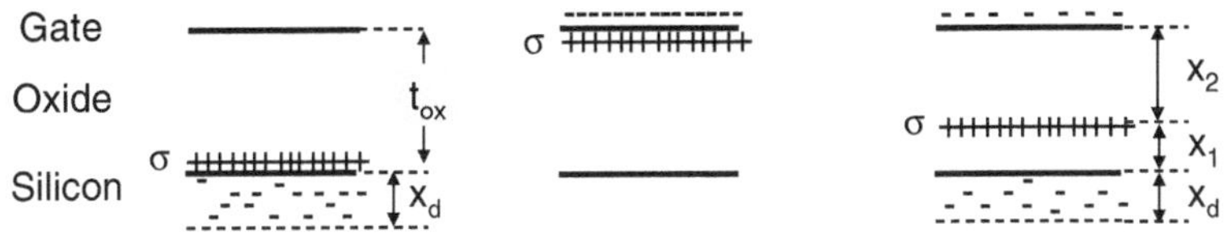

Fig. 4.22 Charge-sheet σ approximations, illustrated for positive charge on p-type silicon.

The effect of an arbitrary distribution of charge $\rho(x)$ within the insulator can be evaluated by integrating the differential charge sheets $dQ(x) = \rho(x)dx$ as

$$Q_s = -\int_0^{t_{ox}} \frac{x}{t_{ox}}\rho(x)dx \quad \mathrm{C/cm^2}, \tag{4.80}$$

where x is the distance from the metal-insulator interface. It is the effective insulator charge $Q_{eff} = -Q_s$ that causes the measured shift in flatband voltage

$$\Delta V_{FB} = -\frac{Q_{eff}}{C_{ox}} = \frac{-\int_0^{t_{ox}}(x/t_{ox})\rho(x)dx}{C_{ox}} \quad \mathrm{V}. \tag{4.81}$$

Since the distribution of insulator charge and location of its centroid is typically not known a priori, the effective charge cannot be calculated. It is, however, extracted from the measured shift in V_{FB}.

It should be noted that insulator charges only cause a parallel shift in the *CV*-plot when the charge is independent of gate voltage. While Q_f is assumed to be gate-voltage independent, Q_{it}, Q_m and Q_{ot} can vary with applied gate voltage, as discussed in the following section.

4.7 Characterization and Parameter Extraction

The *MOS* structure is a powerful tool that is extensively used to extract important process and *MOSFET* parameters [32, 33].

4.7.1 Extraction of Equivalent Oxide Thickness, t_{eq}

In accumulation, the measured capacitance of an *MOS* structure with a polysilicon gate increases asymptotically toward a maximum value C_{max} that consists of three capacitances in series: the physical equivalent oxide capacitance, C_{ox}, the silicon capacitance C_{Si}, and the polysilicon capacitance C_{poly}, where

$$C_{\max} = \left[\frac{1}{C_{ox}} + \frac{1}{C_{Si}} + \frac{1}{C_{poly}}\right]^{-1} \quad \text{F/cm}^2. \tag{4.82}$$

For an oxide dielectric, the physical oxide thickness, $t_{ox\text{-}phys}$, can be directly measured by ellipsometry or high-resolution transmission-electron microscopy (*HRTEM*) [34]. Typically, the Capacitance Equivalent Thickness, *CET*, is larger than $t_{ox\text{-}phys}$ and $C_{max} < C_{ox}$ because of quantum-effects discussed in Sect. 4.3.2.6. For p-type silicon and n-type polysilicon gate, both the gate and silicon are accumulated when the structure is biased in accumulation. The same applies to p-type polysilicon and n-type silicon. If $t_{ox\text{-}phys}$ is considerably larger than the depth of the carrier centroid in both accumulated layers, then $C_{Si} >> C_{ox}$ and $C_{poly} >> C_{ox}$. Thus, the error introduced by approximating $C_{max} \approx C_{ox}$ is not appreciable. In this case, both the high-frequency and low-frequency *CV*-plots give the same value in accumulation

$$CET = \frac{\varepsilon_0 \varepsilon_{ox}}{C_{max}} \approx \frac{\varepsilon_0 \varepsilon_{ox}}{C_{ox}} \quad \text{cm.} \tag{4.83}$$

As the dielectric thickness decreases below about 10 nm, however, the approximation in (4.83) becomes increasingly inaccurate because of the quantum effects in silicon and polysilicon. For $t_{ox\text{-}phys} = 10\,\text{nm}$, the *CET* is about 10% larger than $t_{ox\text{-}phys}$ [32].

The oxide thickness can also be extracted from he quasi-static ("low-frequency") plot in inversion. In this range, the polysilicon gate is depleted when silicon is inverted and the discrepancy increases because of polysilicon depletion. An additional error is introduced if the accumulating field causes appreciable tunneling current through the oxide [35] (Sect. 4.8.1). Polysilicon depletion and quantum-mechanical effects can be simulated and the results used to extract the *CET* [36,37]. Algorithms have been developed, however, to extract the physical oxide thickness without the need for detailed simulations [38,39]. It should be noted that the insulator dielectric "constant" that was so far assumed to be constant can vary with the gate material and its interactions with the dielectric. This is observed, for example, with a tungsten-silicided polysilicon gate on oxide [40]. Changes in dielectric properties are more pronounced with metal gates and high-*K* dielectrics, as will be discussed in Chap. 5.

4.7.2 Workfunction Difference

Equation (4.75) provides a means for separately extracting the gate workfunction from a plot of the flatband voltage as a function of oxide thickness [41]. Provided that the samples are prepared under conditions such that the effective oxide charge is independent of oxide thickness and remains constant during measurement, the plot is then linear with a slope of $-Q_{eff}/\varepsilon_{ox}$. The intercept with the vertical axis is the workfunction difference ϕ_{ms} (Fig. 4.23). One method to fabricate the samples is to grow a thick oxide on the wafer and then gradually thin the oxide by etching to produce the desired samples of different oxide thickness onto which the gate material is deposited

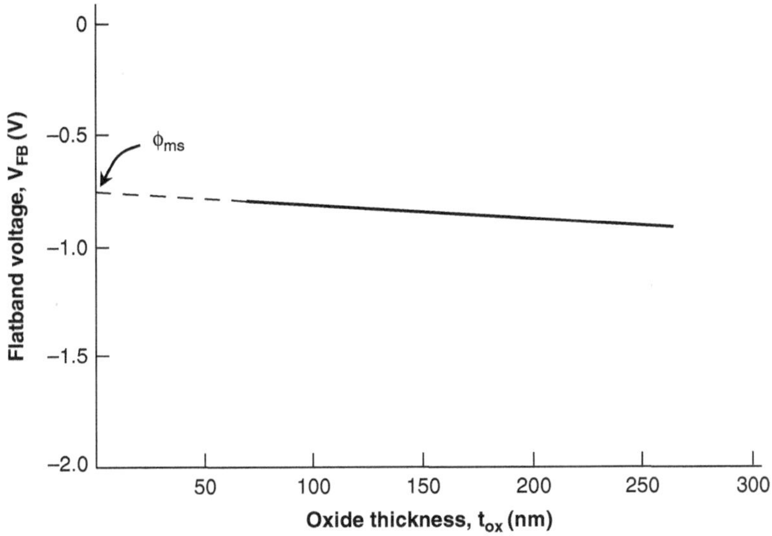

Fig. 4.23 Extraction of work function difference from V_{FB} measurements as a function t_{ox} (Adapted from [40])

4.7.2.1 Fermi-Level Pinning

So far, it was assumed that the interface-state charge density Q_{it} resides totally at the silicon surface. Some combinations of gate and dielectric materials, however, appear to exhibit a high density of electronic states at the gate-dielectric interface, denoted here as Q_{itm}. This has the tendency to pin the gate Fermi level near the charge-neutrality level so that the apparent work function of the gate in contact with the dielectric is fixed and differs from its value in vacuum [42, 43]. Fermi-level pinning was first discussed in Chap. 2 to explain the weak dependence of Schottky-barrier height on metal workfunction (Fig. 2.49). A similar effect is observed with advanced gate-stacks that incorporate high-*K* metal-oxide dielectrics, such as hafnium oxide or aluminum oxide. The high density of interface states is related to the interaction between gate and dielectric materials. Tailoring the workfunction difference is one desired means of adjusting the *NMOS* and *PMOS* threshold voltages without modifying the silicon dopant concentration. Fermi-level pinning practically eliminates this flexibility. Fermi-level pinning is further discussed in Chap. 5.

4.7.3 Extraction of Dopant Concentration

The ionized impurity concentration can be obtained from high-frequency *CV* measurements in depletion.

4.7.3.1 Uniformly Doped Silicon

For uniformly-doped silicon, the extraction of dopant concentration from the high-frequency *CV*-plot is straightforward. The minimum capacitance $C_{Si\text{-}min}$ is found from the maximum and minimum measured capacitances, C_{ox} and C_{min}

$$C_{Si-min} = \frac{C_{ox}C_{min}}{C_{ox} - C_{max}} \quad \mathrm{F/cm^2}. \tag{4.84}$$

Once $C_{Si\text{-}min}$ is known, the dopant concentration can be found from (4.50) and (4.51).

4.7.3.2 NonUniform Profile

For a slow-varying non-uniform profile, the above procedure results in an effective dopant concentration that can be used for process monitoring and to approximate the Debye length at flatband. The ionized-impurity profile can be obtained from the slope of $1/C^2$ versus V_G in depletion, referred to as the inverse-square-capacitance method. Assuming p-type silicon, an incremental charge on the gate induces a charge dQ_s in silicon where, with the depletion approximation

$$dQ_s = -dQ_m = -CdV_G = qN_A(x_d)dx_d \quad C/\mathrm{cm}^2. \tag{4.85}$$

From

$$\frac{1}{C_{Si}} = \frac{x_d}{\varepsilon_0\varepsilon_{Si}} = \frac{1}{C} - \frac{1}{C_{ox}}, \tag{4.86}$$

$$dx_d = \varepsilon_0\varepsilon_{Si} d\left(\frac{1}{C_{Si}}\right) = \varepsilon_0\varepsilon_{Si} d\left(\frac{1}{C_{ox}} + \frac{1}{C_{Si}}\right) = \varepsilon_0\varepsilon_{Si} d\left(\frac{1}{C}\right). \tag{4.87}$$

Substituting in (4.85) gives

$$N_A(x_d) = -\left[q\varepsilon_0\varepsilon_{Si}\frac{d}{dV_G}\left(\frac{1}{C}\right)\right]^{-1}, \tag{4.88}$$

or

$$N_A(x_d) = -\frac{2}{q\varepsilon_0\varepsilon_{Si}}\frac{dV_G}{d(1/C)^2}. \tag{4.89}$$

The above relation shows that at any point, the ionized impurity concentration is inversely proportional to the slope $dV_G/d(1/C^2)$. For acceptors, the slope is positive. The depth is found from (4.85).

4.7.3.3 Limitations of the Technique

If the concentration changes appreciably within a distance comparable to the extrinsic Debye length L_D, the value obtained from (4.89) will be an average of the true impurity profile over the extrinsic Debye length. This is because (4.89) actually measures the majority carrier concentration that is assumed to be approximately equal to the ionized-impurity concentration. Since the variation of majority carrier concentration occurs within a few Debye lengths, changes in dopant concentration over a distance comparable to L_D are not resolved. Equation (4.89) yields a good approximation of dopant profile for x_d larger than approximately $3L_D$. Below this value, majority-carriers begin to affect the results and correction factors must be introduced to extend the profile to the surface.

For a slowly varying gate voltage, the maximum profile depth is limited by inversion carriers. The depth can, however, be extended into deep depletion by pulsing the gate and measuring the capacitance during the pulse, as described in the next section. The maximum depth in this case is limited by avalanche breakdown or tunneling.

4.7.3.4 Pulsed CV Profiling

The pulsed *CV* technique extends the range of concentration versus depth beyond the depth at steady state. The method uses voltage pulses of short duration during which negligible minority carriers can be generated, allowing the structure to go into deep depletion without the interference of minority carriers. The high-frequency capacitance is measured during each pulse. The voltage is typically pulsed from flatband to depletion with incrementally increasing voltage amplitudes. Typical pulse widths range from 1 to 10 ms [32, 33].

4.7.4 Lifetime Measurements

When the gate voltage is pulsed from an initial value V_{G1}, which is chosen at flatband for convenience, to V_{G2} in strong inversion, minority-carriers cannot be instantaneously supplied to establish equilibrium. The structure then goes into deep depletion to "expose" more ions and satisfy neutrality (Fig. 4.24a).

In the absence of light there are two mechanisms for supplying minority carriers to the inversion layer of an *MOS* structure; (a) Thermal generation within the depletion region x_d and drift to the silicon surface; (b) Thermal generation outside the depletion region at a distance of approximately one diffusion length from x_d, followed by diffusion to the depletion boundary and drift through the depletion layer to the surface. For each generated pair, a minority carrier is provided as an inversion carrier to the surface and a majority carrier neutralizes one ionized impurity. At time t_0 in Fig. 4.24a, V_{G2} is kept constant and the depletion depth begins to relax with time from deep-depletion to the equilibrium condition at time t_∞ as minority carriers are provided. Relaxation is best described by a measuring the capacitance versus time, C-t, while keeping V_{G2} constant, as illustrated in Fig. 4.24b [5, 32, 44, 45].

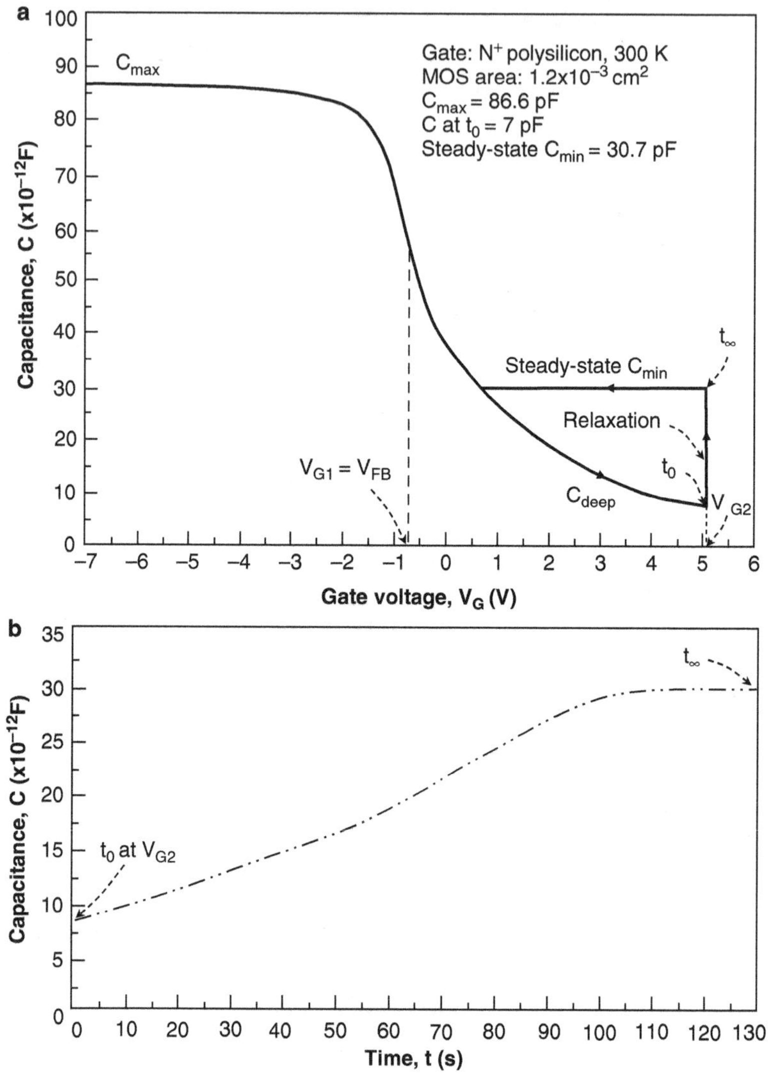

Fig. 4.24 **a** *CV*-plot in deep depletion. Voltage pulsed from V_{G1} to V_{G2}. Capacitance relaxes to equilibrium at constant V_{G2}. **b** *C-t* plot in relaxation. Capacitance relaxes from deep depletion at t_0 (V_{G2} in a) to equilibrium at t_∞ as minority carriers are generated

For p-type silicon, the following relation holds at any time

$$C_{ox}\left[V_G - \psi(t)\right] = -q\left[N_n(t) + \int_0^{x_{d(t)}} N_A(x)dx\right], \tag{4.90}$$

where N_n is the inversion-electron density per unit area. Since V_G is set constant at V_{G2}, differentiating the above relation and rearranging gives

$$\frac{dN_n}{dt} = \frac{C_{ox}}{q}\frac{d\psi_s}{dt} - N_A\left[x_d(t)\right]\frac{dx_d}{dt}. \tag{4.91}$$

The depletion width is related to capacitance through

$$x_d(t) = \frac{\varepsilon_0\varepsilon_{si}(C_{ox} - C_{HF})}{C_{ox}.C_{HF}}, \tag{4.92}$$

where C_{HF} is the high-frequency capacitance in deep depletion. With the depletion approximation, the change of surface potential with time is

$$\frac{d\psi(t)}{dt} = -\frac{qx_d(t)}{\varepsilon_0\varepsilon_{Si}}N_A\left[x_d(t)\right]\frac{dx_d}{dt}. \tag{4.93}$$

Assuming an effective minority-carrier lifetime τ, the rate of increase in inversion-carrier density N_n can be approximated by

$$\frac{dN_n}{dt} = -\frac{n_i\left[x_d(t) - x_{d-inv}\right]}{\tau}, \tag{4.94}$$

where $x_{d\text{-}inv}$ is the steady-state depletion layer depth in inversion that is approached at $t = \infty$. Combining the above equations and using the identity

$$\frac{2}{C_{HF}^3}\frac{dC_{HF}}{dt} = -\frac{d}{dt}\left(\frac{1}{C_{HF}}\right)^2$$

gives the so-called Zerbst relation for uniform N_A

$$-\frac{d}{dt}\left(\frac{C_{ox}}{C_{HF}}\right)^2 \approx -\frac{2n_i}{\tau N_A}\frac{C_{ox}}{C_{inv}}\left(\frac{C_{inv}}{C_{HF}} - 1\right), \tag{4.95}$$

where C_{inv} is the steady-sate inversion capacitance. A plot of the term on the left of (4.95) versus $(C_{inv}/C_{HF} - 1)$ can be approximated by a straight line. The lifetime is obtained from the slope of the straight line [44]. Variants of the above relation are discussed in [32, 45].

A long relaxation time indicates a long lifetime and hence a low density of generation sites, such as heavy metals. The measurement can be used to routinely monitor silicon quality.

4.7.5 Extraction of Interface-State Distribution

Interface traps are represented by a charge Q_{it} associated with charging and discharging traps within the bandgap. The charged-state of a trap depends on whether it is of acceptor or donor type, and on the energy-level of the trap with respect to the surface Fermi-level. Typically, both donor and acceptor levels exist. The position of trap-levels with respect to the Fermi-level and their trap occupancy changes with applied gate voltage. Therefore, as the gate voltage is swept, the Fermi-level "scans" the surface and the trap-distribution above and below the Fermi-level changes, modifying Q_{it}. Several methods have been implemented to extract the trap density and distribution from *MOS* measurements. Among these are the high-frequency capacitance method [1], the combined high- and low-frequency method [46, 47], and the conductance method [48]. Only the first two methods are discussed in this section.

4.7.5.1 High-Frequency Capacitance Method

In the high-frequency method, also referred to as the Terman method [1], the frequency is chosen sufficiently high so that interface traps are not charged or discharged by the signal. Therefore, the capacitance due to interface traps can be assumed to be zero. The voltage sweep, however, changes the position of interface-traps with respect to the Fermi-level and hence the charged state of traps. This modifies the effective charge Q_{eff} seen by silicon and the flatband voltage varies as the V_G is swept, stretching out the plot along the voltage axis (Fig. 4.25, dotted lines). For a uniform distribution of interface traps within the bandgap, the stretch-out is gradual, causing a nonparallel shift of the plot. The stretch-out is distorted by a non-uniform distribution of states in a manner that can be used to measure lateral non-uniformitics of Q_{it} along a MOSFET channel [49]. The relation between the silicon capacitance C_{Si} and surface potential ψ_s is known for a given dopant concentration. For a given oxide thickness, a relation between ψ_s and V_G can be found and a *CV*-plot with $Q_{it} = 0$ established (solid curve in Fig. 4.25). Since C depends only on C_{Si} (ψ_s) and oxide thickness, the same capacitance on all plots in Fig. 4.25 correspond to the same surface potential but not to the same gate voltage. The shift

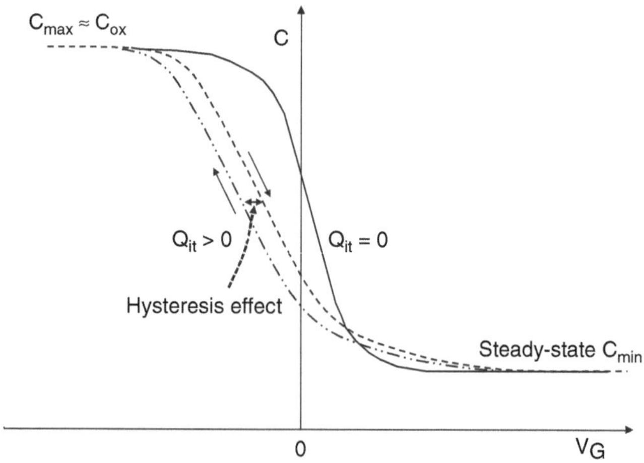

Fig. 4.25 High-frequency plots for $Q_{it} = 0$ and $Q_{it} > 0$. Stretch-out is a measure of surface state density

in gate voltage is caused by a change in Q_{it}. If a ψ_s-V_G plot is generated for a curve with $Q_{it} > 0$ and compared to ψ_s-V_G for the theoretical curve, a difference ΔV_G is found between the two curves at the same surface potential. The interface state density can then be approximated as [5, 32]

$$D_{it} \approx \frac{C_{ox}}{q} \frac{d(\Delta V_G)}{d\psi_s} \quad \text{cm}^{-2}. \tag{4.96}$$

The high-frequency method is considered to be only useful for measuring interface trap densities above $10^{10}\,\text{cm}^{-2}$ and above, mainly because of inaccuracies in capacitance measurements [32]. Other techniques that use a gate-controlled pn junction or a *MOSFET* to measure the interface trap density are described in Chap. 5.

There are cases where sweeping from accumulation to inversion yields a considerably different *CV* plot than sweeping from inversion to accumulation, or vice versa. This "hysteresis" effect is illustrated in Fig. 4.25 with the two arrows indicating the sweep direction. It is observed with or without the "stretching" of the *CV* plot, particularly with new high-*K* dielectric materials [50, 51]. It is typically seen when the field reaches a certain magnitude in each polarity, causing charge-trapping or detrapping within the dielectric and hence a shift in the flatband voltage. The magnitude of hysteresis depends on energy and density of dielectric traps, and dielectric history. For example, a hysteresis of 100–150 mV is observed on HfO_2 annealed in N_2 when V_G was swept ± 3 V, but was negligible when V_G was swept ± 1 V [50] (Chap. 5).

4.7.5.2 Combined High- and Low-Frequency Method [46, 47]

An equivalent circuit of the oxide, interface, and silicon capacitances is shown in the insert of Fig. 4.26. At high frequency, the interface capacitance C_{it} is zero since the traps do not follow the signal, so the measured capacitance is

$$C_{HF} = \frac{C_{ox}\, C_{Si}}{C_{ox} + C_{Si}} \quad \text{F/cm}^2. \tag{4.97}$$

At low frequency, the traps can follow the signal and a capacitance associated with charging and discharging the traps is added in parallel to the silicon capacitance in (4.97)

$$C_{LF} = \frac{C_{ox}\,(C_{Si} + C_{it})}{C_{ox} + C_{Si} + C_{it}} \quad \text{F/cm}^2, \tag{4.98}$$

where C_{LF} is the low-frequency capacitance. Combining the above relations gives

$$C_{it} = \frac{C_{ox}\, C_{LF}}{C_{ox} - C_{LF}} - \frac{C_{ox}\, C_{HF}}{C_{ox} - C_{HF}} \quad \text{F/cm}^2. \tag{4.99}$$

The interface trap-density is related to the interface capacitance by

$$C_{it} = qD_{it}(\psi_s) = \frac{dQ_{it}}{d\psi_s}, \tag{4.100}$$

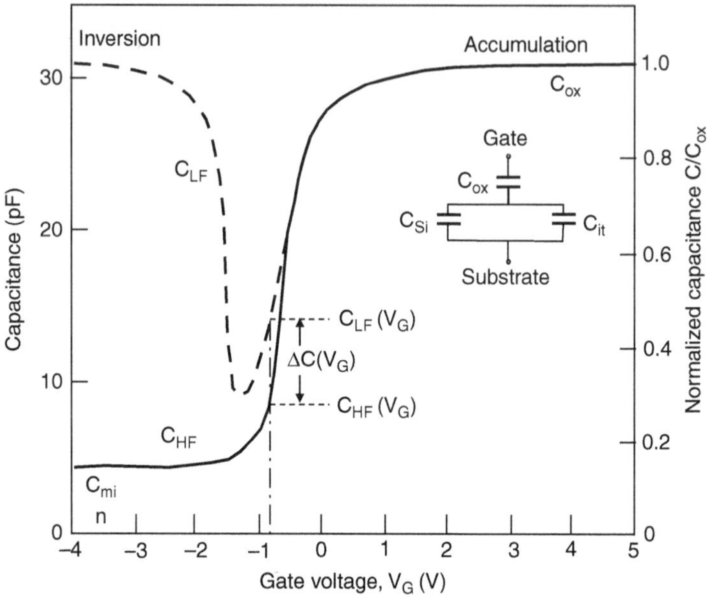

Fig. 4.26 High- and low-frequency *CV*-plots showing the difference $\Delta C(\mathrm{V_G})$ caused by interface traps (Adapted from [46])

where D_{it} is the interface trap density in cm^{-2}.(eV)$^{-1}$ given by [44]

$$D_{it} = \frac{C_{ox}}{q}\left[\frac{C_{LF}/C_{ox}}{1-C_{LF}/C_{ox}} - \frac{C_{HF}/C_{ox}}{1-C_{HF}/C_{ox}}\right] \quad \mathrm{cm^{-2}eV^{-1}}. \tag{4.101}$$

A typical combination of low- and high-frequency plots is shown for n-type silicon in Fig. 4.26.

The interface trap density is determined directly from low- and high-frequency *CV* measurements.

The relation between surfaced potential and gate voltage can be found following a procedure in [52]. The low-frequency capacitance is

$$C(V_G) = \frac{dQ}{dV_G} = C_{ox}\frac{dV_{ox}}{dV_G}. \tag{4.102}$$

Since $dV_G = dV_{ox} + d\psi_s$, (4.102) can be written as

$$C(V_G) = C_{ox}\left(1 - \frac{d\psi_s}{dV_G}\right), \tag{4.103a}$$

or

$$\frac{d\psi_s}{dV_G} = 1 - \frac{C(V_G)}{C_{ox}}. \tag{4.103b}$$

Integrating 4.103b with the flatband as the reference gives the surface potential as

$$\psi_s(V_G) - \psi_s(V_{FB}) = \int_{V_{FB}}^{V_G} \left[1 - \frac{C(V_G)}{C_{ox}}\right] dV_G. \tag{4.104}$$

4.7.6 Extraction of Mobile Ion Concentration

The most common mobile ions are alkali ions, such as Na^+, K^+, and Li^+. They can be moved though the oxide by applying a gate voltage at elevated temperature, typically in the range 200–400 °C. When the gate voltage is positive, the ions drift to the oxide-silicon interface and when the gate is negative they drift back to the gate-oxide interface. Their existence can be detected by a parallel shift in the *CV*-plot or by a peak in the quasi-static $I-V$ plot.

4.7.6.1 Extraction from the Shift in V_{FB}

The shift in the *CV*-plot is determined from the change in V_{FB} when the location of the mobile-charge centroid changes. The measurement is done by biasing the gate at elevated temperature for some time (typically 10–30 min) to allow the ions to move totally to one electrode, cooling the structure with the voltage applied to the gate, and then measuring the flatband voltage. The procedure is repeated with the gate biased in the opposite polarity. The difference between the two flatband voltages ΔV_{FB} gives the mobile ion concentration as

$$N_I = \frac{|\Delta V_{FB}|.C_{ox}}{q}. \tag{4.105}$$

4.7.6.2 Extraction from Quasi-Static *I–V* Measurements [53]

This method is also referred to as the triangular voltage sweep (*TVS*). A linear voltage-ramp is applied to the gate at a very slow ramp-rate *a*, typically in the range 10–50 mV/s, as described in Sect. 4.4.1. The current is measured with the structure held at elevated temperature. In the absence of mobile ions, the measured current is the displacement current. When ions are present, a peak in current is seen superimposed at $V_G \approx 0$ on the displacement current (Fig. 4.27). At the elevated temperature, the capacitance is almost flat because of the large increase in n_i. The ion current is added to the displacement current in (4.64) as

$$j_G = \frac{dQ_G}{dt} = aC_{ox} + j_{Ion} \quad \mathrm{A/cm^2}. \tag{4.106}$$

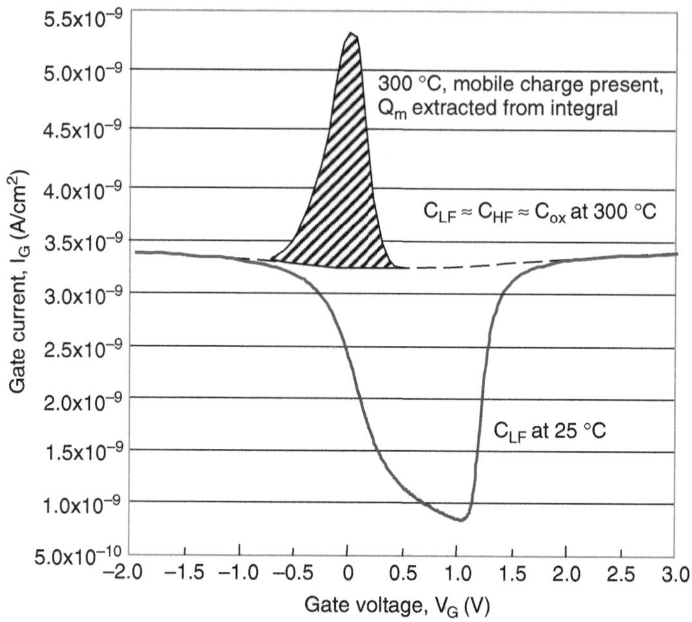

Fig. 4.27 Quasi-static measurement of mobile ions. The ion density is determined from the cross-hatched area

For negligible *DC* current through the oxide, the mobile ion density N_I is determined from the cross-hatched area in Fig. 4.27 which is the integral

$$\int_{-V_G}^{+V_G} (j_G - aC_{ox})dV_G = aqN_I \quad \text{cm}^{-2}, \tag{4.107}$$

where a is the ramp-rate. One of the advantages of the *TVS* method is that it is sensitive to mobile charge densities as low as $10^9\,\text{cm}^{-2}$. It also enables the separation of different mobile charges, for example, Na^+ and K^+, because their peaks occur at different gate voltages.

4.8 Carrier Transport Through the Dielectric

Referring to the band diagram in Fig. 4.17, the barrier heights of 3.2 eV for electrons and 4.6 eV for holes are too large for thermal emission of carriers from silicon into the oxide at room temperature. Thus, assuming an ideal oxide void of charge, traps and defects, for carriers to transit the oxide from the silicon, they must be either excited over the barrier by a high-field avalanche or tunnel through the energy barrier. Tunneling is a quantum-mechanical effect that predicts an increasing probability for carriers to penetrate the oxide as the distance through the oxide is reduced below ~4 nm. For an oxide thickness below 1 nm, the film becomes practically transparent to carriers. Avalanche injection occurs when silicon is driven into deep depletion, increasing the field in silicon and creating hot carriers that can be injected over the barrier into the oxide.

4.8.1 Tunneling Through the Oxide

For tunneling to occur through the oxide, the field in the oxide must be larger than $\sim 1 \times 10^7\,\mathrm{V/cm}$. Tunneling can occur directly through the full oxide width (Fig. 4.28a), or by the Fowler-Nordheim (*FN*) mechanism by which the field reduces the required tunneling distance (Fig. 4.28b).

Both direct and *FN* tunneling occur simultaneously, but for an ultra-thin oxide, direct tunneling dominates. The situation in 4.28b is that the oxide is too thick for appreciable direct tunneling to occur but sufficiently thin that band-bending creates a triangular barrier that reduces the tunneling distance.

4.8.1.1 Fowler-Nordheim Tunneling

In its simplified form, the relation for Fowler-Nordheim tunneling is [54, 55]

$$j_{FN} = AE_{ox}^2 e^{-B/E_{ox}} \quad \mathrm{A/cm^2}, \tag{4.108}$$

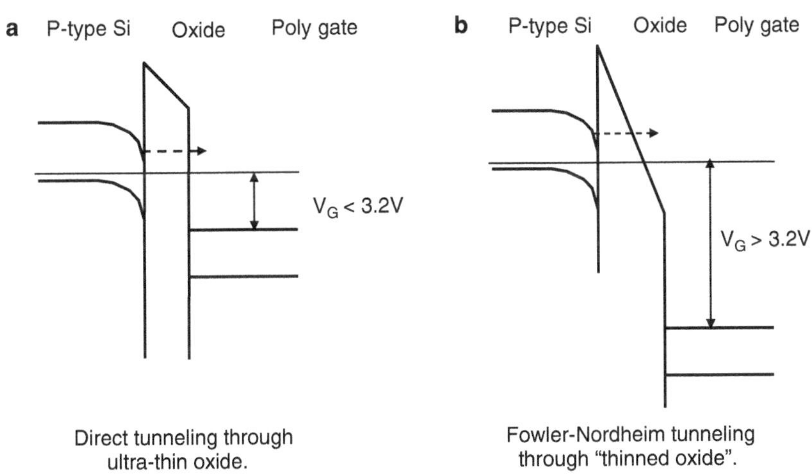

Fig. 4.28 Comparison of Fowler-Nordheim and direct tunneling

where A and B are constants defined as

$$A \approx \frac{q^3}{8\pi h\phi_{ox}} = \frac{1.54x10^{-6}}{\phi_{ox}} \quad \mathrm{A/V^2}, \tag{4.109}$$

$$B \approx \frac{8\pi\sqrt{2m^*\phi_{ox}^3}}{3qh} = 4.823x10^7\phi_{ox}^{3/2} \quad \mathrm{V/cm}, \tag{4.110}$$

where ϕ_{ox} is the barrier height at the silicon- or polysilicon-oxide interface in eV. Without consideration of barrier lowering and quantum effects at the silicon surface [54], the oxide-silicon barrier is $\phi_{ox} \approx 3.2\,eV$ for electrons and $\phi_{ox} \approx 4.6\,eV$ for holes. The effective electron mass in the oxide m^* is assumed to be about 0.4 m_O, where m_O is the free electron mass [56]. A detailed derivation of (4.108) can be found in [32, 54]. This yields $A \approx 4.81\ x\ 10^{-7}\,A/V^2$, $B \approx 2.76\ x\ 10^8\,V/cm$.

4.8.1.2 Direct Tunneling

Assuming the same constants A and B defined in (4.109) and (4.110), the relation for direct tunneling is found as [55]

$$j_{Direct} = AE_{ox}^2 e^{-B^*/E_{ox}} \quad \mathrm{A/cm^2}, \tag{4.111}$$

where

$$B^* = B\left[1 - \left(1 - \frac{qV_{ox}}{\phi_{ox}}\right)^{1.5}\right] \quad \mathrm{V/cm}. \tag{4.112}$$

The gate current $j_G = j_{FN} + j_{Direct}$. Figure 4.29 shows the calculated direct-tunneling and Fowler-Nordheim currents as a function of field in the oxide for $t_{ox} = 3\,nm$ and $t_{ox} = 6\,nm$. The values obtained from (4.108) and (4.111) are in good agreement with the reported results in [55]. It can be seen that in this oxide-thickness range, direct tunneling dominates.

4.8.2 Avalanche Injection [57]

For avalanche injection, the *MOS* structure is driven into deep depletion. With the depletion approximation and for a uniform boron concentration, the deep depletion depth is

$$x_{d-deep} \approx \sqrt{\frac{2\varepsilon_0\varepsilon_{Si}\psi_{s-deep}}{qN_A}} \quad \mathrm{cm}, \tag{4.113}$$

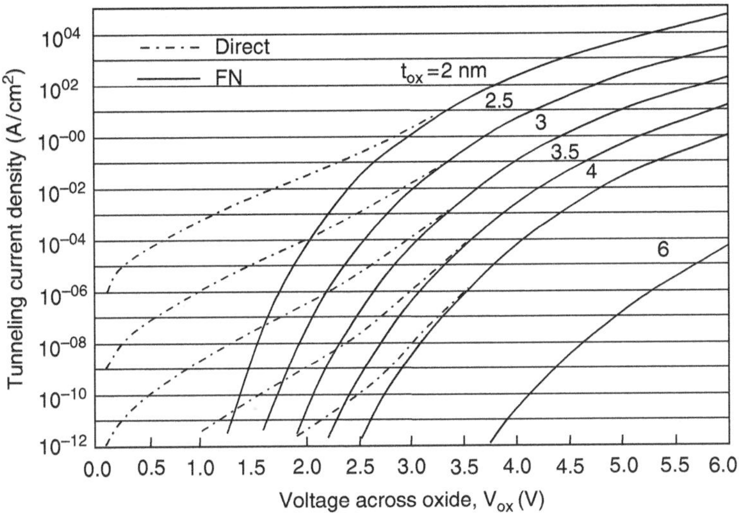

Fig. 4.29 Calculated direct and Fowler-Nordheim (*FN*) currents as a function of gate voltage for $t_{ox} = 2\,\text{nm}$ to 6 nm

In this mode, the surface potential is high and the voltage across the oxide $V_{ox} = V_G - \psi_s$ is low. For a uniformly doped substrate, the peak field is

$$E_{peak} = \frac{qN_A x_{d-deep}}{\varepsilon_0 \varepsilon_{Si}} \quad \text{V/cm}. \tag{4.114}$$

For a non-uniform substrate, the peak field is found by integrating N_A from the silicon surface to $x_{d\text{-}deep}$

$$E_{peak} = \frac{q\int_0^{x_{d-deep}} N_A(x)dx}{\varepsilon_0 \varepsilon_{Si}}. \tag{4.115}$$

When the field in silicon 'reaches a "critical" value where avalanche breakdown occurs, a plasma of hot electrons and holes is created in silicon. A fraction of the electrons gets accelerated toward the oxide-silicon interface. Those electrons with energy sufficient to surmount the oxide barrier are injected into the oxide and transported to the gate, constituting the avalanche injection current, as illustrated in the band-diagram in Fig. 4.30.

The measurement is typically done by applying a sinusoidal signal of large enough amplitude to cause avalanche breakdown. The signal frequency is high so that minority carriers cannot follow and form an inversion layer. Avalanche multiplication occurs during one half of a cycle. Field-enhancements at gate edges and carrier trapping in the oxide and at its interfaces are neglected and will be considered in the following chapter.

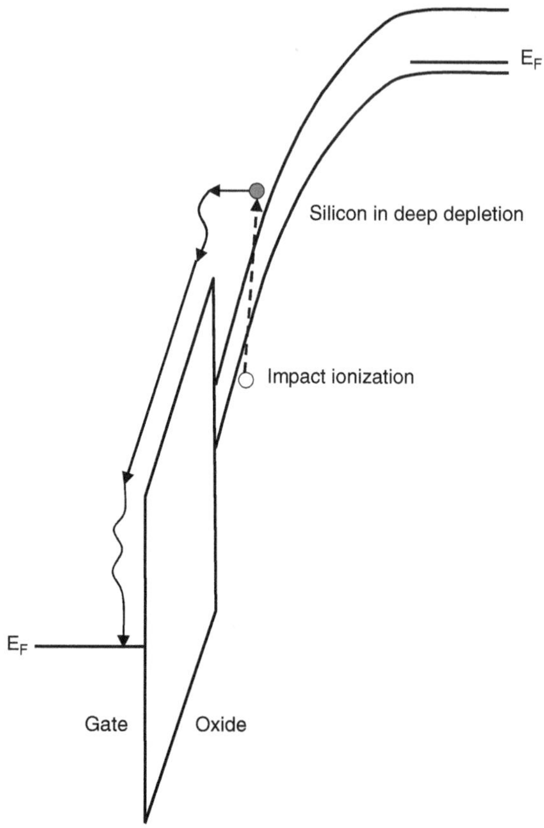

Fig. 4.30 Band diagram of *MOS* structure under avalanche injection (Adapted from [1])

4.9 Problems

(The temperature is 300 K unless otherwise stated)

1. A silicon substrate has a uniform concentration $N_A = 10^{17}\,cm^{-3}$. Prepare a table showing the value of $F(u_s, u_b)$, the electric field, and the space-charge concentration as ψ_s varies from -0.4 to $+0.4\,V$.

2. Show that $F(-u_s, -u_b) = -F(u_s, u_b)$.

3. Show that $\sinh u_b = \frac{\bar{n}-\bar{p}}{2n_i}$, $\cosh u_s = \frac{n_s+p_s}{2n_i}$.

4. Compare the results of Problem 2 with the vales of E_s and Q_s obtained with the depletion approximation.

5. Plot the surface field Es as a function of surface potential in Fig. 4.6, indicating the field-direction.

6. An ideal MOS structure is constructed on a p-type substrate of concentration $N_A = 1.5 \times 10^{16}\,cm^{-3}$. Assume full-ionization and calculate the surface charge and electric field for the following cases: $u_s = u_b$, $u_s = 0$, $u_s = -u_b$. In each case, indicate the charge polarity and direction of the electric field.

7. For the values of u_s and u_b defined below, indicate the biasing conditions and draw the energy-band and block-charge diagrams that characterize the static state of the system. (a) $u_b = 12$, $u_s = 12$; (b) $u_b = -9$, $u_s = 3$; (c) $u_b = 9$, $u_s = 18$; (d) $u_b = 15$, $u_s = -15$; (e) $u_b = -15$, $u_s = 0$.

8. In an MOS structure, the dielectric is a dual insulator consisting of 20-nm oxide and 20-nm silicon-nitride. The substrate is p-type and the gate is a metal biased to $-10\,V$.

(a) From (4.30) and (4.31), derive expressions for the fields in the oxide and nitride.
(b) Calculate the field in the 20-nm oxide and compare the result to the field obtained for an equivalent single oxide-dielectric.
(c) Find the field in silicon.

9. Find the intrinsic capacitance per unit area for silicon at 300 K and 400 K.

10. At what gate voltage will the Fermi-level coincide with the valence band-edge in accumulation and the conduction band-edge in inversion? Assume p-type silicon, quasi-static conditions and 10-nm oxide thickness.

11. Calculate the surface potential at the minimum low-frequency silicon capacitance and the minimum total capacitance for an MOS structure constructed on p-type silicon with $N_A = 10^{14}\,cm^{-3}$ and $N_A = 10^{16}\,cm^{-3}$. Assume 25-nm oxide.

12. Derive an equation for the drift time of alkali ions through an oxide film of thickness tox. Generate plots for the drift time through 50-nm oxide as a function of temperature in the range 0–500 °C.

13. An MOS structure is fabricated on a p-type substrate with $N_A = 10^{17}\,cm^{-3}$. The gate is degenerately-doped polysilicon and the dielectric consists of 10-nm oxide and 10-nm nitride. Assume $Q_{eff} = 0$ and calculate the inversion electron concentration for $V_G = 3\,V$.

14. Consider an MOS structure with uniform substrate of concentration $N_D = 5 \times 10^{16}\,cm^{-3}$ and an equivalent oxide thickness of 25 nm. The structure is driven into deep depletion so that the inversion concentration is zero. For a critical field at avalanche breakdown of $5 \times 10^5\,V/cm$, calculate the depletion depth at onset of avalanche breakdown. Repeat for a Gaussian arsenic profile with its peak at the silicon surface, an implanted dose of $5 \times 10^{12}\,cm^{-2}$, and standard deviation of $0.2\,\mu m$. At what inversion concentration would x_d relax to half of its initial value?

15. A high-frequency MOS CV plot shows $Cmax = 180\,pF$ and $C_{min} = 108\,pF$. Pulsed CV measurements were done on the same structure with the flatband as the voltage reference. The results are given in the table below. The MOS area is $2.53 \times 10^{-3}\,cm^2$. Plot the ionized-impurity concentration versus depth.

$V_G - V_{FB}$ (V)	C (pF)	$V_G - V_{FB}$ (V)	C (pF)	$V_G - V_{FB}$ (V)	i (pF)
0.0	165.0	4.5	95.7	9.0	70.2
0.5	156.5	5.0	91.8	9.5	68.1
1.0	144.5	5.5	88.4	10.0	66.1
1.5	133.1	6.0	85.2	10.5	64.3
2.0	124.2	6.5	82.3	11.0	62.3
2.5	116.6	7.0	79.5	11.5	60.5
3.0	110.3	7.5	76.9	12.0	58.7
3.5	104.7	8.0	75.5	12.5	56.9
4.0	99.9	8.5	72.3	13.0	55.2

16. A silicon surface is doped with $N_A = 10^{16}\,cm^{-3}$. What is the maximum effective charge density that can be present before inversion occurs? For this condition calculate the field in silicon.

References

1. L. M. Terman, "An investigation of surface states at a silicon/silicon oxide interface employing metal-oxide-silicon diodes," Solid-State Electron., 5 (5), 285–299, 1962.
2. K. Lehovec, A. Slobodskoy, and J. L. Sprague, "Field effect-capacitance analysis of surface states on silicon," Phys. Stat. Sol. (b), 3 (3), 447–464, 1963.
3. A. S. Grove, B. E. Deal, E. H. Snow, and C. T. Sah, "Investigation of thermally oxidized silicon surfaces using metal-oxide-semiconductor structures," Solid-State Electron., 8 (2), 145–163, 1965.
4. R. H. Kingston and S. F. Neustadter, "Calculation of the space-charge, electric field, and free carrier concentration at the surface of a semiconductor," J. Appl. Phys., 26 (6), 718–720, 1955.
5. E. H. Nicollian and J. R. Brews, MOS Physics and Technology, John Wiley & Sons, 1982.
6. C. E. Young, "Extended curves of the space charge, electric field, and free carrier concentration of the surface of a semiconductor, and curves of the electrostatic potential inside a semiconductor," J. Appl. Phys., 32 (3), 329–332, 1961.
7. A. Many, Y. Goldstein, and N. B. Grover, Semiconductor Surfaces, North Holland, 1971.
8. F. Stern and W. E. Howard, "Properties of semiconductor surface inversion layers in the quantum limit," Phys. Rev., 163 (3), 816–835, 1967.
9. M. J. van Dort, P. H. Woerlee, and A. J. Walker, "A simple model for quantization effects in heavily-doped silicon MOSFETs at inversion conditions," Solid-State Electron., 37 (3), 411–414, 1994.
10. C.-Y. Hu, S. Banerjee, K. Sandra, G. G. Streetman, and R. Sivan, "Quantization effects in inversion layers of PMOSFETs on Si (100) substrates," IEEE Electron Dev. Lett., 17 (6), 276–278, 1996.
11. S. A. Hareland, S. Krishnamurhty, S. Jallepalli, C.-F. Yeap, K. Hasnat, A. F. Tasch, and C. M. Maziar, "A computationally efficient model for inversion layer quantization effects in deep submicron n-channel MOSFETs," IEEE Trans. Electron Dev., 43.
12. Y. Ohkura, "Quantum effects in Si n-MOS inversion layer at high substrate concentration," Solid-State Electron., 33 (12), 1581–1585, 1990.
13. J. A. López-Villanueva, P. Cartujo-Casinello, J. Banqueri, F. Gámiz, and S. Rodríguez, "Effects of the inversion layer centroid on MOSFET behavior," IEEE Trans. Electron Dev., 44 (11), 1915–1922, 1997.

14. F. Li, H.-H. Tseng, L. F. Register, P. J. Tobin, and S. K. Barnerjee, "Asymmetry in gate capacitance-voltage (C-V) behavior of ultrathin metal gate MOSFETs with HfO_2 gate dielectrics," IEEE Trans. Electron Dev., 53 (8), 1943–1946, 2006.
15. N. Rodriguez, F. Gamiz, and J. B. Roldan, "Modeling of inversion layer centroid and polysilicon depletion effects on ultrathin-gate-oxide MOSFET behavior: The influence of crystallographic orientation," IEEE Trans. Electron Dev., 54 (4), 723–732, 2007.
16. S.-I. Tagaki and A. Toriumi, "Quantitative understanding of inversion-layer capacitance in Si MOSFETs," IEEE Trans. Electron Dev., 42 (12), 2125–2130, 1995.
17. G. Doucet and F. van de Wiele, "Threshold voltage of nonuniformly doped MOS structures," Solid-State Electron., 16 (3), 417–423, 1973.
18. B. E. Deal, "Standardized terminology for oxide charges associated with thermally oxidized silicon," IEEE Trans. Electron Dev., ED-27, 606–608, 1980.
19. F. J. Grunthaner, P. J. Grunthaner, R. P. Velasquez, B. F. Lewis, J. Maserjian, and A. Madhukar, "High-resolution x-ray-photoelectron spectroscopy as a probe of local atomic structure: Application to amorphous SiO_2 and the $Si-SiO_2$ interface," Phys. Rev. Lett., 43, 1683–1685, 1979.
20. F. J. Grunthaner, R. P. Velasquez, and M. H. Hecht, "X-ray photoelectron spectroscopy study of the chemical structure of thermally nitrided SiO_2," Appl. Phys. Lett., 44 (10), 969–971, 1984.
21. E. H. Poindexter, E. R. Ahlstrom, and P. J. Caplan, Proc. Intl. Conf. on the Physics of SiO_2 and its Interfaces, S. T. Pantilides, ed., p. 227, Pergamon Press, N. Y., 1978.
22. N. M. Johnson, D. K. Biegelsen, M. D. Moyer, S. T. Chang, E. H. Poindexter, and P. J. Caplan, "Electronic traps and P_b centers at the Si/SiO_2 interface: Band-gap energy distributions," J. Appl. Phys., 56 (10) 2844–2849, 1984.
23. T. Sakurai and T. Sugano, "Theory of continuously distributed trap states at $Si-SiO_2$ interfaces," J. Appl. Phys., 52 (4), 2889–2896, 1981.
24. G. Van den Bosch, G. Groeseneken, H. E. Maes, R. B. Klein, and N. S. Saks, "Oxide and interface degradation resulting from substrate hot hole injection in metal oxide semiconductor field effect transistors at 295 K and 77 K," J. Appl. Phys., 75, 2073–2080, 1994.
25. B. E. Deal, M. Sklar, A. S. Grove, and E. H. Snow, "Characteristics of the surface-state charge (Qss) of thermally oxidized silicon," J. Electrochem. Soc., 114, 266–274, 1967.
26. E. H. Snow, A. S. Grove, B. E. Deal, and C. T. Sah, "Ion transport phenomena in insulating films," J. Appl. Phys., 36, 1664–1673, 1965.
27. G. F. Derbenwick, "Mobile ions in SiO_2: Potassium," J. Appl. Phys., 48, 1127–1132, 1977.
28. A. G. Tangena, N. F. De Rooij, and J. Middelhoek, "Sensitivity of MOS structures for contamination with H^+, Na^+ and K^+ ions," J. Appl. Phys., 49, (11), 5576–5583, 1978.
29. J. S. Logan and D. R. Kerr, "Migration rates of alkali ions in SiO_2 films," Solid-State Dev. Res. Conf., 1965.
30. G. Greeuw and J. F. Verwey, "The mobility of Na^+, Li^+, K^+ ions in thermally grown SiO_2 films," J. Appl. Phys., 56, 2218–2224, 1984.
31. J. P. Stagg, "Drift mobilities of N^+ and K^+ ions in SiO_2 films," Appl. Phys. Lett., 31 (8), 532–533, 1977.
32. D. K. Schroder, Semiconductor Material and Device Characterization, John Wiley & Sons, 1998.
33. B. El-Kareh and R. J. Bombard, Introduction to VLSI Silicon Devices; Physics, Technology and Characterization, Kluwer Academic Publishers, 1986.
34. K. S. Krisch, J. D. Bude, and L. Manchanda, "Gate capacitance attenuation in MOS devices with thin gate dielectrics," IEEE Electron Dev. Lett., 17 (11), 521–524, 1996.
35. W. K. Henson, K. Z. Ahmed, E. M. Vogel, J. R. Hauser, J. J. Wortman, R. D. Venables, M. Xu, and D. Venables, "Estimating oxide thickness of tunnel oxides down to 1.4 nm using conventional capacitance–voltage measurements on MOS capacitors," IEEE Electron Dev. Lett., 20 (4), 179–181, 1999.
36. D. Vasileska, D. K. Schroder, and D. K. Ferry, "Scaled silicon MOSFETs: Degradation of the total gate capacitance," IEEE Trans. Electron Dev., 44 (4), 584–587, 1997.

37. C. Bowen, C. L. Fernando, G. Klimeck, A. Chatterjee, D. Blanks, R. Lake, J. Hu, J. Davis, M. Kulkarni, S. Hattangady, and I.-C. Chen, "Physical oxide thickness extraction and verification using quantum mechanical simulation," IEDM Tech. Digest, pp. 869–897, 1997.
38. S. Walstra and C. T. Sah, "Thin oxide thickness extrapolation from capacitance-voltage measurements," IEEE Trans. Electron Dev., 44 (7), 1136–1142, 1997.
39. B. J. R. Hauser, "Bias sweep rate effects on quasi-static capacitance of MOS capacitors," IEEE Trans. Electron Dev., 44 (6), 1009–1012, 1997.
40. Y. C. Cherng, Han-Cheng Wulu, F. H. Yang, and C. Y. Lu," Decrease of gate oxide dielectric constant in tungsten polycide gate processes," IEEE Electron Dev. Lett., 14 (5), 243–245, 1993.
41. T. J. Hwang, S. H. Rogers, and B. Z. Li, "Work function measurement of tungsten polycide gate structure," J. Electron. mater., 12, 667–678, 1983.
42. Y.-C. Yeo, P. Ranade, T.-J. King, and C. Hu, "Effects of high-K gate dielectric materials on metal and silicon gate workfunctions," IEEE Electron Dev. Lett., 21 (6), 342–344, 2002.
43. C. C. Hobbs, L. R. C. Fonseca, A. Knizhnik, V. Dhandapani, S. B. Samavedam, W. J. Taylor, J. M. Grant, L, G. Dip, D. H. Triyoso, R. I. Hegde, D. C. Gilmer, R. Garcia, D. Roan, M. L. Lovejoy, R. S. Rai, E. A. Hebert, H.-H. Tseng, S. G. H. Anderson, B. E. White, and P. J. Tobin, "Fermi-Level Pinning at the Polysilicon/Metal Oxide interface – Parts I and II," IEEE Trans. Electron Dev., 51 (6), 971–984, 2004.
44. M. Zerbst, "Relaxation effects at semiconductor-insulator interfaces", Z. Angew. Phys., 22, 30–33, 1966.
45. J. S. Kang and D. K. Schoder, "The pulsed MIS capacitor: A critical review," Phys. Stat. Sol., 89a, 13–43, 1985.
46. R. Castagné and A. Vapaille, "Determination of the $SiO_2 - Si$ interface properties by means of very low frequency MOS capacitance measurements," Surf. Sci., 28, 157–193, 1971.
47. M. Kuhn, "A quasi-static technique for MOS C-V and surface state measurements," Solid-State Electron., 13 (6) 873–885, 1970.
48. E. H. Nicollian and A. Goetzberger, "The Si–SiO_2 interface – electrical properties as determined by the metal-insulator-silicon conductance technique," Bell Syst. Tech. J., 46, 1055–1133, 1967.
49. S. W. Huang and J.-G. Hwu, "Lateral nonuniformity of effective oxide charges in MOS capacitors with Al_2O_3 gate dielectrics," IEEE Trans. Electron Dev., 53 (7), 1608–1614, 2006.
50. B. H. Lee, L. Kang, W.-J. Qi, R. Nieh, Y. Jeon, K. Onishi, and J. C. Lee, "Ultrathin hafnium oxide with low leakage and excellent reliability for alternative gate dielectric application," IEEE IEDM Tech. Digest, pp. 133–137, 1999.
51. C. Leroux, J. Mitard, G. Ghibaudo, X. Garros, G. Reimbold, B. Guillaumot, and F. Martin, "Characterization and modeling of hysteresis phenomena in high K dielectrics," IEDM Tech. Digest, 737–740, 2004.
52. C. N. Berglund, "Surface states at steam-grown silicon-silicon dioxide interfaces," IEEE Trans. Electron Dev., ED-13 (10), 701–705, 1966.
53. M. Kuhn and D. J. Silversmith, "Ionic contamination and transport of mobile ions in MOS structures," J. Electrochem. Soc., 118, 966–970, 1971.
54. M. Depas, B. Vermeire, P. W. Mertens, R. L. van Meirhaeghe, and M. M. Heyns, "Determination of tunneling parameters in ultra-thin oxide layer poly-Si/SiO_2/Si structures," Solid-State Electron., 38 (8), 1465–1471, 1995.
55. M. Lenzlinger and E. H. Snow. "Fowler-Nordheim tunneling into thermally grown SiO_2" J.Appl. Phys., 40, 278–283, 1969.
56. W.-C. Lee and C. Hu, "Modeling CMOS tunneling currents through ultrathin gate oxide due to conduction and valence-band electron and hole tunneling," IEEE Trans. Electron Dev., 48 (7), 1366–1373, 2001.
57. E. H. Nicollian, A. Goetzberger, and C. N. Berglund, "Avalanche injection currents and charging phenomena in thermal SiO_2," Appl. Phys. Lett., 15, 174–176, 1969.

Chapter 5
Insulated-Gate Field-Effect Transistor

5.1 Introduction

A field-effect transistor consists of four terminals: gate, source, drain and body or substrate (Fig. 5.1). The field created by a voltage applied to a gate modulates the resistance of region under the gate between source and drain. The modulated region of the transistor body is referred to as the channel. There are three types of field-effect transistors shown in Fig. 5.1, of which the *MOSFET* is one. In a *MOSFET*, the gate consists of an *MOS* structure that slightly overlaps two junctions located on either side of the gate. The source and drain are of opposite polarity to the body. This type of transistor is also known as an insulated field-effect transistor, *IGFET*, because the gate is separated from the body by an insulator. In a junction field-effect transistor or *JFET*, one or two pn junctions act as the gate that modulates the width of a conductive path between the source and drain. The source and drain are of the same polarity type as the body. A metal-semiconductor field-effect transistor, or *MESFET*, operates similarly to a *JFET*, except that the gate consists of a Schottky-barrier diode formed between a metal in contact with the semiconductor.

This chapter focuses on the *MOSFET* which is by far the most common. The discussion of a JFET follows in the next chapter.

5.2 Qualitative Description of *MOSFET* Operation

There are two major types of *MOSFET*s, n-channel and p-channel. In an n-channel *MOSFET*, or simply *NMOS*, the body is p-type and the source and drain are n-type. In a p-channel *MOSFET*, or simply *PMOS*, the source and drain are p-type and the body is n-type. The *MOSFET* can be normally-on (enhancement mode) or normally-off (depletion-mode). The symbols for the four different types are shown in Fig. 5.2. To simplify the discussion, the source is kept fixed at zero potential and the voltages on gate, drain, and body are referred to the source. Unless otherwise stated, the gate

B. El-Kareh, *Silicon Devices and Process Integration: Deep Submicron and Nano-Scale Technologies*,
© Springer Science+Business Media, LLC 2009

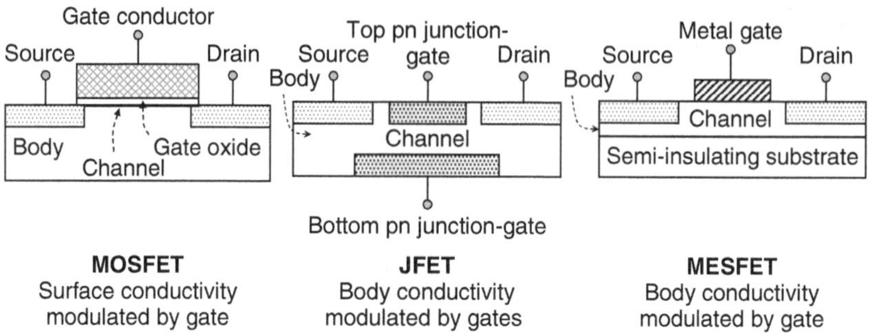

Fig. 5.1 Three field-effect transistor types

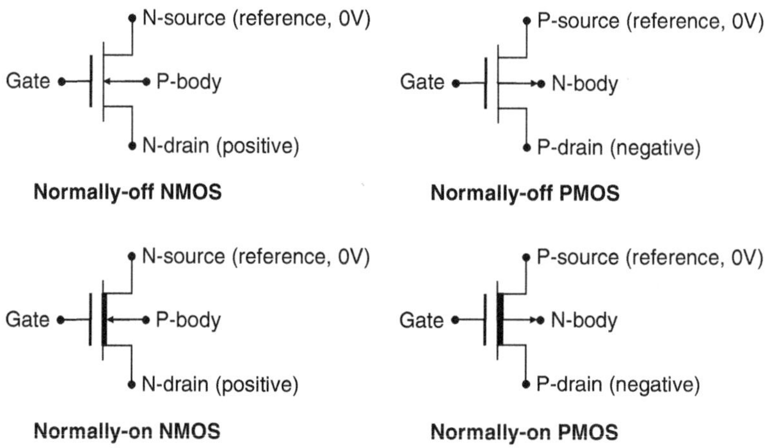

Fig. 5.2 *MOSFET* symbols. Source is chosen as reference at 0 V

voltage V_G, drain voltage V_D, and body voltage V_B refer to the gate-to-source voltage V_{GS}, drain-to-source voltage V_{DS}, and body-to-source voltage V_{BS}.

When the surface under the gate of an *NMOS* is inverted, a thin layer of inversion electrons forms between source and drain, creating a conductive path between the two regions. When the surface of a *PMOS* is inverted, inversion holes form the conductive path. In a normally-off *MOSFET*, the channel region does not conduct at zero gate voltage. It may be slightly depleted, accumulated, or at flatband. A gate voltage must be applied to "enhance" it into inversion and turn it on. In a normally-on *MOSFET*, the channel is initially inverted at $V_G = 0$, and a gate voltage must be applied to turn it off by depleting the conducting film between source and drain.

A typical *MOSFET* layout is shown in the top-view of Fig. 5.3. It consists of two symmetrically arranged gate-controlled junctions on either side of the gate. The channel region is the intersection of gate and active area. L_D and W_D are, respectively, the drawn channel length and width. L_{poly} is the patterned polysilicon gate dimension, and L_{met} the distance between source and drain metallurgical junctions.

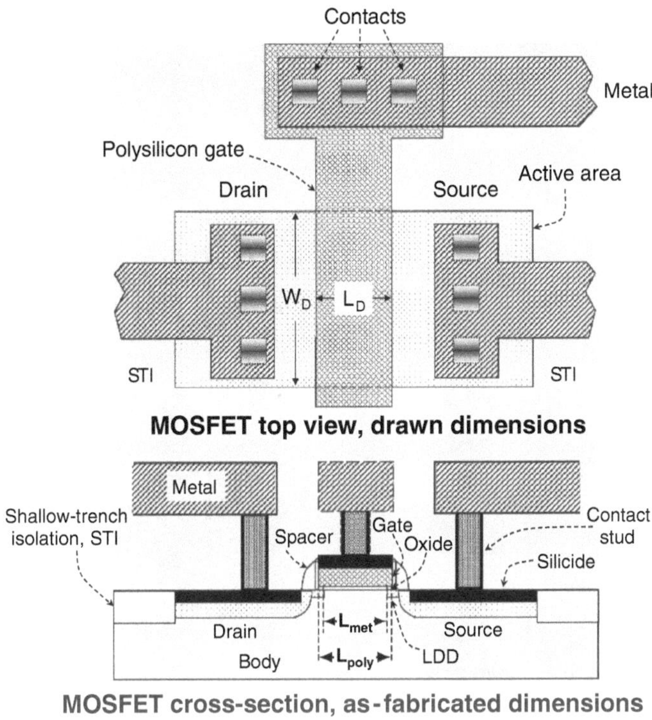

Fig. 5.3 Top-view and cross-section of typical *MOSFET*

Depending on the impurity profiles and applied voltages, the effective (electrical) channel length L_{eff} can be smaller or larger than L_{met}. The structure is illustrated in Fig. 5.3 for shallow-trench isolation (STI), polysilicon gate, oxide gate-dielectric and pn junction source and drain. Other materials and configurations are applicable and will be described in this chapter and in Chap. 7. The main processing steps to construct an *NMOS* are briefly illustrated in Fig. 5.4 for reference.[1]

MOSFETs are typically isolated from each other by thick oxide. Shallow-trench isolation regions are patterned, etched, oxide-filled, and planarized by chemical-mechanical polishing, *CMP* (Fig. 5.4a). The region outside *STI* is called the active area. W_D is the drawn channel-width. It is the distance between isolation edges in a direction parallel to the polysilicon gate (Fig. 5.4b). The effective channel width W_{eff} is the electrically extracted channel width which is typically different than W_D.

For an *NMOS*, boron is typically implanted at multiple energies and doses to form the p-well and tailor its surface (Fig. 5.4a). The process is repeated with arsenic or phosphorus to form the *PMOS* n-well and tailor its surface. The well constitutes the body of the *MOSFET*.

[1] A description of unit processing steps can be found in Fundamentals of Semiconductor Processing Technologies, by B. El-Kareh, Kluwer Academic Publishers, 1997. CMOS and BiCMOS technologies are reviewed in Chap. 7.

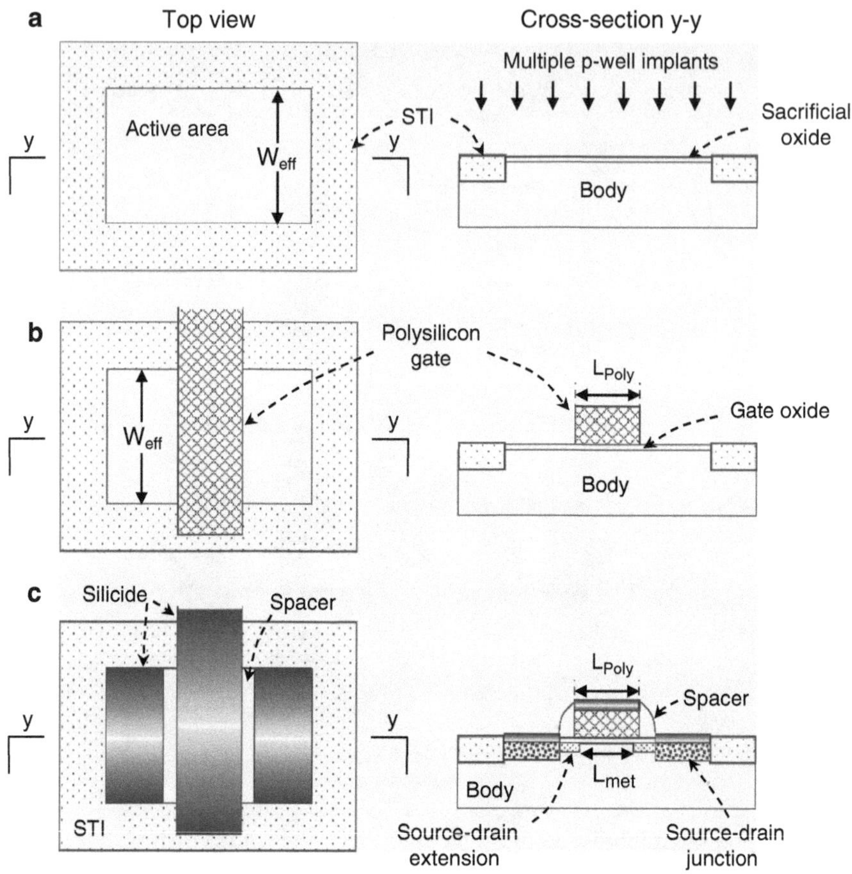

Fig. 5.4 **a** Definition of *NMOS* active area and isolation, W_{eff}, and p-well multiple implants. **b** Gate-stack patterning. **c** Formation of source-drain extensions, spacers, source-drain junctions, and silicidation.

The surface is prepared for gate oxidation. A thin gate oxide is grown and polysilicon deposited and patterned (Fig. 5.4b). The etched polysilicon dimension L_{poly} is the design variable that determines the channel length.

A thin oxide is deposited or grown to form a protective liner on the polysilicon sidewalls. Source-drain extensions (or "lightly-doped drain," *LDD*) are typically implanted at this point at different energy, dose, and angle to tailor the profile at junction edges facing the channel (Fig. 5.4c). The purpose of these implants is to reduce the electric field at junction edges facing some of the channel, and to suppress short-channel effects, as will be discussed in Sect. 5.4.3.

An oxide or nitride film of appropriate thickness is then deposited and etched directionally to form the spacers on polysilicon sides. The contacted source-drain and the gate are heavily doped, typically by implantation. The structure is then annealed and silicided (Fig. 5.4c). The spacers keep the heavily-doped source-drain junctions and their silicides at a "safe" distance from the gate. L_{met} is the distance between source and drain metallurgical junctions along the surface.

Inter-level isolation films are deposited and planarized; contacts are patterned, etched, filled with metal, and planarized to form "contact studs." Metal is then deposited, patterned, etched, and annealed, forming the first metal connections to the *MOSFET* terminals, as illustrated in Fig. 5.3.

The principles developed for an MOS structure in the preceding chapter are directly applicable to the theory of a *MOSFET*. The starting point is a three-terminal device consisting of a body, a pn junction, and a gate that slightly overlaps the junction. The resulting structure, called a gate-controlled pn junction, allows the extraction of several important *MOSFET* parameters by observing the junction characteristics as a function of surface field. The structure is then extended to a *MOSFET* by adding a fourth terminal, a second junction on the other side of the gate. The analysis begins with a channel of sufficiently large dimensions so that edge-effects on transistor characteristics can be neglected and the channel can be assumed to be fully controlled by the gate. The transistor dimensions are then reduced by a procedure known as "scaling," and short- and narrow-channel effects are analyzed, leading to fundamental limits as the *MOSFET* approaches nanoscale dimensions. The chapter concludes with mobility-enhancement techniques and alternative materials for gate dielectric, gate conductor, and junctions.

5.3 Gate-Controlled PN Junction, or Gated Diode

A gate-controlled pn junction, or gated diode, is an *MOS* structure adjacent to a junction with the gate slightly overlapping the junction [1]. It is a three-terminal device, illustrated in Fig. 5.5 for a p-type body and n^+-junction. The following discussion is equally applicable to an n-type body and p^+ junction by appropriate changes in voltage polarities. For simplicity, the body is assumed to be uniformly doped and held at 0 V. Also, the effective oxide charge, Q_{eff}, and workfunction difference between gate and body, ϕ_{ms}, are assumed to be zero. The effects of nonzero Q_{eff} and ϕ_{ms} can be accounted for by a shift in flatband voltage, as discussed in Chap. 4.

5.3.1 Junction at Equilibrium

For $V_G = 0$, the p-surface is at flatband. It is not possible, however, that the same gate material simultaneously satisfies the flatband condition in both p-region and n-regions. For example, when the p-region is at flatband, the n^+-region under the gate is accumulated (See Problem 1). With the n^+-region at ground, there is no current and the structure is at equilibrium (Fig. 5.5a). The Fermi levels in the p-body and n^+-region align laterally and vertically. As the gate voltage is increased above zero, the surface potential increases and the p-region under the gate becomes depleted (Fig. 5.5b). The depletion region under the gate is referred to as the field-induced depletion region to distinguish it from the junction depletion. The

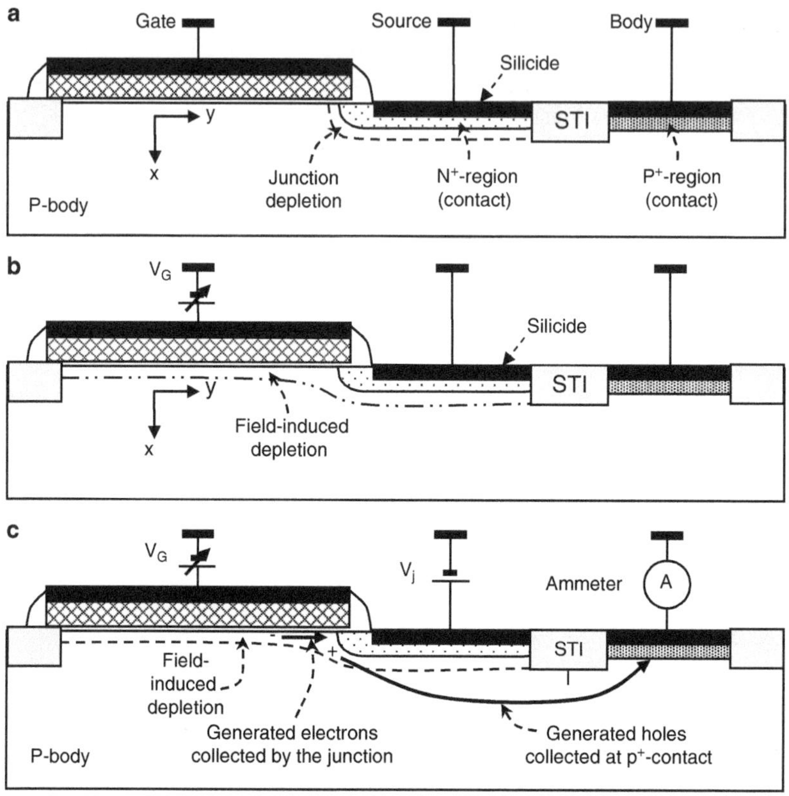

Fig. 5.5 **a** Schematic cross-section of a gate-controlled pn junction with gate, n^+-region, and body at ground. Surface of p-region under the gate is at flatband. **b** Gate-controlled pn junction with surface of p-body in depletion. **c** Measurement of junction reverse current as a function of gate voltage in a gate-controlled pn junction

gate voltage creates a two-dimensional field near the edge of the junction where both depletion regions merge. Fringe-field effects at the edges of the polysilicon gate are not considered in this discussion. The field is positive, that is, points in the x-direction (into silicon) over the p- and n-regions. At the edge of the junction, some field lines from the gate terminate on ionized impurities in the p-region and on excess electrons in the overlapped n-region. The ionized impurities in the p-region near the junction edge are thus shared between n^+-region and gate, reducing the junction barrier. As discussed in the previous chapter, the electron concentration in the field-induced depletion region is initially negligible and the space-charge consists mainly of ionized impurities.

It is important to note that the Fermi-levels remain aligned, vertically and horizontally, since there is no net current in any direction. Band-bending at the surface of the p-region brings the conduction band edge close to the Fermi-level, which determines the electron concentration in the conduction band. In the n^+-region, the Fermi-level practically coincides with the conduction band-edge.

5.3.2 Reverse Biased Junction: Depleting Gate Voltage

The depletion width increases with increasing gate voltage until it reaches a maximum value x_{dmax} at onset of strong inversion when V_G approaches the threshold voltage V_T where $\psi_s \approx 2\ \phi_b$. In the absence of a junction, equilibrium in weak and strong inversion would be established by the slow process of thermal generation of electron-hole pairs (*ehp*). In a gate-controlled junction, however, the barrier under the gate at the edge of the junction is sufficiently reduced to allow injection of electrons into the field-induced depletion region and rapidly establish equilibrium.

When a voltage is applied to the junction, the structure is no longer at equilibrium. The Fermi level splits into a quasi-Fermi level for electrons and a quasi-Fermi level for holes. A positive voltage V_j on the n-region pulls down the quasi-Fermi-level for electrons by V_j and the junction reverse voltage increases from the built-in voltage V_b to $V_b + V_j$ (Chap. 2). Let a small reverse voltage be applied to the junction and the reverse current be measured as a function of V_G (Fig. 5.5c).

In the plot of the current-voltage characteristic in Fig. 5.6, it is assumed that the leakage at the *STI* boundaries of the gate is negligible. When the p-surface is at flatband, the reverse current is essentially that of an isolated pn junction characterized by a depletion layer of thickness x_d and a barrier $V_b + V_j$, as discussed in Chap. 2. The current consists of thermal generation of *ehp* within the junction depletion region, ehp generation outside the depletion region at an average distance of one minority-carrier diffusion length, and surface *ehp* generation at the junction perimeter. The current measured is shown in region *A* of the solid curve in Fig. 5.6. As V_G is increased, a field-induced depletion region forms and thermally generated electrons within and outside the depleted region drift to the junction while the generated holes drift to the p-contact.

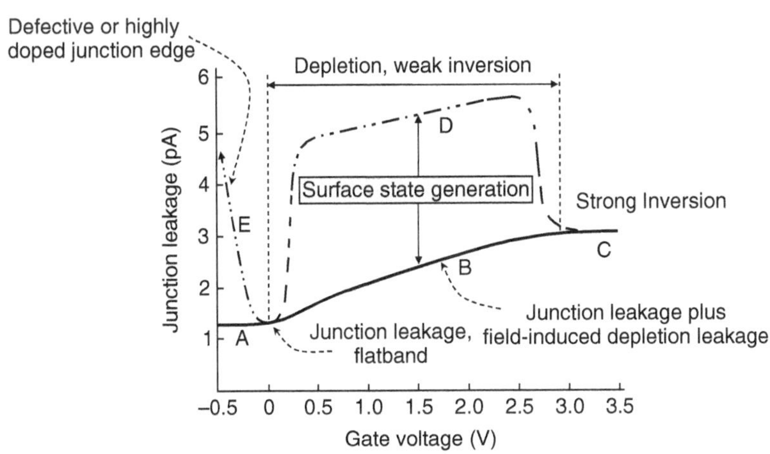

Fig. 5.6 Junction leakage as a function of gate voltage in gate-controlled pn-junction measured for a reverse voltage VR = 1 V

In the absence of interface states, the current measured would increase proportionally to the field-induced depletion depth. This is shown as region B of the solid curve in Fig. 5.6. This current would then saturate at the onset of strong inversion where x_d reaches x_{dmax} (region C of the solid curve in the figure). For an inverting V_G, the steady-state inversion electron concentration is rapidly established by injection of electrons from the n^+-region into the field-induced depletion region. The fraction of inversion-layer electrons coming from thermal generation is typically negligible compared to that injected by the n^+-region. To bring the p-type surface to the onset of strong inversion, however, the surface potential must increase from $\psi_s \approx 2\phi_b$ to $\psi_s \approx 2\phi_b + V_j$, where the inversion layer is essentially at the same potential as the n-region.

With the depletion approximation, the maximum width of the field-induced depletion region becomes

$$x_{d\max} = \sqrt{\frac{2\varepsilon_0 \varepsilon_{Si}(2|\phi_b| + V_j)}{qN_A}} \quad \text{cm}. \tag{5.1}$$

The total bulk charge is then

$$Q_{b\max} = qN_A x_{d\max} = \sqrt{2qN_A \varepsilon_0 \varepsilon_{Si}(2|\phi_b| + V_j)} \quad C/cm^2. \tag{5.2}$$

The threshold voltage is

$$V_T = -\frac{Q_{b\max}}{C_{ox}} + \psi_s = -\frac{Q_{b\max}}{C_{ox}} + 2|\phi_b| + \mathrm{V_j} \quad \text{V}. \tag{5.3}$$

Note that V_T will increase over its value at $V_j = 0$ for two reasons, first by the amount of V_j to compensate for the voltage on the n^+-region and then by $\Delta V_T = \Delta Q_b / C_{ox}$, where ΔQ_b is the difference between Q_{bmax} with reverse voltage and Q_{bmax} at $V_j = 0$. In Fig. 5.6, V_T is ~2.7 V for a reverse voltage of 1 V on the n^+-region. Similarly, V_T would decrease if a small forward voltage (negative V_j) is applied to the n^+-region. For (5.2) to be applicable in forward bias, however, the forward voltage must be sufficiently small that injection of electrons from the n^+-region into the p-region can still be neglected and the depletion approximation holds.

In the presence of interface states, a peak in current is observed superimposed on the junction and field-induced depletion leakage. It is shown as region D in the dashed curve of Fig. 5.6. This current is characterized by a generation rate [1–3]

$$U_s = \sigma v_{th} N_{it} \frac{n_i^2 - p_s n_s}{p_s + n_s + 2n_i} = s_0 \frac{n_i^2 - p_s n_s}{p_s + n_s + 2n_i} \quad cm^{-2}s^{-1}, \tag{5.4}$$

where

$$s_0 = \sigma\, v_{th} N_{it} \quad cm/s \tag{5.5}$$

is referred to as the surface generation velocity.

In (5.4) and (5.5), σ is the average capture cross-section ($\approx 10^{-15}\,\text{cm}^2$), assumed to be the same for electrons and holes; $\mathrm{v_{th}}$ the thermal velocity ($\approx 10^7\,\text{cm/s}$); p_s and $\mathrm{n_s}$, the surface hole and electron concentrations; $\mathrm{n_i}$ the intrinsic carrier concentration ($\approx 1.4 \times 10^{10}\,\text{cm}^{-3}$ at 300 K); and N_{it} the effective density of interface states per unit area, assumed to be located at mid-gap. When interface states are distributed across the bandgap, only those states that are approximately within one kT of mid-gap are effective generation centers. In this case, the effective density of interface states can be thought of as a value N_{it} that would give the same surface generation velocity as the distributed states.

For a fully depleted surface, p_s and n_s are negligible compared to n_i and (5.4) reduces to

$$U_s = \frac{1}{2} n_i s_o \quad cm^{-2}/s^{-1}, \tag{5.6}$$

and the current component due to interface-state generation becomes

$$I_{it} = \frac{1}{2} q n_i s_o A_s \quad A, \tag{5.7}$$

where A_s is the depleted surface area. In strong inversion, $\mathrm{n_s}$ increases rapidly and becomes much larger than n_i. From (5.4), it can be seen that an increase in $\mathrm{n_s}$ or p_s reduces the generation rate and hence I_{it}. This is observed in Fig. 5.6 as a rapid drop in current at onset of strong inversion. The drop can be visualized by considering that in strong inversion, the probability that states are occupied by electrons is very high so the states are no longer effective as generation sites.

5.3.3 Reverse Biased Junction: Accumulating Gate Voltage

In some cases, an increase in current is observed when the gate voltage is negative with respect to the n^+-junction. This is shown as region E of the dashed curve in Fig. 5.6. While the negative gate voltage accumulates the surface of the p-region, it depletes the surface of the overlapped n-region. An increase in V_j further depletes the n-region. Four mechanisms can be responsible for the rise in current as the surface of the n-region depletes: thermal generation, defect-induced leakage, impact ionization, and tunneling.

5.3.3.1 Thermal Generation

This current component is attributed to transport-limited thermal-generation of electron-hole pairs within the depleted layer of the heavily-doped n-region [4]. The current is found to increase exponentially as the negative gate voltage is increased, and to be limited by diffusion of minority-carriers along the depleted n-surface. Its mechanism can be distinguished from impact-ionization and band-to-band tunneling by measuring the temperature dependence of reverse current. While impact ionization and tunneling have weak temperature dependence, thermally-generated current is practically reduced to zero at cryogenic temperature [4].

5.3.3.2 Defect-Induced Leakage

Defects can be created in the highly-doped n-region in the form of dislocations, stacking faults and precipitates by contamination or mechanical stress. Such defects

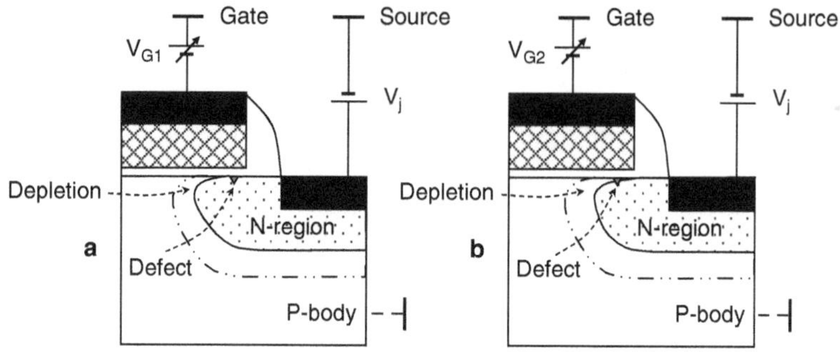

Fig. 5.7 Surface defect near junction edge. **a** Negative V_{G1}, p-region accumulated, n-region depleted, defect away from depletion boundary, low defect-induced leakage. **b** $V_{G2} < V_{G1}$, depletion in n-region extends to defect, rapid increase in leakage

can become centers of high generation rate. If a defect is located at a distance larger than approximately one minority-carrier diffusion length from the depletion boundary, its contribution to leakage is negligible. For example, while the contact to the n^+-region is highly defective, its impact on reverse leakage is minimized by keeping its boundary in silicon at a "safe" distance from the junction. Defects can be randomly distributed and fall anywhere in the p- or n-region. Figure 5.7 illustrates the case where a surface defect falls near the edge of the n-region. In Fig. 5.7a, the boundary of the depleted n-region is shown sufficiently far from the defect so that no defect-induced increase in leakage is observed. When $V_j - V_G$ increases, the depletion layer expands laterally toward the defect (Fig. 5.7b), eventually approaching it and causing a rapid increase in leakage (Region E of the dashed curve in Fig. 5.6).

5.3.3.3 Impact Ionization and Junction Breakdown

So far, the reverse characteristics of the gate-controlled junction were described for a sufficiently small reverse bias on the junction where impact ionization was negligible. As the reverse voltage is increased, the probability for impact ionization increases and, when a "critical field" is reached, avalanche breakdown occurs. In the absence of a gate or surface charge, the breakdown voltage is that of an isolated junction, depending on the impurity profile and curvature at junction corners and edges where the field is highest. The curvature effect is shown for a step junction as a function of background concentration in Fig. 2.40 of Chap. 2. The two-dimensional field created by an overlapping gate at the junction edge can strongly affect the junction breakdown voltage. A detailed analysis of the field distribution at the junction edge requires two-dimensional computer simulations. One can, however, get an insight into the breakdown mechanism by making some simplifying assumptions. Consider, for example, the case where V_j is increased while the gate and body are kept at ground (Fig. 5.8).

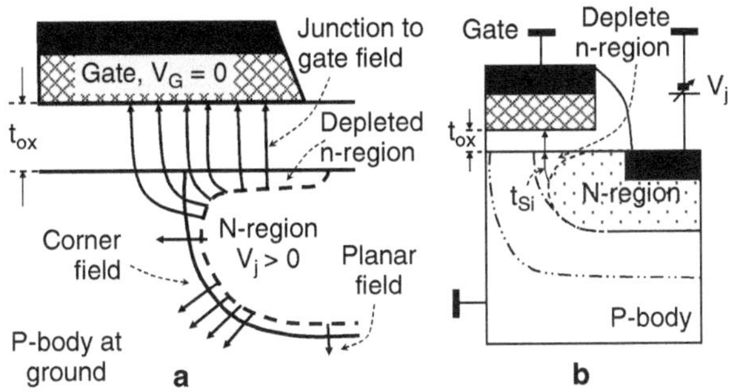

Fig. 5.8 Junction field configuration with gate and body grounded. **a** Corner and planar fields. **b** Selected *Si* path t_{Si}

The surface of the p-region is at flatband and the overlapped surface of the n-region is depleted. Since the junction is reverse-biased and the gate is at ground, an inversion layer will not form on the surface of the n-region. Instead, the surface of the n-region goes into deep depletion as the surface potential increases. The field points away from the silicon surface. t_{Si} is a field line in silicon, and t_{ox} the thickness of the oxide that is assumed to be ideal. From the continuity of the displacement vector, the fields in silicon and oxide are related by

$$E_{ox} = \frac{\varepsilon_{Si}}{\varepsilon_{ox}} E_{Si} \approx 3E_{Si}. \tag{5.8}$$

As V_j increases, the fields increase in the silicon and the oxide. For a low to moderately-doped p-region ($N_A <\approx 5 \times 10^{17}\,\text{cm}^{-3}$), the field along t_{Si} is larger than in other regions of the junction.

Figure 5.9 compares approximate calculations of the surface field at the junction corner to the field at the junction floor as a function of V_j. For this comparison, it is assumed that $N_D = 5 \times 10^{17}\,\text{cm}^{-3}$ along t_{Si}, $t_{ox} = 10\,\text{nm}$, the junction depth $x_j = 300\,\text{nm}$, and $N_A = 10^{17}\,\text{cm}^{-3}$ at the junction floor. The comparison is simplified by making step-junction approximations with uniform dopant concentrations.

Since the impact ionization rate depends exponentially on electric field (2.131), it follows that for a given V_j, impact-ionization in silicon occurs at a considerably higher rate along a path t_{Si} than in other junction regions. Thermally-generated carriers traveling through the field-induced depletion region are hence multiplied by the high field along t_{Si}, whereby secondary electrons drift to the n-region and secondary holes to the p-region. The net is an increase in reverse current due to impact ionization. The multiplication factor is given by (2.132) as

$$M = \frac{1}{1 - \int_0^{t_{Si}} \alpha_i(E)dx}, \tag{5.9}$$

where α_i (E) is the field-dependent ionization rate shown in Fig. 2.26.

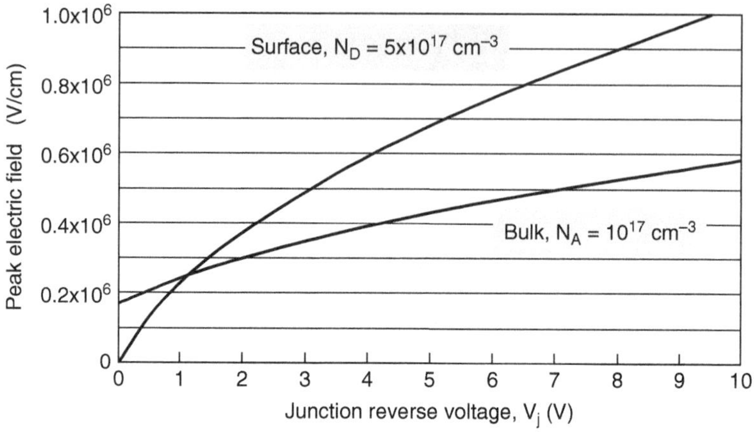

Fig. 5.9 Comparison of peak surface field in field-induced depletion region in n-region to peak field in bulk depletion region

Avalanche breakdown occurs at a "critical field" where the integral in the denominator of (5.9) approaches unity and the multiplication factor M tends to infinity. The critical field increases with dopant concentration [5, 6]. From (2.136), it is approximated as $6 \times 10^5\,\mathrm{V/cm}$ for $N_A = 10^{17}\,\mathrm{cm}^{-3}$, and $8.2 \times 10^5\,\mathrm{V/cm}$ for $N_D = 5 \times 10^{17}\,\mathrm{cm}^{-3}$. An approximate ionization rate $\alpha_i \approx 1.5 \times 10^5\,\mathrm{cm}^{-1}$ can be extrapolated for electrons and holes from Fig. 2.36. This gives an approximate path-length in silicon of $t_S \approx 65\,\mathrm{nm}$ at the breakdown field. From (5.8), the field in the oxide is $2.4 \times 10^6\,\mathrm{V/cm}$. The breakdown voltage is then approximated as the sum of voltages across t_{ox} and t_{Si}:

$$BV \approx |E_{ox}|\,.t_{ox} + |E_{Si}|\,.t_{Si} \approx 6.3V.$$

This can be compared to the breakdown voltage of $\approx$10 V obtained for an isolated step junction with background concentration $N_A = 10^{17}\,\mathrm{cm}^{-3}$ and $x_j = 300\,\mathrm{nm}$ (2.138).

When a negative gate voltage V_G is applied, the surface of the p-region becomes accumulated. Detailed analysis and experimental results show that for low to moderate concentrations N_A in the p-region and $t_{ox} \ll x_d$, the junction breakdown voltage depends linearly on $V_j - V_G$ and is practically independent of N_A and x_j for a wide range of background concentrations and curvatures [1, 7].

A positive V_G, reduces $|V_j - V_G|$ and causes a field-induced depletion region at the surface of the p-region (Fig. 5.5b). The field at the junction edge is therefore reduced, and the breakdown voltage increases as V_G increases, until it reaches the value of the planar junction breakdown [1, 7].

Several simplifying assumptions were made in the above analysis. In actual structures, the impurity concentrations are nonuniform and higher than assumed. Also, the ionization rates are position-dependent since the electric field in silicon varies with distance. The results should therefore be only considered as rough approximations.

5.3.3.4 Band-to-Band Tunneling

Avalanche breakdown occurs when a critical field is reached where the ionization integral in the denominator of (5.9) approach unity. The integral depends on both the field-dependent ionization rate and path-length. The critical field, however, increases with increasing dopant concentration in the depleted n-region. This is because, for a given surface field, the ionization-rate decreases and the depletion region narrows as N_D increases at the surface of the n-region, reducing the probability for impact ionization. As the depletion width narrows below ~20 nm, the probability for band-to-band tunneling increases rapidly. The mechanism is similar to tunneling in a Zener diode discussed in Chap. 2. For band-to-band tunneling to occur in silicon, the field must be higher than $\sim 10^6\,\mathrm{V/cm}$ and band-bending must exceed the energy gap of ~1.2 eV [8,9]. The band-diagrams in Fig. 5.10 illustrate the generation of electron-hole pairs by tunneling of electrons from the valence band into the conduction band in the heavily-doped, gate-overlapped n- or p-region. As ψ_s increases above 1.2 V, the valence band and conduction band overlap on the energy scale, so that filled states and empty states appear opposite each other, separated by the thin depletion region. The tunneling current density from the filled to the empty states follows the relation [8,9]

$$j_T = AE_{Si}e^{-B/E_{Si}}, \tag{5.10}$$

where A is a pre-exponential constant that depends on the effective mass, energy gap, phonon scattering and geometry; $B = 27.6\,\mathrm{MV/cm}$, and E_{Si} the field in silicon at the oxide-silicon interface. E_{Si} depends on the difference between junction and gate potentials, V_{JG}, and on dopant concentration in the region where tunneling occurs. The field can be approximated under similar assumptions as for impact ionization. The following relation holds for nonzero workfunction difference and effective oxide charge

$$V_{JG} - V_{FB} = V_{ox} + \psi_s = V_{ox} + \frac{E_g}{q} \approx V_{ox} + 1.2 \approx E_{ox}t_{ox} + 1.2, \tag{5.11}$$

where V_{FB} is the flatband voltage and, from (5.8), $E_{ox} = 3E_{Si}$.

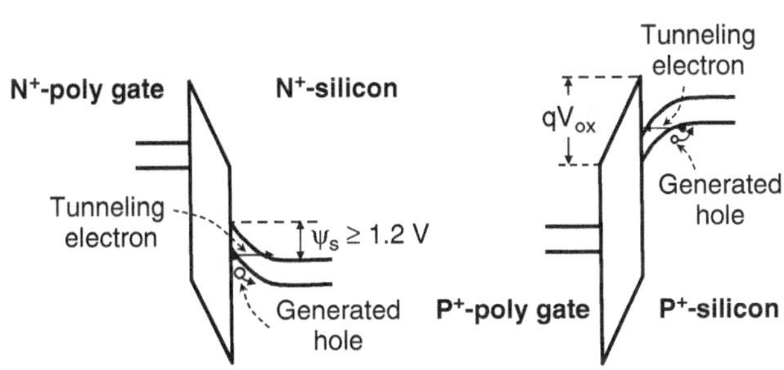

Fig. 5.10 Energy band diagrams illustrating band-to-band tunneling in silicon in the overlapped n- and p-regions. Generated holes drift to the substrate [10–12]

For $V_{FB} = 0$, (5.11) simplifies to [11, 12]

$$E_{Si} \approx \frac{V_{JG} - 1.2}{3t_{ox}}. \tag{5.12}$$

The gate-induced tunneling current increases as the oxide thickness is reduced and the dopant concentration in the gate-overlapped junction increased. It is important to control this current component in leakage-sensitive applications, such as *DRAM* and *CMOS* logic, as will be discussed in Chap. 8.

5.4 MOSFET Characteristics

This section discusses the characteristics of an *NMOS*. The discussion is equally applicable to a *PMOS* by appropriate changes in polarities. For a symmetrically arranged *MOSFET* shown in Fig. 5.3, either of the junctions can act as the source or drain. The source and drain are distinguished only by the applied voltages. The source is the junction at the lower potential in an *NMOS* and at the higher potential in *PMOS*. For simplicity, the source is fixed at 0 V and all other voltages are referred to the source. Therefore, the drain is the junction that is biased positive in *NMOS* and negative in *PMOS*.

5.4.1 Long and Wide Channel

There is no sharp boundary between a long and a short channel, or between a wide and narrow channel. Short and narrow-channel effects are gradual and, as will be discussed in Sects. 5.4.3 and 5.4.4, become more pronounced when the channel length or width is reduced to the order of the field-induced depletion depth. For now, it suffices to say that a channel is considered long if the drain bias has negligible effect on the threshold voltage and effective channel length. It is considered wide if edge effects have little impact on channel characteristics. The following discussion focuses on a long and wide channel.

For simplicity, the p-body is assumed to be uniformly doped and held at ground. For $V_G = V_{FB}$, the channel surface is at flatband and the source and drain behave like two independent pn junctions that are characterized by a built-in voltage V_b, as described in Chap. 2. Let both junctions be initially at ground. As shown for an *MOS* structure, threshold voltage V_T is the gate voltage at onset of strong inversion. This is the gate voltage at which the surface potential $\psi_s \approx 2\phi_b$, where ϕ_b is the Fermi potential defined as

$$\phi_b = \frac{kT}{q} \ln \frac{N_A}{n_i} \quad \text{V}, \tag{5.13a}$$

and

$$V_T = -\frac{Q_{b\max}}{C_{ox}} + V_{FB} + \psi_s = \frac{\sqrt{2\varepsilon_0 \varepsilon_{Si} q N_A \psi_s}}{C_{ox}} + V_{FB} + \psi_s \quad \text{V}. \tag{5.13b}$$

From the above relations, it follows that V_T decreases in magnitude as temperature increases (See Prob. 5.6).

The inversion layer constitutes a conducting film that connects the two junctions along the surface. Assuming a constant effective surface mobility for electrons μ_{eff}, the inversion-layer conductivity is

$$G = \frac{W_{eff}}{L_{eff}} \int_0^{x_i} \sigma(x) dx = -\frac{W_{eff}}{L_{eff}} \mu_{eff} \int_0^{x_i} q\, n(x) dx = -\frac{W_{eff}}{L_{eff}} \mu_{eff} Q_n \quad \text{S}. \tag{5.14}$$

where W_{eff}, L_{eff} are, respectively, the effective channel width and length, $n(x)$ the electron concentration as a function of depth from the surface, x_i the depth at which silicon is intrinsic, that is, at which $n = n_i$, and Q_n the inversion electron charge density. It follows that the channel sheet resistance is

$$R_S = \frac{1}{\mu_{eff}\, Q_n} \quad \text{Ohm/square} \tag{5.15}$$

The channel resistance is then

$$R_{Ch} = \frac{L_{eff}}{W_{eff}} \frac{1}{\mu_{eff}\, Q_n} \quad Ohm. \tag{5.16}$$

5.4.1.1 Current-Voltage Characteristic: Linear Mode

If a very small voltage $V_D \leq 50\,\text{mV}$ is applied to the drain, the drain current can be approximated as

$$I_D = \frac{V_D}{R_{Ch}} \cong \frac{W_{eff}}{L_{eff}} \mu_{eff}\, Q_n\, V_D \quad A. \tag{5.17}$$

The dependence of drain current on gate voltage is simplified by assuming that at $V_G = V_T$, the electron charge density $Q_n \approx 0$ and the surface charge Q_s consists of only ionized impurity charge, that is, $Q_s \approx Q_{bmax}$. The electron charge can then be related to the gate voltage above V_T, the "gate overdrive" as

$$Q_n \cong (V_G - V_T) C_{ox} \quad C/cm^2. \tag{5.18}$$

Substituting in (5.17) gives the drain current I_D as

$$I_D \cong \frac{W_{eff}}{L_{eff}} \mu_{eff}\, C_{ox} (V_G - V_T)\, V_D \quad A. \tag{5.19}$$

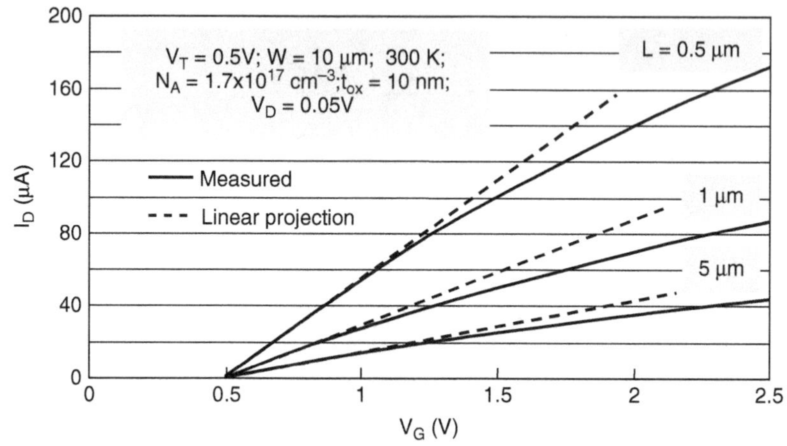

Fig. 5.11 *MOSFET* characteristics in the linear mode. The $I_D - V_G$ plots are not strictly linear because of mobility degradation by the gate field

According to (5.19), I_D increases linearly with V_D for a constant V_G, and linearly with V_G for a constant V_D. The *MOSFET* is therefore said to operate in the linear mode (Fig. 5.11). As will be shown later, however, the characteristics are not strictly linear. This is because of the degradation in effective mobility as V_G increases, and the reduction in Q_n at the drain edge as V_D increases. Note that at the source edge, the channel potential V_{Ch} is 0 and at the drain edge $V_{Ch} = V_D$. The threshold voltage therefore increases along the channel from $V_{T\text{-}source}$ at the source to $V_{T\text{-}drain}$ at the drain because Q_{bmax} increases due to the greater depletion-layer width. Q_n is lower at the drain than at the source for two reasons: first because the gate-overdrive $(V_G - V_{T\text{-}drain})$ decreases by V_D, second because $V_{T\text{-}drain} > V_{T\text{-}source}$. Neglecting the small increase in V_T, the gate overdrive is $(V_G - V_T)$ at the source and $[V_G - (V_T + V_D)]$ at the drain. For a small V_D, an average gate-overdrive can then be approximated as $[V_G - (V_T + V_D/2)]$. Substituting in (5.19) gives a better approximation of the linear mode as

$$I_D \cong \frac{W_{eff}}{L_{eff}} \, \mu_{eff} \, C_{ox} \left(V_G - V_T - \frac{V_D}{2} \right) V_D \quad \text{A}. \tag{5.20}$$

where V_T is the threshold voltage given by (5.13b).

5.4.1.2 The Gradual Channel Approximation [13, 14]

A more exact expression can be obtained by use of the gradual channel approximation which accounts for the dependence of V_T on position along the channel. Consider an elemental channel-section dy of resistance dR. The voltage drop in this section is

$$dV = I_D dR = \frac{I_D dy}{W \mu_{eff} |Q_n(y)|}, \tag{5.21}$$

where due to current continuity, I_D is independent of position y. At a distance y from the source, the total surface charge Q_s is the sum of ionized-impurity charge Q_b and inversion-layer charge Q_n

$$Q_s(y) = Q_b(y) + Q_n(y). \tag{5.22}$$

The surface potential at a distance y from the source is

$$\psi_s(y) = 2\phi_b + V_{Ch}(y) \quad V, \tag{5.23}$$

where $V_{Ch}(y)$ is the position dependent voltage along the channel. At the drain edge, $V_{Ch} = V_D$ and at the source, $V_{Ch} = 0$.

The total surface charge is

$$|Q_s(y)| = [V_G - \psi_s(y)]\, C_{ox} = [V_G - 2|\phi_b| - V_{Ch}(y)]\, C_{ox} \quad C/cm^2. \tag{5.24}$$

The bulk charge is

$$Q_b(y) = \sqrt{2\varepsilon_0 \varepsilon_{Si} q N_A [2|\phi_b| + V_{Ch}(y)]} \quad C/cm^2. \tag{5.25}$$

The inversion charge density Q_n is the difference between total and bulk charge:

$$Q_n(y) = Q_s(y) - Q_b(y) \quad C/cm^2. \tag{5.26}$$

Substituting (5.24)–(5.26) into (5.21) and integrating from source ($y = 0$, $V = 0$) to drain ($y = L$, $V = V_D$) gives

$$I_D = \beta \left\{ \left[V_G - 2|\phi_b| - \frac{V_D}{2} \right] V_D - K \left[(V_D + 2|\phi_b|)^{3/2} - (2\phi_b)^{3/2} \right] \right\} \quad A, \tag{5.27}$$

where β and K are defined as

$$\beta = \frac{W_{eff}}{L_{eff}} \mu_{eff} C_{ox}, \tag{5.28}$$

$$K = \frac{2}{3} \frac{\sqrt{2\varepsilon_0 \varepsilon_{Si} q N_A}}{C_{ox}}. \tag{5.29}$$

Equation (5.27) is only valid for $V_D < V_G - V_T$, that is, below the saturation condition discussed in the next section. A Taylor expansion of (5.27) for very small V_D reverts again to (5.20). Figure 5.12 illustrates the $I_D - V_D$ characteristics obtained from (5.27) (solid curve) and 5.20 (dashed curve). As can be seen, there is good agreement for very small V_D. The solid curve is more accurate and lower than the dashed curve because (5.27) accounts for the higher V_T near the drain.

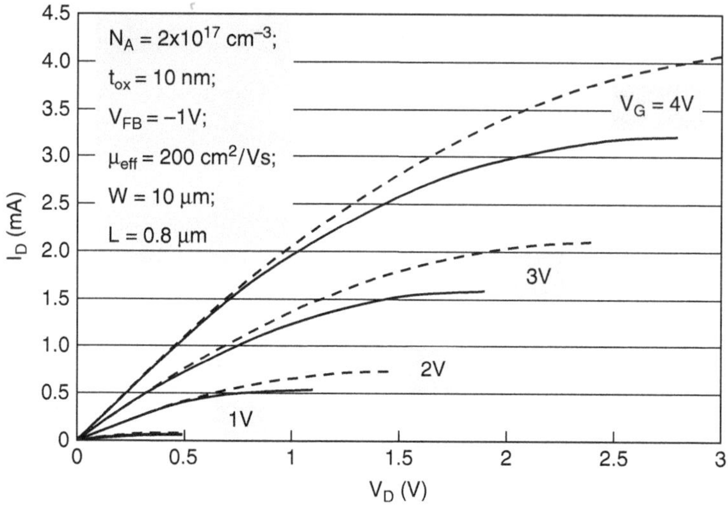

Fig. 5.12 Approximation of *NMOS* $I_D - V_D$ characteristic in the linear mode. Solid curve: (5.27); dashed curve: (5.20)

5.4.1.3 Current–Voltage Characteristic: Saturation Mode

When V_D increases, the gate overdrive decreases at the drain edge and, as V_D approaches $(V_G - V_T)$, the gate overdrive drops to near zero and Q_n becomes negligible at the drain edge when compared to the rest of the channel. The channel is said to be pinched-off at the drain boundary. The current is now limited by the voltage drop between source and the pinch-off point $V_P = (V_G - V_T)$. Substituting $V_G - V_T$ for V_D in (5.20) gives

$$I_{Dsat} \cong \frac{W_{eff}}{2L_{eff}} \mu_{eff} C_{ox}(V_G - V_T)^2 \quad A. \tag{5.30}$$

The above relation shows that, for a given V_G, the drain current saturates and becomes independent of drain voltage when $V_D \geq V_G - V_T$. The corresponding drain voltage is referred to as V_{Dsat}. The *MOSFET* is said to operate in the saturation mode (or in the parabolic mode since I_{Dsat} increases with the square of gate overdrive). V_{Dsat} can be obtained from (5.24–5.26) under the assumption that $Q_n \approx 0$ at y = L where $V_{Ch} = V_{Dsat}$

$$V_{Dsat} \cong V_G - V_{FB} - 2\phi_b + \frac{\varepsilon_0 \varepsilon_{Si} q N_A}{C_{ox}^2}\left[1 - \sqrt{1 + \frac{2C_{ox}^2(V_G - V_{FB})}{\varepsilon_0 \varepsilon_{Si} q N_A}}\right] \quad V. \tag{5.31}$$

The threshold voltage at the drain edge is found by substituting $(\psi_s + V_{Dsat})$ for ψ_s in (5.13b):

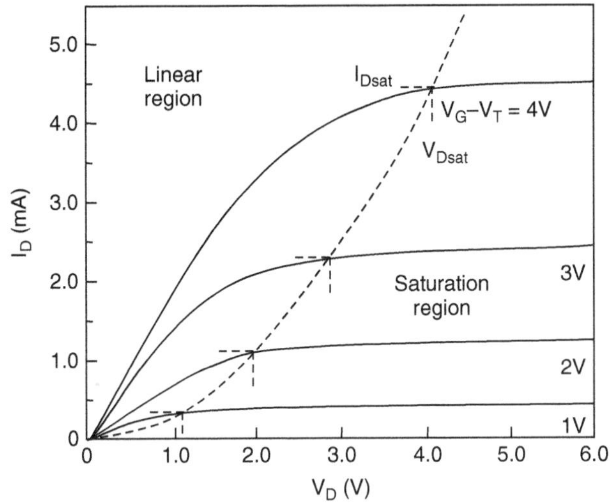

Fig. 5.13 Illustration of *NMOS* $I_D - V_D$ characteristics with gate overdrive as parameter

$$V_{T-drain} = \frac{\sqrt{2\varepsilon_0\varepsilon_{Si}qN_A(2|\phi_b| + V_{Dsat})}}{C_{ox}} + V_{FB} + 2|\phi_b| + V_{Dsat} \quad \text{V}. \tag{5.32}$$

The value obtained from (5.32) should be used for the threshold voltage in (5.30). The effect of drain bias on V_T in (5.30) is, however, frequently ignored for simplicity. Figure 5.13 illustrates the current–voltage characteristics for a long-channel *NMOS* in the "linear" and saturation modes. The dotted parabolic curve is the locus of I_{Dsat} versus V_{Dsat}.

5.4.1.4 Field and Charge Distribution at and above Pinch-Off

At pinch-off, the drain to source voltage drops along the channel, from V_{Dsat} at the drain edge to 0 at the source edge. The channel potential V_{Ch} and hence the threshold voltage is a function of position y between source and drain. Figure 5.14 approximates the channel voltage and electric fields as a function of distance y from source for the pinch-off condition. The corresponding charge densities are shown in Fig. 5.15. The surface potential increases while the *vertical* fields in the oxide and silicon decrease from source to drain. With the depletion approximation, the bulk charge Q_b increases with the square root of surface potential while the inversion electron charge Q_n and total surface charge Q_s decrease from source to drain.

As V_D increases above V_{Dsat}, the pinch-off point, defined as the point P where $V_{Ch}(y) = V_G - V_T(y)$, moves from the drain edge toward the source (Fig. 5.16).

The voltage between P and source is $[V_G - V_T(y)]$ and the voltage between drain and P is $V_D - [V_G - V_T(y)]$. The distance δL between P and drain can be obtained by making a step-junction approximation for the pinch-off region:

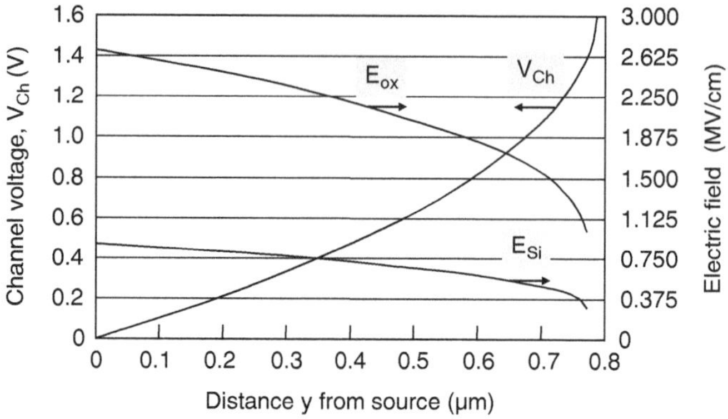

Fig. 5.14 Approximated channel voltage and electric fields for the *NMOS* in Fig. 5.11 and $V_D = V_G - V_T = V_{Dsat} = 1.6\,\text{V}$

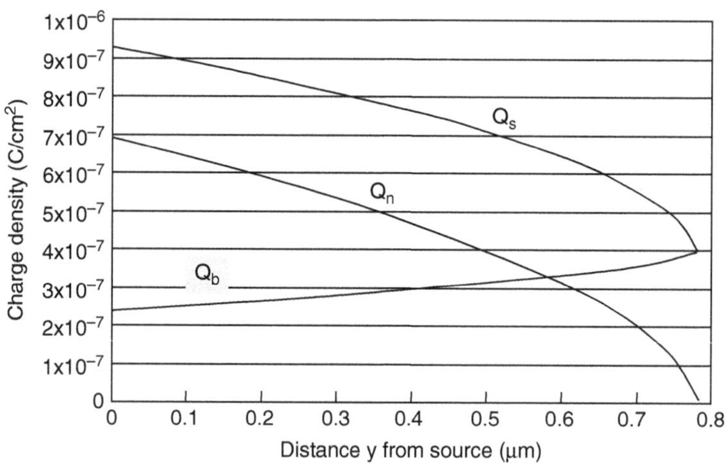

Fig. 5.15 Approximated charge densities for the *NMOS* in Fig. 5.11 and $V_D = V_G - V_T = V_{Dsat} = 1.6\,\text{V}$

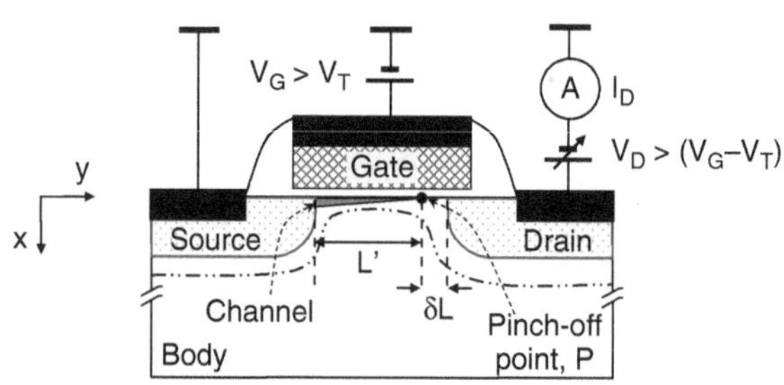

Fig. 5.16 Exaggerated channel cross-section illustrating the displacement of the pinch-off point P along the channel toward the source for $V_G > V_T$ and $V_D > V_G - V_T$. The potential at P is $V_G - V_T$

$$\delta L \cong \sqrt{\frac{2\varepsilon_0\varepsilon_{Si}[V_D - (V_G - V_T)]}{qN_A}} \quad cm. \tag{5.33}$$

The channel between source and P can be treated as a resistor of effective length $L' = L - \delta L$ having a varying sheet-resistance (varying Q_n) along the surface.

For a long channel, the lateral field in the channel is small. Inversion electrons therefore drift from source to P at approximately uniform mobility, μ_{eff}. As can be inferred from (5.33), the pinched-off section is a region of high field at which electrons drift at velocity saturation, $v_{sat} \approx 10^7$ cm/s. At any point y of the channel, the drain current is

$$I_D = W_{eff}\ Q_n(y)\ v_n(y) \quad A, \tag{5.34}$$

where $v_n(y)$ is the electron drift velocity in the y-direction at a position y of the channel. Between source and P, the drift velocity is

$$v_n(y) = \mu_{eff}\ E_y = \mu_{eff}\frac{\partial V(y)}{dy} \quad cm/s. \tag{5.35}$$

Since I_D is the same crossing any plane normal to the channel and Q_n decreases from source to drain, $v_n(y)$ must increase. At the pinch-off point P, Q_n becomes small but does not strictly go to zero. Instead, it must have a finite value at P, obtained from (5.34) as

$$Q_n(P) = \frac{I_D}{W_{eff}\ v_{sat}} \quad C/cm^2. \tag{5.36}$$

5.4.1.5 Channel Conductance, g_d

The channel conductance is defined as

$$g_D = \left.\frac{\partial I_D}{\partial V_D}\right|_{V_G} \quad S. \tag{5.37}$$

In the linear mode, the conductance is approximated from (5.20) as

$$g_{Dlin} = \frac{W_{eff}}{L_{eff}}\mu_{eff}\ C_{ox}(V_G - V_T) \quad S. \tag{5.38}$$

According to (5.30), the drain current in saturation should be independent of drain voltage and the saturation conductance, g_{Dsat}, should ideally be zero, that is, the slope in the $I_D - V_D$ characteristics should be zero in saturation. In real structures, an increase in drain voltage increases δL in (5.33), reducing the effective channel length from L to L' (Fig. 5.16). This is referred to as the channel-length modulation. As can be seen from (5.30), as L decreases with increasing V_D the drain current increases, resulting in a finite slope in the $I_D - V_D$ characteristics. In long channels where the fraction $\delta L/L$ is very small, the channel conductance is near zero. As the channel length decreases, however, $\delta L/L$ increases and the conductance can no longer be neglected. Channel-length modulation will be further detailed in Sect. 5.4.3.1.

5.4.1.6 Channel Transconductance, g_m

The channel transconductance is the change of drain current with gate voltage, defined as

$$g_m = \left.\frac{\partial I_D}{\partial V_G}\right|_{V_D} \quad S. \tag{5.39}$$

The linear transconductance g_{mlin} per unit width is found from (5.20) as

$$\frac{g_{mlin}}{W_{eff}} \cong \frac{1}{L_{eff}}\, \mu_{eff}\, C_{ox}\, V_D \quad S/cm. \tag{5.40}$$

From (5.30), the saturation transconductance per unit width g_{msat} is

$$\frac{g_{msat}}{W_{eff}} \cong \frac{1}{L_{eff}}\, \mu_{eff}\, C_{ox}\, (V_G - V_{T-drain}) S/cm. \tag{5.41}$$

The transconductance per unit width is increased by reducing the channel length and increasing the effective surface mobility.

5.4.1.7 Body-Bias Effect on Threshold Voltage

With applied body-bias, the threshold voltage is

$$V_T = \frac{\sqrt{2\varepsilon_0\varepsilon_{Si} q N_A (2\phi_b + V_B)}}{C_{ox}} + V_{FB} + 2\phi_b \quad V, \tag{5.42}$$

where V_B is the source-to-body bias. V_T can be increased by applying a positive V_B (reverse bias) and decreased with negative V_B.

The sensitivity of V_T to body-bias is found by differentiating (5.42) with respect to V_B

$$\frac{dV_T}{dV_B} = C_{ox}\frac{dQ_b}{dV_B} = \frac{1}{2C_{ox}}\sqrt{\frac{2\varepsilon_0\varepsilon_{Si} q N_A}{2\phi_b + V_B}} \quad V. \tag{5.43}$$

It follows that the sensitivity can be reduced by decreasing N_A and increasing V_B. Figure 5.17 approximates $I_D - V_G$ characteristics for $V_D > (V_G - V_T)$ and varying V_B.

Circuit applications in which the *MOSFET* source-to-body junction is reverse or slightly forward biased are common. Such a situation is shown for a Dynamic Random Access Memory (DRAM) cell in Fig. 5.18. The gate of the *NMOS* is connected to the array "word-line." The *NMOS* operates bi-directionally with one junction connected to the memory cell and the other to the array "bit-line." The figure

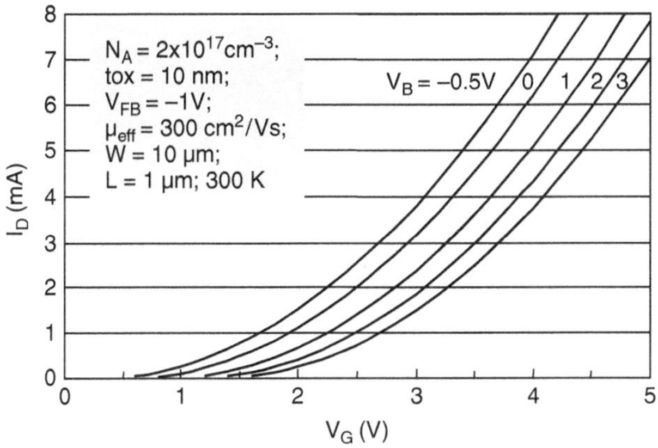

Fig. 5.17 $I_D - V_G$ characteristics in saturation mode, reflecting an increase in V_T for reverse V_B and decrease in V_T for forward V_B

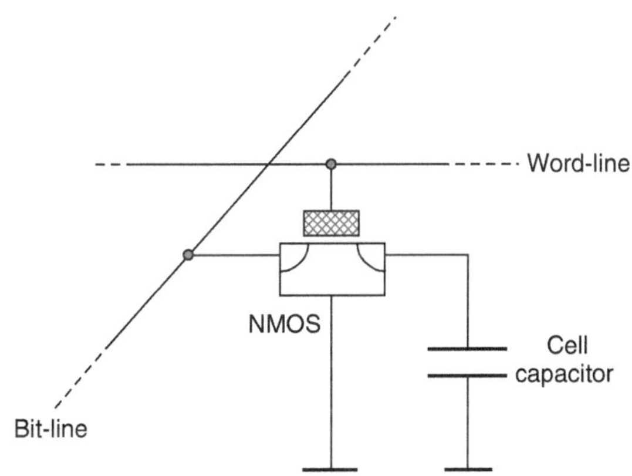

Fig. 5.18 Charging of capacitor in a *DRAM* memory cell. With high voltage applied to the drain (bit-line) and gate (word-line), the capacitor charges, which raises the source potential. Threshold voltage at source increases; gate overdrive decreases; and channel pinches off at source

describes a "write" operation where the word-line and bit-line are brought to a high potential, charging the cell and raising the source potential. This increases the body-to-source bias and hence the threshold voltage. Eventually, the channel pinches-off at the source. Thus, when the source voltage reaches one V_T below the drain voltage, the node potential reaches its maximum value found from (5.31) as

$$V_{Node} \cong V_G - V_{FB} - 2\phi_b + \frac{\varepsilon_0 \varepsilon_{Si} q N_A}{C_{ox}^2}\left[1 - \sqrt{1 + \frac{2C_{ox}^2 (V_G - V_{FB})}{\varepsilon_0 \varepsilon_{Si} q N_A}}\right] \quad V. \tag{5.44}$$

The gate voltage required to support the maximum node voltage is

$$V_G = V_T + V_{Node} = \frac{\sqrt{2\varepsilon_0 \varepsilon_{Si} q N_A (2\phi_b + V_{Node})}}{C_{ox}} + V_{FB} + 2\phi_b + V_{Node}. \tag{5.45}$$

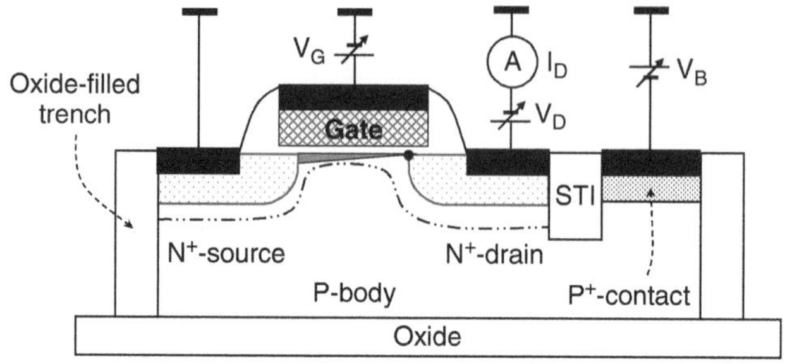

Fig. 5.19 V_T adjustment by body-bias V_B in an isolated *NMOS*

For a given gate voltage, the maximum node voltage can be increased by reducing the body-bias effect.

The body-bias effect can be utilized to locally adjust V_T. An isolated *MOSFET* or group of *MOSFET*s is best suited for this purpose. A dielectrically isolated structure that allows independent biasing of the *MOSFET* body is shown in Fig. 5.19. By applying an appropriate reverse or forward bias to the source-to-body junction, V_T can be locally increased or decreased.

5.4.1.8 Subthreshold Characteristics

The approximation that for $V_G \leq V_T$, the inversion electron charge Q_n is zero and the surface charge consists solely of ionized impurity charge Q_b was made to simplify the derivation of the current-voltage relations. This assumption was justified because, while the electron concentration increases exponentially with increasing surface potential ψ_s, it remains negligible compared to the ionized-impurity concentration when $V_G \leq V_T$. Q_n and I_D, however, do not drop abruptly to zero at threshold. The drain current has a finite value at $V_G = V_T$. It decreases exponentially when V_G is reduced below threshold, with a slope on the log-linear scale that is inversely proportional to the thermal voltage kT/q. In this region, the *MOSFET* is said to operate in the subthreshold mode.

A *MOSFET* operating in the subthreshold mode is similar to a bipolar transistor operating in the active mode. In an *NMOS*, the gate voltage increases the surface potential, reducing the barrier at the source edge, and locally forward-biasing the source to body junction. As the barrier is reduced, a fraction of electrons in the n^+-source gains enough thermal energy to overcome the barrier and get injected into the p-region. Similarly, holes are injected from the p-region into the source. The gate voltage therefore controls minority-carrier injection near the source. The injection mechanism is identical to that of a pn junction in forward bias discussed in Chap. 2. Since the source is heavily doped, the injection of holes is very small and can be neglected in this discussion. The concentration of excess minority carriers near the source increases exponentially with increasing surface potential, following the Boltzmann approximation

$$n_{s0} = \bar{n}_{p0}\, e^{q\psi_s/kT} \quad cm^{-3}. \tag{5.46}$$

n_{s0} is the electron surface concentration near the source, and $\bar{n}_{p0}$ the equilibrium concentration of minority-electrons in the p-region. At the drain boundary, the electron concentration is lower than at the source because of the reverse bias applied to the drain. It is found as

$$n_{sL} = \bar{n}_{p0} e^{q(\psi_s - V_D)/kT} \quad cm^{-3}. \tag{5.47}$$

n_{sL} is the electron surface concentration at the drain at distance L from the source. Assuming no recombination within the channel, the slope in electron density is

$$\frac{dn}{dy} = \frac{n_{s0} - n_{sL}}{L} = \frac{\bar{n}_p e^{q\psi_s/kT}\left(1 - e^{-qV_D/kT}\right)}{L}. \tag{5.48}$$

The drain current is the sum of drift and diffusion current. Initially, the surface potential does not change appreciably from source to drain and the drift current is negligible. The diffusion current therefore dominates, following the relation

$$I_D = qAD_n \frac{dn}{dy} \quad A, \tag{5.49}$$

where A is the cross-sectional area of the inversion layer, taken normal to the surface and parallel to the source and drain, dn/dy is the gradient of the inversion-carrier concentration along the surface, and D_n is the electron diffusivity, related to the mobility by the Einstein relation

$$D_n = \frac{kT}{q}\mu_n \quad cm^2/s. \tag{5.50}$$

The channel cross-sectional area can be approximated as $W \cdot x_i$, where x_i is the depth below the surface where silicon is intrinsic. For a uniformly doped p-region, the band-bending as a function of depth x is

$$\psi(x) = \psi_s \left(1 - \frac{x_d(x)}{x_{d\,max}}\right)^2, \tag{5.51}$$

where

$$x_{d\,max} = \sqrt{\frac{2\varepsilon_0 \varepsilon_{Si}(2\phi_b)}{qN_A}}. \tag{5.52}$$

Since at x_i, $\psi(x_i) = \phi_b$, one finds $x_i \approx 0.293\ x_{dmax}$. The cross-sectional area and gradient in (5.49) are then

$$A = W\sqrt{\frac{0.34\varepsilon_0 \varepsilon_{Si}\phi_b}{qN_A}}, \tag{5.53a}$$

$$\frac{dn}{dy} = \frac{n_i^2 e^{q\psi_s/kT}\left(1 - e^{-qV_D/kT}\right)}{L.N_A}. \tag{5.53b}$$

For $V_D > 2kT/q$, (5.53b) simplifies to

$$\frac{dn}{dy} \approx \frac{n_i^2 e^{q\psi_s/kT}}{L.N_A}. \tag{5.53c}$$

The drain current can now be approximated as

$$I_D \approx I_0 e^{q\psi_s/kT} \quad A, \tag{5.54}$$

where

$$I_0 = q\frac{W}{L}\sqrt{\frac{0.34\varepsilon_0\varepsilon_{Si}\phi_b}{qN_A}}\mu_{eff}\frac{kT}{q}\frac{n_i^2}{N_A} \quad A.$$

Without interface states, the relation between ψ_s and gate voltage is

$$\psi_s \approx \frac{C_{ox}}{C_{ox}+C_{Si}}V_G = \frac{V_G}{n}, \tag{5.55}$$

where C_{Si} is the silicon capacitance at the source edge, and n is referred to as the ideality factor defined as

$$n = \frac{C_{ox}+C_{Si}}{C_{ox}} = \left(1+\frac{\varepsilon_{Si}t_{ox}}{\varepsilon_{ox}x_{d(channel)}}\right). \tag{5.56a}$$

When interface states are present, the additional capacitance associated with Q_{it} must be taken into account, and n increases to

$$n = \frac{C_{ox}+C_{Si}+C_{it}}{C_{ox}}. \tag{5.56b}$$

The ideality factor n typically ranges from 1–1.3. It varies slightly with body-bias because of the dependence of C_{Si} on V_B. Figure 5.20 shows the $I_D - V_G$ plots in the subthreshold and above threshold mode, illustrating the decrease in subthreshold slope at low temperature [15]. Threshold voltage can be defined as the gate voltage at which the drain current departs from its exponential dependence. If I_{D0} is the drain current at V_T when $\psi_s \approx 2\phi_b$. Then from (5.54) the current is

$$I_{D0} \approx I_0 e^{q(2\phi_b)/kT}. \tag{5.57}$$

The subthreshold current then decreases with decreasing gate voltage as shown in Fig. 5.21. It follows the relation

$$I_D \approx I_{D0} e^{q(V_G-V_T)/nkT}, \tag{5.58}$$

can be found from (5.57), (5.55), and (5.54). The subthreshold current at $V_G = 0$ is referred to as the off-current, I_{off}. It must be controlled and kept as small as possible in applications such as dynamic memory where it directly affects the time that the cell retains information, and in logic designs where the stand-by power must be kept small, as further discussed in Chap. 8.

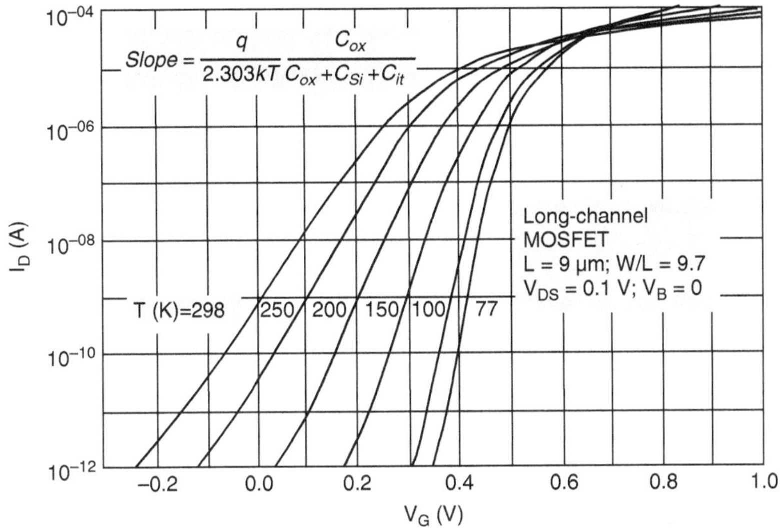

Fig. 5.20 $I_D - V_G$ characteristics for a long-channel *NMOS* in the subthreshold and above threshold modes with temperature as parameter [15]

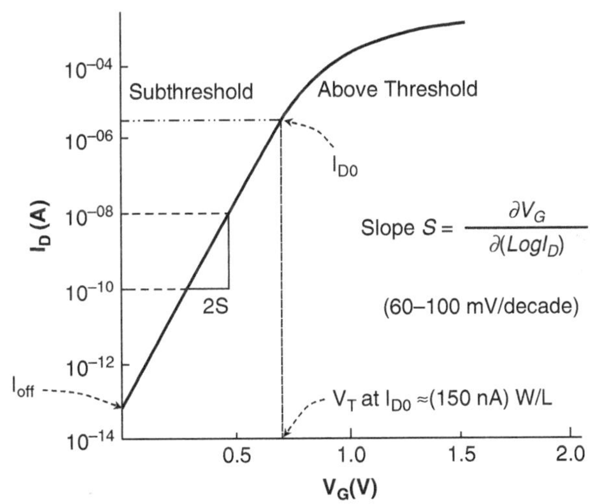

Fig. 5.21 $I_D - V_G$ characteristic for a long-channel *MOSFET* defining the subthreshold slope and off-current. I_{DO} estimated as $150nA \times W/L$

Below V_T, the structure is in weak inversion and electrons flow by diffusion. In strong inversion above V_T, the drain voltage drops along the channel and creates a lateral field in which carriers flow predominantly by drift. In the transition region around V_T, both drift and diffusion should be taken into account.

5.4.1.9 Inversion Carrier Mobility

Inversion carriers are confined to a very thin surface-layer of about 10 nm where quantization effects become important, particularly at high surface fields (Sect. 5.8). As a result of confinement, carriers are subjected to additional scattering mechanisms near the surface to those in the bulk. Therefore, the carrier mobility is expected to be smaller at the surface than in the bulk and to decrease with increasing electric field. An effective surface mobility, μ_{eff}, was introduced in the preceding sections to account for the difference in mobility between surface and bulk. Several theories and models have been suggested to describe the surface mobility [16–21]. The effective mobility depends on lattice scattering, ionized-impurity scattering, and additional surface scattering such as increased acoustical and optical phonon scattering [22, 23], surface roughness scattering [24, 25], and interface charge scattering [22]. Because of the complexity of scattering mechanisms, however, no simple equation has been found for the dependence of μ_{eff} on electric field that is applicable to all *MOSFET* conditions. Therefore, the dependence of mobility on surface field is best described by extracting the mobility from measured $I_D - V_G$ data and combining the results with well-established values of inversion-layer charge [26].

The purpose of this section is not to compare the theories, but to highlight the relations between μ_{eff} and vertical field, temperature, and process conditions. The effect of lateral field on mobility is neglected here since for long channels the lateral field is small and, for a given vertical field, the mobility can be assumed to be constant along most of the surface.

Experimental results show that, for the same impurity concentration and temperature, the inversion-carrier mobility is two to three times smaller at the surface than in the bulk [25]. The surface mobility increases with decreasing temperature because of the reduced phonon scattering. At a given temperature, the mobility for a thermally oxidized surface is found to have a unique dependence on vertical field, independent of oxide thickness, as shown for a long-and wide-channel *MOSFET* above threshold in Fig. 5.22 [27]. The plots in Fig. 5.22 are referred to as universal mobility curves and used as benchmarks to compare mobility enhancements or degradations when new structures and materials are introduced (Sect. 5.5).

The effective vertical field is averaged over the electron distribution in the inversion layer and defined as [27]

$$E_{eff} = \frac{Q_b + \eta Q_n}{\varepsilon_0 \varepsilon_{Si}} \quad V/cm, \tag{5.59}$$

where for a (100) surface, the averaging factor $\eta = 1/2$ for electrons [27, 28], and $\eta = 1/3$ for holes [29]. An expression for the dependence of mobility on effective vertical field is of the form

$$\mu_{eff} = \frac{\mu_0}{1 + \left[\alpha\left(E_{eff} - E_{ref}\right)\right]^{\beta}} \quad cm^2/Vs, \tag{5.60}$$

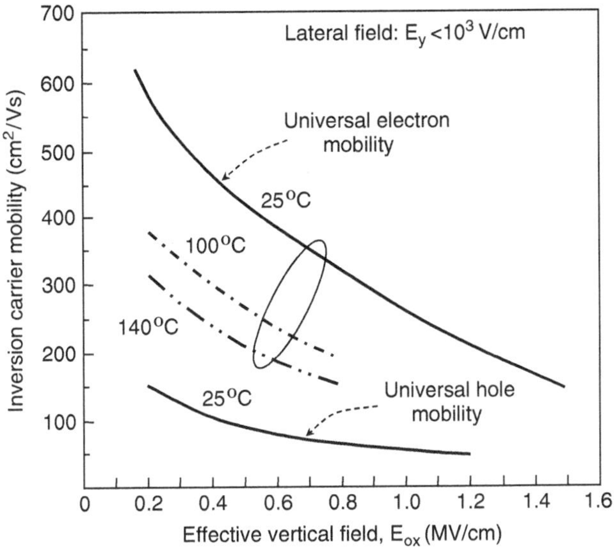

Fig. 5.22 Universal inversion-layer mobility plots for electrons [27] and holes [28]

where E_{ref} and μ_0 are, respectively, the effective field and surface mobility for $Q_n \approx 0$, and α, β are constants to be extracted from measurements. A commonly used empirical relation for the dependence of effective mobility on gate voltage is [29]

$$\mu_{eff} = \frac{\mu_0}{1+\theta\,(V_G - V_T)} \quad cm^2/Vs, \tag{5.61}$$

where θ varies with oxide thickness and channel concentration. θ is obtained from a plot of μ_0/μ_{eff} versus $(V_G - V_T)$, and is typically <0.1.

5.4.1.10 Intrinsic and Extrinsic Resistances

The intrinsic resistance is the resistance of the channel R_{Ch} defined by (5.16). It is determined by the effective channel dimensions, the effective surface mobility, and the inversion layer charge and hence the gate voltage. Resistances in series with the channel, including source, drain and contact resistances, are extrinsic. Figure 5.23 is a schematic of current contours, summarizing the main extrinsic-resistance components on the source-side of a typical structure. Similar resistance components are found on the drain side. For long channels and symmetrically-arranged source and drain, they can be assumed the same as for the source side.

The arrows on the current lines indicate the direction of electron flow. Because carriers are constrained to flow laterally near the surface, the current lines crowd at several locations of dissimilar conductivities along the current path. Current-lines crowd at contact-1 between the contact-stud and silicide, at contact-2 between silicide and n-type source, and at the transition between the inversion layer and the

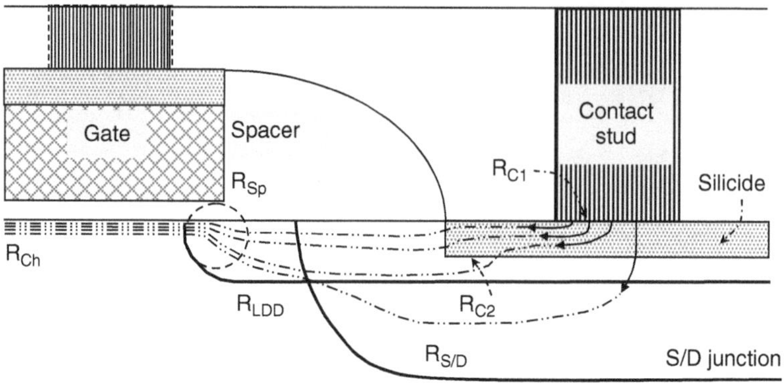

Fig. 5.23 Schematic current contours identifying the main extrinsic-resistance components on the source-side of an *NMOS*

lightly doped source/drain extension. The latter gives rise to a spreading resistance similar to the resistance of a point contact to silicon [30–32]. Since the inversion layer is very thin, it can be compared to a sharp wedge from which the current lines spread-out into the source-drain extension.

The total *MOSFET* resistance can be approximated as

$$R = R_{Ch} + R_{Ext} + R_{wire} \quad Ohm, \tag{5.62}$$

where $R_{Ext} = 2R_{Sp} + 2R_{C1} + 2R_{C2} + 2R_{S/D} + 2R_{LDD}$ Ohm,

R_{Ch} = channel resistance,
R_{wire} = Wiring or interconnect resistances (not shown),
R_{Sp} = spreading resistance,
R_{C1} = contact resistance between stud and silicide,
R_{C2} = contact resistance between silicide and silicon,
$R_{S/D}$ = series resistance of source or drain region,
R_{LDD} = series resistance of lightly-doped extension regions.

The spreading resistance depends on the local resistivity where spreading occurs. It is therefore sensitive to the gate voltage and a complex function of the carrier concentration in the region overlapped by the gate. At the source edge, and the drain edge for a small drain voltage, this region becomes accumulated with electrons and hence less resistive, while the p-type surface gets inverted. Under the simplifying assumption of an idealized uniform resistivity ρ at the source edge, the spreading resistance can be approximated as [31–34]

$$R_{Sp} \approx \frac{0.64\,\rho}{W} \ln \frac{\kappa\, x_j}{x_{Ch}} \quad \text{Ohm}, \tag{5.63}$$

where W is the channel width, x_j the source junction depth, x_{Ch} the channel depth, and κ a factor found as 0.58, 0.75, and 0.90, respectively in [31], [33], and [34].

Since the ratio x_j/x_{Ch} is very large, the exact value of κ is of minor importance. A relation for a nonuniform profile at the source-edge is derived in [32].

The is no spreading resistance at the drain in saturation since the carriers drift to the drain at velocity saturation away from the surface as the drain edge becomes depleted.

The impact of extrinsic resistance on *MOSFET* characteristics can be seen as a reduction in drain voltage, gate overdrive and transconductance. Let V_D' and V_G' be the externally applied drain and gate voltage. The actual voltages seen by the structure are

$$V_D = V_D' - I_D(R_{ext-S} + R_{ext-D}), \tag{5.64a}$$

$$V_G = V_G' - I_D R_{ext-S}. \tag{5.64b}$$

In (5.64), $R_{ext\text{-}S}$ and $R_{ext\text{-}D}$ are, respectively, the source and drain extrinsic resistances. The effective gate to source voltage decreases by $I_D R_{ext\text{-}S}$. There is also an increase in threshold voltage ΔV_T due to the body effect that can be found by substituting $(\psi_s + I_D R_{ext\text{-}S})$ for ψ_s in (5.13). The ratio of intrinsic (zero R_{ext}) to measured saturation transconductance can be approximated as

$$\frac{g_{m-Int}}{g_{m-Meas}} = \left[\frac{V_G' - V_T}{V_G' - V_T - \Delta V_T - I_D R_{ext-S}}\right]^2. \tag{5.65}$$

The effects of extrinsic resistances on transistor performance become increasingly important as the channel length is reduced and the channel resistance R_{Ch} decreases.

5.4.1.11 Transit Time and Cutoff Frequency

The transit time τ is the average time for inversion carriers to travel from source to drain. For a long channel in the linear mode, the lateral field is

$$E_y = \frac{V_D}{L_{eff}} \quad V/cm, \tag{5.66}$$

and, for a long channel where $E_y < 5 \times 10^3$ V/cm, the drift velocity is

$$v_d = \mu_{eff}\, E_y \quad cm/s. \tag{5.67}$$

The transit time is then

$$\tau = \frac{L_{eff}}{v_d} = \frac{L_{eff}^2}{\mu_{eff}\, V_D} \quad s. \tag{5.68}$$

The *MOSFET* gain is defined as the ratio of the amplitude of small-signal drain current to the amplitude of small-signal gate current:

$$Gain = \frac{\partial I_D}{\partial V_G \omega (C_{GS} + C_{GD})} = \frac{g_{mlin}}{\omega (C_{GS} + C_{GD})}. \tag{5.69}$$

where ω is the frequency, g_{mlin} the linear transconductance, C_{GS} the gate to source capacitance, including fringe and channel, and C_{GD} the gate to drain capacitance, including fringe.

The cutoff frequency (or gain bandwidth product) f_T is a good *MOSFET* figure of merit. It is the frequency at which the gain is 1. It is also the frequency where the carriers just follow the signal. From (5.68), f_T in the linear mode is

$$f_T = \frac{\omega_T}{2\pi} = \frac{1}{2\pi\tau} = \frac{\mu_{eff}\, V_D}{2\pi L_{eff}^2} \quad Hz. \tag{5.70}$$

From (5.69) with gain = 1,

$$f_T = \frac{g_{mlin}}{2\pi (C_{GS} + C_{GD})} \quad Hz. \tag{5.71}$$

In the saturation mode, (5.68) becomes

$$\tau = \frac{L_{eff}}{v_d} = \frac{L_{eff}^2}{\mu_{eff}\,(V_G - V_T)} \quad s, \tag{5.72}$$

and the cutoff frequency

$$f_T = \frac{\mu_{eff}(V_G - V_T)}{2\pi L_{eff}^2} = \frac{g_{msat}}{2\pi (C_{GS} + C_{GD})} \quad Hz. \tag{5.73}$$

Another *MOSFET* figure of merit of particular importance for analog designs is the frequency at which the ratio of output to input power (unilateral power gain) goes to 1. This is called the maximum oscillation frequency f_{max} defined for a long channel as

$$f_{\max} = \frac{f_T}{\sqrt{8\pi f_T R_G C_{GD}}} \quad Hz, \tag{5.74}$$

where R_G is the gate resistance and C_{GD} the gate to drain capacitance. f_{max} is increased by increasing f_T, reducing the gate sheet resistance and gate to drain capacitance.

5.4.2 Scaling to Small Dimensions

Scaling refers to the methodology needed to relate device dimensions to the appropriate electrical and material parameter changes to obtain equivalent device functions as dimensions are reduced. *MOSFET* dimensions are scaled down in size to reduce cost and improve performance. Scaling is made possible by advanced processes such as lithography, etch, oxidation, and doping techniques, and flexibility in choice of materials (Chap. 7). This enables the reduction in minimum feature size, that is, the minimum line-width or line-to-line space that can be printed on the die, to deep submicron and nanoscale dimensions. For the same number of circuits, this means reduced die-size, higher productivity and lower cost per die, lighter die, and smaller package. For the same die-size, it means more circuits and functions per die. Scaling lowers the applied voltage, reduces capacitances, reduces power density and potentially increases circuit speed.

5.4.2.1 Scaling Considerations

The full integrated process can be divided in roughly three areas, the front-end of the line (*FEOL*), the back-end of the line (*BEOL*), and packaging. *FEOL* consists of all processing steps from silicon wafer to silicidation of gate, source and drain. *BEOL* includes contacts, wiring and other steps between silicidation and bond- or solder-pads. The silicide is a common layer to *FEOL* and *BEOL*. Packaging is the process of connecting the bond- or solder-pads to an appropriate chip- or system-package. The bond- or solder-pads constitute the interface to *BEOL* and packaging.

Several considerations must be taken into account when scaling *MOSFET*s to smaller dimensions. These not only include the front-end of the line, but also the wiring strategy and packaging. The discussion here focuses on the front-end of the line. Scaling considerations related to the back-end of the line are discussed in Chap. 7. Packaging is beyond the scope of this book.

The three most important scaling considerations are: the minimum feature that can be printed with the available processing capabilities, the maximum allowable vertical and horizontal electric fields, and the expected component lifetime. Figure 5.24 shows the actual and projected minimum feature size as a function of time [35].

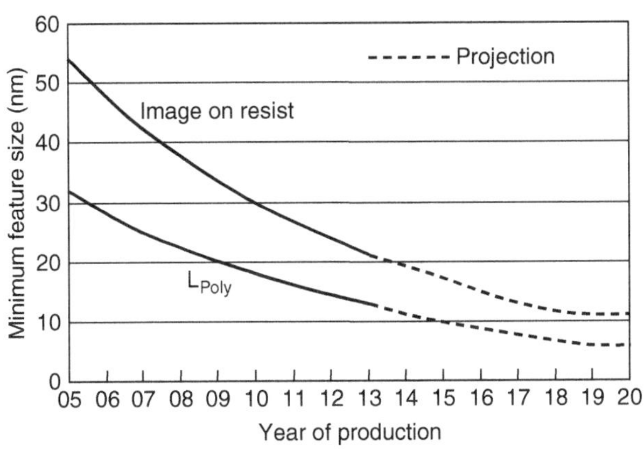

Fig. 5.24 Minimum feature size versus year of production [35]

Table 5.1 Important scaling considerations

Parameter	Considerations
Channel L, W, x_{jlat}	E_y, I_{off}, snap-back, *SCE, RSCE, NCE, RNCE*, impact ionization, *DIBL*, punch-through.
W_{Metal}, $W_{Contact}$, W_{Via}	Reliability, contact/Via resistance.
t_{eq}, x_j, $x_{silicide}$, t_{Poly}, t_{Metal}	E_{ox}, E_{Si}, I_G (tunneling), silicide penetration, current density.
V_D	I_{off}, E_y, reliability, snap-back, *SCE, RSCE*, impact ionization, *DIBL*, punch-through, chip power, self-heating.
V_G	E_{ox}, E_{Si}, *GIDL*, I_G (tunneling).
Current	Reliability, self-heating.
N_{Ch}	V_T, *SCE, GIDL, DIBL*, μ_{eff}, impact ionization, C_j

Glossary: L length, W width, E_y lateral field at surface, I_{off} off-current, E_{ox} gate dielectric field, E_{Si} silicon field, E_x vertical field in silicon, I_G gate current, *DIBL* Drain-induced barrier lowering, *GIDL* gate-induced drain leakage, μ_{eff} effective surface mobility, C_j junction capacitance, x_{jlat} lateral extent of junction, N_{Ch} Effective channel dopant concentration, *SCE* short-channel effects, *RSCE* reverse short-channel effects, *NCE* narrow-channel effects, *RNCE* reverse narrow-channel effects

The minimum polysilicon width L_{poly} is typically smaller than the minimum photo-resist feature because of a bias created by etching. The metallurgical channel length L_{met} is smaller than L_{Poly} due to polysilicon sidewall oxidation steps and lateral extent of source and drain. The minimum on-chip electrical channel length L_{eff} is typically different than L_{met}. Table 5.1 summarizes important scaling considerations. Short- and narrow-channel effects, *SCE, NCE*, and reverse short- and narrow-channel effects are discussed in the following sections.

5.4.2.2 Constant-Field Scaling

If the vertical and lateral electric fields are kept the same, a scaled and nonscaled *MOSFET* will have essentially the same electric characteristics. This is the principle of constant field scaling. Let V_{DD} denote the maximum supply voltage. If V_{DD}, t_{ox}, L_{eff}, W_{eff}, x_j, and V_T are scaled down from a given initial device by a factor k, and the channel dopant concentration scaled up by the same factor, then the scaled and initial *MOSFET*s will behave approximately the same. Figure 5.25 illustrates scaled *MOSFET* parameters. Factors of important constant-field scaling are given in Table 5.2 [36].

5.4.2.3 Issues with Scaling

Several parameters do not scale at the same rate as the minimum feature size because of practical limits imposed on them. Scaling limits of key parameters are shown in Table 5.3. Their impact on transistor performance is detailed in the following sections.

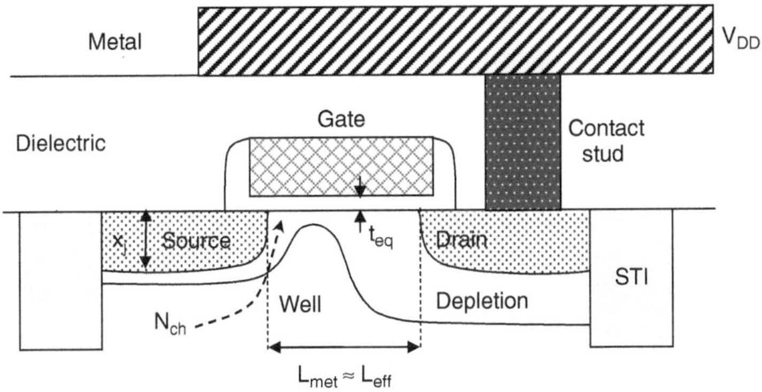

Fig. 5.25 Illustration of *MOSFET* scaled parameters

Table 5.2 Simplified constant field scaling, s > 1 [36]

Parameter	Initial	Scaled
Channel length	L	L/s
Channel width	W	W/s
Gate area	A_G	A_G/s^2
Junction depth	x_j	x_j/s
Well dopant concentration	N_C	$N_C \cdot s$
Power supply voltage	V_{DD}	V_{DD}/s
Equivalent oxide thickness	t_{eq}	t_{eq}/s
Dielectric field	E_{ox}	E_{ox}
Threshold voltage	V_T	$V_T/\sqrt{s}$
Silicon field at threshold	E_{Si}	$E_{Si}\cdot\sqrt{s}$
Lateral field in channel	E_y	E_y
Gate capacitance	C_G	C_G/s
Drive current	I_{Dsat}	I_{Dsat}/s
Inverter delay	$\tau = CV_{DD}/I_{Dsat}$	τ/s
Power dissipation/circuit	$P = V_{DD}.I_{Dsat}$	P/s^2
Power density	P/A	P/A

Table 5.3 Scaling limits of key parameters

Parameter	Scaling limit
V_T	Off-current, active and standby power
L	Off-current, active and standby power, punch-through
E_{Si}	Mobility, reliability, power
x_j	Silicide spiking, sheet resistance
R_C, R_{VIA}	Contact, via resistivity, area
R_{ext}	x_j, reliability, spacers
J_{Wire}	Electromigration
t_{ox}	Reliability, quantum-mechanical tunneling [37]
N_{Ch}	μ_{eff}, GIDL, V_T – variability

The total power dissipation is the sum of active (switching) and standby power (Chap. 8):

$$P \approx n\ f\ C_L V_{DD}^2 + V_{DD} I_{Leak} W. \tag{5.75}$$

where n is the number of active circuits, C_L the load capacitance that must be charged and discharged in each clock cycle, f the clock frequency, and I_{Leak} the total leakage current of which I_{off} is a large fraction. The standby power due to I_{off} is

$$P_{\text{Standby}} = W_{Total}\ V_{DD}\ I_{off} = W_{Total}\ V_{DD}\ I_{D0} e^{qV_T/nkT}\, W, \tag{5.76}$$

where W_{Total} is the total on-chip *MOSFET* width in µm, and I_{off} the off-current in A/µm. The circuit delay is

$$\tau \approx \frac{C_L\ V_{DD}}{I_{Dsat}} s, \tag{5.77}$$

where C_L is the load capacitance to be charged and discharged. Thus, increasing I_{Dsat} reduces the delay.

5.4.2.4 Aggressive Scaling

Selective aggressive scaling is applied where high performance is needed and an increase in power consumption is acceptable. An example of aggressive scaling to improve current drive and speed is shown in Table 5.4 [38, 39]. An increase in vertical field by a factor $\varepsilon > 1$ is justified by an improvement in dielectric reliability as the oxide thickness is reduced.

Figure 5.26 shows how the off-current increases when V_T is reduced. The *MOSFET* is optimized for either high-performance and power, or low-power applications.

Table 5.4 Example of non-constant field scaling [38, 39] ($s > 0$, $\theta > 0$)

Parameter	Initial	Scaled
Channel length	L	L/s
Channel width	W	W/s
Gate area	A_G	A_G/s^2
Junction depth	x_j	x_j/s
Channel dopant concentration	N_C	$N_C \theta s$
Power supply voltage	V_{DD}	$V_D \theta/s$
Equivalent oxide thickness	t_{eq}	t_{eq}/s
Dielectric field	E_{ox}	$E_{ox}\ \theta$
Threshold voltage	V_T	$V_T \sqrt{\theta/s}$
Silicon field	E_{Si}	$E_{Si}\sqrt{\theta s}$
Lateral field	E_y	$E_y.\theta$
Gate capacitance	C_G	C_G/s
Drive current	I_{Dsat}	$I_{Dsat}\theta^2/s$
Inverter delay	$\tau = CV_{DD}/I_{Dsat}$	$\tau/\theta.s$
Power dissipation/circuit	$P = V_{DD}.\ I_{Dsat}$	$P\ \theta^3/s^2$
Power density	P/A	$(P/A).\theta^3$

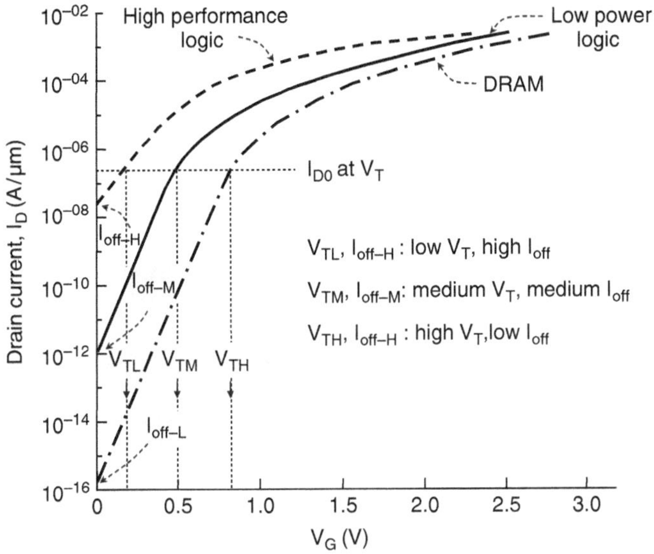

Fig. 5.26 Trade-off between power and performance

For high-performance (*HP*) digital applications, such as microprocessors, the transistor is aggressively scaled for high drive current I_{Dsat} to reduce circuit delay. A high-performance transistor is therefore designed with a low threshold voltage, high gate overdrive, and small channel length, resulting in high I_{Dsat} and high off-current $I_{off\text{-}H}$, hence high active and standby power. The increase in subthreshold slope in the top curve Fig. 5.26 is due to a short-channel effect discussed in the following section.

Applications in the intermediate range with medium threshold-voltage can be grouped into roughly two categories: low-operating power (*LOP*) and low standby power (*LSTP*). *LOP* transistors are not scaled as aggressively as *HP* transistors. They exhibit therefore lower drive-current, hence lower performance, and considerably lower off-current than *HP*. They are applicable to, for example, mobile computing such as notebooks. *LSTP* transistors are scaled conservatively to maintain very low off-current for mobile electronics with limited battery capacity.

In a *DRAM* cell (Fig. 5.18), where low leakage, hence long cell retention-time is more important than transistor speed, the transistor is optimized with a sufficiently high threshold voltage V_T and long channel to ensure a very low off-current ($I_{off\text{-}L}$) [40].

Power dissipation in high-performance applications limits the maximum V_{DD} that can be applied to the *MOSFET* and how far V_T and L_{eff} can be scaled-down. Figure 5.27 illustrates how the scaling limit on V_T can degrade circuit delay.

Assume, for example, that V_{DD} is scaled from 1.0 V to 0.7 V and, because of power constraints, V_T is kept at a nominal minimum of 0.3 V. Since I_{Dsat} is proportional to approximately the square of gate-overdrive

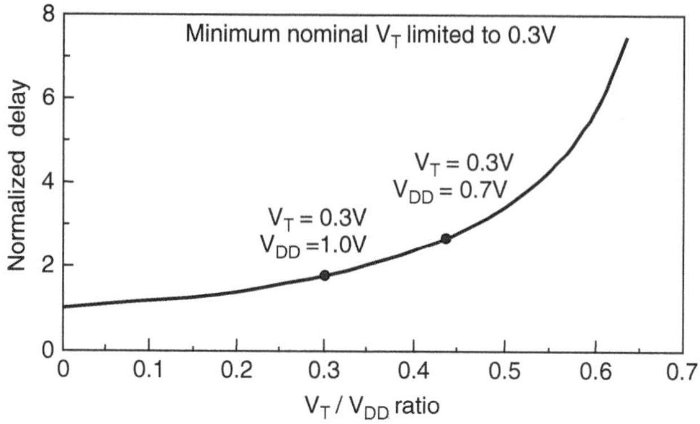

Fig. 5.27 Circuit delay versus V_T/V_{DD} ratio [40]. V_{DD} is the maximum voltage applied on gate and drain

$$I_{Dsat} \propto (V_G - V_T)^2,$$

the ratio of I_{Dsat} at $V_{DD} = 0.7\,\text{V}$ to I_{Dsat} at $V_{DD} = 1.0\,\text{V}$ is 0.33, increasing the delay time in (5.77) by 33%.

Fluctuations in processes, such as lithography, etch diffusion, and oxidation, cause variations in device parameters that require specification of tolerances around the nominal parameter values (Chap. 7). In particular, tolerances on threshold voltage, ΔV_T, and channel length, ΔL, have a profound effect on deep submicron and nanoscale *MOSFET* performance, power, reliability, and yield. For a specification of $V_T \pm \Delta V_T$ and $L \pm \Delta L$, the nominal values are typically defined by taking into account the minimum absolute values as the "worst-case" condition for I_{off}, power, and reliability. The maximum values then determine the worst-case performance.

The temperature dependence of V_T should also be considered (Problem 2). Since chips typically operate at an elevated temperature that can range from 65°C to 125°C, depending on application, the lowest nominal V_T should be defined at the maximum operating temperature.

Scaling to ultra-shallow junctions ($x_j < 100nm$) reduces the lateral junction-spread x_{jlat} and depletion width x_{dlat}, necessitating a smaller drawn channel length for a given L_{eff} than nonscaled junctions. Shallower source-drain junctions also reduce short-channel effects, as described in the following section. As x_j is reduced, however, the source-drain sheet resistance and hence R_{ext} increases. This requires novel doping techniques to increase the source-drain active dopant concentration without appreciably increasing $x_j$or x_{jlat}. Also, integration of silicide and contacts becomes more difficult because of possible increase in junction leakage due to metal penetration. Metal penetration should also be considered when thinning the polysilicon gate (Chap. 7).

Contact and via resistances become dominant at deep submicron and nanoscale dimensions. For a given contact or via resistivity, a reduction of the minimum

Table 5.5 Fundamental parameters that do not scale

Parameter	*MOSFET* parameters affected
Temperature, *kT/q*	Subthreshold slope, n_i, qV_{DD}/kT shrinks
ϕ_m, V_b	Contact potential
v_{sat}	I_{Dsat} at high field[1]

[1] Velocity improves by overshoot, quasi-ballistic and ballistic transport in deep submicron and nanoscale dimensions, as discussed later.

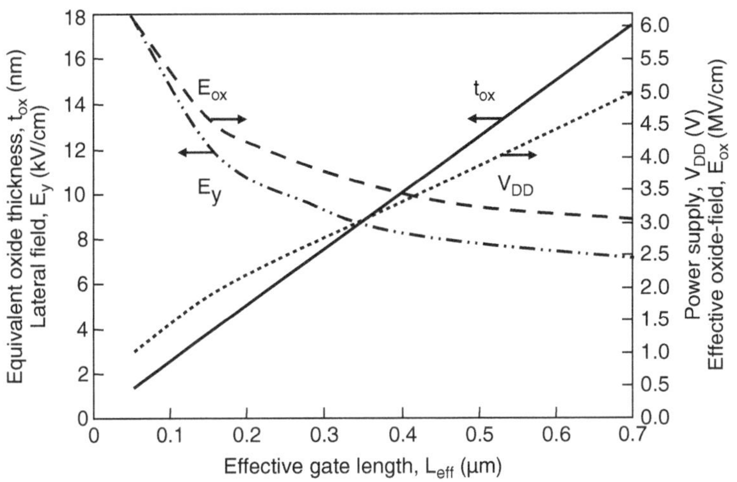

Fig. 5.28 Trends in oxide thickness t_{ox}, power supply voltage V_{DD}, oxide field E_{ox}, and lateral field E_y for high performance logic

contact width from 0.25 μm to 0.05 μm can increase resistance by a factor of 10–100, depending on current direction and contact geometry.

There are also fundamental physical parameters that directly affect transistor characteristics and do not scale. Among them are the Boltzmann factor *kT/q*, the workfunction and contact potential, and the saturation velocity (Table 5.5). The interface-state density Q_{it} is a process and material parameter that does not scale. It can degrade the surface mobility and modify the threshold voltage.

For high-performance logic, trends in equivalent oxide thickness t_{ox}, power supply voltage V_{DD}, oxide field E_{ox}, and lateral field E_y are shown in Fig. 5.28. The trend in oxide thickness is approximated as $t_{ox} \approx L_{eff}/45$ where L_{eff} is the minimum channel length for which long-channel subthreshold behavior is observed [38, 41]. The trend in V_{DD} is taken from [42, 43]. When calculating the maximum oxide field, a 10% tolerance was added to V_{DD}, and ~0.6 nm was added to the oxide thickness to account for the inversion layer thickness and polysilicon-gate depletion. The oxide field increases from ~3.5 MV/cm at $L_{eff} = 0.7\,\mu m$ to ~5.5 MV/cm at $L_{eff} = 0.1\,\mu m$. The lateral field, approximated as $E_y \approx V_{DD}/L_{eff}$, increases from ~70 kV/cm at $L_{eff} = 0.7\,\mu m$ to ~180 kV/cm at $L_{eff} = 0.1\,\mu m$.

5.4.3 Short-Channel Effects, SCE

The gradual-channel approximation in Sect. 5.4.1.2 applies to a long channel in which the surface potential can be assumed to be uniform over most of the channel, the lateral field is small compared to the vertical field, the mobility along the channel is constant, and the drain bias has negligible effect on threshold voltage and effective channel length. In long channels, a one-dimensional analysis can be made under the assumption that the entire channel is in "full control" of the gate. This assumption, however, becomes increasingly inaccurate as the channel length is reduced, where several deviations from long-channel behavior are observed that require two-dimensional or even three-dimensional analysis to accurately simulate the *MOSFET* behavior. Several approximations and simplifying assumptions are made to describe, in simple terms, important short-channel effects (*SCE*) and their relations to device structure and impurity profiles.

5.4.3.1 Channel-Length Modulation

For long channels, the saturation conductance defined by (5.37) is near zero since it can be assumed that the spread in lateral depletion width at the drain boundary, δL in (5.33), is negligible compared to the total channel length. In short channels, δL constitutes an appreciable fraction of the channel length. Since inversion carriers travel through the depleted region at saturation velocity, the contribution of δL to the total channel resistance is negligible. The effective (electrical) channel length, L_{eff}, is hence essentially the length of the region between the source metallurgical junction and pinch-off point. Since δL increases with increasing drain bias, L_{eff} decreases and I_{Dsat} increases appreciably as V_D increases. This manifests itself as an increase in the slope of the $I_D - V_D$ characteristics at constant V_G, (Fig. 5.29). Channel-length modulation in *MOSFETs* is similar to base-width modulation in bipolar transistors. It is characterized by an Early voltage V_A which is the intercept of the extended $I_D - V_D$ characteristic with the V_D axis, as shown in Fig. 5.29 for $V_G = 2$ V. The slope of the line is approximately

$$g_D = \left.\frac{\partial I_D}{dV_D}\right|_{V_G} = \frac{I_D'}{|V_D'| + |V_A|}.$$

V_A is then defined as

$$|V_A| = \frac{I_D}{g_D} - \left|V_D^I\right| \quad V. \tag{5.78}$$

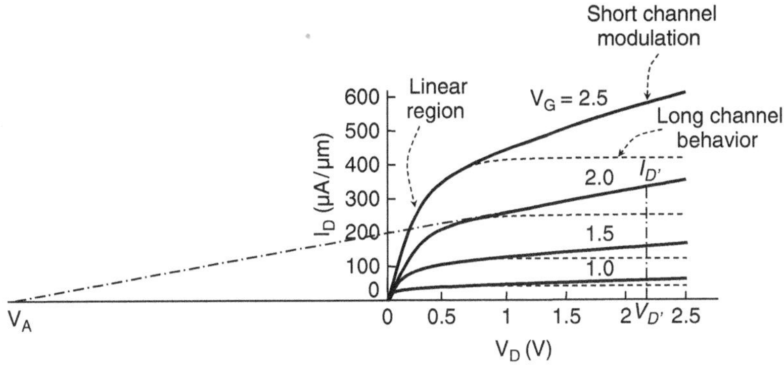

Fig. 5.29 Increase in I_D with increasing V_D in saturation due to channel -length modulation. V_A is the Early voltage

The Early voltage is particularly important to analog applications where linearity is essential.

5.4.3.2 Dependence of V_T on Channel Length and Drain Bias

In long channels, the threshold voltages in the linear and saturation modes are practically identical and independent of channel length and drain bias. As the effective channel length L_{eff} is reduced, the linear V_T decreases and the saturation V_T decreases further. The spread between both threshold voltages becomes larger with decreasing L_{eff} and increasing V_D. A simple model to explain the dependence of V_T on channel length is illustrated for an *NMOS* in Fig. 5.30 [44].

Neglecting fringe-fields, the positive charge on the gate, Q_m, in an isolated *MOS* structure is uniformly neutralized by negative charge of ionized acceptors in silicon, Q_b (Fig. 5.30a). The charge on the gate per unit width (normal to the paper) is

$$Q'_m = -Q_b L \quad C/cm,$$

where L is the length of the structure. In an isolated pn junction with zero surface charge, the positive ionized donor charge in the n^+-region is fully neutralized by negative acceptor charge at the junction floor and perimeter (Fig. 5.30b).

In a *MOSFET* with the gate biased toward inversion, the charge at the source and drain edges is partially neutralized by ionized acceptor charge in the p-body and partially compensated by the gate in the overlapped regions. This means that a fraction of the bulk charge in the channel region is neutralized by source and drain and less charge in the field-induced depletion region is associated with the gate. The bulk charge per unit width is approximated by a trapezoidal shape, as shown in Fig. 5.30c. The gate charge per unit width can then be expressed as

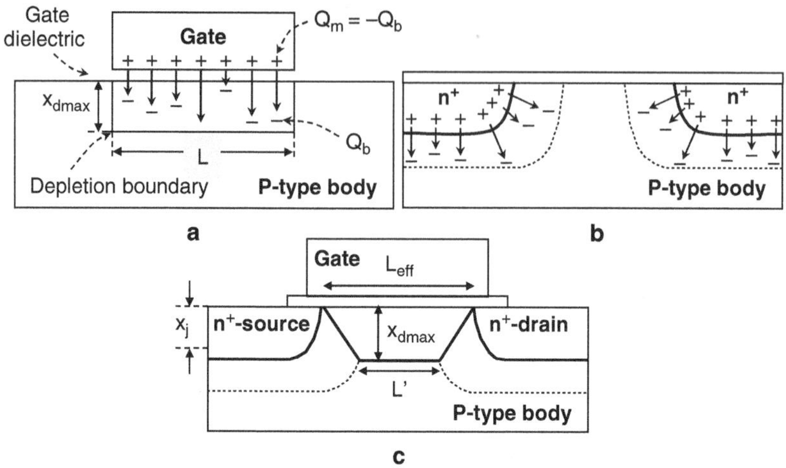

Fig. 5.30 Charge-sharing. **a** Isolated *MOS* capacitor. **b** Isolated junctions. **c** Charge-sharing in *MOSFET* [44]

$$Q_m' = -Q_b \frac{L_{eff} + L'}{2} \quad C/cm. \tag{5.79}$$

This is smaller than the charge per unit width Q_b. L_{eff} that would be induced in the absence of source and drain. The net result is a reduced threshold voltage. In a long channel, the difference between the value in (5.79) and Q_b L_{eff} is negligible and V_T is not appreciably affected. Since charge-sharing at the source and drain boundaries is essentially independent of channel length, the difference becomes larger as the channel gets shorter, decreasing V_T. The difference is found to increase with increasing source and drain junction depth [43–45].

Figure 5.31 shows the decrease, or "roll-off," of linear and saturation *NMOS* V_T as the channel length is reduced. There is more roll-off in the saturation mode than in the linear mode because of the larger spread in depletion at high drain bias. While the charge-sharing model and its variants describe the dependence of linear V_T on channel length with reasonable accuracy [44–47], it is not well-suited for the saturation mode where two dimensional analyses become essential. Approximate two-dimensional analytical solutions have been successfully implemented to analyze the *MOSFET* in saturation [48–51]. The models show that, as the drain voltage is increased in a short channel, the field lines emanating from the drain extend toward the source, reducing the barrier for minority-carrier injection at the source, hence reducing V_T. This is referred to as drain-induced barrier lowering, DIBL [49–51]. More accurate results and a better understanding of the dependence of threshold voltage on channel length and applied voltage can be obtained from numerical two-dimensional *MOSFET* analysis [52–56]. Figure 5.32 shows the effect of drain voltage on the barrier V_b as obtained from numerical simulations of an *NMOS* [56].

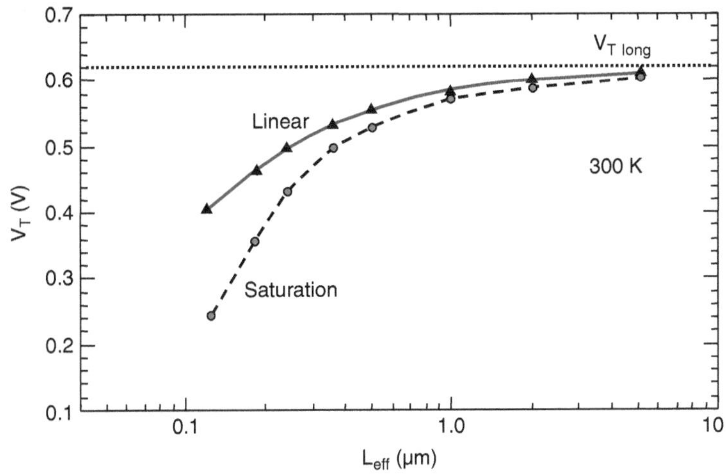

Fig. 5.31 Roll-off of *NMOS* V_T with decreasing channel length and increasing drain voltage

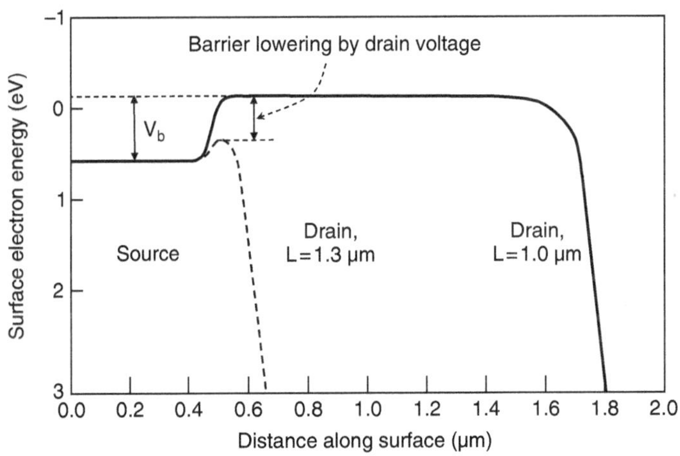

Fig. 5.32 Surface electron energy versus lateral position along the channel for two channel lengths

The source is a "reservoir" of electrons at ground, and the drain a deep potential well. The source-body junction is characterized by a barrier V_b to electron injection and, to turn-on the device, the gate voltage is increased to lower the barrier. In the long channel, the barrier is flat over most of the channel and the drain field does not affect the barrier at the source. Any mechanism, other than the gate voltage, that lowers the barrier reduces V_T. When the channel-length is sufficiently reduced, drain field-lines begin to reach the barrier region, sharing Q_b and lowering the barrier for electron injection from the source into the channel. The current increases then exponentially with drain voltage and less gate voltage is needed to turn-on the device. When the depletion regions merge, barrier-lowering at the source becomes approximately a linear function of drain bias.

The voltage needed to turn-on the device can be approximated by the linear relation

$$V_{T-short} = V_{T-long} - \frac{\gamma}{L} V_D \quad V, \tag{5.80}$$

where γ is a constant obtained from measurements.

5.4.3.3 Velocity Saturation

The lateral field in deep submicron *MOSFET*s can exceed the "critical field" at which the carrier velocity reaches its scattering-limited velocity, v_{sat}. For example, in an *NMOS* with $L_{eff} = 0.1\,\mu\text{m}$ operating at $V_G = V_D = 1\,\text{V}$, the lateral field is $\sim 10^5\,\text{V/cm}$. In this case, the transit time in (5.68) can be expressed as

$$\tau = \frac{L_{eff}}{v_{sat}} \quad s, \tag{5.81}$$

and the drain current in (5.34) becomes

$$I_D = \frac{W_{eff}\, L_{eff}\, \bar{Q}_n}{\tau} = \bar{Q}_n\, W_{eff}\, v_{sat} \quad A, \tag{5.82}$$

independent of channel length! In deep submicron *MOSFET*s, the drain current is, however, found to increase beyond the value given by (5.82) due to a mechanism known as velocity overshoot, as discussed in Sect. 5.5.1.3.

5.4.3.4 Punch-Through

The same mechanism that reduces the magnitude of threshold voltage causes punch-through. The difference between threshold-voltage lowering and punch-through lies in the role of the gate. As V_T is reduced, I_{off} increases, as expected from the exponential dependence of I_{off} on V_T in (5.58). When the channel length is reduced and the drain bias increased, positive charges in the drain begin to be imaged directly on electrons in the source, inducing transport of electrons from source to drain without the contribution of the gate. This gives rise to an additional current component superimposed on the "normal" subthreshold current and referred to as punch-though current. It manifests itself as an increase in subthreshold slope and I_{off} (Fig. 5.33). Since the barrier-lowering mechanism is a two-dimensional drain-field effect, the dependence of slope and I_{off} on V_D in Fig. 5.33 can only be reproduced numerically.

Figure 5.34 shows the variation of measured *NMOS* and *PMOS* subthreshold slope as a function of channel length [57]. As can be seen, the slope increases as the channel length is reduced. In the extreme case of punch-through ($V_D = 6\,\text{V}$ in Fig. 5.33), the drain current becomes practically independent of gate voltage.

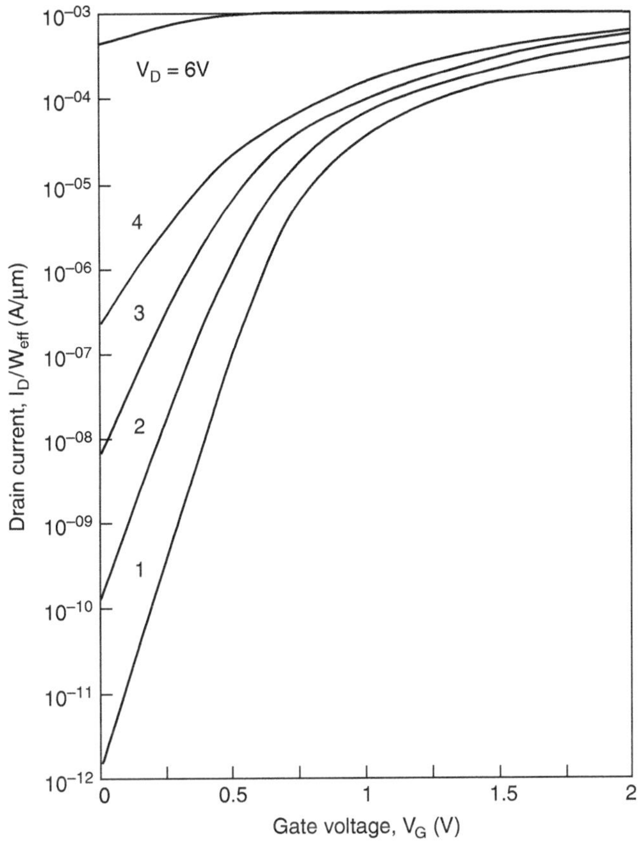

Fig. 5.33 Subthreshold characteristics of a non-optimized *NMOS* exhibiting punch-through

The onset of punch-through can be defined as the bias condition where the drain and source depletion regions merge. In a one-dimensional approximation (Fig. 5.35a), this is when

$$x_{dS} + x_{dD} = L_{eff} \ ,$$

where x_{dS} and x_{dD} are, respectively, the source and drain depletion widths. For a uniformly doped body, the punch-through voltage can be approximated as

$$V_{PT} \approx L_{eff}^2 \frac{qN_A}{2\varepsilon_0\varepsilon_{Si}} - 1.8 \cong 7.72\, L_{eff}^2 N_A - 1.8 \quad V, \tag{5.83}$$

where L_{eff} is in μm and N_A in cm^{-3}. The 1.8 term is the approximate sum of the built-in voltages at the source and drain.

In *MOSFET*s, punch-through typically occurs below the channel, where there is less control by the gate and the body dopant concentration is lower than at near the

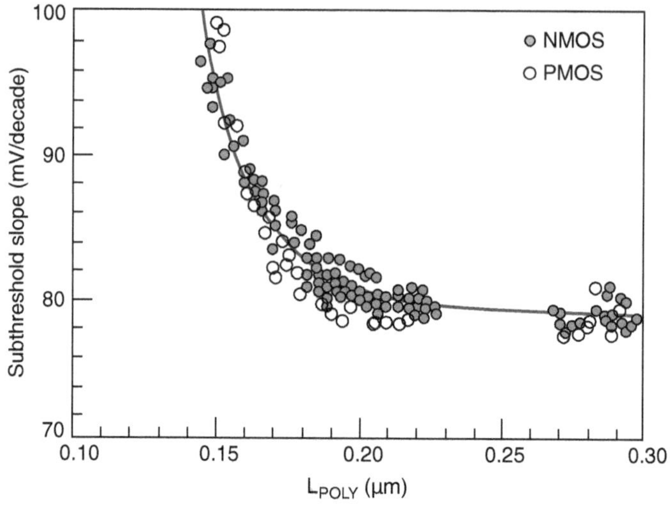

Fig. 5.34 Measured *NMOS* and *PMOS* subthreshold slope as a function of channel length (Adapted from [57])

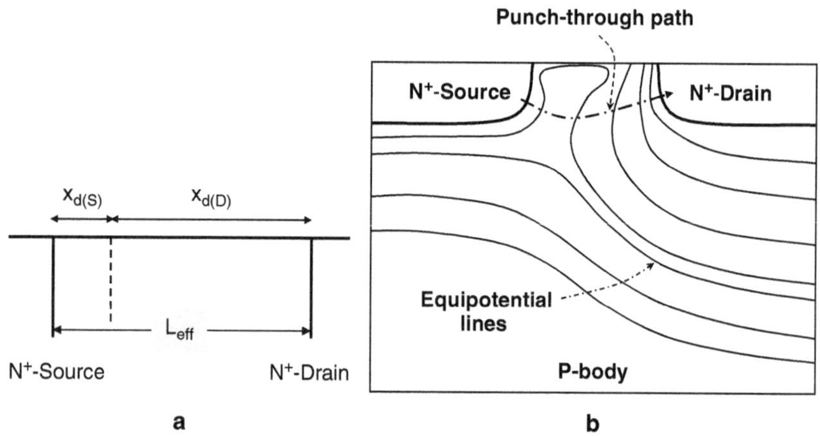

Fig. 5.35 Punch-through condition. **a** One-dimensional approximation. **b** Equipotential lines and approximate punch-through path

surface. Figure 5.35b shows two-dimensional contours of equipotential lines and the approximate path of punch-through current.

At punch-through, the drain bias effectively forward-biases the source-to-body junction. The current then increases exponentially with increasing drain bias above punch-through:

$$I_{PT} \approx I_{PT0} e^{aq(V_D - V_{PT})/kT} \quad A, \tag{5.84}$$

where a is a fitting parameter and I_{PT0} is the drain current at onset of punch-through.

The V_T roll-off and punch-through current are short-channel effects that can be reduced by increasing the dopant concentration in the channel region from the surface to approximately the depth of the contacted junction to minimize the spread of the depletion layers at the source and drain. This is done, however, at the cost of increasing the body-effect and junction capacitance and reducing the mobility and junction breakdown voltage. Methods to optimize the channel profile are discussed in Chap. 7.

5.4.4 Reverse Short-Channel Effects, RSCE

As *MOSFET* process technologies advanced, a new short-channel phenomenon was observed as an initial increase in V_T prior to the "normal" roll-off as the channel length was reduced (Fig. 5.36). For lack of a model, this effect, first observed in 1981 [58], was called "Anomalous short-channel effect" or "Delayed short-channel effect." The effect is now commonly referred to as "Reverse short-channel effect, *RSCE*."

Several models have been proposed to explain *RSCE* [59–70]. They can be arranged in roughly two groups, one based on a laterally nonuniform dopant profile in the channel and the other on nonuniform interface or oxide charge distribution along the channel.

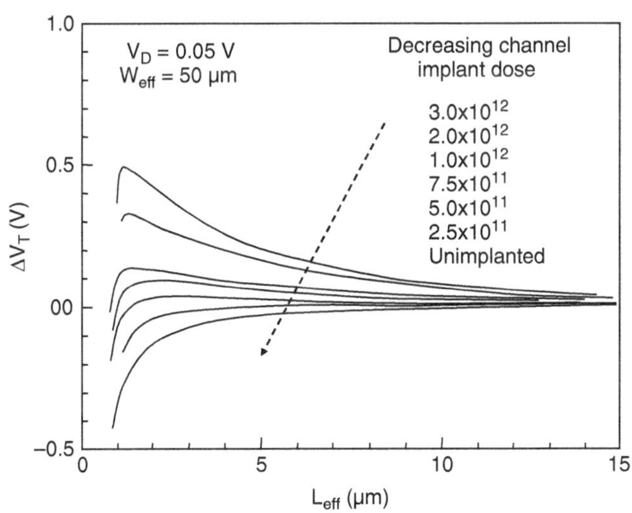

Fig. 5.36 Change in V_T as a function of effective channel length for various implant doses (Adapted from [58])

5.4.4.1 Nonuniform Dopant Distribution along the Channel

The results in [58] are reproduced in [59, 60] and explained by oxidation-enhanced diffusion (*OED*) of boron in the channel during the reoxidation step after polysilicon-gate patterning.[2] To increase the punch-through voltage, a multiple channel implant is implemented that includes an "anti-punch-through" implant beneath the channel where punch-through would occur (Fig. 5.35b). During the reoxidation step, silicon interstitial point defects are injected in the junction region and diffuse laterally into the channel with an average diffusion length of $\lambda_i = 1.4\,\mu\text{m}$ obtained by fitting the data. The two-dimensional coupled interstitial-boron diffusion from the "anti-punch-through" region toward the surface results in a laterally nonuniform increase in surface boron concentration. For $L_{Poly} \gg \lambda_i$, the concentration increases near the edges, causing a "small" increase in V_T. For $\text{L}_{\text{Poly}} \leq 2\lambda_i$, the increased concentration from the source and drain sides overlap, further increasing the threshold voltage for short channels [59, 60]. The effect is also observed for *PMOS* with an anti-punch-through phosphorus implant, but to a smaller extent.

RSCE is also attributed to enhanced boron diffusion during self-aligned silicidation [61]. The model is based on the diffusion of silicon into the silicide creating vacancies that are injected into the channel. As in the previous model, this results in enhanced boron diffusion towards the channel surface. Enhanced diffusion is, however, attributed to vacancies rather than silicon interstitials. The effect is not observed for *PMOS* (N-body).

Excess channel dopant concentration near source and drain edges are also observed to contribute to *RSCE* in *MOSFET*s with channels down to deep submicron dimensions [62–67]. This is particularly important to explain *RSCE* for structures which have a uniform vertical profile in the channel rather than a graded anti-punch-through implant profile examined in [59, 60]. The proposed mechanism is based on transient-enhanced diffusion, *TED*, caused by point-defects generated by the source-drain ion implantation (Fig. 5.37). As a result, boron diffusion is enhanced near junctions where it more readily moves to the surface, causing accumulation near the surface adjacent to the junctions and a gradient in the lateral boron profile from edge to center [62–65]. The threshold voltage therefore initially increases as the channel length is reduced.

The model is supported by measurement of temperature dependence of *RSCE* showing that the effect decreases from 300 K to 77 K, an observation that can only be explained by enhanced boron concentration at the source and drain edges [66]. In addition, *MOSFET*s fabricated on silicon-on-insulator (*SOI*) with thick buried oxide (*BOX*) exhibit a decrease in *RSCE* as the top silicon thickness is reduced, a trend explained by a reduced lateral distribution of silicon interstitials due to their high recombination velocity in the buried oxide that acts as a sink [67].

[2] For more details on unit processes, the reader is referred to Fundamentals of Semiconductor Processing Technologies, by B. El-Kareh, Kluwer Academic Publishers, 1995. Process integration is discussed in Chap. 7.

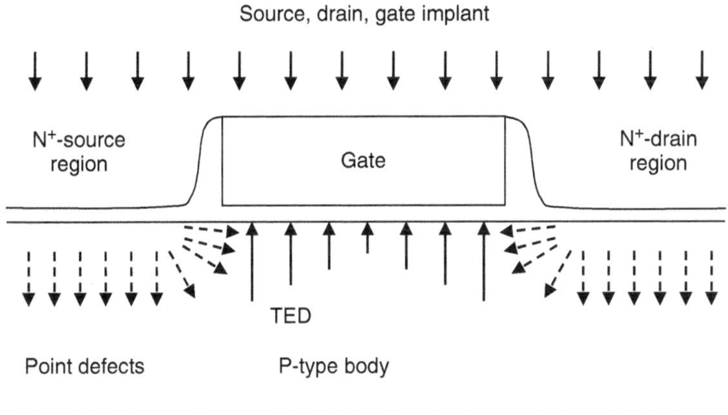

Fig. 5.37 Conceptual diagram of local interstitial point-defect injection during source-drain implantation, causing TED

5.4.4.2 Nonuniform Interface Charge Distribution in Channel

The lateral distribution of interstitials is also believed to create negatively-charged interface states at the edges of the source and drain, that locally increase the magnitude of V_T and account for *RSCE* in *NMOS* as the channel length is reduced [68,69]. There is no mention, however, of how this localized charge affects *PMOS*. A model for *RSCE* in *PMOS* is suggested, based on boron penetration from the polysilicon gate through the gate oxide [70–72]. In *MOSFET*s with very thin oxide, boron can penetrate into the channel surface, creating negatively charged ions. Boron penetration is more severe with BF_2^+-implanted or fluorine co-implanted gates [70, 71]. A positive shift in surface charge increases the magnitude of the *PMOS* threshold voltage. In this model, it is believed that sidewall oxidation depletes boron from the edges so that V_T is more negative in the center than at the edges, explaining the initial increase in V_T as the channel length is reduced [72].

In deep submicron *MOSFET*s, the channel concentration is locally increased near the source-drain edges to reduce the V_T roll-off and punch-through current while minimizing the impact on body-effect and inversion-carrier mobility. For example, boron can be implanted at an angle with respect to the *NMOS* polysilicon-gate sidewall at an appropriate energy and dose to increase the concentration beneath the channel near the source-drain edges and reduce the spread of source and drain depletion layers [73, 74] (Fig. 5.38). The resulting profile is called a "halo" or "pocket" (Chap. 7). Depending on the implant conditions, the halo can encroach into the surface and locally increase the threshold voltage near source and drain and cause RSCE. This effect can be utilized to balance the V_T roll-off and RSCE and achieve a more uniform V_T in short channels by optimizing the channel and halo implant profiles [75]. The halo concentration is typically negligible when compared to the polysilicon-gate concentration. At the polysilicon-oxide boundary, however, there may be some noticeable compensation of arsenic in polysilicon by the boron

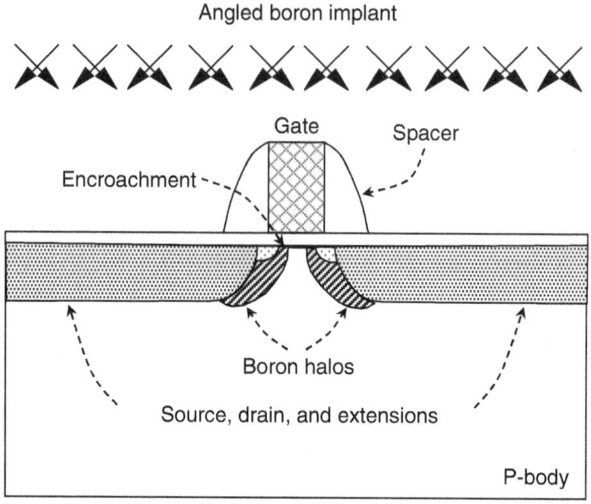

Fig. 5.38 Large tilt-angle implant to form boron "halos" in deep submicron *NMOS* [72, 73]

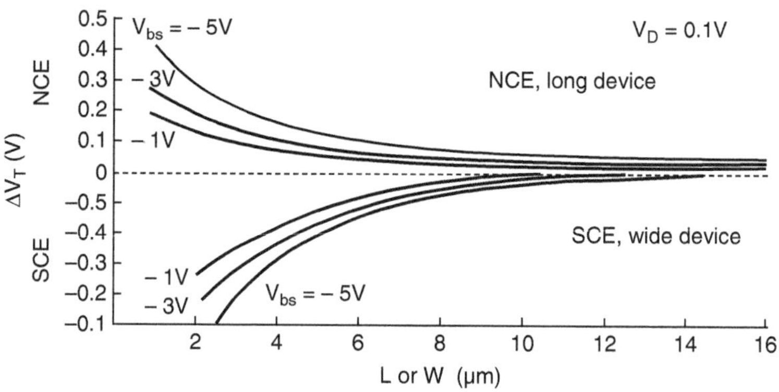

Fig. 5.39 Change in threshold due to short and narrow-channel effects (Adapted from [77])

halo implant. This can enhance the polysilicon depletion effect and locally increase the equivalent gate-oxide thickness, hence the threshold voltage near the source and drain edges, causing RSCE [76].

5.4.5 *Narrow Channel Effects, NCE*

In contrast to the short-channel effect where $|V_T|$ decreases as the length L is reduced, the narrow-channel effect, NCE, is an increase in $|V_T|$ as the width W is reduced. Figure 5.39 compares the short- and narrow-channel effects for a

LOCOS-isolated *NMOS* with an implanted channel having a peak boron concentration of $\sim 6 \times 10^{16}\,\text{cm}^{-3}$ (Chap. 7) [76]. By adjusting L and W, the short- and narrow-channel effects can be made to cancel each other.

The narrow-channel effect was first observed in 1975 when extracting the channel vertical impurity profile from measurement of threshold voltage as a function of substrate to source reverse voltage (body-bias effect) [77]. The threshold voltage was found to increase not only with reverse body to source voltage but also with decreasing channel width as the gate-width is reduced to the same order of magnitude as the field-induced depletion under the gate.

A simple geometrical model to explain the narrow-width effect is illustrated in Fig. 5.40 where the depletion region is shown, looking from source to drain, to extend beyond the gate-width. The threshold voltage increases, owing to the extra bulk charge ΔQ_b at the edges of the fringe field-induced depletion region outside the metal gate. While the contribution of ΔQ_b to the total Q_b is negligible in wide channels, it becomes increasingly important as the channel width is reduced.

Figure 5.41 is a schematic cross-section of an older-generation NMOS fabricated with just four masking steps, looking from source toward drain. To form the structure, openings are first patterned in the thick field oxide to define the active area, including source, drain and channel. Only the channel of patterned width W is shown in the figure.

After source and drain are formed, the thin gate oxide is grown, contacts are etched and aluminum is deposited and patterned. The thick field oxide isolates *MOSFET*s laterally from each other. The taper is caused by isotropic wet etching,

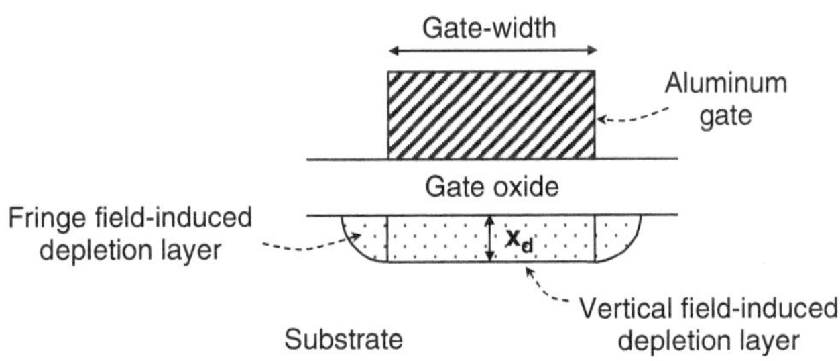

Fig. 5.40 Geometrical model to explain the narrow-channel effect. The schematic cross-section is shown looking from source toward drain. The depletion layer extends outside the gate width [37]

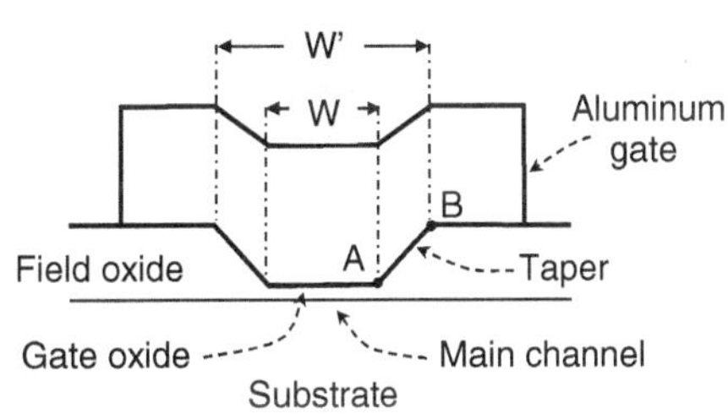

Fig. 5.41 Schematic cross-section of tapered field-oxide, shown looking from source toward drain [78]

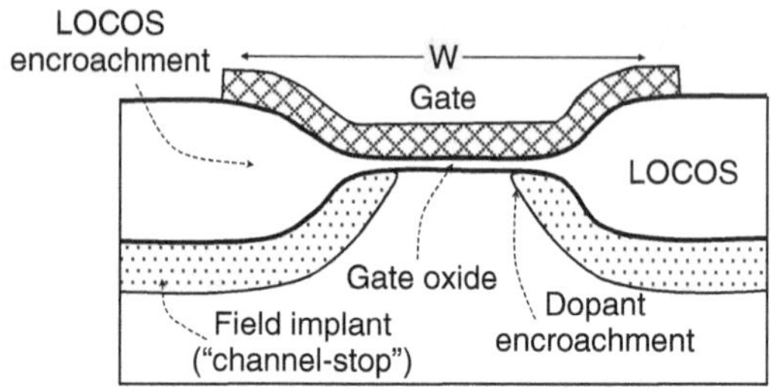

Fig. 5.42 Schematic cross-section of LOCOS-implanted *MOSFET*, looking from source toward drain

increasing the width from W at the bottom of the opening to W' at the top. This creates a position-dependent oxide thickness and hence threshold voltage that increases along the taper. As the gate voltage is increased, the inversion layer initially forms under the thin gate oxide of width W and then expands laterally under the tapered regions, widening the effective (electrical) channel width, W_{eff} [78]. For any width W, the tapered edges of the channel invert at a higher gate voltage than the center thin-oxide region and the threshold voltage of the entire channel (thin oxide and tapered edges) increases with decreasing channel width for small W.

In more advanced technologies, the thick field oxide in grown locally and partially recessed by a process commonly known as local oxidation of silicon, or *LOCOS* (Fig. 5.42). Local oxidation is achieved by patterning a silicon-nitride film, leaving nitride over active areas where it acts as a barrier to oxidizing species. Oxidation proceeds vertically and laterally, resulting in some lateral encroachment into the active area and a taper in the oxide. To increase the *LOCOS* threshold voltage, dopants of the same type as in the *MOSFET* body are implanted into the *LOCOS* regions to form a "channel-stop" under the field oxide [79]. Since this is typically done at an early stage of the process, subsequent thermal cycles can cause dopants to diffuse toward the surface and encroach into the channel boundaries, locally increasing the threshold voltage in those regions. The threshold voltage increases from channel center to channel edges because of the encroachment of both the channel-stop implant and *LOCOS*. Therefore, as the gate voltage is increased, the inversion layer initially forms in the channel center and gradually expands to the edges, increasing the effective channel width [80]. For a given channel width, the combined effects of oxide taper and dopant encroachment are found to increase the threshold voltage with decreasing width for small W [81, 82].

A quantitative analysis of the narrow-channel effect as a function of oxide taper and lateral impurity profile near the channel edges requires two- or three-dimensional solutions. A schematic cross-section of an *NMOS* in strong inversion is shown in Fig. 5.43, looking from source toward drain. A uniformly doped substrate at ground and an abrupt transition from thin to thick oxide are assumed for simplicity.

In the middle of the channel, away from the oxide edges, the depletion depth is x_{dmax} and the surface potential ψ_s is uniform. In strong inversion, $\psi_s \approx 2\phi_b$, (ϕ_b: bulk Fermi potential). The lateral field E_y is zero and Poisson equation can be solved in

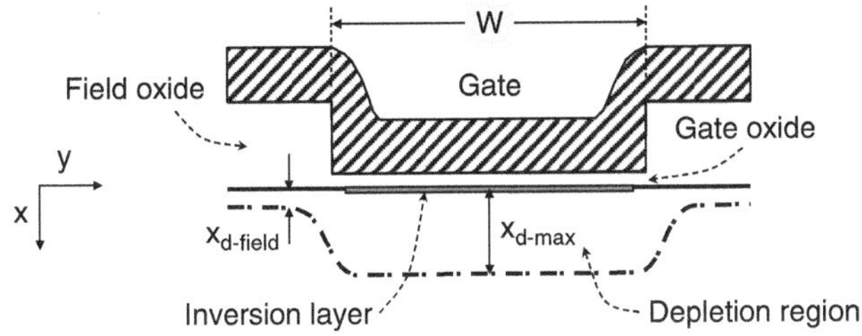

Fig. 5.43 Schematic cross-section of wide-channel *NMOS* in inversion, looking from source toward drain

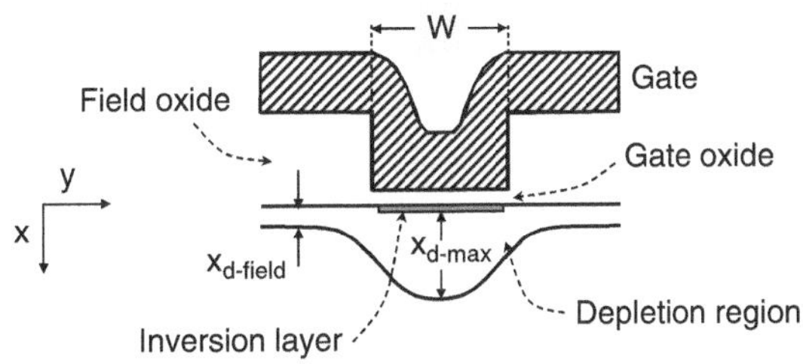

Fig. 5.44 Schematic cross-section of narrow-channel *NMOS* in inversion, looking from source toward drain

one-dimension to calculate the field and potential distribution within the depletion region

$$\frac{dE_x}{dx} = -\frac{qN_A}{\varepsilon_0 \varepsilon_{Si}}, \tag{5.85}$$

where N_A is the bulk acceptor concentration. Under the field oxide, away from the edge, the surface potential $\psi_{s\text{-}field}$ is smaller than $2\phi_b$ and $x_{d\text{-}field} < x_{dmax}$ because of the larger gate-voltage drop across the thick field oxide. Near the edges, there is a transition from $\psi_{s\text{-}field}$ to $\approx 2\phi_b$ and hence $x_{d\text{-}field}$ to x_{dmax}, where a lateral field component $E_y(x)$ exists and Poisson's equation must be solved in two dimensions

$$\frac{\partial E_x}{\partial x} + \frac{\partial E_y}{\partial y} = -\frac{qN_A}{\varepsilon_0 \varepsilon_{Si}}. \tag{5.86}$$

The lateral field component can be visualized by considering that part of the field-lines emanating from positive charge at the gate edges end laterally on negatively ionized impurities under the field oxide. This constitutes an additional charge that must be imaged by the gate at turn-on, increasing the threshold voltage. The channel can be described by two regions in parallel, a center region with a target threshold voltage and edge regions with a threshold voltage higher than V_T. For a wide channel, this additional edge-charge is negligible compared to the total bulk charge under the thin oxide, and the increase in V_T is negligible. As the channel width is reduced and becomes comparable to x_{dmax} (Fig. 5.44), the additional edge charge becomes more important, increasing the overall *NMOS* V_T.

Numerical two-dimensional simulations have been used to demonstrate the effect of channel width on threshold voltage for the structures shown in Figs. 5.42–5.44

[83, 84]. As previously stated, V_T is found to increase with decreasing width. One reason for this increase is the additional depletion charge that must be induced by the gate in the transition regions. Another reason is related to the gradient in lateral field: as the width is further reduced, the fringe-field at the edges begins to reach the center of the channel and the depletion depth at the center becomes a function of the total device width [84, 85]. The effects are exaggerated by increasing the source to body bias because of the increase in the ratio of $x_{d\text{-}max}$ to $x_{d\text{-}field}$ [84].

5.4.6 Reverse Narrow-Channel Effects, RNCE

LOCOS is widely used to isolate *MOSFET*s of channel dimensions above $\sim$0.6 µm. As dimensions are scaled down to deep-submicron and nanoscale dimensions, the process becomes increasingly inefficient. One important reason is the lateral encroachment of *LOCOS* into the active area (Fig. 5.42), the so-called "bird's beak," that consumes part of the patterned active area and hence reduces circuit density. This limitation led to the development of fully recessed, box-shaped oxide isolation, commonly known as shallow-trench isolation, or *STI*. The shallow-trench isolation process eliminates the "bird's beak" and substantially improves planarity of the structure. One observed drawback of *STI* is, however, a *decrease* in threshold voltage as the width is reduced, referred to as the reverse narrow channel effect, *RNCE*. To simplify the description of this effect, only the two extreme cases of nonrecessed and fully-recessed isolation are considered. Figure 5.45 compares the two structures and the field distribution under the edges of their gates at turn-on [86, 87].

In the nonrecessed structure of Fig. 5.45a, the two-dimensional field creates an additional charge stored at the edges of the channel, resulting in an increase in threshold voltage. This is the "conventional" narrow-channel effect discussed in the preceding section. In the *STI* structure of Fig. 5.45b, the field distribution at the corners enhances the depletion under the gate, thus reducing the threshold voltage in those regions [86–89]. The threshold voltage therefore decreases from center to edge. The effect is aggravated by any mechanism that contributes to depleting

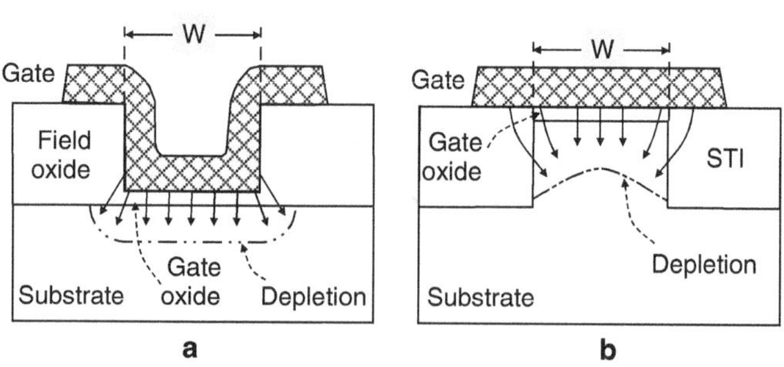

Fig. 5.45 Comparison of field distribution at turn-on of **a** Non-recessed field oxide isolation, **b** Fully-recessed shallow-trench isolation, *STI* (Adapted from [87]

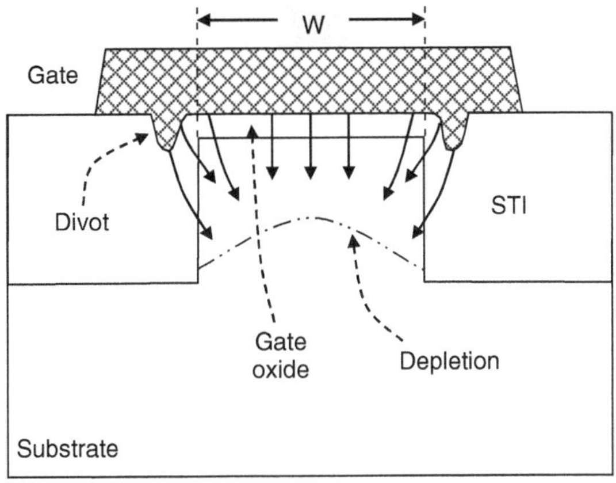

Fig. 5.46 Enhanced corner field caused by a divot with the gate conductor wrapping around the gate

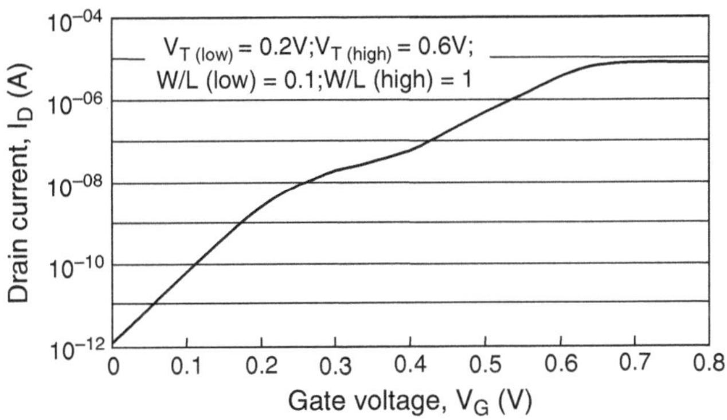

Fig. 5.47 Simulated "double-hump" in the $I_D - V_G$ characteristic of edge and center transistors in parallel

the channel near the edges. Among these are the depletion of channel dopants near the *STI* edges [90], interface charge at the channel-*STI* boundary [91, 92], and the presence of a "divot" that can be caused by stress-enhanced etch rate at the *STI* corners [93], allowing the gate conductor to wrap around the corners and intensify the two-dimensional field (Fig. 5.46). In this case, the entire channel region can be approximated by transistors of varying threshold voltage in parallel. Figure 5.46 illustrates this effect for two transistors, one in the center region with an assumed $V_T = 0.6\,\text{V}$ and $\text{W}/\text{L} = 1$ and the other near the edges with $V_T = 0.2\,\text{V}$ and $\text{W}/\text{L} = 0.1$. The resultant I_D–V_G characteristic exhibits a "double-hump," as shown in Fig. 5.47 [94, 95].

As the low-V_T edge-transistor reaches the threshold voltage, the higher-V_T center transistor is still in the subthreshold mode. The edge transistor is assumed to have a smaller subthreshold slope $\partial V_G/\partial(LogI_D)$ than the center transistor because the silicon capacitance is smaller at the edge [87, 88]. A similar edge effect is found

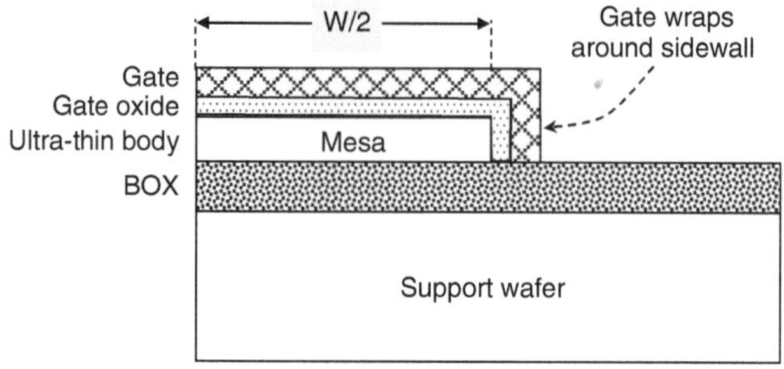

Fig. 5.48 Schematic cross-section of mesa isolated transistor with gate-oxide and gate wrapping around the body edge [96]

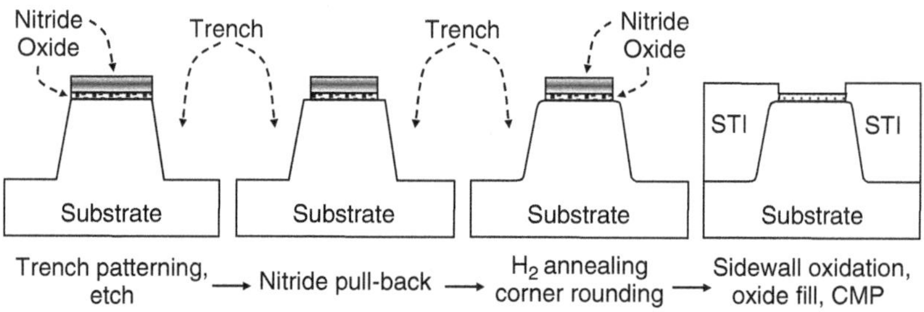

Fig. 5.49 Main process steps to round the *STI* corners with H_2-anneal. The nitride mask is pulled back before annealing [95]

on mesa-isolated silicon on insulator (*SOI*) transistors with the gate oxide and gate wrapping around the *MOSFET* body (Fig. 5.48) [96].

The narrow-channel effect can change from "conventional," that is, an increase in V_T and the width is reduced, to "inverse," where V_T decreases with decreasing width, as the recess-depth of the field oxide is reduced [97]. In fully-recessed isolation, *RNCE* can be suppressed by increasing the silicon concentration near the channel sidewall [98] and reducing dopant segregation at the channel sidewalls [99–102]. *RNCE* can also be suppressed by reducing the two-dimensional field at the *STI* corner. One method is to round the corner by high-temperature oxidation, for example, by rapid-thermal oxidation (*RTO*), immediately after trench etch [94], or after the *STI* chemical-mechanical polishing (*CMP*) [103]. Corner rounding can also be achieved after etching the trench by recessing the masking nitride and annealing in a hydrogen ambient at optimized temperature and pressure conditions [95, 104]. Hydrogen is found to enhance the migration of silicon at corners, resulting in a rounded edge. The main steps for this process are shown in Fig. 5.49.

Another method of corner rounding is shown in Fig. 5.50. A polysilicon buffer is placed between the *STI* masking nitride and oxide pad. When the trench sidewall is oxidized, a self-aligned oxide and "bird's beak" are formed at the *STI* corner, reducing the two-dimensional field [105].

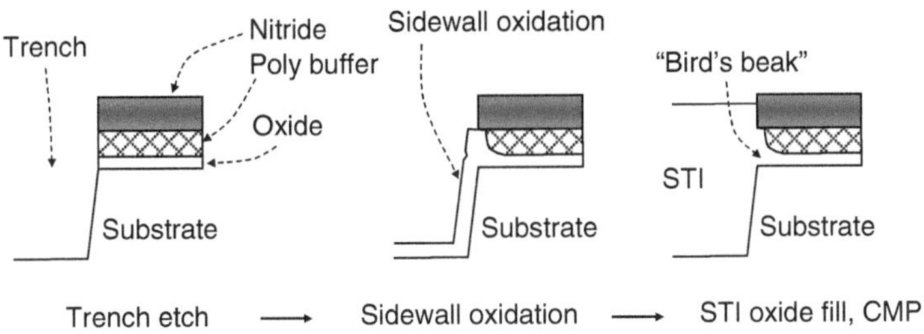

Fig. 5.50 Main process steps to round the *STI* corners with a polysilicon buffer mask [105]

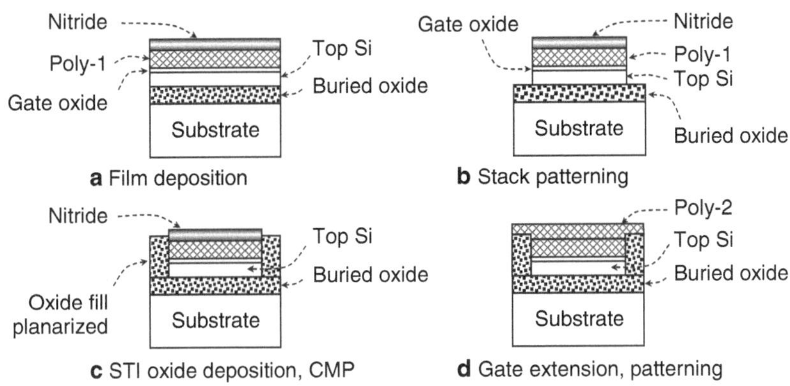

Fig. 5.51 Main process steps to elevate the *STI* above the silicon surface [106, 107]

Raising the *STI* above the silicon surface substantially improves the reverse narrow channel effect. This can be achieved by, for example, isolating the structure after gate-oxide and polysilicon deposition, as illustrated for an SOI *MOSFET* in Fig. 5.51 [106, 107].

Another *STI* -edge problem is found when the gate oxide thickness is reduced to sub-2.5 nm. The oxide is found to thin at the corner, reducing the threshold voltage and oxide reliability along the *STI* edges [108]. Thinning is attributed to local stress that reduces the oxidation rate during the initial stages of oxidation. Oxidizing in a nitrogen ambient using radical-*N* is found to improve oxide uniformity around *STI* corners, suppressing *RNCE* and improving oxide reliability [109, 110].

5.4.7 Small-Size Effects

Small-size effects include both short- and narrow-channel effects and are three-dimensional in nature. For simplicity, they have been treated as separate two-dimensional short-channel and narrow-channel effects. As both channel dimensions are reduced, however, it has become evident that short- and narrow-channel effects are coupled in modulating the depletion width so that three-dimensional models have become necessary to improve accuracy [87, 97, 111, 112].

Another issue with reducing the *MOSFET* size is the increase in the channel dopant fluctuation. Following the conventional scaling strategy, the channel dopant concentration must increase to suppress short-channel effects as the channel length is reduced. For nearly uniform channel profiles, the channel dopant concentration has actually increased from about $10^{16}\,\text{cm}^{-3}$ in micron-scale *MOSFETs* to above $10^{18}\,\text{cm}^{-3}$ in deep submicron and nanoscale technologies. A reduction in channel dimensions accompanied by an increase in dopant concentration can cause substantial fluctuations in threshold voltage due to statistical variation in dopant concentration as predicted in [113,114]. Experiments and simulations show that for a uniform channel impurity profile, the contribution of dopant fluctuation to the spread in V_T can be approximated by [115–117]

$$\sigma_{VT} \approx \frac{\sqrt[4]{q^3 \varepsilon_0 \varepsilon_{Si} \phi_b}}{\sqrt{2\varepsilon_0 \varepsilon_{ox}}} \frac{t_{ox} \sqrt[4]{N_{Ch}}}{\sqrt{L_{eff}\, W_{eff}}} \quad V, \tag{5.87}$$

where σ_{VT} is the standard deviation in V_T, t_{ox}, is the oxide thickness, N_{Ch} is the average channel dopant concentration, ϕ_b is the Fermi potential, and L_{eff}, W_{eff} are, respectively, the effect channel length and width. It is concluded that the variation in threshold voltage increases with increasing dopant concentration and decreasing channel dimensions. The variability in V_T is predicted to be dominated by the variation in dopant concentration in deep submicron and nanoscale technologies where $6\sigma_{VT}$ can reach about 250 mV. The spread in V_T also causes an increase in mismatch between identically designed adjacent *MOSFETs* [118–122]. This can become a serious problem in analog designs.

5.5 Mobility Enhancement

As the channel length is reduced below 100 nm, the *MOSFET* current drive capability increases, however, at the cost of increasing the off-state current, I_{off}, and hence the stand-by power. Enhancing the inversion-carrier mobility is a means of improving *MOSFET* performance without the need to reduce the channel length. As found in Chap. 1 (1.72), the one-dimensional low-field drift velocity is

$$v_{dy} = \frac{q\tau}{m_y^*} E_y = \mu_y E_y \quad cm/s, \tag{5.88}$$

where y is the direction from source to drain, v_{dy}, the drift velocity, τ the mean-free time between collisions, m_y^* the carrier effective mass, μ_y the carrier mobility, and E_y the lateral field. In the high-field region, above $\sim 10^4\,\text{cm/s}$, carriers lose their energy to the lattice predominantly by emission of optical phonons. In each optical phonon scattering event, a carrier loses an energy E_p to the crystal. From the energy balance, the drift velocity saturates to [8]

$$v_{sat} \approx \sqrt{\frac{E_p}{m*}} \cong 10^7 \quad \mathrm{cm/s}, \tag{5.89}$$

where $E_p \approx 0.063\,\mathrm{eV}$ is the optical-phonon energy. Carriers lose energy to the lattice through collisions at the same rate as they gain energy from the field. Thus, their average drift velocity no longer increases. The low-field drift velocity and hence carrier mobility can be increased by increasing the mean-free time between collisions. Both the low and high-field velocity can be increased by reducing the effective mass.

5.5.1 Mean-Free Time Between Collisions, τ

The mean-free time between collisions is limited by carrier scattering events. These include phonon (lattice) scattering, Coulomb scattering due to ionized impurities, interface charge scattering, and surface roughness scattering. Any mechanism that reduces the collision probability increases τ and hence μ.

5.5.1.1 Low Temperature

Phonon scattering can be decreased by reducing the operating temperature. Figure 5.52 shows the measured temperature dependence of the effective low-field electron and hole mobilities in the range 50–300 K, as extracted from the linear characteristics of long and wide channel *NMOS* [15, 123], and *PMOS* [124]. The effective mobility follows approximately the $T^{-1.5}$ dependence of lattice scattering. For a short-channel *MOSFET* in saturation, the inversion-carrier mobility is position-dependent and a function of a two-dimensional field. The drift velocity increases from source to drain where it reaches its saturation value of $\sim 10^7\,\mathrm{cm/s}$ at 300 K. The mobility is hence difficult to define, but an effective mobility can be extracted from the saturation *MOSFET* characteristics. The effective mobility in saturation is found to increase by a factor of 1.7 as the temperature is reduced from 297 K to 77 K [15]. A similar behavior is obtained for the inversion-hole mobility in *PMOS*, as shown for low and high vertical fields Fig. 5.52 [124].

While reducing the temperature to cryogenic levels substantially improves the carrier mobility and other important *MOSFET* parameters, such as subthreshold slope (Fig. 5.20), series resistances, and power dissipation, the need for a complex cooling medium makes this approach unattractive for most applications.

5.5.1.2 Channel Profile Optimization

There are two surface scattering mechanisms related to ionized impurities: Coulomb scattering and surface roughness scattering due to the surface field caused by the ionized impurity charge. Thus, an increase in channel dopant concentration degrades the surface mobility and *MOSFET* current drive. Several techniques have

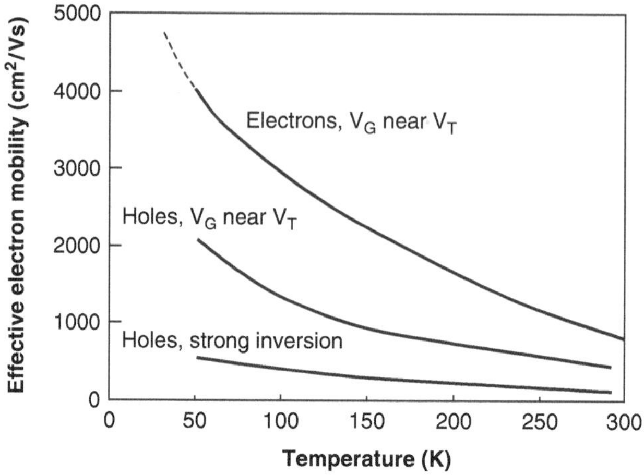

Fig. 5.52 Variation of effective electron mobility with temperature. Adapted from [15, 123] for electrons, from [124] for holes

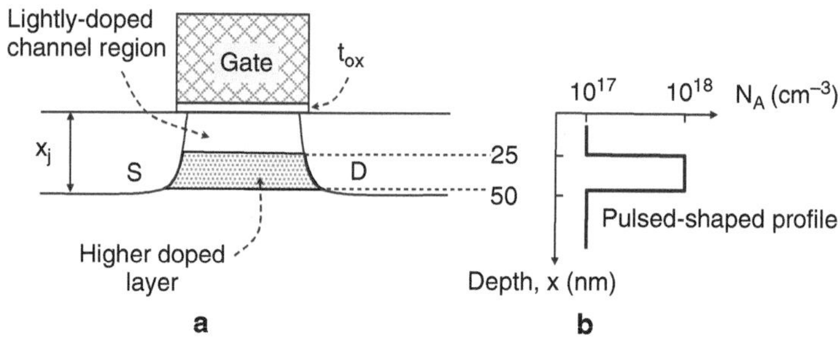

Fig. 5.53 Example of channel profile optimization. **a** Cross-section of lightly doped channel region over a heavier-doped layer; **b** Pulsed-shaped channel vertical profile (Adapted from [129])

been developed to achieve a channel of low dopant concentration, reducing the impact on surface mobility and dopant fluctuation, while suppressing short channel effects. Figure 5.53 illustrates an example of a pulsed-shaped profile of a low-doped channel and an underlying region of higher body concentration.

The lightly doped channel can be formed by low-temperature or ultra-high vacuum epitaxial growth of an undoped silicon layer on top of a higher doped body-region [125–127], by implanting a higher-doped pulsed-shape layer beneath a lightly-doped channel [128–130], or by implanting low-diffusivity dopants, such as indium for *NMOS* and antimony for *PMOS*, to form a super-steep retrograde well beneath the lightly-doped channel, as illustrated in Fig. 5.54 [131–133]. The vertical profile that is achieved is found to substantially improve the *MOSFET* current drive in the linear regime. Less improvement is found, however, in saturation because of the increased body-effect at the drain boundary and the resulting reduction in the saturation drain voltage, V_{Dsat} [134, 135].

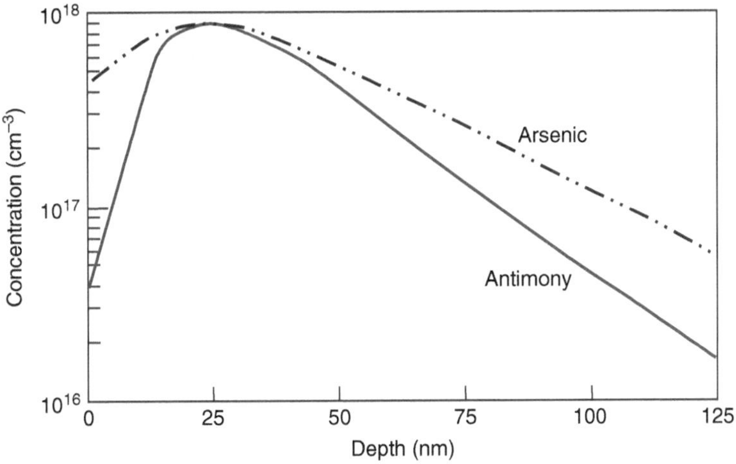

Fig. 5.54 Example of super-steep retrograde *PMOS* well. For the same peak concentration, the profile is steeper with antimony than with arsenic, resulting in lower surface concentration [131]

5.5.1.3 Velocity Overshoot

When carriers are injected from the source into the channel, they experience an abrupt change in electric field, from near zero within the source, where electrons are "cold," to an average field in the channel that can exceed 10^5 V/cm in deep submicron *MOSFET*s. The carriers "heat-up" in the channel as they are accelerated and gain energy from the lateral field. Whenever carriers travel through an abrupt increase in electric field, they exceed their steady-state drift velocity for a short period of time τ or a short distance λ, after which they gradually return to their equilibrium velocity as a result of collisions with the lattice. Similarly, when the field is instantaneously turned-off at a time $t = 0$, the average drift velocity will gradually approach zero and the decay will follow the expression given in (1.72)

$$<v_{dy}(t)> = <v_{dy}(0)>e^{-t/\tau},$$

where $<v_{dy}(t)>$ is the average drift velocity along the channel at time t, and $<v_{dy}(0)>$ is the average drift velocity at time $t = 0$. The quantity τ is called the relaxation time, with τ typically in the range 10^{-13} s to 10^{-12} s. If the channel length becomes comparable to the mean-free path between collisions λ, or if the transit time across the channel becomes comparable to the relaxation time τ, carriers injected at the source may not reach their steady-state velocity before being collected at the drain but can traverse the channel without collisions and thus attain a higher velocity than the steady-state velocity before relaxation effects take place [136,137]. This effect is referred to as velocity overshoot. The analysis of velocity overshoot is very complex, requiring detailed numerical simulations [136–141]. Results obtained from the Monte Carlo method are shown for electrons at 300 K in Fig. 5.55 [136].

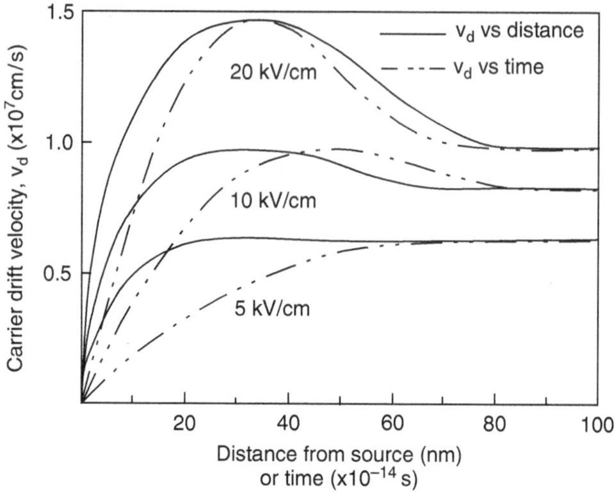

Fig. 5.55 Simulated electron drift velocity as a function of distance and time for an abrupt change in lateral field $\Delta Ey = 5\,\mathrm{kV/cm}$, $10\,\mathrm{kV/cm}$ and $20\,\mathrm{kV/cm}$ (adapted from [136])

The dashed lines show the time dependence of the average drift velocity of electrons injected "cold" at the source and drifting into the channel of high lateral field E_y, where E_y is assumed to be uniform for simplicity. The solid lines show the drift velocity as a function of distance from the source. It is evident from the plots that for a channel length of $0.1\,\mu\mathrm{m}$ or shorter, the average carrier drift velocity and hence the drain current can substantially exceed the value predicted without velocity overshoot. Velocity overshoot is also observed for any channel length in the drain region where there is a large field gradient [138]. The effect on the average drift velocity becomes, however, only appreciable in short- or ultra-short channels.

Electron velocity overshoot has been experimentally observed by extracting the drift velocity from measurements of transconductance on *MOSFETs* with $0.1\,\mu\mathrm{m}$ or shorter gate length at cryogenic temperatures. As the temperature decreases, the phonon mean-free path increases above its value at 300 K, increasing the probability for velocity overshoot. An increase in the average drift velocity by a factor of 1.8 to 2 from 300 K to 77 K for $L_{eff} = 0.1\,\mu\mathrm{m}$ [142, 143].

5.5.2 *Effective Mass*

In Chap. 1, it was shown that a carrier behaves in the crystal as if it had a mass of effective magnitude equal to

$$m* = \frac{\hbar^2}{d^2E/dk^2}, \tag{5.90}$$

where $\hbar = h/2\pi$ is the reduced Planck constant, E the energy and k the wave vector related to the carrier momentum p by $p = \hbar k$. The effective mass is determined by the curvature d^2E/dk^2 that depends on crystallographic orientation [18]. Thus, the mobility is anisotropic [144, 145]. A means to enhance the mobility is to constrain inversion carriers to flow in the direction of lower effective mass. This can be done by an appropriate choice of channel orientation, by applying an appropriate strain to the crystal to reduce the effective mass of electrons or holes in the direction of carrier flow, or by a combination of both.

5.5.2.1 Crystal Orientation

MOSFETs constructed on *(100)* surfaces in *<110>* directions exhibit the highest electron mobility and lowest interface charge density [144, 146]. These are the main reasons for the choosing the wafer orientation shown in Fig. 5.56. This orientation, however, does not optimize the hole mobility. Measurements show that in this orientation a *PMOS* exhibits about half the drive current of an *NMOS* of the same geometry under similar bias conditions.

Rotating the wafer in Fig. 5.56 by 45°, that is, changing the azimuthal direction of inversion carrier flow from *<110>* to *<100>*, improves the hole mobility by about 40% [147, 148]. An even greater improvement in the hole mobility is achieved on (110) wafers with the channel oriented in the <110> direction [149–157]. Figure 5.57 illustrates measured electron and hole mobilities as a function of

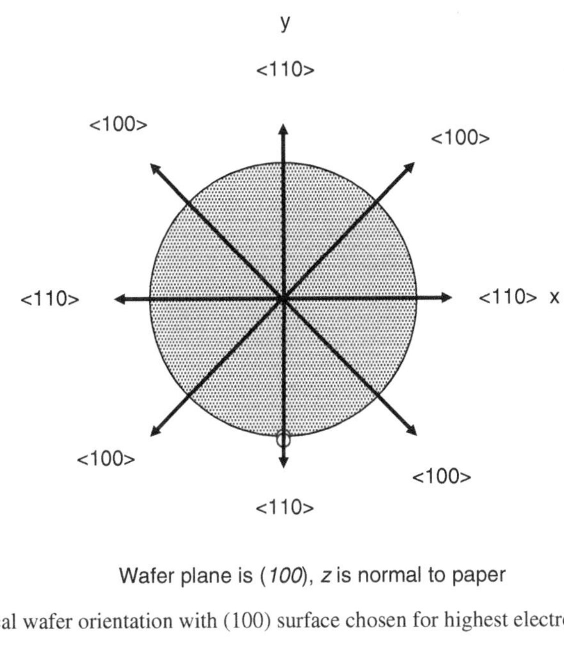

Fig. 5.56 Typical wafer orientation with (100) surface chosen for highest electron mobility

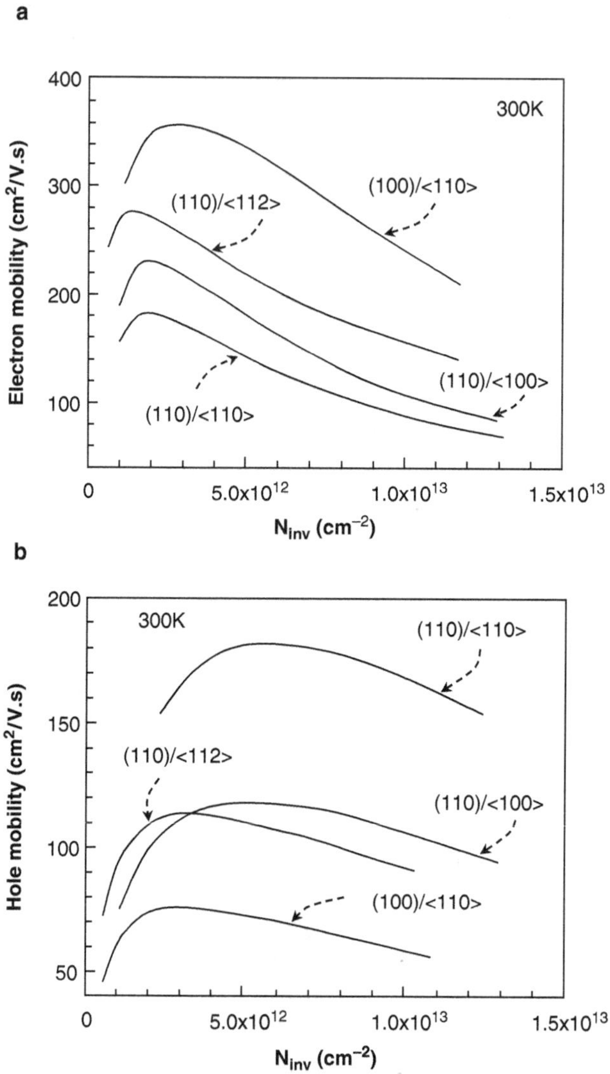

Fig. 5.57 Measured electron and hole mobilities as a function of inversion carrier density N_{inv} for different combinations of crystallographic planes *(hkl)* and channel orientation <uvw> (Adapted from [149])

inversion carrier density N_{inv} in long and wide *MOSFETs* (small lateral field) fabricated on bulk silicon with different combinations of wafer planes *(hkl)* and channel orientations *<uvw>*.

The electron mobility is found to be highest in the (100) plane (Fig. 5.57a). The optimum surface orientation for holes is the (110) plane and the <110> channel orientation (Fig. 5.57b) [149]. A method to combine both orientations in one integrated wafer will be described in Chap. 7.

The low mobility at small inversion carrier concentrations is attributed to Coulomb scattering, mainly by ionized impurities [26]. The effect becomes more pronounced as the channel dopant concentration increases [158]. The physical reason for the initial increase in the effective mobility with N_{inv} is believed to be due to an increase in screening of the ionized-impurity charge and interface charge by the inversion layer [158–162]. As N_{inv} increases, the mobility reaches a peak and then decreases again through an increase in surface roughness scattering as a direct result of the fluctuating potential caused by the imperfect interface which is only a small distance from the inversion layer [163, 164]. The mobility due to surface roughness scattering alone has been shown to vary with the transverse field as $\sim E^{-\gamma}$, where γ ranges from 2–2.6 [158, 164, 165].

5.5.2.2 Strained Silicon

The change in resistance with mechanical stress is referred to as the piezoresistance effect. It is due to a slight change in the crystal lattice constant as a result of a small strain. The effect was first measured on silicon and germanium in 1954 [166]. Figure 5.58 shows the essential features of the measurements. Uniaxial tensile stress X was applied to single crystal rods of different crystal orientations by hanging a weight on a string. The resistance was measured from the IR drop as shown in the figure. The change $(1/X)$ $(d\rho/\rho)$ was then extracted from $(1/X)$ $(\delta R/R)$ and from the geometry of the rod, where ρ is the resistivity of the rod and X the stress in Pa. The piezoresistance coefficients were extracted from the three measurement configurations.

For n-type silicon, a decrease in ρ with increased tensile strain was observed in configuration (a) of Fig. 5.58. This is attributed to an increase in electron mobility [166, 167]. As discussed in Chap. 1, the constant energy surfaces consist of six

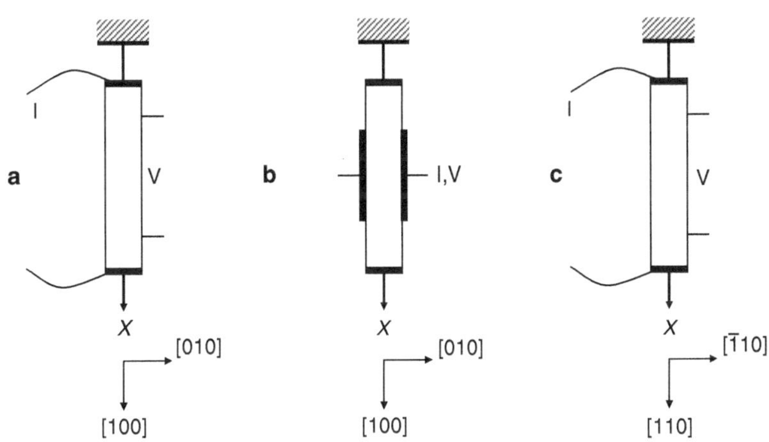

Fig. 5.58 Schematic diagram showing the stress configurations, crystallographic orientations, and resistance measurements [166]

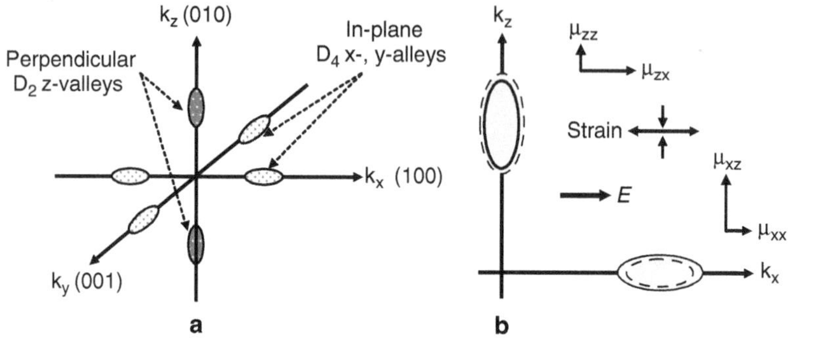

Fig. 5.59 Schematic constant-energy surface diagram. **a** Six ellipsoids of revolution along cube axis; **b** Enlarged view to illustrate the effect of strain in the first configuration of Fig. 5.58. The strain is tensile along the x-axis and compressive along the *z*-axis as indicated by the arrows [166]

ellipsoids of revolution along the cube axes in momentum space (Fig. 5.59a). Thus, there are six groups of electrons located in the ellipsoid energy minima. Because the surfaces are ellipsoidal, the electron effective mass in a given group is anisotropic. In a direction along the axis, it has the value $m^* = 0.98\,m_o$, and a direction vertical to the axis the value $m^* = 0.19\,m_o$, where m_o is the electron rest mass [167]. Hence, the electron mobility in each group is anisotropic. It is small in a direction along the axis and large in a direction perpendicular to the axis (Fig. 5.59b). In the cubic crystal the average mobility is the same in all direction and expressed as

$$\bar{\mu} = \frac{\mu_l + 2\mu_h}{3}, \tag{5.91}$$

where μ_l is the low mobility along the axis and μ_h the high mobility normal to the axis.

In the absence of stress, the two energy valleys shown as solid-line ellipsoids in Fig. 5.59b are equally populated with electrons, but the two groups have different mobilities in the [100] direction. Elastic strain in silicon destroys the cubic symmetry and shifts the energy minima in opposite directions shown by the dashed ellipsoids in Fig. 5.59b. A tensile strain in the x-direction results in contraction in the z-direction. For a strain of 0.01, the shift in energy between D4 and D2 in the figure is approximately (15 eV)$\boldsymbol{e}$, where $\boldsymbol{e}$ is the tensile strain (Fig. 5.60) [167]. For $\boldsymbol{e} = 0.01$, the shift is 0.15 eV which is ~5 kT. As a consequence, practically all electrons will transfer from the higher energy x-valley to the lower-energy z-valley.

Thus, if a rectangular *NMOS* channel is placed in the (100) plane with the source-drain axis oriented in the *x*-direction and a tensile stress applied along the source-drain, a compressive stress would result in the *z*-axis. Essentially all inversion electrons would then be transferred from the higher-energy in-plane *x*-valleys to the lower-energy out-of-plane *z*-valleys. The carriers then move from source to drain, perpendicular to the z-axis which is the direction of low effective mass and hence high mobility. In addition, the probability of an electron to be scattered from a

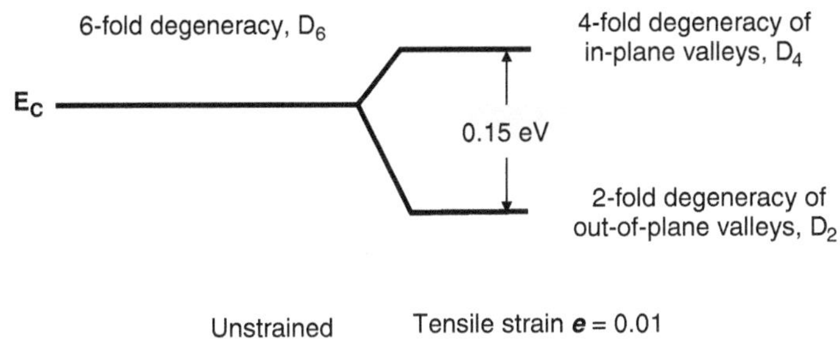

Fig. 5.60 Illustration of energy band splitting under strain. Essentially, all electrons are transferred from in-plane to out-of-plane valleys

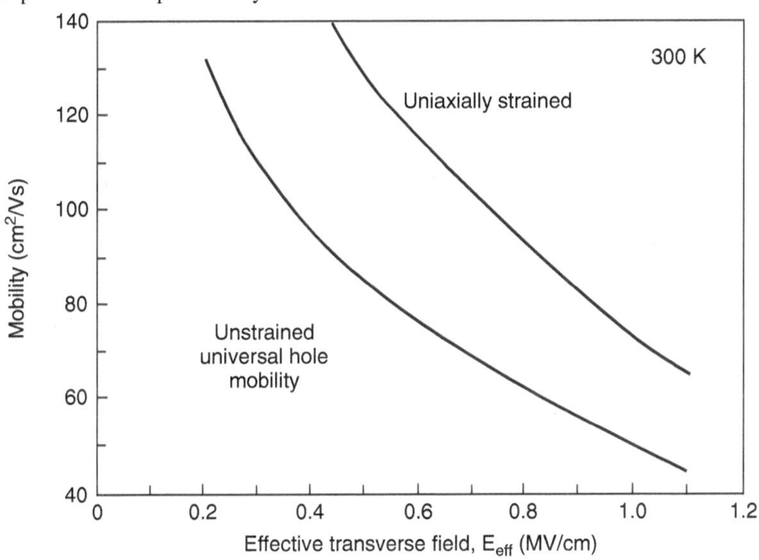

Fig. 5.61 Increase in hole mobility in *PMOS* under uniaxial compressive strain [173]

z-valley into an x-valley decreases because the number of final states in other valleys is reduced. This further increases the electron mobility [167]. An overall increase in strain-induced electron mobility by a factor larger than 1.7 has been experimentally verified [168–171]. Similar *NMOS* results are obtained by applying compressive out-of-plane stress instead of a tensile in-plane stress [172].

Strain in silicon also improves the hole mobility. By applying a uniaxial compressive strain in the source-drain direction, a substantial increase in the hole mobility is obtained in the low- and high vertical field regimes as shown in Fig. 5.61 [173]. The valence band structure along the <100> axis in *Si* is shown for unstrained silicon in Fig. 5.62a (Chap. 1). The offset band *V3* is too deep in energy to be appreciably populated by holes. The light-hole band *V2* has a higher curvature and hence smaller effective mass than the heavy-hole band *V1*. The two bands coincide at $k = 0$. A proposed valence band structure for strained Si is shown in Fig. 5.62b [174]. In this

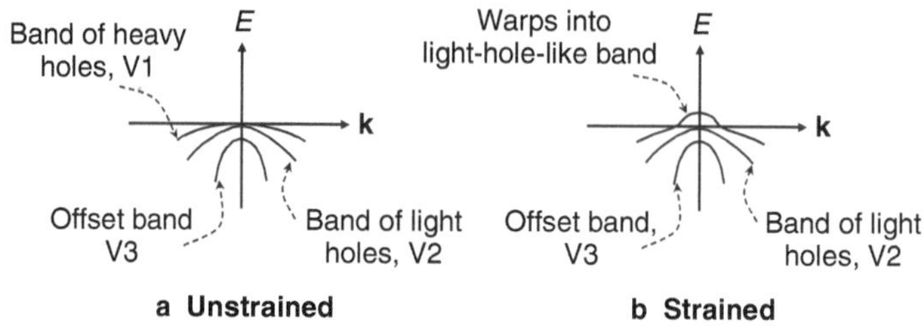

Fig. 5.62 Valence band diagram. **a** Unstrained; **b** Suggested strained [173, 174]

model, the strain lifts the degeneracy of the light- and heavy-hole bands at k = 0. Also, the heavy-hole band drops in energy (moves up) and becomes "warped," exhibiting a curvature similar to that of the light-hole band [173, 174]. As a consequence, most of the inversion holes populate the upper band ($V1$) and move from source to drain with a smaller effective mass and hence higher mobility than without strain.

Several methods have been demonstrated to simultaneously induce tensile strain in *NMOS* and compressive strain in *PMOS*. They will be detailed in Chap. 7.

5.6 Ultrathin Oxide and High-K Dielectrics

A silicon-dioxide film of thickness less than ~4 nm is referred to as ultrathin oxide. Assuming a maximum allowable oxide-field of 6×10^6 V/cm, the maximum voltage that can be applied across a 4-nm oxide is ~2.4 V. In small devices, when the voltage across the oxide drops below 3.2 V, the shape of the tunneling barrier is trapezoidal rather than triangular, and the current through the oxide is mainly due to direct tunneling (Chap. 4). The direct tunneling current increases rapidly as the oxide thickness is reduced below 4 nm. To maintain the steady increase in performance and circuit density while suppressing short-channel effects, the ratio of channel length to gate oxide thickness must remain above ~45 [38,41]. Thus, for an effective gate length of 70 nm, the silicon-dioxide thickness must be ~1.5 nm. For such a thickness, the gate current can increase above ~10 A/cm^2 (Chap. 4). Because of the associated standby power dissipation, this can become prohibitive, even in power-tolerant high-performance logic applications.

The primary motivation for the development of high dielectric-constant gate insulators is to avoid an increase in the gate leakage current associated with the rapid increase in tunneling current through silicon-dioxide as the SiO_2 thickness is reduced. Figure 5.63 compares the *NMOS* gate current density, measured at 300 K for a fixed gate voltage $V_G = V_{FB} - 1$ V on three dielectrics of $K = 3.9$ (SiO_2), $K = 5$, and $K = 18$ as a function of equivalent oxide thickness, t_{eq} or *EOT* (Chap. 4). For the same t_{eq}, higher-K materials are thicker than SiO_2, providing a longer tunneling distance and hence considerably smaller gate current.

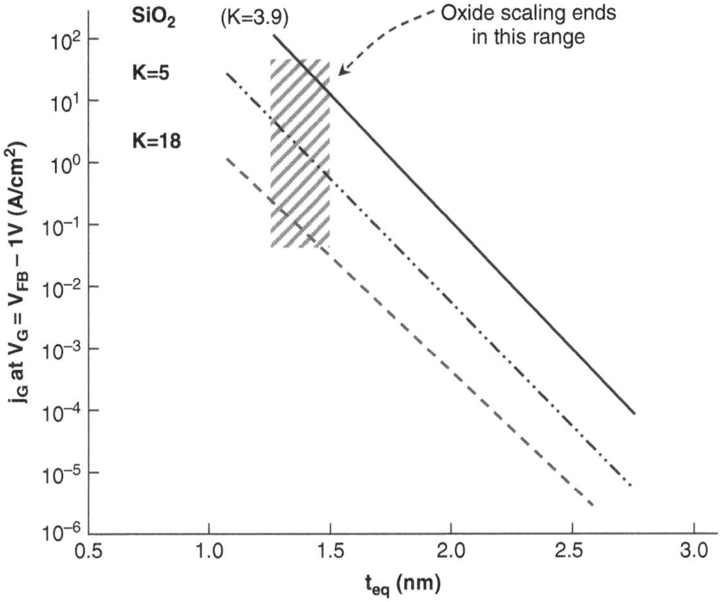

Fig. 5.63 Comparison of tunneling current for three gate dielectrics of different dielectric constants and same equivalent oxide thickness (Adapted from [175])

Example: for t_{eq}=2 nm, the physical dielectric thickness is, respectively, 2 nm, $2 \times (5/3.9) = 2.56$ nm, and $2 \times (18/3.9) = 9.2$ nm for SiO_2, $K=5$, and $K=18$.

Figure 5.64 shows how the standby power dissipation decreases with increasing dielectric constant. The plot for SiO_2 of t_{ox} = 1.5 nm is based on measurements [176]. The curve for the high-K dielectric shows the potential reduction in power dissipation for a high-K material with the same t_{eq} value [177].

5.6.1 High-K Dielectric Requirements

One of the present challenges in integrating high-K gate-dielectrics into a *CMOS* process is to achieve the high gate dielectric-constant and low gate leakage while maintaining surface and interface properties, and long-term reliability equivalent to those of silicon-dioxide.

Table 5.6 summarizes important high-K dielectric requirements. A minimum dielectric constant of about 15 is needed to achieve an equivalent oxide thickness of $\leq$1 nm while maintaining an acceptable level of gate leakage. The bandgap must be greater than about 5 eV to provide sufficiently large conduction and valence band offsets (barrier heights) to limit thermionic (Schottky) electron and hole emission, as illustrated for electrons in Fig. 5.65.

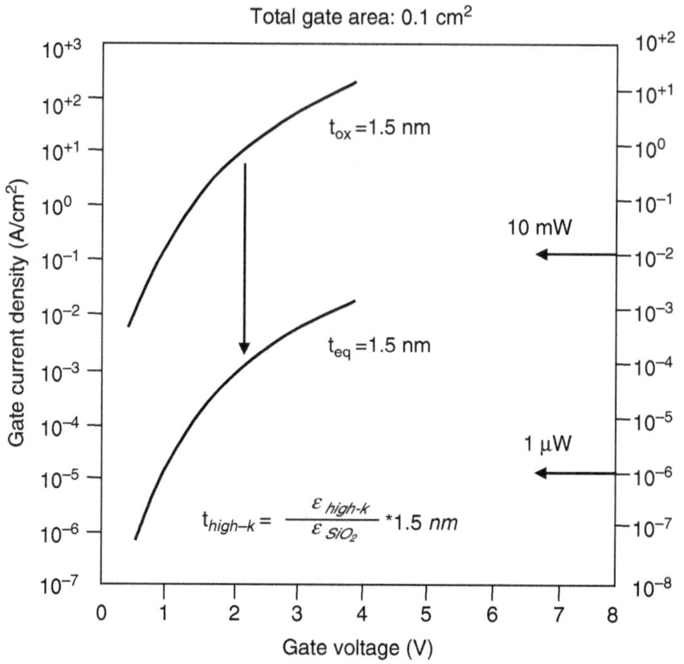

Fig. 5.64 Gate leakage and power dissipation for 1.5-nm thick SiO_2 compared to that of a high-*K* material of same equivalent oxide thickness [177]

Table 5.6 Important high-*K* dielectric requirements

Dielectric constant	>15
Bandgap	>5 eV
Barrier height for electrons, ΔE_c	>1 eV
Barrier height for holes, ΔE_v	>1 eV, satisfied with most materials
Thermal stability	Similar to SiO_2
Chemical stability	Similar to SiO_2
Interface trap density, Si	$<10^{11}/cm^2$
Interface trap density, gate	Low, to avoid Fermi-level pinning
V_T stability, hysteresis	10 mV hysteresis
Inversion-carrier mobility	95% of SiO_2
Long-term reliability	>10 year

For gate-first processing, the material must withstand high processing temperatures, in particular source-drain annealing temperatures, without reacting with silicon or exhibiting changes in its properties. The interface-state density, Q_{it}, on both the dielectric-silicon and dielectric-gate interfaces must be kept small. An increase in Q_{it} at the dielectric-silicon interface can degrade the inversion-carrier mobility and modify the threshold voltage. A high density of interface charge at the gate-insulator interface can result in Fermi-level pinning, limiting the flexibility of adjusting the gate workfunction (Chap. 4).

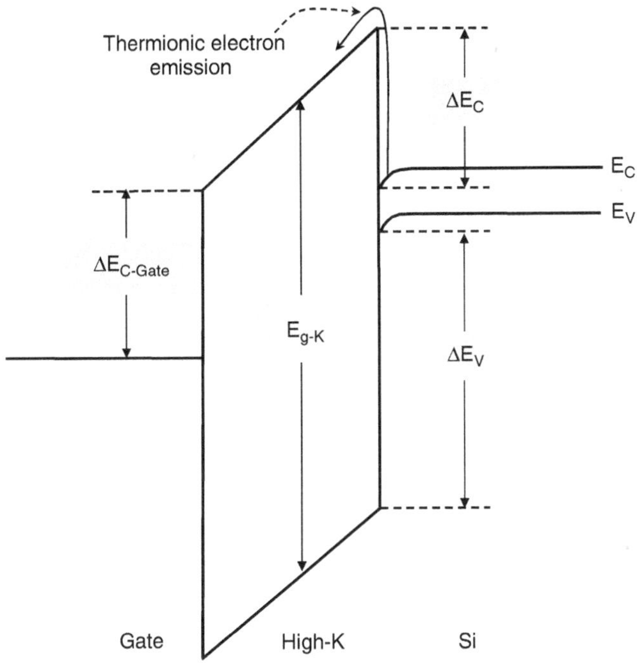

Fig. 5.65 Band diagram illustrating the barriers (band offsets) for electron and hole injection in a conductor-insulator-Si system

5.6.2 High-K Materials

The need for insulators of higher dielectric constant than oxide is not limited to high-performance logic. In *DRAM* applications, for example, an increase in capacitance density reduces the cell capacitor area required for a given cell capacitance (Chap. 8). Also, a larger capacitance density is needed to reduce the size of on-chip decoupling capacitors in the nF range. A variety of high-K materials are introduced, ranging from forming oxynitrides with $K = 4$–7 to the development of ferroelectric insulators of $K > 500$.

5.6.2.1 Oxynitrides

The dielectric constant of oxynitrides (*SiON*) can range from ~4 to 7, depending on the composition, thickness, and deposition conditions. Ultrathin oxynitride films of 1.8–2.8 nm thickness and $K > 5.7$ have been formed by nitridizing silicon in an ammonia (NH_3) atmosphere at 800–900°C to grow an ultra-thin film of silicon nitride (Si_3N_4), and then rapid-thermal annealing (*RTA*) in diluted nitrous oxide (N_2O) at 800–1000°C to reoxidize silicon. The thin interfacial oxide reduces the interface-state and fixed charge densities by eliminating the $N-H$ bonds associated with the

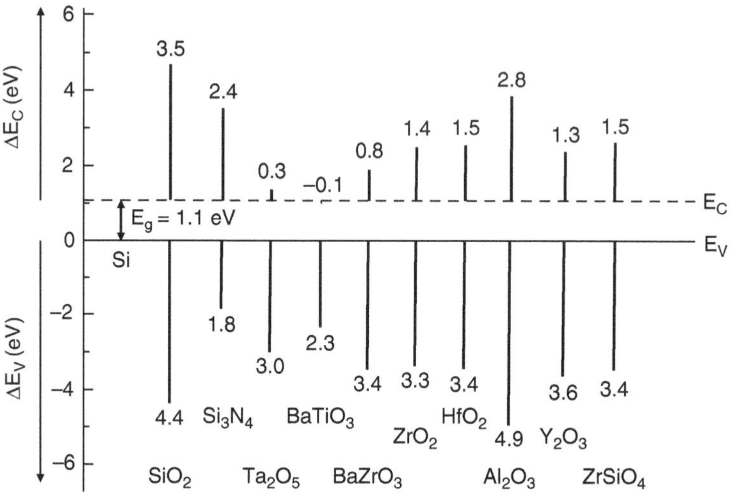

Fig. 5.66 Calculated conduction band and valence band offsets of various dielectrics on silicon (Adapted from [185])

Si_3N_4 in direct contact with silicon [178–184]. Silicon-dioxide and oxynitrides have, however, reached the limit of 1.1–1.5 nm due to the rapid increase in gate current.

5.6.2.2 Hafnium-Based Dielectrics

Figure 5.66 compares calculated band offsets of various high-K dielectrics on silicon [185]. Because of its sufficiently wide bandgap of ~6.0 eV and band offsets $\Delta E_C \approx 1.5\,\text{eV}$, $\Delta E_V \approx 3.4\,\text{eV}$, and good thermal stability in contact with silicon [186], hafnium-dioxide appears to be the most promising gate-insulator for high-performance applications. When in direct contact with silicon, however, hafnium-dioxide and other hafnium-based dielectrics, such as *HfON*, *HfSiO*, *HfSiON*, and *HfTaON* degrade the surface mobility to considerably below 80% of the value obtained with silicon dioxide [187–189]. The mobility degradation is associated with coulomb scattering with charges at the silicon-dielectric interface or within the insulator, surface roughness, and phonon scattering [190, 191]. Inserting an ultrathin SiO_2 layer between the high-K dielectric and silicon improves carrier mobility, but at the cost of reducing the effective dielectric constant of the composite film [188, 192].

5.6.2.3 Other High-K Materials

A comprehensive discussion of high-K materials that are being explored for digital, memory and analog applications is beyond the scope of this book. For a review, the reader is referred to [177]. The dielectrics shown in Fig. 5.66 constitute only one part of the materials that are being explored. In addition to Al_2O_3 and ZrO_2, other binary metal oxides, such as TiO_2 and La_2O_3 have also been demonstrated. The main motivation for exploring them is their predicted thermal stability with silicon, wide bandgap and band offsets.

Aluminum-oxide gate-dielectrics have been demonstrated with $t_{eq} < 1.5\,\text{nm}$, gate leakage of 0.4 A/cm2 at 1 V gate-overdrive, interface-state density $<10^{11}\,\text{cm}^{-2}$, and good thermal stability in contact with silicon [193–195]. The dielectric constant of 10–11 is, however, too low for high-speed *MOSFETs* at nanoscale dimensions. In addition, high bulk charge levels have been observed.

Metal-insulator-metal (*MIM*) capacitors of $HfO_2 - Al_2O_3$ laminates have also been demonstrated for RF and mixed-signal applications [196]. The capacitors exhibit low leakage ($10^{-9}\,\text{A/cm}^2$) at 3.3 V and 125 °C, a capacitance density of $3.1\,\text{fF}/\mu\text{m}^2$, and linear and quadratic voltage coefficients of capacitance (*VCC*) of, respectively, $-80\,\text{ppm/V}$ and $100\,\text{ppm/V}^2$ (Chap. 6).

Tantalum pentoxide (Ta_2O_5) exhibits a high dielectric constant of typically ~25 [197]. Its conduction band offset to silicon is, however, low in the range 0.6–0.8 eV [198]. In ultrathin films, this can result in excessive gate leakage at operating voltages. The material is therefore not attractive for high-speed digital applications. Ta_2O_5 of thickness >3 nm has been used as a node-dielectric in *DRAM* cells, increasing the cell-capacitance density while maintaining an acceptable retention time [199]. The Ta_2O_5 dielectric constant has been enhanced to ~50 by depositing Ta_2O_5 epitaxially on a ruthenium electrode under conditions to form a hexagonal crystal-symmetry (as opposed to a conventional orthorhombic symmetry), making the material attractive to *DRAM* designs in the Gbit range [200].

Because of its very high dielectric constant ($K > 100$), Barium-Strontium-Titanate, *BST* $[(Ba_xSr_y)TiO_3]$ is very attractive for Gbit-scale *DRAM* cells. Since *BST* cannot withstand high-temperature processes, it is only used for stacked-capacitor cells that are constructed after all high-temperature cycles are completed [201] (Chap. 8). *BST* films of equivalent oxide thickness below 1 nm and leakage current in the nA/cm^2 range have been successfully deposited on a ruthenium- or platinum-base electrode, demonstrating its potential for Gbit *DRAM* stacked-capacitor cells [201–204].

5.7 Gate Stack

The first *NMOS* structure that was fabricated about four decades ago had an aluminum gate. Because of the low melting point of aluminum, the source and drain were diffused before metal deposition. As a consequence, the gate was not self-aligned to source and drain. This was not a serious issue since the channel was several microns long and the misalignment was a small fraction of the total channel length. As the channel length was reduced to micron and submicron dimensions, self-alignment between gate and source-drain became an absolute necessity. Since polysilicon and refractory metals can withstand high-temperature processes, they can be deposited and patterned before high-temperature source-drain processes, allowing source and drain to be implanted and annealed after gate patterning and be self-aligned to the gate.

The main motivation for the development of refractory-metals to replace polysilicon gates is: the increase in the contribution of polysilicon depletion to the electrical equivalent gate oxide thickness; the increase in polysilicon gate resistance as the *MOSFET* horizontal and vertical dimensions are reduced; and secondary benefits of a metal gate might include reduction or elimination of boron diffusion into and through the gate dielectric. Another important advantage of refractory metals that will become apparent from the discussion below is the ability to tune the gate workfunction and independently adjust the *NMOS* and *PMOS* threshold voltages without the need to adjust the channel dopant concentration.

5.7.1 Polysilicon Workfunction

The polysilicon gate must be degenerately doped to minimize gate depletion and the resulting increase in the capacitance equivalent thickness, CET, when the *MOSFET* is turned on (Chap. 4). In modern CMOS, a dual polysilicon gate-workfunction is implemented with the *NMOS* gate heavily doped n-type and the *PMOS* gate heavily doped p-type. The polysilicon-gate workfunction is therefore approximately equal to the silicon affinity $\chi(\sim 4.15\,\mathrm{V})$ in *NMOS* and to $\chi + E_g/q$ ($\sim 5.25\,\mathrm{V}$) in *PMOS* (Chap. 4). Assuming zero effective oxide charge and intrinsic silicon, the room temperature threshold voltage would be $\sim -0.55\,\mathrm{V}$ for *NMOS* (n^+-poly) and $+0.55\,\mathrm{V}$ for *PMOS* (p^+-poly), that is, both *MOSFETs* would be "normally-on," as illustrated for *NMOS* in Fig. 5.67. To adjust the threshold voltage and also suppress short-channel effects, the channel is doped with increasing concentration as the channel length is reduced. An increase in channel dopant concentration results in an increase in vertical field and in ionized-impurity scattering, reducing the effective inversion-carrier mobility. The impact on mobility can be reduced by adjusting the workfunction instead of dopant concentration, and confining the inversion

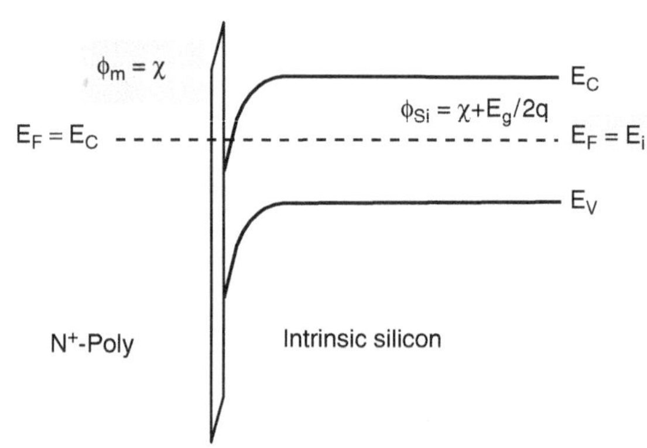

Fig. 5.67 Energy-band diagram for *NMOS* showing the effect of workfunction difference between n^+-poly and intrinsic silicon. The workfunction difference is $\phi_{ms} = \phi_m - \phi_{Si} = E_g/2q \approx -0.55\,\mathrm{V}$

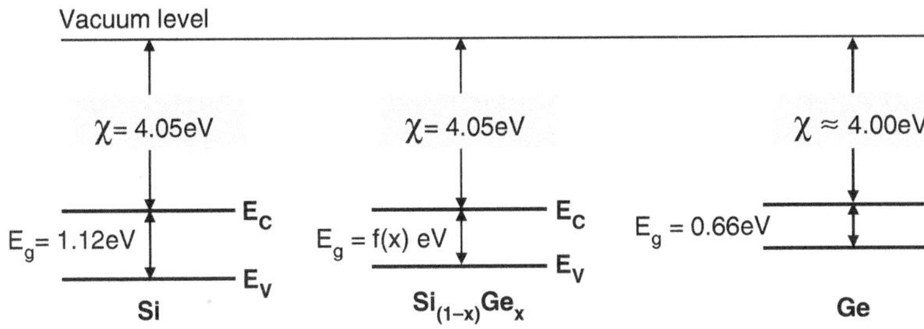

Fig. 5.68 Energy-band diagrams showing the reduction in polysilicon workfunction by incorporating germanium [206]

carriers to a region of low concentration, such as with a super-steep retrograde profile (Fig. 5.54).

The polysilicon workfunction can be adjusted for *PMOS* by incorporating germanium into the heavily doped p-type polysilicon [205–208]. Figure 5.68 compares the conduction-band and valence-band energy levels in single-crystal *Si*, $Si_{1-x}Ge_x$, and *Ge* materials [206].

The electron affinities of *Si* and *Ge* are comparable, but *Ge* has a considerably smaller energy gap. Thus, the valence band is about 0.45 eV higher in *Ge* than in *Si*. Incorporating *Ge* into polysilicon is found to raise the valence band and hence reduce the *SiGe* workfunction by an energy that increases with the fraction x of *Ge*. For example, for $x \approx 0.45$, the p^+-poly-SiGe workfunction decreases by $\sim$0.4 V while the n^+-poly-*SiGe* workfunction decreases only slightly. A large fraction of the reduction in energy gap is related to strain induced by the *SiGe* [206]. Thus, the workfunction difference between p^+-poly and the n-type *PMOS* body decreases by $\sim$0.4 V, requiring considerably less channel dopant concentration to achieve a specific threshold voltage. This results in reduced normal field and ionized-impurity scattering, hence an increase in inversion-carrier mobility. A single workfunction p^+-poly-SiGe gate for *NMOS* and *PMOS* was demonstrated with improved current drive by adjusting the *Ge* fraction [209].

Another advantage of adding Ge to p^+-poly is the increase in the active dopant concentration and hole mobility as found from Hall-measurements [205–207]. Also, the activation temperature of boron in p^+-poly-*SiGe* is reduced. The p^+-gate resistivity is found to decrease by a factor of 4 at a *Ge* fraction of 0.45. The n^+-poly-*SiGe* resistivity is not appreciably affected for $x \leq 0.45$ but begins to increase sharply at higher *Ge* concentrations. [205–207].

5.7.2 Metal Gates

The main advantages of refractory metals over polysilicon are the reduced sheet resistance and elimination of gate depletion effects. The workfunction of metal-gates must, however, satisfy both the *NMOS* and *PMOS* requirements as did the n^+-poly and p^+-poly gates. This has been initially achieved by choosing a single metal of midgap workfunction, that is, a workfunction approximately equal to that of intrinsic silicon. As the channel concentration is increased, however, to suppress short-channel effects, the *NMOS* and *PMOS* threshold voltages obtained with a midgap gate become too large, $\sim \pm 0.5$. The need for considerably lower threshold voltages in *MOSFET*s at nanoscale dimensions prompted the development of metal gates that are separately optimized for *NMOS* and *PMOS*.

5.7.2.1 Midgap Metal Gates

A midgap metal gate has a workfunction $\phi_m \approx \chi + E_g/2q \approx 4.71\,\text{V}$, approximately equal to that of intrinsic silicon. It allows the design of symmetrical *NMOS* and *PMOS* threshold voltage characteristics with a single gate material, simplifying the process in addition to solving the polysilicon problems mentioned above. Tungsten (*W*), *TiN*, and stacked *W/TiN* gates have a workfunction near 4.8 V and are hence appropriate materials as midgap gates [210–215]. Structures with a *W/TiN* gate-stack combine the high *TiN/Si* interface quality with the low tungsten resistivity.

The main disadvantage of midgap gates can be described as follows: As the channel length is reduced, the channel dopant concentration is increased to suppress short-channel effects. At nanoscale dimensions, the channel concentration can reach levels above $10^{18}\,\text{cm}^{-3}$ where the Fermi potential approaches $-0.5\,\text{V}$ for *NMOS* and $+0.5\,\text{V}$ for *PMOS*. Unless the dopant concentration is modified, the threshold voltage is fixed at about $\pm 0.5\,\text{V}$. Nanoscale *MOSFET*s require V_T in the range $\pm 0.2\,\text{V}$ to 0.25 V to ensure an adequate gate overdrive. The threshold voltage can be adjusted to a magnitude lower than 0.5 V by counter-doping in a very thin region under the surface, that is, forming a thin "buried n-channel" in *NMOS* and "buried p-channel" in *PMOS*, of appropriate dose. The obvious disadvantage of this approach is the reduced surface mobility and increased susceptibility to short-channel effects [216–219]. A lightly-doped channel well obtained with a step channel profile or super-steep retrograde well reduces the impact on mobility but is not to adequate to reduce V_T to the required level. Also, this approach exacerbates the short-channel effect.

5.7.2.2 Dual Workfunction Metal Gates

The limitations of single-metal midgap gates, such as W and TiN, prompted the development of dual workfunction *NMOS* and *PMOS* metal gates for nanoscale *MOSFET*s. To satisfy low-voltage applications, the metal workfunction must be about $\chi + 0.2\,\text{V}$ (*NMOS*), and about $\chi + E_g/q - 0.2\,\text{V}$ (*PMOS*), that is, about 4.35 V for *NMOS* and 5.07 V for *PMOS* [220]. The metals are hence said to have band-edge workfunctions (Fig. 5.69).

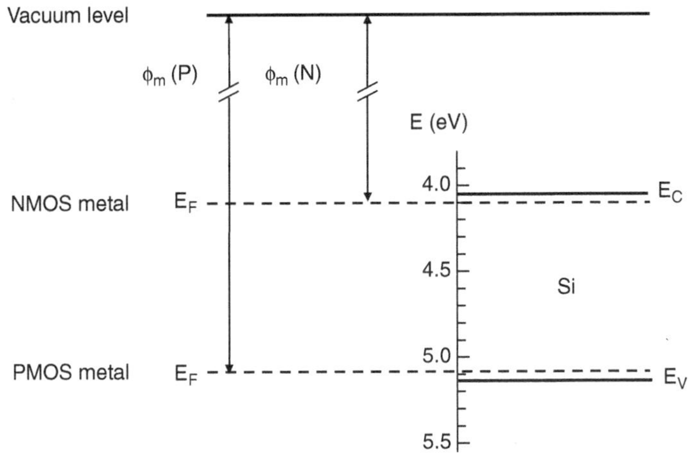

Fig. 5.69 Optimized dual workfunction metal gates. The Equilibrium Fermi level lies 0.2 V below the conduction band-edge (*NMOS*) and 0.2 V above the valence band-edge (*PMOS*)

Several refractory metals have been investigated as a replacement of polysilicon to form dual-workfunction gates. Among them are molybdenum (*Mo*), platinum (*Pt*), Tantalum (*Ta*), Titanium (*Ti*), Tantalum Nitride (*TaN*), Tantalum Silicon Nitride ($TaSi_xN_y$) and fully silicided gate (*FUSI*). Key properties of some of the materials are briefly described here. Details on processes to deposit the metals and tune their workfunction can be found in Chap. 7 and references therein.

To be viable for nanoscale *CMOS*, the metal gates must have band-edge workfunctions as described in Fig. 5.69. They must exhibit electrical, thermal, and chemical stability with underlying thin gate dielectrics, in particular with high-*K* dielectrics. Metals exhibit anisotropy in their workfunction [221]. For example, single-crystal molybdenum varies from ~4.4 V in the (112) plane to ~4.95 V in the (100) plane [222]. The metal workfunction depends on the deposition condition, annealing and dielectric material upon which it is deposited. For example, the *Mo* workfunction is found as 5.05 V over SiO_2 and 4.95 V over ZrO_2, a difference of 100 mV [223]. To simplify the process, it would be advantageous to deposit a single gate material and then tune its workfunction independently for *NMOS* and *PMOS*. For example, by selectively implanting inert ions, such as Ar^+ or N^+ into molybdenum, the microstructure of *Mo* can be amorphized and its workfunction reduced to near 4 V, suitable for *NMOS*, while the nonimplanted *Mo* film maintains a higher workfunction of 4.95 V that is suitable for *PMOS* [224].

Ruthenium-base gate materials also appear promising as dual-workfunction metal gates. *Ru* is found to have a *PMOS* workfunction of 5 V while a *Ru-Ta* alloy can be tuned to an *NMOS* workfunction as low as 4.2 V [225]. The material is also found to be compatible with *Hf*-base high-*K* dielectrics [226].

Other integration processes to form dual workfunction metal gates include doping hafnium nitride (*HfN*) with lanthanum (*La*) to form an *NMOS* gate, and doping

TaN with *Al* to tune the workfunction for a *PMOS* gate [227], or tuning *TiN* for *PMOS* and *TaSiN* for *NMOS* [228, 229].

5.7.2.3 Fully Silicided Gate, FUSI

Full gate silicidation (*FUSI*) is found to be an excellent method to integrate a metal gate into a *CMOS* process, and tune its workfunction to an *NMOS* or *PMOS* value [230]. Most of the *FUSI* work focuses on nickel and its silicides. The work function of *NiSi FUSI* depends on whether it is formed on arsenic-, boron- or undoped polysilicon [231, 232]. The workfunction is found to be 4.58 V on arsenic-doped, 5.1 V on boron-doped and 4.87 V on undoped polysilicon. The difference is attributed to pile-up of impurities at the NiSi-SiO_2 interface, which is more pronounced for arsenic than for boron.

Tuning of *FUSI* on high-*K HfSiON* has been achieved by selectively reducing the polysilicon height in *PMOS* to obtain a nickel-rich silicide, thus modifying the phase of full nickel-silicidation from *NiSi* of workfunction ~4.44 V on the *NMOS* gate to $Ni_{31}Si_{12}$ of workfunction ~5.0 V on the *PMOS* gate [233]. More detail on this process is given in Chap. 7.

5.7.2.4 Fermi-Level Pinning

The extracted gate workfunction on high-*K* dielectrics is frequently found to differ appreciably from its value on SiO_2 or vacuum. This is observed on both metal and polysilicon gates on high-*K* dielectrics [223, 234–242]. As for Schottky-barriers discussed in Chap. 2, this is attributed to a high density of electronic states at the interface between the gate and high-*K* dielectric, resulting in Fermi-level pinning (Fig. 5.70) [243].

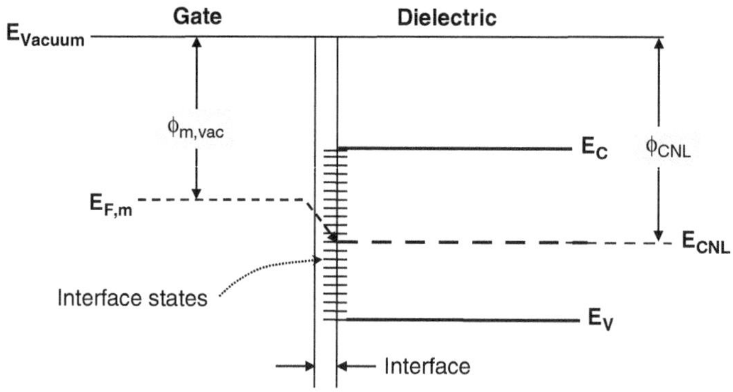

Fig. 5.70 Illustration of Fermi-level pinning caused by high electronic-state density at the interface between gate and high-K dielectric. E_{CNL} is the charge neutrality level

For polysilicon on HfO_2 and Al_2O_3, interface states are believed to be related to Si-Hf and Si-O-Al bonds [241]. For metal- high-*K* interfaces, interfaces are related to metal-induced gap states in the insulator (*MIGS*) [243]. Interface states are assumed to be continuously distributed throughout the band gap. The density of acceptor-type states is found to increase toward the conduction band edge and the density of donor type states to increase toward the valence band edge [234]. Acceptor states are negatively charged when below the Fermi level and neutral above the Fermi level. Donor states are neutral below the Fermi level and positively charged above the Fermi level. If the Fermi level coincides with the charge-neutrality level, E_{CNL}, the surface is neutral. Figure 5.70 illustrates the case where the metal Fermi level is initially above E_{CNL}. In this case, band alignment occurs by electron transfer from the metal to interface states. An interface dipole is created, pulling the metal workfunction down toward E_{CNL} [234]. Because of the high interface state density, the metal Fermi level becomes pinned near E_{CNL} and a higher effective metal workfunction is measured.

Fermi level pinning is undesirable because it limits the flexibility in tuning the gate workfunction. Understanding the nature of interface states and controlling their level is very important.

5.7.2.5 Line-Edge Roughness, LER

The variability of *MOSFET* electrical parameters, such as off-current, on-current, threshold voltage, matching, reliability, and speed is caused primarily by fluctuations in effective channel length, gate-dielectric thickness, and channel dopant fluctuations. The parameters are therefore specified at nominal values, allowing for a variability of 3σ to 6σ, where σ is the standard deviation of a particular parameter. The variation of gate line-width is the main contributor to fluctuations in effective channel length. The worst-case I_{off} is then specified for the minimum gate line-width ("3σ to 6σ low"), and the worst-case I_{Dsat} for the maximum gate line-width ("3σ to 6σ high"). Patterned gates have typically rough edges because of their granularity and variability caused by lithography and etch. They exhibit an additional variability in the line-width within the *MOSFET*, referred to as line-edge roughness, *LER* [244] (Fig. 5.71).

For a channel length larger than ~80 nm, *LER* is found to have negligible impact on *MOSFET* parameters. As the channel length is reduced below 80 nm, however, the contribution of *LER* to channel length variability becomes increasingly important [245–247]. Simulations of 32-nm to 50-nm *MOSFET*s show that to limit the increase in I_{off} below three times the value predicted without *LER*, the maximum 3σ variation on *LER* must be 3 nm [248, 249].

The *LER* can be smoothened by implant and diffusion processes. Using scanning tunneling microscopy (*STM*), the *LER* is found to depend strongly on implanted dose of source-drain extensions, halos, and co-implanted species [250].

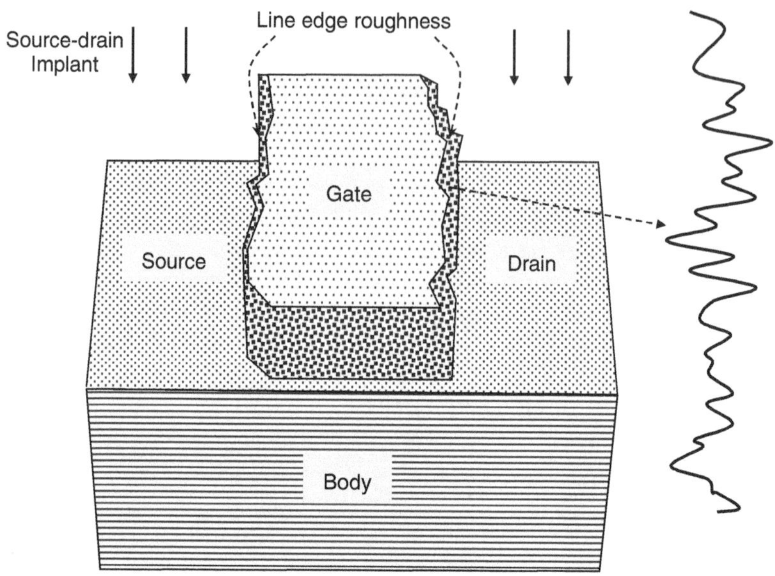

Fig. 5.71 Simulated *MOSFET*-gate line-edge roughness (Adapted from [244])

5.8 Three-Dimensional Structures, *FinFETS*

A *FinFET* belongs to the family of nonplanar multi-gate structures, whereby the gate is wrapped around a thin silicon pillar, or "fin," which constitutes the body of the *MOSFET*. The channel is typically formed on two or three sides of the fin. Figure 5.72 illustrates a FinFET constructed on silicon-on-insulator (*SOI*), with a channel formed on the sides and top of the fin, referred to as a tri-gate *FinFET*.

As with the planar *MOSFET*, the channel length is defined by the polysilicon or metal line-width and source-drain extensions. The channel width of the structure in Fig. 5.72 is equal to two times the fin-height plus the top fin-width. Thus, the *FinFET* occupies a smaller horizontal area per unit channel-width than the planar *MOSFET*. Another advantage of *FinFET*s is the control of the channel by the wrap-around gate, requiring considerably less dopant concentration in the *MOSFET* body to suppress short-channel effects. This results in a smaller subthreshold slope and higher inversion-carrier mobility. The gate can be made to wrap around multiple fins in parallel to increase the current-carrying capability. *FinFET*s are thus very attractive for nanoscale *CMOS*.

Process enhancements to planar *MOSFET*s can also be applied to *FinFET*s. A *FinFET* with molybdenum gate and hafnium-oxide high-K dielectric of t_{eq} in the range 1.75–1.95 nm has been demonstrated [251]. By tuning the molybdenum workfunction with nitrogen implant, the *NMOS* and *PMOS* threshold voltages were reduced to +0.28 V and −0.17 V, respectively. The corresponding subthreshold slopes were near ideal, 62.5–67.5 mV/decade. A metallic wrap-around gate was also obtained by full nickel silicidation and a dual workfunction achieved by impurity

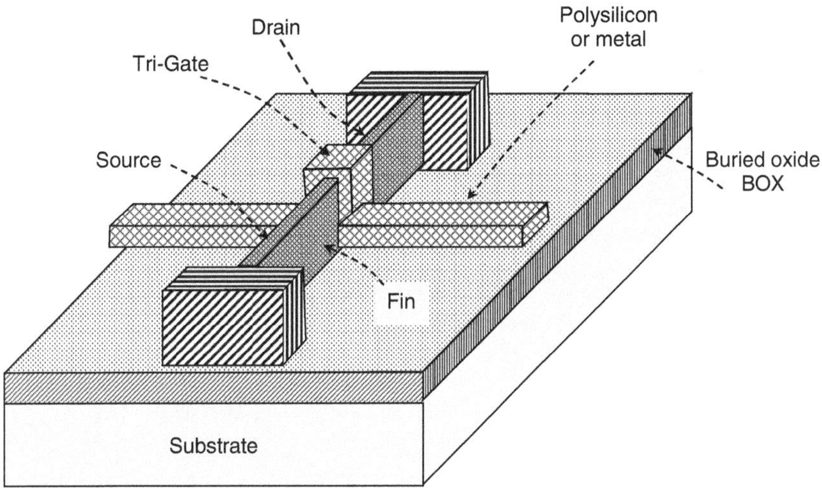

Fig. 5.72 Illustration of tri-gate *FinFET* constructed on silicon-on-insulator (*SOI*)

segregation, as discussed in the preceding section [252]. The *FinFET* channel mobility can also be enhanced by appropriate crystal orientation and strain [253–255]. Another performance enhancement can be obtained by full source-drain silicidation to form Schottky-barriers of appropriate barrier height, and diffusing source-drain extensions from the silicide [256]. The technique results in low source-drain resistance and ultra-short source-drain extensions.

Processes related to *FinFETs* are further discussed in Chap. 7.

5.9 Problems

(The temperature is 300 K unless otherwise stated.)

1. In a gate-controlled pn junction, the body is p-type of concentration $5 \times 10^{16}\,\mathrm{cm}^{-3}$. The gate dielectric has an equivalent oxide thickness of 25 nm and the gate overlaps the n-region that has an effective dopant concentration of $10^{18}\,\mathrm{cm}^{-3}$ in the overlap region. The gate, body and n-region are at zero potential. Find the surface potential in the p-region and in the overlapped n-region for an effective oxide charge $Q_{eff} = 0$ and for $Q_{eff} = +10^{12}$ charges/cm^2.

2. In a gate-controlled pn junction, the body is p-type, uniformly doped with $N_A = 2 \times 10^{15}\,\mathrm{cm}^{-3}$. The source is heavily-doped n^+, maintained at ground potential. The surface of the body is tailored by implanting boron which, at the end of all processes, can be approximated by a Gaussian profile with a peak concentration of $5 \times 10^{16}\,\mathrm{cm}^{-3}$, located at the surface, and a straggle $\Delta R_p = 0.1\,\mu\mathrm{m}$. Calculate the threshold voltage and body-bias effect for $V_{FB} = -0.6\,\mathrm{V}$, $t_{eq} = 25\,\mathrm{nm}$, and a body-to-source reverse bias $V_B = 1\,\mathrm{V}$.

3. The body of an *NMOS* is uniformly doped with $N_A = 2 \times 10^{16}\,\text{cm}^{-3}$, the gate is degenerately doped n-type polysilicon, and the equivalent gate oxide thickness is $t_{eq} = 10\,\text{nm}$. Calculate the effective oxide charge necessary to turn-on the *NMOS* for zero gate voltage and zero body bias.

4. Positive electrostatic charge is delivered to the gate of an *NMOS* at a rate of 0.5 pC/s. The gate dielectric is SiO_2. How long does it take to breakdown the oxide? Assume the effective gate dimensions to be $W_{eff} = 20\,\mu\text{m}$ and $L_{eff} = 5\,\mu\text{m}$.

5. The gate of the *NMOS* in Problem 4 is connected to a pn junction. What would be the time to breakdown for a junction leakage of 200 fA? 600 fA?

6. Derive a relation for the temperature dependence of the long-channel threshold voltage. Plot the threshold voltage as a function of temperature in the range −20°C to 140°C for an *NMOS* having the following properties: Uniform channel with $N_A = 10^{17}\,\text{cm}^{-3}$, $t_{ox} = 10\,\text{nm}$, N^+-polysilicon gate, L = 10 µm, $Q_{eff} = +10^{11}$ charges/cm^2, and source and body at ground.

7. Show that for a uniformly dopant *MOSFET* body, the body-bias effect can be expressed as

$$\frac{dV_T}{dV_B} = \frac{C_{Min}}{C_{Max} - C_{Min}},$$

where V_B is the source to body voltage.

8. The gate dielectric of a *PMOS* consists of 10-nm SiO_2 and 20-nm Si_3N_4. The effective channel length and width are, respectively, 2 µm and 5 µm and the channel is uniformly doped with $N_D = 5 \times 10^{16}\,\text{cm}^{-3}$. The source is at ground and the drain at −0.1 V. Assume that electrons are trapped at the oxide-nitride interface and estimate the total number of electrons necessary to bring the *MOSFET* to the onset of strong inversion for the following three cases:

(a) Degenerately-doped n^+-polysilicon gate at ground potential,
(b) Degenerately-doped p^+-polysilicon gate at ground potential,
(c) No gate.

9. Given: Long-wide channel *NMOS*; $t_{eq} = 10\,\text{nm}$; body uniformly doped with $N_A = 10^{17}\,\text{cm}^{-3}$; source to body bias = 1 V. How much must the gate voltage change to change the inversion layer concentration by one decade in the middle of the weak inversion regime?

10. A long- and wide-channel *NMOS* is biased as shown in the figure below. Given: $V_{FB} = -0.7\,\text{V}$,

Equivalent gate oxide thickness $t_{eq} = 25\,\text{nm}$,
Uniform body concentration: $N_A = 4 \times 10^{16}\,\text{cm}^{-3}$,
Subthreshold ideality factor: n = 1.3,
V_T defined at $I_D = I_o = 80\,\text{nA}$,
Node to body leakage 1 pA.

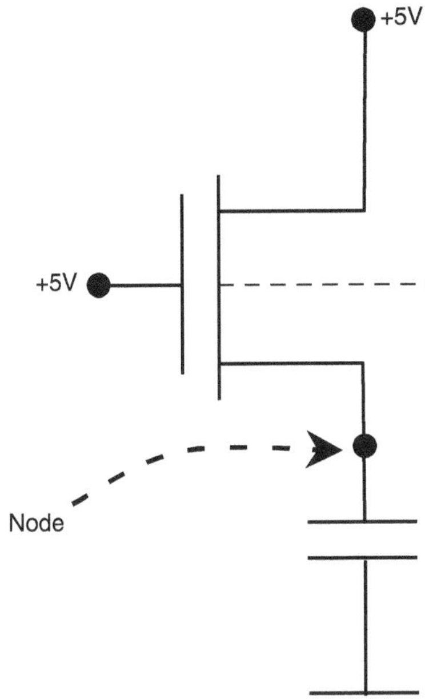

Find the maximum node voltage under steady-state conditions.

11. To maximize circuit speed, a *MOSFET* is designed to the minimum possible channel length. Short-channel constraints, however, require that 3σ low V_T for the shortest *MOSFET* be above 0.3 V. The dependence of V_T on channel length is found empirically as

$$V_T = 1.1 - 0.6/L$$

where L is μm. The channel length tolerance is $\pm 0.2\,\mu\text{m}$. Estimate:

(a) The shortest *nominal* channel length allowable,
(b) The nominal V_T for the channel length in a),
(c) The 3σ high V_T for the channel in a).

12. Consider an *NMOS* where $t_{eq} = 5\,\text{nm}$, $V_T = 0.5\,\text{V}$, $V_G = 2.5\,\text{V}$, $V_D = 0.1\,\text{V}$, $V_S = V_B = 0\,\text{V}$, and a total extrinsic resistance of 600Ohm-μm. The inversion electron mobility at $V_G = V_T$ is $650\,\text{cm}^2/\text{Vs}$. Estimate the channel length at which the intrinsic and extrinsic resistances are equal.

13. For the box-shaped body boron profile shown in the figure below, $t_{eq} = 10\,\text{nm}$, $Q_{eff} = 10^{11}\,\text{q}\,\text{C/cm}^2$, degenerately doped n-type polysilicon gate, and source, drain and body are grounded. Find the threshold voltage, the maximum depletion depth, and the surface field at threshold.

14. With the source and body at ground, the threshold voltage of a long-channel, isolated enhancement-mode *PMOS* is $-0.15\,\text{V}$. The threshold voltage is adjusted to

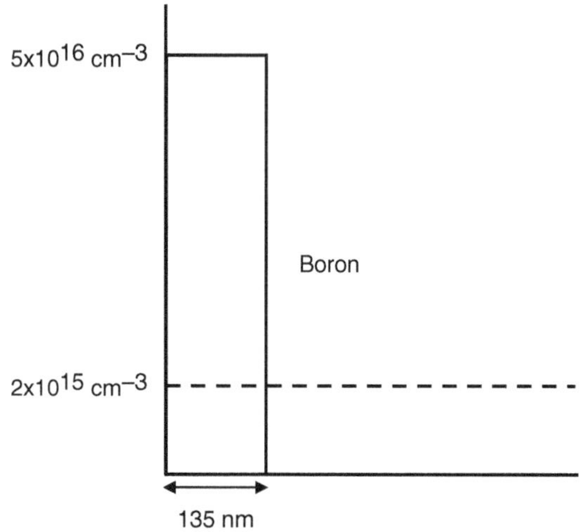

$-0.3\,\text{V}$ by applying a body-to-source reverse voltage, V_B. Find V_B, assuming $t_{eq} = 5\,\text{nm}$ and a uniform body concentration of $N_D = 10^{17}\,\text{cm}^{-3}$.

15. The channel length of a wide *PMOS* is $2\,\mu\text{m}$. Given: Hole mobility at $V_T = 200\,\text{cm}^2/\text{Vs}$, $\text{V}_\text{T} = -0.6\,\text{V}$, $V_G = 2\,\text{V}$, $\text{t}_\text{eq} = 5\,\text{nm}$, and body uniformly doped at $N_D = 4 \times 10^{17}\,\text{cm}^{-3}$. Estimate the time of flight of holes from source to drain.

References

1. A. S. Grove and D. J. Fitzgerald, "Surface effects on pn junctions – Characteristics of surface space-charge regions under non-equilibrium conditions," Solid-State Electron., 9 (8), 783–806, 1966.
2. R. N. Hall, "Electron-hole recombination in germanium," Phys. Rev., 87 (2), 387–394, 1952.
3. W. Shockley and W. T. Read, "Statistics of recombination of holes and electrons," Phys. Rev., 85 (5), 835–842, 1952.
4. W. P. Noble, S. H. Voldman, and A. Bryant, "The effects of gate field on the leakage characteristics of heavily doped junctions," IEEE Trans. Electron Dev., 36 (4), 720–726, 1989.
5. W. Shockley, "Problems related to p-n junctions in silicon," Solid-State Electron., 2 (1), 35–67, 1961.
6. J. F. Verwey, R. P. Kramer, and B. J. de Maagt, "Mean free path of hot electrons at the surface of boron-doped silicon," J. Appl. Phys., 46 (6), 2612–2619, 1975.
7. B. El-Kareh, "Effect of surface field on junction avalanche breakdown," IEDM Tech. Digest., 11–14, 1972.
8. J. L. Moll, Physics of Semiconductor Devices, McGraw Hill, Chap. 9, 1964.
9. E. G. Kane, "Theory of tunneling," J. Appl. Phys., 32 (1), 83–91, 1961
10. C. Chang and J. Lien, "Corner-field induced drain leakage in thin oxide MOSFETs," IEDM Tech. Dig., 714–717, 1987.
11. T. Y. Chan, J. Chen, P. K. Ko, and C. Hu, "The impact of gate-induced drain leakage current on MOSFET scaling," IEDM Tech. Dig., 718–721, 1987.
12. J. Chen. T. Y. Chan. I. C. Chen, P. K. Ko, and C. Hu, "Sub-breakdown drain leakage current in MOSFET," IEEE Electron. Dev. Lett., 8 (11), 515–517, 1987.

13. H. K. J. Ihantola and J. L. Moll, "Design theory of a surface field-effect transistor," Solid-State Electron., 7 (6) 423, 1964.
14. C. T. Sah, "Characteristics of the metal-oxide-semiconductor transistors," IEEE Trans. Electron Dev., ED-11 (7) 324–345, 1964.
15. F. H. Gaensslen, V. L. Rideout, E. J. Walker, and J. J. Walker, "Very small MOSFETs for low-temperature operation," IEEE Trans. Electron Dev., ED-24 (3), 218–229, 1977.
16. J. R. Schriefer, "Effective carrier mobility on surface space charge layers," Phys. Rev., 97, 641–646, 1955.
17. F. Fang and S. Triebwasser, "Carrier surface scattering in silicon inversion layers," IBM J. Res. Dev., 8, 410–415, 1964.
18. F. Stern and W. E. Howard, "Properties of semiconductor surface inversion layers in the electric quantum limit," Phys. Rev. B, 163, 816–835, 1967.
19. F. F. Fang and A. B. Fowler, "Transport properties of electrons in inverted silicon surface," Phys. Rev., 169, 616–631, 1968.
20. V. G. K. Reddi, "Majority carrier surface mobilities in thermally oxidized silicon," IEEE Trans. Electron Dev., ED-15 (3), 151–160, 1968.
21. C.-T. Sah, T. H. Ning, and L. L. Tschopp, "The scattering of electrons by surface oxide charge and by lattice vibrations," Surf. Sci., 32, 561–575, 1972.
22. T. Nishida and C.-T. Sah, "A physically based mobility model for MOSFET numerical simulations," IEEE Trans. Electron Dev., 34 (2), 310–320, 1987.
23. H. Shin, G. M. Yeric, A. F. Tasch, and C. M. Maziar, "Physically based models for effective mobility and local field mobility of electrons in MOS inversion channels," Solid-State Electron., 34 (6), 545–552, 1991.
24. S.-I. Tagaki, A. Toriumi, M. Iwase, and H. Tango, "On the universality of inversion layer mobility in silicon MOSFETs Part I – Effect of substrate impurity concentration," IEEE Trans. Electron. Dev., 41 (12), 2357–2362, 1994.
25. O. Leistiko, A. S. Grove, and C. T. Sah, "Electron and hole mobilities in inversion layers on thermally oxidized silicon surfaces," IEEE Trans. Electron. Dev., ED-12 (5), 248–254, 1965.
26. J. R. Hauser, "Extraction of experimental mobility data for MOS devices," IEEE Trans. Electron. Dev., 43 (11), 1981–1988, 1996.
27. A. G. Sabnis and J. T. Clemens, "Characterization of the electron mobility in the inverted <100> surface," IEEE IEDM Tech. Dig., 18–21, 1979.
28. S. C. Sun and J. D. Plummer, "Electron mobility in inversion and accumulation layers on thermally oxidized silicon surfaces," IEEE Trans. Electron. Dev., ED-27 (8), 1497–1508, 1980.
29. J. T. Watt and J. D. Plummer, "Universal mobility-field curves for electrons and holes in MOS inversion layers," VLSI Tech. Dig., 81–82, 1987.
30. K. Y. Fu, "Mobility degradation due to the gate field in the inversion layer of MOSFETs," IEEE Electron. Device Lett., 3 (10), 292–293, 1982.
31. G. Baccarani and G. A. Sai-Halasz, "Spreading resistance in submicron MOSFETs," IEEE Electron. Device Lett., 4 (2), 27–29, 1983.
32. K. K. Ng and W. T. Lynch, "Analysis of the gate-voltage-dependent series resistance of MOSFETs," IEEE Trans. Electron. Dev., 33 (7), 965–972, 1986.
33. K. K. Ng, R. J. Bayruns, and S. C. Fang, "The spreading resistance of MOSFETs," IEEE Electron. Device Lett., 6 (4), 195–197, 1985.
34. J. M. Pimbley, "Two-dimensional current flow in the MOSFET source-drain," IEEE Trans. Electron. Dev., ED-33 (7), 986–996, 1986.
35. International Technology Roadmap for Semiconductors – 2006 update (www.itrs.net).
36. R. H. Dennard, F. H. Gaensslen, V. L. Rideout, E. Bassous, and A. R. LeBlanc, "Design of ion-implanted MOSFET's with very small physical dimensions," IEEE J. Solid-State Circuits, SC-9, 256–268, 1974.
37. S.-H. Lo, D. A. Buchanan, Y. Taur, and W. Wang, "Quantum-mechanical modeling of electron tunneling current from the inversion layer of ultra-thin-oxide NMOSFETs," IEEE Electron. Device Lett., 18 (4), 209–211, 1997.

38. Y. Taur, S. Wind, Y. J. Mii, Y. Lii, D. Moy, K. A. Jenkins, C. L. Chen, P. J. Coane, D. Klaus, J. Bucchignano, M. Rosenfield, M. G. R. Thompson, and M. Polcari, "High performance 0.1 μm CMOS devices with 1.5 V Power Supply," IEEE IEDM Tech. Dig., 127–130, 1993.
39. J. R. Brews, W. Fichtner, E. H. Nicollian, and S. M. Sze, "Generalized guide for MOSFET miniaturization," IEEE Electron. Device Lett., EDL-1 (1), 2–4, 1980.
40. W.-H. Lee, T. Osakama, K. Asada, and T. Sugano, "Design methodology and size limitations of submicrometer MOSFETs for DRAM applications," IEEE Trans. Electron. Dev., 35 (11), 1876–1884, 1988.
41. K. K. Ng, S. A. Eshraghi, and T. D. Stanik, "An improved generalized guide for MOSFET scaling," IEEE Trans. Electron. Device, 40 (10), 1895–1897, 1993.
42. Y. Taur, "CMOS design near the limit of scaling," IBM J. Res. Dev., 46 (2/3), 213–222, 2002.
43. B. Davari, "CMOS technology scaling, 0.1 μm and beyond," IEEE IEDM Tech. Dig., 555–558, 1996.
44. L. D. Yau, "A simple theory to predict the threshold voltage of short-channel IGFETs," Solid-State Electron., 17 (10), 1059–1063, 1974.
45. G. W. Taylor, "The effects of two-dimensional charge sharing on the above-threshold characteristics of short-channel IGFETs," Solid-State Electron., 25 (8), 22 (8), 701–717, 1979.
46. O. Jaentsch, "A geometrical model of the threshold voltage of short and narrow-channel MOSFETs," Solid-State Electron., 25 (1), 59–61, 1982.
47. G. Merckel, "A simple model of the threshold voltage of short and narrow channel MOSFETs," Solid-State Electron., 23 (12), 1207–1213, 1980.
48. R. R. Troutman, "VLSI limitations from drain-induced barrier lowering," IEEE Trans. Electron, Device, ED-26 (4), 461–469, 1979.
49. T. H. Nguyen and J. D., Plummer, "Physical mechanisms responsible for short channel effects in MOS devices," IEEE IEDM Tech. Dig., 596–599, 1981.
50. C.-Y. Wu and S.-Y. Yang, "An analytic and accurate model for the threshold voltage of short channel MOSFETs in VLSI," Solid-State Electron., 27 (7), 651–658, 1984.
51. S. C. Jain and P. Balk, "A unified analytical model for drain-induced barrier lowering and drain-induced high electric field in a short-channel MOSFET," Solid-State Electron., 30 (5), 503–511, 1987.
52. P. E. Cottrell and E. M. Buturla, "Steady state analysis of field effect transistors via the finite element method," IEEE IEDM Tech. Dig., 51–54, 1975.
53. S. Selberherr, "MINIMOS – A two-dimensional MOS transistor analyzer," IEEE Trans. Electron Dev., ED-27 (8), 1440–1560, 1980.
54. J. A. Greenfield and R. W. Dutton, "Nonplanar VLSI device analysis using the solution of Poisson's equation," IEEE Trans. Electron Dev., ED-27 (8), 1520–1532, 1980.
55. C. L. Wilson and J. L. Blue, "Two-dimensional finite element charge-sheet model for a short-channel MOS transistor," Solid-State Electron., 25 (6), 461–477, 1982.
56. W. P. Noble, "Short channel effects in dual gate field effect transistors," IEEE IEDM Tech. Digest., 483–486, 1978.
57. M. T. Bohr and Y. A. El-Mansy, "Technology for advanced high-performance microprocessors," Trans. Electron Dev., 45 (3), 620–625, 1998.
58. M. Nishida and H. Onodera, "An anomalous increase in threshold voltages with shortening the channel lengths for deeply boron-implanted n-channel MOSFETs," IEEE Trans. Electron Dev., ED-28 (9), 1101–1103, 1981.
59. M. Orlowski, C. Mazuré, and F. Lau, "Submicron short channel effects due to gate reoxidation induced lateral interstitial diffusion," IEEE IEDM Tech. Digest, 632–635, 1987.
60. C. Mazuré and M. Orlowski, "Guidelines for reverse short-channel behavior," IEEE Electron Dev. Lett., 10 (12), 556–558, 1989.
61. C.-Y. Hu and J. M. Sung, "Reverse short-channel effects on threshold voltage in submicron silicide devices," IEEE Electron Dev. Lett., 10 (10), 446–448, 1989.
62. T. Kunikiyo, K. Mitsui, M. Fujinaga, T. Uchida, N. Kotani, and Y. Akasaka, "Numerical modeling of processes and devices for integrated circuits," NUPAD IV workshop, May, 1992.

63. C. S. Rafferty, H.-H. Vuong, S. A. Esharghi, M. D. Giles, M. R. Pinto, and S. J. Hillenius, "Explanation of reverse short channel effect by defect gradients," IEEE IEDM Tech. Digest, 311–314, 1993.
64. H. Brut, A. Juge, and G. Ghbaudo, "Physical model of threshold voltage in silicon MOS transistors including reverse short channel effects," Electron Lett., 31 (5), 411–412, 1995.
65. K. Nishi, H. Matsuhashi, T. Ochini, K. Kasai, and T. Nishikawa, "Evidence of channel profile modification due to implantation damage studied by new method, and its implication to reverse short channel effect of nMOSFETs," IEEE IEDM Tech. Digest, 993–995, 1995.
66. B. Szelag, F. Balestra, and G. Ghibaudo, "Comprehensive analysis of reverse short-channel effect in silicon MOSFETs from low-temperature operation," IEEE Electron Dev. Lett., 19 (12), 511–513, 1998.
67. D. Tsoulakas, C. Tsamis, D. N. Kouvatsos, P. Revva, and E. Tsoi, "Reduction in the reverse short channel effect in thick SOI MOSFETs," IEEE Electron Dev. Lett., 18 (3), 90–92, 1997.
68. N. D. Arora and M. S. Sharma, "Modeling the anomalous threshold voltage behavior in submicrometer MOSFETs," IEEE Electron Dev. Lett., 13 (2), 92–94, 1992.
69. H. Jacobs, A. v. Schwerin, D. Scharfetter, and F. Lau, "MOSFET reverse short channel effect due to silicon interstitial capture in gate oxide," IEEE IEDM Tech. Digest, 307–310, 1993.
70. J. M. Sung, C. Y. Lu, M. L. Chen, and S. J. Hillenius, "Fluorine effect on boron diffusion of p^+ gate devices," IEEE IEDM Tech. Digest, 447–450, 1989.
71. J. R. Pfiester, L. C. Parrillo, and F. K. Baker, "A physical model for boron penetration through thin gate oxides from p^+ polysilicon gates," IEEE Electron Dev. Lett., 11 (6), 247–249, 1990.
72. C.-Y. Chang, C.-Y. Lin, J. W. Chou, C. C.-H. Hsu, H.-T. Pan, and J. Ko, "Anomalous reverse short-channel effect in p^+ polysilicon gated p-channel MOSFET," IEEE Electron Dev. Lett., 15 (11), 437–439, 1994.
73. T. Hori and K. Kurimoto, "A new half-micron p-channel MOSFET with LATIPS (Large-Tilt-Angle-Implanted-Punchthrough-Stopper)," IEEE IEDM Tech. Digest, 394–397, 1988.
74. H. Wakabayashi, M. Ueki, M. Narihiro, T. Fukai, N. Ikezawa, T. Matsuda, K. Yishida, K. Takeuchi, Y. Ochiai, T. Mogami, and T. Kunio, "Sub-50-nm physical gate length CMOS technology and beyond using steep halo," IEEE IEDM Tech. Digest, 89–93, 2002.
75. R. Qwoziecki, T. Skotnicki, P. Bouillon, and P. Gentil, "Optimization of Vth roll-off in MOSFETs with advanced channel architecture – Retrograde doping pockets," IEEE Trans. Electron Dev., 46 (7), 1551–1561, 1999.
76. A. Sadovnikov, A. Kalnitsky, A. Bergemont, and P. Hopper, "The effect of polysilicon doping on the reverse short-channel effect in sub-quarter micron NMOS transistors," IEEE Trans. Electron Dev., 48 (2), 393–395, 2001.
77. P. P. Wang, "Device characteristics of short-channel and narrow-width MOSFETs," IEEE Trans. Electron Dev., ED-25 (7), 779–786, 1978.
78. K. O. Jeppson, "Influence of the channel width on the threshold voltage modulation in m.o.s.f.e.t.s," Electron. Lett., 11 (14), 997–299, 1975.
79. H. N. Kotecha and K. E. Beilstein, "Current and capacitances in narrow width MOSFET structures," IEEE IEDM Tech. Digest, 47–50, 1975.
80. J. D. Sansbury, "MOS field threshold increase by phosphorus-implanted field", IEEE Trans. Electron Dev., ED-20 (5), 473–476, 1973.
81. M. B. Bandali and T. C. Lo, "On the modeling of the self-aligned field implanted MOS devices with narrow width," IEEE IEDM Tech. Digest, 573–576, 1975.
82. H. Kotecha and W. P. Noble, "Interaction of IGFET field design with narrow channel device operation," IEEE IEDM Tech. Digest, 724–727, 1980.
83. L. A. Akers, M. M. E. Beguwala, and F. Z. Custode, "A model of a narrow-width MOSFET including tapered oxide and doping encroachment," IEEE Trans. Electron Dev., ED-28 (12), 1490–1495, 1981.
84. K. E. Kroell and G. K. Ackermann, "Threshold voltage of narrow channel field effect transistors," Solid-State Electron., 19 (1), 77–81, 1976.
85. W. P. Noble and P. E. Cottrell, "Narrow channel effects in insulated gate field effect transistors," IEEE IEDM Tech. Digest, 582–586, 1976.

86. C.-R. Ji and C. T. Sah, "Analysis of the narrow gate effect in submicrometer MOSFETs," IEEE Trans. Electron Dev., ED-30 (12), 1672–1677, 1983.
87. N. Shigyo, M. Konaka, R. L. M. Dang, "Three-dimensional simulation of inverse narrow-channel effect," Electron. Lett., 18 (6), 274–275, 1982.
88. N. Shigyo, S. Fukuda, T. Wada, K. Hieda, T. Hamamoto, H. Watanabe, K. Sunouch, and H. Tango, "Three-dimensional analysis of subthreshold swing and transconductance for fully recessed oxide (trench) isolated 1/4-μm-width MOSFETs," IEEE Trans. Electron Dev., 35 (7) 945–950, 1988.
89. L. A. Akers, "The inverse-narrow-width effect," IEEE Electron Dev. Lett., EDL-7 (7), 419–421, 1986.
90. S. S.-S. Chung and T.-C. Li, "An analytical threshold-voltage model of trench-isolated MOS devices with nonuniformly doped substrates," IEEE Trans. Electron Dev., 39 (3), 614–622, 1992.
91. K. Ohe, Y. S. Kugo, H. Umimoto, and S. Odanaka, "The inverse-narrow-width effect of LOCOS isolated n-MOSFET in a high-concentration p-well," IEEE Electron Dev. Lett., 13 (12), 636–638, 1992.
92. E. Herbert, K. M. Hong, Y. C. Cheng, and K. Y. Chan, "The narrow-channel effect in MOSFETs with semi-recessed oxide structures," IEEE Trans. Electron Dev., 37 (3), 692–670, 1990.
93. L. A. Akers, M. Sugino, and J. M. Ford, "Characterization of the inverse-narrow-width effect," IEEE Trans. Electron Dev. ED-34 (12), 2476–2484, 1987.
94. T. Osishi, K. Shiozawa, A. Furukawa, Y. Abe, and Y. Tokuda, "Isolation edge effect depending on gate length of MOSFETs with various isolation structures," IEEE trans. Electron Dev., 47 (4), 822–827, 2000.
95. A. H. Parera, J.-H. Lin, Y.-C. Ku, M. Azrak, B. Taylor, J. Hayden, M. Thompson, and M. Blackwell, "Trench isolation for 0.45 μm active pitch and below," IEEE. IEDM Tech. Digest, 679–682, 1995.
96. S. Masuda, T. Sato, H. Yoshimura, A.Sudo, I. Mizushima, Y. Tsunashima, and Y. Toyoshima, "Novel corner rounding process for shallow-trench isolation utilizing MSTS (Micro-Structure Transformation of Silicon)," IEEE IEDM Tech. Digest, 137–140, 1998.
97. J. B. Kuo, Y. G. Chen, and K. W. Su, "Sidewall-related narrow-channel effect in mesa-isolated fully-depleted ultra-thin SOI NMOS devices," IEEE Electron Dev. Lett., 16 (9), 379–381, 1995.
98. B. Agrawal, V. K. De, and J. D. Meindl, "Three-dimensional analytical subthreshold models for bulk MOSFETs," IEEE Trans. Electron Dev., 42 (12), 2170–2180, 1995.
99. K. Ohe, S. Odanaka, K. Moriyama, T. Hori, and G. Fuse, "Narrow-width effects of shallow trench-isolated CMOS with n^+-polysilicon gate," IEEE Trans. Electron Eev., 36 (6), 1110–1116, 1989.
100. A. Ono, R. Ueno, and I. Sakai, "TED control technology for suppression of reverse narrow channel effect in 0.1 μm MOS devices," IEEE IEDM Tech. Digest, 227–230, 1997.
101. C.-Y. Chang, S.-J. Chang, T.-S. Chao, S.-D. Wu, and T.-Y. Huang, "Reduced reverse narrow channel effect in thin SOI nMOSFETs," IEEE Electron Dev. Lett., 21 (9),460–462, 2000.
102. S.-J. Chang, C.-Y. Chang, C. Chen, J.-Y. Chou, T.-S. Chao, and T.-Y. Huang, "An anomalous crossover in Vth toll-off for indium-doped nMOSFETs," IEEE Electron Dev. Lett., 21 (9), 457–459, 2000.
103. J. Kim, T. Kim, J. Park, W. Kim, B. Hong, and G. Yoon, "A shallow trench isolation using nitric oxide (NO)-annealed wall oxide to suppress inverse narrow width effect," IEEE Electron Dev. Lett., 21 (12), 575–577, 2000.
104. C. P. Chang, C. S. Pai, F. H. Baumann, C. T. Liu, C. S. Rafferty, M. R. Pinto, E. J. Lloyd, M. Bude, F. P. Klemens, J. F. Miner, K. P. Cheung, J. I. Colonell, W. Y. C. Lai, H. Vaidya, S. J. Hillenius, R. C. Liu, and J. T. Clemens, "A highly manufacturable corner rounding solution for 0.18 μm shallow trench isolation," IEEE IEDM Tech. Digest, 661–664, 1997.
105. T. Sato, I. Mizushima, J.-I. Iba, M. Kito, Y. Takegawa, A. Sudo, and Y. Tsunashima, "Trench transformation technology using hydrogen annealing for realizing highly reliable device structure with thin dielectric film," VLSI Tech. Digest, 206–207, 1998.

106. K. Horita, T. Kuroi, Y. Itoh, K. Shiozawa, K. Eikyu, K. Goto, Y. Inoue, and M. Inuishi, "Advanced shallow trench isolation to suppress the inverse narrow channel effects for 0.24 μm pitch isolation and beyond," VLSI Tech. Digest, 179–180, 2000.
107. W. P. Noble, A. K. Ghatalia, and B. El-Kareh, "MOSFET with raised STI isolation self-aligned to the gate stack," US patent 5,539,229, Dec. 28, 1994.
108. J.-W. Lee, Y. Saitoh, R. Koh, and T. Mogami, "Elevated field insulator (ELFIN) process for device isolation of ultrathin SOI MOSFETs with top silicon film less than 20 nm," IEEE Electron Dev. Lett., 32 (8), 467–469, 2002.
109. C. T. Liu, F. H. Baumann, A. Ghetti, H. H. Vuong, C. P. Chang, K. P. Cheung, J. I. Colonell, W. Y. C. Lai, E. J. Lloyd, J. F. Miner, C. S. Pai, H. Vaidya, R. C. Liu, and J. T. Clemens, "Severe thickness variation of sub-3 nm gate oxide due to Si surface faceting, poly-Si intrusion, and corner stress," VLSI Tech. Digest, 94–95, 1999.
110. M. Togo, K. Watanabe, M. Terai, T. Fukai, M. Narihiro, K. Arai, S. Koyama, N. Ikezawa, T. Tatsumi, and T. Mogami, "Impact of radical oxynitridation on characteristics and reliability of sub-1.5 nm-thick gate dielectric FETs with narrow channel and shallow-trench isolation," IEEE IEDM Tech. Digest, 813–816, 2001.
111. M. Togo, K. Watanabe, M. Terai, T. Yamamoto, T. Fukai, T. Tatsumi, and T. Mogami, "Improving the quality of sub-1.5-nm-thick oxynitride gate dielectric for FETs with narrow channel and shallow-trench isolation using radical oxygen and nitrogen," IEEE Trans. Electron Dev., 49 (10), 1736–1741, 2002.
112. K. H.-L. Hsueh, J. J. Sanchez, T. A. Demassa, and L. A. Akers, "Inverse-narrow-width and small-geometry MOSFET threshold voltage model," IEEE Trans. Electron Dev., 35 (3), 325–338, 1988.
113. B. Hoeneisen and C. A. Mead, "Fundamental limitations in microelectronics – I. MOS technology," Solid-State Electron., 15 (7), 819–829, 1972.
114. R. W. Keyes, "Effect of randomness in the distribution of impurity ions on FET thresholds in integrated electronics," IEEE J. Solid-State Circuits, 10 (4), 245–247, 1975.
115. T. Mizuno, J.-I. Okamura, and A. Toriumi, "Experimental study of threshold voltage fluctuations due to statistical variation of channel dopant number in MOSFETs," IEEE Trans. Electron Dev., 41 (11), 2216–2221, 1994.
116. P. A. Stolk and D. B. M. Klaassen, "The effect of statistical dopant fluctuation on MOS device performance," IEEE IEDM, Tech. Digest, 627–630, 1996.
117. P. A. Stolk, F. P. Widdershoven, and D. B. M. Klaassen, "Modeling statistical dopant fluctuations in MOS transistors," IEEE Trans. Electron Dev., 45 (5), 1960–1971, 1998.
118. H.-S. Wong and Y. Taur, "Three-dimensional 'atomistic' simulation of discrete random dopant distribution effect in sub-0.1 μm MOSFETs," IEEE IEDM, Tech. Digest, 705–708, 1993.
119. K. R. Lakshimikumar, R. A. Hadaway, and M. A. Copeland, "Characterization and modeling of mismatch in MOS transistors for precision analog design," IEEE J. Solid-State Circuits, 21 (12), 1057–1066, 1986.
120. M. J. M Pelgrom, A. C. J. Duinmaijer, and A. P. G. Welbers, "Matching properties of MOS transistors," IEEE J. Solid-State Circuits, 24 (10), 1433–1440, 1989.
121. P. G. Drennan and C. C. McAndrew, "A comprehensive MOSFET mismatch model," IEDM Tech. Digest, 167–170, 1989.
122. H. Yang, V. Macary, J. L. Huber, W.-G Min, B. Baird, and J. Zuo, "Current Mismatch due to local dopant fluctuations in MOSFET channel," IEEE Trans. Electron Dev., 50(11), 2248–2254, 2003.
123. S.-C. Liu, J. B. Kuo, K.-T. Huang, and S.-W. Sun, "A closed-form back-gate-bias related inverse narrow-channel effect model for deep-submicron VLSI CMOS devices using shallow trench isolation,", IEEE Trans. Electron Dev., 47 (4), 725–733, 2000.
124. S. K. Tewksbury, "N-channel enhancement-mode MOSFET characteristics from 10 to 300 K," IEEE Trans. Electron Dev., ED-28 (12), 1519–1529, 1981.
125. M. Aoki, T. Ishii, T. Yoshimura, Y. Kiyota, S. Iijima, T. Yamanaka, T. Kure, K. Ohyu, T. Nishida, S. Okazaki, K. Seki, and K. Shimohigashi, "0.1 μm CMOS devices using low-impurity channel transistors (LICT)," IEEE IEDM, Tech. Digest, 939–941, 1990.

126. K. Noda, T. Tatsumi, T. Uchida, K. Nakajima, H. Miyamoto, and C. Hu, "A 0.1-μm delta-doped MOSFET fabricated with post-low-energy implanting selective epitaxy," IEEE Trans. Electron Dev., 45 (4), 809–814, 1998.
127. A. Hori, T. Hirai, M. Tanaka, H. Nakaoka, H. Umimoto, and M. Yasuhira, "A 0.1-μm CMOS with a step channel profile formed by ultra high vacuum CVD and in-situ doped ions," IEEE IEDM Tech. Digest, 909–911, 1993.
128. A. Asenov and S. Saini, "Suppression of random dopant-induced threshold voltage fluctuations in sub-0.1-μm MOSFETs with epitaxial and δ-doped channels," IEEE trans. Electron Dev., 46 (8), 1718–1724, 1999.
129. R.-H. Yan, A. Ourmazd, and K. F. Lee, "Scaling of Si MOSFET: from bulk to SOI to bulk," IEEE Trans. Electron Dev., 39 (7), 1704–1710, 1992.
130. K. F. Lee, R. H. Yan, D. Y. Jeon, G. M. Chin, Y. O. Kim, D. M. Tennant, B. Razavi, H. D. Lin, Y. G. Wey, E. H. Westerwick, M. D. Morris, R. W. Johnson, T. M. Liu, M. Tarsia, M. Cerullo, R. G. Swartz, and A. Ourmazd, Room temperature 0.1 μm CMOS technology with 11.8 ps gate delay," IEEE IEDM Tech. Digest, 131–133, 1993.
131. N. Kawakami, K. Egusa, and K. Shibahara, "Reduction of threshold voltage fluctuation of p-MOSFETs by antimony super steep retrograde well channel," Second International Workshop on Junction Technology, 7–10, 2001.
132. J. B. Jacobs and D. Antoniadis, "Channel profile engineering for MOSFETs with 100 nm channel length," IEEE Trans. Electron Dev., 42 (5), 870–875, 1995.
133. J.-H. Lee, J. Lee, S. Talwar, Y. Wang, D. Weon, S. Hahn, C. Kang, T. Hong, Y. Kim, H. Lee, S. Lee, J. Rob, D. Kang, and J. Park, "Laser thermal annealed SWSR well prior to epi-channel growth (LASPE) for 70 nm nFETs," IEEE IEDM Tech. Digest, 441–444, 2000.
134. S. Venkatesan, J. W. Lutze, C. Lage, and W. J. Taylor, "Device drive current degradation observed with retrograde channel profiles," IEEE IEDM Tech. Digest, 419–422, 1995.
135. S. E. Thompson, P. A. Packan, and M. T. Bohr, "Linear versus saturated drive current: tradeoffs in super steep retrograde well engineering," VLSI Tech. Digest, 154–255, 1996.
136. J. G. Ruch, "Electron dynamics in short channel field-effect transistors," IEEE Trans. Electron Dev., ED-19 (5), 652–654, 1972.
137. D. K. Ferry, J. R. Barker, and H. L. Grubin, "Hot-carrier constraints on transient transport in very small semiconductor devices," IEEE Trans. Electron Dev., ED-38 (8), 905–911, 1981.
138. Y.-J. Park, T.-W. Tang, and D. H. Navon, "Monte Carlo surface scattering simulation in MOSFET structures," IEEE Trans. Electron Dev., 30 (9), 1110–1116,1983.
139. S. E. Laux and M. V. Fischetti, "Monte Carlo simulation of submicrometer Si n-MOSFETs at 77 and 300 K," IEEE Electron Dev. Lett., 9 (9), 467–469, 1988.
140. G. Baccarani and M. R. Wordeman, "An investigation of steady-state velocity overshoot in silicon," Solid-State Electron., 28 (4), 407–416, 1985.
141. T. Kobayashi and K. Saito, "Two-dimensional analysis of velocity overshoot effects in ultrashort-channel Si MOSFETs," IEEE Trans. Electron Dev., ED-33 (4), 788–792, 1985.
142. G. A. Sai-Halasz, M. R. Wordemann, D. P. Kern, S. Rischton, and E. Ganin, "High transconductance and velocity overshoot in NMOS devices at the 0.1-μm gate-length level," IEEE Electron Dev. Lett., 9 (9), 464–466, 1988.
143. S. Y. Chou, D. A. Antoniadis, and H. I. Smith, "Observation of electron velocity overshoot in sub-100-nm-channel MOSFETs in silicon," IEEE Electron Dev. Lett., 6 (12), 665–667, 1985.
144. T. Sato, Y. Takeishi, and H. Hara, "Mobility anisotropy of electrons in inversion layers on oxidized silicon surfaces," Phys. Rev. B., 4 (6), 1950–1960, 1971.
145. C. T. Sah, J. R. Edwards, and T. H. Ning, "Observation of mobility anisotropy of electrons on (110) silicon surfaces at low temperatures," Physica Status Solidi (a), 10 (1), 153–160, 1972.
146. S.-i. Tagaki, A. Toriumi, M. Iwase, and H. Tango, "On the universality of inversion layer mobility in Si MOSFETs: Part II – effects of surface orientation," IEEE Trans. Electron Dev., 41 (12), 2362–2368, 1994.
147. D. Colman, R. T. Bate, and J. P. Mize, "Mobility anisotropy and piezoresistance in silicon p-type inversion layers," J. Appl. Phys., 39 (4) 1823–1831, 1968.

148. H. Sayama, Y. Nishida, H. Oda, T. Oishi, S. Shimizu, T. Kunikiyo, S. Sonoda, Y. Inoue, and M. Inuishi," Effect of <100> channel direction for high performance SCE immune pMOSFET with less than 0.15 μm gate length," IEEE IEDM Tech. Digest, 657–660, 1999.
149. M. Yang, V. W. C. Chan, K. K. Chan, L. Shi, D. M. Fried, J. H. Sathis, A. I. Chou, E. P. Gusev, J. A. Ott, L. E. Burns, M. V. Fischetti, and M. Ieong, "Hybrid-orientation technology (HOT): Opportunities and challenges," IEEE Trans. Electron Dev., 53 (5), 965–978, 2006.
150. M. Yang, E. P. Gusev, M. Ieong, O. Gluschnkov, D. C. Boyd, K. K. Chan, P. M. Kozlowski, C. P. D'Emic, R. M. Sicina, P. C. Jamison, and A. I. Chou, "Performance dependence of CMOS silicon substrate orientation for ultrathin oxynitride and HfO_2 gate dielectrics," IEEE Electron Dev. Lett., 24 (5), 339–341, 2003.
151. H. S. Momose, T. Ohguro, K. Kojima, S.-I. Nakamura, and Y. Toyoshima, "1.5-nm gate oxide CMOS on (100) surface-oriented Si substrate," IEEE Trans. Electron Dev., 50 (4), 1001–1007, 2003.
152. M. Kinugawa, M. Kakumu, T. Usami, and J. Matsugana, "Effects of silicon surface orientation on submicron CMOS devices," IEEE IEDM Tech. Digest, 581–584, 1985.
153. T. Mizuno, N. Sugiyama, T. Tezuka, Y. Moriyama, S. Nakaharai, T. Maeda, and S. Takagi, "Physical mechanism for high hole mobility in (110)-surface strained- and unstrained-MOSFETs," IEEE IEDM Tech. Digest, 809–812, 2003.
154. K. Onishi, C. S. Kang, R. Choi, H.-J. Cho, Y. H. Kim, S. Krishnan, M. S. Akbar, and J. C. Lee, "Performance of polysilicon gate HfO_2 MOSFETs on (100) and (111) silicon substrate," IEEE Electron Dev. Lett., 24 (4) 254–256, 2003.
155. L. Chang, M. Ieong, and M. Yang, "CMOS circuit performance enhancement by surface orientation optimization," IEEE Trans. Electron Dev., 51 (10), 1621–1627, 2004.
156. H. Nakamura, T. Ezaki, T. Ewamoto, M. Togo, T. Ikezawa, N. Ikarashi, M. Hane, and T. Yamamoto, "Effects of selecting channel direction in improving performance of sub-100 nm MOSFETs fabricated on (110) surface Si substrate," Jap. J. Appl. Phys., 43 (4B), 1723–1728, 2004.
157. M. Aoki, K. Yano, T. Masuhara, and K. Shimohigashi, "Fully symmetric cooled CMOS on (110) plane," IEEE Trans. Electron Dev., 36 (8), 1429–1433, 1989.
158. M. J. van Dort, P. H. Woerlee, A. J. Walker, C. A. H. Juffermans, and H. Lifka, "Influence of high substrate doping levels on the threshold voltage and the mobility of deep-submicrometer MOSFETs," IEEE Trans. Electron Dev., 39 (4), 932–938, 1992.
159. S. Tagaki, M. Iwase, and A. Toriumi, "On the universality of inversion-layer mobility," IEEE IEDM Tech. Digest, 398–401, 1988.
160. A. Hiroki, S. Odanaka, K. Ohe, and H. Esaki, "A mobility model for submicrometer MOSFET simulations including hot-carrier-induced device degradation, IEEE Trans. Electron Dev., 35 (9), 1487–1493, 1988.
161. F. Gamiz, J. B. Roldan, H. Kosina, and T. Grasser, "Improving strained-Si on Si1-xGex deep submicron MOSFETs performance by means of stepped doping profile," IEEE Trans. Electron Dev., 48 (9), 1878–1884, 2001.
162. W. P. Soppa and H.-G. Wagemann, "Investigation and modeling of the surface mobility of MOSFETs from -25 to $+150°C$," IEEE Trans. Electron Dev., 35 (7), 970–977, 1988.
163. D. K. Ferry, "Effects of surface roughness in inversion layer transport," IEEE IEDM Tech. Digest, 605–608, 1984.
164. Y. C. Cheng and E. A. Sullivan, "On the role of scattering by surface roughness in silicon inversion layers," Surface Science, 34 (3), 717–731, 1973.
165. K. Sonoda, K. Taniguchi, and C. Himaguchi, "Analytical device model for submicrometer MOSFETs," IEEE Trans. Electron Dev., 38 (12), 2662–2668, 1991.
166. C. S. Smith, "Piezoresistance effect in germanium and silicon," Phys. Rev. B., 94 (1), 42–46, 1954.
167. R. W. Keyes, "High-mobility FET ins strained silicon," IEEE Trans. Electron Dev., ED-33 (6), 853, 1986.
168. J. Wesler, J. L. Hoyt, S. Takagi, and J. F. Gibbons, "Strain dependence of the performance enhancement in strained-Si-n-MOSFETs," IEEE IEDM Tech. Digest, 373–376, 1994.

169. K. Rim, J. L. Hoyt, and J. F. Gibbons, "Fabrication and analysis of deep submicron strained-Si N-MOSFETs," IEEE Trans. Electron Dev., 47 (7), 1406–1415, 2000.
170. K. Rim, S. Koester, M. Hatgrove, J. Chu, P. M. Mooney, J. Ott, T. Kanarsky, P. Ronsheim, M. Ieong, A. Grill, and H.-S. P. Wong, "Strained Si NMOSFETs for high-performance CMOS technology," VLSI Tech. Digest, 50–51, 2001.
171. H. M. Nayfey, C. W. Leitz, A. J. Pitera, E. A. Fitzgerald, J. L. Hoyt, and D. A. Antoniadis, "Influence of high channel doping on the inversion layer electron mobility in strained silicon n-MOSFETs," IEEE Electron Device Lett., 24 (4), 248–250, 2003.
172. C.-H. Ge, C.-C. Lin, C.-H. Ko, C.-C. Huang, B.-W. Chan, B.-C. Perng, V.-C. Sheu, P.-Y. Tsai, L.-G. Yao, C.-L. Wu, T.-L. Lee, C.-J. Chen, C.-T. Wang, S.-C. Lin, Y.-C. Yeo, and C. Hu, "Process-strained Si (PSS) CMOS technology featuring 3D strain engineering," IEEE IEDM Tech. Digest, 73–76, 2003.
173. S. E. Thompson, M. Armstrong, C. Auth, M. Alavi, M. Buchler, R. Chau, S. Cea, T. Ghani, T. Hoffman, C.-H. Jan, C. Kenyon, J. Klaus, K. Kuhn, Z. Ma, B. Mcintyire, K. Mistry, A. Murthy, B. Obradovic, R. Nagisetty, P. Nguyen, S. Sivakumar, R. Shaheed, L.. Shifren, B. Tufts, S. Tyagi, M. Bohr, and Y. El-Masry, "A 90-nm logic technology featuring strained-silicon," IEEE Trans. Electron Dev., 51 (11), 1790–1796, 2004.
174. M. L. Lee and E. A. Fitzgerald, "Hole mobility enhancements in nanometer-scale strained-silicon heterostructures grown on Ge-rich relaxed $Si_{1-x}Ge_x$," J. Appl. Phys., 94 (4), 2590–2596, 2003.
175. K. Ishimaru, M. Takayanagi, T. Watanabe, S. Inaba, M. Fujiwara, and D. Matsushita, "Scaled CMOS with SiON and high-k," 209th ECS Meeting, Silicon Materials Science and Technology, Denver, Colorado, 2006.
176. B. Brar, G. D. Wilk, and A. C. Seabaugh, "Direct extraction of the electron tunneling effective mass in ultrathin SiO_2," Appl. Phys. Lett., 69 (18), 2728–2730, 1996.
177. G. D. Wilk, R. M. Wallace, and J. M. Anthony, "High-k gate dielectrics: Current status and material properties considerations," Appl. Phys. Rev., 89 (10), 5243–5275, 2001.
178. T. M. Pan, T. F. Lei, H. C. Wen, and T. S. Chao, "Characterization of ultrathin oxynitride (18–21 Å) gate dielectrics by NH_3 nitridation and N_2O RTA treatment," IEEE Trans. Electron Dev., 48 (5), 907–912, 2001.
179. T. M. Pan, T. F. Lei, and T. S. Chao, "Robust ultrathin oxynitride dielectrics by NH_3 nitridation and N_2O RTA treatment," IEEE Electron Dev. Lett., 21 (8), 178–180, 2000.
180. S. C. Song, H. F. Luan, Y. Y. Chen, M. Gardner, J. Fulford, M. Allen, and D. L. Kwong, "Ultra Thin ($<20\AA$) CVD Si_3N_4 gate dielectric for deep-sub-micron CMOS devices," IEEE IEDM Tech. Digest, 373–376, 1998.
181. T. P. Ma, "Making silicon nitride film a viable gate dielectric," IEEE Trans. Electron Dev., 45 (3), 680–690, 1998.
182. S. C. Song, H. F. Luan, C. H. Lee, A. Y. Mao, S. J. Lee, J. Gelpey, S, Marcus, and D. L. Kwong, "Ultra thin high quality stack nitride/oxide gate dielectrics prepared by in-situ rapid thermal N_2O oxidation of NH_3-nitrided Si," Digest VLSI Technol. Syst. Applic., 78–81, 1999.
183. V. J. Kapoor, R. S. Bailey, and H. J. Stein, "Hydrogen-related memory traps in thin silicon nitride films," J. Vac. Sci. Technol., A, 1 (2), 600–603, 1983.
184. K. Allaert, A, Van Calster, H. Loos, and A. Lequesne, "A comparison between silicon nitride films made by PCVD on N_2-SiH_4/Ar and N_2-SiH_4/He," J. Electrochem Soc., 132 (7), 1763–1766, 1985.
185. J. Robertson, "Band offsets of wide-band-gap oxides and implications for future electronics," J. Vac. Sci. Technol., B 18 (3), 1785–1791, 2000.
186. K. J. Hubbard and D. G. Schlom, "Thermodynamic stability of binary oxides in contact with silicon," J. Mater. Res., 11 (11), 2757–2761, 1997.
187. C. Hobbs, T. Tseng, K. Reid, B. Taylor, L. Dip, L. Hebert, R. Garcia, R. Hegde, J. Grant, D. Gilmer, A. Franke, V. Dhandapani, M. Azrak, L. Prabhu, R. Rai, S. Bagchi, J. Conner, S. Backer, F. Dumbuya, B. Nguyen, and P. Tobin, "80 nm poly-Si gate CMOS with HfO_2 gate dielectric," IEEE IEDM Tech. Digest, 651–654, 2001.

188. X. Yu, M. Yu, and C. Zhu, "A comparative study of HfTaON/SiO_2 and HfON/SiO_2 gate stacks with TaN metal gate for advanced CMOS applications," IEEE Trans. Electron Dev., 54 (2), 284–290, 2007.
189. A. L. P. Rotondaro, M. R. Visokay, J. J. Chambers, A. Shanware, R. Khamankar, H. Bu, R. T. Laaksonen, L. Tsung, M. Douglas, R. Kuan, M. J. Bevan, T. Grider, J. McPherson, and L. Colombo, "Advanced CMOS Transistors with a Novel HfSiON Gate Dielectric," VLSI Tech. Digest, 148–149, 2002.
190. T. Yamaguchi, R. Iijima, T. Ino, A. Nishiyama, H. Satake, and N. Fukushima, "Additional scattering effects for mobility degradation in Hf-silicate gate MISFETs," IEEE IEDM Tech. Digest, 621–624, 2002.
191. Z. Ren, M. V. Fischetti, E. P. Geusev, E. A. Cartier, and M. Chudzik, "Inversion channel mobility in high-k high performance MOSFETs," IEDM Tech. Digest, 793–796, 2003.
192. A. Morioka, H. Watanabe, M. Miyamura, T. Tatsumi, M. Saitoh, T. Ogura, T. Iwamoto, T. Ikarashi, Y. Saito, Y. Okada, H. Watanabe, Y. Mochiduki, and T. Mogami, "High mobility MISFET with low trapped charge in HfSiO films,"VLSI. Tech. Digest, 165–166, 2003.
193. A. Chin, C. C. Liao, C. H. Lu, W. J. Chen, and C. Tsai, "Device and reliability of high-K Al_2O_3 gate dielectric with good mobility and low D_{it}," VLSI. Tech. Digest, 135–136, 1999.
194. A. Chin, Y. H. Wu, S. B. Chen, C. C. Liao, and W. J. Chen, "High-quality La_2O_3 and Al_2O_3 gate dielectrics with equivalent oxide thickness 5–10 Å," VLSI. Tech. Digest, 16–17, 2000.
195. D. A. Buchanan, E. P. Gusev, E. Cartier, H. Okorn-Schymidt, K. Rim, M. A. Gribelyuk, A. Mokuta, A. Ajmera, M. Copel, S. Guha, N. Bojarczuk, A. Callegari, C. D'Emic, P. Kozlowski, K. Chan, R. J. Fleming, P. C. Jamison, J. Brown, and R. Arndt, "80-nm poly-silicon gated n-FETs with ultra-thin Al_2O_3 gate dielectric for ULSI applications," IEEE IEDM Tech. Digest, 223–226, 2000.
196. S.-J. Ding, H. Hu, C. Zhu, M. F. Li, S. J. Kim, B. J. Cho, D. S. H. Chan, M. B. Yu, A. Y. Du, A. Chin, and D.-L. Kwong, "Evidence and understanding of ALD HfO_2-Al_2O_3 laminate MIM capacitors outperforming sandwich counterparts," IEEE Electron Dev. Lett., 25 (10), 681–683, 2004.
197. C. Kizilyalli, R. Y. S. Huang, and P. K. Roy, "MOS transistors with stacked SiO_2-Ta_2O_5-SiO_2 gate dielectrics for giga-scale integration of CMOS technologies," IEEE Trans. Electron Dev., 19 (11), 423–425, 1998.
198. B. C. Lai, J.-C. Yu, and J. Y.-M. Lee, "Ta_2O_5/silicon barrier height measured from nMOSFETs fabricated with Ta_2O_5 gate dielectric," IEEE Electron Dev. Lett., 22 (5), 221–223, 2001.
199. K. W. Kwon, I. S. Park, D. H. Han, E. S. Kim, S. T. Ahn, and M. Y. Lee, "Ta_2O_5 capacitors for 1 Gbit DRAM and beyond," IEDM Tech. Dig., 835–838, 1994.
200. M. Hiratani, T. Hamada, S. Iijima, Y. Ohji, I. Asano, N. Nakanishi, and S. Kimura, "A heteroepitaxial MIM-Ta_2O_5 capacitor with enhanced dielectric constant for DRAMs of G-bit generation and beyond," VLSI Tech. Digest, 41–42, 2001.
201. A. Nitayama, Y. Kohyama, and K. Hieda, "Future directions for DRAM memory cell technology," IEEE IEDM Tech. Digest, 335–338, 1998.
202. B. T. Lee, K. H. Lee, C. S. Hwang, W. D. Kim, H. Horii, H.-W. Kim, H.-J. Cho, C. S. Kang, J. H. Chung, S. I. Lee, and M. Y. Lee, "Integration of (Ba, Sr)$Ti0_3$ capacitor with platinum electrodes having $Si0_2$ spacer," IEEE IEDM Tech. Digest, 249–252, 1997.
203. M. Hiratani, T. Nabatame, Y. Matsui, Y. Shimamoto, Y. Sasago, Y. Nakamura, Y. Ohji, I. Asano, and S. Kimura, "A conformal ruthenium electrode for MIM capacitors in Gbit DRAMs using the CVD technology based on oxygen-controlled surface reaction," VLSI Tech. Digest, 102–103, 2000.
204. H. Horii, B. T. Lee, H. J. Lim, S. H. Joo, C. S. Kang, C. Y. Yoo, H. B. Park, W. D. Kim, S. I. Lee, and M. Y. Lee, "A self-aligned stacked capacitor using novel Pt electroplating method for 1 Gbit DRAMs and beyond," VLSI Tech. Digest, 103–104, 1999.
205. T.-J. King, J. R. Pfiester, J. D. Schott, J. P. McVittie, and K. C. Saraswat, "A polycrystalline-$Si_{1-x}Ge_x$-gate CMOS technology," IEEE IEDM tech. Digest, 253–256, 1990.

206. T.-J. King, J. P. McVittie, K. C. Saraswat, and J. R. Pfiester, "Electrical properties of heavily doped polycrystalline silicon-germanium films," IEEE Trans. Electron Dev., 41 (2), 228–232, 1994.
207. Y. V. Ponomarev, P. A. Stolk, C. Salm, J. Schmitz, and P. H. Woerlee, "High-performance deep submicron CMOS technologies with polycrystalline-SiGe gates," IEEE Trans. Electron Dev., 47 (4), 848–855, 2000.
208. P.-E. Hellberg, S.-L. Zhang, and C. S. Petersson, "Work function of boron-doped polycrystalline Si_xGe_{1-x} films," IEEE Electron Dev. Lett., 18 (9), 456–458, 1997.
209. V. Z.-Q. Li, M. R. Mirabedini, R. T. Kuehn, j. J. Wortman, and C. Ozturk, "Single gate 0.15 mm CMOS devices fabricated using RTCVD in-situ boron doped $Si_{1-x}Ge_x$ gates," IEEE IEDM Tech. Digest, 833–836, 1997.
210. N. Kasai, N. Endo, and A. Ishitani, "Deep-submicron tungsten gate CMOS technology," IEEE IEDM Tech. Digest, 242–245, 1988.
211. D. H. Lee, S. W. H. Joo, G. H. Lee, J. Moon, T. E. W. Shim, and J. G. Lee, "Characteristics of CMOSFETs with sputter-deposited W/TiN stack gate," VLSI Tech., 119–120, 1995.
212. J. C. Hu, H. Yang, R. Kraft, A. L. P. Rotondaro, S. Hattangady, W. W. Lee, R. A. Chapman, C.-P. Chao, A. Chatterjee, M. Hanratty, M. Rodder, and I.-C. Chen, "Feasibility of using W/TiN as metal gate for conventional 0.13 μm CMOS technology and beyond," IEEE IEDM Tech. Digest, 825–828, 1997.
213. B. Maiti, P. J. Tobin, C. Hobbs, R. I. Hegde, F. Huang, D. L. O'Meara, D. Jovanovic, M. Mendicino, J. Chen, D. Connelly, O. Adetutu, J. Mogab, J. Candelaria, and L. B. La, "PVD YiN gate NMOSFETs on bulk silicon and fully depleted silicon-on-insulator (FD-SOI) substrates for deep sub-quarter micron CMOS technology," IEEE IEDM Tech. Digest, 781–784, 1998.
214. R. Li and Q. Xu, "Damascene E/TiN gate MOSFETs with improved performance for 0.1-μm regime," IEEE Trans. Electron Dev., 49 (11), 1891–1896, 2002.
215. J.-M. Hwang and G. Pollack, "Novel polysilicon stacked-gate structure for fully-depleted SOI/CMOS," IEEE IEDM Tech. Digest, 345–348, 1992.
216. I. De, D. Johri, A. Srivastava, and C. M, Osburn, "Impact of gate workfunction on device performance at the 50 nm technology node," Solid-State Electron., 44 (8), 1077–1080, 2000.
217. Y. Abe, T. Oishi, K. Shiozawa, Y. Takuda, and S. Satoh, "Simulation study of comparison between metal gate and polysilicon gate for sub-quarter-micron MOSFETs," IEEE Electron Dev. Lett., 29 (12), 632–634, 1999.
218. E. Josse and T. Stotnicki, "Polysilicon gate with depletion – or – metallic gate with buried channel: what evil is worse?," IEEE IEDM Tech. Digest, 661–664, 1999.
219. K. Maitra and V. Misra, "A simulation study to evaluate the feasibility of midgap workfunction metal gates in 25 nm bulk CMOS," IEEE Electron Dev. Lett., 24 (11), 707–709, 2003.
220. H. Zhong, G. Heuss, and V. Misra, "Electrical properties of RuO_2 gate electrodes for dual metal gate Si-CMOS," IEEE Electron Dev. Lett., 21, (12), 593–595, 2000.
221. H. B. Michaelson, "The work function of the elements and its periodicity," J. Appl. Phys., 48 (11), 4729–4733, 1977.
222. Z. B. Zhaolin Wei, P. Grange, and B. Delmon, "XPS and XRD studies of fresh and sulfided MO_2N," Appl. Surf. Sci., 135, 107–114, Sept. 1998.
223. Q. Lu, R. Lin, P. Ranade, Y. C. Yeo, X. Meng, H. Takeuchi, T.-J. King, C. Hu, H. Luan, S. Lee, W, Bai, C.-Ho. Lee, D.-L. Kwong, X. Guo, X. Wang, and T. P. Ma, "Molybdenum metal gate MOS technology for post-$Si0_2$ gate dielectrics," IEEE IEDM Tech. Digest, 641–644, 2000.
224. H. Wakabayashi, Y. Saito, K. Takeuchi, T. Mogami, and T. Kunio, "A dual-metal gate CMOS technology using nitrogen-concentration-controlled TiN_x film," IEEE Trans. Electron Dev. 48 (10), 2363–2369, 2001.
225. V. Misra, H. Zhong, and H. Lazar, "Electrical properties of Ru-based alloy gate electrodes for dual metal gate Si-CMOS," IEEE Electron Dev. Lett., 23 (6), 354–356, 2002.
226. J. H. Lee, Y.-S. Suh, H. Lazar, R. Jha, J. Gurganus, Y. Lin, and V. Mistra, "Compatibility of dual metal gate electrodes with high-K dielectrics for CMOS," IEEE IEDM Tech. Digest, 323–326, 2003.

227. X. P. Wang, A. E.-J. Lim, H. Y. Yu, M.-F. Li, C. Ren, W.-Y. Loh, C.-X. Zhu, A. Chin, A. D. Trigg, Y.-C. Yeo, S. Biesemans, G.-Q. Lo, D.-L. Kwong, "Work function tunability of refractory metal nitrides by lanthanum or aluminum doping for advanced CMOS devices," IEEE Trans. Electron Dev. 54 (11), 2871–2877, 2007.
228. S. B. Samavedam, L. B. La, J. Smith, S. Dakshina-Murthy, E. Luckowski, J. Schaeffer, M. Zavala, R. Martin, V. Dhandapani, D. Triyoso, H. H. Tseng, P. J. Tobin, D. C. Gilmer, C. Hobbs, W. J. Taylor, J. M. Grant, R. I. Hedge, J. Mogab, C. Thomas, P. Abramowitz, M. Moosa, J. Conner, J. Jiang, V. Arunachalam, M. Sadd, B.-Y. Nguyen, and B. White, "Dual-metal gate CMOS with HfO_2 gate dielectric," IEEE IEDM Tech. Digest, 433–436, 2002.
229. Y. Nara, S. Inumiya, F. Ootsuka, and Y. Ohgi, "Integration of dual-metal-gate CMOS with sub-0.9-nm EOT HfSiON gate dielectrics," 3rd Intnl. Symp. On Adv. Gate Stack Tech., Sept. 27, 2006.
230. J. Kedzierski, E. Nowak, T. Kanarski, Y. Zhangt, D. Boyd, R. Carru8thers, C. Cabral, R. Amos, C. Lavoie, R. Roy, J. Newbury, E. Sullivan, J. Benedict, P. Saunders, K. Wong, D. Canapair, M. Krishnan, K.-L. Lee, B. A. Rainey, D. Fried, P. Cottrell, H.-S. P. Wong, M. Ieong, and W. Haensch, "Metal-gate FinFET and fully-depleted SOI devices using total gate silicidation," IEEE IEDM Tech. Digest, 247–250, 2002.
231. J. H. Sim, H. C. Wen, J. P. Lu, and D. L. Kwong, "Dual work function metal gates using full nickel silicidation of doped poly-Si," IEEE Electron Dev. Lett., 24 (10), 631–633, 2003.
232. J.Kedzierski, D. Boyd, P. Sonsheim, S. Zafar, J. Newbery, J. Ott, C. Cabral Jr., M. Ieong, and W. Haensch, "Threshold voltage control in NiSi-gated MOSFETs through silicidation induced impurity segregation (SIIS)," IEEE IEDM Tech. Digest, 315–318, 2003.
233. A. Veloso, T. Hoffmann, A. Lauwers, S. Brus, J.-F. de Marneffe, S. Locorotondo, C. Vrancken, T. Kauerauf, A. Shickova, B. Sijmus, H. Tigelaar, M. A. Pawlak, H. Y. Yu, C. Demeurisse, S. Kubicek, C. Kerner, T. Chiarella, O. Richard, H. Bender, M. Niwa, P. Absil, M. Jurczak, S. Biesemans and J. A. Kittl, "Dual work function phase controlled Ni-FUSI CMOS (NiSi NMOS, Ni_2Si or $Ni_{31}Si_{12}$ PMOS): Manufacturability, Reliability & Process Window Improvement by Sacrificial SiGe cap," VLSI Tech. Digest, Paper 12.2, 2006.
234. Y.-C. Yeo, T.-J. King, and C. Hu, "Metal-dielectric band alignment and its implications for metal gate complementary metal-oxide-semiconductor technology," J. Applied Phys. 92 (12) 7266–7271, 2002.
235. Y.-C. Yeo, P. Ranade, T.-J. King, and C. Hu, "Effects of high-κ gate dielectric materials on metal and silicon gate workfunctions," IEEE Electron Dev. Lett., 23 (6), 342–344, 2002.
236. G. Brown and V. Misra, "Evaluation of Fermi level pinning in low, midgap and high workfunction metal gate electrodes on ALD and MOCVD HfO_2 under high temperature exposure," IEEE IEDM Tech. Digest, 295–298, 2004.
237. S. B. Samavedam, L. B. La, P. J. Tobin, B. White, C. Hobbs, L. R. C. Fonseca, A. A. Demkov, J. Schaeffer, E. Luckowski, A. Martinez, M Raymond, D. Triyoso, D. Roan, V. Dhandapani, R. Garcia, S. G. H. Anderson, K. Moore, H. H. Tseng, C. Capasso, O. Adetutu, D. C. Gilmer, W. J. Taylor, R. Hedge, and J. Grant, "Fermi level pinning with sub-monolayer MeO_x and metal gates," IEEE IEDM Tech. Digest, 307–31-, 2003.
238. C. Hobbs, L. Fonseca, V. Dhandapani, S. Samavedam, B. Taylor, J. Grant, L. Dip, D. Triyoso, R. Hedge, D. Gilmer, R. Garcia, D. Roan, L. Lovejoy, R. Rai, L. Hebert, H. Tseng, B. White, and P. Tobin, "Fermi level pinning at the polySi/metal oxide interface," VLSI Tech. Digest, 9–10, 2003.
239. K. Shirashi, K. Yamada, K. Torri, Y. Akasaka, K. Nakajima, M. Khono, T. Chikyo, H. Katajima, and T. Arikado, "Physics of Fermi level pining at the polySi/Hf-based high-k oxide interface," VLSI Tech. Dig., 108–109, 2004.
240. M. Koyama, Y. Kamimuta, T. Ino, A. Kaneko, S. Inumiya, K. Eguchi, M. Takayanagi, and A. Nishiyama, "Careful examination of the asymmetric Vfb shift problem for poly-Si/HfSiON gate stack and its solution by the Hf concentration control in the dielectric near the poly-Si interface with small EOT expense," IEEE IEDM Tech. Digest, 499–452, 2004.
241. C. Hobbs, L. Fonseca, A. Knizhnik, V. Dhandapani, S. Samavedam, W. J. Taylor, J. M. Grant, L. Dip, D. H. Triyoso, R. I. Hedge, D. C. Gilmer, R. Garcia, D. Roan, M. L. Lovejoy,

R. S. Rai, E. A. Hebert, H. H. Tseng, S. G. H. Anderson, B. E. White, and P. Tobin, "Fermi level pinning at the polySi/metal oxide interface – Part I," IEEE Trans. Electron Dev., 51 (6), 971–977, 2004.

242. ibid. Part II, 978–984.
243. J. Tersoff, "Schottky barrier heights and the continuum of gap states," Phys. Rev. Lett., 52 (6), 465–468, 1984.
244. S. Xiong and J. Bokor, "A simulation study of gate line edge roughness effects on doping profiles of short-channel MOSFET devices," IEEE Trans. Electron Dev., 51 (2), 228–232, 2004.
245. J. A. Croon, G. Storms, S. Winkelmeier, I. Pollentier, M. Ercken, S. Decoutere, W. Sansen, and H. E. Maes, "Line edge roughness: characterization, modeling and impact on device behavior," IEEE IEDM Tech. Digest, 307–310, 2002.
246. S. Xiong, J. Bokor, Q. Xiang, P. Fisher, I. Dudley, P. Rao, H. Wang, and B. En, "Is gate line edge roughness a first-order issue in affecting the performance of deep sub-micron bulk MOSFET devices?," IEEE Trans. Semicon. Manuf., 17 (3), 357–361, 2004.
247. T. Linton, M. Chandhok, B. J. Rice, and G. Schrom, "Determination of the line edge roughness specification for 34 nm devices," IEEE IEDM Tech. Digest, 303–306, 2002.
248. C. H. Díaz, H.-J. Tao, Y.-C. Ku, A. Yen, and K. Young, "An experimentally validated analytical model for gate line-edge roughness (LER) effects on technology scaling," IEEE Electron Dev. Lett., 22 (6), 287–289, 2001.
249. H.-W. Kim, J.-Y. Lee, J. Shin, S.-G. Woo, H.-K. Cho, and J.-T. Moon, "Experimental Investigation of the impact of LWR sub-100-nm device performance," IEEE Trans. Electron Dev., 51 (12), 1984–1988, 2004.
250. H. Fukutome, Y. Momiyama, T. Kubo, Y. Tagawa, T. Aoyama, and H. Arimoto "Direct evaluation of gate line edge roughness impact on extension profiles in sub-50-nm n-MOSFETs," IEEE Trans. Electron Dev., 53 (11), 2755–2763, 2006.
251. D. Ha, H. Takeuchi, Y.-K. Choi, T.-J. King, D.-L. Kwong, A. Agrawal, and M. Ameen, "Molybdenum-gate HfO2 CMOS FinFET technology," IEEE IEDM Tech. Digest, 643–646, 2004.
252. J. Kedzierski, M. Ieong, T. Kanarsky, Y. Zhang, and H.-S. P. Wong, "Fabrication of metal gated FinFETs through complete gate silicidation with Ni," IEEE Trans. Electron Dev., 51 (12) 2115–2120, 2004.
253. W. Xiong, C. Rinn Cleavelin, P. Kohli, C. Huffman, T. Schulz, K. Schruefer, G. Gebara, K. Mathews, P. Patruno, Y.-M. Le Vaillant, I. Cayrefourcq, M. Kennard, C. Mazure, K. Shin, and T.-J. King Liu, "Impact of Strained-Silicon-on-Insulator (sSOI) substrate on FinFET Mobility," IEEE Electron Dev. Lett., 27 (7), 612–614, 2006.
254. K.-M. Tan, T.-Y. Liow, R. T. P. Lee, C.-H. Tung, G. S. Samudra, W.-J. Yoo, and Y.-C. Yeo, "Drive-current enhancement in FinFETs using gate-induced stress," IEEE Electron Dev. Lett., 27 (9), 769–771, 2006.
255. K.-M. Tan, T.-Y. Liow, R. T. P. Lee, K. M. Hoe, C.-H. Tung, N. Balasubramanian, G. S. Samudra, and Y.-C. Yeo, "Strained p-channel FinFETs with extended Π-shaped silicon–germanium source and drain stressors," IEEE Electron Dev. Lett., 28 (10), 905–908, 2007.
256. B.-Y. Tsui and C.-P. Lin, "A novel 25-nm modified Schottky-Barrier FinFET with high performance," IEEE Electron Dev. Lett., 430–432, 2004.

Chapter 6
Analog Devices and Passive Components

6.1 Introduction

Analog devices allow the design of circuits whose inputs and outputs are continuously varying quantities, such as resistance, capacitance, current, and voltage. The measured analog signal has an infinite number of possible values. The information is conveyed by the instantaneous value of the signal. Initially, analog circuits were designed primarily with bipolar transistors (Chap. 3). Analog *MOSFETs*, however, have become increasingly important because of their higher packing density, lower cost, high input impedance, and performance that has gradually approached that of bipolar transistors.

While several component parameters can be simultaneously optimized for digital and analog applications, there are specific analog requirements that are different from digital. In particular, the trend in high-performance, high-density digital *MOSFETs* is to reduce the size to deep submicron and nanoscale dimensions and operate at supply voltages as low as about 0.8 V, while analog devices typically require higher voltages and hence larger dimensions, particularly to ensure a sufficiently large signal-to-noise ratio. Other parameters, such as high transistor cut-off frequency, high maximum oscillation frequency, and high Early voltage, low bipolar base resistance, low component mismatch, and low noise, are of particular importance to analog applications. The relative importance of these parameters depends on application. When mixed analog and digital components are designed on the same die, such as in system on a chip (SoC), there is a trade-off between simultaneously optimizing the two sets of components and manufacturing cost.

The first section of this chapter discusses analog components, their key parameters and requirements. Many analog circuits use matched component pairs of intended identical characteristics. This is particularly important in the design of converters and operational amplifiers. Also, current and voltage signals fluctuate around their nominal values. These fluctuations, referred to as noise, interfere with signals and limit their accuracy. Mismatch and noise have become increasingly

B. El-Kareh, *Silicon Devices and Process Integration: Deep Submicron and Nano-Scale Technologies*,
© Springer Science+Business Media, LLC 2009

important as device dimensions and operating voltages are reduced. Sources of mismatch and noise are reviewed in the last two sections.

6.2 Analog Devices

Analog devices can be categorized into active and passive. The most important active analog devices are the *NPN* and *PNP* transistors and their complementary arrangement, sometimes referred to as *CBIP*. Since bipolar transistors have already been discussed in detail in Chap. 3, they will not be covered in this chapter. Other active devices are the junction field-effect transistors, *JFET*, and the analog MOSFET. Passive devices are precision resistors, precision capacitors, varactors, and inductors.

6.2.1 Junction Field-Effect Transistor, JFET

A Junction Field-Effect Transistor consists of a semiconductor channel, whose thickness and hence resistance can be varied by narrowing or widening one or two pn junction depletion regions. The basic principles of a *JFET* were demonstrated by Lilienfeld and O'Heil [1–3], almost 20 years before the first bipolar transistor was invented, and then developed by Shockley [4].

As in MOSFETs, there are two types of *JFET* s, the n-channel *JFET (NJFET)* and the p-channel *JFET (PJFET)*. Each type of *JFET* can be normally-on (depletion mode) or normally-off (enhancement mode). The most common type of *JFET* is, however, normally on. Figure 6.1 illustrates a normally-on double-gated *NJFET* with no voltage applied.

The channel consists of an n-type region, assumed uniformly doped for simplicity. The heavily doped contacts on both sides of the channel are the *JFET* source and drain. The gate is shown with two pn junctions, a top gate and a bottom gate, that

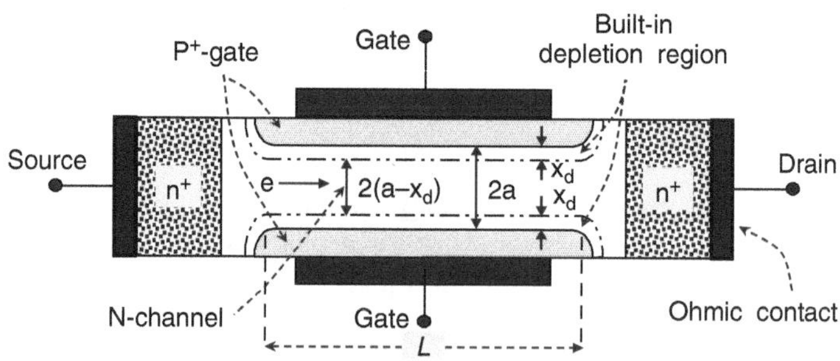

Fig. 6.1 Schematic of an n-channel *JFET* at thermal equilibrium

are typically tied together and used to modulate the conducting width of the channel. The metallurgical width of the channel is the distance $2a$ between the top and bottom metallurgical junctions. The width of the conducting channel is $2(a - x_d)$, where x_d is the width of the built-in junction depletion region in the channel. In the absence of applied voltage and for a uniformly-doped channel, the width of the depletion region is uniform. The length L of the channel is approximately the length of the plane portion of the gate. The channel width is in a direction normal to the page. The channel resistance is:

$$R_{Ch} = \frac{\rho L}{S} = \frac{\rho L}{2W(a - x_d)} \tag{6.1}$$

where ρ is the channel resistivity, L the channel length, S the channel cross-sectional area, W the channel width, $2a$ the metallurgical channel thickness, and x_d the junction depletion width.

6.2.1.1 JFET at Turn-Off

The following assumptions are made to simplify the discussion:

1. The gate to channel junction is a one-sided abrupt junction and the channel is uniformly doped.
2. The structure is symmetric with respect to center axis.
3. Top and bottom gate are tied together.
4. The source is grounded; V_G and V_D are, respectively, the gate and drain voltage with respect to source.
5. The channel is long, that is, the lateral field is very small compared to vertical field.

When a bias voltage V_G is applied to the gate such that the channel is fully depleted at the source, the channel is turned-off. This is when x_d increases to the value a at the source (Fig. 6.2).

The width of the depletion region at turn-off is

$$x_d = \sqrt{\frac{2\varepsilon_0 \varepsilon_{\mathrm{Si}}(|V_b| \pm |V_G|)}{qN_D}} \cong \sqrt{\frac{1.3 \times 10^7(|V_b| \pm |V_G|)}{N_D}} = a, \tag{6.2}$$

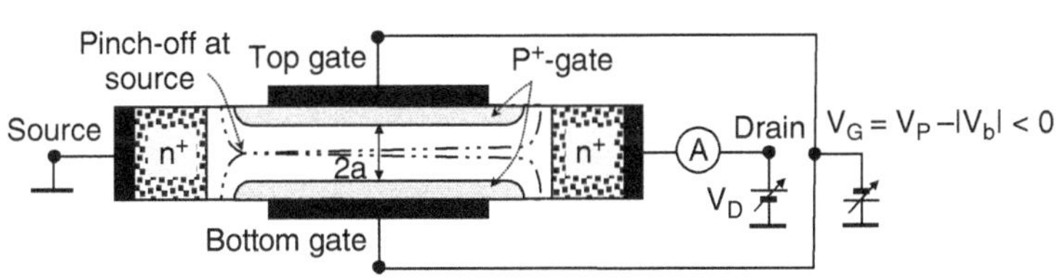

Fig. 6.2 Schematic of an n-channel *JFET* at turn-off

where V_b is the built-in voltage, V_G is the applied gate voltage, and N_D is the channel dopant concentration, assumed uniform. The plus sign is used for a reverse-biased gate, where the depletion region expands, and the minus sign for a forward-biased gate, where the depletion region narrows. The reverse gate voltage at turn-off, also referred to as the threshold voltage V_T is

$$V_T = |V_b| - \frac{qN_D a^2}{2\varepsilon_{Si}\varepsilon_0} = |V_b| - V_P \cong |V_b| - 7.72 \times 10^{-8} N_D a^2 \quad \text{V}. \tag{6.3}$$

where V_P is the pinch-off voltage defined as

$$V_P = \frac{qN_D a^2}{2\varepsilon_0 \varepsilon_{Si}} \quad \text{V}. \tag{6.4}$$

For a depletion-mode *NJFET*, the threshold voltage is negative. The built-in voltage for a one-sided step-junction, can be approximated as (Chap. 2)

$$V_b \cong 0.55 + \frac{kT}{q} \ln \frac{N_D}{n_i} \quad \text{V}. \tag{6.5}$$

Example: For $N_D = 1.2 \times 10^{16}\,\text{cm}^{-3}$, $a = 0.5\,\mu\text{m}$, and 300 K, $n_i \approx 1.4 \times 10^{10}\,\text{cm}^{-3}$, $V_b \approx 0.90\,\text{V}$, and $V_T \approx -1.41\,\text{V}$.

6.2.1.2 Linear Mode

For a given gate voltage of magnitude below threshold, the total reverse voltage seen at the drain end of the channel is $|V_b| + V_D \pm V_G$ where V_D is the positive drain voltage. As V_D increases, the gate to channel depletion region widens at the drain boundary, decreasing the thickness of the conducting channel, as illustrated in Fig. 6.3 for $V_G = 0$. For small V_D, in the range 10–100 mV, the increase in x_d remains small and has negligible impact on the overall channel resistance. In this case, the drain current is a linear function of drain voltage. Neglecting extrinsic source and drain resistances, the drain current is found as

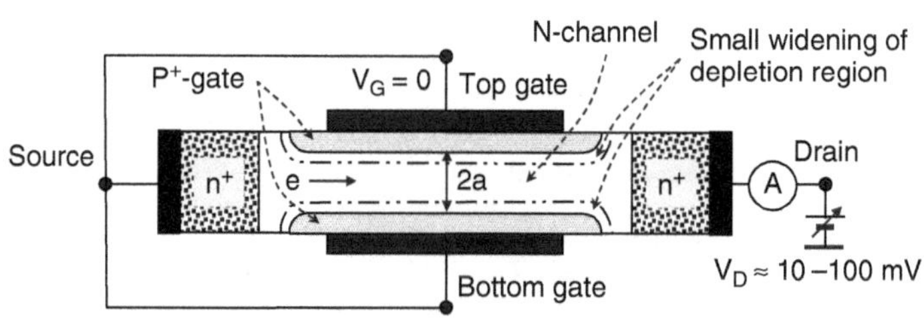

Fig. 6.3 Schematic of *NJFET* biased in the linear mode

$$I_D = \frac{V_D}{R_{Ch}} = \frac{2W(a-x_d)}{\rho L}V_D = \frac{2Wq\mu_n N_D}{L}(a-x_d)V_D \quad \text{A}. \tag{6.6}$$

Combining (6.2) and (6.6) gives

$$I_D \approx \frac{2Wq\mu_n N_D}{L}\left[a - \sqrt{\frac{2\varepsilon_0\varepsilon_{Si}(|V_b| \pm |V_G|)}{qN_D}}\right]V_D \quad \text{A}, \tag{6.7}$$

or

$$I_D \approx \frac{2Wq\mu_n N_D a}{L}\left[1 - \sqrt{\frac{2\varepsilon_0\varepsilon_{Si}(|V_b| \pm |V_G|)}{qN_D a^2}}\right]V_D \quad \text{A}. \tag{6.8}$$

Defining

$$G_0 = \frac{2Wq\mu_n N_D a}{L}, \tag{6.9}$$

and using the definition of pinch-off voltage in (6.4) gives

$$I_D = G_0\left[1 - \sqrt{\frac{(|V_b| \pm |V_G|)}{|V_P|}}\right]V_D \quad \text{A}. \tag{6.10}$$

The drain current is maximum when the gate is at zero potential. It decreases as the reverse gate to source voltage increases and goes to zero when $|V_G| + |V_b| = |V_P|$.

The channel linear conductance $g_{D\text{-}lin}$ is

$$g_{D-lin} = \left.\frac{\partial I_D}{\partial V_D}\right|_{V_G} = G_0\left[1 - \sqrt{\frac{|V_b| \pm |V_G|}{|V_P|}}\right] \quad \text{S}, \tag{6.11}$$

and the linear transconductance $g_{m\text{-}lin}$ is

$$g_{m-lin} = \left.\frac{\partial I_D}{\partial V_G}\right|_{V_D} = \frac{G_0}{2V_P}\sqrt{\frac{|V_P|}{|V_b| \pm |V_G|}}\,V_D \quad \text{S}. \tag{6.12}$$

6.2.1.3 Transition Mode

The channel voltage drops from V_D at the drain boundary to zero at the source. The reverse gate-to-channel voltage is therefore position dependent. For very small V_D, the channel voltage can be assumed uniform and approximated as $V_b \pm V_G$. As V_D increases, this approximation becomes inaccurate and the position-dependence channel voltage must be taken into account. At any point in the channel, the reverse voltage is $V_b + V_{Ch}(y) \pm |V_G|$, where y is the distance from the source. The reverse voltage is $\mathrm{V_b} \pm |\mathrm{V_G}|$ at the source and increases to $V_b + V_D \pm |V_G|$ at the drain. The depletion width and hence channel thickness is therefore position dependent. The channel thickness decreases from source to drain (Fig. 6.4). Thus, an elemental section dy of the channel would have a higher resistance near the drain

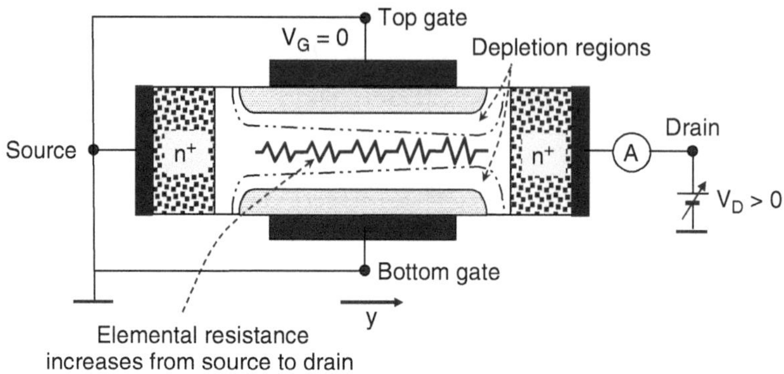

Fig. 6.4 Schematic of *NJFET* biased in the transition mode

than near the source. As for the *MOSFET*, the *JFET* current-voltage characteristic in this mode can best be analyzed by the method of gradual channel approximation in which the lateral field is very small compared to the vertical field and the channel depth is assumed to be solely dependent on gate voltage.

The resistance of an elemental section *dy* of the channel is

$$dR(y) = \frac{dy}{2Wq\mu_n N_D[a - x_d(y)]} \quad \text{Ohm.} \tag{6.13}$$

From Ohm's law, $dV(y) = I_D \cdot dR(y)$, where I_D is the same at any plane crossing the channel. Integrating from source, where V = 0, y = 0, to drain, where $V = V_D$, $y = L$, yields

$$2q\mu_n N_D W \int_0^{V_D} \left\{ \left[a - \sqrt{\frac{2\varepsilon_0\varepsilon_{Si}}{qN_D}(|V_G| + |V_b| + V(y))} \right] \right\} dV(y) = I_D \int_0^L dy$$

and

$$I_D = G_0\, V_P \left\{ \frac{V_D}{V_P} - \frac{2}{3}\left[\left(\frac{|V_G| + |V_b| + V_D}{V_P} \right)^{3/2} \right] + \frac{2}{3}\left[\left(\frac{|V_G| + |V_b|}{V_P} \right)^{3/2} \right] \right\}, \tag{6.14}$$

where V_P and G_0 are defined in (6.4) and (6.8). Figure 6.5 shows the $I_D - V_D$ characteristics in the linear and transition regions obtained from (6.14). The gradual-channel approximation is only valid up to the point *P* after which the current saturates, as will be discussed below.

A Taylor expansion of (6.14) for very small V_D results in (6.10). As V_D further increases, the overall channel resistance increases and the rate of increase of drain current with drain voltage departs from linearity, as shown in the transition regions in Fig. 6.5.

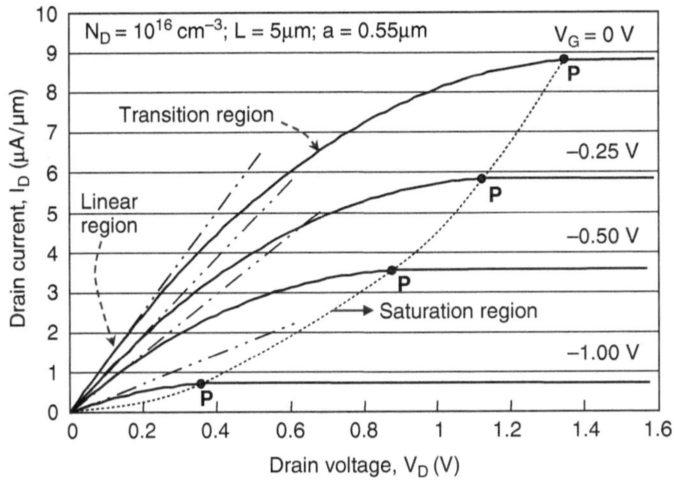

Fig. 6.5 I_D versus V_D plots showing the linear, transition and saturation regions

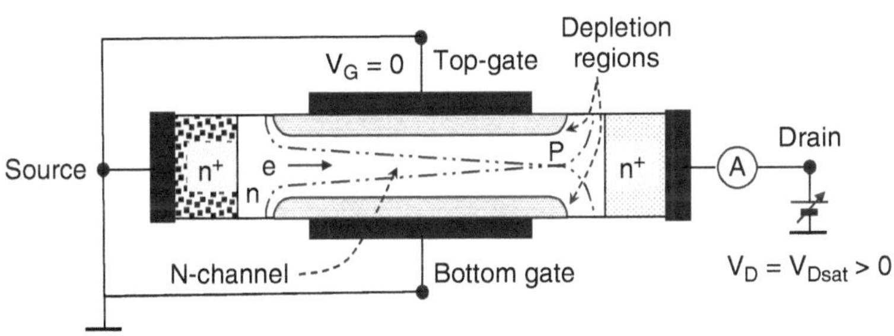

Fig. 6.6 Schematic of an *NJFET* illustrating the pinch-off condition at the drain for $V_G = 0$

6.2.1.4 Onset of Saturation, Pinch-Off at Drain

For each of the points P in Fig. 6.5, there is a value of V_D where the drain to gate reverse voltage is sufficiently large to pinch-off, that is, fully deplete the channel at the drain, while the rest of the channel is still conductive (Fig. 6.6). The drain voltage for this condition is referred to as V_{Dsat} to indicate that, for a long channel, the current saturates at this point and does not further increase as the drain voltage is increased.

At onset of pinch-off, the total drain to gate reverse voltage is

$$V_{Dsat} + V_b + |V_G| = \frac{qN_D a^2}{2\varepsilon_{Si}\varepsilon_0} = V_P \tag{6.15a}$$

or

$$V_{Dsat} = V_P - |V_G| - |V_b| \tag{6.15b}$$

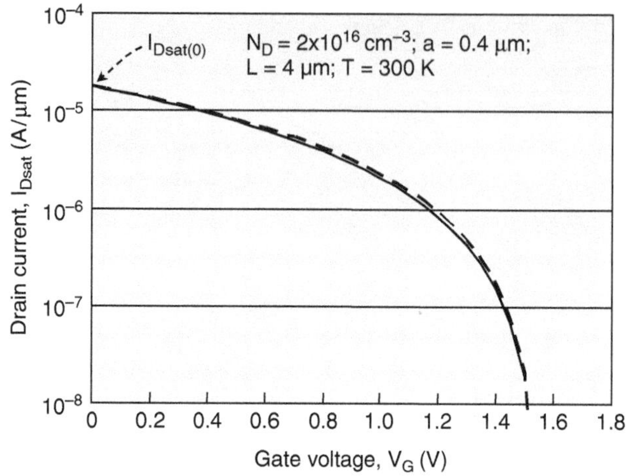

Fig. 6.7 Comparison of I_{Dsat} versus V_G obtained from (6.16) and (6.17) (*solid line*) with the square-law characteristic in (6.18) (*dashed line*)

Substituting (6.15a) in (6.14) gives

$$I_{Dsat} = G_0 \left\{ \frac{V_P}{3} - (|V_G| + |V_b|) \left[1 - \frac{2}{3}\sqrt{\frac{|V_G| + |V_b|}{V_P}} \right] \right\} \quad \text{A.} \tag{6.16}$$

The drain current reaches its maximum when $V_G = 0$, that is

$$I_{Dsat(0)} = G_0 \left\{ \frac{V_P}{3} - |V_b| \left[1 - \frac{2}{3}\sqrt{\frac{|V_b|}{V_P}} \right] \right\} \quad \text{A.} \tag{6.17}$$

A good approximation is found as

$$I_{Dsat} \approx I_{Dsat(0)} \left(1 - \frac{|V_G|}{V_P - |V_b|} \right)^2 \quad \text{A.} \tag{6.18}$$

Figure 6.7 compares the $I_{Dsat} - V_G$ characteristic found from (6.16) and (6.17) with the square-law relation in (6.18).

6.2.1.5 Saturation Conductance and Transconductance

Ideally, the saturation conductance should be zero. As V_D increases above pinch-off, the depletion region between drain and channel expands, displacing the pinch-off point P toward the source (Fig. 6.8).

The potential at P remains essentially constant as (6.15a)

$$V_P = V_{Dsat} + V_b + |V_G| = \frac{qN_D a^2}{2\varepsilon_{Si}\varepsilon_0}.$$

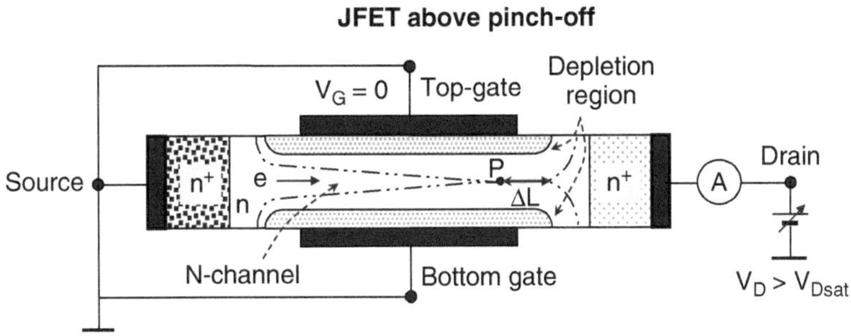

Fig. 6.8 Illustration of *NJFET* above pinch-off

The displacement of the pinch-off point is approximated as

$$\Delta L = x_{d-\text{lateral}} = \sqrt{\frac{2\varepsilon_0\varepsilon_{Si}(V_D - V_P)}{qN_D}} \quad \text{cm}. \tag{6.19}$$

For a long channel, ΔL is negligible compared to the channel length and the drain current remains essentially constant at the value of I_{Dsat} defined by (6.16) and (6.17), independent of V_D. In this case, the saturation conductance is near ideal, that is, the saturation resistance appears infinite (Fig. 6.5). To visualize the flow of carriers above pinch-off, consider the non-depleted channel between source and P separately from the depleted region within ΔL. Since P is positive with respect to source, the voltage across the non-depleted channel region induces a field E which, for a long channel, is sufficiently small that Ohm's law applies. This gives rise to a drift electron current component from source to point P of density (Chap. 1)

$$j_n = qnv_n = qn\mu_n E \quad \text{A/cm}^2.$$

Since the current through any plane normal to the channel is the same, the electron velocity must increase from source to drain as the non- depleted channel cross-section decreases. The voltage difference between drain and point P drops across a narrow region ΔL, creating a sufficiently high field to accelerate electrons to the saturated velocity in that region. Carriers reaching P are hence swept to the drain at saturation velocity. Pinch-off therefore means that the free carrier concentration is considerably reduced but not zero. The *JFET* drain can be compared to the collector of an *NPN* transistor.

For short channels, ΔL can become a sizable fraction of the total channel length so that the effective channel length decreases as V_D increases (Fig. 6.9). This is referred to as *channel length modulation*. Consequently, the channel resistance is reduced and I_{Dsat} increases as V_D increases. Thus, the conductance becomes finite in saturation, as illustrated in Fig. 6.9. The short-channel drain current can be modeled by adding a modulating factor λ as

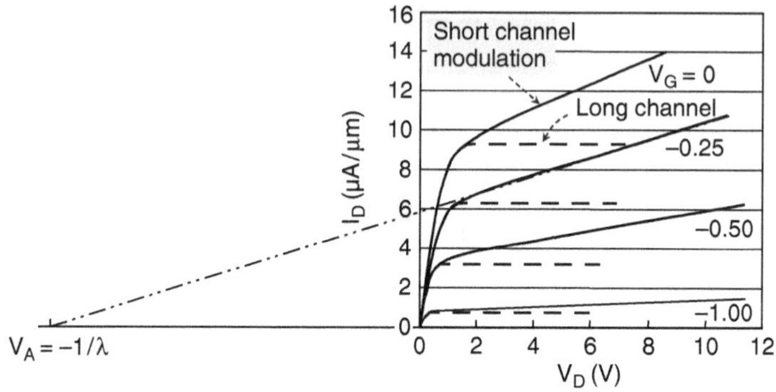

Fig. 6.9 Illustration of channel length modulation resulting in finite conductance in short-channel *JFET*s. V_A is the Early voltage

$$I_{Dsat} \approx I_{Dsat(0)} \left(1 - \frac{|V_G|}{V_P - |V_b|}\right)^2 (1 + \lambda V_D) \quad \text{cm.} \tag{6.20}$$

When $V_D = -1/\lambda$, $I_{Dsat} = 0$. The value of V_D at this point is called the Early voltage (Fig. 6.8).

The saturation transconductance g_{msat} is found from (6.16) as

$$g_{msat} = \frac{\partial I_D}{\partial V_G} = G_0 \left[1 - \sqrt{\frac{|V_b| \pm |V_G|}{V_P}}\right] \quad \text{S.} \tag{6.21}$$

This is identical to the linear drain conductance in (6.11). A good approximation is obtained from (6.18) as

$$g_{msat} = \frac{2I_{Dsat(0)}}{V_P - |V_b|} \left(1 - \frac{|V_b| \pm |V_G|}{V_P}\right) \quad \text{S.} \tag{6.22}$$

The maximum transconductance is

$$g_{m-\max} = \frac{2I_{Dsat(0)}}{V_P - |V_b|} \quad \text{S.} \tag{6.23}$$

6.2.1.6 AC Response

A figure of merit is the frequency at which the *JFET* gain drops to 1. This is the frequency f_T at which the small-signal current through the input gate capacitance is equal to the small-signal drain current. The small-signal gain is

$$\text{Gain} = \frac{\partial I_D}{\partial Q_G\, \omega} = \frac{\partial I_D}{\partial V_G\, \omega\, C_G} = \frac{g_m}{2\pi f\, C_G}, \tag{6.24}$$

where

$$C_G = C_{GS} + C_{GD} + C_{GCh}.$$

C_{GS}, C_{GD}, and C_{GCh} are, respectively, the gate to source, gate to drain and gate to channel capacitances. The frequency at unity gain is

$$f_T = \frac{g_m}{2\pi\, C_G} \quad \text{Hz.} \tag{6.25}$$

6.2.1.7 Effect of Extrinsic Resistances

So far, extrinsic resistances outside the channel were not taken into account and only the voltages seen at the junction depletion boundaries were considered. The voltages seen at the contacts must, however, include voltage drops across parasitic series resistances at the source and drain. Because of the very low gate current, voltage drop across the gate series resistance can be neglected. Thus, the terminal drain to source voltage, denoted as V_D', is

$$V_D' = V_D + I_D(R_S + R_D) \quad \text{V,} \tag{6.26}$$

where R_S and R_D are, respectively, the extrinsic source and drain series resistance, including contact resistances, and V_D is the voltage seen at the drain depletion boundary. The terminal gate to source voltage is

$$V_G' = V_G + I_D R_S \quad \text{V,} \tag{6.27}$$

where V_G is the voltage seen at the gate depletion boundary. The measured transconductance and conductance are

$$g_m' = \frac{g_m}{1 + R_S g_m + (R_S + R_D) g_d} \quad \text{S,} \tag{6.28}$$

and

$$g_d' = \frac{g_d}{1 + R_S g_m + (R_S + R_D)\, g_d} \quad S, \tag{6.29}$$

where g_m, g_d are, respectively, the intrinsic transconductance and conductance, that is, without accounting for extrinsic series resistances.

The measured transconductance and conductance are thus reduced by extrinsic series resistances.

6.2.1.8 Gate Leakage Sources

Gate leakage currents reduce the *JFET* input impedance and contribute to noise. The main contributors to gate leakage are:

- Thermal generation within the gate junction depletion region,
- Thermal generation outside the gate junction depletion region,

– Thermal generation at junction surface intercepts,
– Impact ionization near the drain.

Thermal generation within and outside the junction depletion region and generation at junction surface intercepts are discussed in Chap. 2.

Impact ionization is caused by the high field at the drain. As the drain voltage increases, carriers entering the depletion region at the drain gain sufficient energy from the field to create electron-hole pairs (ehp) by impact. This applies to thermally generated carriers and channel electrons coming from the source. Defining the ionization rate α_i as the number of electron-hole pairs (ehp) generated within a path length of one centimeter, the total number of ehp generated in the high-field drain region is (Chap. 2)

$$\int_{x_{dn}}^{x_{dp}} \alpha_i(E)dx,$$

where x_{dn} and x_{dp} are, respectively, the depletion boundaries at the drain and gate, and $\alpha_i\ (E)$ the field-dependent ionization rate, taken to be of the form as

$$\alpha_i(E) = ae^{-b/|E|} \quad \text{cm}^{-1}.$$

Approximate values for a and b are given in Chap. 2, Table 1.

Incident carriers are thus multiplied by impact ionization with a multiplication factor.

$$M = \frac{1}{1 - \int_{x_{dn}}^{x_{dp}} \alpha_i(E)dx}. \tag{6.30}$$

Avalanche multiplication, and hence the drain to gate junction breakdown BV_{DG} occurs when the integral in the denominator of (6.30) approaches unity and M tends to infinity. For a long channel, breakdown occurs when the voltage between channel and gate reaches a critical value and the drain and gate current increase sharply. Typical multiplication and breakdown characteristics are shown for four different *PJFET* channel-lengths L in Fig. 6.10.

Initially, the gate current is negligible compared to the drain current. As the drain voltage increases, the probability for impact ionization and multiplication increases and the gate current increases almost exponentially with drain voltage. This is shown in Fig. 6.10 for a gate current above 1 pA. The fraction of thermally generated current is small compared to the drain current. Thus, the current that is multiplied is predominantly the drain current I_D. Since electrons and holes are generated in pairs, the drain current must increase at the same rate as the gate current. In the multiplication range, however, the increase in drain current is a very small fraction of I_D and hence not discernable. As avalanche multiplication is approached, the gate and drain currents become comparable.

The impact-ionization gate current I_G is a function of incident drain current and the two- or three-dimensional field near the drain boundary. For the same bias conditions, I_G increases as the channel length is reduced (Fig. 6.10). The breakdown

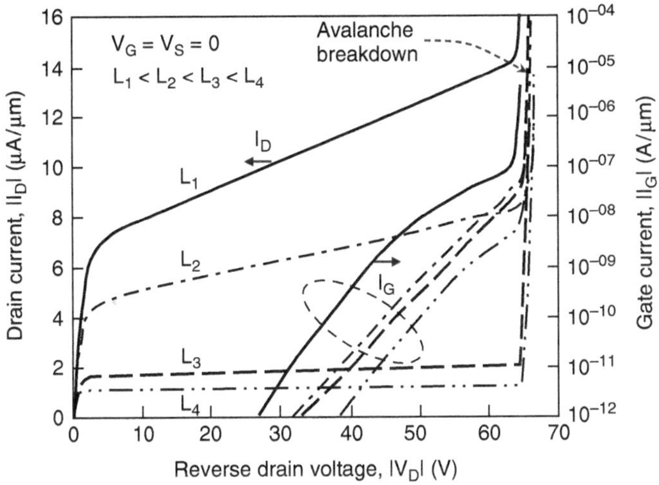

Fig. 6.10 Drain and gate currents per unit channel width versus drain voltage characteristics for four *PJFET* channel lengths L. Absolute values are shown

voltage is essentially the same for the four channel lengths shown in the figure. For long channels, the breakdown as measured with the source open, BV_{DGO}, is essentially the same as that measured with the source tied to the gate, BV_{DGS}. As the channel length is reduced, however, a point is reached where punch-through, that is, merger of drain and source depletion regions, occurs at a voltage lower than avalanche breakdown. In this case, BV_{DGS} is smaller than BV_{DGO}. Figure 6.10 shows the $I_D - V_D$ and $I_G - V_D$ characteristics for $V_G = V_S = 0$. If a reverse gate voltage is applied, the drain voltage at breakdown is reduced by the same amount since, for long channels, the breakdown field depends on the total drain to gate reverse voltage.

6.2.2 Analog/RF MOSFETs

Several requirements on analog and radio-frequency (*RF*) *MOSFETs* differ from those of digital *MOSFETs* [5–8]. Among them are the stronger focus on high transconductance, high-frequency response, high dynamic range (signal to noise ratio), small low-frequency noise, high output impedance, and small mismatch between adjacent devices.

When only analog circuits are present, *MOSFETs* can be independently optimized to meet specific circuit needs. Analog and radio-frequency (*RF*) *MOSFETs* are, however, typically integrated with memory on the same die, mainly for cost reduction. The trend in deep submicron and nanoscale *CMOS* results in conflicting technology requirements that necessitate a trade-off between simultaneously

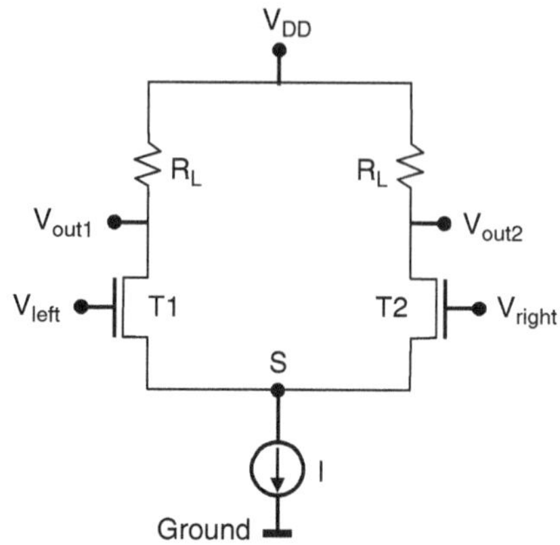

Fig. 6.11 Basic NMOS differential amplifier

optimizing both sets of devices and manufacturing cost. In addition, isolation between the two groups of circuits becomes mandatory to avoid propagation of noise from digital switching to the "quiet" analog circuits. This section discusses special requirements for key analog *MOSFET*s.

The importance of high transconductance, high Early voltage (Chap. 5, Fig. 5.29), low mismatch between adjacent devices, and low noise can be best understood by considering a simple differential amplifier shown in Fig. 6.11 [9].

The two *NMOS* devices have a common source to which a constant current source is connected. The *NMOS* drains are each connected to a load resistor R_L that is connected to the supply voltage V_{DD}. If the input voltages V_{left} and V_{right} are exactly the same, then transistors T1 and T2 will have the same gate to source voltage $V_G - V_S$, where V_S is the voltage across the current source. Thus, T1 and T2 will have the same drain currents. If the V_{left} is increased by δV_G and V_{right} decreased by δV_G, the current in T1 will increase by δI_D and the current in T2 will decrease by δI_D. V_{out1} will decrease by $\delta I_D \cdot R_L$ and V_{out2} will increase by $\delta I_D \cdot R_L$. Thus, the differential gain from V_{left} to V_{out1} is [9]

$$A_{\text{diff}} = -\frac{2\partial I_D.R_L}{2\partial V_G} = -\frac{\partial I_D.R_L}{\partial V_G}. \tag{6.31}$$

The term $\delta I_D/\delta V_G$ is the transistor transconductance g_m. The gain is therefore proportional to transconductance as

$$A_{\text{diff}} = -g_m R_L. \tag{6.32}$$

The above relations were derived under the assumption of zero *MOSFET* conductance g_D, that is, infinite transistor output resistance. If g_D is finite, the parallel combination of R_L and the transistor output resistance $1/g_D$ results in a degraded gain as

$$A_{\text{diff}} = -\frac{g_m R_L}{1 + g_D R_L}. \tag{6.33}$$

Thus high transconductance g_m and low conductance g_D (high Early voltage) are of particular importance to analog designs. Another consideration in differential amplifiers is the minimum differential voltage that can be detected. Mismatch between transistors T1 sand T2 produce an offset in the differential voltages at V_{out1} and V_{out2} that cannot be distinguished from the signal being amplified. This is a *DC* error that limits the resolution of the system. Therefore, reducing mismatch-induced offsets is extremely important to the design of most analog circuits. Input voltage noise also results in an upper limit of useful amplifier gain because the noise is amplified at V_{out1} and V_{out2}. The main issue in mixed digital-analog designs is the reduction in operating voltage with each digital generation to reduce power and increase density. To maintain the same signal-to-noise ratio as the power supply voltage is reduced, the noise level must be lowered. In the limiting case where the digital voltage becomes too low for analog circuits, the structures can be optimized for a dual power-supply voltage, for example, by fabricating *MOSFET*s with two different gate oxides, a thin oxide for core digital circuits and a thicker oxide to sustain a higher analog voltage. A thicker gate oxide, however, limits the scalability of analog *MOSFET*s because of related short-channel effects (Chap. 5).

6.2.2.1 Lateral Channel Profile Optimization

To minimize short-channel effects as the channel length is reduced, a "halo," also known as a "pocket," is typically implanted to locally increase the body concentration around source and drain (Chaps. 5 and 7). A pocket around the drain, however, often degrades analog performance in terms of transconductance, Early voltage, and threshold voltage mismatch. Transconductance directly affects the minimum voltage-noise (Sect. 6.4), and the operating frequency (bandwidth). As discussed in Chap. 5, the transconductance per unit width can be increased by reducing the channel length, increasing the mobility, and reducing the threshold voltage. With a symmetrical pocket implant, the transconductance degrades because of the increase in local threshold voltage and decrease in local mobility associated with the higher channel concentration at the source and drain. The *MOSFET* Early voltage is reduced because an increase in drain voltage above saturation laterally depletes part of or all pocket implant, reducing the local threshold voltage and resulting in an increase in drain current and hence higher conductance g_D. The effect is similar to that of channel-length modulation in short channels except that it also occurs in long channels. The threshold voltage mismatch increases because of the increased contribution of pocket dopant fluctuation to mismatch.

One method to alleviate the above problems is to implant the pocket only at the source, that is, asymmetrically as illustrated in Fig. 6.12b [10–18]. This approach, also referred to as lateral asymmetric channel (*LAC*), was initially introduced to reduce the threshold-voltage roll-off and the lateral electric field as the channel

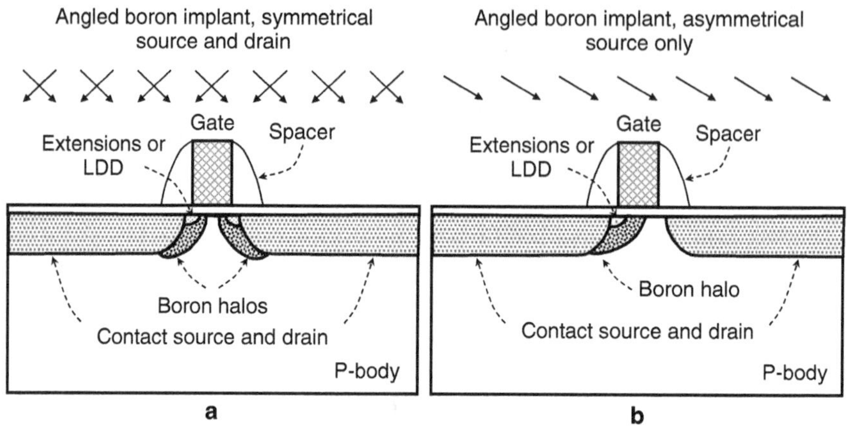

Fig. 6.12 Comparison of symmetrical (a) and asymmetrical (b) pocket (halo) implant

length is decreased [10] (Chap. 5). A high lateral field increases the inversion carrier temperature (Chap. 1). Hot carriers incident on the silicon surface and interface with the gate dielectric are known to create surface states and dielectric traps, degrading the channel conductivity and causing local shifts in threshold voltage, a serious reliability concern. The lower dopant concentration near the drain results in superior hot-carrier reliability than with symmetrical halo structures because of the reduced electric field. The lateral profile near the source can be optimized to increase the average inversion carrier mobility [11–18]. Figure 6.13 compares the conventional laterally-uniform channel profile with a typical asymmetrical lateral profile [11]. The lateral gradient in concentration near the source creates a field and field-gradient such that electrons injected from the source into the channel are accelerated along the channel toward the drain, enhancing velocity overshoot, hence increasing the average mobility and *MOSFET* transconductance [11–18] (Chap. 5).

The combination of low dopant concentration at the drain and high concentration at the source also reduces the dependence of Early voltage and subthreshold characteristics on channel length. In the subthreshold regime, the transport of carriers is similar to that of a bipolar transistor with the source acting as the emitter, the body as the base and the drain as the collector of the transistor (Chaps. 3 and 5). The density of minority carriers that are injected from the source into the channel and subsequently diffuse to the drain is inversely proportional to the lateral Gummel number in the channel. In asymmetrical structures, the Gummel number is determined mainly by the laterally integrated dopant concentration near the source and less dependent on the remaining part of the channel. This results in an almost channel-length independent subthreshold current [19].

In asymmetrically halo-doped structures, the average channel concentration away from the source can be made lower than in symmetrical structures, resulting in a smaller gate capacitance when the MOSFET is driven into saturation. This increases f_T, f_{max} (Chap. 5), and *RF* performance [16]. A significant improvement

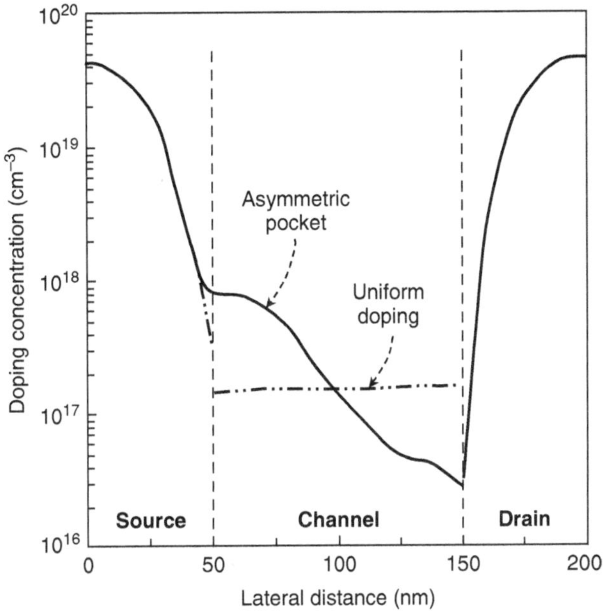

Fig. 6.13 Lateral impurity profile along the channel surface comparing uniform doping and asymmetric pocket implant in an NMOS (Adapted from [11])

in the ratio of transconductance-to-drain current, g_m/I_D, is obtained in the subthreshold mode [20–22]. The transconductance in the subthreshold mode can be derived from (5.53)–(5.57) in Chap. 5 as

$$g_m = \frac{qI_D}{nkT} \quad \text{S}. \tag{6.34}$$

In subthreshold, g_m depends linearly on drain current as opposed to its dependence on the square-root of I_D in strong inversion. Also, the ratio $g_m/I_D = q/(nkT)$ in (6.34) is independent of device geometry and considerably higher than in the saturation mode, resulting in higher gain (Fig. 6.14). An even higher g_m/I_D ratio is found in an optimized asymmetrical structure by reducing the angle of halo implant at the source and reducing the width of high dopant concentration [22].

6.2.3 Integrated Passive Components

Passive components include resistors, capacitors, varactors, and inductors. They can be connected to the die as discrete elements or integrated within the die. An example of integrating an inductor in a CMOS base process is shown in Fig. 7.7 of Chap. 7. This section discusses integrated resistors, capacitors and varactors in a *CMOS* or *BiCMOS* process.

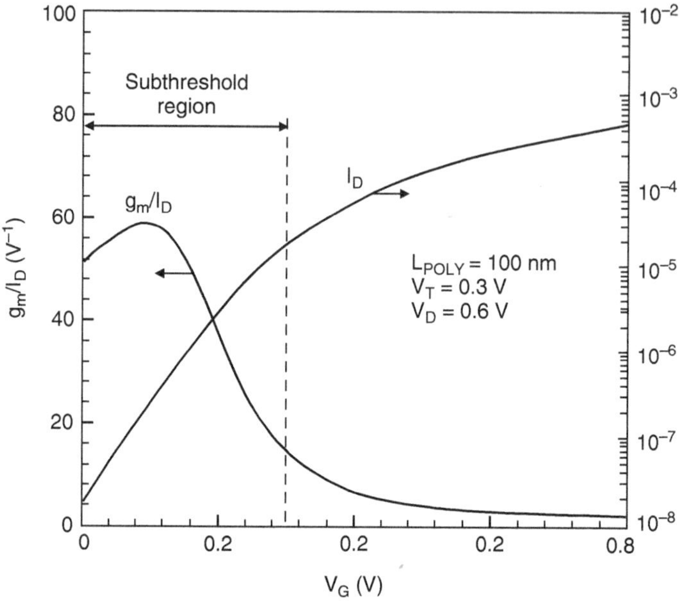

Fig. 6.14 Variation of I_D and g_m/I_D as a function of V_G in a 1.2V technology (Adapted from [22])

6.2.3.1 Resistors

Integrated resistors can be diffused in silicon or deposited as thin films (Chap. 7). In a *CMOS* technology, available diffused resistors are the *NMOS* source-drain, *PMOS* source-drain, n-well, and p-well. The *NMOS* or *PMOS* gate conductor, typically polysilicon, can serve as a thin-film resistor. Additional resistors are available in a *BiCMOS* technology. Among them are the *NPN* and *PNP* base and emitter polysilicon, and buried-layer resistors. In all cases, additional masking and implantation steps may be needed to block silicidation and tailor the sheet resistance. Additional precision thin-film resistors, such as nickel-chromium (*NiCr*) and silicon-chromium (*SiCr*) may also be required.

Resistor Design

A typical straight-line diffused or polysilicon resistor is shown in plan view in Fig. 6.15a. It consists of resistor body of the desired sheet resistance, and two highly-doped connecting end regions. A thin dielectric, typically SiO_2 or Si_3N_4, is patterned to block silicidation over the resistor body and allow doping the resistor ends at high concentration to reduce contact resistances. Schematic cross-sections of typical diffused and polysilicon resistors are shown, respectively in Fig. 6.15b and c.

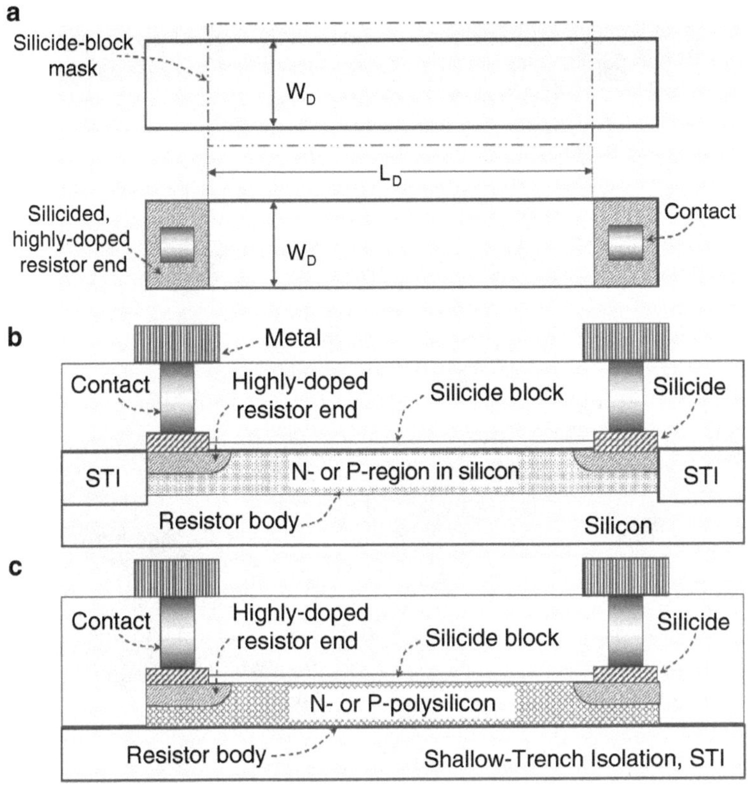

Fig. 6.15 **a** Plan view of a diffused or polysilicon straight-line resistor; **b** Schematic cross-section of a diffused resistor; **c** Schematic cross-section of a polysilicon resistor

The drawn resistor width W_D is the width of the patterned resistor body, and the drawn length L_D is defined by the length of the silicide block film. The electrical resistor dimensions L_E and W_E can be smaller or larger than the drawn dimensions. The difference between designed and electrical dimensions are defined as

$$\Delta L = L_D - L_E, \tag{6.35a}$$

$$\Delta W = W_D - W_E. \tag{6.35b}$$

Resistor Parameters

The most important resistor parameters are the sheet resistance, matching of resistor pairs (Sect. 6.3), temperature coefficient of resistance *TCR*, voltage coefficient of resistance *VCR*, linearity, and parasitics. The resistance of a straight-line resistor has the value

$$R = nR_S + 2R_{\text{end}}, \tag{6.36}$$

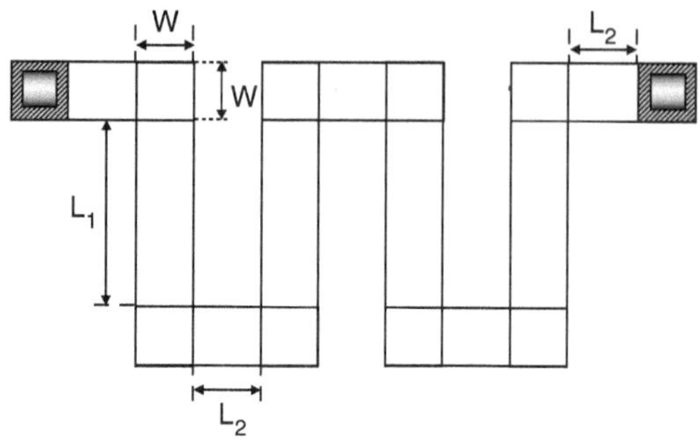

Fig. 6.16 Schematic of a meander resistor

where n is the number of squares defined as L_E/W_E. R_S is the body sheet resistance and R_{end} the end-resistance that comprises contact, series, and edge resistance of the heavily-doped region and is specified in Ohm-µm.

The choice of sheet resistance is a trade-off between size and performance. For example, given a resistor value, the higher the sheet resistance the smaller the area occupied by the resistor. As the area is reduced, however, the mismatch between identical resistor pairs increases (Sect. 6.3). For convenience and layout efficiency, the resistor can be designed in a meander form as shown in Fig. 6.16. For the resistor geometry in Fig. 6.16, the effective number of squares is

$$n = \frac{4L_1 + 5L_2}{W_E} + 8x0.559. \tag{6.37}$$

The second term in (6.37) is the contribution of 8 corners to n; the value 0.559 is obtained by conformal mapping techniques [23].

The total resistance is given by (6.36). The end resistance is the sum of contact resistance between metal and silicide R_C, silicide series resistance $R_{Silicide}$, and edge resistance R_{edge} that comprises the transition resistance between silicide and silicon:

$$R = 2(R_C + R_{\text{Silicide}} + R_{\text{Edge}}/W_E) + nR_S. \tag{6.38}$$

The edge resistance can be dominant because of the uptake of dopants by the silicide and the depletion of dopants at the silicon interface resulting in a barrier between silicide and silicon. R_{edge} is expressed in Ohm-µm.

The difference between drawn and electrical resistance length ΔL is usually negligible because L_E is defined by the thin silicide block mask that is typically precisely patterned by *RIE*, resulting in very small ΔL compared to L_D. The difference in width, ΔW can be extracted from measurement on straight-line resistors with the same drawn width and varying drawn lengths, that is, with different number of squares $n = L_D/W_D$. From (6.35b), one gets

$$\Delta W = W_D - R_S \frac{L_{D2} - L_{D1}}{R_2 - R_1}, \tag{6.39}$$

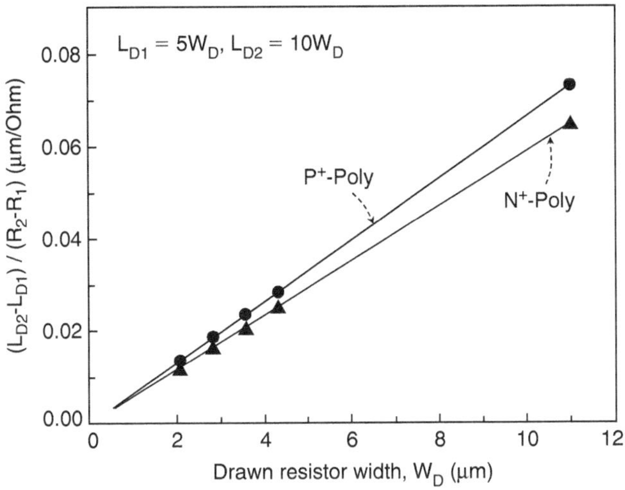

Fig. 6.17 Linear dependence of $(L_{D2} - L_{DI})/(R_2 - R_I)$ on resistor width (Adapted from [24])

where R_S is the sheet resistance of the resistor body, L_{DI}, L_{D2} the drawn lengths of two resistors (Fig. 6.15a), and R_I, R_2 the total resistance measured on the resistor pair. Figure 6.17 shows the linear dependence of the ratio $(L_{D2} - L_{DI})/(R_2 - R_I)$ on W_D for $L_{DI} = 5W_D$ and $L_{D2} = 10W_D$ [24]. When extracting ΔW from (6.38), it is assumed that for a fixed width, R_{end} and R_S are the same for the two resistors. ΔW is obtained from the intercept of the curve-fit with the horizontal axis and R_S from the slope.

The contact resistance can be estimated from measurements on a structure as in Fig. 2.61 and the silicide resistance from silicided resistor pairs of the same width but different lengths. The edge resistance can then be approximated from (6.38).

Polysilicon resistors are typically integrated in very small sizes with a wide range of values from tens of Ohms to Mega-Ohms by controlling the dopant type and concentration. The transport of carriers through polysilicon is different than through single-crystal silicon. Polysilicon consists of crystallites, called grains, joined together by grain boundaries. A grain boundary consists of a few atomic layers of disordered atoms [25]. The average grain size depends on deposition conditions and dopant concentration. The crystallites are typically arranged at large angles between adjoining grains. Inside the grain the atoms are arranged periodically as in single crystal silicon. Thus, the transport of carriers through the grain can be treated in the same manner as described in Chap. 1. The transport of carriers from grain to grain is, however, more complex. Figure 6.18a shows the average free-carrier concentration obtained from Hall measurements versus average boron concentration in a p-type polysilicon film, and Fig. 6.18b shows measured and calculated film resistivity in the same dopant range [25]. Calculations can be made in one dimension by making simplifying assumptions on grain size, trap density and dopant concentration. At low dopant concentration, the hole concentration is only a very

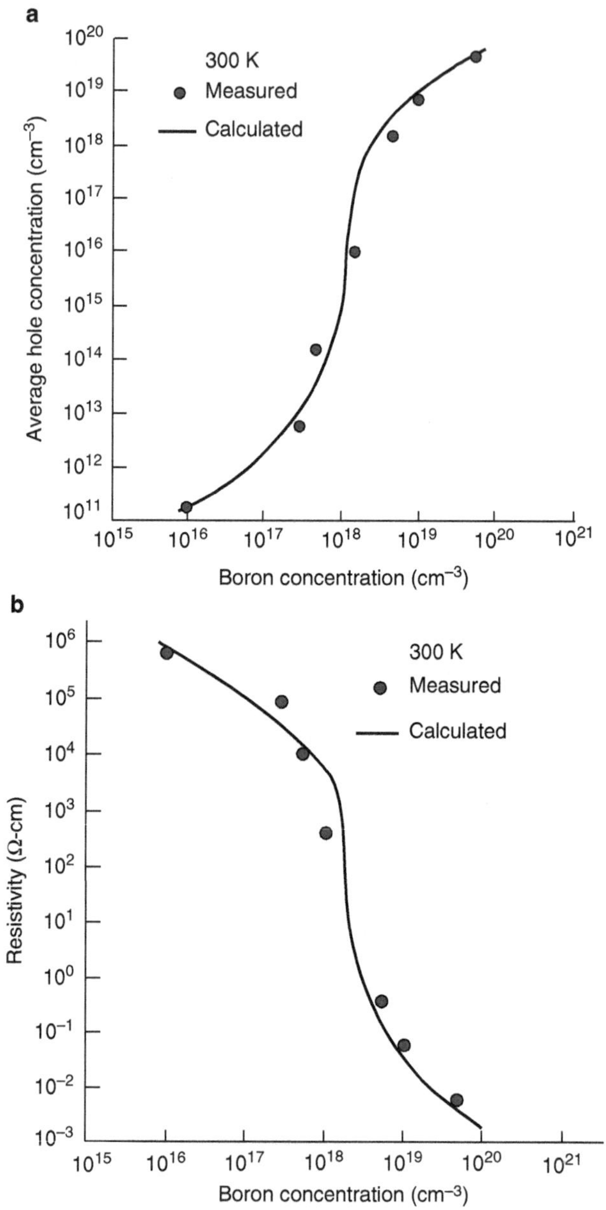

Fig. 6.18 a Hall-measured and calculated hole concentration versus boron concentration. **b** Measured and calculated resistivity of p-type polysilicon (Adapted from [25])

small fraction of the boron concentration. When the dopant concentration reaches about $5 \times 10^{17}\,\text{cm}^{-3}$, the concentration of holes increases rapidly and approaches that of boron at higher dopant concentrations.

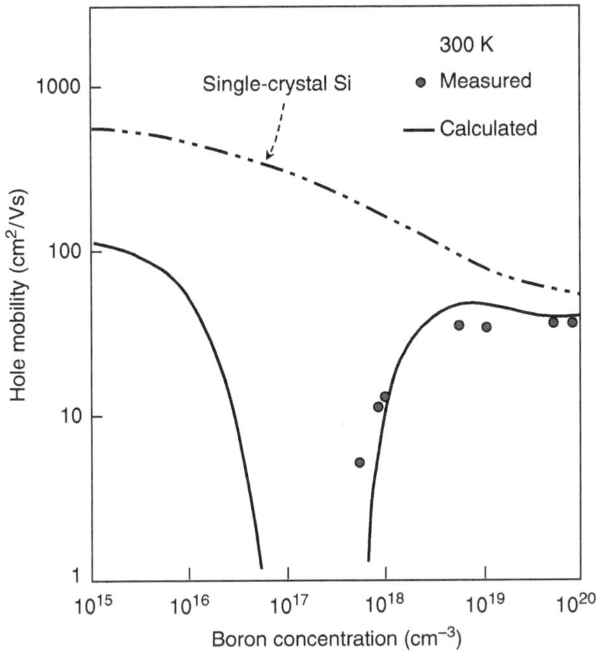

Fig. 6.19 Measured and calculated hole mobility versus dopant concentration of polysilicon film with a grain size of 122 nm (Adapted from [26])

The hole mobility is found to have a minimum in the dopant concentration range $10^{17}\,\mathrm{cm}^{-3}$ to $10^{18}\,\mathrm{cm}^{-3}$ [71, 72] and to increase at higher or lower dopant concentration as shown for a boron doped polysilicon film in Fig. 6.19.

The hole mobility is also found to increase with increasing grain size and film thickness [25–29]. Since the resistivity is a function of mobility and carrier concentration, it will depend on the polysilicon deposition conditions and grain size. One of the major difficulties in high resistivity polysilicon resistors is its large sensitivity to dopant concentrations in the range $10^{16}\,\mathrm{cm}^{-3}$ to $10^{18}\,\mathrm{cm}^{-3}$ (Fig. 6.18b). For example, over a dopant range of 5×10^{17} to 5×10^{18} a resistivity change of about five decades has been observed in polysilicon compared to only one decade change in single crystal silicon [25–29].

The effect of grain boundaries on the electrical properties of polysilicon has been described by two models. One model is based on the segregation of dopants at grain boundaries where they become inactive [30, 31]. The second model assumes that the disordered atoms at the boundary are sites of incomplete bonding, creating interface states that trap and immobilize free carriers, thus reducing the free carrier concentration within the grain [25, 29, 32]. The charged traps create a space charge surrounding the grain that constitutes a barrier to carrier flow from grain to grain, which reduces the carrier mobility. This is schematically illustrated in a simplified two-dimensional structure in Fig. 6.20.

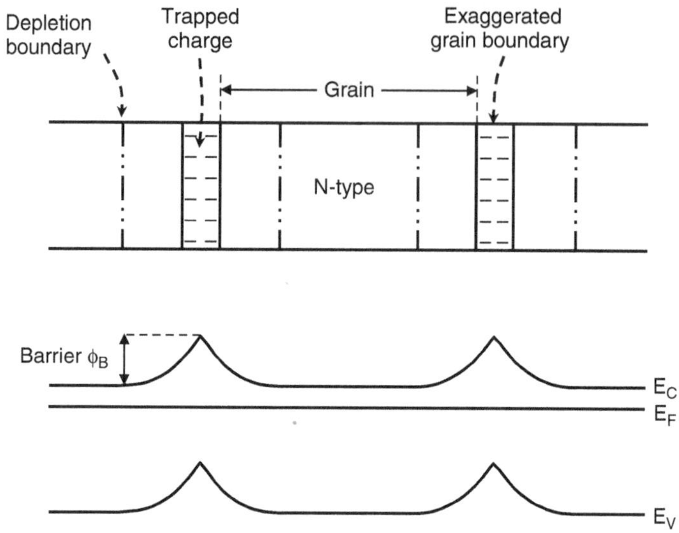

Fig. 6.20 Energy band diagram for polysilicon crystallites [25]

The grain boundary is found to be of negligible thickness compared to the grain size. Interface states are assumed to be initially neutral. In n-type polysilicon, the states are below the Fermi level and hence negatively charged by trapping electrons. In p-type polysilicon, the states are above the Fermi level, hence positively charged after trapping holes.

The transport of carriers across the grain boundary is mainly by thermionic emission, that is, by carriers that possess enough thermal energy to surmount the barrier ϕ_B (Chap. 2). For p-type polysilicon and a grain size L, the conductivity is found as [25]

$$\sigma = Lq^2\bar{p}\left(\frac{1}{2\pi m^*kT}\right)^{1/2} e^{-q\phi_B/kT}, \tag{6.40}$$

where m^* is the carrier effective mass and $\bar{p}$ the average hole concentration. The effective mobility can be obtained from the relation $\sigma = qp\mu_{\text{eff}}$ as

$$\mu_{\text{eff}} = Lq\left(\frac{1}{2\pi m^*kT}\right)^{1/2} e^{-q\phi_B/kT}. \tag{6.41}$$

It follows that the effective mobility of carriers traveling across the grain boundary increases with increasing temperature.

The mobility minimum is observed when the barrier reaches its maximum value which is when the grain is fully depleted. The condition of full-depletion depends on dopant concentration, grain size, and interface trap density. For typical polysilicon films at room temperature, the mobility minimum is found in the concentration range of $10^{17}\,\text{cm}^{-3}$ to $10^{18}\,\text{cm}^{-3}$ [25, 26].

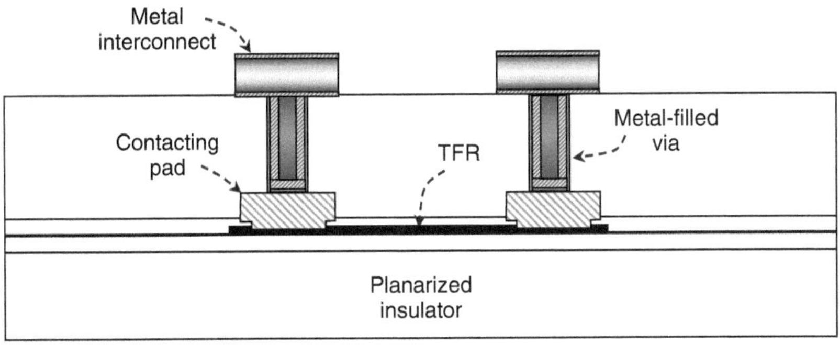

Fig. 6.21 Schematic cross-section of a thin-film resistor (*TFR*)

A modified grain-boundary trapping model is suggested in [33], whereby both the segregation and trapping mechanisms influence the polysilicon film conductivity. The trapping mechanism also explains the reduction in polysilicon resistivity after hydrogen annealing. The adsorption of hydrogen at grain boundaries is found to reduce the interface trap density by passivation of dangling bonds [34].

The term "Thin-Film Resistor" (*TFR*) is reserved for resistors made of thin metal or metal alloys, such nickel-chromium *NiCr* [35, 36], aluminum-doped *NiCr* [37], silicon-chromium *SiCr* [38], titanium-nickel-chromium *TiNiCr* [39], and tantalum-nitride *TaN* [40]. Their geometrical shape is similar to those of diffused and polysilicon resistors but their fabrication is different. Figure 6.21 is a schematic cross-section of a *NiCr TFR*.

The film is deposited at a thickness of about 5 nm–20 nm, typically by sputtering from a target of the same composition onto a planarized insulator surface. After patterning the *TFR*, a film of appropriate etch-selectivity to *NiCr*, such as a titanium-tungsten alloy, is deposited and patterned to form end contacting pads.

Thin film resistors can have a sheet resistance from about 30 Ohm/square to 2000 Ohm/Square. The resistance can be adjusted by a process known as "laser trimming," a procedure that removes parts of the resistor to increase the number of squares n, thus incrementally increasing the resistance [41–43]. The resistance is monitored during trimming until the desired value is obtained. A 3σ resistance tolerance of 0.1% or less can be achieved by laser trimming. The trim factor is the ratio of final to initial film resistance.

Temperature Coefficient of Resistance, TCR

The resistivity can be expressed in terms of the carrier mean-free path λ, which is the average distance traveled by the carrier between collisions (Chap. 1). The mean-free path in typical diffused resistors is considerably smaller than the resistor dimensions, so that the resistor exhibits bulk properties whereby the carrier mobility is essentially limited by phonon and impurity scattering. The temperature dependence

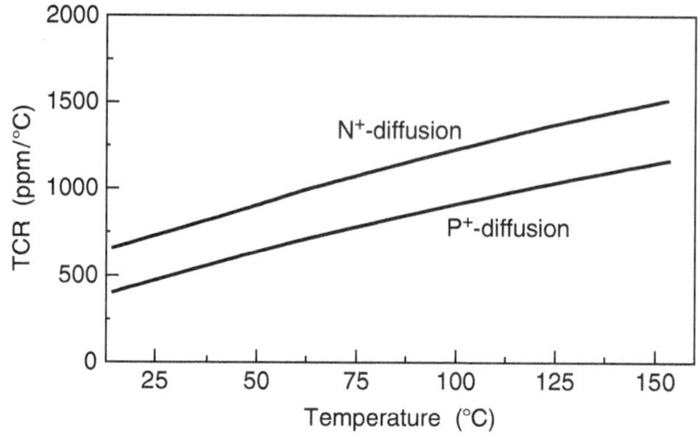

Fig. 6.22 Temperature coefficient of resistance (TCR) measured on diffused resistors of width 0.6 μm and length 3 μm (Adapted from [44])

of sheet resistance in the range $-50\,^{\circ}\mathrm{C}$ to $+125\,^{\circ}\mathrm{C}$ can then be approximated by the linear relation

$$R_S(T) = R_{S0}\left[1 + \alpha(T - T_0)\right], \tag{6.42}$$

where α is the temperature coefficient of resistance (TCR) expressed in parts per million per °C (ppm/°C) as

$$TCR = \alpha = 10^6 \frac{1}{R_{S0}} \frac{dR_S}{dT} \quad \mathrm{ppm}/^{\circ}\mathrm{C}, \tag{6.43}$$

and R_{S0} is the sheet resistance at temperature T_0. Thus, diffused resistors exhibit a positive TCR as shown for source-drain resistors in Fig. 6.22 [44].

The conduction in polysilicon resistors differs from that in diffused resistors due to the presence of grain boundaries. The conduction within the grain is similar to that in single crystal silicon with a positive TCR, while the conduction across the grain-boundary is dominated by thermionic emission that exhibits a negative TCR as can be extracted from (6.40). Thus, the TCR of polysilicon resistors is a combination of both mechanisms and will depend on temperature range, grain size distribution, and dopant type and concentration [25, 28–33, 44–47]. The polysilicon resistor TCR is found to be adjustable to near zero by tailoring the dopant type and concentration, for example, by arsenic implantation into phosphorus-doped polysilicon [46], or by adjusting the boron concentration in p-type polysilicon, as shown in Fig. 6.23 [47].

The capability of adjusting the polysilicon TCR to near zero is one of the major advantages of polysilicon over diffused resistors.

The advantages of thin-film resistors (TFR) over polysilicon resistors are their higher precision after trimming, lower TCR across a wide range of temperatures, and the higher flexibility in adjusting the sheet resistance for specific applications.

The conduction mechanism in a TFR can be categorized according to the ratio of film thickness d to electron mean-free path λ. The electron mean-free path in metals is ~30 nm [38, 48–50]. For $d \gg \lambda$, the resistor exhibits bulk-like behavior,

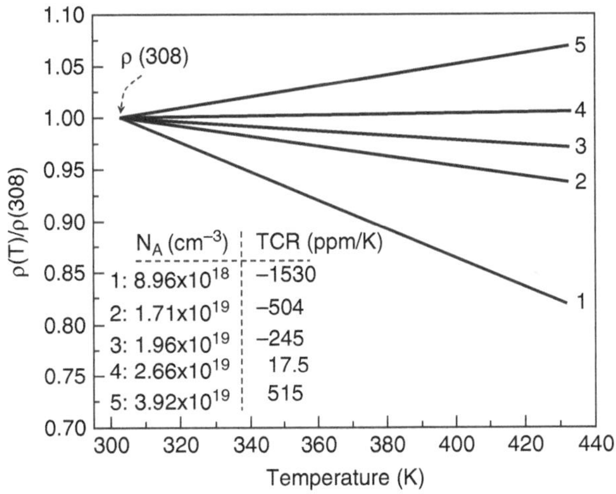

Fig. 6.23 Normalized resistivity $\rho(T)/\rho(308K)$ and extracted *TCR* for boron-implanted polysilicon (Adapted from [47])

and (6.42) and (6.43) apply. For $d \approx \lambda$, there are additional inelastic scattering events at the resistor surface that add another component to the film resistivity as

$$\rho(T) = \rho_{\text{Surf}}(T) + \rho_0[1 + \alpha(T - T_0)], \tag{6.44}$$

where ρ_{Surf} is the surface-scattering related resistivity and ρ_0 is the bulk resistivity at T_0. Since the mean-free path decreases with increasing temperature, the contribution of surface scattering also decreases. This negative trend leads to a *TCR* that is still positive but smaller than the bulk *TCR* [48]. For ultra-thin films with $d \ll \lambda$, there is evidence of agglomeration of the film into an array of small individual islands with boundaries between islands. For electron conduction to occur, electrons must be transferred from one particle to the next across the boundary. It is the mechanism for this transfer that determines the film resistance [49]. The sheet resistance is found to greatly increase and exhibit a negative temperature coefficient of resistance. The negative *TCR* suggests a thermally activated process which initially was attributed to thermionic emission but later found to be caused by a thermally assisted tunneling mechanism [49]. With an assumed activation-energy E_a for carriers to cross the boundary, the sheet resistance decreases exponentially with inverse temperature as [48]

$$R_S(T) \propto e^{E_a/kT}, \tag{6.45}$$

with a negative *TCR* given by

$$TCR = -\frac{E_a}{kT^2}. \tag{6.46}$$

Voltage Coefficient of Resistance, VCR

The voltage coefficient of resistance *VCR* is a measure of the sensitivity of resistance to the applied voltage between its terminals

$$VCR = 10^6 \frac{1}{R_{S0}} \frac{dR_S}{dV} \quad \text{ppm/V}. \tag{6.47}$$

For polysilicon and thin-film resistors that are typically deposited over a thick insulator, the *VCR* is negligible for all voltages that do not cause excessive current which results in an increase in resistor temperature, and are sufficiently low to avoid high-field transport effects and impact ionization. The *VCR* of diffused resistors is appreciable, particularly when the resistor forms pn junctions with its surroundings. This is because as the applied reverse voltage is increased, the depletion width spreads into the resistor body, reducing the conductive path and hence increasing the sheet resistance. Since the voltage drops from one resistor-end to the other, the depletion width becomes position-dependent causing non-uniform sheet resistance.

Resistor Linearity

A resistor is linear if its magnitude does not change with current or applied voltage. If, however, the power generated in the resistor is sufficiently high to cause an increase in resistor temperature, referred to as self-heating, there will be a nonlinearity in the current–voltage characteristics that is closely related to the *TCR* of the resistor in the same temperature range. The temperature rise due to self-heating depends on how efficiently the heat is removed from the resistor and its surroundings, that is, on the thermal conductivity of the layers surrounding the resistor. For example, for the same power dissipation, a diffused resistor in silicon will exhibit less self-heating than a polysilicon or *TFR* deposited on a thick insulator because the thermal conductivity of typical insulators is about two orders of magnitude smaller than that of silicon. Similarly, a polysilicon resistor or *TFR* placed higher up above silicon will exhibit more self-heating than resistors placed closer to the substrate. Thus, to minimize self-heating, a maximum allowable current density must be specified. It is typically in the range $10^4 - 5 \times 10^4\,\text{A/cm}^2$ (0.1–$0.5\,\text{mA}/\mu\text{m}^2$).

Application of a large voltage across a polysilicon resistor that results in a current density exceeding a threshold value, typically in the range 5–10 $\text{mA}/\mu\text{m}^2$, may cause an irreversible change in resistor value. This effect has been utilized to trim polysilicon resistors that deviate from the target value due to slight variations in process conditions [51–53]. A simple electrical trimming technique was demonstrated on heavily doped polysilicon films grown by Chemical-Vapor Deposition (*CVD*) [51]. By applying pulses of current density above the threshold level, the resistance decreased steadily with increased pulse amplitude, pulse width, and number of pulses. A total resistance change greater than 50% was observed. The average temperature, however, did not rise appreciably because of the low duty cycle of the pulses. The mechanism of trimming is believed to be related to a reduction in

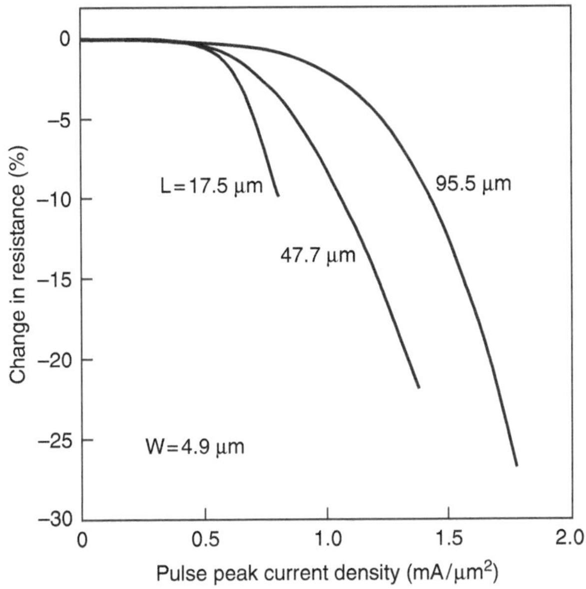

Fig. 6.24 Percent change in resistance versus current pulse amplitude for phosphorus-doped polysilicon resistors of the same width and different lengths (Adapted from [53])

barrier height at the grain boundary and the resulting increase in mobility. Since grain boundaries have a considerably higher resistance than grains, most of the power of each pulse is dissipated at grain boundaries, modifying the barrier. By applying current pulses of amplitude larger than the threshold level but lower than the amplitude of trimming pulses, the resistance was found to "recover" to its initial value, or even to higher than its initial value.

Results of trimming polysilicon resistors doped with $7.5 \times 10^{19}\,cm^{-3}$ phosphorous are shown for three resistors of same width and different lengths in Fig. 6.24 [53]. A reduction in resistance up to 27% was observed after three pulses of 10 μs duration at intervals of 300 μs. The dependence of trimming on resistor length is not well understood.

Parasitic Capacitance

Parasitic capacitances between a resistor and surrounding conductors are illustrated in Fig. 6.25 for a resistor placed on the same plane as the first metal. The capacitance per unit area down to silicon is the equivalent of three capacitances in series, C_{PMD}, C_{STI} *and* C_{Si}. C_{PMD} is the pre-metal dielectric capacitance defined as $\varepsilon_0\varepsilon_{ox}/t_{eq}$, where t_{eq} is the equivalent oxide thickness of the inter-level dielectric separating the resistor from the planarized wafer surface. C_{STI} depends on the *STI* thickness t_{STI} as $\varepsilon_0\varepsilon_{ox}/t_{STI}$, and C_{Si} is capacitance between the silicon surface and

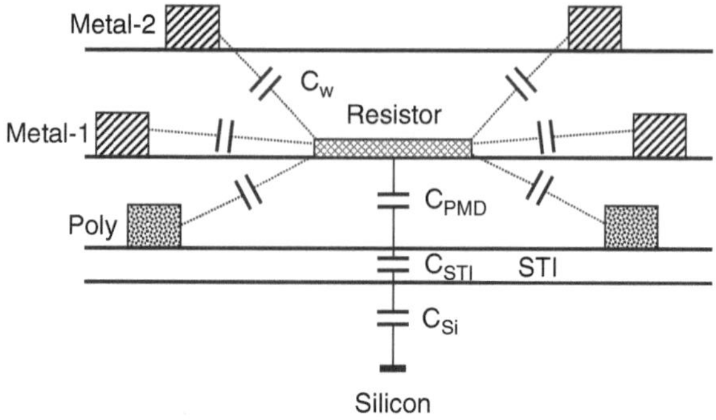

Fig. 6.25 Illustration of the main parasitic capacitances for a resistor placed at the first metal level

bulk. C_{Si} depends on the surface dopant concentration and resistor bias conditions with respect to silicon (Chap. 4).

Wiring capacitance, C_w is the capacitance to other conductors in the vicinity of the resistor. C_w depends on the distance and dielectric constant of the insulator between resistor and conductor and, to some extent, on the resistor thickness. A diffused resistor will typically have a larger parasitic capacitance than a polysilicon or thin-film resistor because it is embedded in conductive silicon. To reduce the parasitic capacitance, the resistor area should be reduced, for example, by using a higher sheet resistance, and the resistor placed higher-up above silicon, and at a large distance from other conductors. Also, decreasing the insulator dielectric constant reduces C_w and decreasing the silicon surface dopant concentration reduces silicon capacitance C_{Si}.

6.2.3.2 Capacitors

The selection of integrated capacitors is based on several criteria, such as high capacitance per unit area (capacitance density), low voltage coefficient of capacitance (*VCC*), low temperature coefficient of capacitance (*TCC*), low mismatch between capacitor pairs (Sect. 6.3), low leakage at the operating voltage and temperature, low parasitic capacitance, low trap density, and low additional cost.

On-chip decoupling capacitors are required, in addition to intrinsic junction and dielectric circuit capacitances, to reduce power-supply noise. They are most efficient when placed close to switching circuits and hence they occupy valuable area. Thus, the most important criteria of decoupling capacitors is the high capacitance density to reduce the area consumed by the capacitor, and the low leakage through the dielectric at the operating voltage. An estimate of the total decoupling capacitance C_{decap} that is required to keep the circuit voltage "bounce" within a specified value is given in [54]. In a time interval Δt, the circuit switching charge ΔQ is

$$\Delta Q = \frac{\bar{I}}{2f} \quad \text{C}, \tag{6.48}$$

where $\bar{I}$ is the average current during switching and f is the clock frequency. The voltage "bounce" across the connected decoupling capacitor is

$$\Delta V = \frac{\Delta Q}{C_{\text{decap}}} \quad \text{V}. \tag{6.49}$$

C_{decap} must be sufficiently large to keep ΔV below a specified fraction r of the power supply voltage V_{DD}, that is

$$C_{\text{decap}} > \frac{\bar{I}}{2frV_{DD}} = \frac{\bar{P}}{2frV_{DD}^2} \quad \text{F}, \tag{6.50}$$

where $\bar{P} = \bar{I}V_{DD}$ is the average power dissipated in the circuit. For a typical ratio $r \approx 0.05$, (6.50) becomes

$$C_{\text{decap}} > \frac{10\bar{P}}{fV_{DD}^2} \quad \text{F}. \tag{6.51}$$

Equation (6.51) gives a reasonable estimate of C_{decap} [54]. Techniques to increase the capacitance density and reduce the area are described below.

High precision capacitors are key elements for advanced analog *CMOS* technology, particularly in the area of A/D converters and switched capacitor filters. In addition to the high capacitance density and low leakage, they must exhibit low *VCC* and *TCC*, low mismatch between capacitor pairs, low noise, low parasitic capacitance, low trap density, and a high quality factor Q.

The variation of capacitance with voltage in capacitors of the type in Fig. 6.26b–d can be typically fitted to a parabola of the form (Fig. 6.27)

$$C = C_0(1 + \alpha_1 V + \alpha_2 V^2), \tag{6.52}$$

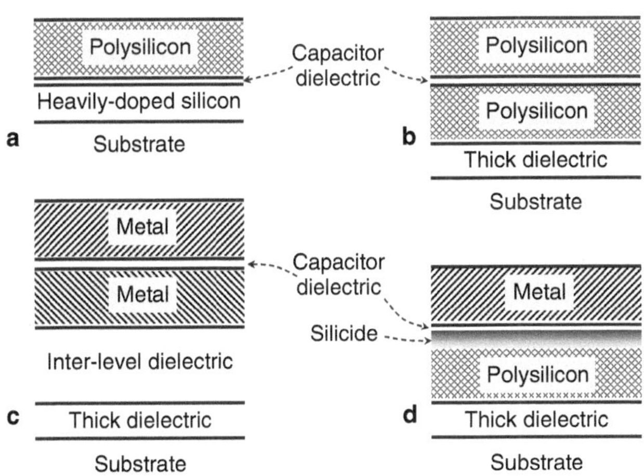

Fig. 6.26 Schematic cross-sections of typical precision capacitors. **a** Polysilicon-Insulator-Silicon; **b** Polysilicon-insulator-Polysilicon; **c** Metal-Insulator-Metal (MIM); **d** Metal-Insulator-Silicide-Polysilicon

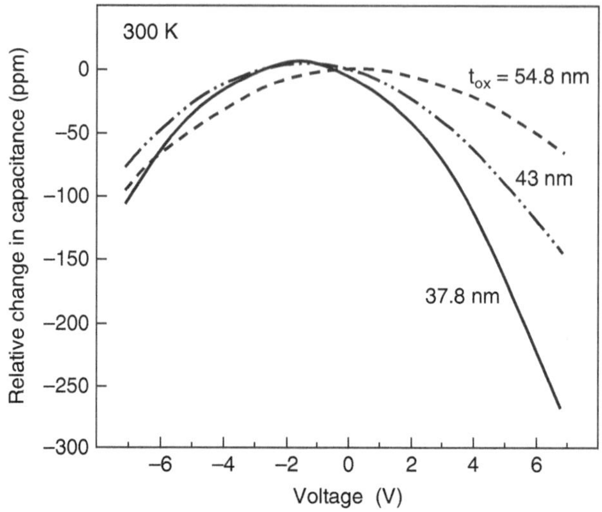

Fig. 6.27 Relative change in capacitance versus applied voltage between plates of n^+-polysilicon/oxide/n^+-silicon capacitor for different oxide thicknesses (Adapted from [55])

where α_1 is referred to as the linear *VCC* and α_2 the quadratic *VCC*. The temperature coefficient *TCC* represents the fractional rate of change of total capacitance per °C and is expressed in ppm/°C as

$$TCC = 10^6 \times \frac{1}{C}\frac{\partial C}{\partial T} \quad \text{ppm/°C}, \tag{6.53}$$

where *C* is the total capacitance.

Matching between capacitor pairs and capacitor noise are discussed detailed in Sects. 6.3 and 6.4. The discussion of parasitic capacitance for a resistor applies also to a capacitor. A low trap density is needed to reduce relaxation effects discussed later in this section.

Junction capacitors are not attractive for precision analog applications because of their inherent high *VCC, TCC*, parasitic capacitance, and series plate resistance. The most widely used precision capacitors are shown schematically in Fig. 6.26. A polysilicon-insulator-silicon capacitor of the type shown in Fig. 6.26a is reported in [55, 56]. The capacitor consists of an n^+-polysilicon plate patterned over silicon-dioxide grown over a heavily-doped collector sinker of an *NPN* transistor in a *BiCMOS* technology [55]. The voltage-dependence of capacitance for this structure is shown in Fig. 6.27 for different oxide thicknesses and silicon phosphorus surface concentration of 10^{20} cm^{-3}.

The total capacitance is a combination of dielectric capacitance and space-charge capacitances in silicon and polysilicon. A positive voltage on polysilicon with respect to silicon depletes polysilicon and reduces the total capacitance. When the voltage polarity is reversed, the silicon surface is depleted and the total capacitance drops again.

The temperature coefficient of capacitance can be resolved into three components [57]

$$TCC = TCC_{(\text{thermal})} + TCC_{(SC)} + TCC_{(OX)}. \tag{6.54}$$

The first term represents the change in plate area and dielectric thickness due to thermal expansion

$$TCC_{(\text{thermal})} = 10^6 \times \left(\frac{1}{A}\frac{dA}{dT} - \frac{1}{t_{eq}}\frac{dt_{\text{eq}}}{dT} \right) \quad \text{ppm}/^\circ\text{C}. \tag{6.55}$$

The second term represents the temperature dependence of the space charge (surface depletion) capacitance

$$TCC_{(SC)} = 10^6 \times \left(\frac{C_{ox}}{C_{Si}^2}\frac{dC_{Si}}{dT} \right) \quad \text{ppm}/^\circ\text{C}. \tag{6.56}$$

The third term corresponds to the temperature dependence of the dielectric constant

$$TCC_{(OX)} = 10^6 \times \left(\frac{1}{\varepsilon_{ox}}\frac{d\varepsilon_{ox}}{dT} \right) \quad \text{ppm}/^\circ\text{C}. \tag{6.57}$$

The *TCC* of an n^+-polysilicon/oxide/n^+-silicon capacitor is found to be negligible [56], and to range from 30 ppm/°C to 60 ppm/°C [55]. Most of the temperature dependence is attributed to vertical and horizontal thermal expansions [57]. N^+-polysilicon/oxide/n^+-polysilicon capacitors of the type shown in Fig. 6.26b exhibit the same trend in *VCC* and *TCC* as for the n^+-polysilicon/oxide/n^+-silicon capacitors [56]. The parasitic capacitance is, however, smaller because the bottom polysilicon plate in Fig. 6.26b is placed over thick oxide, resulting in a smaller capacitance to the substrate than the junction capacitance in Fig. 6.26a. The *VCC* can be further reduced by forming plates with metallic characteristics (Fig. 6.26d). For example, a *TiN*-oxide-silicide capacitor with $t_{ox} = 50\,\text{nm}$ exhibits an average linear *VCC*, α_1 in (6.52), of about $-2.1\,\text{ppm/V}$ and a quadratic *VCC*, $\alpha_2 \approx -9.1\,\text{ppm/V}^2$ while the *TCC* is found negligible [58, 59]. The very low *VCC* is due to the metallic characteristics of *TiN* and silicide. The sheet resistance of *TiN* and silicide are, however, large when compared to aluminum or copper. A "true" metal-insulator-metal (*MIM*) capacitor not only exhibits a lower plate sheet-resistance but can also be placed higher up above silicon reducing the coupling to the substrate. The low plate sheet-resistance and low coupling to the substrate, combined with a high resistivity of the substrate region underneath the capacitor, are important to ensure a high quality factor Q of the passive element for high precision analog and *RF* designs.[1] Such a capacitor is shown in Fig. 4.26c. It has been demonstrated in a *SiGe BiCMOS* technology by sandwiching 50-nm oxide between the top two aluminum levels, resulting in near zero *VCC* and high Q [60]. A *MIM* structure that uses a 20-nm thick

[1] The quality factor is defined as: $Q = 2\pi \frac{\text{Energy stored}}{\text{Energy dissipated per cycle}}$. It is a dimensionless parameter that, for example, describes how much amplitude is lost in one cycle of an oscillating system. An oscillating mechanical system with low frictional forces would have a high Q. In a viscous medium, the system would have a low Q

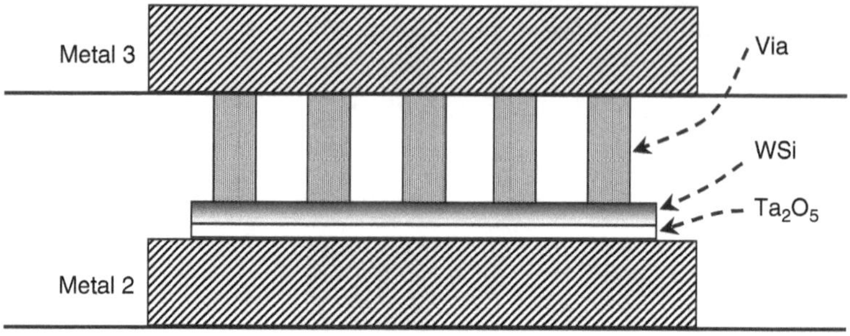

Fig. 6.28 Schematic cross-section of *MIM* capacitor with a 20-nm thick tantalum-oxide (Ta_2O_5) dielectric, a top tungsten-silicide (*WSi*) plate and multiple tungsten-filled Vias to reduce the plate resistance (Adapted from [61])

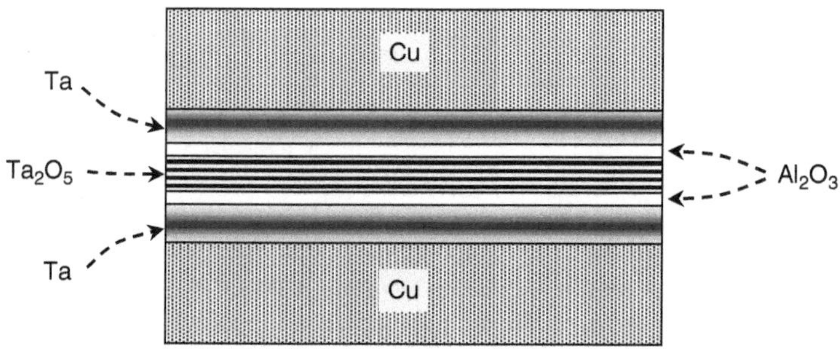

Fig. 6.29 Schematic cross-section of *Cu-*Ta_2O_5*-Cu MIM* capacitor with Al_2O_3 barriers (Adapted from [62])

amorphous tantalum-oxide (Ta_2O_5, $K \approx 20$) as a capacitor dielectric is shown in Fig. 6.28 [61]. Tungsten-silicide is used at the top plate and multiple vias are patterned to reduce the plate resistance. The high capacitance density considerably reduces the area occupied by the capacitor. Without proper treatment of interfaces, however, the leakage, *VCC* and *TCC* in tantalum-oxide capacitors may be too high for precision analog applications [62]. A *MIM* capacitor that uses tantalum-oxide as a dielectric, copper plates, and Al_2O_3/Ta barriers above and below the dielectric to protect against oxygen and copper diffusion is reported in [63] (Fig. 6.29). The structure exhibits a capacitance density of $4.4\,\text{fF}/\mu\text{m}^2$ for a Ta_2O_5 thickness of 40 nm, linear VCC $\alpha_l = 150\,ppm/V$, quadratic VCC $\alpha_2 = 400\,ppm/V^2$, and $TCC = 200\,ppm/K$ [62]. The coefficients are higher than with *MIM* capacitors with an insulator of lower dielectric constant and are believed to be due to changes in the Ta_2O_5 dielectric properties with temperature and voltage. Other high-*K* *MIM* capacitors for decoupling and *RF* applications are reported in [63–66].

Capacitors are frequently used as memory elements in analog circuits. For example, the capacitor *C* in Fig. 6.30 is charged to a voltage that corresponds to the analog information to be stored. The information is read with a high-impedance amplifier. If, prior to storing the information, the capacitor still "remembers" a fraction of the

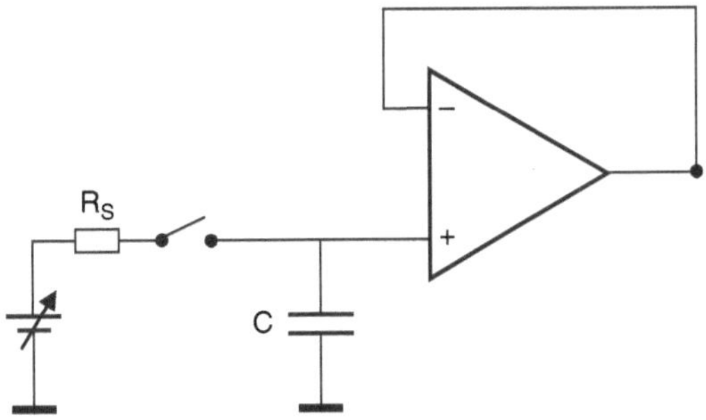

Fig. 6.30 Analog memory [67]

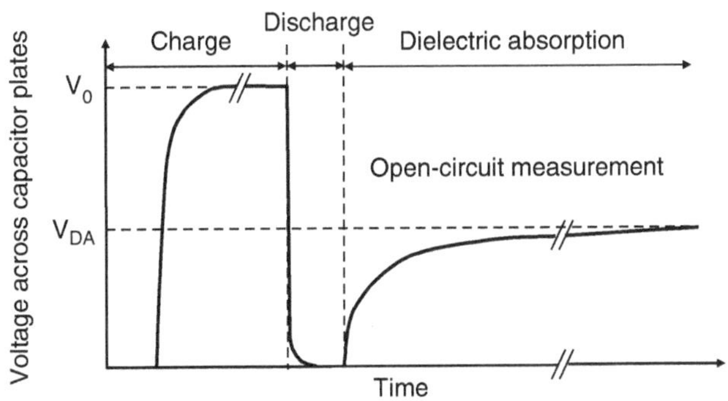

Fig. 6.31 Example of capacitor voltage versus time, normalized to V_0. Open-circuit V_{DA} is a measure of dielectric absorption (Adapted from [69])

voltage that was applied to it in a previous operation, an error in the read-out occurs [67]. The inability of a charged capacitor to discharge completely to zero volts after shorting its electrodes is called dielectric absorption, *DA* [67–69]. This effect is also known as “capacitor soakage” or “capacitor memory.” It is caused by slow relaxation of residual polarization (dipoles aligned to the field) or trapped charge after the voltage across the plates is removed. It is best understood by considering the following experiment:

The capacitor is initially charged for a sufficiently long time until a voltage V_0 is obtained across its plates. The plates are then shorted for a very short time and the open-circuit voltage across the plates measured directly after the discharge (Fig. 6.31). Depolarization and trap-depopulation will give rise to a transient current I_{DA} to the plates, charging the plates and raising the voltage across the plates asymptotically from zero to V_{DA}, as shown in Fig. 6.31 [69]. V_{DA} is a measure of dielectric absorption and is defined as

$$V_{DA} = \frac{1}{C}\int I_{DA}dt \quad \text{V.} \tag{6.58}$$

6.2.3.3 Varactors

The term varactor is an abbreviation for variable reactor that refers to the variation of capacitance with applied voltage to a pn junction, a Schottky-barrier diode, or an *MOS* structure. It is used, for example, in the tuning stage of a radio receiver or a voltage-controlled oscillator (*VCO*) [70–72].

Junction Varactor

For a reversed-biased junction, the capacitance depends on reverse voltage as (Chap. 2)

$$C_j \propto V_R^{-n}, \tag{6.59}$$

where $n = 1/2$ for a one-sided abrupt junction, $n = 1/3$ for a linearly-graded junction, and V_R is the sum of built-in voltage V_b and applied voltage V_a. Figure 6.32 shows a schematic of a varactor formed between the p-base and n-collector of an *NPN* transistor in an *SOI BiCMOS* technology [73]. The p-base contact receives the *PMOS* source-drain implants and the contacts to collector sinkers receive the *NMOS* source-drain implants. Thus, the structure consists of a p^+-n junction, in which the depletion region expands almost fully into the n-region. A similar varactor formed between base and specially-designed collector in a bulk *BiCMOS* technology is proposed in [74].

The profile in the n-region can be expressed as

$$N_D(x) = G\left(\frac{x}{x_0}\right)^m, \tag{6.60}$$

where G and x_o are constants, $m = 0$ for a one-sided abrupt junction, $m = 1$ for a linearly-graded junction. As the depletion region expands into the n-region by an increment dx, the field increases by an increment

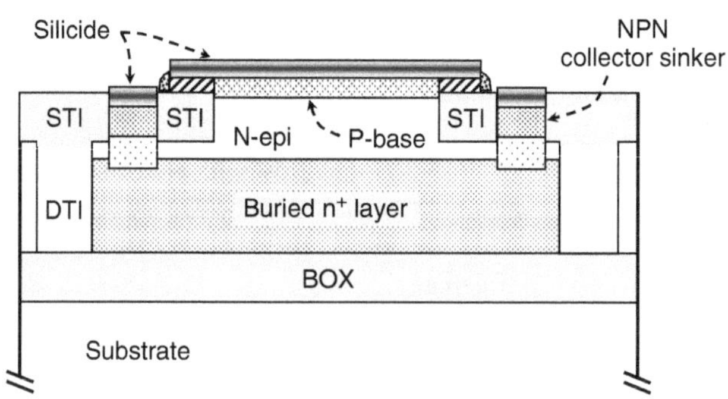

Fig. 6.32 Schematic of a varactor formed between base and collector in an *SOI BiCMOS* technology [73]

$$dE = \frac{qN_D}{\varepsilon_0 \varepsilon_{Si}}(x)dx = G'\left(\frac{x}{x_0}\right)^m dx. \tag{6.61}$$

The capacitance–voltage relation is obtained by solving Poisson's equation

$$\frac{d^2V}{dx^2} = -G'\left(\frac{x}{x_0}\right)^m. \tag{6.62}$$

Integrating Poisson's equation with appropriate boundary conditions gives the dependence of depletion layer width as a function of reverse voltage as [75, 76]

$$x_{dn} \propto V_R^{1/(m+2)}. \tag{6.63}$$

Thus, the capacitance can be expressed as

$$C_j = \frac{\varepsilon_0 \varepsilon_{Si}}{x_{dn}} \propto V_R^{-1/(m+2)}. \tag{6.64}$$

For a one-sided abrupt junction, $m = 0$ and, from (6.59) $n = 1/2$. For a linearly-graded junction $m = 1$ and $n = 1/3$ (Fig. 6.33). The capacitance-voltage characteristic of a one-sided p^+n step-junction is shown in Fig. 6.34.

A junction is said to be hyperabrupt if $n > 1/2$, in which case m must be negative [75,77]. In a special case where $m = -3/2$ and $n = 2$ the capacitance is proportional to V_R^{-2}. When such a capacitor is used with an inductor L in a resonant circuit, the resonant frequency varies linearly with applied reverse voltage V_a. For $V_a \gg V_b$

$$\omega = \frac{1}{\sqrt{LC_j}} \propto \frac{1}{\sqrt{V_a^{-2}}} \propto V_a. \tag{6.65}$$

By choosing different values of m, one can obtain a wide variety of C_j versus V_a characteristics for specific applications.

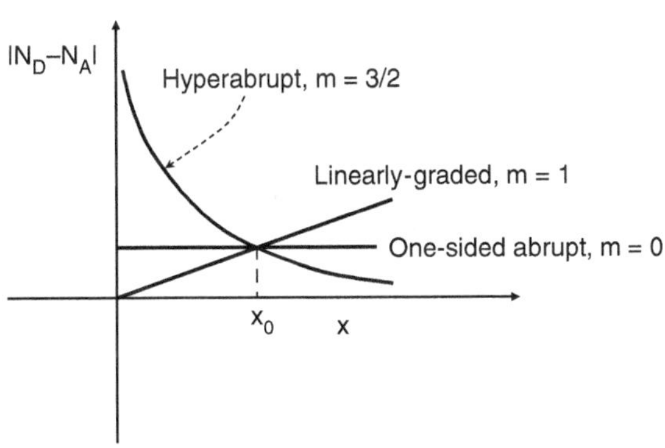

Fig. 6.33 Illustration of impurity profiles in the n-region of a p^+n junction with varying values of the exponent in (6.60)

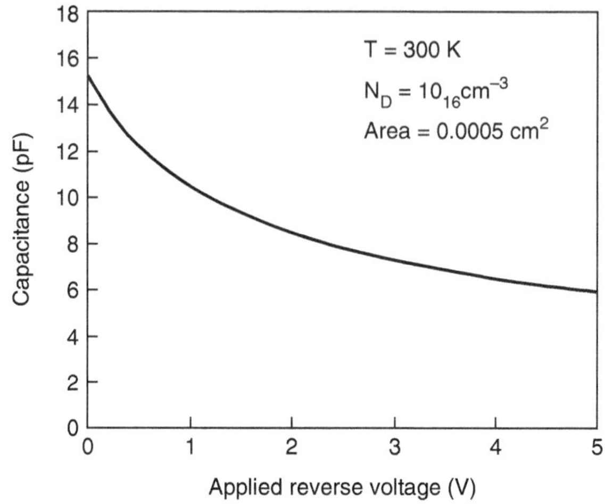

Fig. 6.34 Capacitance-voltage characteristic of a p^+n one-sided step junction

The most important varactor parameters are the capacitance per unit area, capacitance sensitivity, tuning range, quality factor, low-frequency noise, and breakdown voltage. The sensitivity is defined as [75, 76]

$$s = -\frac{dC}{C}\frac{V}{dV} = \frac{-d(\log C)}{d(\log V)} = \frac{1}{m+2}. \tag{6.66}$$

For an abrupt junction, $s = 1/2$, and for a linearly-graded junction, $s = 1/3$. The highest variation with biasing voltage is for a hyperabrupt junction with $m = -3/2$ and $s = 2$.

The tuning range is determined largely by the ratio of the maximum to the minimum varactor capacitances in the voltage range of operation and is reduced by parasitics. It is expressed as

$$C_{\text{ratio}} = \frac{C_{\text{max}}}{C_{\text{min}}} = \frac{C_{\text{int-max}} + C_{\text{par}}}{C_{\text{int-min}} + C_{\text{par}}}, \tag{6.67}$$

where C_{int} is the varactor intrinsic variable capacitance and C_{par} the parasitic capacitance assumed for simplicity to be voltage independent. The parasitic capacitance of the varactor deteriorates the tuning range and hence the frequency tuning range of, e.g, the *VCO*.

The quality factor Q is a measure of energy stored in the circuit element to energy lost by, for example, Eddy currents in the substrate. It is defined as

$$Q = \omega\frac{\text{Energy stored}}{\text{Power lost}} = 2\pi\frac{\text{Energy stored}}{\text{Energy lost per cycle}}, \tag{6.68}$$

where ω is the angular frequency. A simplified equivalent circuit of a varactor is shown in Fig. 6.35, where R_S is the series resistance and R_P the parallel equivalent resistance all junction leakage-current components

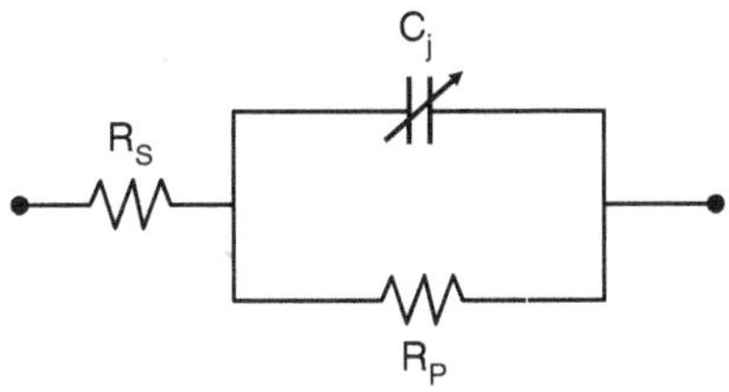

Fig. 6.35 Equivalent circuit of a varactor. R_S is the series resistance and R_P the parallel equivalent resistance of all junction leakage components

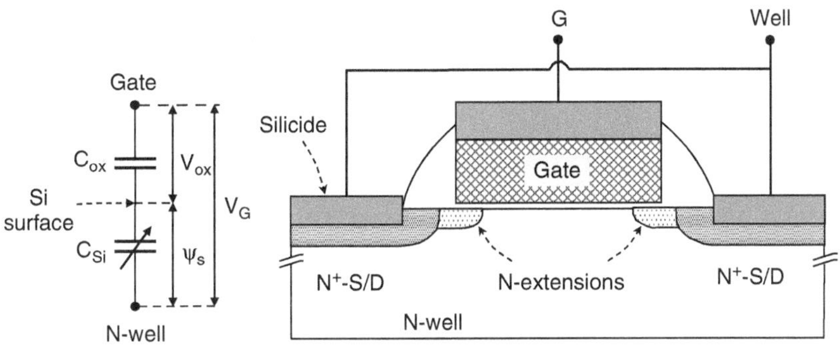

Fig. 6.36 Cross-section and equivalent circuit of an enhancement-mode varactor

The maximum varactor quality factor is defined as [78]

$$Q_{\max} \approx \left(\frac{R_P}{4R_S}\right)^{1/2}. \tag{6.69}$$

Q_{max} increases as the junction leakage decreases and the series resistance increases. A Schottky diode would have an almost perfect one-sided abrupt junction, but typical Schottky diodes exhibit high reverse leakage that increases rapidly with increasing reverse voltage, thus reducing the quality factor.

MOS Varactor

An *MOS* varactor can be formed between the gate and well in a standard *CMOS* process. Figure 6.36 illustrates a varactor formed on the n-well of an accumulation-mode *PMOS*, operating in the accumulation and depletion regions of the *CV* characteristic (Chap. 4) [79–81]. The choice of an n-well rather than a p-well is made because of the ease of isolation in bulk *CMOS* and the higher electron mobility. For an applied gate voltage with respect to n-well far above flatband, the surface is in strong accumulation. As the gate voltage decreases and becomes negative, the capacitance decreases from a maximum value C_{ox} to a minimum value C_{min} (Chap. 4).

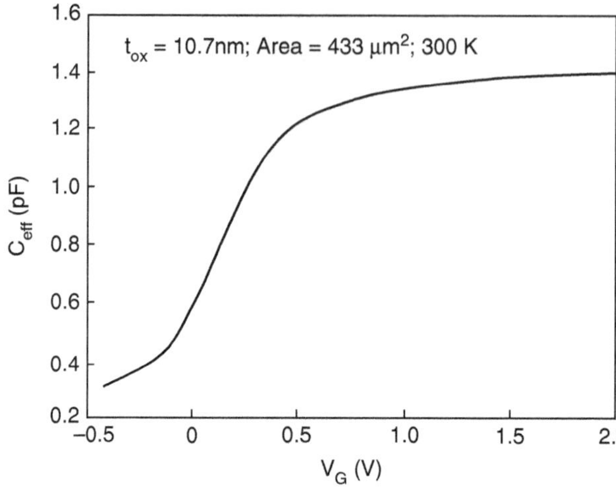

Fig. 6.37 *CV* plot obtained on a varactor formed by an array of structures similar to that of Fig. 6.36

Figure 6.37 shows a measured *CV* plot at 2.5 GHz on a varactor formed by MOS structures similar to that in Fig. 6.36, arranged in an array of 14 gates of 1.95 μm effective channel length and 15.85 μm effective channel width [79].

The quality factor of the polysilicon-oxide-well varactor can be increased by reducing the gate length and hence the series resistance from the n^+-regions to the region under the gate. The tuning range, however, decreases due to the increasing fraction of parasitic gate overlap and fringe capacitances [80].

The two-terminal structure in Fig. 6.36 can be extended to a three-terminal varactor by placing a p^+-junction in the same n-well in direct contact with the region under the gate as in a *PMOS*, allowing a wider varactor tuning range [81]. The p^+- and n^+-regions can be separated from each other by a silicide-block film, as shown in Fig. 6.38. A varactor of the desired capacitance can be formed by an array of structures shown in Fig. 6.38 [81].

The *MOS* capacitance reaches its maximum value in strong accumulation (gate positive). As the voltage on the gate is reduced and changes polarity, the *MOS* capacitance decreases and reaches its minimum value C_{min} when $V_G \approx V_T$. In the absence of a p^+-region or with the p^+-region floating, C_{min} can only be established with an adequate supply of electron-hole pairs, for example, by thermal generation. Otherwise, the structure goes into deep depletion as the gate voltage is decreased below V_T (Chap. 4). With the p^+-region floating, however, the capacitance in deep depletion is undetermined and increases with time as more inversion holes are supplied. By applying a negative bias on the p^+-region, generated holes are swept-away from under the gate and the structure remains at a fixed surface potential in deep depletion. Thus, the *MOS* capacitance is lower than C_{min}, increasing the tuning range. A varactor of this type constructed in a 0.35 μm technology yields a tuning range of 3:1 with the p^+-region floating, while it increases to 5:1 with the p^+-region at -3 V.

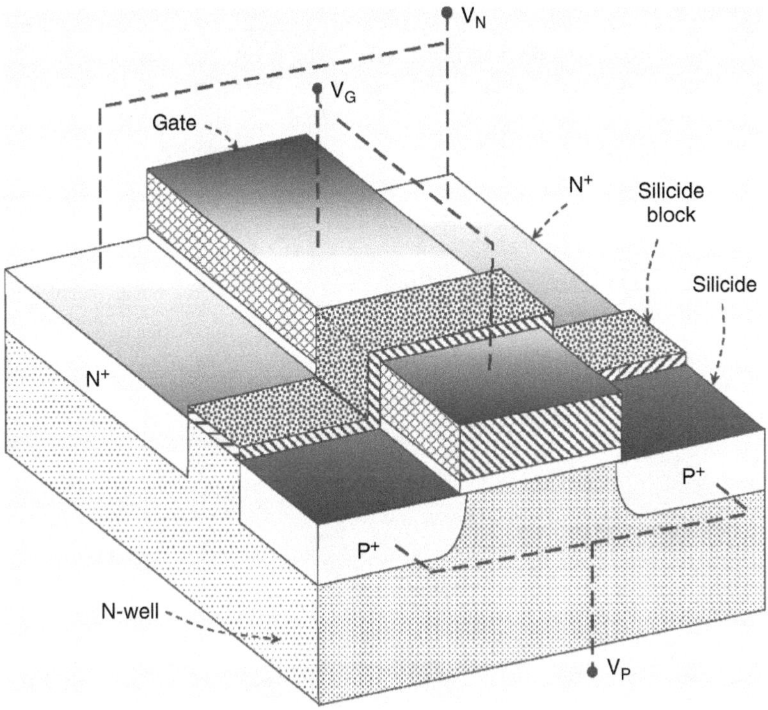

Fig. 6.38 Three-terminal accumulation-mode *PMOS* varactor (Adapted from [81])

A three-terminal varactor that combines pn junction and *MOS* variable capacitances to achieve a high *Q*-factor is also demonstrated in [82, 83]. A *Q*-factor of 108 at 2.5 GHz on a varactor of 723 fF total capacitance has been reported in [83].

The accumulation mode n-well varactor is found to exhibit considerably higher low-frequency noise than pn junction varactors. The *MOS* noise is found to strongly depend on bias conditions [84].

6.3 Matching

Identically designed devices that are processed and biased under the same conditions are expected to have identical electrical parameters. In reality, however, there are fluctuations in device dimensions and impurity concentrations that cause global and local variability in parameters. Global variation accounts for the total variation in the value of a parameter across a die, a wafer, a batch, or from batch to batch. Local variation is a microscopic effect that causes mismatch in parameters between adjacent devices on the same die [85, 86]. Since analog circuits are more sensitive to differences and ratios of parameters rather than their absolute values, parametric mismatch fluctuations between adjacent, identically designed and biased devices is critical for precision analog and mixed-signal applications. It is also important to digital designs. For any electrical parameter P, the mismatch between two adjacent devices 1 and 2 is the difference $\Delta P = P_2 - P_1$, with P_1 and P_2, respectively, the parameter values for devices 1 and 2. The differences are assumed to be random and have a normal distribution with zero mean and variance $\sigma_{\Delta P}^2$.

6.3.1 MOSFET Mismatch

In Chap. 5, the *MOSFET* current-voltage relationship was given in the linear region as

$$I_D = \beta \left(V_G - V_T - \frac{V_D}{2} \right) V_D \quad \text{A}, \tag{6.70}$$

and in the saturation region as

$$I_D = \frac{\beta}{2} (V_G - V_T)^2 \quad \text{A}, \tag{6.71}$$

where V_T is the threshold voltage, V_G the gate to source voltage, V_D the drain to source voltage and β the so-called current factor defined as

$$\beta = \mu_{\text{eff}} C_{\text{ox}} \frac{W_{\text{eff}}}{L_{\text{eff}}} = \mu_{\text{eff}} \frac{\varepsilon_0 \varepsilon_{\text{ox}}}{t_{eq}} \frac{W_{\text{eff}}}{L_{\text{eff}}}. \tag{6.72}$$

6.3.1.1 Inverse-Area Law

The main sources of mismatch in *MOSFET* drain current at a fixed gate voltage or in gate voltage at a fixed drain current are the differences in V_T and β between the two structures [86–90]. These random differences have a normal distribution with zero mean and variance that depends on gate area, $W_{eff}\, L_{eff}$, and distance d between the centers of identical structures as [87]:

$$\sigma_{\Delta VT}^2 = \frac{A_{VT}^2}{\overline{W}_{\text{eff}} \overline{L}_{\text{eff}}} + S_{VT}^2 d^2, \tag{6.73}$$

$$\frac{\sigma_{\Delta\beta}^2}{\beta^2} = \frac{A_\beta^2}{\overline{W}_{\text{eff}} \overline{L}_{\text{eff}}} + S_\beta^2 d^2, \tag{6.74}$$

where A_{VT}, A_β, S_{VT}, and S_β are technology-dependent factors and the bars in the denominators mean average values. Equations (6.73) and (6.74) are sometimes referred to as the Pelgrom laws [87]. For closely-spaced structures of area less than

$100\,\mu\text{m}^2$, where the separation between devices is of the same order as the device dimensions, the second terms in the above equations can be ignored [86–88]. A_{VT} is expressed in the convenient unit of $\text{mV}\cdot\mu\text{m}$ and found to decrease with technology generation from 30-35 mV μm in a 2.5 μm technology to ∼5 mV μm in a 0.18 μm technology [88]. Similarly, A_β is best expressed in $\%\cdot\mu\text{m}$. It is also found to decrease from 2.3–3.2% μm to ∼1% μm in the same technology range [88].

The inverse area dependence of A_{VT} and A_β can be explained by considering the different terms that affect the variability [86]. Consider, for example, the *NMOS* threshold defined as (Chap. 5):

$$V_T = -\frac{Q_{b\max}}{C_{\text{ox}}} - \frac{Q_{\text{eff}}}{C_{\text{ox}}} + \phi_{ms} + \psi_s \quad \text{V}, \tag{6.75}$$

where ψ_s is the surface potential at onset of strong inversion given as

$$\psi_s = 2\phi_b = 2\frac{kT}{q}\ln\frac{N_A}{n_i} \quad \text{V}, \tag{6.76}$$

and the bulk charge Q_{bmax} defined as

$$Q_{b\max} = \frac{\sqrt{2\varepsilon_0\varepsilon_{Si}qN_A\psi_s}}{C_{ox}} + \frac{q\phi_I}{C_{ox}} \quad \text{C/cm}^2. \tag{6.77}$$

The last term in (6.77) allows for an additional threshold adjust implant dose ϕ_I which is assumed to have a delta function profile at the surface. The variance in V_T is determined from the variance in the individual terms on the right-side of (6.75). Since the workfunction difference $\phi_{ms} = \phi_m - \phi_b$ and surface potential ψ_s have a logarithmic dependence on dopant concentrations, they do not contribute appreciably to mismatch and can be approximated as constants. The gate oxide thickness and dielectric constant are typically well-controlled, so C_{ox} can also be assumed constant. This assumption may not, however, apply to high-*K* and composite dielectrics for which no adequate data is available. The variance in V_T is thus mainly related to that of the effective oxide charge Q_{eff} and integrated bulk dopant concentration Q_{bmax}. Under the assumption that Q_{eff} follows a Poisson distribution [91, 92], its variance can be approximated as inversely proportional to gate area

$$\sigma^2_{Qeff} = \frac{qQ_{eff}}{\bar{W}_{eff}\bar{L}_{eff}}. \tag{6.78}$$

There is also a fluctuation in the total dopant concentration due to the random placement of impurity atoms [93–95]. This causes a variance in the mismatch of Q_{bmax} and hence in V_T. Assuming a channel-length independent gate-controlled depletion width x_{dmax}, and that the dopant ions follow a Poisson distribution, the variance in Q_{bmax} due to dopant fluctuations an be expressed as [86]

$$\sigma^2_{Qb\max} = \frac{Q^2_{b\max}}{4\bar{W}_{\text{eff}}\bar{L}_{\text{eff}}x_{d\max}\bar{N}_{\text{eff}}} \propto \frac{Q^2_{b\max}}{\bar{W}_{\text{eff}}\bar{L}_{\text{eff}}}. \tag{6.79}$$

The variance in (6.79) becomes more significant as the channel dimensions are reduced and the number of dopant atoms contained in the channel depletion layer decreases [93–98]. Dopant fluctuation can be considerably reduced by forming a nearly intrinsic channel, such as with thin-film *SOI, FinFETs*, or with a super-steep well profile that confines the channel to a thin, lightly doped region (Chap. 5).

A similar analysis is made for $\sigma_{\Delta\beta}$ by considering that the local variability of t_{ox} is typically negligible and, for long and wide channels $\sigma_{\Delta\beta}$ is essentially dependent on the variability of effective mobility.

Another cause of V_T mismatch is believed to be related to the polysilicon grain-size distribution and the difference in dopant diffusivity between grain and grain-boundary (Fig. 6.39) [99]. Immediately after implanting the gate, source and drain, dopants are distributed within the top part of the polysilicon gate, as illustrated schematically in Fig. 6.39a (possible channeling along grain boundaries is disregarded). During the initial stages of anneal, diffusion proceeds rapidly along the grain boundaries (Fig. 6.39b) [99–108]. Upon further annealing, crystallites become almost uniformly doped, however, leaving randomly distributed lightly-doped regions close to the oxide interface (Fig. 6.39c). It is those regions where polysilicon depletion is more pronounced, resulting in a local increase in V_T and in $\sigma_{\Delta VT}$.

A similar mechanism is observed for arsenic and phosphorus doped polysilicon [99, 100]. It is found that $\sigma_{\Delta VT}$ induced by polysilicon grain-size distribution will only follow an inverse channel area relationship when the channel dimensions are large compared to the polysilicon grain size, which is in the range of 0.2–0.3 μm [99]. When the grains are sufficiently wide to cover the whole channel area, the effect of rapid diffusion along grain boundaries disappears and, if the thermal budget is insufficient to fully dope the grain, a large fraction of the polysilicon gate remains lightly doped resulting in higher V_T. Thermal treatment at a higher thermal budget, however, increases the probability for boron penetration into silicon from highly-doped polysilicon, locally reducing V_T in *PMOS* (Fig. 6.39d). Boron penetration is assumed to be a microscopic random effect and to result in additional fluctuations in V_T, hence an increase in $\sigma_{\Delta VT}$ [100].

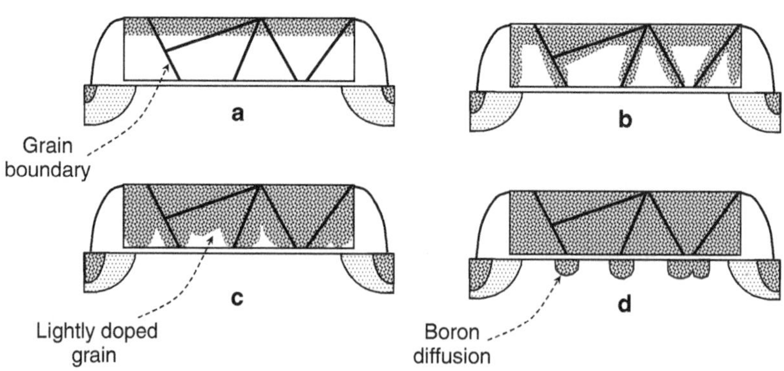

Fig. 6.39 Schematic illustration of diffusion along grain boundaries and within grains. **a** Directly after implant; **b** Fast diffusion along grain boundary; **c** Islands of lightly doped grains; **d** Local boron penetration (Adapted from [99])

6.3.1.2 Validity of the Inverse Area Law

Although the dependence of V_T mismatch on inverse gate area is reported to be valid down to 180 nm dimensions [88], and even to structures of dimensions as small as $W_{\text{eff}}/L_{\text{eff}} = 1\,\mu\text{m}/50\,\text{nm}$ [109], the mismatch does not always follow the Pelgrom-Lakshmikumar law. This is because several mechanisms that were not taken into account when developing (6.73) and (6.74) become significant when channel dimensions are reduced. Among them are the dependence of V_T and I_D on channel dimensions due to short- and narrow-channel effects and their inverse, line-edge roughness, enhanced fluctuation in dopant number, Fermi-level pinning at the polysilicon grain boundaries, voltage drops across source and drain series resistances, field-dependent mobility, and body-factor mismatch.

As discussed in Chap. 5, line edge roughness, *LER*, is a local fluctuation of up to 5 nm in line width caused by the granularity in photoresist and variations in the photon or electron beams [110–112]. *LER* causes a random variation of *MOSFET* gate length along the gate width direction, resulting in additional variance in the mismatch of threshold voltage, drive current, and off-state leakage. As channel dimensions are reduced below approximately 100 nm, the sensitivity of V_T, I_D and I_{off} to *LER* increases. Thus, the amount of *LER* that can be tolerated shrinks with every technology generation [113–116]. The maximum *LER* that can be tolerated is specified as 3 nm for the range 34-nm to 50-nm *MOSFET*s [114, 115].

For small-size *MOSFET*s, the depletion depth x_d in (6.79) can no longer be assumed uniform but is a function of L_{eff}, W_{eff} and V_D (Chap. 5). Fluctuations in channel length and width are thus found to add new terms to the variance of V_T [117]. The total charge within x_d that is controlled by the gate is found to depend on L_{eff} and W_{eff} approximately as [117].

$$\frac{Q_{b\max}^2}{Q_{b\max(0)}^2} \approx 1 - \frac{\text{k1}}{\overline{\text{L}}_{\text{eff}}} + \frac{\text{k2}}{\overline{\text{W}}_{\text{eff}}}, \tag{6.80}$$

where $Q_{bmax(0)}$ is the depletion charge density in long and wide channels and *k1, k2* are process and geometry dependent constants. The signs on the right side of (6.80) are related to short- and narrow-channel effects whereby the total depletion charge controlled by the gate decreases as the channel length is reduced, but increases as the channel width is reduced. For inverse short- and narrow-channel effects, the signs must be interchanged. By substituting (6.79) in (6.78) and retaining the first-order term of a Taylor expansion, the variance in the average L_{eff} and W_{eff} is found to add two terms to the variance in V_T mismatch in small-size *MOSFET*s [117].

$$\sigma_{\Delta VT}^2 \approx \frac{\text{A}_{1\Delta\text{VT}}^2}{\overline{\text{W}}_{\text{eff}}\overline{\text{L}}_{\text{eff}}} + \frac{\text{A}_{2\Delta\text{VT}}^2}{\overline{\text{W}}_{\text{eff}}\overline{\text{L}}_{\text{eff}}^2} - \frac{\text{A}_{3\Delta\text{VT}}^2}{\overline{\text{W}}_{\text{eff}}^2\overline{L}_{\text{eff}}}, \tag{6.81}$$

where A_i in the numerators denote process and geometry-dependent constants. The signs in the last two terms in (6.80) must again be interchanged for inverse short- and narrow-channel effects.

Several mismatch models have been suggested with varying degrees of complexity [118–122]. In [118], it is assumed that the *MOSFET* mismatch is caused by four parameters that affect the ratio $\Delta I_D/I_D$: the threshold-voltage mismatch ΔV_T, the current factor mismatch $\Delta\beta$, the source-drain resistance mismatch ΔR_{SD}, and the body factor mismatch $\Delta\sigma$. The current factor β is defined in (6.72). σ is a factor that reflects the effect of body-bias on I_D. *LER* is taken into account in the calculation of ΔV_T and ΔI_D. It is concluded that in cross-coupled *MOSFET*s, $\sigma(\Delta V_T)$ varies as $(W_{eff}L_{eff})^{-3/4}$ and $\sigma_{\Delta\beta}$ varies as $(W_{eff}L_{eff})^{-1/2}$.

In a suggested model by [119, 120], variations in process parameters such as flatband voltage V_{FB}, equivalent oxide thickness t_{eq}, effective surface mobility μ_{eff}, channel dopant number N, and extrinsic source-drain resistances R_{SD}, are taken into account in addition to variations in channel size dimensions. Correlations, for example, between V_T and β through t_{ox} are also considered. The model analyzes the contributions to mismatch of individual process parameters and allows the identification of the most significant contributors [119].

The local variance in channel length is found to depend on the channel width as

$$\sigma_L^2 \propto \frac{1}{\overline{\mathrm{W}}_{\mathrm{eff}}}, \tag{6.82}$$

implying that as the channel width increases, line edge roughness "averages out" and the mismatch in L_{eff} decreases [120].

Similarly, the local variance of channel width is found to depend on length as.

$$\sigma_W^2 \propto \frac{1}{\overline{L}_{\mathrm{eff}}}. \tag{6.83}$$

For small-size *MOSFET*s, the simple equations (6.82) and (6.83) do not necessarily apply. The local variance of source-drain resistance, channel dopant number, mobility, and oxide thickness are, however, found to depend on the inverse of channel area [120].

6.3.1.3 Other Mechanisms Affecting Mismatch

In addition to dopant non-uniformities created in silicon and in polysilicon by the polysilicon granularity, the grain boundaries are believed to represent surfaces of high interface-state density that cause Fermi-level pinning at different crystallite orientations with respect to the silicon surface (Chap. 5) [123]. Simulations suggest that Fermi-level pinning causes a variance in ΔV_T that increases as the gate-dielectric thickness is reduced. The variance can be comparable to that caused by dopant non-uniformities [123].

Another mechanism affecting mismatch is related to metal upper-level metal coverage of *MOSFET*s that can result in incomplete passivation of interface states [124]. This is demonstrated on a specially designed test structure to measure mismatch, in which only one of the transistors of a matched pair is covered by first or second metal. Figure 6.40 illustrates the impact of metal coverage on mismatch in current which is defined as

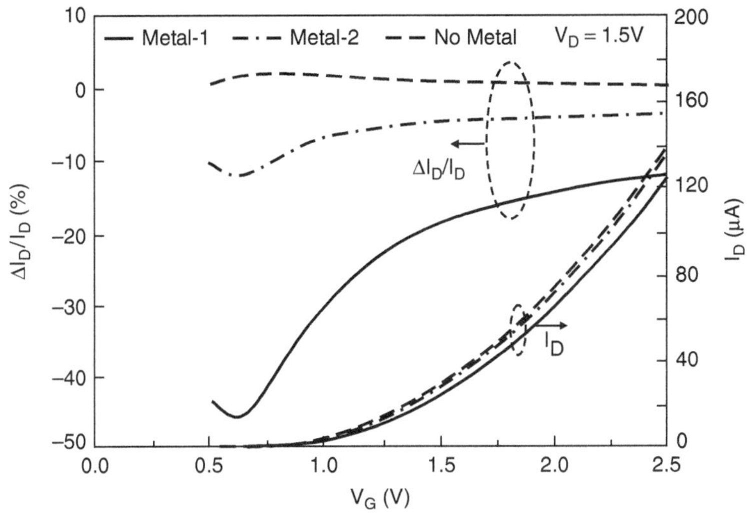

Fig. 6.40 Mismatch of a transistor pair with W/L = 10/10. Dashed line: Both transistors not covered with metal; Solid line: Left transistor covered with metal-1; Dashed-dot line: Left transistor covered with metal-2; body at 0 V (Adapted from [124])

$$\frac{\Delta I_D}{I_D} = 100\,\frac{I_{D-\text{left}} - I_{D-\text{right}}}{I_{D-\text{right}}}\quad\%. \tag{6.84}$$

A large mismatch in drain current is observed when, for example, the left transistor is covered with metal and the right transistor is left uncovered. The mismatch is attributed to incomplete annealing of interface states when a metal cover is present, and to result from a combination of V_T-mismatch and β-mismatch. Independent measurements show that the covered transistor exhibits higher V_T and smaller β than the transistor without metal coverage. Also, the mismatch is higher when the transistor is covered with first-level metal than with second-level metal. This is attributed to the difference in efficiency of annealing interface states. A smaller mismatch is observed when both transistors are symmetrically covered with metal. The mismatch can be considerably reduced by an additional anneal, if compatible with the structure, that is believed to passivate a larger fraction of interface states.

The results in [124] are obtained for zero bias between plate and transistor. During operation, however, there may be a potential difference between plate and transistor causing shifts in V_T and β, resulting in an additional mismatch between the two transistors even when the plates are symmetrically arranged. The shift can be accelerated by the elevated operating temperature, typically in the range 65 °C-125 °C.

6.3.1.4 Extraction of V_T Mismatch

The algorithm used to extract threshold voltage mismatch is another important factor to consider. The maximum-slope linear extrapolation method described in Chap. 5

can be ambiguous since the position and magnitude of transconductance peak depends on several factors, including mobility and source-drain resistance [125]. The fixed current per channel-square method is more precise, provided the fixed current is defined near the onset of strong inversion. The threshold voltage is then the gate voltage where the drain current reaches the value (Chap. 5)

$$I_{D0} = \frac{W_{\text{eff}}}{L_{\text{eff}}} I_x \quad \text{A}, \tag{6.85}$$

where I_x, the drain current per channel square, ranges from 80–400 nA for NMOS, 40–200 nA for *PMOS*, depending on technology.

6.3.2 Bipolar Transistor Mismatch

The mechanisms for bipolar transistor mismatch fluctuations are similar to the local microscopic effects discussed for MOSFETs. In Chap. 3, the collector current of an *NPN* transistor was given at low-level injection and negligible multiplication as

$$I_C = \frac{qA_E}{(kT/q)\int_0^{W_B}(N_A(x)dx)/(\mu_n(x)\,.\,n_i^2(x))}(e^{qV_{BE}/kT}-1) \quad \text{A}, \tag{6.86}$$

where A_E is the area of the emitter of width W_E and length L_E, $N_A(x)$, $\mu_n(x)$, and $n_i(x)$ are, respectively, the position-dependent base dopant concentration, electron mobility and intrinsic carrier concentration in the base, W_B is the base width, and V_{BE} the base-emitter forward voltage. A similar relation applies to a *PNP* transistor. The exact low-level injection biasing conditions vary with emitter area and transistor geometry and profiles, but V_{BE} typically ranges from 0.6 to 0.8 V. Disregarding the effects of emitter, base, and collector resistances, the collector-current ratio of two identically biased transistors is

$$\frac{I_{C1}}{I_{C2}} = \frac{I_{SC1}}{I_{SC2}} = \frac{A_{E1}\int_0^{W_{B2}}(N_A(x))/(\mu_n(x).n_i^2(x))dx}{A_{E2}\int_0^{W_{B1}}(N_A(x))/(\mu_n(x).n_i^2(x))dx}, \tag{6.87}$$

where I_{SC1}, I_{SC2} are, respectively, the collector saturation currents for transistors 1 and 2. Ideally, for identically designed transistors the ratio in (6.87) should be 1. There are, however, variances in several process parameters that cause a mismatch between the two transistors. Among them are: the variance in emitter size, in intrinsic-base dopant profile, and in base width. The base dopant profile and base-width can be monitored by measuring the base pinch resistance, that is, the resistance of the active base [126]. In silicon-germanium (*SiGe*) transistors there are additional variances in the *Ge* profile and the *Ge* concentrations at the emitter and collector depletion boundaries that strongly impact transistor gain and Early voltage.

For a forward voltage in the range 0.6–0.8 V, the *NPN* base current can be assumed to consist predominantly of injected holes into the emitter. This assumption is only valid for conditions where recombination-generation and impact ionization contributions to base current are negligible. The ratio of base currents is then

$$\frac{I_{B1}}{I_{B2}} = \frac{I_{SB1}}{I_{SB2}}, \tag{6.88}$$

where I_{SB1}, I_{SB2} are the base saturation currents of transistor 1 and 2. The mechanisms that contribute to I_{SB} are, however, complicated by the presence of an interface oxide that can be contiguous or "broken" and variations in the emitter-polysilicon morphology that can affect both the emitter junction depth and profile, representing other sources of fluctuation.

The collector and base currents are correlated through the emitter area since they both depend on A_E. The fluctuation in their mismatch depends on emitter area A_E and distance d between transistor centers as [86, 87, 127, 128]

$$\frac{\sigma_{\Delta IC}^2}{\overline{I}_C^2} = \frac{b_{1C}^2}{\overline{W}_E \overline{L}_E} + b_{2C}^2 d^2, \tag{6.89}$$

$$\frac{\sigma_{\Delta IB}^2}{\overline{I}_B^2} = \frac{b_{1B}^2}{\overline{W}_E \overline{L}_E} + b_{2B}^2 d^2, \tag{6.90}$$

where b_i are process and device dependent factors. When defining the variance in current gain $\beta = I_C/I_B$, however, the correlation between I_C and I_B through A_E must be considered [86, 87, 129]. The variance in β is then expressed as

$$\frac{\sigma_{\Delta\beta}^2}{\overline{\beta}^2} = \frac{\sigma_{\Delta IC}^2}{\overline{I}_C^2} + \frac{\sigma_{\Delta IB}^2}{\overline{I}_B^2} - r\frac{\sigma_{\Delta IC}}{\overline{I}_C}\frac{\sigma_{\Delta IB}}{\overline{I}_B}, \tag{6.91}$$

where r is the correlation coefficient between the mismatches in I_C and I_B. The smaller the r, the less the variance in device parameters depends on emitter size fluctuations.

A plot of mismatch versus emitter current shows a larger variance at very low currents where leakage components and variability in barrier heights are significant. The variability in mismatch of I_C and I_B also increases at high emitter current densities, typically $>100\,\mu\text{A}/\mu\text{m}^2$ because of the variability in emitter, base, and collector series resistance across which voltage drops become significant [130]. Between the two extreme regions, mismatch variability is almost constant over several orders of magnitude of the active operating region.

6.3.3 Resistor Mismatch

Mismatch between resistor pairs is caused by variability in sheet resistance, resistor width and length, and end resistance, including contact and edge resistance. The resistance can be expressed as [131]

$$R \approx R_S \frac{(L_D + \Delta L)}{(W_D + \Delta W)} + \frac{2R_{\text{end}}}{(W_D + \Delta W)} + 2R_C, \tag{6.92}$$

where R_S is the sheet resistance of the resistor body, L_D, W_D, respectively, the drawn resistor length and width, ΔL, and ΔW the difference between drawn and effective length and width, and R_{end} the end resistance. For typical resistors, the end resistance is dominated by the spreading resistance at the edge, that is, the transition from silicide to silicon, so that (6.92) can be approximated by adding a δL term to the resistor length as

$$R \approx R_S \frac{(L_D + \Delta L + \delta L)}{(W_D + \Delta W)} = R_S \frac{L_{\text{eff}}}{W_{\text{eff}}}. \tag{6.93}$$

As for a transistor, the variance of resistor mismatch is the sum of local microscopic variability and global systematic variability and can be expressed for wide resistors as [86–89]

$$\frac{\sigma^2_{\Delta R}}{\bar{R}^2} = \frac{A^2_{\Delta R}}{W_{\text{eff}} L_{\text{eff}}} + S^2_{\Delta R} d^2, \tag{6.94}$$

where $A_{\Delta R}$ and $S_{\Delta R}$ are process-related constants and d is the distance between the resistor centers. For closely spaced resistors, the second term in (6.94) becomes negligible. Figure 6.41 shows matching results obtained on a phosphorous-doped polysilicon resistor-pair of 1.1 kOhm/Square sheet resistance, measured at room temperature at three operating points: constant voltage across resistor and constant

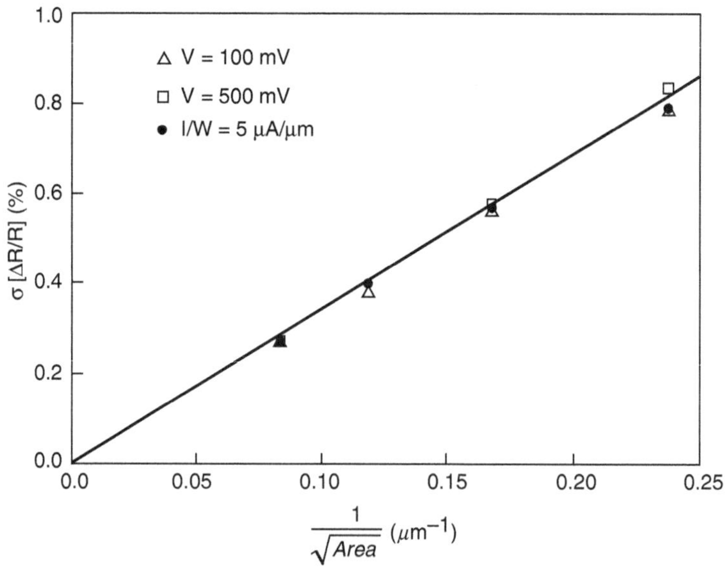

Fig. 6.41 Mismatch variance of a 1.1 kOhm/Square polysilicon resistor as a function of inverse resistor area obtained for three operating points (Adapted from [132])

current per unit width [132]. Matching is found to be independent of operating point in this range. A considerably lower mismatch variance ($<0.1\%$) is found for thin-film resistors (TFR) than for polysilicon resistors [133].

6.3.4 Capacitor Mismatch

The mismatch of integrated capacitor pairs can be approximated by [134]

$$\frac{\sigma_{\Delta C}^2}{\overline{C}^2} \approx \frac{A_{\Delta C}^2}{W_{\text{eff}} L_{\text{eff}}} + \frac{\sigma_W^2}{W_{\text{eff}}^2} + \frac{\sigma_L^2}{L_{\text{eff}}^2}, \tag{6.95}$$

where, for an oxide dielectric, $A_{\Delta C}$ is estimated as 1 nm but can be larger for other dielectrics due to the larger variability in thickness and dielectric constant. σ_W and σ_L represent, respectively, the line-edge variations due to roughness caused lithography and etch and have typically the same values. A curve-fit to measured data is shown in Fig. 6.42. While the mismatch variance of square-shaped capacitors, where the last two terms in (6.95) have an equal effect on area, follows the first-order $1/\sqrt{WL}$-law, an appreciable departure is observed on capacitors with different W and L [134].

Direct measurement of small capacitances can be time-consuming and increasingly inaccurate as the dimensions are reduced. A simple, accurate and fast method to measure small capacitances using the principle of capacitance voltage-division was developed in [135–137], and then later refined to measure the mismatch between capacitor pairs [138–140]. The basic principles of the method are described in Figs. 6.43 and 6.44.

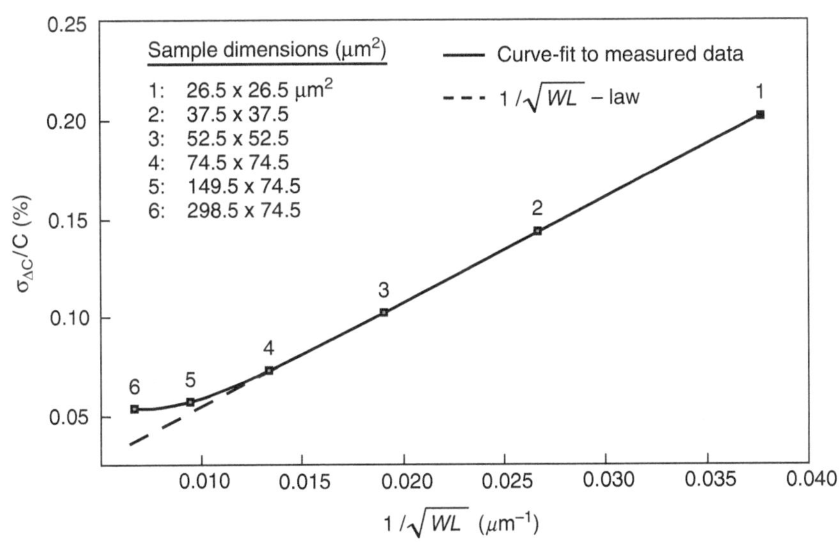

Fig. 6.42 Measured mismatch variance of capacitors versus inverse square-root of area. A departure from the $1/\sqrt{WL}$-law is observed on non-square-shaped capacitors (Adapted from [134])

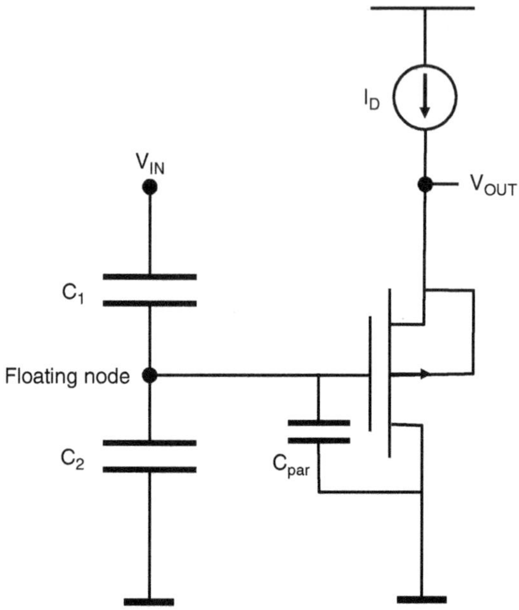

Fig. 6.43 The floating gate capacitance measurement method (Adapted from [137, 139]). C_{par} is the parasitic capacitance

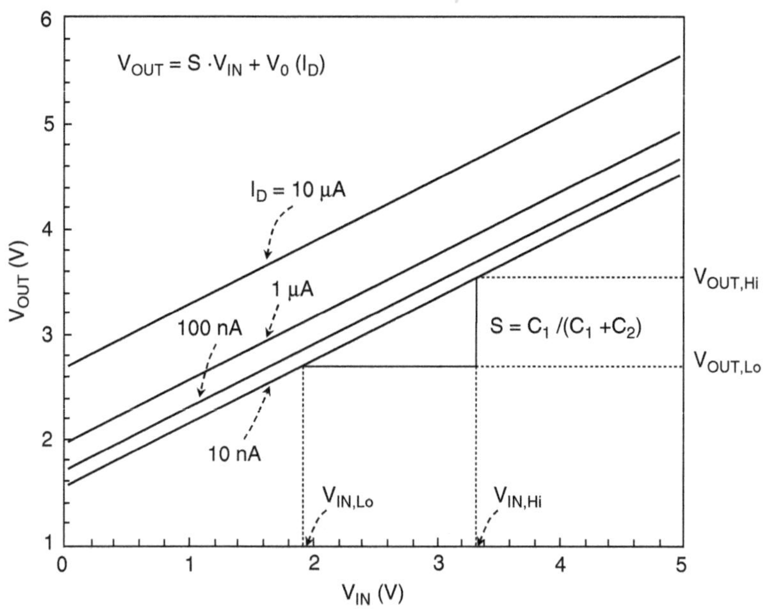

Fig. 6.44 Example of output characteristics of the floating-gate capacitance measurement method and definition of the slope S (Adapted from [137])

The capacitor C_1 to be measured is connected in series with a reference-capacitor of known value C_2 and their common node connected to a floating gate of a "sense-*MOSFET*" operating in a source-follower configuration. The *MOSFET* drain and the second plate of the reference capacitor are grounded. A voltage V_{IN} is applied to the second plate of the capacitor to be measured. A constant current I_D is forced through the *MOSFET*, establishing a fixed gate to source potential that depends on the current amplitude. The voltage on the floating node depends on the capacitance ratio of the two capacitors. In the source-follower configuration, a change in V_{IN} induces the same change in source potential so that, when parasitic capacitances can be neglected, a plot of V_{OUT} versus V_{IN} yields a straight-line of slope (Fig. 6.44)

$$S = \frac{C_1}{C_1 + C_2}, \tag{6.96}$$

from which the unknown capacitance can be extracted as

$$C_1 = C_2 \frac{S}{1 - S}. \tag{6.97a}$$

The floating-gate method was applied to the measurement of matching by determining the difference between two identically designed capacitors instead of the ratio between a known and an unknown capacitor [138, 139]. The relative capacitor mismatch is

$$\frac{\Delta C}{\overline{C}} = \frac{C_1 - C_2}{\overline{C}}, \tag{6.97b}$$

where

$$\overline{C} = \frac{C_1 + C_2}{2}. \tag{6.98}$$

The capacitor mismatch can then be extracted from the slope S as

$$\frac{\Delta C}{\overline{C}} = 4(S - 0.5). \tag{6.99}$$

When the capacitor mismatch drops below about 0.05%, the impact of the parasitic capacitance C_{par} on accuracy becomes noticeable. To improve the measurement accuracy, a double-slope measurement is implemented in which the slope is measured twice: first with C_1 connected to V_{IN} (C_2 at ground), resulting in a slope S_1, and then with C_2 connected to V_{IN} and C_1 to ground giving a slope S_2 [4.139]. Assuming that the parasitic capacitance C_{par} is the same for both configurations,

$$S_1 = \frac{C_1}{C_1 + C_2 + C_{\mathrm{par}}}, \tag{6.100}$$

and

$$S_2 = \frac{C_2}{C_1 + C_2 + C_{\mathrm{par}}}. \tag{6.101}$$

The mismatch is then

$$\frac{\Delta C}{\overline{C}} = 2\frac{S_1 - S_2}{S_1 + S_2} = 2\frac{C_1 - C_2}{C_1 + C_2}, \tag{6.102}$$

which is independent of C_{par}. A more sophisticated differential floating-gate mismatch measurement technique which is believed to further improve the resolution is detailed in [140].

6.4 Noise

Noise in electronic circuits is defined as the random fluctuation in signal current or voltage (Fig. 6.45). It is the main limitation of the accuracy of a measuring device and ultimately sets a lower limit on signals that can be detected and processed [141, 142]. The average value of current noise i_n obtained by integrating over sufficient time is zero and therefore not useful for noise analysis. Instead, the mean square of noise current $\overline{i_n^2}$ or voltage $\overline{v_n^2}$ is used to describe the noise power. The noise power is found to vary with frequency, particularly in the low and very high frequency ranges. A measure of how it is distributed over frequency is the power spectral density defined as

$$S_i = \frac{\overline{i_n^2}}{\Delta f} \quad \mathrm{A^2/Hz}, \tag{6.103a}$$

or

$$S_v = \frac{\overline{v_n^2}}{\Delta f} \quad \mathrm{V^2/Hz}, \tag{6.103b}$$

where Δf is the bandwidth, that is, the frequency range over which noise is measured.

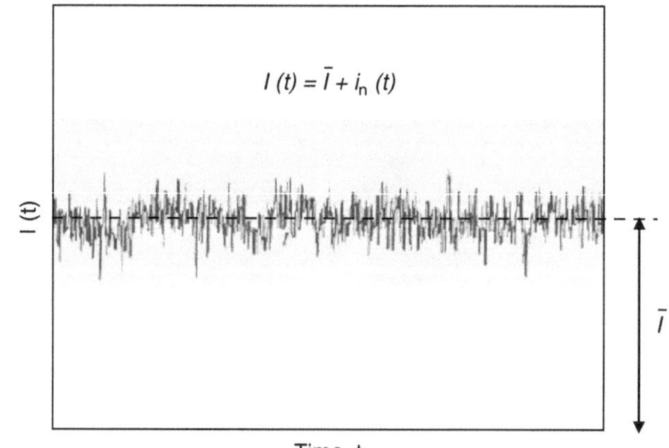

Fig. 6.45 Illustration of current noise, $i_n(t)$, superimposed on an average signal current $\bar{I}$

6.4.1 Classification of Noise

Noise sources are classified as thermal noise, shot noise, generation-recombination noise, and flicker noise also known as *1/f* noise.

6.4.1.1 Thermal Noise

Thermal noise, also called Nyquist or Johnson noise, is due to the random motion of carriers in a conductive medium of resistance R. The average thermal carrier velocity at a temperature T is (Chap. 1)

$$\overline{v_{th}} = \sqrt{\frac{3kT}{m^*}} \approx 10^7 \text{cm/s} \quad \text{at} \quad 300\,\text{K}, \tag{6.104}$$

where m^* is the carrier effective mass and $k = 8.62 \times 10^{-5}\,\text{eV/K} = 1.38 \times 10^{-23}\,\text{J/K}$. In the absence of an electric field, on the average, the centroid of carriers does not move in a specific direction. In short periods of time, however, more carriers can move in one direction than the other, causing the noise. When a low to medium electric field is applied, a current is induced by drifting carriers along the field but the drift velocity is much smaller than the thermal velocity. Thus, thermal noise is unaffected by the presence or absence of direct current and is directly proportional to temperature [141]

$$S_i = \frac{\overline{i_n^2}}{\Delta f} = \frac{4kT}{R} \quad \text{A}^2/\text{Hz}, \tag{6.105a}$$

or

$$S_V = \frac{\overline{v_n^2}}{\Delta f} = 4kTR \quad \text{V}^2/\text{Hz}. \tag{6.105b}$$

For example, the thermal noise spectral density measured in a resistor of $R =$ 1 KOhm at 300 K is

$$\sqrt{\overline{v_n^2}} = \sqrt{S_V \Delta f} = \frac{4nV}{\sqrt{Hz}}$$

From (6.105a) and (6.105b), it is concluded that the thermal-noise spectral-density is independent of frequency and hence often called white noise. This is found to be the case up to a frequency of 10^{13} Hz [141]. The noise level is sometimes specified by a temperature equivalent as

$$T_N = \frac{\overline{v_n^2}}{4kR\Delta f}, \tag{6.106}$$

or a resistance-equivalent

$$R_N = \frac{\overline{v_n^2}}{4kT_0\Delta f}, \tag{6.107}$$

where $T_0 = 290\,\mathrm{K}$ is the standard noise temperature. If only thermal noise is present then T_N and R_N are the actual temperature and resistance extracted from (6.105). Otherwise, they can be higher.

6.4.1.2 Shot Noise

Shot noise is a random event related to energy of individual carriers that are transported over a barrier and is present in diodes, *MOSFETs* and bipolar transistors. In contrast to thermal noise, shot noise is always associated with direct current and depends on the carrier energy and velocity directed toward the barrier. Thus, the current I that appears to be at a steady level actually consists of random current pulses. The fluctuation in I is called shot noise of spectral power density

$$\frac{\overline{i^2}}{\Delta f} = 2q\bar{I} \quad \mathrm{A}^2/\mathrm{Hz}, \tag{6.108}$$

where $\bar{I}$ is the average current. For a pn junction, (6.108) holds as long as the frequency f is lower than $1/\tau$ where τ is the carrier transit time through the depletion region. Since typically $\tau < 10\,\mathrm{ps}$, the shot noise remains frequency-independent (white spectrum) up to frequencies in the higher GHz range [141, 142].

6.4.1.3 Generation-Recombination Noise

The Shockley-Read-Hall (*SRH*) generation-recombination process was described in Chap. 1. It was found to be associated with localized electronic states, called "traps," with energies in the forbidden gap either at the surface or in the bulk. The traps can randomly capture or emit carriers. Trapping and detrapping charge can cause fluctuations in current (or voltage) as a result of fluctuation in the number of carriers and their mobility. Consider, for example, the inversion layer of an *NMOS* and assume for simplicity that the drain voltage is very small and the *MOSFET* operates in the linear mode. The average inversion resistance per unit channel width is (Chap. 5):

$$\bar{R} = \frac{L}{q\overline{N\mu}} \quad \mathrm{Ohm/square}, \tag{6.109}$$

where N is the number of electrons per unit area and μ is the individual carrier mobility. A small drain field E across the resistor produces an average current

$$\bar{I} = q\overline{N\mu}\,E \quad \mathrm{A/cm}. \tag{6.110}$$

Both the number of carriers and the carrier mobility can fluctuate as [142]

$$N(t) = \bar{N} \pm \Delta N(t), \tag{6.111a}$$

$$\mu(t) = \bar{\mu} \pm \Delta\mu(t), \tag{6.111b}$$

The generation-recombination (GR) noise is discussed in terms of number fluctuations and mobility fluctuations. Trapping and detrapping of carriers can cause fluctuations in the number of an ensemble of carriers by modifying the threshold voltage, and fluctuations in carrier mobility by modifying the vertical field [143]. Thus, the fluctuations in number of carriers and their mobility are correlated [144]. A similar reasoning applies to *JFET*s, bipolar transistors and passive devices.

6.4.1.4 Random-Telegraph Signal Noise

Random telegraph signal (RTS) noise, also known as "popcorn" or "burst" noise is a special case of generation-recombination noise, observed at low frequencies [142–150]. In *MOSFET*s of small channel dimensions, only a few traps are present. Individual events of trapping and detrapping of carriers can then be observed to cause the current to switch randomly between two levels, spending an average time τ_h in the high state and τ_l in the low state [151]. This results in a waveform similar to that of a random telegraph signal, displayed as a function of time in Fig. 6.46b.

When multiple interface trap-levels are "active," the RTS exhibits more than two current levels. A single such random-telegraph-signal has a frequency-dependent Lorentzian-shaped power spectral density as shown in Fig. 6.47 [141, 142, 148, 150, 152, 153]

$$S_i = \frac{\overline{i^2}}{\Delta f} \approx A \frac{\overline{I^\gamma}}{1+(f/f_c)^2} \quad \mathrm{A^2/Hz}, \tag{6.112}$$

where A is a frequency-independent process constant, $\gamma = 0.5$–2, and

$$f_c = \frac{1}{2\pi}\left(\frac{1}{\tau_h}+\frac{1}{\tau_l}\right) \quad \mathrm{Hz}. \tag{6.113}$$

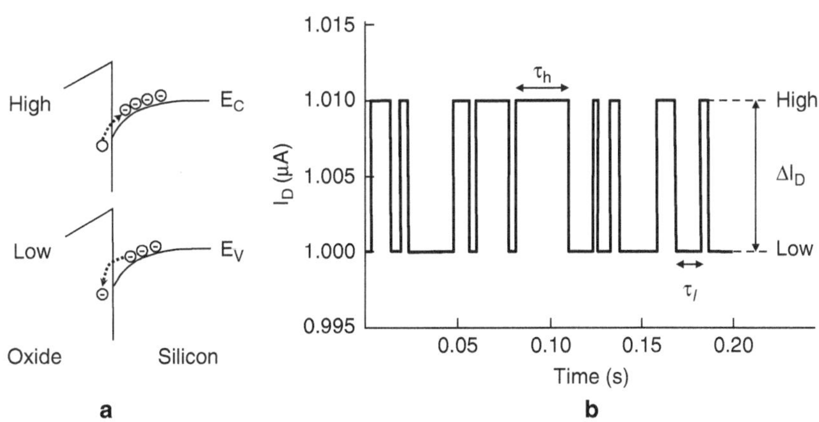

Fig. 6.46 Schematic representation of **a** electron trapping and detrapping, **b** RTS-like noise in NMOS drain current (Adapted from [142])

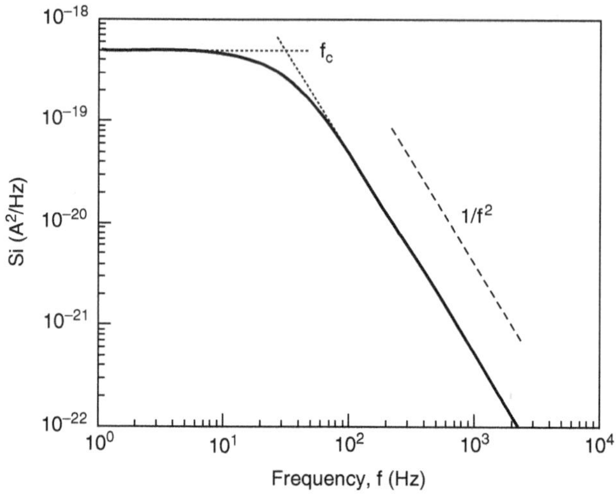

Fig. 6.47 Lorentzian-shaped power spectral density for RTS-like noise in Fig. 6.46 [141]

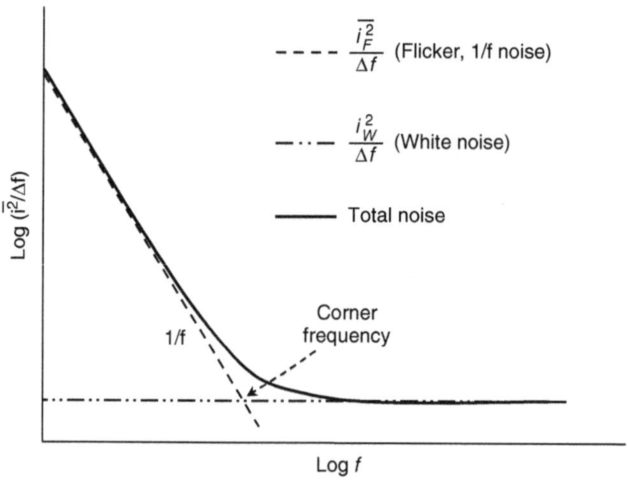

Fig. 6.48 Schematic representation of flicker noise spectral density versus frequency

6.4.1.5 Flicker or 1/f Noise

Flicker noise is a low-frequency noise that is known to result from carrier trapping and detrapping by electronic states within the forbidden gap in the semiconductor bulk and at the surface. The flicker-noise spectral density follows a $1/f^{\alpha}$-law where $\alpha \approx 1$ (Fig. 6.48).

Flicker noise is always associated with direct current and has a spectral density of $1/f$ noise of a general form [153, 154]

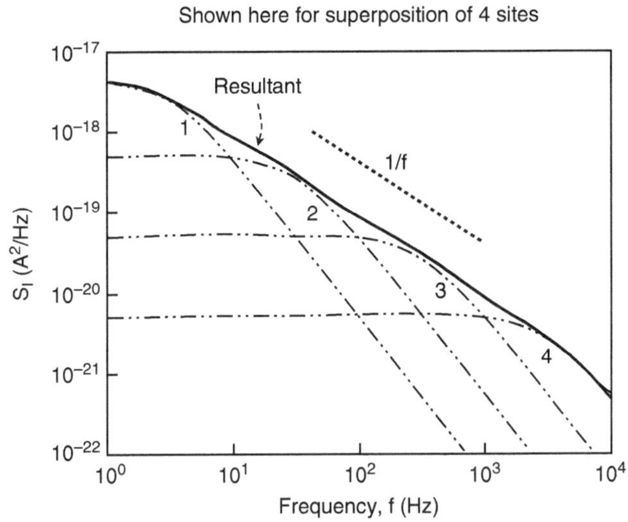

Fig. 6.49 Superposition of four "Lorentzians" resulting in a power spectral density that approximately follows a 1/f frequency dependence (Adapted from [142])

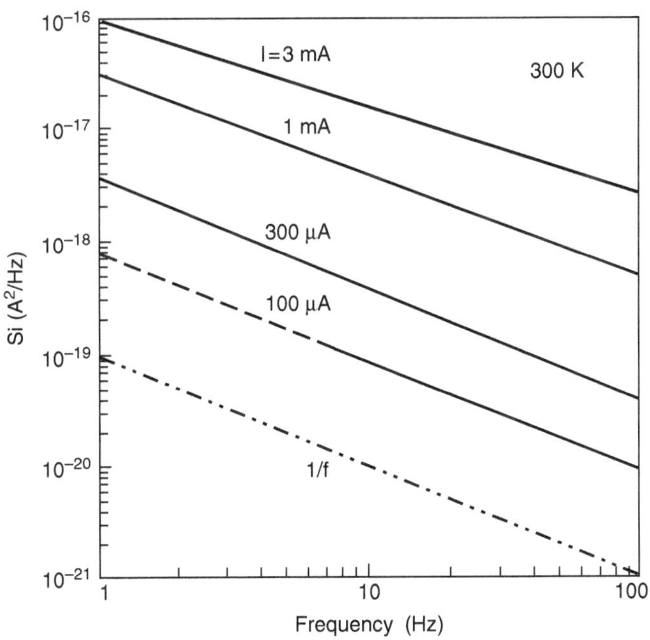

Fig. 6.50 Current noise spectra for high-sheet polysilicon resistor for several currents and $W/L = 4/7.1$. Solid lines are curve fits to actual data (Adapted from [155])

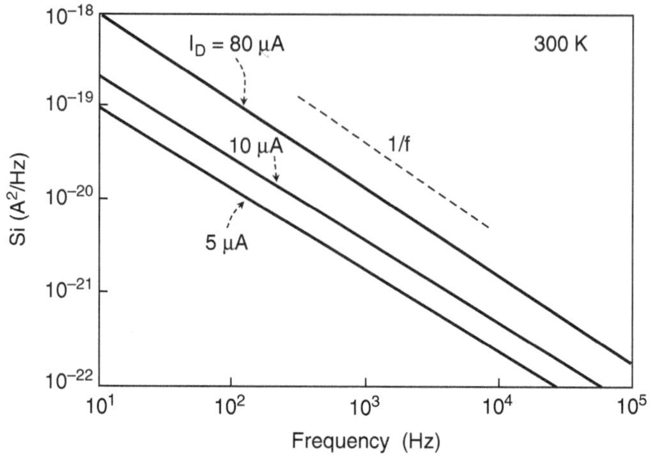

Fig. 6.51 Low-frequency noise results on an *NMOS* with $W/L = 4/8$ for three drain currents. Solid lines are curve fits to actual data and follow the $1/f$ trend (Adapted from [149])

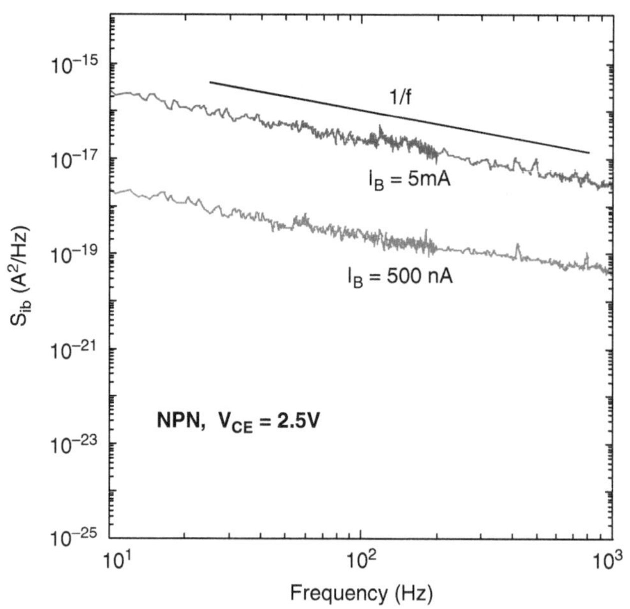

Fig. 6.52 Low-frequency noise measurements on an *NPN SiGe*-base transistor for two base currents I_B [156]

$$\frac{\overline{i^2}}{\Delta f} = K_1 \frac{I^a}{f^b} \quad A^2/\text{Hz}, \tag{6.114}$$

where a is a constant ranging form 0.5–2, b $\approx$ 1, and K_l is a process-dependent parameter.

The information gained from *RTS*-like noise is very valuable to modeling the flicker noise [144]. It is now believed that the $1/f$ characteristic of low-frequency noise is the result of superposition of multiple "Lorentzians" of the type shown in Fig. 6.47 with distributed time-constants [142, 149], as illustrated in Fig. 6.49 for the superposition of 4 "Lorentzians" of different time-constants. Figures 6.50–6.52 are examples of measured flicker-noise on a polysilicon resistor [155], a *MOSFET* [149], and a bipolar transistor [156].

The value K_l in (6.114) typically varies from one device to another and across one wafer. This is because the flicker noise strongly depends on the trap density, in particular the surface state density. Therefore, to reduce the flicker noise, it is important to minimize the interface trap density and contamination.

6.5 Problems

The temperature is 300 K unless otherwise stated.

1. The concentration in a double-gated *JFET* channel is uniform at $2 \times 10^{16}\,\text{cm}^{-3}$. Assume one-sided abrupt junctions and calculate the distance between the gate metallurgical junctions that will yield a threshold voltage $|V_T| = 1.4\,\text{V}$.

2. The channel doping profile of a double-gated *JFET* can be approximated by two half-Gaussian distributions as shown in the figure below. Assume one-sided abrupt gate junctions. Calculate the pinch-off voltage.

3. For a channel length $L_{eff} = 5\,\mu\text{m}$ in problem 1, calculate the channel resistance for $V_G = V_S = 0\,\text{V}$ and V_D at onset of pinch-off.

4. The lateral channel profile in an asymmetrical NMOS of effective channel length $L_{eff} = 0.1\,\mu\text{m}$ can be approximated by a superposition of a Gaussian boron distribution and a uniform boron concentration of $10^{17}\,\text{cm}^{-3}$. The Gaussian profile peaks at $10^{18}\,\text{cm}^{-3}$ at the source and drops to $10^{17}\,\text{cm}^{-3}$ in the middle of the channel. Assume a one-sided abrupt source-channel junction and plot the built-in field in the channel as a function of distance from the source.

5. Estimate the amount of decoupling capacitance required to keep the voltage noise of a switching circuit to less than 5% of the power supply voltage V_{DD}. Assume 100 W circuit power dissipation, a clock-frequency of 1 GHz, and $V_{DD} = 1.4\,\text{V}$. What is the required capacitance per unit area to fit the decoupling capacitor within an area of $2\,\text{cm}^2$?

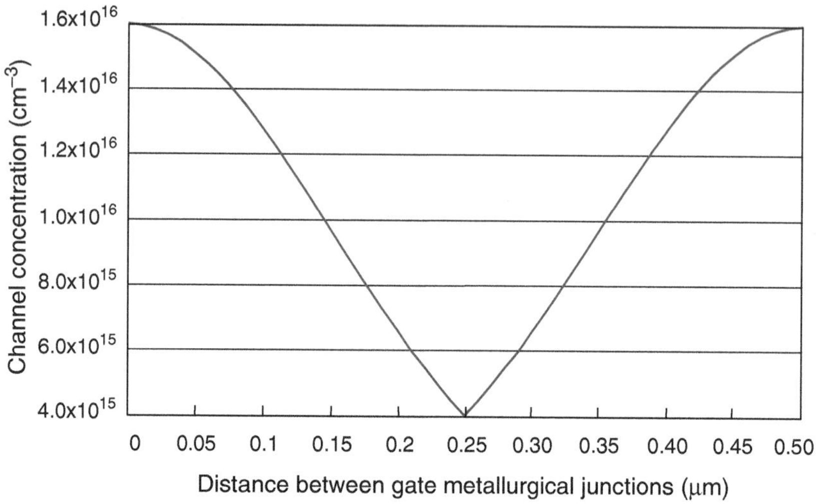

6. The figure below is a schematic of a precision analog capacitor with SiO_2 as a dielectric, designed in a cross-coupled arrangement. Show that this arrangement reduces the voltage coefficient of capacitance when compared to a single equivalent capacitor having the same dielectric.

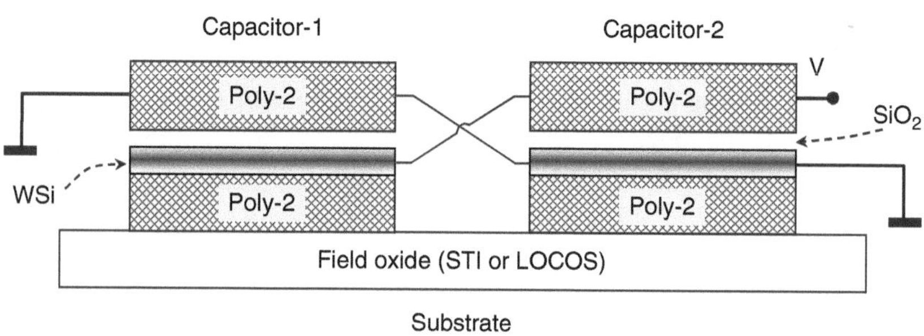

References

1. J. E. Lilienfeld, "Method and Apparatus for Controlling Electric Current," US patent #1,745,175, issued January 28, 1930. Filed: Canadian application October 1925, US application October 1926.
2. J. E. Lilienfeld, "Device for Controlling Electric Currents," US patent 1,900,018, issued March 7, 1933, filed March 1928.
3. O. Heil, "Improvements in or Relating to Electrical Amplifiers and other Control Arrangements," U.K. patent 439, 457, issued December 1935, filed March 1935.

4. W. Shockley, "A unipolar 'field-effect' transistor," Proc. IEEE, 40 (11), 1365–1376, 1952.
5. J. C. Guo, "Halo and LDD engineering for multiple V_{TH} high performance analog CMOS devices," IEEE Trans. Semcon. Manuf., 20 (3), 313–322, 2007.
6. E. A. Vittoz, "Future of analog in the VLSI environment," Proc. ISCASM, 1372–1375, 1990.
7. K. Bult, "Analog broadband communication circuits in pure digital deep sub-micron CMOS," IEEE ISSCC Tech. Dig., 76–78, 1999.
8. J. J. P. Bruines, "Process outlook for analog and RF applications," Microelectron. Eng., 54, 35–48, 2000.
9. N. H. E. West and K. Eshraghian, Principles of CMOS VLSI Design, Addison-Wesley Publishing Company, 1993.
10. T. N. Buti, S. Ogura, N. Rovedo, and K. Tobimatsu, "A new asymmetrical halo source GOLD drain (HS-GOLD) deep sub-half-micrometer n-MOSFET design for reliability and performance," IEEE Trans. Electron Dev., 38 (8), 1757–1764, 1991.
11. H. Shin and L. Lee, "A 0.1-μm asymmetric halo by large-angle-tilt implant (AHLATI) MOSFET for high performance and reliability," IEEE Trans. Electron Dev., 46 (4), 820–822, 1999.
12. B. Cheng, V. R. Rao, and J. C. S. Woo, "Exploration of velocity overshoot in a high-performance deep sub-0.1-μm SOI MOSFET with asymmetric channel profile," IEEE Electron Dev. Lett., 20 (10), 538–540, 1999.
13. D. G. Borse, K. N. M. Rani, N. K. Jha, A. N. Chandorkar, J. Vasi, V. R. Rao, B. Cheng, and J. C. S. Woo, "Optimization and realization of sub-100-nm channel length single halo p-MOSFETs," IEEE Trans. Electron Dev., 49 (6), 1077–1079, 2002.
14. H. V. Deshpande, B. Cheng, and J. C. S. Woo, "Channel engineering for analog device design in deep submicron CMOS technology for system on chip applications," IEEE Trans. Electron Dev., 49 (9), 1558–1565, 2002.
15. K. Narasimhulu, D. K. Sharma, and V. R. Rao, "Impact of lateral asymmetric channel doping on deep submicrometer mixed-signal device and circuit performance," IEEE Trans. Electron Dev., 50 (12), 2481–2488, 2003.
16. K. Narasimhulu, M. P. Desai, S. G. Narendra, and V. R. Rao, "The effect of LAC doping on deep submicrometer transistor capacitances and its influence on device RF performance," IEEE Trans. Electron Dev., 51 (9), 1416–1422, 2004.
17. J. H. Song, Y. J. Park, and H. S. Min, "Drain current enhancement due to velocity overshoot effects and its analytical modeling," IEEE Trans. Electron Dev., 43 (11), 1870–1974, 1996.
18. M. Lundstrom, "Elementary scattering theory of the Si MOSFET," IEEE Electron Dev. Lett., 18 (7), 361–363, 1997.
19. H. S. Shin, C. Lee, S. W. Wang, B. G. Park, and H. S. Min, "Channel length independent subthreshold characteristics in submicron MOSFETs," IEEE Electron Dev. Lett., 19 (4), 137–139, 1998.
20. E. Vittoz and J. Fellrath, "CMOS analog integrated circuits based on weak inversion operation," IEEE J. Solid-State Circuits, SC-12 (3), 224–231, 1977.
21. B. C. Paul, A. Raychowdhury, and K. Roy, "Device operation for digital subthreshold logic operation," IEEE Trans. Electron Dev., 52 (2), 237–247, 2005.
22. S. Chakraborty, A. Mallik, C. K. Sarkar, and V. R. Rao, "Impact of halo doping on the subthreshold performance of deep-submicron CMOS devices and circuits for ultra-low power analog/mixed-signal applications," IEEE Trans. Electron Dev., 54 (2), 241–247, 2007.
23. R. W. Berry, P. M. Hall, and M. T. Harris, Thin Film Technology, Van Nostrand, 1968.
24. W.-C. Liu, K.-B. Thei, H.-M. Chuang, K.-W. Lin, C.-C. Cheng, Y.-S. Ho, C.-W. Su, S.-C. Wong, C.-H. Lin, and C. H. Diaz, "Characterization of polysilicon resistors in sub-0.25 μm CMOS ULSA applications," IEEE Electron Dev. Lett., 22 (7), 318–320, 2001.
25. J. Y. Seto, "The electrical properties of polycrystalline silicon films," J. Appl. Phys., 46 (12), 5247–5254, 1975.
26. N. C.-C. Lu, L. Gerzberg, C.-Y. Lu, and J. D. Meindl, "Modeling and optimization of monolithic polycrystalline silicon resistors," IEEE Trans. Electron Dev., ED-28 (7), 818–830, 1981.
27. M. M. Mandurah, K. C. Saraswat, and T. I. Kamins, "Phosphorus doping of low pressure chemically vapor-deposited silicon films," J. Electrochem. Soc., 126 (8), 1019–1023, 1979.

28. N. C. C. Lu, L. Gerzberg, and J. D. Meindl, "A quantitative model of the effect of grain size on the resistivity of polycrystalline silicon resistors," Electron Dev. Lett., EDL-1 (3), 38–41, 1980.
29. T. I. Kamins, "Hall mobility in chemically deposited polycrystalline silicon," J. Appl. Phys., 42 (11), 4357–4365, 1971.
30. M. E. Cowher and T. O. Sedgwick, "Chemical vapor deposited polycrystalline silicon," J. Electrochem. Soc., 119 (11), 1565–1570, 1972.
31. A. L. Fripp, "Dependence of resistivity on the doping level of polycrystalline silicon," J. Appl. Phys., 46 (3), 1240–1244, 1975.
32. P. Rai-Choudhury and P. L. Hower, "Growth and characterization of polycrystalline silicon," J. Electrochem. Soc., 120 (12), 1761–1766, 1971.
33. G. Baccarani and B. Riccò, "Transport properties of polycrystalline silicon films," J. Appl. Phys., 49 (11), 5565–5570, 1978.
34. M. Nakabayshi, M. Ikegami, and T. Daikoku, "Influence of hydrogen on electrical characteristics of poly-Si resistor," Jpn. J. Appl. Phys., 32, Part 1 (9A), 3734–3738, 1993.
35. F. Hegner, "The industrial production of high quality nickel-chromium resistors with controlled temperature coefficient of resistance," Thin Solid Films, 57 (2), 359–362, 1979.
36. G. Nocerino and K. E. Singer, "The electrical and compositional structure of thin Ni-Cr films," Thin Solid Films, 57 (2), 343–348, 1979.
37. M. A. Bayne, "Al-doped Ni-Cr for temperature coefficient of resistance control in hybrid thin-film resistors," J. Vac. Sci. Technol. A4 (6), 3142–3145, 1986.
38. F. Wu, A. W. McLaurin, K. E. Henson, D. G. Managhan, and S. L. Thomasson, "The effects of the process parameters on the electrical and microstructure characteristics of the CrSi thin resistor films: part I," Thin Solid Films, 332, 418–422, 1998.
39. D. Nachrodt, U. Pachen, A. Ten Have, and H. Vogt, "Ti/Ni(80%)Cr(20%) thin-film resistor with a near zero temperature coefficient of resistance for integration in a standard CMOS process," IEEE Electron Dev. Lett., 29 (3), 212–214, 2008.
40. P. Zurcher, P. Alluri, P. Chu, A. Duvallet, C. Happ, R. Henderson, J. Mendonca, M. Kim, M. Petras, M. Raymond, T. Remmel, D. Roberts, B. Steimle, J. Stipanuk, S, Straub, T. Sparks, M. Tarabbia, H. Thibieroz, and M. Miller, "Integration of thin film MIM capacitors and resistors into copper metallization based RF-CMOS and Bi-CMOS technologies," IEEE IEDM Tech. Dig., 153–156, 2000.
41. P. Fehlhaber, "Laser trimming of SiCr thin-film resistors," IEEE IEDM Tech. Dig., 9–10, 1969.
42. R. H. Wagner, "Functional laser trimming: An overview," Laser Processing of Semiconductors and Hybrids, SPIE Proceedings, 611, 8–17, 1986.
43. M. J. Mueller and W. Mickanin, "Functional laser trimming of thin film resistors on silicon IC," Laser Processing of Semiconductors and Hybrids, SPIE Proceedings, 611, 70–83, 1986.
44. H.-M. Chuang, K.-B. Thei, S.-F. Tsai, and W.-X. Liu, "Temperature-dependent characteristics of polysilicon and diffused resistors," IEEE Trans. Electron Dev., 50 (5), 1413–1415, 2003.
45. W. A. Lane and G. T. Wrixon, "The design of thin-film polysilicon resistors for analog IC applications," IEEE Trans. Electron Dev., 36 (4), 738–744, 1989.
46. D. W. Lee, T. M. Roh, H. S. Park, J. Kim, J. G. Koo, and D. Y. Kim, "Fabrication technology of polysilicon resistors using novel mixed process for analogue CMOS application," IEEE Electron Lett., 35 (7), 803–804, 1999.
47. M. S. Raman, T. Kifle, E. Bhattacharya, and K. N. Bhat, "Physical model for the resistivity and temperature coefficient of resistivity in heavily doped polysilicon," IEEE Trans. Electron Dev., 53 (8), 1885–1892, 2006.
48. P. Steinmann, S. M. Stuart, R. Higgins, "Controlling the TCR of thin film resistors," Euro. Dev. Res. Conf., 451–453, 2000.
49. C. A. Neugebauer and M. B. Webb, "Electrical conduction mechanism in ultrathin, evaporated metal films," J. Appl. Phys., 33, 74–82, 1962.
50. J. R. Sambles and T. W. Preist, "The effects of surface scattering upon resistivity," J. Phys. F: Met. Phys., 12, 1971–1987, 1982.

51. Y. Amemiya, T. Ono, and K. Kato, "Electrical trimming of heavily doped polycrystalline silicon resistors," IEEE Trans. Electron Dev., ED-26 (11), 1738–1742, 1979.
52. K. Kato, T. Ono, and Y. Amemiya, "A physical mechanism of current-induced resistance decrease in heavily doped polysilicon resistors," IEEE Trans. Electron Dev., ED-29 (8), 1156–1161, 1982.
53. S. Das and S. K. Lahiri, "Electrical trimming of ion-beam-sputtered polysilicon resistors by high current pulses," IEEE Trans. Electron Dev., 41 (8), 1429–1434, 1994.
54. C. T. Black, K. W. Guarini, Y. Zhang, H. Kim, J. Benedict, E. Sikorski, I. V. Babich, and K. R. Milkove, "High-capacity, self-assembled metal-oxide-semiconductor decoupling capacitor," IEEE Electron Dev. Lett., 25 (9), 622–624, 2004.
55. T.-I. Liou and C.-S. Teng, "n+-poly-to-n+-silicon capacitor structure for single poly analog CMOS and BiCMOS process," IEEE Trans. Electron Dev., 36 (9), 1620–1628, 1989.
56. S. A. St Onge, S. G. Franz, A. F. Puttlitz, A. Kalinoski, B. E. Johnson, and B. El-Kareh, "Design of precision capacitors for analog applications," IEEE Trans. Comput. Hybrids Manuf. Tech., 15 (6), 1064–1070, 1992.
57. J. L. McCreary, "Matching properties, and voltage and temperature dependence of MOS capacitors," IEEE J. Solid-State Circuits, SC-14 (6), 608–616, 1987.
58. C. Kaya, H. Tigelaar, J. Peterson, M. De Wit, J. Fattaruso, D. Hester, S. Kiriaki, K.-S. Tan, and F. Tsay, "Polycide/metal capacitors for high precision A/D converters," IEEE IEDM Tech. Dig., 782–785, 1988.
59. J. A. Babcock, S. G. Balster, A. Pinto, C. Dirnecker, P. Steinmann, R. Jumpertz, and B. El-Kareh, "Analog characteristics of metal-insulator-metal capacitors using PECVD nitride dielectrics," IEEE Trans. Electron Dev., 22 (5), 230–232, 2001.
60. K. Stein, G. Hueckel, E. Eld, T. Bartush. R. Groves, N. Greco, and D. Harame, "High reliability metal insulator metal capacitors for silicon germanium analog applications," IEEE BCTM Tech. Dig., 191–194, 1997.
61. T. Yoshitomi, Y. Ebuchi, H. Kimijima, T. Ohguro, E. Morifuji, H. S. Momose, K. Kasai, K. Ishimaru, F. Matsuoka, Y. Katsumata, M. Kinugawa, and H. Iwai, "High performance MIM capacitor for RF BiCMOS/CMOS LSI," IEEE BCTM Tech. Dig., 133–136, 1999.
62. T. Ishikawa, D. Kodama, Y. Matsui, M. Hiratani, T. Furusawa, and D. Hisamoto, "High-capacitance $Cu/Ta_2O_5/Cu$ MIM structure for SoC applications featuring a single-mask add-on process," IEEE IEDM Tech. Dig., 940–942, 2002.
63. C. Zhu, H. Hu, X. Yu, S. J. Kim, A. Chin, M. F. Li, B. J. Cho, and D.-L. Kwong, "Voltage and temperature dependence of capacitance of high-K HfO_2 MIM capacitors: A unified understanding and prediction," IEEE IEDM Tech. Digest, 879–882, 2003.
64. H. Hu, S.-J. Ding, H. F. Lim, C. Zhu, M. F. Li, S. J. Kim, X. F. Yu, J. H. Chen, Y. F. Yong, B. J. Cho, D. S. H. Chan, S. C. Rustagi, M. B. Yu, C. H. Tung, A. Du, D. My, P. D. Foo, A. Chin, and D.-L. Kwong, "High performance ALD HfO_2-Al_2O_3 laminate MIM capacitor for RF and mixed signal IC applications," IEEE IEDM Tech. Dig., 379–382, 2003.
65. S. J. Kim, B. J. Cho, M. B. Yu, M.-F. Li, Y.-Z. Xiong, C. Zhu, A. Chin, and D.-M. Kwong, "Metal-insulator-metal RF bypass capacitor using niobium oxide (Nb_2O_5) with HfO_2/Al_2O_3 barriers," IEEE Electron Dev. Lett., 26 (9), 625–627, 2005.
66. K. C. Chiang, C.-C. Huang, G. L. Chen, W. J. Chen, H. L. Kao, Y.-H. Wu, A. Chin, and S. P. McAlister, "High performance $SrTiO_3$ MIM capacitors for analog applications," IEEE Trans. Electron Dev., 53 (9), 2312–2319, 2006.
67. K. Hyyppä, "Dielectric absorption in memory capacitors," IEEE Trans. Instrum. Meas., 21 (1), 53–56, 1972.
68. J. C. Kuenen and G. C. M. Meijer, "Measurement of dielectric absorption of capacitors and analysis of its effects on VCO," IEEE Trans. Instrum. Meas., 45 (1), 89–97, 1996.
69. C. Iorga, "Compartmental analysis of dielectric absorption in capacitors," IEEE Trans. Dielectrics, 7 (2), 187–192, 2000.
70. P. Andreani and S. Mattisson, "On the use of MOS varactors in RF VCOs," IEEE J. Solid-State Circuits, 35 (6), 905–910, 2000.

71. J. Maget, R. Kraus, and M. Tiebout, "A physical model of a CMOS varactor with high capacitance tuning range and its application to simulate integrated VCOs," Solid-State Electron, 46, 1609–1615, 2002.
72. C.-S. Chang, C.-P. Chao, J. G. J. Chern, and J. Y.-C. Sun, "Advanced CMOS technology portfolio for RF IC applications," IEEE Trans. Electron Dev., 52 (7), 1324–1334, 2005.
73. B. El-Kareh, S. Balster, W. Leitz, P. Steinmann, H. Yasuda, M. Corsi, K. Dawoodi, C. Dirnecker, P. Foglietti, A. Haeusler, P. Menz, M. Ramin, T. Scharnagl, M. Schiekofer, M. Schober, U. Schulz, L. Swanson, D. Tatman, M. Waitschull, J. W. Weijtmans, and C. Willis, "A 5V complementary-SiGe BiCMOS technology for high-speed precision analog circuits," IEEE BCTM, 211–214, 2003.
74. Y. Morandini, J.-F. Larchanchel, and C. Gaquiere, "Evaluation of SiGeC HBT varactor using different collector access and base-collector junction configuration in BiCMOS technologies," IEEE BCTM Tech. Dig., 246–249, 2007.
75. R. A. Moline and G. F. Foxhall, "Ion-implanted hyperabrupt junction voltage variable capacitors," IEEE Trans. Electron Dev., ED-19 (2), 267–273, 1972.
76. S. M. Sze, Properties of Semiconductor Devices, John Wiley & Sons, 1981.
77. P. J. Kannam, S. Ponczak, and J. Olmstead, "Design considerations of hyperabrupt varactor diodes," IEEE Trans. Electron Dev., ED-18 (3), 109–115, 1971.
78. M. H. Norwood and E. Shatz, "Voltage variable capacitor tuning: A review," Proc. IEEE, 56 (5), 788–798, 1968.
79. T. Soorapanth, C. P. Yue, D. K. Shaeffer, T. H. Lee, and S. S. Wong, "Analysis and optimization of accumulation-mode varactor for RF ICs," IEEE Symp. VLSI Circuits Tech. Dig., 32–33, 1998.
80. F. Svelto, P. Erratico, S. Manzini, and R. Castello, "A metal-oxide-semiconductor varactor," IEEE Electron Dev. Lett., 20 (4), 164–166, 1999.
81. F. Svelto, S. Manzini, and R. Castello, "A three terminal varactor for RF IC's in standard CMOS technology," IEEE Trans. Electron Dev., 47 (4), 893–895, 2000.
82. W. M. Y. Wong, P. S. Hui, Z. Chen, K. Shen, J. Lau, P. C. H. Chan, and P.-K. Ko, A wide tuning range gated varactor," IEEE J. Solid-State Circuits, 35 (5), 773–779, 2000.
83. J.-H. Gau, R.-T. Wu, Steven Sang, C.-H. Kuo, T.-L. Chang, H.-H. Chen, A. Chen, and J. Ko, "Gate-assisted high-Q-factor junction varactor," IEEE Electron Dev. Lett., 26 (9), 682–683, 2005.
84. Y.-J. Chan, C.-F. Huang, C.-C. Wu, C.-H. Chen, and C.-P. Chao, "Performance consideration of MOS and junction diodes for varactor application," IEEE Trans. Electron Dev., 54 (9), 2570–2573, 2007.
85. J.-B. Shyu, G. C. Temes, and F. Krummenacher, "Random error effects in matched MOS capacitors and current sources," IEEE J. Solid-State Circuits, SC-17, 1070–1076, 1982, and SC-19 (6), 948–955, 1984.
86. K. R. Lakshmikumar, R. A. Hadaway, and M. A. Copeland, "Characterization and modeling of mismatch in MOS transistor5s for precision analog design," IEEE J. Solid-State Circuits, SC-21 (6), 1057–1066, 1986.
87. M. J. M. Pelgrom, A. C. J., Duinmaijer, and A. P. G., Welbers, "Matching properties of MOS transistors," IEEE J. Solid-State Circuits, 24 (5), 1433–1440, 1989.
88. P. R. Kinget, "Device mismatch and tradeoffs in the design of analog circuits," IEEE J. Solid-State Circuits, 40 (6), 1212–1224, 2005.
89. M. J. M. Pelgrom, H. P. Tuinhout, and M. Vertregt, "Transistor matching in analog CMOS applications," IEEE IEDM Tech. Dig., 915–918, 1998.
90. S. Lovett, M. Welten, A. Mathewson, and B. Mason, "Optimizing MOS transistor matching," IEEE J. Solid-State Circuits, 33 (1), 147–150, 1998.
91. G. Baccarani, M. Severi, and G. Soncini, "A new method for the determination of the interfaced-state density in the presence of statistical fluctuation of the surface potential," Appl. Phys. Lett., 23 (5), 265–267, 1973.
92. R. Castagne and A. Vapaille, "Apparent interface state density introduced by the spatial fluctuations of surface potential in an M.O.S. structure," Electron Lett., 6 (22), 691–693, 1970.

93. R. W. Keyes, "Physical limits in digital electronics," Proc. IEEE, 740–768, 1975.
94. B. Hoeneisen and C. A. Mead, "Fundamental limits in microelectronics – I. MOS technology," Solids-State Electron., 819–829, 1972.
95. K. Takeuchi, T. Tatsumi, and A. Furukawa, "Channel engineering for the reduction of random-dopant-placement-induced threshold voltage fluctuation," IEEE IEDM Tech. Dig., 841–844, 1997.
96. P. A. Stolk and D. B. M. Klaassen, "The effect of statistical dopant fluctuations on MOS device performance," IEEE IEDM Tech. Dig., 627–630, 1996.
97. T. Mizuno, J.-I. Okamura, and A. Toriumi, "Experimental study of threshold voltage fluctuation due to statistical variation of channel dopant number in MOSFETs," IEEE Trans. Electron Dev., 41 (11), 2216–2221, 1994.
98. A. Asenov and S. Saini, "Polysilicon gate enhancement of the random dopant induced threshold voltage fluctuations in sub-100 nm MOSFETs with ultrathin gate oxide," IEEE Trans. Electron Dev., 47 (4), 805–812, 2000.
99. H. P. Tuinhout, A. H. Montree, and P. A. Stolk, "Effects of gate depletion and boron penetration on matching of deep submicron CMOS transistors," IEEE IEDM Tech. Dig., 631–634, 1997.
100. R. Difrenza, J. C. Vildeuil, P. Llinares, and G. Ghibaudo, "Impact of grain number fluctuations in the MOS transistor gate on matching performance," IEEE ICMTS, 244–249, 2003.
101. H. Ryssel, H. Iberl, M. Bleier, G. Prine, K. Haberger, and H. Kranz, "Arsenic-implanted polysilicon layers," Appl. Phys., 24 (3), 197–200, 1981.
102. B. Swaminathan, K. C. Saraswat, and R. W.. Dutton, "Diffusion of arsenic in polycrystalline silicon," Appl. Phys. Lett., 40 (9), 795–798, 1982.
103. M. Arienzo, Y. Komem, and A. E. Michel, "Diffusion of arsenic in bilayer polycrystalline silicon films," J. Appl. Phys., 55 (2), 365–369, 1984.
104. H. Schaber, R. V. Criegern, and I. Weitzel, "Analysis of polycrystalline diffusion source by secondary ion mass spectroscopy," J. Appl. Phys., 58 (11), 4036–4042, 1985.
105. J. M. C. Stork, M. Arienzo, and C. Y. Wong, "Correlation between the diffusive and electrical barrier properties of the interface in polysilicon contacted n^+-p junctions," IEEE Trans. Electron Dev., 32, 1766–1770, 1985.
106. J. L. Hoyt, E. F. Crabbé, R. F. W. Pease, J. F. Gibbons, and A. F. Marshall, "Lateral uniformity of $n+/p$ junctions formed by arsenic diffusion from epitaxially aligned polycrystalline silicon on silicon", J. Electrochem. Soc., 135 (7), 1773–1779, 1988.
107. S. Nédèle, D. Mathiot, and M. Gaunneau, "Diffusion of boron on polycrystalline silicon," ESSDERC Tech. Dig., 153–156, 1996.
108. A. Wang and K. C. Saraswat, "A strategy for modeling of variations due to grain size in polycrystalline thin-film transistors," IEEE Trans. Electron Dev., 47 (5), 1035–1043, 2000.
109. J. T. Horstmann, U. Hilleringmann, and K. F. Goser, "Matching analysis of deposition defined 50-nm MOSFETs," IEEE Trans. Electron Dev., 45 (1), 299–306, 1998.
110. S. Winkelmeier, M. Sarstedt, M. Ereken, M. Goethals, and K. Ronse, "Metrology method for the correlation of line edge roughness for different resists before and after etch," Microelectron. Eng., 57–58, 665–672, 2001.
111. S. Xiong and J. Bokor, "A simulation study of gate line edge roughness effects on doping profiles of short-channel MOSFET devices," IEEE Trans. Electron Dev., 51 (2), 228–232, 2004.
112. L. H. A. Leunissen, M. Ercken, and G. P. Patsis, "Determining the impact of statistical fluctuations on resist line edge roughness," Microelectron. Eng., 78–79, 2–10, 2005.
113. C. H. Diaz, H.-J. Tao, Y.-C. Ku, A. Yen, and K. Young, "An experimentally validated analytical model for gate line-edge roughness (LER) effects on technology scaling," IEEE Electron Dev. Lett., 22 (6), 287–289, 2001.
114. T. Linton, M. Chandhok, B. J. Rice, and C. Schrom, "Determination of the line edge roughness specification for 34 nm devices," IEEE IEDM Tech. Dig., 303–306, 2002.
115. J. A. Croon, G. Storms, S. Winkelmeier, I. Pollentier, M. Ercken, S. Decoutere, W, Sansen, and H. E. Maes, "Line edge roughness: Characterization, modeling and impact on device behavior," IEEE IEDM Tech. Dig., 307–310, 2002.

116. G. Declerck, "A look into the future of nanoelectronics," Symp. VLSI Tech. Dig., 6–10, 2005.
117. M. Steyart, J. Bastos, R. Roovers, P. Kinget, W. Samsen, B. Graindourze, A. Pergoot, and Er. Janssens, "Threshold voltage mismatch in short-channel MOS transistors," Electron. Lett., 30 (18), 146–148, 1994.
118. S.-C. Chyi, K.-H. Pan, and D.-J. Ma, "A CMOS mismatch model and scaling effects," IEEE Electron Dev. Lett., 18 (6), 261–263, 1997.
119. P. H. Drennan and C. C. McAndrew, "A comprehensive MOSFET mismatch model," IEEE IEDM Tech. Dig., 167–170, 1999.
120. P. G. Brennan and V. C. McAndrew, "Understanding MOSFET mismatch for analog design," IEEE J. Solid-State Circuits, 38 (3), 450–456, 2003.
121. T. Serrano-Gotarredona and B. Linares-Barranco, "A new five-parameter MOS transistor mismatch model," IEEE Electron Dev. Lett., 21 (1), 37–39, 2000.
122. H. Klimach, A. Arnaud, C. Galup-Montoro, and M. C. Schneider, "MOSFET mismatch modeling: A new approach," IEEE Des. Test Comput., 23 (1), 20–29, 2006.
123. A. R. Brown, G. Roy, and A. Asenov, "Poly-Si-gate-related variability in decananometer MOSFETs with conventional architecture," IEEE Trans. Electron Dev., 54 (11), 3036–3063, 2007.
124. H. Tuinhout, M. Pelgrom, R. Penning de Vries, and M. Vertregt, "Effects of metal coverage on MOSFET matching," IEEE IEDM Tech. Dig., 735–738, 1997.
125. J. A. Croon, H. P. Tuinhout, R. Difrenza, J. Knol, A. J. Moonen, S. Decoutere, H. E. Maes, and W. Sansen, "A comparison of extraction techniques for threshold voltage mismatch," Proc. IEEE 2002 Conf. Microelectronics Test Structures, 15, 225–240, 2002.
126. P. G. Drennan, C. C. McAndrew, J. Bates, and D. Schroder, "Rapid evaluation of the root causes of BJT mismatch," Proc. International Conf. on Microelectronic Test Structures (ICMTS), 122–127, 2000.
127. P. G. Drennan, C. C. McAndrew, and J. Bates, "A comprehensive vertical BJT mismatch model," IEEE BCTM Tech. Dig., 83–86, 1998.
128. H. P. Tuinhout, "Improving BiCMOS technologies using BJT parametric mismatch characterization," IEEE BCTM Tech. Dig., 163–170, 2003.
129. C. C. McAndrew, J. Bates, T. T. Ida, and P. Drennan, "Efficient statistical BJT modeling, why β is more than I_C/I_B," IEEE BCTM Tech. Dig., 28–31, 1997.
130. S. Bordez, S. Danaie, R. Difrenza, J.-C. Vildeuil, and G. Morin, "Study of bipolar matching at high current level with various test configurations leading to a new model approach," IEEE BCTM Tech. Dig., 62–65, 2005.
131. P. G. Drennan,"Diffused resistor mismatch modeling and characterization," IEEE BCTM Tech. Dig., 27–30, 1999.
132. R. Thewes, R. Brederlow, C. Dahl, U. Kollmer, C. G. Linnenbank, B. Holzapfl, J. Becker, J. Kissing, S. Kessel, and W. Weber, "Explanation and quantitative model for the matching behavior of poly-silicon resistors," IEEE IEDM Tech. Dig., 771–774, 1998.
133. H. Thibieroz, P. Shaner, and Z. C. Butler, "Mismatch and flicker noise characterization of tantalum nitride thin film resistors for wireless applications," IEEE ICMTS, 14, 207–212, 2001.
134. U. Grünebaum, J. Oehm, and K. Schumacher, "Mismatch Modeling and Simulation – A Comprehensive Approach," Analog Integrated Circuits and Signal Processing, Kluwer Academic Publishers, 29, 165–171, 2001.
135. H. Iwai and S. Kohyama, "On-chip capacitance measurement circuits in VLSI structures," IEEE Trans. Electron Dev., ED-29 (10), 1622–1626, 1982.
136. B. Eitan, "Channel-length measurement technique based on a floating-gate device," IEEE Electron Dev. Lett., 9 (7), 340–342, 1988.
137. C. Kortekaas, "On-chip quasi-static floating-gate capacitance measurement method," IEEE ICMTS, 3, 109–113, 1990.
138. H. P. Tuinhout, H. Elzinga, J. T. Brugman, and F. Postma, "Accurate capacitor matching measurements using floating gate test structures," IEEE ICMTS, 8, 133–137, 1995.
139. H. P. Tuinhout, H. Elzinga, J. T. Brugman, and F. Postma, "The floating gate measurement technique for characterization of capacitor matching," IEEE Trans. Semicon. Manuf., 9 (1), 2–8, 1996.

140. J. Hunter, P. Gudem, and S. Winters, "A differential floating gate capacitance mismatch measurement technique," IEEE ICMTS, 13, 142–147, 2000.
141. A. van der Ziel, Noise in Solid State Devices and Circuits, John Wiley & Sons, 1986.
142. M. von Haartman and M. Östling, Low-Frequency Noise in Advanced NOS Devices, Springer, 2007.
143. C. Surya and T. Y.Hsiang, "Surface mobility fluctuations in metal-oxide-semiconductor field-effect transistors," Phys. Rev. B., 35 (12), 6343–6347, 1987.
144. K. K. Hung, P. K. Ko, C. Hu, and Y. C. Cheng, "Random telegraph noise of deep-submicrometer MOSFETs," IEEE Electron Dev. Lett., 11 (2), 90–92, 1990.
145. K. S. Ralls, W. J. Skocpol, L. D. Jackel, R. E. Howard, L. A. Fetter, R. W. Epworth, and D. M. Tennant, "Discrete resistance switching in submicrometer silicon inversion layers: Individual interface traps and low-frequency (1/f?) noise," Phys. Rev. Lett., 52 (3), 228–231, 1984.
146. M. J. Uren, D. J. Day, and M. J. Kirton, "1/f and random telegraph noise in silicon metal-oxide-semiconductor field-effect transistors," Appl. Phys. Lett., 47 (11), 1195–1197, 1985.
147. M. J. Kirton, M. J. Uren, and S. Collins, "Individual interface states and their implication for low-frequency noise in MOSFETs," Appl. Surf. Science, 30 (1–4), 148–152, 1987.
148. Y. F. Lim, Y. Z. Xiong, N. Singh, R. Yang, Y. Jiang, D. S. H. Chan, W. Y. Loh, L. K. Bera, G. Q. Lo, N. Balasubramanian, and D. -L. Kwong, "Random telegraph signal noise in gate-all-around Si-FinFET with ultra-narrow body," IEEE Trans. Electron Dev., 77 (9), 765–768, 2006.
149. S.-R. Li, W. McMahon, Y.-L. R. Lu, and Y.-H. Lee, "RTS noise characterization in flash cells,"IEEE Electron Dev. Lett., 29 (1), 106–108, 2008.
150. C. M. Compagnoni, R. Gusmeroli, A. S. Spinelli, and A. Visconti, "RTN V_T instability from the stationary trap-filling condition: An analytical spectroscopic investigation," IEEE Trans. Electron. Dev., 55 (2), 655–661, 2008.
151. R. H. Howard, W. J. Skocpol, L. D. Jackel, P. M. Mankiewich, L. A. Fetter, D. M. Tennant, R. Epworth, and K. S. Ralls, "Single electron switching events in nanometer-scale Si MOSFETs," IEEE Trans. Electron Dev., ED-32 (9), 1669–1674, 1985.
152. S. Machlup, "Noise in semiconductors: Spectrum of a two-parameter random signal," J. Appl. Phys., 25, 241–243, 1954.
153. R. C. Jaeger and A. J. Broderson, "Low-frequency noise sources in bipolar junction transistors," IEEE Trans. Electron Dev., ED-17 (2), 128–134, 1970.
154. J. L. Plumb and E. R. Chenette, "Flicker noise in transistors," IEEE Trans. Electron Dev., 10 (5), 304–308, 1963.
155. O. Roux dit Buisson and G. Moria, "Flicker noise characteristics of polysilicon resistors in submicron BiCMOS technologies," IEEE ICMTS, 10, 49–51, 1997.
156. E. Zhao, R. Krithivasan, A. K. Sutton, Z. Jin, J. D. Cressler, B. El-Kareh, S. Balster, and H. Yasuda, "An investigation of low-frequency noise in complementary SiGe HBTs," IEEE Trans. Electron Dev., 53 (2), 329–338, 2006.

Chapter 7
Enabling Processes and Integration

7.1 Introduction

Integrating a process on a chip requires a thorough and, throughout the development cycle, continuous understanding of how it will be applied. This includes the definition of a set of required components, component parameters and their tolerances, range of operating temperature, reliability expectations, set of available tools and their limitations at the time of production, and overall cost. The integration cycle begins with the definition of a full process flow. Process and device simulators are utilized to optimize the flow and ensure that it will satisfy nominal values for all parameters of the required components. Experimental short-loops are then run to evaluate individual process modules. These loops can require special test structures that may be part of a full test-die. Simulations and experiments are run in parallel to ensure the feasibility of a full process flow. A complete test die is then designed to be utilized for a full process-flow. The test-die typically contains structures that address component parameters and their tolerances, reliability structures, yield structures, process monitors, failure analysis structures, modeling structures, and sub-circuits.

The chapter begins with a description of a typical *CMOS* logic process flow, applicable to *MOSFET*s of ~0.18 μm channel length. This will be referred to as the "conventional process." A conventional *BiCMOS* process follows, that is applicable to analog/mixed-signal designs. The flow utilizes several unit-process concepts that are described in [1] and the references therein. Only a brief review of selected "conventional" unit processes is presented in this chapter. The second part of the chapter focuses on advanced enabling processes, applicable to structures of nanoscale dimensions.

7.2 A Conventional CMOS Logic Process Flow

A conventional *CMOS* logic flow is schematically described in Figs. 7.1a–7.1m. The starting material is a <100> oriented p^+-wafer with a typical boron concentration of about $10^{19}\,cm^{-3}$ upon which a p-type epitaxial layer is grown at a low

B. El-Kareh, *Silicon Devices and Process Integration: Deep Submicron and Nano-Scale Technologies*,

© Springer Science+Business Media, LLC 2009

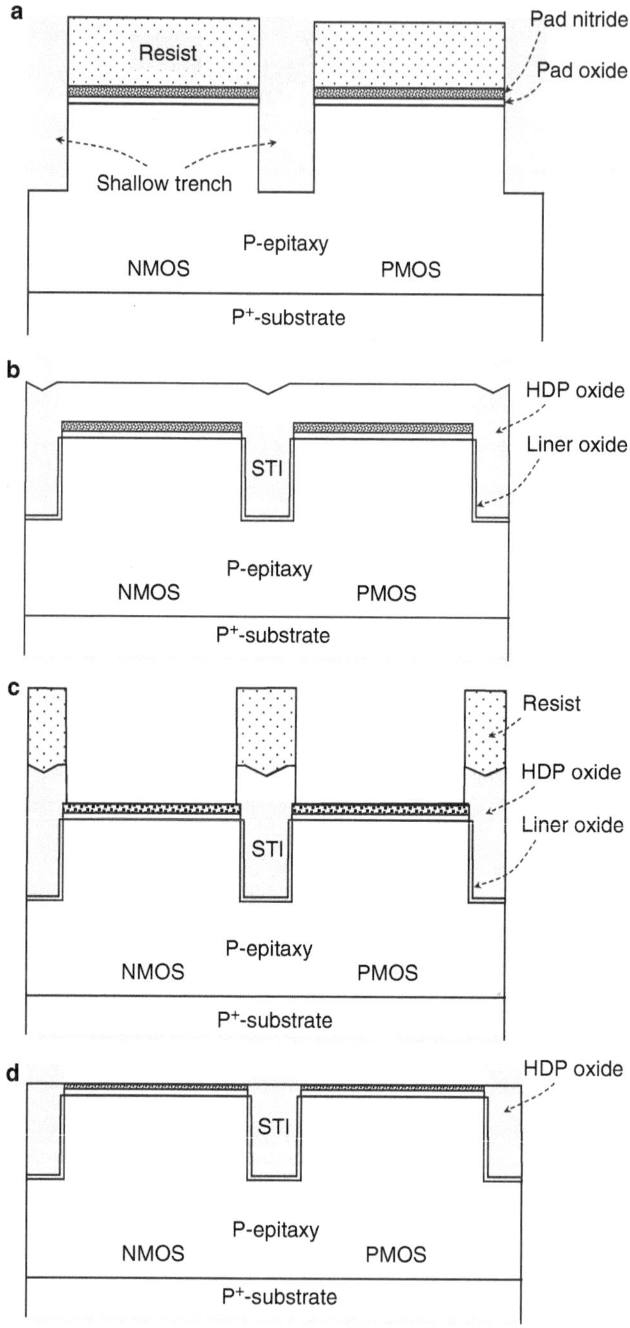

Fig. 7.1 A conventional CMOS process flow. **a** Patterning of shallow-trench isolation (STI) regions. **b** STI oxide liner and oxide fill. **c** Patterning of STI oxide-fill. Oxide left on STI regions. **d** Oxide-fill surface after planarization by CMP.

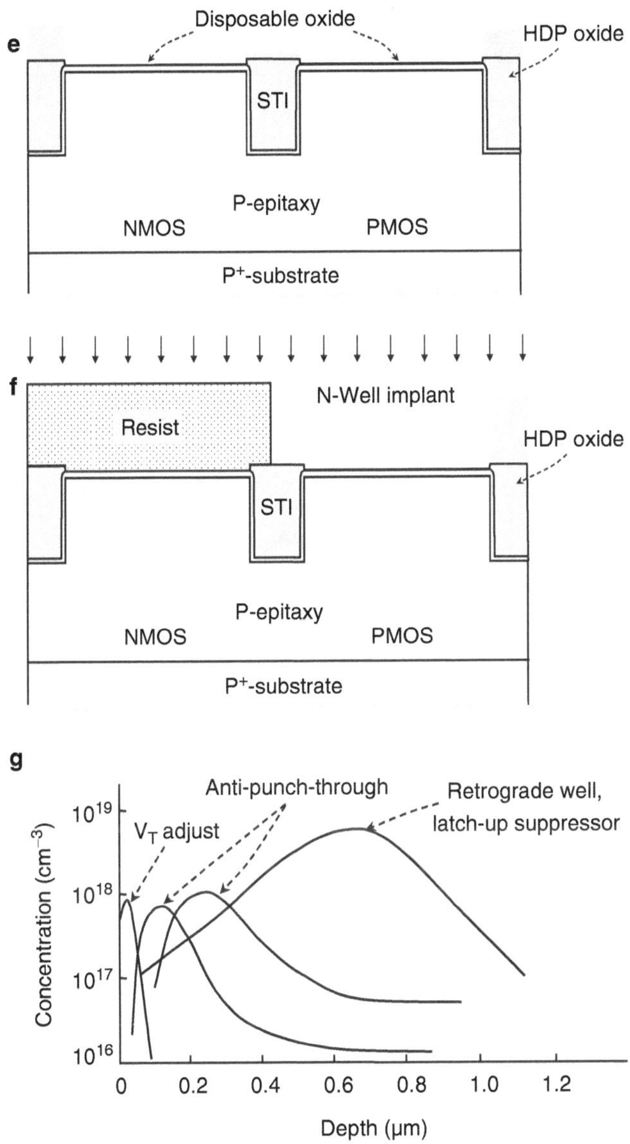

Fig. 7.1 (continued) A conventional CMOS process flow. **e** Growth of disposable oxide after removal of pad nitride and pad oxide. **f** Pattering and implantation of PMOS n-well implants. The *NMOS* p-well is formed in a similar process (not shown). **g** Example of multiple p-well implant profiles. Similar profiles apply to n-well implants.

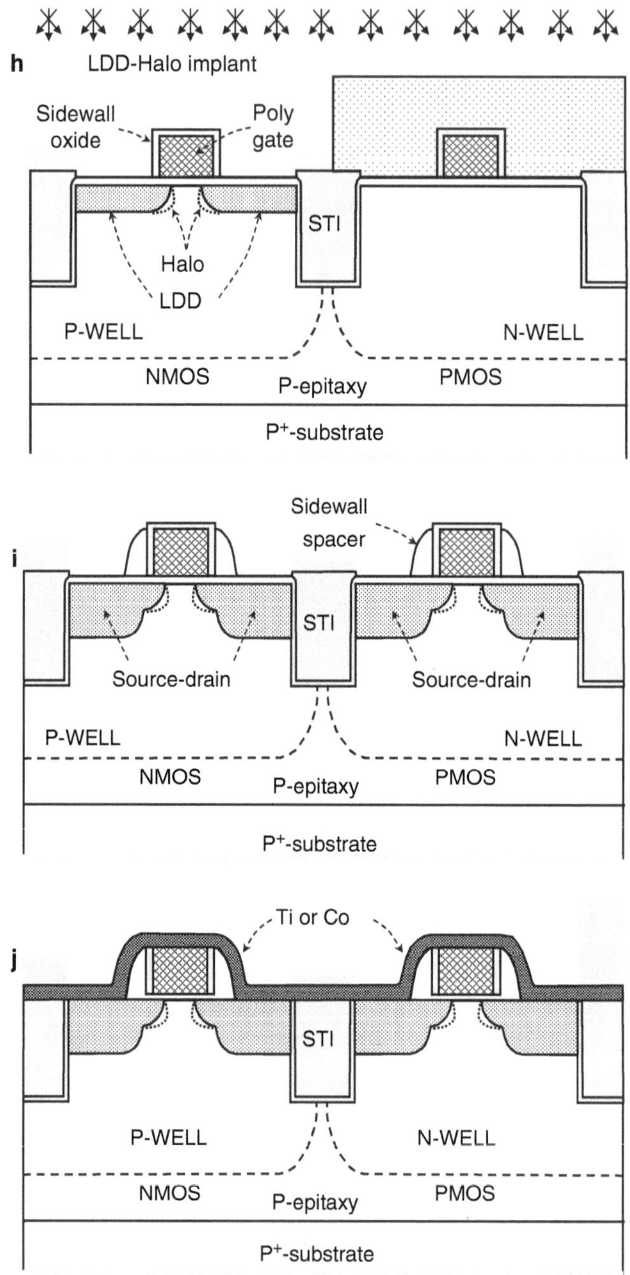

Fig. 7.1 (continued) A conventional CMOS process flow. **h** Pattering and implantation of NMOS LDD or source-drain extensions, and halo. A similar process is repeated for PMOS. **i** MOSFETs after nitride spacer directional etch. **j** Deposition of titanium or cobalt on the entire wafer.

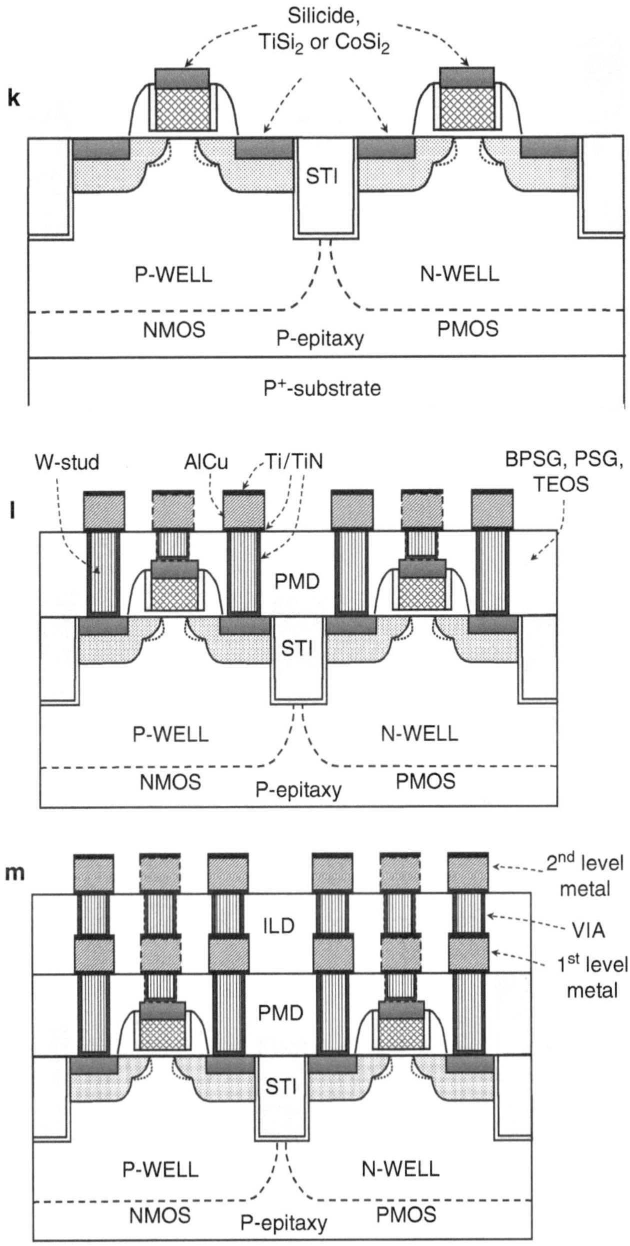

Fig. 7.1 (continued) A conventional CMOS process flow. **k** Formation of self-aligned silicide. **l** Pre-metal dielectric (PMD), contacts, and first metal. **m** Inter-level dielectric (ILD), vias and second-level metal. A similar process applies to higher metal levels

boron concentration, typically in the range 10^{15}–10^{16} cm^{-3}, to allow tailoring the surface and subsurface to the desired dopant profile. Another advantage of epitaxy is that it is essentially void of oxygen [1]. The heavily doped substrate serves as a low-resistance path to suppress latch-up (Chap. 8). A 10–20 nm pad oxide (SiO_2) is grown and covered with a deposited 100–150 nm pad nitride (Si_3N_4). Shallow trench isolation (*STI*) regions are patterned and etched (Fig. 7.1a). The trenches can be etched with the patterned pads acting as a stencil ("hard-mask") or, alternatively, the photoresist (simply, resist) mask can be left in place to etch both the nitride-oxide stack and the silicon trench. The *STI* depth typically ranges from 0.25 to 0.5 μm.

The resist is removed and a thin, typically 10-15 nm, liner-oxide film is grown on the trench sidewalls and floor. A silicon-dioxide layer is deposited by, for example, high-density plasma (*HDP*) enhanced *CVD*, at an appropriate thickness to fill the *STI* of varying width across the wafer (Fig. 7.1b).

The oxide layer is patterned by covering the *STI* regions with resist and dry-etching the oxide layer from most of the active regions (Fig. 7.1c). This is done to reduce the mass of oxide that must be removed by chemical-mechanical polishing (*CMP*) in the next step.

The resist is removed and *CMP* with high-selectivity slurry is used to planarize the surface and remove the excess trench oxide over the *STI* regions. The nitride layer serves as an "etch-stop" (Fig. 7.1d). A high temperature oxidation may follow to round the trench corners and relieve the stress created during *STI* processing, and to suppress the reverse narrow-channel effect and subthreshold kink (Chap. 5).

The pad nitride and pad oxide are removed by wet or reactive-ion etching and a screen oxide, referred to as sacrificial or disposable oxide, is grown (Fig. 7.1e). The film serves as an amorphous layer for implantation in the next step, and also as a screen layer to protect the surface from extraneous contamination and particles during implantation. It also allows placing the peak of the implanted concentration near the silicon surface.

The *NMOS* regions are covered with resist and the *PMOS* n-well and channel are implanted at multiple energies and doses (Fig. 7.1f). The process is repeated for the *NMOS* p-well and channel implants (not shown). An example of multiple p-well profile is shown in Fig. 7.1g.

A high-energy and medium- to high-dose boron implant forms the retrograde profile, that is, the impurity concentration drops-off toward the surface. The main purpose of this implant is to reduce the well sheet-resistance to suppress latch-up. Medium-energy, medium-dose boron implants increase the sub-surface concentration in regions susceptible to punch-through. Both the retrograde and anti-punch-through profiles must be shaped to avoid excessive encroachment into the surface and impact on threshold voltage. A low-energy shallow, low-dose boron or indium implant adjusts the threshold voltage to the desired value.

The resist and disposable oxide are removed. The surface is cleaned and the gate dielectric, typically 3–4 nm oxide, is grown, followed by the deposition of undoped polysilicon without breaking the vacuum. The polysilicon is pattered by directional reactive ion etching (*RIE*). Because of the high etch-selectivity between silicon and oxide, etching essentially stops on the underlying gate-oxide (Fig. 7.1h)

A liner-oxide, typically 10–15 nm thick, is grown or deposited conformal over polysilicon. It serves to remove the *RIE* damage, as an offset spacer on the polysilicon sidewalls, and as a protective film during phosphorus/arsenic implantation of the *NMOS* lightly-doped drain (*LDD*), or source-drain extensions (Fig. 7.1h). Boron halos may also be implanted during this step or after sidewall-spacer formation in the next step. The purpose of the lightly-doped drain or extension is to reduce the electric field at the drain boundary near the channel surface. Halos are implanted at large tilt to increase the channel concentration in regions susceptible to punch-through and, in some designs, to provide reverse short-channel effect (RSCE) behavior (Chap. 5). A similar process is applied to *PMOS* arsenic/phosphorus halos and boron *LDD* or source-drain extensions.

The resist is removed and a 100-150 nm nitride film typically deposited by *CVD* to form the polysilicon-gate spacers. When etched directionally, nitride spacers are left along the polysilicon sidewalls (Fig. 7.1i).

The purpose of the spacers is to keep the highly-doped source-drain regions and their silicides (next step) at a sufficiently large distance from the channel and gate. This is to avoid an increase in electric field at the drain, and silicide encroachment into gate and channel regions. The selectivity between nitride and oxide allows etching to stop within the underlying oxide film.

The *PMOS* is covered with resist, as in Fig. 7.1h, and the *NMOS* source, drain, and gate simultaneously implanted, typically with arsenic at a low energy and high dose. A similar process is used for the *PMOS* boron source, drain and gate implants. It should be noted that in this dual gate-workfunction process, the implanted dose and final impurity profile in the polysilicon gate are limited by the requirements on source-drain junction depth and lateral extent, and the associated thermal budget, such as activation anneal cycles. A single workfunction, for example, in-situ phosphorus-doped polysilicon for both *NMOS* and *PMOS* would allow more heavily doped polysilicon, independent of source-drain, but would require a boron buried channel in *PMOS* to re-adjust V_T (Chap. 5).

The oxide is removed from source, drain, and gate regions and a refractory metal, typically 20–30 nm titanium, cobalt, or nickel (Sect. 7.3.5) is sputter-deposited on the entire surface (Fig. 7.1j).

The wafer is subjected to a two-step rapid-thermal anneal (*RTP*) to form the silicide, $TiSi_2$ or $CoSi_2$, in exposed silicon regions, as described in [1] and in Sect. 7.4.4.4. Silicide does not form on dielectrics, such as oxide and nitride. The unreacted metal is removed by wet-etching, leaving self-aligned silicides in source, drain and gate regions (Fig. 7.1k).

Figure 7.1*l* encompasses several steps. A dielectric, sometimes referred to as pre-metal dielectric (*PMD*), is deposited to isolate the first interconnect metal from polysilicon and silicon. The dielectric can consist of boro-phospho-silicate glass (*BPSG*), phospho-silicate glass (*PSG*), oxide deposited by the decomposition of tetra-ethyl-orthosilicate (*TEOS*), or a combination of glass and *TEOS*. A thin *CVD* silicon-nitride or undoped oxide film (not shown) typically precedes *PSG* or *BPSG* to act as a barrier for dopant diffusion from the glass into silicon, and also as an etch-stop when patterning contacts.

The glass is reflowed at 700 °C–800 °C, depending on the boron and phosphorus concentration. It is then "softly" planarized by *CMP*, in preparation for contact patterning. Contacts are patterned and etched directionally down to the silicide on polysilicon, source, drain and wells. A thin, about 25-nm thick, *Ti* film is deposited, typically by sputtering (physical-vapor deposition, *PVD*), followed by a 10–50 nm *PVD TiN* film. The *Ti* film lowers the contact resistance and improves adhesion with silicide, and the *TiN* film serves as an adhesion layer for the subsequent *W* and as a barrier to prevent reaction of tungsten with silicon and fluorine penetration at the bottom of the contact. For high aspect-ratio (ratio of depth to width) contacts, ionized *PVD* (*IPVD*) achieves higher sidewall and floor uniformity (Sect. 7.4.3.2). Tungsten is deposited by *CVD* through a reduction of tungsten hexafluoride (WF_6) by silane (SiH_4) or hydrogen (H_2), and then planarized by *CMP*. The first-level metal typically consists of sputter-deposited 15–20 nm *Ti*, followed by 25–40 nm *TiN* as a barrier layer, 300–600 nm *Al-0.5%Cu* alloy and 15 nm *Ti* with 25–40 nm *TiN* on the top. Alloying aluminum with Cu and the presence of top and bottom *Ti/TiN* layers considerably extend the electro- and stress migration lifetime of interconnects.[1] *TiN* also serves as anti-reflective coat. The first metal level is patterned by directional etching using photoresist as a mask. Figure 7.1*l* shows the structure after first-metal level patterning.

An inter-level dielectric (*ILD*) layer, typically 0.5–1.0 μm *TEOS* oxide, is deposited and planarized by *CMP* in preparation for patterning vias to form connections between first and second level metal. As for contacts, the vias are lined with *Ti/TiN*, filled with tungsten and planarized. The second-level metal follows a similar process as for the first level (Fig. 7.1m). The sequence is essentially repeated for the third through the nth level. The metal thickness may, however, hierarchically increase at upper levels to allow a higher current-carrying capability and to offset the degradation of wiring *RC* time delays associated with scaling of wiring. More advanced patterning and planarization processes, known as "damascene" and "dual damascene," are described in [1] and in Sect. 7.5.1.1.

7.3 A BiCMOS Process Flow

A variant of an analog/RF BiCMOS process that integrates CMOS, bipolar transistors (Chap. 3), and passive components (Chap. 6) on the same die is described in Figs. 7.2–7.7 [2]. Integration is best achieved by modularly adding bipolar modules to a base-line CMOS process (Fig. 7.2).

Figure 7.3 shows schematic cross-sections of completed *NPN* and *PNP* structures up to silicidation [2]. The numbers in the figure refer to the different transistor regions defined in the bottom of the figure. The substrate is a high resistivity

[1] Electromigration is the movement of metal ions under the influence of an "electron wind" at high current densities. This causes voids at the cathode side, locally increasing the resistance, and hillocks at the anode side resulting in shorts. Stress-migration is the movement of metal ions under the influence of mechanical stress.

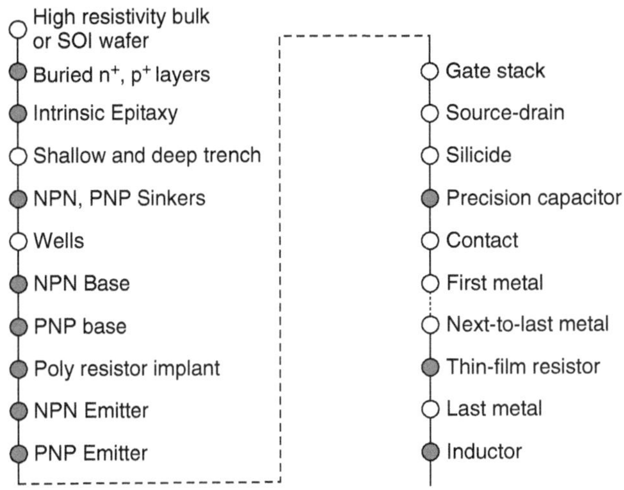

Fig. 7.2 A typical modular *BiCMOS* flow. Bipolar/analog-specific modules are highlighted in dark

SOI wafer for ease of isolation and to achieve a high quality-factor for analog passive devices. The top silicon thickness is about 500 nm and the buried-oxide (*BOX*) thickness typically ranges from 0.1 to 1 μm (Sect. 7.4.1.5). A 15–20 nm screen-oxide layer is grown and a n^+-buried layer (*NBL*) implanted with arsenic or antimony at low energy and high dose, using photoresist as a mask. The p^+-buried layer (*PBL*) is formed in a similar process.. The heavily-doped buried layers serve as low-resistance collector paths for the *NPN* and *PNP* transistors. A 0.4–0.6 μm intrinsic epitaxial film is grown, upon which the bipolar transistors and *CMOS* will be built. shown in region 3 in the figure. Shallow-trench isolation (*STI*) regions and deep trench isolation (*DTI*) regions are formed by etching and oxide-filling *STI* and *DTI* in either sequence. The surface is planarized with *CMP* after both regions are filled with oxide. Shallow-trench, deep-trench, and buried oxide fully encapsulate bipolar and *CMOS* structures with oxide. A disposable oxide film is grown in preparation for implanting collector "sinkers" to reduce the collector "down-resistance" between collector-contact and buried-layer, and for implanting *CMOS* wells. *NPN* sinkers are implanted, typically with a combination of phosphorus and arsenic at multiple energies, using resist for masking. The process is repeated for implanting *PNP* sinkers with boron. N-well and p-well regions are masked and implanted through the same disposable oxide, as described in Fig. 7.1f.

The gate-stack is removed from the *NPN* regions and a boron-doped silicon-germanium (*SiGe*) *NPN* base is grown epitaxially to a thickness of typically 100–200 nm. Germanium and boron are incorporated in-situ by adding germane (GeH_4) and diborane (B_2H_6) to the gas stream. The desired *Ge* and boron profiles in the base are achieved by adjusting the GeH_4 and B_2H_6 concentration during growth. Figure 7.4 shows a *SIMS* profile obtained for a nearly flat *Ge* concentration in the base [3]. The *SiGe* layer is capped with Si grown at 625 °C by atmospheric

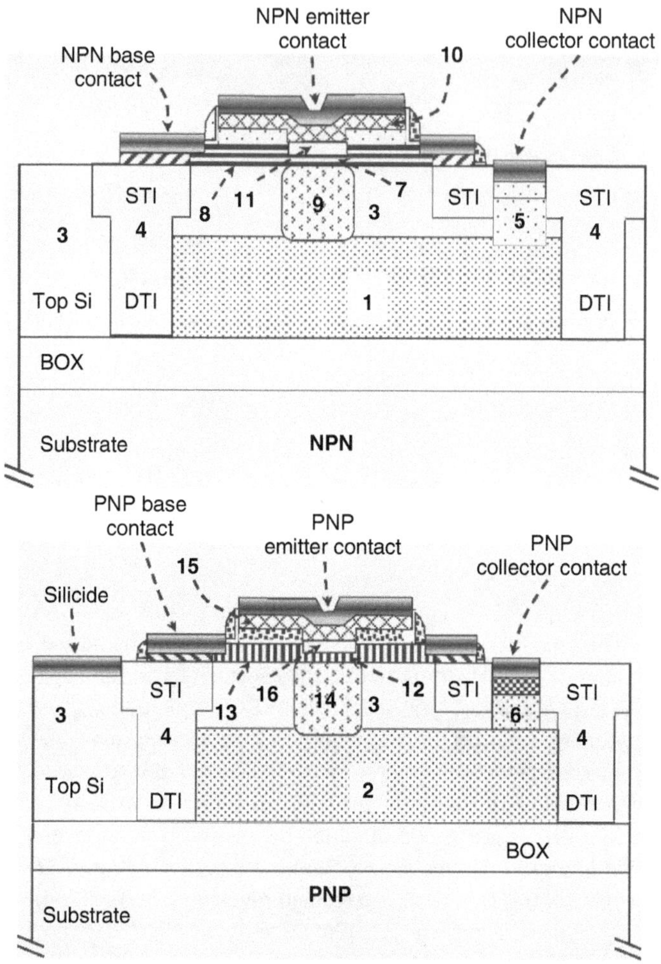

1: N^+ Buried Layer, NBL
2: P^+ Buried Layer, PBL
3: Intrinsic epitaxy
4: Shallow and deep trenches
5: NPN sinker
6: PNP sinker
7: NPN intrinsic base
8: NPN extrinsic base
9: NPN SIC
10: NPN poly emitter
11: NPN silicon emitter
12: PNP intrinsic base
13: PNP extrinsic base
14: PNP SIC
15: PNP poly emitter
16: PNP silicon emitter

SIC: Selectively Implanted Collector
STI: Shallow-Trench Isolation
DTI: Deep Trench Isolation

Fig. 7.3 *NPN* and *PNP* structures, shown after silicidation

pressure *CVD*. The *Ge* transitions at the *SiGe/Si* interfaces are exceptionally steep. Other *Ge* profiles, such as triangular and trapezoidal, can also be obtained. A thin silicon capping layer is grown over the *SiGe* film to serve as a buffer between the emitter and base (Chap. 3). The base layer grows as single-crystal silicon on silicon

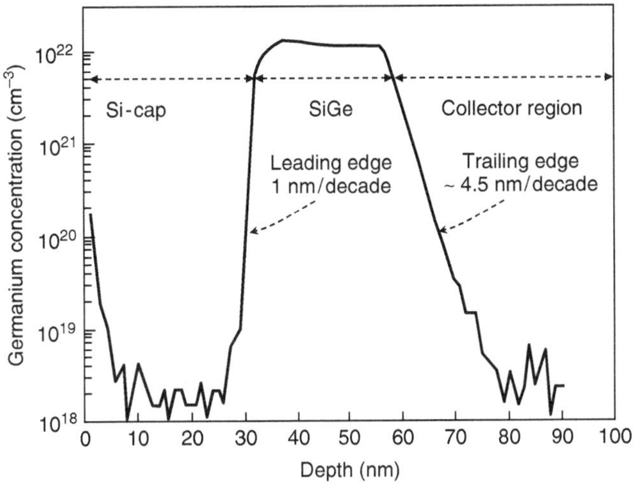

Fig. 7.4 Reconstruction of *SIMS* profile of a *SiGe* layer (Adapted from [3])

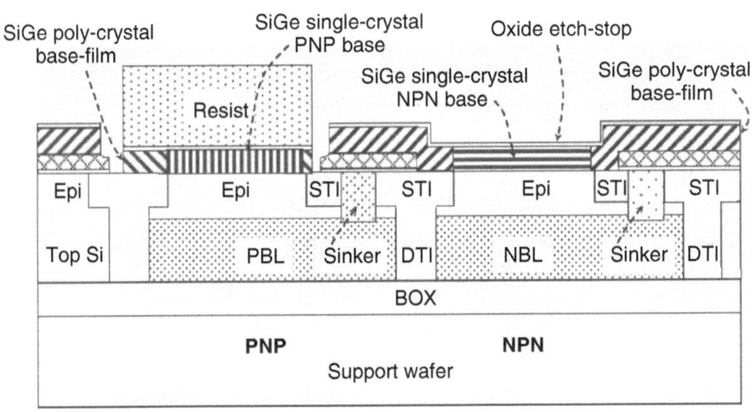

Fig. 7.5 Bipolar structures at *PNP* base patterning

and as polycrystalline silicon on polysilicon and insulators. This is shown in different shadings in Fig. 7.3. A 20–30 nm oxide etch-stop is deposited on the entire surface.

The *NPN* oxide and base are removed from the *PNP* region. The *PNP SiGe* base is deposited with the desired *Ge* profile. The *PNP* base is covered with 15–20 nm deposited oxide and the base implanted with arsenic or phosphorus. The base is doped by implantation rather than in-situ because of possible profile interactions between dopants and germanium. The *PNP* base is patterned as shown in Fig. 7.5.

At this point, polysilicon resistors may be masked and implanted to the desired sheet resistance. A dielectric stack of typically 20–30 nm thermal SiO_2 and 50–100 nm deposited Si_3N_4 serves as an insulator between the emitter polysilicon (deposited in the next step) and the extrinsic base. The *NPN* emitter region

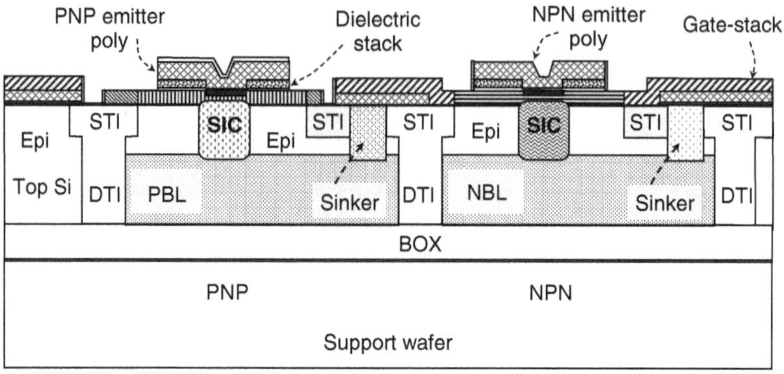

Fig. 7.6 Bipolar structures after *NPN*, *PNP* emitter polysilicon patterning

is patterned and dry-etched through the nitride. Phosphorus is implanted through the opening at a low-dose (10^{12}–10^{13} cm^{-2}) and medium energy (100–200 keV) to form the *NPN* selectively implanted collector (*SIC*). The purpose of *SIC* is to delay the onset of the Kirk effect (Chap. 3, Sect. 3.4.2). After a short oxide-wet-etch, an interfacial oxide (IFO), ~0.5 nm thick Sect. 3.3.4.1, is formed by low-temperature rapid-thermal oxidation, followed by deposition of 150–200 nm *NPN* polysilicon emitter and a thin screen oxide. The emitter polysilicon is implanted with arsenic at medium low-energy and a dose of about 2×10^{16} cm^{-2} and subjected to rapid-thermal annealing at about 1000°C to form the active emitter in silicon (11 in Fig. 7.3). The *NPN* emitter is patterned and etched directionally, stopping on the dielectric-stack. A nitride film is deposited to serve as a *PNP* polysilicon emitter etch-stop. The *PNP* emitter process module is similar to that of the *NPN* structure. Boron is implanted through the *PNP* emitter opening at low-dose (10^{12}–10^{13} cm^{-2}) and medium energy (50–100 keV) to form the *PNP SIC* (14 in Fig. 7.3). After the *PNP IFO* growth and emitter polysilicon deposition, boron is implanted into the polysilicon at low energy and a dose in the range 5×10^{15}–8×10^{15} cm^{-2}. The *NPN* and *PNP* structures are shown in Fig. 7.6 after *PNP* emitter-polysilicon patterning.

The *CMOS* gate-stack, polysilicon resistors and capacitor are simultaneously patterned and directionally etched, followed by silicidation as described in Sect. 7.1. The *NMOS* source and drain serve as contacts to the *PNP* base, *NPN* collector, polysilicon resistor and capacitor contacts. The *PMOS* source and drain serve as contacts to the *NPN* base and *PNP* collector contacts. An insulator, such as the nitride spacer deposited over the polysilicon gate, is patterned over polysilicon resistors to serve as a block during silicidation of source, drain, gate, and capacitor plate. Cross-sections of polysilicon resistors and capacitors are shown in Figs. 6.15 and 6.26 in Chap. 6.

Silicidation is followed by interconnect process modules similar to those described in Sect. 7.1. A thin-film *SiCr* or *NiCr* resistor may be inserted between the upper-level metals, as shown in Fig. 6.21 in Chap. 6. Also, a spiral inductor can be formed when patterning the last metal or a combination of two or three metal levels to reduce the inductor line resistance, as shown in Fig. 7.7.

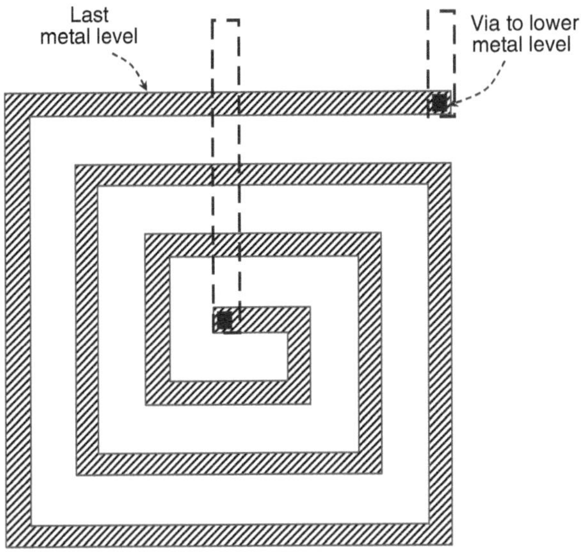

Fig. 7.7 Example of a spiral inductor

7.4 Advanced Enabling Processes

The purpose of this section is to briefly review conventional processes and to describe new enabling processes for deep submicron and nanoscale technologies.

7.4.1 Crystal Growth and Wafer Preparation

The first step in wafer preparation is to grow a nearly defect-free silicon crystal with the desired dopant type and concentration. There are essentially two methods for growing single-crystal silicon: The Czochralski (*CZ*) crystal pull method and the Float Zone (*FZ*) method [1].

7.4.1.1 Czochralski Crystal-Pull Method[2]

The Czochralski crystal-pull method to grow single-crystal silicon is illustrated in Fig. 7.8 [4]. It is the method of choice for high volume production of large (300 to 400 mm diameter) wafers. Ultra-high purity pieces of polysilicon are melted in an inert atmosphere (typically argon) in a quartz crucible. The crucible is kept at a

[2] Discovered 1916 by Polish Professor Jan Czochralski who accidentally dipped his pen into a tin-melt rather than in the ink-pot while writing his report. When he quickly pulled it out, he noticed a tin-crystal that was grown on the tip of his pen.

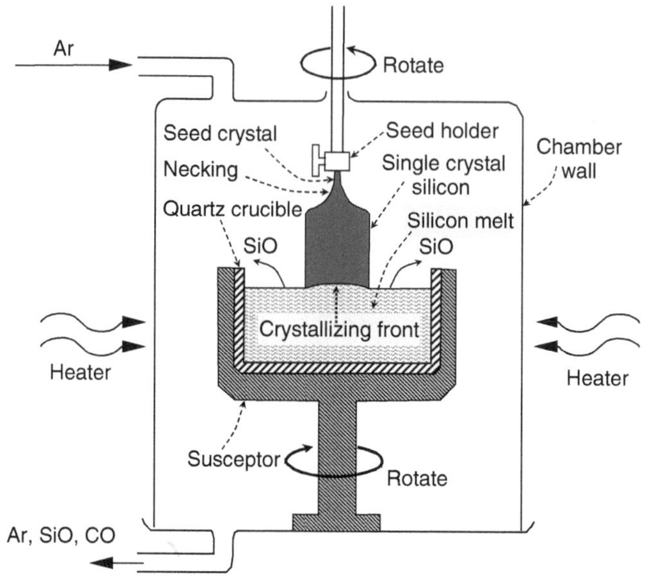

Fig. 7.8 Schematic illustration of key elements of the Czochralski (*CZ*) crystal-pull method (heat shields not shown)

temperature slightly above the silicon melting point of 1412 °C using surrounding heaters and heat shields to precisely control radial and axial temperature gradients in the melt. Doping is achieved by adding controlled amounts of elements, such as boron, phosphorus or arsenic into the melt. A seed crystal of the desired orientation is partially lowered into the melt at a temperature near the solidification point to initiate the growth. It is then withdrawn while the seed and crucible rotate in opposite directions to improve homogeneity. The seed must not only sustain the weight of the ingot but also the rotation torque. The seed is pulled at a varying rate, starting at about 3 cm per minute to grow a crystal with a diameter of a few mm, and then gradually slowing down the pull-rate to increase the crystal diameter to the desired value. The growing crystal replicates the crystalline orientation of the seed crystal. The gradual decrease in pull-rate is referred to as the Dash process [5]. It ensures that the crystal will be dislocation free even though the seed crystal may contain dislocations caused by the thermal shock with the melt.

To end the growth process, the pull-rate is gradually increased to slowly reduce the crystal diameter and form an end-cone similar to the seed cone to avoid forming dislocations. Since the solubility of impurities is higher in the melt than in the crystal, the ratio of impurity concentration in the crystal C_S to its concentration in the melt C_L is called the segregation coefficient k_{seg}

$$k_{seg} = \frac{C_S}{C_L} \tag{7.1}$$

For most impurities, k_{seg} is smaller than one. Thus, the melt concentration increases as the crystal is grown and the melt is consumed. This results in an increase in crystal concentration from the seed-end to the tail-end of the crystal. The rotation rate, pull-rate, and temperature are computer controlled and adjusted to minimize radial and axial concentration inhomogeneities.

Since a reaction between the *Si* melt and the quartz (SiO_2) crucible is unavoidable, oxygen will dissolve in the melt and be initially incorporated in the crystal primarily as interstitials (O_i). The oxygen concentration and distribution can be controlled by adjusting the rotation, pull-speed and other growth parameters. When controlled, the crystal oxygen concentration can range from about $1.2\times10^{18}\,\mathrm{cm}^{-3}$ to $1.9\times10^{18}\,\mathrm{cm}^{-3}$ [6, 7]. Thermal convection due to density gradients as a result of temperature gradients plays an important role in the incorporation of oxygen in the crystal by causing the melt to stream along the crucible walls and dissolve more silicon dioxide [6]. A better control is achieved by growing the crystal with an axial magnetic field to the melt. The method, referred to as magnetic Czochralski (*MCZ*), suppresses convection, resulting in a reduction of the oxygen content by one order of magnitude compared to non-magnetic growth [8]. An improved control of oxygen concentration can be achieved by regulating the melt rotation with an electromagnetic force rather than crucible rotation (Fig. 7.9) [9].

The presence of oxygen can be beneficial or detrimental, depending on its total concentration. If maintained at an optimum concentration between $\sim10^{18}\,\mathrm{cm}^{-3}$ and $1.8\times10^{18}\,\mathrm{cm}^{-3}$, it has the advantage of enabling intrinsic defect-gettering, as described below [10]. At an initial concentration above $1.8\times10^{18}\,\mathrm{cm}^{-3}$, however, oxygen can precipitate out and cause stacking faults, dislocations, and wafer warpage [11–17].

The crystal is then cut into slices which, after polishing, results in wafers used as the starting material for chip production. The polished wafers must satisfy stringent requirements with respect to impurity concentration, defect density, thickness, flatness, diameter, crystal orientation, and their tolerances.

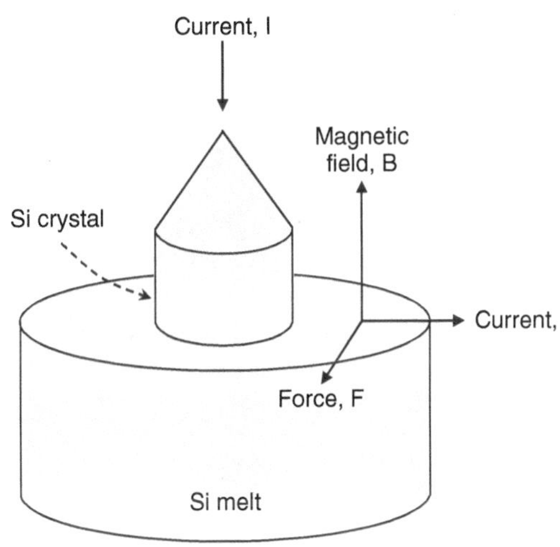

Fig. 7.9 Schematic diagram of the principle of melt rotation by electromagnetic force. The current passes axially through the crystal and radially through the melt. The resulting force F rotates the melt [9]

7.4.1.2 Intrinsic Gettering [10]

Gettering is the process by which mobile impurities, such as heavy metals are removed from active device regions of the wafer. The metals and point defects (vacancies, interstitials) diffuse through the crystal and are trapped by defective regions, such as dislocations or oxide precipitates that are intentionally created away from the device region. It is an important step in yield enhancement that can be performed during wafer preparation or subsequent processing steps. There are essentially two types of gettering, extrinsic and intrinsic. One example of extrinsic gettering is the bombardment of the wafer backside with argon ions to produce a damaged region that can act as a sink for mobile impurities and point defects. Deposition of a polysilicon layer on the wafer backside has also been used for external gettering by providing grain boundaries that can act as traps for mobile impurities.

Intrinsic gettering removes mobile metal impurities, particularly metals that degrade the minority carrier lifetime, from the active device region of the wafer through oxygen precipitates forming bulk micro-defects (*BMD*) that are intentionally created in a region deep in the bulk of the wafer [10]. The requirement for intrinsic gettering is that the initial interstitial oxygen content (O_i) be over the O solubility limit, for example, in the range of $\sim 10^{18}\,\text{cm}^{-3}$ and $1.8\times 10^{18}\,\text{cm}^{-3}$. A typical gettering thermal cycle is the so-called high-low-high sequence that consists of three steps (Fig. 7.10): (a) A high temperature ($>1100\,^{\circ}$C for 2–4 h) heat treatment in an inert atmosphere, such as *Ar*, N_2, or H_2, to out-diffuse oxygen from the top (and bottom) region of the wafer, leaving a zone that is denuded of oxygen and essentially free of oxygen-induced defects; (b) A low temperature anneal (600 °C–750 °C for $\sim$2 h) to enhance nucleation of oxygen precipitates beneath the denuded zone; (c) A high temperature cycle (1000 °C–1100 °C for $\sim$2 h) to getter heavy metals and point defects. The depth of the denuded zone typically ranges from 5 to 40 μm, depending on the level of oxygen concentration, temperature, thermal-cycle time, and ambience [6, 10]. The cycle-time for the first high-temperature treatment can be shortened by rapid-thermal annealing (*RTA*) as demonstrated on germanium-doped *CZ* wafers in [18].

7.4.1.3 High-Resistivity Czochralski Wafers

Wafers of resistivity above $\sim$100 Ohm-cm are required for special applications, such as components with high quality-factor[3] for *RF* applications. High-resistivity crystals, of the order of 1 kOhm-cm, can be produced with magnetic Czochralski, *MCZ* [19]. In the presence of interstitial oxygen, however, the resistivity can be modified during processing thermal cycles in the range 350 °C–550 °C [20–23]. At an oxygen concentration larger than $\sim 10^{17}\,\text{cm}^{-3}$, significant amounts of shallow-donor type oxygen aggregates, referred to as thermal donors, can be formed at

[3] See footnote 1, Chap. 6, page 401.

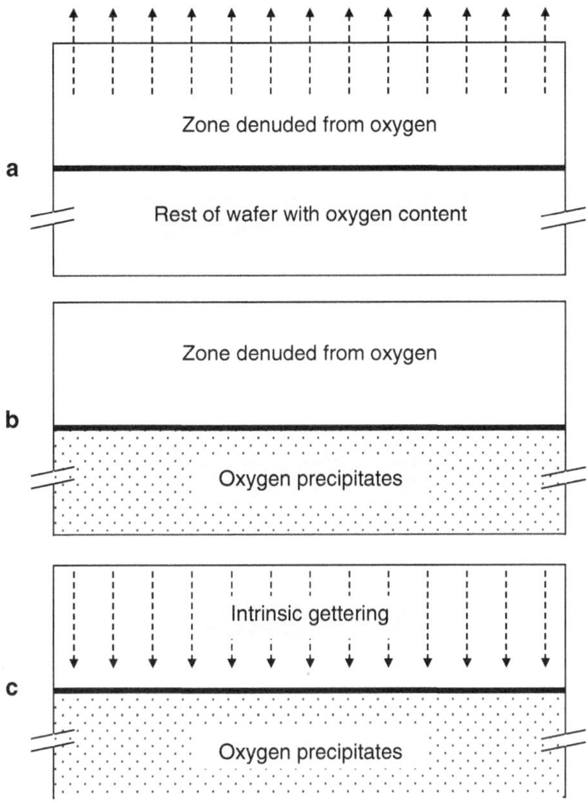

Fig. 7.10 Illustration of intrinsic gettering. **a** Thermal cycle at >1100 °C for oxygen to escape; **b** Low-temperature anneal at 600 °C–750 °C to enhance nucleation of oxygen precipitates; **c** High-temperature intrinsic gettering-anneal at 1000 °C–1100 °C

a sufficiently high concentration to critically modify the wafer resistivity. The thermal donor concentration can be considerably reduced by a special annealing sequence [20, 24].

7.4.1.4 Float-Zone Crystal Growth

The principle of float-zone (*FZ*) crystal growth is illustrated in Fig. 7.11. A seed crystal with the desired orientation is clamped at the starting end of the ingot that typically consists of a polysilicon rod. A molten zone is formed by, for example, an *RF* heating coil and moved slowly from the seed to the ingot end. The molten zone is stabilized by surface tension and magnetic levitation (not shown). At the freezing interface, there will be a distribution of impurities between the melt and the crystal due to segregation. Since k_{seg} is smaller than one, impurities will be driven from the seed-side to the opposite end side. A high-purity crystal can be produced with

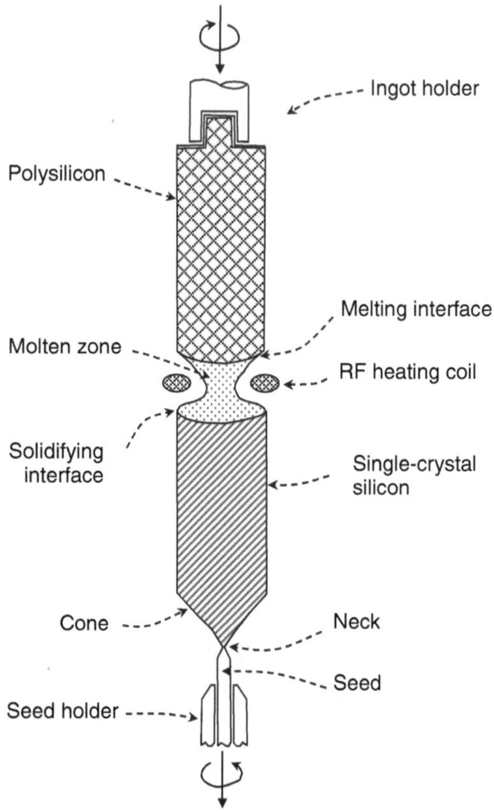

Fig. 7.11 Schematic illustration of key elements of the float-zone (*FZ*) method

multiple zone-passes along the ingot, a technique known as zone-refining. Since the melt does not touch a crucible, a crystal with very low interstitial oxygen concentration ($<10^{15}\,\mathrm{cm}^{-3}$) can be grown. Thus, crystals of very high resistivity can be obtained.

Dopants can be introduced into the molten zone by means of a controlled gas mixture. Doping the crystal can also be performed by neutron transmutation doping (*NTD*), for example, by converting silicon (Si_{30}) to phosphorus (P_{31}) to form high-resistivity n-type wafers [25–27].

The main drawbacks of float-zone crystal growth are the limited wafer size ($\leq$200 mm) and cost. *FZ*-wafers are typically used for special application, such as power transistors, solar cells, and very high-resistivity (in excess of 1 kOhm-cm) wafers for *RF* designs where the component's quality-factor is important [28].

7.4.1.5 Silicon-On-Insulator, SOI

A silicon-on-insulator wafer is shown in Fig. 7.12. It consists of a support wafer, an insulator, typically a buried oxide (*BOX*), and a top single-crystal. Depending

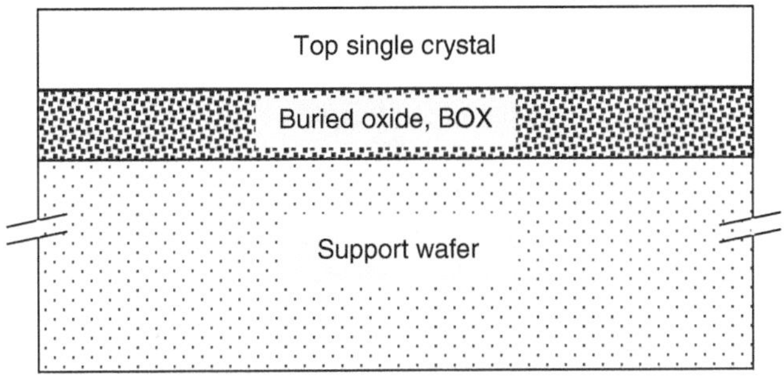

Fig. 7.12 Illustration of silicon on insulator (SOI)

on application, the top silicon-crystal, referred to as body, can range from less than 50 nm to several μm. If needed, it can be thinned to the desired thickness by oxidation and etch, or thickened by epitaxial growth. In high-performance *MOSFET* designs, the source and drain abut the *BOX*. If the field-induced depletion region under the gate abuts the *BOX*, the structure is referred to as fully-depleted (*FD*) *SOI*. Otherwise, the structure is called partially-depleted (*PD*). The *BOX* thickness typically varies from about 100 to 1 μm and depends primarily on the method of *SOI* production and circuit requirements. In most applications, the type and resistivity of the support wafer are chosen to reduce parasitic effects on device operation. For example, a very high-resistivity wafer is chosen for analog/mixed-signal applications to reduce losses by Eddy-currents in the wafer.

There are several advantages of *SOI* wafers over "bulk" wafers. Among them are:

(a) Reduced junction capacitance and hence higher performance and lower power dissipation when junctions abut the *BOX*.
(b) Formation of a quantum well in ultra-thin body (<50 nm) SOI *MOSFET*s, improving speed.
(c) Elimination of latch-up in *CMOS* (Chap.8).
(d) Reduction in *MOSFET* body-effect, particularly in fully-depleted *SOI* due to the pinning of the field-induced depletion region at the *BOX*.
(e) Reduction in subthreshold slope, particularly in *FD SOI* due to the decreased silicon capacitance at the source-end of the channel.
(f) Full isolation when used in conjunction with lateral dielectric isolation, for example, oxide-filled deep trenches. In mixed-signal applications, this allows the separation of analog from digital circuits to suppress noise coupling. In non-fully-depleted SOI *MOSFET*s, it also allows individual adjustment of threshold voltage by applying an appropriate body-bias (referred to as dynamic threshold voltage).
(g) Construction of three-dimensional structures, such as *FinFET*s, as discussed in Chap. 5.
(h) High-voltage bipolar transistor, *MOSFET, JFET* applications, as discussed in Chaps. 3, 5, and 6.

The main disadvantages of *SOI* wafers are:

(a) The starting wafer cost. This can, however, be offset by process simplifications enabled by *SOI*, such as elimination of wells.
(b) Floating body-effect in high-performance *MOSFETs*. When the body is not tied to a fixed potential, impact ionization at the drain charges the body and reduces the threshold voltage, or even induces a bipolar effect that leads to voltage snap-back between source and drain. Several methods have been suggested to suppress this parasitic effect [29–32].
(c) Increase in thermal resistance. Since the thermal conductivity of silicon dioxide is over two orders of magnitude lower than that of silicon, the dissipation rate of power into the silicon wafer is reduced, leading to an increase in component temperature due to self-heating. This can be reduced by thinning the *BOX*.

The most common methods to produce *SOI* wafers are *SIMOX* (Separation by Implanted Oxygen), Bond and Etch-back (BESOI), and the Smart Cut.

Forming a silicon-dioxide layer by oxygen-ion implantation was first reported in [33] and further investigated in [34, 35]. It was first applied to device fabrication in [36]. Initially, the implantation conditions to form stoichiometric buried SiO_2 were an ion energy of 150 keV and dose $1.2\times10^{18}\,\mathrm{cm}^{-2}$ followed by an anneal cycle at 1150°C for 2 h and longer, resulting in a buried oxide thickness of about 200 nm. Because of the high-cost related to the required long oxygen implant and anneal times, the techniques was modified to *SIMOX* at essentially a lower dose of about $4\times10^{17}\,\mathrm{cm}^{-3}$, decreasing the buried-oxide thickness below ~100 nm [37,38].

Bonding of two wafers can be achieved by oxidizing one or both wafers and bringing them in intimate contact at elevated temperature [39]. This was initially accomplished by applying a voltage between the wafers, generating an electrostatic force in the gap between them that pulls the wafers together (Fig. 7.13) [40]. A more efficient method is pressing the oxidized surfaces of the two wafers together and subjecting them to an oxidizing atmosphere [41]. It is believed that in this process a partial vacuum is created between the wafers, bringing them in intimate contact and creating *Si-O-Si* bonds between them.

A controlled thinning of the device wafer is required to form the top silicon film with the desired thickness. This is achieved by grinding and polishing, or by etch-back with one or two highly selective "etch stops" built into the device wafer [39, 42]. The complexity of incorporating etch-stops and the additional cost of consuming two "bulk" wafers to produce one *SOI* wafer in *BESOI* are reduced by combining wafer bonding with hydrogen implantation to precisely define the top silicon depth [43, 44]. This technique, dubbed as "Smart Cut," can be described in six basic steps illustrated in Fig. 7.14 [44]:

(a) Thermal oxidation of the device wafer. This will form the buried oxide (*BOX*) when the *SOI* wafer is completed.
(b) Hydrogen implantation at a dose that can range from about $10^{16}\,\mathrm{cm}^{-2}$ to $10^{17}\,\mathrm{cm}^{-2}$ and an energy that defines the peak of hydrogen concentration (the projected range). Hydrogen creates micro-cavities parallel to the

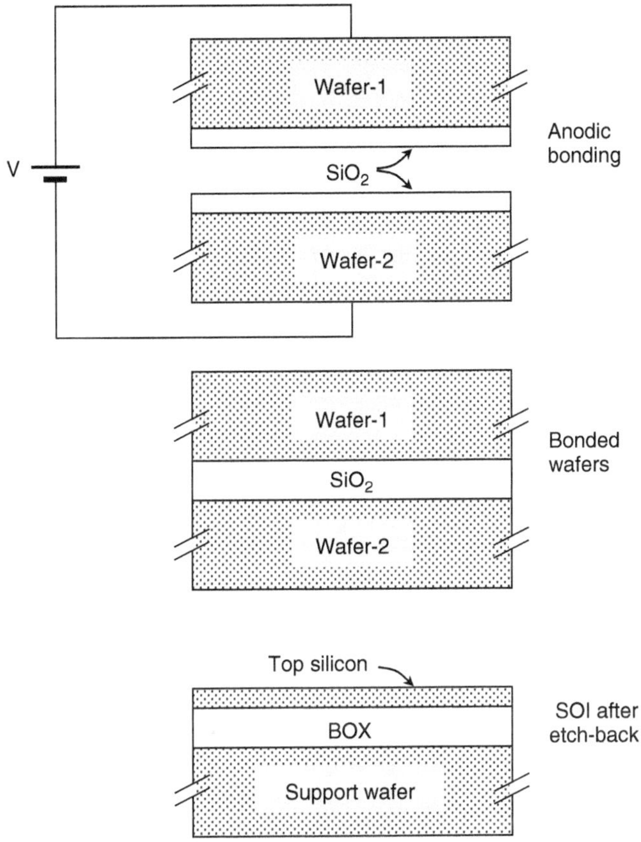

Fig. 7.13 Illustration of earlier anodic bonding of silicon on insulator and etch-back

wafer surface and, after bonding to a second wafer and thermal treatment at 400 °C–600 °C [steps (c) and (d)], induces an in-depth splitting over the whole wafer at the implanted projected range [44].

(c) Bonding the oxidized wafer to a second wafer that will be referred to as the support wafer.
(d) First thermal treatment at 400 °C–600 °C, resulting in propagation of cracks parallel to the bonded surface and splitting of the top silicon interface.
(e) High temperature treatment (1050 °C–1100 °C) to stabilize the bonded interface.
(f) Touch-up polishing to smooth the surface.

The split-wafer that was initially oxidized can now be used as the new support wafer when forming the next *SOI* wafer, eliminating the need for two "bulk" wafers to produce one *SOI* wafer.

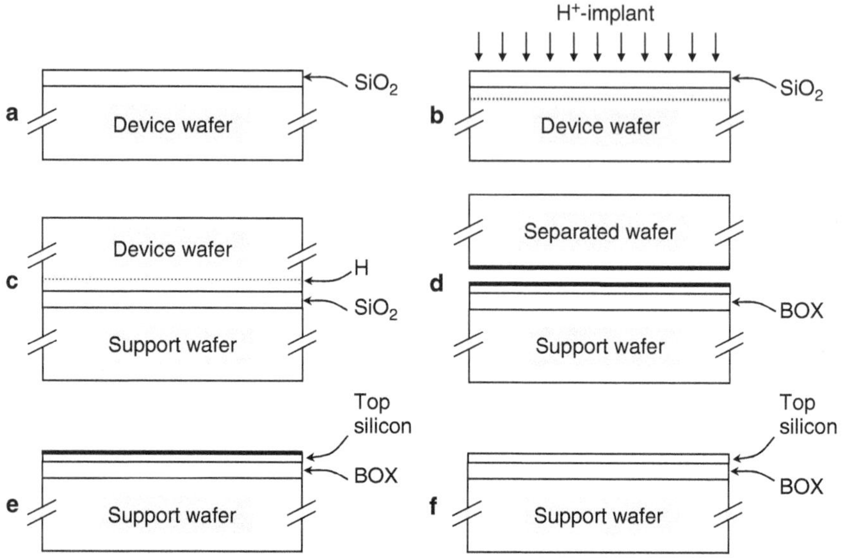

Fig. 7.14 The "Smart Cut" process sequence. **a** Oxidation of device wafer; **b** Hydrogen implant; **c** Bonding of oxidized device wafer to support wafer; **d** First thermal treatment to induce split; **e** Second thermal treatment to stabilize bond; **f** Touch-up polish (Adapted from [44])

7.4.2 Short-Duration Thermal Processes

Thermal processes are an integral part of process integration. They are required for several steps, such as activation of implanted species and annealing of implant damage, dielectric growth, gettering, silicidation, glass-reflow, passivation of the silicon-surface, and rapid-thermal deposition. Junctions such as the *MOSFET* source and drain and the emitter and base of bipolar transistors play a dominant role in device size and performance. It is therefore important to minimize and control the junction vertical and lateral extent, particularly at submicron, deep submicron, and nanoscale dimensions. This is accomplished by optimizing the total temperature-time product, referred to as the thermal budget. The objective is to achieve a high degree of implanted dopant activation, that is, to provide dopants with sufficient thermal energy and time to move them from interstitial to substitutional sites and to remove implant damage, while minimizing the vertical and lateral diffusion of dopants.

7.4.2.1 Rapid-Thermal Processes, RTP

Because of their large heat capacity, furnaces with large boat-loads of wafers require anneal-temperature cycles of tens of minutes and are hence not suitable for short-duration thermal processing. This led to the development of rapid-thermal-

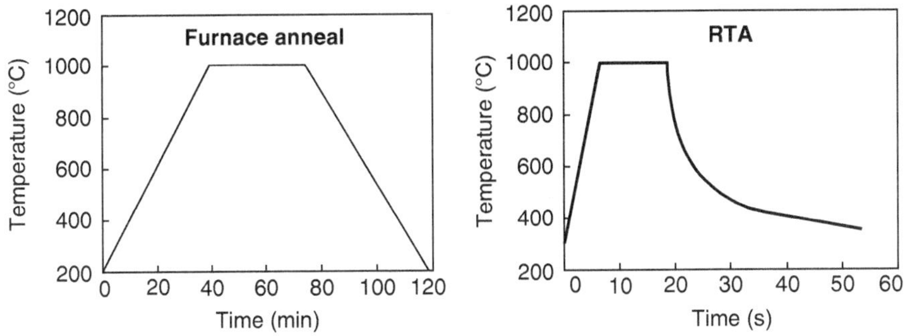

Fig. 7.15 Comparison of typical furnace and *RTA* anneal cycles (Adapted from [47])

processing (*RTP*) tools that typically consist of heating the entire wafer by tungsten halogen lamps for a wide range of temperatures and durations as short as a few seconds [45–48]. Figure 7.15 compares typical rapid-thermal anneal (*RTA*) and furnace-anneal temperature cycles. The *RTA* cooling rate is determined by radiative cooling from the wafer and hence is much slower than the heat-up.

Rapid-thermal anneal enables the reduction of junction depth while maintaining high dopant activation and low contact and series resistances [49–53]. In combination with $^{49}BF_2{}^{+}$-preamorphized silicon, *RTA* enables the formation of junctions as shallow as about 100 nm [49, 50]. The shallower $^{49}BF_2{}^{+}$-implanted junction compared to the boron-implanted junction is a result of two factors: amorphizing the surface impedes channeling, and fluorine reduces boron transient-enhanced-diffusion (*TED*) caused by the presence of implant-induced point defects [49].

High-speed silicon bipolar transistors with a polysilicon emitter acting as a diffusion source have also been obtained by highly activating the extrinsic base and simultaneously annealing the emitter and base junction depths with *RTA* [54]. A similar process simplification has also been demonstrated for high-speed analog/mixed-signal silicon-germanium (*SiGe*) *BiCMOS* processes by simultaneous *RTA* of the bipolar emitter and base and *MOSFET* source-drain junctions [55, 56].

7.4.2.2 Spike and Flash Anneal

As the *MOSFET* channel size is scaled down below 90-nm, ultra-shallow source-drain junctions of depth x_j far below 100 nm with surface concentrations above $10^{20}\,\text{cm}^{-3}$ become necessary to reduce short-channel effects while maintaining low contact and spreading resistances and reducing implant-induced defects (Chap. 5). Similarly, the depth of source-drain junction extensions and their sheet resistance R_S must be optimized. In addition, the silicide thickness must also be reduced below 40 nm and ultra-thin oxides of thickness less 5 nm must be controllably grown. These requirements have necessitated the extension of rapid-thermal processing tools to fast ramp-rates and peak-temperature durations of about 1–2 s, referred to as spike-anneal [57–63]. Further reduction in channel-size below about 45 nm requires

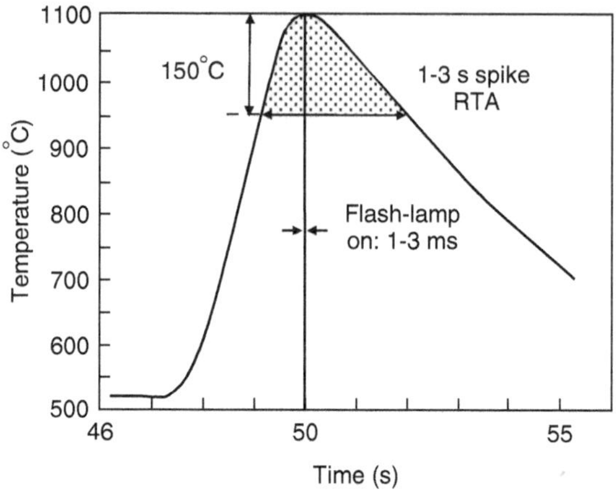

Fig. 7.16 Comparison of temperature-time profiles for spike and millisecond flash-lamp anneals

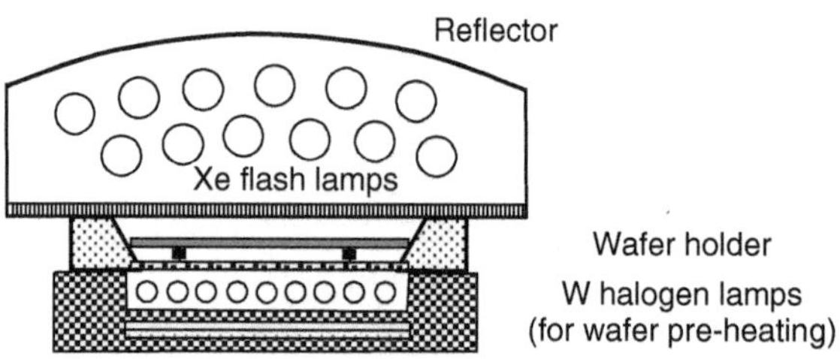

Fig. 7.17 Schematic diagram of flash-lamp anneal system (Adapted from [65, 68])

the development of extremely short heating and cooling rates and peak-temperature durations in the millisecond range, referred to as flash-lamp anneal (*FLA*) [64–72], or laser-spike anneal (*LSA*) [73–84]. Figure 7.16 compares the temperature versus time for *RTA*, spike and flash anneals. Spike anneal can be considered as an extension of *RTA* with shorter dwell times at high temperature. A typical spike anneal cycle consists of ramping up the temperature from a pre-anneal temperature of about 500 °C to 1000–1100 °C at a typical rate of 150 °C/s with 1–2 s dwell-time at peak temperature (e.g., above 950 °C), and then ramping down at a similar fast rate [57, 58, 61, 63]. A millisecond flash-lamp anneal system is shown schematically in Fig. 7.17. The top heat source is typically white-light *Xe* flash lamps assisted by bottom *RTA* with near infrared tungsten halogen lamps [65, 68]. The wafer is typically pre-heated with RTA prior to flash annealing. For example, after reaching an intermediate temperature of 700 °C to 750 °C, the wafer is flashed to peak temperatures of 1275 °C to 1325 °C [72]. The shorter wavelength of *FLA* and *LSA* causes the energy to be absorbed in a thin layer at the surface. The bulk of the wafer acts as a (solid) thermal heat sink for the heated surface layer. Thus, the cool-down in *FLA* and *LSA* is very rapid. The flash ramp rate is typically of the order of 10^6 °C/s for both ramp-up and ramp-down [64].

7.4.2.3 Laser Anneal

Laser annealing systems utilize *XeCl* or *KrF* high-intensity excimer *UV* laser pulses of nanosecond duration to heat the wafer surface either above the silicon melting point [74–84], or anneal the surface in a non-melting mode [85–91]. In the melt-anneal mode, the surface is pre-amorphized to a specific depth by implanting, for example, *Ge* or *Si* and then rapidly heated to the melting point by intense laser irradiation. The melt-temperature of amorphous silicon is about 250 °C below that of crystalline silicon, giving the process enough latitude to control the size of the molten zone [92–94]. Also, because the dopant diffusivity in molten silicon is about eight orders of magnitude higher than in the solid, dopants distribute almost uniformly in the melt and diffusion stops abruptly at the liquid-solid interface [74]. Melting and recrystallization occur in very short times, $<100\,\text{ns}$, due to the short laser pulses that cause only a very thin layer of silicon to melt so that the heat rapidly dissipates into the silicon bulk [76]. Thus, ultra-shallow abrupt junctions with dopant concentrations above the solid-solubility can be achieved [76, 78]. The main issues with laser melt-anneal is the residual defect density at the amorphized-crystalline boundary (referred to as end-of-range defects), the need for an amorphous layer, the impact of irradiation on the gate-stack and gate-dielectric integrity, and deactivation of dopant impurities during subsequent thermal cycles [95].

Non-melt laser anneals typically utilize 15–30 ns pulses at $0.4–0.6\,\text{J/cm}^2$ whereby the temperature reaches $\sim$1200 °C–1400 °C at an absorption depth in silicon of about 7 nm, while the silicon bulk remains essentially unheated [85]. At ultra-low boron implant energies of $\leq$500 eV, the projected range is about 2.5 nm and the entire boron profile is expected to be heated during the pulse. After the pulse, the heat dissipates into the bulk and the surface cools down in less than 1 ms [85, 86]. A high degree of dopant activation with negligible dopant movement can be achieved with non-melt laser anneal, enabling the design of high-performance *MOSFET*s with sub-5 nm channel length and reduced series and contact resistances [87, 89–91]. Figure 7.18 illustrates the improvement of sub-5 nm *MOSFET* channel halo control with the low thermal-budget millisecond (or laser spike) anneal when compared to conventional RTP.

An important consideration with *RTA*, *FLA* and *LSA* is the temperature control and uniformity, particularly in the presence of multiple patterns of varying layer stacks and optical properties that can cause significant local temperature variations. These pattern effects can be severe limitations of laser spike anneal and flash-lamp anneal [73].

7.4.2.4 Rapid Thermal Oxidation and Nitridation, RTO, RTN

The most common thermally-grown dielectrics on silicon are silicon-dioxide (SiO_2), and dielectrics incorporating nitrogen in SiO_2. The grown films serve primarily as gate-dielectrics [97–104], *DRAM* cell capacitor dielectrics, precision analog capacitor dielectrics discussed in Chap. 6 [105, 106], emitter-base interface

a
b
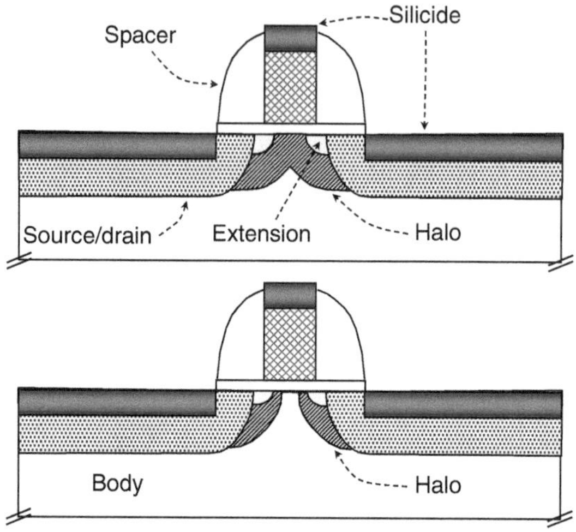

Fig. 7.18 Extension of halo design with millisecond annealing; (**a**) Conventional *RTA*; (**b**) Millisecond anneal (Adapted from [90, 96])

oxide discussed in Chap. 3 [56, 107, 108], and device isolation. This section presents examples where it is more advantageous to grow dielectrics by a rapid-thermal process than by a conventional furnace process.

Reduction of Thermal Budget with RTO

The main advantage of rapid-thermal oxidation over conventional furnace oxidation is the ability to grow SiO_2 at high temperature to achieve high-quality oxide while minimizing the thermal budget. For example, it is well known that adding chlorine-containing species, such as HCl, Cl_2, or $C_2H_2Cl_2$, to an oxidizing ambient greatly improves the electrical stability of oxide films and their interface with silicon by gettering metal ions and creating *Si-Cl* bonds at the surface [109–113]. This is especially important to gate oxides. To obtain good gate-oxide bulk and interface properties, however, the oxide film must be subjected to high temperatures, in the range 950 °C–1100 °C, during oxide growth or post-oxidation annealing [109, 110]. Rapid-thermal oxidation was introduced to reduce the overall thermal budget while achieving the high-quality thin thermal-oxide films of thickness 10 nm or below [97–100]. *RTO* films are typically grown at 1050–1150 °C in controlled oxygen ambient [98, 99].

Another example is the oxidation of silicon-germanium alloys. In some strained-*MOSFET* configurations, a small amount of germanium is added to silicon near the surface. Conventional furnace oxidation of the strained surface can cause the *SiGe* structure to relax with the formation of defects (Chap. 5). To avoid strain relaxation, the gate oxide is typically formed by a low-temperature process, but the quality of the oxide is poor when compared to conventional thermal oxidation. Also, oxidation of *SiGe* alloys reveals selective oxidation of silicon, rejecting *Ge* from the oxide layer and thus results in pileup of metallic-Ge at the oxide-silicon interface, which degrades the dielectric quality [102–104]. Application of rapid thermal oxidation *RTO* to *SiGe* strained layers is particularly attractive for minimization of thermal budget to avoid strain relaxation. In addition, *RTO* offers higher flexibility in reducing the *Ge* pile-up at the surface and better control over oxide growth in the thin oxide regime compared with furnace oxidation [102].

Suppression of Boron Penetration with Oxynitride

One of the potential problems with thin SiO_2 is boron penetration from the *PMOS* gate into the oxide and channel. A slight modification of pure SiO_2 by incorporating small mount of nitrogen into the oxide is found to not only reduce boron penetration but also to increase the dielectric constant and improve the dielectric field-strength and oxide-silicon interface stability [114–129]. The improvements are mainly attributed to the "pile-up" of controlled amounts of nitrogen at the silicon-oxide interface [118, 120]. The lightly nitrided oxides, referred to as oxynitrides, are formed by growing SiO_2 at high temperature in either furnace or *RTP* in an atmosphere containing a mixture of oxygen and ammonia (NH_3) [114–116, 121], nitric oxide (*NO*) [123, 128], or nitrous oxide (N_2O) [117–120, 124, 124, 129]. Rapid-thermal oxidation and nitridation offer higher flexibility in forming high-quality oxynitride films while maintaining a low thermal budget. For example, a thin oxide film can be grown in dry oxygen by *RTO* at 1100 °C, the oxide then nitrided by *RTN* in an ammonia atmosphere, and the nitrided-oxide reoxidized by *RTO* in dry oxygen at 900 °C–1150 °C. The reoxidized nitrided-oxide is referred to as *ONO* [116]. Ammonia was initially used as a nitrogen source because it would dissociate more readily than N_2 and provide more nitrogen at the $Si - SiO_2$ interface, resulting in a more efficient barrier to boron penetration. The resulting presence of large concentrations of hydrogen, however, causes an increase in charge trapping and *MOSFET* instability [121]. Because of those deleterious effects, other sources such as N_2O, NO, or a mixture of NO/N_2O are now more commonly used [117–120].

Trench Sidewall Oxidation

Oxide-filled shallow-trench isolation, *STI*, remains the preferred choice for device isolation in nanoscale technologies. A typical *STI* module consists of growing 15–25 nm "pad-oxide" and depositing 150–200 nm pad silicon-nitride, transferring the trench pattern to the silicon-nitride that acts as a stencil, and directionally etching silicon to a depth of 200–400 nm, depending on the technology node. The trench surface is then thermally oxidized to remove the etch-damage and provide a protective 10–20 nm thick liner prior to oxide-fill deposition and planarization.

Oxidation of the trench surface is an important step in the isolation module. It is typically done at a temperature above 1000 °C for a short duration of time to provide

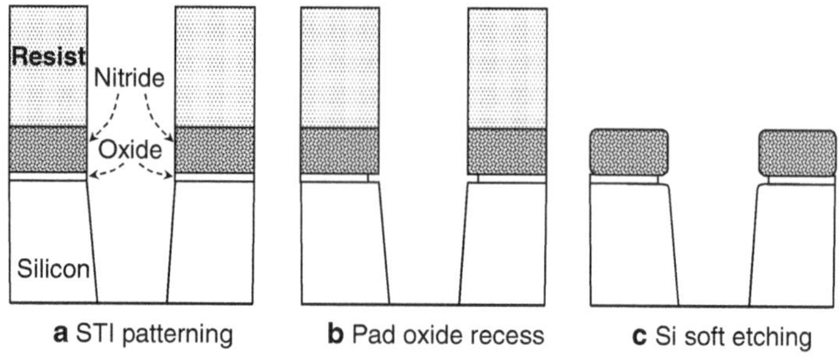

Fig. 7.19 Schematic of top *STI* corner rounding. **a** Silicon nitride and silicon etch; **b** Pad oxide recess in highly diluted *HF* solution; **c** Corner rounding by silicon soft etching in $O_2 + CF_4$ plasma (Adapted from [132])

a high-quality side-wall interface with silicon and some rounding of the trench corners. One important consideration with *STI* is the shape of the corner at the surface boundary between *STI* and *MOSFET* channel just before forming the gate stack (gate dielectric and conductor). Rounding the corner is necessary to suppress gate-oxide thinning and gate-polysilicon wrap-around the corner, both causing a reverse narrow-channel effect, *RNCE*, as described in Chap. 5. Corner rounding from high-temperature oxidation alone is found inadequate to suppress *RNCE* because of the small extent of rounding and the presence of a "divot" as described in Chap. 5. This has prompted the development of alternate techniques to suppress *RNCE* while minimizing stress-induced defects and encroachment into the channel [128–136]. One method consists of recessing the nitride stencil to expose the silicon corners, and then annealing in hydrogen to round the corners, as described in Fig. 5.49 of Chap. 5 [130]. This so-called two-step method was found to produce a corner with radius of curvature up to 35 nm [131]. Another method undercuts the pad oxide in a slow etchant, such as a 1:99 *HF* solution or a low-concentrated ammonium-fluoride solution, after *STI* etch (Fig. 7.19) [130, 131]. Silicon is then subjected to "soft," isotropic etching in an $O_2 + CF_4$ plasma mixture, resulting in the desired corner rounding [132].

The annealing ambient is also found to play a role in gate-oxide thinning at the *STI* corners. Annealing in an argon atmosphere after growing the sacrificial oxide greatly reduces gate-oxide thinning at the *STI* corners when compared to annealing in nitrogen [134]. The more enhanced gate-oxide thinning in nitrogen ambient is attributed to N_2 pile-up at the silicon-oxide interface at the *STI* corner, retarding oxidation [134].

7.4.3 Thin-Film Deposition

Thin-films can be formed by physical-vapor deposition, *PVD*, such as sputtering, or by chemical vapor deposition, *CVD* [1]. While *CVD* typically exhibits superior conformality in high aspect-ratio features, that is, depth or height of feature divided by its width, sputtering is preferred wherever applicable because of its simplicity and lower cost. This section describes two advanced deposition methods that have become increasingly important in deep submicron and nanoscale technologies as the aspect ratio of features, such as trenches or contacts, is further increased: Atomic Layer Deposition, *ALD* and Ionized Physical Vapor Deposition, *IPVD*.

7.4.3.1 Atomic Layer Deposition, ALD

Atomic layer deposition, *ALD*, initially called atomic layer epitaxy, *ALE*, was developed in the 70s [137]. It was initially introduced as a variant of *PVD* for epitaxial film growth of *ZnS* films for use as electroluminescent films. The process forming the film by sequential evaporation from two separate sources is illustrated in Fig. 7.20 [138]. The elements *Zn* and S_x were pulsed-evaporated in time from two separate sources to grow *ZnS* on a substrate at an adjusted temperature, using shutters, so that the substrate was exposed to one element at a time. The deposition conditions allow a single layer of sulfur to grow on a single layer of zinc, that is, the *Zn-S* bond ensures monoatomic *S* coverage [138]. Similarly, a single layer of zinc grows on a single layer of sulfur.

The process has exceptional features when compared to other deposition techniques [138]:

(a) It is self-limiting and the film thickness does not depend on the rate of reactions, provided that the dose in each reaction step is sufficiently high to ensure full

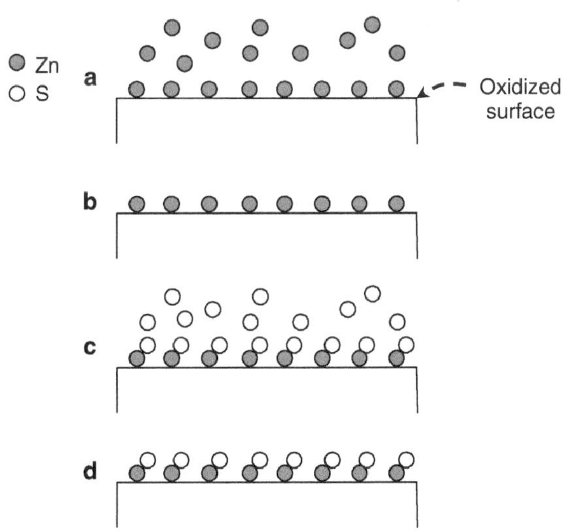

Fig. 7.20 *ALD (ALE)* one-cycle sequence for *ZnS* with separately evaporated *Zn* and S_x. **a** Oxidized wafer surface exposed to *Zn* vapor and bond formed between *Zn* and *O*; **b** Extra *Zn* is purged; **c** *Zn* surface exposed to S_x vapor and *Zn-S* bond ensures monoatomic *S* coverage; **d** Extra *S* purged. Cycle can be repeated to form another layer of *ZnS* (Adapted from [138, 139]

monolayer coverage. The self-limiting nature of the sequence is the foundation of *ALD* [141].

(b) The wafer temperature can be adjusted to result in re-evaporation from the surface of any loosely bonded atoms. In the absence of vapor-phase reactions, this results in a highly stable stoichiometric film.

(c) Process conditions allow only two-dimensional nucleation, resulting in very uniform layers even in ultra-thin films.

The *PVD*-based method described in [137] has found very limited practical use. Instead, *ALD* methods based on chemically reactive molecular precursors has become the dominant path [138–146]. The method is illustrated in Fig. 7.21 for the growth of *ZnS* films from $ZnCl_2$ and H_2S [138, 142].

Many *ALD* reactions use two volatile reactive chemicals, called precursors. The precursors are pulsed separately into the substrate and react sequentially with the surface one-at-a-time. Between the pulses, the reactor is purged with an inert gas or evacuated. In the first reaction step, the precursor is chemisorbed at the substrate surface, and during subsequent purging, excess precursors are removed from the reactor. In the second step, another precursor is introduced on the substrate and the desired film growth reaction takes place. Byproducts and excess precursors are then purged from the reactor, completing one reaction cycle.

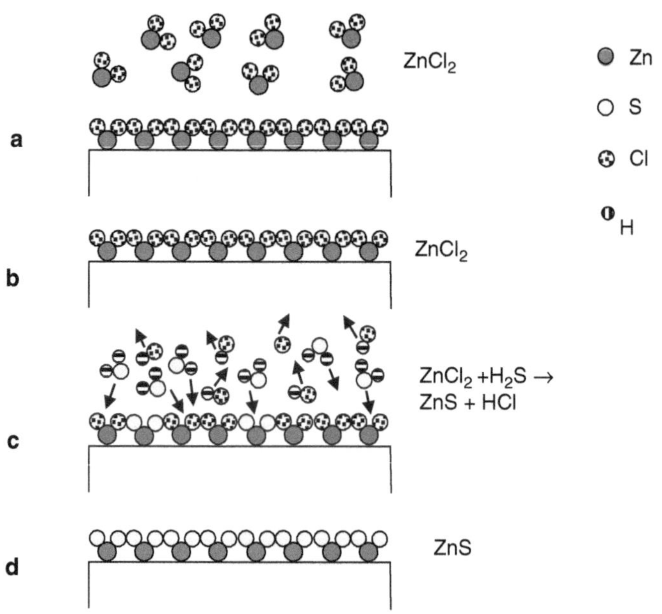

Fig. 7.21 Simplified schematic diagram illustrating the growth of *ZnS* from $ZnCl_2$ and H_2S [138, 142]. **a** Reactant vapor of $ZnCl_2$ pulsed onto wafer, $ZnCl_2$ chemisorbed on the surface; **b** Excess $ZnCl_2$ purged from the chamber; **c** Reactant vapor of H_2S pulsed onto wafer, reacts with $ZnCl_2$; **d** Excess reactants and byproduct *HCl* purged from chamber

The key steps of an *ALD* cycle can be summarized as follows:

1. A precursor gas is pulsed so that an atomic layer can be chemisorbed,
2. The precursor gas is flushed with an inert gas or evacuated,
3. The precursor is thermally reacted, often in the presence of another gas, for example, H_2O or O_2, to form a layer of the desired film,
4. The remainder of the precursor molecules is flushed away so that the entire cycle can be repeated.

The amount of material deposited in one *ALD* reaction cycle is referred to as growth per cycle, *GPC* [138, 142, 143]. The growth per cycle is typically much less than a molecular layer because the large physical size of the precursor prevents chemisorption of a complete atomic layer of the desired species.

The excellent control of film deposition, unmatched uniformity and conformality on surfaces of planar and three-dimensional structures have earned *ALD* widespread acceptance for depositing ultra-thin, conformal layers required for many nanoscale structures. This includes high-K gate dielectrics [144–153], metal gates [148, 154], *DRAM* cell capacitor-dielectric layers [155], *MIM* capacitor dielectric layers [156], diffusion barrier liners [157, 158], and copper seed layers [159]. Barrier layers and the copper seed layer are discussed in Sect. 7.5.1.

High-K dielectrics include hafnium-based dielectrics, such as HfO_2 and $Hf_xSi_{1-x}O$ [144–148], aluminum oxide Al_2O_3 [147, 152], zirconium dioxide ZrO_2 [145, 152, 153], tantalum oxide Ta_2O_5 [149], titanium oxide $\mathrm{TiO_2}$ [150], and strontium-titanate $SrTiO_3$ [151]. The films are typically formed from two precursors as described above. For example, *ALD* hafnium-dioxide can be grown using hafnium-tetrachloride ($HfCl_4$) and water as precursors [144, 146]. An ultra-thin silicon-dioxide film is typically grown as an interface oxide between high-K dielectrics and silicon to minimize interface traps and mobility degradation [145]. HfO_2 is grown at about 300 °C and 1.5 Torr. One growth cycle consists of a short pulse (<1 s) of H_2O followed by a short pulse of $HfCl_4$, separated from the first pulse by several seconds. The final reaction is [144, 146]:

$$HfCl_4 \uparrow +2H_2O \uparrow \Rightarrow HfO_2 \downarrow +4HCl \uparrow .$$

In *ALD*, the above reaction is divided into two successive half-reactions, the first with $HfCl_4$ and the second with $\mathrm{H_2O}$. Similarly, *ALD* aluminum oxide can be formed from trimethyl aluminum and water [147, 152].

7.4.3.2 Ionized Physical Vapor Deposition, IPVD

Physical vapor deposition by sputtering has the advantage of simplicity and low cost. Sputtering, however, results in the emission of mainly neutral atoms from the bombarded target surface. Thus, the emission profile depends roughly on the cosine of the emission angle with respect to the normal to the surface [160], the size of the sputter-target relative to the wafer-size, and on possible collisions of sputtered species with the residual gas in the chamber. This results in a varying angle

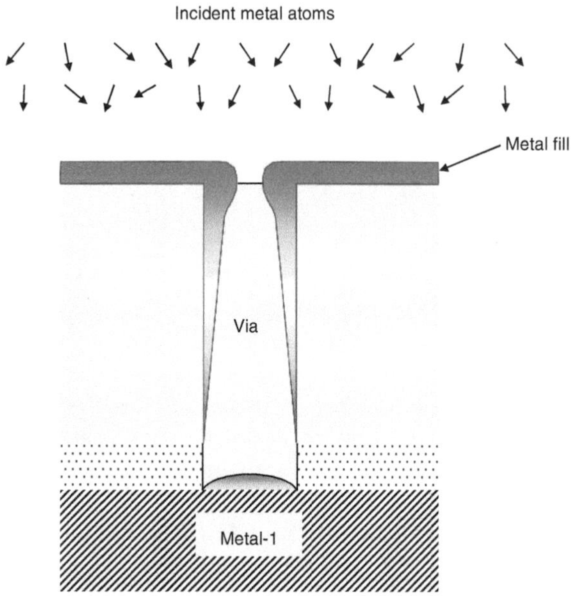

Fig. 7.22 Schematic of sputter-deposited film profile onto a medium aspect-ratio feature (Adapted from [161])

of incidence onto the wafer and a highly non-conformal film-profile along the surface of the opening, as illustrated in Fig. 7.22. Film conformality at any point of the trench surface is defined as the ratio of the deposited film thickness at that point to the thickness on the top flat surface.

To improve conformality, physical collimation of the sputtered flux can be achieved by a honeycomb filter between target and wafer that screens-out sputtered species with large-angle trajectories [161–163]. The collimator, however, significantly reduces the deposition rate and hence the through-put. These limitations have prompted the development of ionized physical vapor deposition, *IPVD*, which is an enhanced deposition technique that allows direct control of the sputtered flux by ionizing the sputtered material and subsequently directing a fraction of the ions toward the wafer to improve the floor-uniformity of the deposited film [164]. The method consists of placing a secondary plasma, typically an argon plasma, with a high electron density ($n_e \gg 10^{11}\,\text{cm}^{-3}$) between target and wafer to ionize the sputtered metal Fig. 7.23 [165, 166]. Metal is sputtered into high-density inductively-coupled plasma (*ICP*) or electron cyclotron resonance (*ECR*) plasma.

Metal atoms are ionized by electron impact and diffuse toward the wafer where they are collimated by the plasma sheath and directionally deposited. The percentage of sputtered flux that is ionized typically varies from 50% to 90% [166]. The method is primarily used to improve the uniformity of deposited metal films on the sidewalls and floor of high aspect-ratio vias [166–170]. Thin films of metals such as *TiN* and *TaN* act as diffusion barriers (liners) and adhesion promoters for aluminum or copper metallization (Sect. 7.4.1) [167–170]. Since the resistivity of

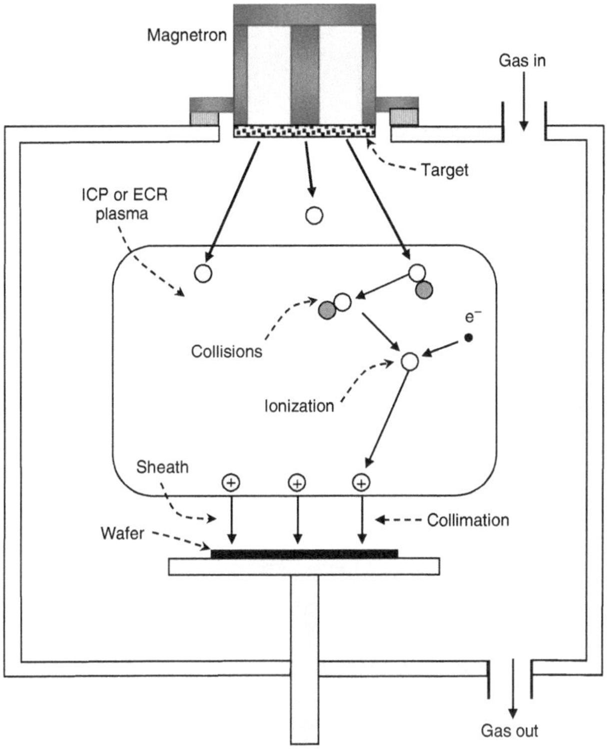

Fig. 7.23 Schematic of a typical ionized *PVD (IPVD)* reactor (Adapted from [165, 166])

barrier metals is typically higher than the resistivity of aluminum or copper, it is important to achieve good sidewall coverage of the barrier metal while minimizing the volume occupied by the barrier.

One method to improve sidewall coverage is to adjust the bias voltage applied on the wafer to control the directionality and bombardment energy of metal and argon ions incident upon the wafer. With sufficiently high ion-bombardment energy, metal can re-sputter from the bottom of the opening and re-deposit onto the sidewalls, improving conformality (Fig. 7.24) [167–171]. The film can also be deposited in two steps. In the first step, the metal is deposited by *IPVD* without applying a high bias on the wafer. This ensures adequate via floor coverage. In the second step, the wafer-bias is increased to accelerate ions to sufficiently high energies and re-sputter metal from the trench-floor onto the sidewalls [170, 171].

7.4.3.3 Lithography Enhancements

So far, technology cycles have been described by a single number directly related to the *DRAM* half metal-pitch in the array, that is, half the distance between the centerlines of adjacent first-level metal lines. This number has been referred to as

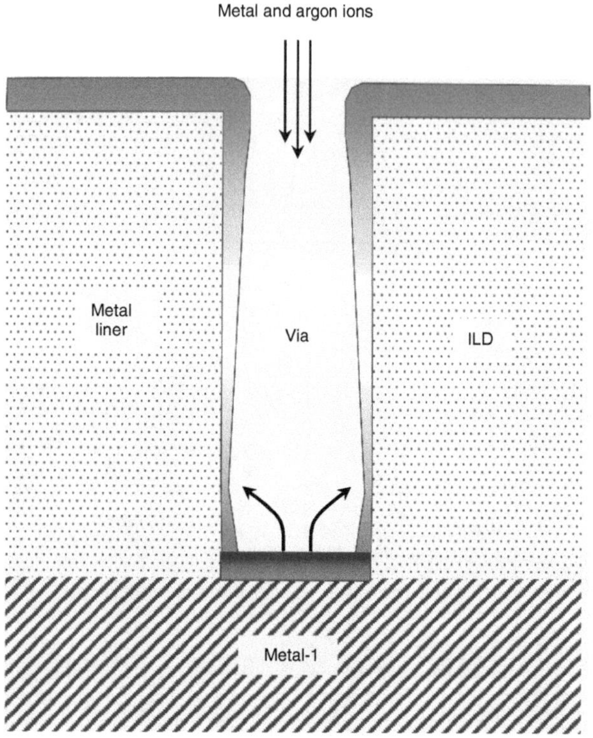

Fig. 7.24 Ionized physical vapor deposition, *IPVD*, with re-sputter from via or trench floor and re-deposition onto sidewalls to improve coverage [167]

the "technology node."[4] For example, the 130-nm process technology refers to an exposure system that can resolve a metal half-pitch of $\sim$130 nm (0.13 μm). The smallest line-width or line-to-line space that can be printed on a chip, called the critical dimensions *CD*, will be smaller than the *DRAM* half-pitch. From diffraction theory, the resolution of a projection imaging system is given by [1]

$$d = k_1 \frac{\lambda_0}{n \sin\theta} = k_1 \frac{\lambda_0}{NA}, \tag{7.1}$$

where d is the resolution, λ_0 the wavelength in vacuum, n is the index of refraction of the medium between lens and photoresist ($n = 1$ in vacuum), θ the aperture angle and NA is the numerical aperture. k_1 is a dimensionless factor that depends on the exposure system and resist. The resolution d can be decreased by reducing λ_0 and k_1, and by increasing NA. For example, the resolution has been extended

[4] The 2005 revision of the International Technology Roadmap for Semiconductors removes the concept of "Technology Node" and replaces it with the "Technology Trend Cycle" specific to a particular product. For continuity, the DRAM half-pitch is used here to describe the technology generation.

to sub-100 nm minimum features by reducing λ_0 from 240 nm, obtained from *KrF* excimer lasers, to 193 nm with *ArF* lasers. For patterns smaller than 65 nm, extension to a still shorter wavelength of 157 nm (F_2 excimer lasers) may be considered. The Raleigh limit of k_1 is 0.5, but the actual value typically ranges from 0.57 to 0.87 [172]. The 0.5 limit can be approached with resolution-enhancement techniques (*RET*), such as phase-shift masks and off-axis illumination [1], and optical proximity corrections (*OPC*) [173–179]. When combining multiple *RET*s, k_1 values near 0.25 are expected. The numerical aperture *NA* typically ranges from 0.6 to 0.8 in a *KrF* system and approaches 0.9 in an *ArF* system [172]. *NA* can be increased to values considerably above 1 by increasing the index of refraction *n*, as discussed below.

7.4.3.4 Optical Proximity Correction, OPC

Optical proximity effects are variations of a shape that depend on the proximity of the shape itself, for example, corners and ends to the line width, to other features that affect patterns [173]. Examples of proximity effects are the difference in printed line-width between an isolated line and a line in a dense array of lines of equal width and space, line-end foreshortening, and corner rounding of rectangular-shape images. Corner rounding can be qualitatively explained by considering that at corners, the intensity of exposure light and density of etch-reactants are shared between two edges, so that exposure and etch proceed slower at the corner itself than at edges away from the corner. Optical proximity correction (*OPC*) is a technique to introduce assist features to compensate for proximity effects. Automatic programs are available to place the correct size of assist features on a given layout [176, 177].

Regions of Different Line Density

Examples of regions of different line density are the center and edge of an array, or nested and isolated lines (Fig. 7.25). Without *OPC*, isolated lines typically print with a smaller width than densely packed lines. Also, the process window for printing small isolated features is typically smaller than for dense features because of the smaller depth of focus [173]. *OPC* is performed by selectively applying adjustments to the drawn line-width, or by placing sub-resolution lines (that do not fully print a pattern on the resist), called scattering bars, adjacent to resolvable lines (Fig. 7.25) [173, 174]. Scattering bars make isolated lines appear like dense lines.

Corner Rounding, Line-End Foreshortening

Examples of designs where line-end foreshortening becomes an issue are the polysilicon gate overlap on field oxide (Fig. 7.26). The drawn polysilicon-gate overlap of the active area must be sufficiently large to ensure that the gate is patterned

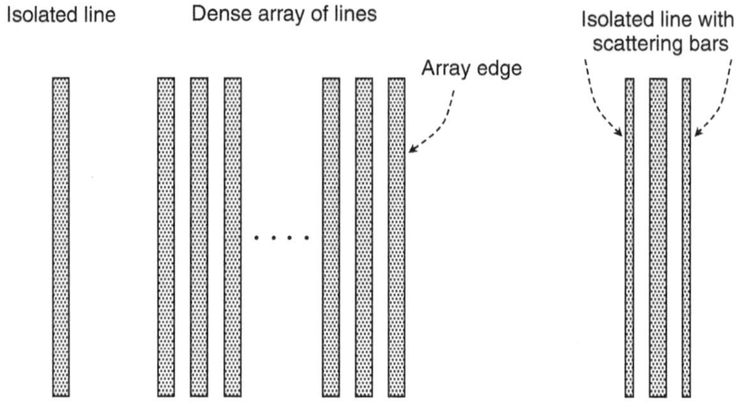

Fig. 7.25 Design of scattering bars to match the widths of isolated lines to the width of dense lines (Adapted from [174])

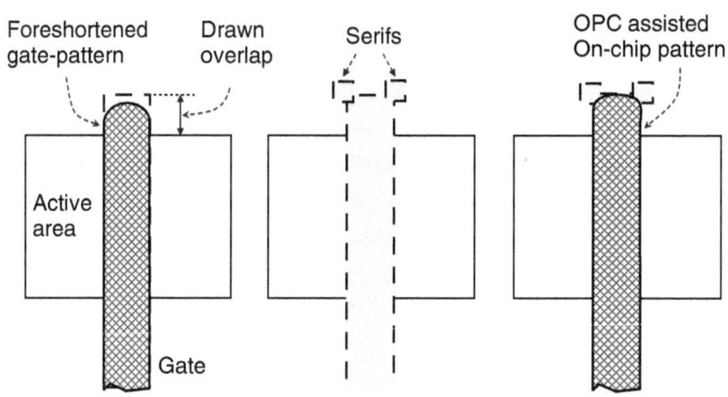

Fig. 7.26 Illustration of line-end foreshortening and correction by adding serifs at the end corners (Adapted from [176])

over the field-oxide area. Without *OPC*, line-end foreshortening would require additional overlap of the drawn shape to meet this requirement. Foreshortening is considerably reduced by adding serifs or "hammerhead" features to the gate pattern, as shown in Fig. 7.26.

Figure 7.27 illustrates how corner rounding can considerably reduce the effective area of a drawn rectangular shape, for example, the emitter of a bipolar transistor. Adding serifs to the corners results in a pattern that is close to the drawn shape.

7.4.3.5 Double Exposure and Double Patterning

Double exposure and double patterning have been introduced to reduce the effective factor k_{1-eff} (or $k_{1-pitch}$) for printing a pitch without the need to reduce the actual k_1 or *NA* of the exposure system (7.1). The methods consist of splitting the design into two masks to relax the minimum pitch and patterning the masks separately.

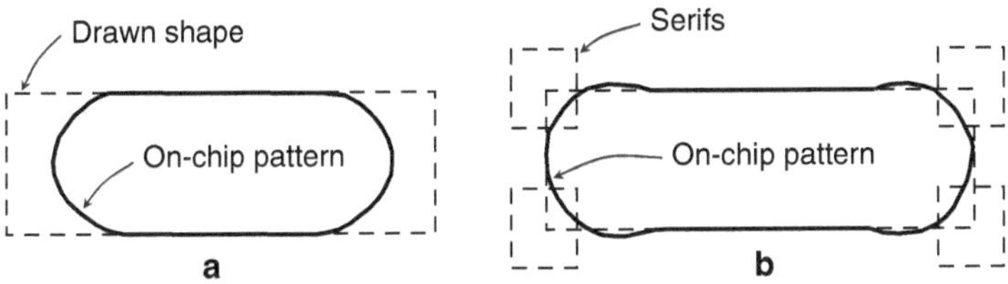

Fig. 7.27 27Illustration of line-end foreshortening and *OPC*. **a** Corner rounding and shape-end foreshortening without *OPC*; **b** Serifs added to the shape corners bring the printed image closer to the drawn shape (Adapted from [176])

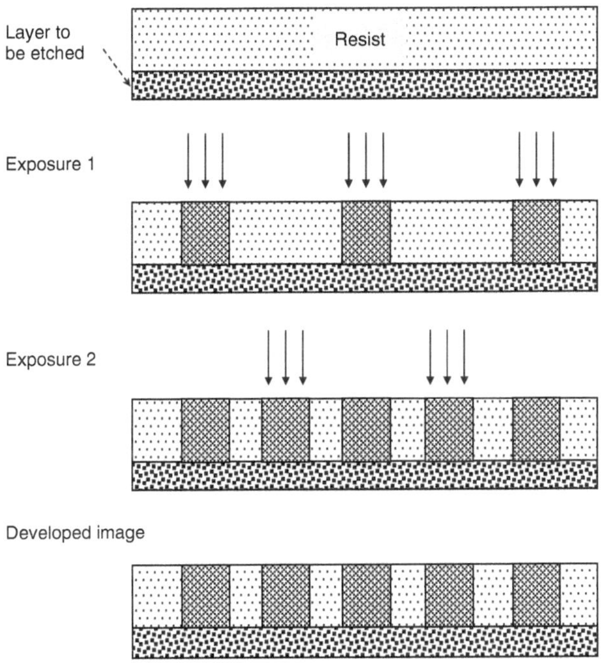

Fig. 7.28 Illustration of double exposure sequence

In double exposure, two same or different masks and two same or different illumination settings are sequentially used to print the desired pattern on the same photoresist layer which is subsequently developed and etched into the substrate Fig. 7.28 [180]. This technique is commonly applied to printing patterns of different dimensions and densities or pitches in the same layer, allowing independent optimizations of exposure conditions for each set of patterns [180–185]. The two exposures may, for example, each consist of lines which are oriented in one or the other of two typically perpendicular directions [181]. In this case, double exposure or patterning allows the decomposition of two-dimensional patterns into two one-dimensional patterns which are easier to print.

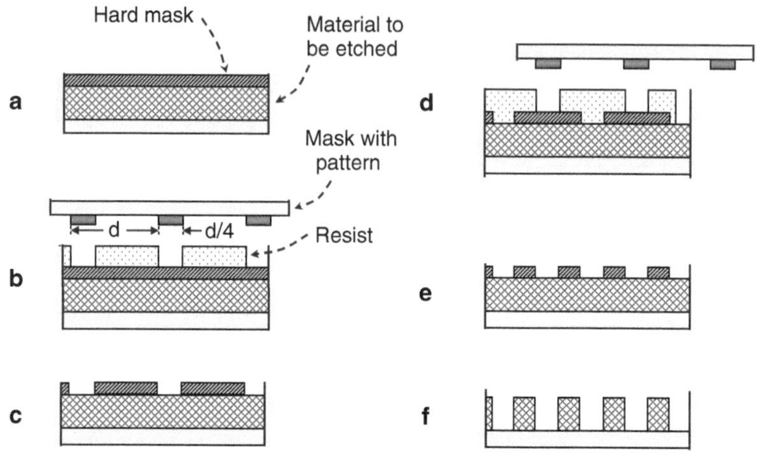

Fig. 7.29 Illustration of double patterning by pitch-splitting with hard mask [180]. **a** Hard-mask application; **b** First trench patterning with pitch = d and CD = d/4; **c** Hard-mask etch and resist stripping; **d** Resist coating and trench patterning shifted ½ pitch; **e** Second hard-mask etch and resist stripping; **f** Final pattern

Double patterning is distinct from double exposure inasmuch as sequential exposures are made on different resist layers, with each exposure followed by one or more etching steps to transfer the resist patterns into the substrate [180]. This avoids the interaction of images on the same resist, resulting in higher resolution than with double exposure alone. Several double patterning schemes have been proposed. Among those are the double-exposure-double-etch scheme shown in Fig. 7.29 [180], and the self-aligned spacer-patterning shown in Fig. 7.30 [186]. The sequence in Fig. 7.29 (shown for positive resist) makes the line-width *CD* independent of overlay [180].

Double exposure and double patterning have several concerns that have delayed introduction into manufacturing [180]. Overlay is a major concern since the second feature must be precisely positioned with respect to the first. Cost is another concern because of the increased complexity in mask design, imaging and process integration, and reduced yield.

7.4.3.6 Immersion Lithography

Immersion microscopy has been well-known for over a century.[5] By placing a medium, such as mineral oil or water, with an index of refraction $n > 1$ between objective and object, the wavelength in the medium is reduced to $\lambda = \lambda_0/n$, effectively increasing the numerical aperture and increasing the resolution by a factor n (7.1).

[5] Ernst Abbe, a German Physicist, was the first to discover late in 1870 that the maximum ray slope entering a lens from an axial point on an object could be increased by a factor equal to the refractive index of the medium between lens and object.

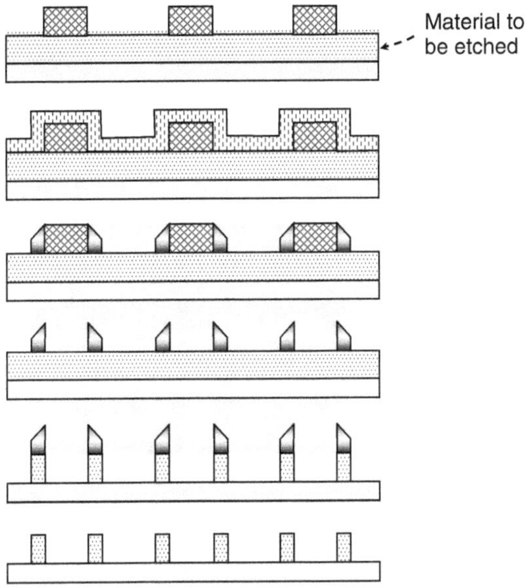

Fig. 7.30 Illustration of double pattering with self-aligned spacer-etch scheme [186, 187]

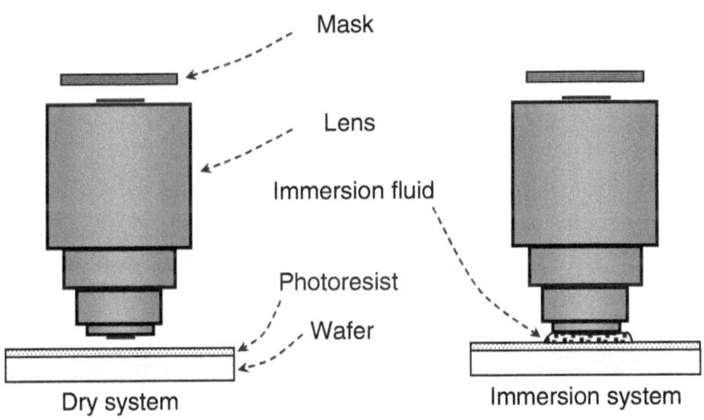

Fig. 7.31 Schematic representation of dry and immersion systems (Adapted from [189])

The method was applied to optical lithography, using a microscope objective as the projection lens [188]. It is referred to as an immersion lithography system (Fig. 7.31). It offers a cost-effective attractive alternative to extending the capability of projection systems by reducing the illuminating wavelength because it does not require the development of new masks, lens and resist material [189]. The requirements on the immersion fluid are uniform high index of refraction, transparency, chemical stability and compatibility with the chosen resist, mechanical stability and bubble-free so it can be used in step-and-scan systems with high-speed lens motion [190].

A numerical aperture $NA = 1.4$ was demonstrated on a commercial optical microscope with high-index oil-immersion ($n \approx 1.5$) optical lithography using 453-nm wavelength. The smallest feature produced was a 230-nm isolated line [191]. Oil immersion was also predicted to extend the *ArF* 193-nm system to $NA = 1.05$ and 125-nm minimum feature at $k_1 = 0.68$ [192]. Other immersion liquids were investigated. Among them are cyclic-hydrocarbons with index of refraction $n \approx 1.5$ at $\lambda_0 = 257\,\text{nm}$ [193, 194], and perfluoropolyethers with $n \approx 1.36$ at $\lambda_0 = 157\,\text{nm}$ (F_2 excimer laser) [194], and water $n \approx 1.44$ at $\lambda_0 = 193\,\text{nm}$ [189, 195, 196]. Water is most attractive because of its ease of use and high index of refraction, effectively reducing the wavelength to $\lambda = \lambda_0/1.44 = 134\,\text{nm}$. This represents an improvement in resolution by 43% [196].

The depth of focus (*DOF*) for immersion imaging is defined based on the effective reduction in wavelength as

$$DOF = k_2 \frac{\lambda_0}{nNA^2}. \tag{7.2}$$

This is significant since the effective *DOF* scales linearly with $1/n$ compared to quadratically with $1/NA^2$ [196, 197].

7.4.3.7 Extreme Ultraviolet and Nanoimprint Lithography

Extreme ultraviolet lithograph (*EUV* or *EUVL*) and nanoimprint lithography (*NIL*) are among the techniques considered for the extension of printing capabilities to 32-nm and smaller dimensions.

EUV systems typically contain several high-reflectivity condenser and projection *Mo/Si* multilayer mirrors [198–200]. The *EUV* wavelength of choice has been selected as 13.5 nm (92 eV photon-energy) because *Mo/Si* mirrors reach a high reflectivity of about 72% at this wavelength [198]. With eight mirrors in the path from source to mask, the optics would then absorb about 93% of the available EUV light. Thus, the light source, typically from plasmas generated by laser or pulse discharges, must be sufficiently bright to compensate this loss. The 13.5-nm wavelength can be produced by transitions of lithium, xenon, or tin plasma-ions, with more focus on the latter two ions. Because of the high radiation energy, *EUV* systems require high vacuum to avoid absorption and ionization along the path, that is, the system must be evacuated after each wafer exposure, reducing the throughput. Other considerations are damage to mirrors and resolution limits similar to those of e-beam and x-ray lithography.

The principle of nanoimprint lithography (*NIL*) is shown schematically in Fig. 7.32 [201]. A mold with the nanostructures is pressed into a thin resist film, for example, polymethymethacrylate (PMMA) that is heated above its glass transition temperature. This creates a thickness contrast pattern in the resist that duplicates the mold nanostructures.

After removal of the mold, anisotropic etching removes the compressed resist regions in the resist, resulting in the desired resist pattern. Silicon dioxide and silicon were initially used as mold materials. The mold was patterned with electron-beam lithography and reactive-ion etching (*RIE*). Sub-10 nm imprints were demonstrated with this technique [202]. Other mold materials are being considered.

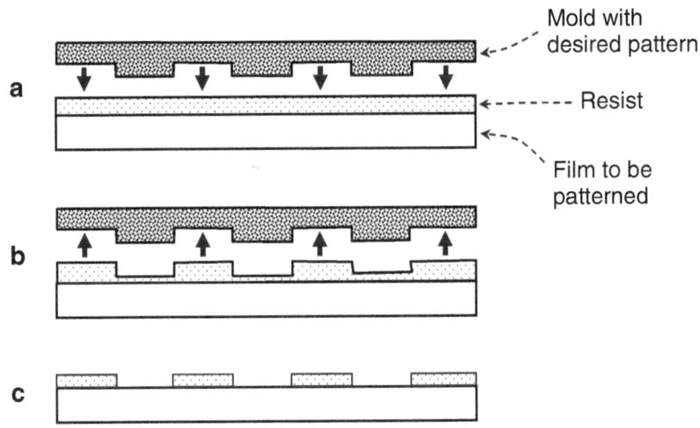

Fig. 7.32 Schematic of a nanoimprint process. **a** Imprint with mold to create a thickness contrast in resist; **b** Mold removal; **c** Pattern transfer with *RIE* to remove residual resist in compressed areas (Adapted from [30])

The major issue with *NIL* is the defect density created by the contact between mold and resist. As a result, *NIL* has been limited so far to designs with less stringent requirements on defects and throughput than deep submicron and nanoscale *CMOS*. Among these are storage media, *MEMS*, flat-panel displays, and biomedical devices [203].

7.4.4 Integration of Ultra-Shallow Junctions

Junctions with a depth x_j<100 nm are referred to as ultra-shallow junctions, *USJ*. Techniques to reduce the junction while maintaining high sheet resistance are discussed in this section.

7.4.4.1 Low-Energy Ion Implantation

Several factors affect the implanted profile in conventional beam-line ion implantation. Among them are the ion energy, mass, charge and dose, the silicon crystallographic orientation with respect to beam, and the degree of surface amorphization [1]. The implantation energy must be reduced to meet the requirements on scaling vertical and horizontal dimensions to deep submicron and nanoscale dimensions. In particular, ultra-shallow $(x_j < 100\,\text{nm})$, low-leakage source-drain and source-drain extension junctions must be formed at ultra-low implant energies of 500 eV or less, and high dose to maintain the required low sheet resistances [204, 205]. This can be done in conjunction with pre-amorphization with silicon or germanium [204–207], and rapid-thermal [207–210], or spike-anneal [211].

Ultra-low implant-energies are particularly important to form ultra-shallow p^+n junctions because of the low boron mass and transient-enhanced diffusion (*TED*). An alternative approach to ultra-low implant energy to achieve ultra-shallow boron profiles is to use boron-containing molecular ions, such as BF_2^+ of molecular mass 49 [1], BCl_2^+ of molecular mass 81 [211], or decaborane $B_{10}H_{14}$ of molecular mass 124 (that ionizes as $B_{10}H_x^+$) [212, 213], which compares to a mass of 11 for boron alone. Thus, the ratio of effective boron energy to the acceleration energy of the molecules is $\sim 11/49$ for BF_2^+, 11/81 for BCl_2^+, and 11/124 for decaborane. For example, when decaborane is implanted at energy 56 keV, the effective boron implant energy is about 5 eV.

Conventional beam-line ion implantation at low energy suffers, however, from fundamental limits that include space charge limitations and inability to efficiently and uniformly dope three-dimensional structures, such as deep trenches and *FinFETs*. The space charge limited current is illustrated for a cathode ray tube in Fig. 7.33. A fraction of the field lines end on the emitted electrons, that is, the space charge of electrons distorts the field. When the field is low, a point is rapidly reached where the cathode is totally shielded from the anode field by the charge of electrons, resulting in zero-field at the cathode and the current saturates. In this case, the current is said to be space-charge limited to

$$j = \frac{4\varepsilon_0}{9}\sqrt{\frac{2q}{m}}\frac{V^{3/2}}{d^2} \quad A/cm^2, \tag{7.3}$$

where j is the current density, ε_0 is permittivity of free space, q the electron charge, m the electron mass, V the voltage between anode and cathode, and d the distance between anode and cathode. Equation (7.3) is referred to as the Schottky-Langmuir space-charge law. A similar situation occurs in an ion implanter operating at a low accelerating voltage, that is, low ion energy. The ion beam current is reduced and the transport of the beam from the ion source to the target becomes more difficult,

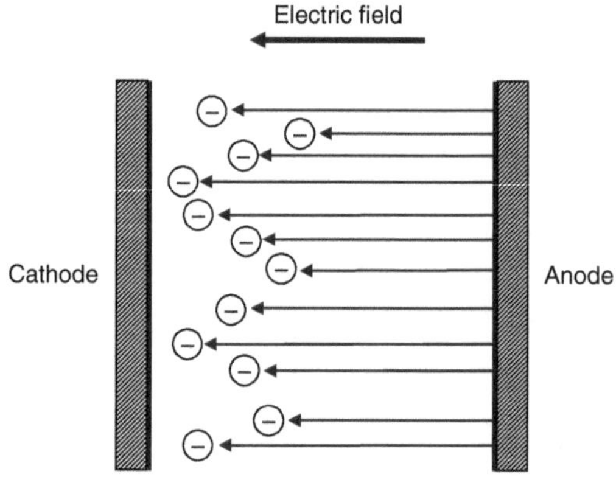

Fig. 7.33 Effect of electron space charge on the electric field between anode and cathode

leading to a strong degradation of throughput and uniformity when the beam energy is decreased below 1 keV [214].

Space-charge limited current is also observed when carriers are injected into an insulator, such as oxide, or a nearly intrinsic semiconductor. In this case, the space-charge limited current law is referred to as Child's law.

Another limitation of beam-line implantation is its inefficiency to uniformly dope high-density, high-aspect-ratio three-dimensional structures, such as deep trenches and *FinFETs*. This is because of very large tilt-angle requirements to implant side-walls of trench- and fin-structures, and shadowing effects on the ion beam which cause non-uniform doping of sidewalls.

7.4.4.2 Plasma Immersion Ion Implantation. PIII

To solve the above technical limitations, plasma immersion ion implantation, *PIII*, also know as plasma doping, *PLAD*, or pulsed plasma doping P^2LAD has been proposed as an alternative to beam-line ion implantation [215–225]. The basic technique was first demonstrated in [226]. A typical configuration of a pulsed plasma doping system is shown schematically in Fig. 7.34 [224]. In this configuration, the plasma is continuous and the ion source is readily available when the voltage pulse is applied.

Plasma immersion ion implantation offers many inherent advantages over conventional beam-line implantation [216]. Among them are: higher throughput than

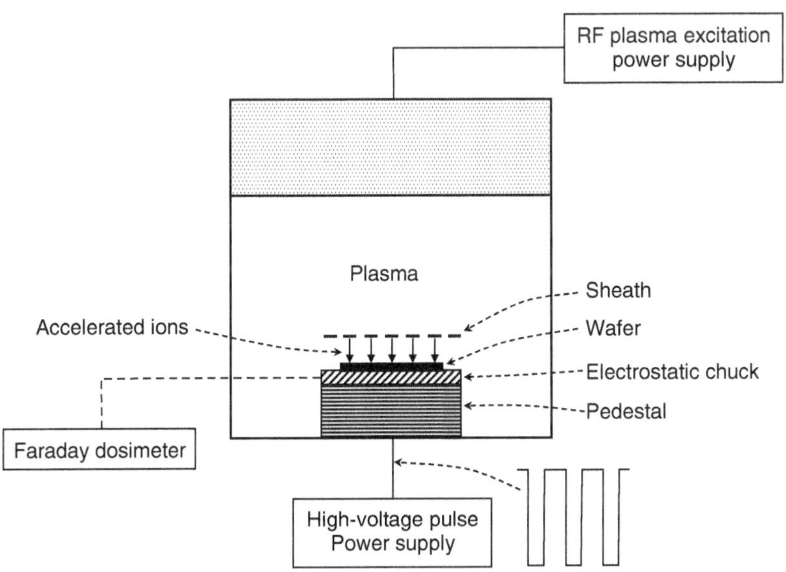

Fig. 7.34 Schematic configuration of a pulsed RF-excited continuous P^2LAD system (Adapted from [224])

those attainable with conventional accelerator beam-lines due to the higher current densities possible with plasma sources [214, 218–220, 224, 225], the higher doping uniformity over larger areas [218–220], the capability of uniformly-doped three-dimensional structures [214, 215, 223], system simplicity and lower cost [218, 224]. As a result, a better trade-off between sheet resistance R_S and junction depth x_j can be achieved with *PIII* than with beam-line implantation. It is also claimed that a lower R_S can be obtained for the same x_j or a smaller x_j for the same R_S [214, 221]. *PIII* has, however, serious limitations in comparison with beam-line implantation [225]. There is no systematic ion-mass separation - all positive ions in the plasma are implanted to some extent. Also, ions are not strictly mono-energetic because the distribution of their energies depends on several factors including gas pressure, pulse shape and plasma density. So far, the technology has not been adopted for manufacturing.

7.4.4.3 Suppression of Diffusion

Diffusion of impurities in silicon is fundamentally related to the interaction between impurities and point defects. Native point defects are silicon vacancies and self-interstitials. Boron and phosphorus diffuse essentially by the mechanism of interstitials, while arsenic and antimony diffuse by the vacancy mechanism [1]. Thus, the junction profile can be strongly influenced by incorporating "neutral" species that affect the density of point defects and hence the dopant diffusivity in the vicinity of the junction. Among these are carbon, fluorine and nitrogen. Figure 3.48 in Chap. 3 illustrates how placing substitutional carbon at a concentration of about 1% in the vicinity of a region doped with boron, phosphorus, arsenic, or antimony, can retard the diffusion of boron and phosphorus and accelerate the diffusion of arsenic and antimony [227]. The plots are shown again in Fig. 7.35 for convenience.

The retardation mechanism is attributed to the creation of silicon interstitials by the so-called kick-out mechanism in which substitutional carbon is replaced by interstitial silicon, creating a flux of silicon interstitials from the doped region toward the carbon-rich region and retarding diffusion of boron and phosphorus, and suppressing transient-enhanced diffusion, *TED*. Also, interstitial carbon can be formed by a dissociative reaction whereby substitution carbon and vacancies are created. The flux of vacancies toward the doped region accelerates the diffusion of antimony and arsenic [227]. Carbon can be incorporated in the desired region epitaxially by adding a carbon-containing compound, such as CH_4-SiH_3, to the gas mixture [228–234], or by co-implantation [235–238]. In both cases, carbon is typically placed just below regions doped with boron or phosphorus where retardation of diffusion and suppression of *TED* is critical, such as the bipolar base and *MOSFET* channel, source-drain, source-drain extensions and halos.

Transient enhanced diffusion of boron is also found to be suppressed by co-implantation of fluorine [90, 239]. The suggested model is that fluorine traps self-interstitials, retarding boron diffusion, and also forms B-F complexes, deactivating a fraction of substitutional boron [239].

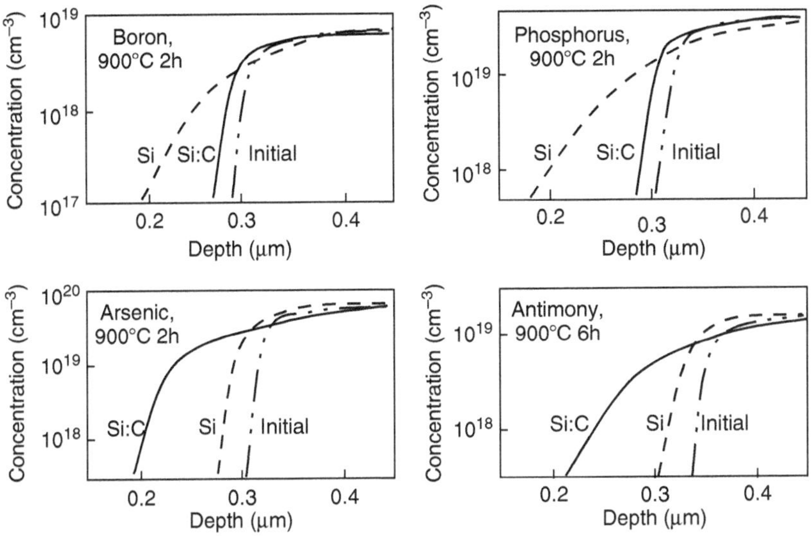

Fig. 7.35 Diffusion of B, P, As, and Sb from highly doped substrates into Si and Si:C layers with substitutional carbon at a concentration of 1% placed in the vicinity of the doped region (Adapted from [227])

7.4.4.4 Junction Silicides

As the source-drain junction depth is reduced to ultra-shallow dimensions, the silicide thickness must also be reduced to maintain low junction leakage. Thus, the optimal silicide thickness becomes a trade-off between sheet resistance, contact resistance, uniformity, and junction leakage [240]. In deep submicron technologies, this trade-off becomes more difficult to achieve with titanium silicide ($TiSi_2$) when the source-drain junctions are silicided simultaneously with the polysilicon gate. The limitation of titanium silicide is the difficulty in transforming the thermodynamic metastable high-resistivity (60–90 μΩ-cm) *C49* phase to the stable low-resistivity (12–20 μΩ-cm) *C54* phase as the polysilicon line-width is decreased from about 1.5 to 0.1 μm. Films composed of the *C49* phase require higher temperature to form the C54 structure when the lateral and vertical dimensions of silicide features are reduced. This is attributed to the decreased number of nucleation sites for the *C54* phase in narrow lines. Higher temperature annealing to form the *C54* phase results in agglomeration and an increase in sheet resistance [241–243]. Agglomeration can be suppressed by depositing thicker titanium and forming a thicker silicide film. However, a thicker silicide film consumes more of the junction and results in increased junction leakage in ultra-shallow junctions. The observation of a line-width dependence of the transformation of $TiSi_2$ to its low-resistance phase was the main reason that cobalt silicide was considered as an alternative to titanium silicide for line-widths below ~0.25 μm [244]. The cobalt-silicide sheet resistance was found to remain constant for n^+-polysilicon and p^+-polysilicon even for line-widths as narrow as 0.08 μm [245].

Nickel mono-silicide (*NiSi*) offers several advantages over titanium and cobalt di-silicides [246–248]: (a) About 30% less silicon consumption than $TiSi_2$ and $CoSi_2$ for the same sheet resistance, which is important when optimizing sheet resistance and leakage in ultra-shallow junctions; (b) Low silicidation temperature; (c) No line-width dependence of sheet-resistance down to sub-50 nm line-widths [247–250], while agglomeration of $CoSi_2$ has been observed for line-widths below ~50 nm [247]; (d) Lower silicidation temperature almost eliminates dopant uptake by the silicide that would cause non-ohmic contact-behavior. Dopant pile-up at the silicide interface and a lower silicide to silicon barrier height for holes is observed in *PMOS*, which further reduces the contact resistivity to *PMOS* source and drain [246, 250].

Nickel mono-silicide is typically formed by sputter-depositing a 9-nm to 30-nm *Ni* film, depending on junction depth, and annealing by *RTA* at 400 °C–600 °C in an inert atmosphere [246–250]. The film is typically capped with a *Ti* or *TiN* layer to avoid oxidation of the silicide-silicon interface [251]. If heated between 200 °C and 300 °C, a higher-resistivity Ni_2Si film begins to form. The film becomes unstable at around 300 °C and converts to the *NiSi*, which is of lower sheet resistance [246]. One important difference between nickel-silicide and cobalt- or titanium-silicide is that silicon diffuses out to the *Ti* or *Co* film to form the silicide while *Ni* diffuses in toward the silicon interface to form *NiSi*, thereby eliminating bridging of the silicide over dielectric films on the *STI* or spacers [246].

In the presence of excess *Si*, which is typical when contact is formed on silicon, thin *NiSi* films react with *Si* to form $NiSi_2$ during post-silicidation processing temperatures of about 700 °C–750 °C. Adding platinum (*Pt*) to *Ni* can improve the thermal stability of *NiPtSi* by delaying the formation of $NiSi_2$ [252, 253]. There is another important advantage of alloying *Pt* with *Ni* that is related to silicon-germanium (*SiGe*) polysilicon gates. Germanium is incorporated in the source and drain of *PMOS* to induce a lateral compressive stress that enhances the hole mobility (Chap. 5). In typical *MOSFET*s, *SiGe* is also present in the gate. The *Ge* in polysilicon SiGe gates tends to segregate out of the *SiGe* during silicidation reaction, so that the amount of *Ge* in the resulting germanosilicide $NiSi_{1-x}Ge_x$ is considerably less than in the starting film. This segregation becomes more pronounced as the *Ge* fraction increases. This is attributed to the higher temperature required to form *NiGe* compared to *NiSi* and it exacerbates the tendency of the germanosilicide to agglomerate [254]. Adding 5–10% *Pt* to *Ni* results in greater Ge incorporation in the germanosilicide thereby reducing the agglomeration and improving its thermal stability [255]. A contact resistivity r_C lower than 10^{-8} Ohm-cm^2 was measured for *NiPt* germanosilicide [256]. The integration of $Ni_{0.9}Pt_{0.1}SiGe$ contacts was demonstrated on tri-gate *FinFET*s with an increase of 18% in *PMOS* drive current when compared to *NiSiGe* [257]. This is attributed to the lower source-drain sheet and contact resistances.

7.4.5 Gate Stack Module

The MOSFET gate stack consists of multiple layers that form the gate dielectric, the gate conductor, and their interfaces. Figure 7.36 illustrates the transition from a stack of silicided, doped polysilicon gates on SiO_2 as the gate dielectric (Fig. 7.36a) to full metal gates on a high-K dielectric (Fig. 7.36b). This section discusses high-K dielectrics and metal gates, including fully-silicided (FUSI) polysilicon gates.

7.4.5.1 High-K Dielectrics

Because of their excellent interface and bulk properties, silicon-dioxide (SiO_2) and oxynitrides (*SiON*) have served as primary gate dielectrics down to nanoscale $(\sim 100\,\text{nm})$ dimensions. As the channel length is reduced, the physical thickness of SiO_2 must also be reduced to maintain adequate gate control of the inversion layer and reduced short-channel effects. At a channel length of about 65 nm and below, however, the physical thickness of SiO_2 must be reduced to less than 1.2 nm where the rapid increase in gate leakage due to direct tunneling makes further scaling impractical (Chap. 5). This fundamental limit prompted the development of insulators of higher dielectric constant than SiO_2, called high-*K* dielectrics, that can be deposited physically thicker than SiO_2, resulting in a significant reduction in gate leakage due to tunneling for the same equivalent oxide thickness, teq (*EOT*). Among these are hafnium-based dielectrics [258–267], sputter-deposited zirconium-oxide (ZrO_2) [268], thermally formed, evaporated, or atomic-layer deposited aluminum oxide (Al_2O_3) [269, 270], and plasma-enhanced *CVD* tantalum oxide (Ta_2O_5) with TaF_5 as a precursor in an $O_2 + H_2$ atmosphere [271].

Hafnium-based dielectrics, such as HfO_2, *HfSiON*, *HfAlON*, and hafnium silicates have emerged as primary choices of gate dielectrics because they satisfy most *CMOS* key dielectric requirements [261–267]. Among these requirements are the dielectric constant, bandgap, band alignment to silicon and thermodynamic stability [260]. Hafnium dioxide can be deposited by reactive co-sputtering of *Hf* in an $Ar + O_2$ atmosphere [258, 262], chemical-vapor deposition (*CVD*) [260], or atomic-layer deposition (*ALD*) with $HfCl_4$ and H_2O as precursors [144]. Other dielectric compositions, such as *HfSiON* [265, 266] and *HfAlON* [267] can be deposited by co-sputtering *Hf* and *Si* targets, or *Hf* and *Al* targets in an $Ar + O_2 + N_2$ atmosphere. In most cases, a very thin $(\sim 0.6\,\text{nm})$ interfacial oxide or oxynitride is formed between

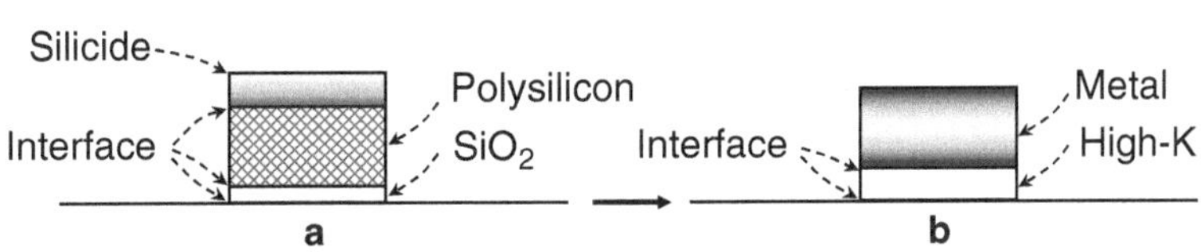

Fig. 7.36 Gate-stack migration from silicided, doped polysilicon on silicon dioxide to metal gates on high-K dielectric

high-*K* dielectrics and silicon to avoid an increase in surface-state trap densities created by direct contact of high-*K* dielectrics with silicon [258,260–267]. The interfacial oxide can be grown, for example, by *RTO*, remote plasma oxidation, or In-Situ Steam Generated (*ISSG*) SiO_2 prior to deposition of the high-*K* dielectric [267]. Its thickness must be optimized to achieve SiO_2-like interface with silicon while minimizing the impact on the overall dielectric constant. For example, a higher dielectric constant can be obtained with *HfSiON* on an oxynitride interface *SiON* than on SiO_2.

7.4.5.2 Metal Gates

Polysilicon gate conductors, even when heavily doped, are known to become depleted to a depth of ~1.2 nm, increasing the capacitance-equivalent-thickness (*CET*) in inversion by about 0.4 nm (Chap. 5). This increase becomes a larger fraction of the total equivalent oxide thickness as device dimensions are scaled down. Thus, it has become necessary to develop metal gates in conjunction with high-*K* dielectrics to eliminate the impact of gate depletion on the *CET*. Additional objectives are also to reduce the gate sheet resistance and eliminate boron penetration observed with heavily-doped polysilicon *PMOS* gates. As for polysilicon gates, there is a strong trade-off between process complexity and performance when choosing between a single-workfunction, near midgap metal gate, and dual-workfunction, band-edge metal gates for nanoscale technologies (Chap. 5). Several single, near midgap workfunction metal gates have been demonstrated. Among them are tungsten (*W*), deposited by sputtering [272] or metal-organic *CVD* [272], *CVD WSi* [274] and W_xN [275], reactively sputtered *TiN* [276–278] and *TaSiN* [280]. Patterning is performed by conventional *CMOS* or a "replacement-gate" process where the gate is formed after source and drain as illustrated in Fig. 7.37 [273, 274, 276, 280]. In this scheme, the *MOSFET* isolation source, drain and gate are completed following a conventional *CMOS* process (Fig. 7.37a). An oxide film is deposited and planarized by *CMP*, exposing the polysilicon (Fig. 7.37b). The polysilicon gate and underlying thin oxide film are removed by wet-etching techniques (Fig. 7.37c). A new gate dielectric that consists of oxide, oxynitride or high-*K* material is deposited followed by the deposition of a thin metal layer, such as *TiN* (Fig. 7.37d). The metal gate, *W* or *Al*, is deposited (Fig. 37e) and planarized (Fig. 7.37f), resulting in the desired metal gate *MOSFET* [273, 274, 280].

The advantages of midgap workfunction metal-gates over polysilicon have been shown for fully-depleted ultra-thin undoped SOI where the threshold voltage for *NMOS* and *PMOS* were simultaneously satisfied without additional doping of the top silicon film. This is because, in ultra-thin *SOI* structures, the threshold voltage and short-channel effects can be independently controlled with the thickness of the top silicon film [273–275, 278, 279]. In bulk silicon, a midgap gate conductor would result in *NMOS* and *PMOS* threshold voltages that are too high for nanoscale applications if not adjusted with counter-doping similar to that of a buried channel. Counter-doping with a buried channel, however, degrades mobility and short-channel characteristics. A single workfunction metal gate becomes less effective as *MOSFET* dimensions are scaled down, nullifying the advantages of metal gates

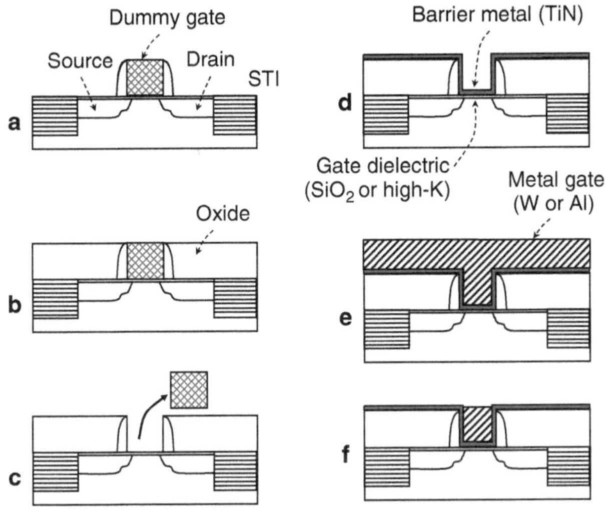

Fig. 7.37 Replacement gate process (Adapted from [276, 280])

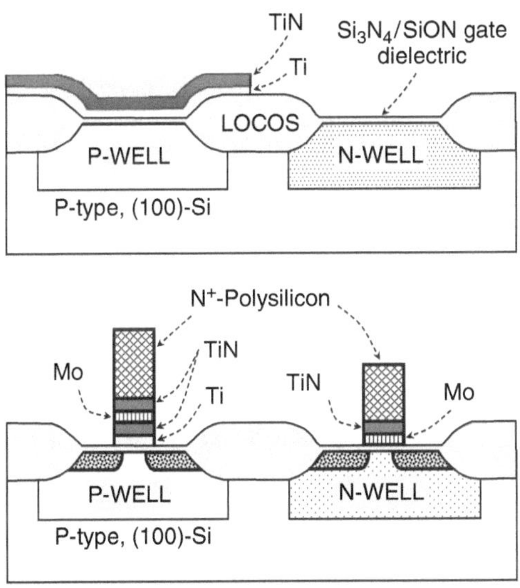

Fig. 7.38 Illustration of dual workfunction metal gate process (Adapted from [287])

over polysilicon. Thus, metal gate conductors with dual workfunction, 4.0–4.3 eV for *NMOS* and 4.8–5.1 eV for *PMOS*, are required to achieve low threshold voltages (Chap. 5) [281–286]. Figure 7.38 illustrates a gate-stack flow that demonstrates the formation of a *CMOS* dual-workfunction metal gate using titanium as the gate electrode for *NMOS* and molybdenum as the gate electrode for *PMOS* [287]. The structure is processed in conventional *CMOS*, including the formation of a 5-nm thick $Si_3N_4/SiON$ gate dielectric stack. This is followed by sputter-deposited 20-nm *Ti*

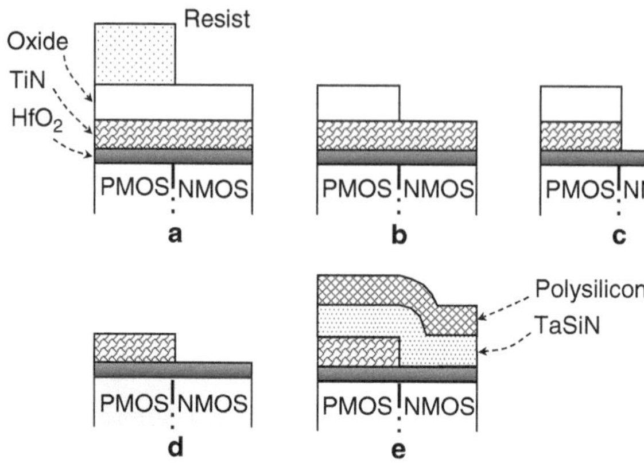

Fig. 7.39 Illustration of dual workfunction metal gate process (Adapted from [289])

and 10-nm *TiN* to form the *NMOS* metal gate. The *NMOS* regions are then covered with photo-resist while the *TiN/Ti* stack over the n-well was etched. 20-nm molybdenum is then sputter-deposited and annealed. This is followed by sputter-deposited 10-nm *TiN*, and 100-nm LPCVD, in-situ n-doped polysilicon as the top electrode. The *TiN* films serve as barrier layers to prevent possible reactions among the metals and the polysilicon during subsequent high temperature processes. The process is completed in conventional *CMOS*.

One example of integrating a dual-metal gate *CMOS* on an *ALD* HfO_2 gate dielectric is reported in [289]. After forming the gate dielectric, *TiN* is deposited by reactive sputtering to form the *PMOS* metal-gate, followed by depositing an oxide film and patterning the *NMOS* regions in photoresist (Fig. 7.39a). The resist pattern is transferred to the oxide by etching the oxide over *NMOS* regions, stopping on *TiN*. The resist is removed and the oxide used as a stencil, or "hard mask," to pattern *TiN* (Fig. 7.39b). *TiN* is wet-etched (Fig. 7.39c), the oxide is wet-etched (Fig. 7.39d), and a *TaSiN* film deposited by reactive sputtering to form the *NMOS* gate electrode, followed by a heavily doped polysilicon film deposited by *CVD* (Fig. 7.39e). The gates are patterned and the process completed in conventional *CMOS* (Not shown). A similar process uses *TaSiN* for *NMOS* and Ruthenium (*Ru*) for *PMOS* [290].

Fully silicided polysilicon (*FUSI*) can also serve as metal gates. This was first demonstrated for single workfunction cobalt-silicide ($CoSi_2$) in a 100-nm *CMOS* technology [33], and for dual-workfunction nickel-silicide gates [291–294]. *MOSFET*s with fully *Ni*-silicided dual workfunction gates were fabricated on *Hf-SiON* using nickel phase-controlled *FUSI* [292]. The workfunction was tuned by modifying the nickel to silicon ratio: for a silicon-rich composition with a ratio $Ni/Si = 0.6$ $\phi_m \approx 4.5\,\text{eV}$, suitable for *NMOS*, and for a nickel-rich film with a ratio $Ni/Si = 1.7$ $\phi_m \approx 4.85\,\text{eV}$, which is suitable for *PMOS* [292]. The flow is shown schematically in Fig. 7.40.

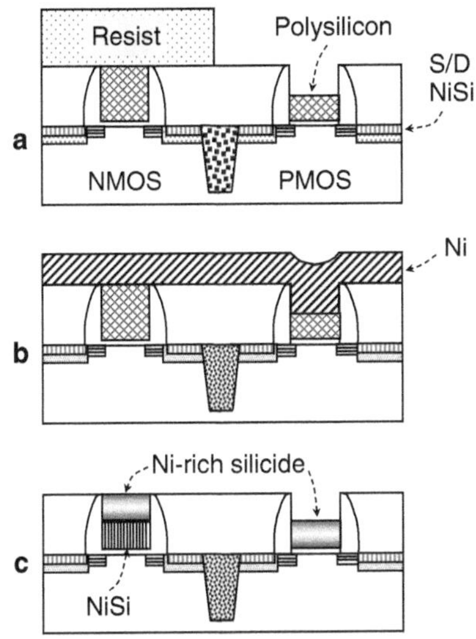

Fig. 7.40 Illustration of dual workfunction *FUSI* gate process (Adapted from [292])

The process follows a conventional *CMOS* flow up to polysilicon sidewall formation, oxide deposition and planarization by *CMP*. Nickel-silicidation of source and drain is done independently of gates. The *NMOS* polysilicon gate is protected with photoresist while the PMOS polysilicon gate is recessed by etch-back (Fig. 7.40a). The resist is removed and a nickel film deposited by, for example, sputtering (Fig. 7.40b). Simultaneous silicidation of *NMOS* and *PMOS* is done by using an optimized 2-step *Ni FUSI* process. The *Ni/Si* ratio is controlled by limiting the first *RTP* step without fully consuming polysilicon. Excess *Ni* is then selectively etched and a second *RTP* step at higher temperature completes silicidation [288].

7.5 Advanced Interconnects

The drive to advanced interconnects is motivated by the need to reduce the wiring *RC* delay,[6] the power supply noise caused by *IR* drops, and the cross-talk noise caused by capacitive coupling between adjacent wires.[7] Thus, both the wiring resistance *R* and capacitance *C* must be reduced. A simplified representation of wire dimensions and capacitances is shown in Fig. 7.41. The wire resistance is

$$R_{wire} = \rho_m \frac{L_m}{W_m t_m} \quad \text{Ohm}, \tag{7.4}$$

[6] The RC product has the unit time: R = Voltage/Current = (Voltage × Time)/Charge; C = Charge/Voltage.

[7] Cross-talk refers to a signal on one line affecting the signal on another line in close proximity.

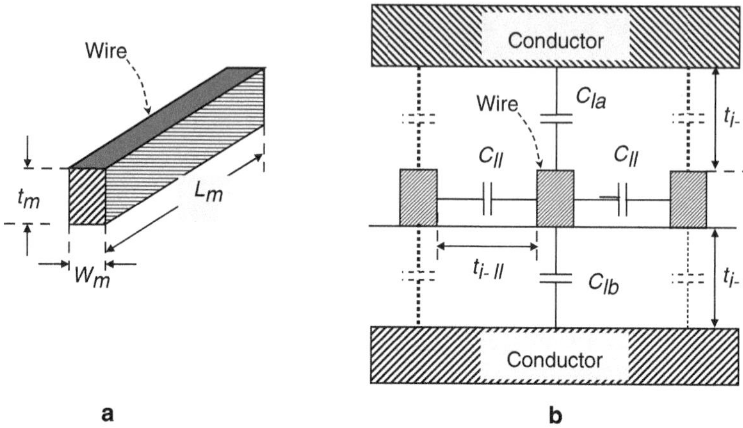

Fig. 7.41 Simplified representation of **a** wire dimensions and **b** wire capacitances

where ρ_m is the metal resistivity, L_m, W_m the wire length and width, and t_m is the metal thickness (Fig. 7.41a).

The wire capacitance C_{wire} is the sum of line to above conductor capacitance (C_{la}), line to below conductor capacitances (C_{lb}) and line to line capacitance (C_{ll}), as shown in Fig. 7.41b. Neglecting fringe-effects and assuming for simplicity that the conductors above and below the wire are continuous equipotential planes rather than individual metal lines, C_{la} and C_{lb} can be expressed as

$$C_{la} = \frac{L_m W_m \varepsilon_0 \varepsilon_i}{t_{i-a}} \quad F, \tag{7.5a}$$

$$C_{la} = \frac{L_m W_m \varepsilon_0 \varepsilon_i}{t_{i-b}} \quad F. \tag{7.5b}$$

where ε_i is the insulator dielectric constant, $t_{i\text{-}a}$ the thickness of the dielectric between the wire and the conductor above and $t_{i\text{-}b}$ is the dielectric thickness between the wire and the conductor below. Similarly, the line to line capacitance can be approximated for each side of the wire as

$$C_{la} = \frac{L_m t_m \varepsilon_0 \varepsilon_i}{t_{i-ll}} \quad F, \tag{7.6}$$

where $t_{i\text{-}ll}$ is the line to line space. Thus, assuming a uniform dielectric material around the wire and neglecting fringe-effects, the RC delay τ_{RC} for the center wire shown in Fig. 7.41b can be approximated from (7.2–7.4) as

$$RC = \tau_{RC} = \frac{\rho_m \varepsilon_0 \varepsilon_i L_m^2}{t_m} \left(\frac{1}{t_{i-a}} + \frac{1}{t_{i-b}} + \frac{2}{t_{i-ll}} \frac{t_m}{W_m} \right) \quad s. \tag{7.7}$$

The dielectric below the first metal level is referred to as the pre-metal dielectric, *PMD*. The dielectric between consecutive metal levels is called inter-level dielectric, *ILD*. The dielectric between metal lines at the same level will be referred to as the intra-level dielectric, *ILLD*.

It follows from (7.7) that for a given wire length and width, the *RC* delay can be reduced by decreasing the conductor resistivity and insulator dielectric constant, and by increasing the dielectric and metal thickness. As the *PMD* and *ILD* thickness increases, however, so does the aspect ratio (depth to width ratio) of contacts and vias, which increases the complexity of etching and filling the openings. An increase in line to line space results in a wider metal pitch (distance between centers of adjacent metal lines) and hence a lower pattern density. Finally, an increase in metal thickness results in higher line to line capacitance and hence cross-talk between adjacent lines. Thus, when optimizing the interconnect design and process, there is a tradeoff between capacitance, resistance, delay, cross-talk, circuit density, and process complexity. This trade-off is alleviated with the introduction of lower resistance copper wiring and low-K intra-level and inter-level dielectrics.

7.5.1 Copper Interconnects

Copper has replaced aluminum in deep submicron and nanoscale interconnects, primarily because of its lower resistivity, higher electromigration resistance, and higher melting point (Table 7.1). The resistivity of copper is less than 2/3 that of aluminum. This property allows either a reduction in interconnect resistance for the same metal thickness or a reduction in metal thickness and hence line to line capacitance for the same wire resistance. The higher melting point is one of the main reasons copper interconnects exhibit a longer electromigration lifetime than aluminum operating at the same current density. The lower copper resistance further improves electromigration and lifetime by reducing Joule heating for the same current.

Table 7.1 Properties of interconnect and contact metals

Property	Al	Cu	Ti	TiN	Co	Ni	W
ρ	2.7	1.67	42.0	20[a]	5.25	6.84	5.65
TCR	0.071	0.130	e	0.0014[b]	0.531	0.692	0.411
TC	222	394	21.9	15[c]	69.0	82.9	166
CTE	23.8	16.5	8.41	d	13.8	13.3	17.4
T_m	660	1085	1670	2150	1495	1455	3387

ρ Resistivity ($\mu\Omega - \text{cm}$), *TCR* Thermal coefficient of resistance (K^{-1}), *TC* Thermal conductivity ($Wm^{-1}K^{-1}$), *CTE* Coefficient of thermal expansion ($\times 10^{-4}K^{-1}$), T_m Melting point (°C)
[a]Minimum value reported in [294], depends strongly on deposition conditions
[b]Minimum value reported in [295], depends strongly on film-stress
[c]Measured on bulk [295]
[d] Not available
[e]Magnitude and polarity depend strongly on film thickness

Two obstacles delayed the introduction of copper interconnects. 1. The inability to pattern the metal by anisotropic reactive ion etching (*RIE*) due to the non-volatility of the by-products; 2. The lack of an appropriate barrier film to be placed around the metal to inhibit Cu migration into the surrounding insulator and silicon, and inhibit oxidizing species to diffuse into copper, corroding the metal. Copper diffuses rapidly into and through insulators and silicon. In silicon, Cu can segregate in active regions, creating near mid-gap states that increase junction leakage. The migration of Cu in insulators can degrade the dielectric integrity and also cause leakage paths between interconnects, particularly along interfaces. The patterning issue was resolved with a metallization process known as "damascene." Novel and efficient barrier liners have been developed for encapsulating the copper lines.

7.5.1.1 Damascene and Dual Damascene

Damascene refers to a decorative art of inlaying different metals into one another, for example, gold into an oxidized steel background. In the damascene interconnect scheme, the metal is "inlayed" into trenches patterned in an insulator [296]. Figure 7.42 illustrates this process for aluminum interconnects and compares it to "conventional" metal patterning. In the conventional process, applicable to aluminum metallization, the metal films are first deposited on a planarized surface

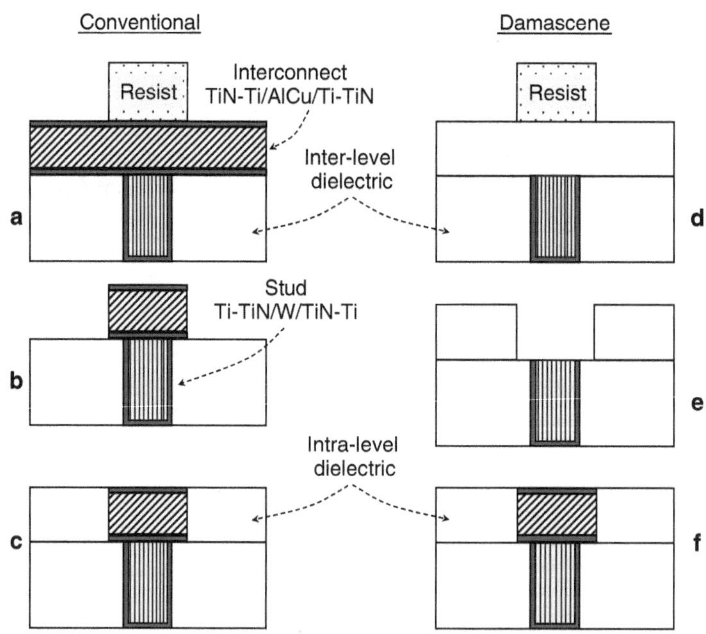

Fig. 7.42 Comparison of "conventional" to damascene patterning, illustrated for aluminum-based interconnects [296]

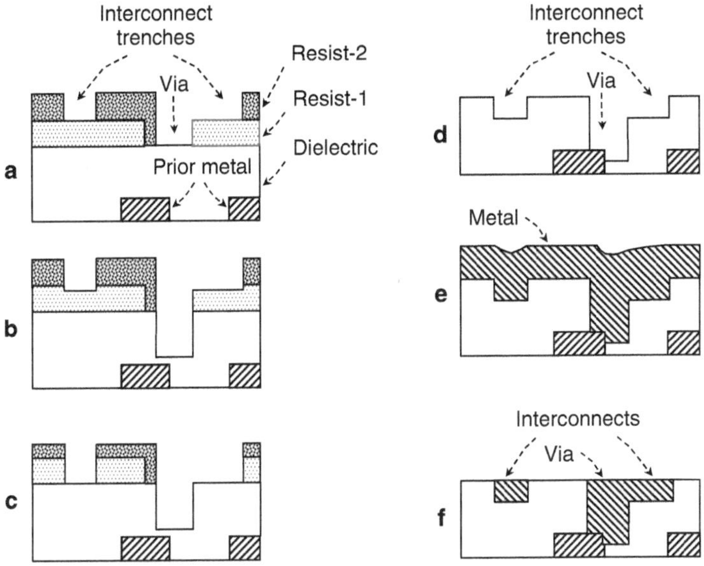

Fig. 7.43 Schematic of dual-damascene process flow, illustrated for two resists of dissimilar properties (Adapted from [296])

and patterned with resist (Fig. 7.42a). The metal is directionally etched by *RIE* (Fig. 7.42b) and then covered with a dielectric and planarized by *CMP* (Fig. 7.42c). In the damascene process, the planarized dielectric is first patterned with resist (Fig. 7.42d), and trenches reactively ion-etched into the dielectric (Fig. 7.42e). The metal films are deposited, filling the trenches, and then planarized down to the surface of the dielectric by *CMP*, leaving embedded metal in the desired wiring pattern (Fig. 7.42f) [296]. The damascene process replaces dry-etching the metal with etching the dielectric, which is simpler.

Dual damascene is an extension of the damascene process whereby via and interconnect patterns are sequentially printed on top of each other in two masking steps (Fig. 7.43) [296, 297]. Resist-1 and resist-2 in Fig. 7.43a have dissimilar exposure, development and etch properties. Other materials can be used for this purpose. Vias are first patterned in resist-1, followed by patterning trenches connected to vias, and "stand-alone" trenches in a second masking step (Fig. 7.43a).

The intersection of the two mask openings defines the self-aligned via-stud. The stud shape is selectively etched in the dielectric almost to completion, with minimal attack on the resists (Fig. 7.43b). The *RIE* reagents are then changed in-situ for patterning the trench images in resist-1 (Fig. 7.43c). The etch-chemistry is changed again and, with both resist-patterns in place, the insulator is etched to the desired interconnect metal depth. Vias are simultaneously etched down to the prior metal level and the resists removed (Fig. 7.43d). The via and prior metal are intentionally shown misaligned to emphasize the process-flexibility. The desired metal is deposited, filling the vias and trenches (Fig. 7.43e). It is then planarized by CMP (Fig. 7.43f).

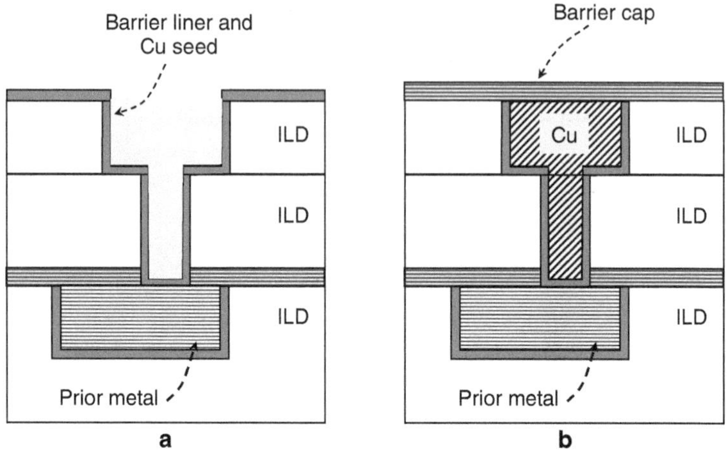

Fig. 7.44 Copper dual-damascene process sequence. **a** Deposition of a thin barrier liner and copper seed after etching via and trench in the *ILD*; **b** Copper plating, *CMP*, and barrier-cap deposition

The steps are repeated for multi-level metals. The surface planarity is maintained throughout the process.

While both patterning by dry-etching and damascene or dual-damascene can be applied to aluminum-based interconnects, only the damascene or dual-damascene process is applicable to copper because of the afore-mentioned non-volatility of copper dry-etch by-products. The copper process sequence is illustrated for dual-damascene in Fig. 7.44 [298]. It consists of etching vias and trenches in the *ILD*, as described in Figs. 7.42 and 7.43. A thin barrier liner, for example, an *ALD* TaN film, is deposited, followed by a thin copper seed-layer (Fig. 7.44a). A copper film is electroplated and planarized by *CMP*. A barrier-cap layer is then deposited over the entire surface (Fig. 7.44b), or selectively over copper only.

7.5.1.2 Barrier Layers

One of the main concerns with copper interconnects is the stability of the *Cu* interface with the surrounding dielectric. Copper does not adhere well to most insulator materials, diffuses rapidly into and through the insulator and silicon, and corrodes rapidly in an oxidizing ambient. It is therefore necessary to encapsulate copper with an efficient barrier to *Cu* and oxygen migration (Fig. 7.44) [299–309]. There are two types of barriers in a copper damascene process, a barrier on the sides and floor of trenches and vias (Fig. 7.44a), and a barrier-cap on top of copper prior to the next inter-level dielectric deposition (Fig. 7.44b) [302]. The key purpose of barrier liners is to prevent copper from diffusing into the *ILD* and oxygen from diffusing into copper, and to promote adhesion with both the interlayer dielectric (*ILD*) and copper.

The most common barrier liners are tantalum nitride (*TaN*), or a combination of *TaN* and *Ta* [302–305]. The films can be deposited by *CVD* [299, 301], PVD [298, 300, 305], or *ALD* [302–304]. Since liner resistivity is considerably higher than that of copper and the liner does not carry appreciable current, the liner must be sufficiently thin to minimize the copper volume displaced by the liner to keep the effective interconnect resistivity at a value of about 2.2 μΩ-cm [300, 302]. The effective resistivity is defined as

$$\rho_{eff} \cong \rho_{Cu} \frac{\text{Volume of Cu} + \text{liner}}{\text{Volume of Cu}} \quad Ohm - nm. \tag{7.6}$$

The effective resistivity approaches the copper resistivity as the liner thickness is decreased. As the interconnect line-width is reduced, metal nitride liners, such as *TaN, TiN* and *WN*, deposited by *ALD* become more efficient since they can form ultra-thin (1–2 nm) barriers with excellent conformality [302–304, 310].

A "self-formed" MnO_x barrier is reported in [306–308]. A thin copper-manganese alloy (*Cu-2%Mn*) is deposited as a copper seed layer by sputtering. Copper is electroplated to the desired thickness, followed by an anneal cycle at 300°C. During annealing *Mn* reacts with SiO_2 to form a thin (≈2 nm) $MnSi_xO_y$ barrier. Excess Mn migrates to the top *Cu* surface and reacts with oxygen to form MnO_x, a key feature of this process [306]. The top MnO_x is removed during copper *CMP*. No barrier is formed at the via-bottom, resulting in intimate contacts between the top and bottom copper layers. A barrier cap is then deposited on the surface. The structure is extended to a bi-layer of an ultra-thin *Ta* film followed by a *CuMn* seed film and copper electro-plating [308]. The structure is annealed after deposition of a barrier cap, whereby a thin *MnO* film forms on all boundaries and encapsulates copper. The ultra-thin *Ta* minimizes the required *Mn* concentration, reducing the interconnect resistance.

Other barrier liner compositions, such as *Ru/TaN* have also been investigated [309].

7.5.1.3 Copper Deposition and CMP

As has been shown, copper is typically deposited by electro-plating. A seed layer is deposited on the barrier liner to facilitate the plating process. It provides an initial low-resistance path for the plating current. Adhesion to the barrier layer can be improved by adding a "glue layer" such as *Ru* [311]. For plating, the front-side is contacted near the wafer edge to provide a uniform current. The film resistance decreases rapidly as the *Cu* layer grows.

Chemical-mechanical polishing is the enabling process for damascene *Cu* interconnects. At every metal level, the copper film is planarized by *CMP*, typically down to the surface of the surrounding insulator (Fig. 7.44b). A measure of the insulator mechanical strength is its Young's modulus [312]. Since low-*K* and ultra-low-*K* dielectrics have a lower Young's modulus than silicon-dioxide (Sect. 7.5.2), polishing copper on the material would require the polishing pad pressure to be greatly

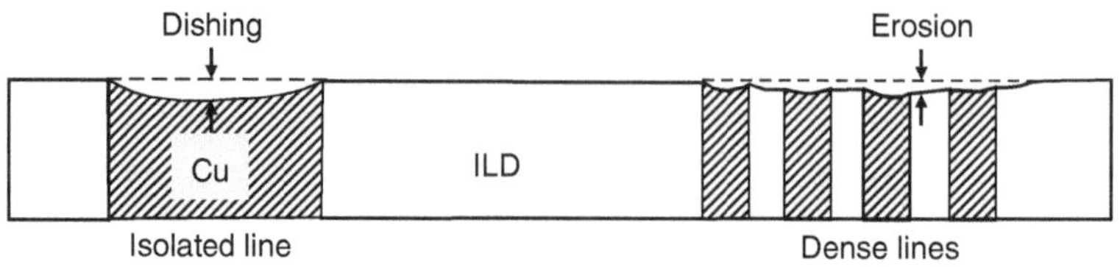

Fig. 7.45 Illustration of copper dishing and dielectric erosion in a damascene CMP process (Adapted from [315]). Dishing and erosion are the result of different polishing properties of the metal and insulator. Erosion depends strongly on pattern density

reduced to avoid cracking, delamination, and excessive metal dishing and dielectric erosion (Fig. 7.45) [313–315]. Since the rate of removal of copper is directly related to the down-force on the pad, this would mean a longer time and hence lower throughput for polishing.

The electrochemical mechanical planarization process, *ECMP*, was developed to alleviate the *CMP* problems, particularly in nanoscale technologies [316]. The wafer is immersed in a specially-designed electrolyte and a voltage applied. A passivation film forms on copper where it prevents the *Cu* dissolution in the electrolyte. A rotating polishing pad with almost zero down pressure ($<0.5\,\text{psi}$) gently removes the passivation film wherever it touches the wafer, allowing Cu to dissolve there [317, 318]. The dissolution of copper in the electrolyte involves oxidation of Cu to Cu^{2+} ions and subsequent Cu ion diffusion

$$Cu \rightarrow Cu^{2+} + 2e. \tag{7.7}$$

The removal rate does not depend on the pad down-pressure but is controlled by the applied voltage.

The *ECMP* system typically consists of three platens [316–318]. In the first platen, most of the *Cu* is removed by electro-chemical mechanical polishing at a high-rate by the applied electric charge, independent of down-force. In the second platen, the remaining thin *Cu* film is cleared in electro-chemical mechanical polishing combined with a precision charge-controlled end-point [317]. The barrier-liner and an eventual hard mask material are removed in a polishing step in the third platen [318]. When compared to "conventional" *CMP, ECMP* removes copper at a considerably higher speed while minimizing damage to the low-*K ILD*.

7.5.1.4 Barrier Cap

The cap layer must seal the top, prevent *Cu* from migrating along the surface, protect the copper from corrosion during subsequent patterning steps, act as a polishing stop and, in case of dielectric caps, the cap must act as etch-stop for vias that do not fully land on the underlying metal [302].

Typical dielectric caps are *PECVD* silicon-nitride (SiN_x) and silicon carbide (*SiC*) which have *K* values of about 5 to 7. The disadvantage of both materials

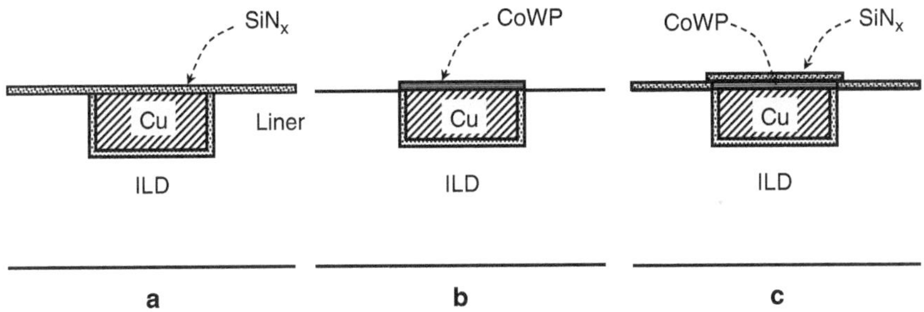

Fig. 7.46 Schematic of cap configurations. **a** SiN_x alone; **b** $CoWP$ alone; **c** $CoWP + SiN_x$

is the relatively high dielectric constant that increases effective K of the ILD. Another drawback is the poor adhesion with copper, allowing Cu to migrate along the dielectric-Cu interface, causing voids in the metal and leakage between metal lines [319, 320]. The problems with dielectric caps can be resolved with self-aligned metal caps, such as selective deposition of 5–20 nm cobalt-tungsten phosphide ($CoWP$) [320–323], or cobalt-tungsten boride ($CoWB$) [324] by electroless plating. Metallic barrier caps are found to be more efficient in suppressing Cu migration and protecting the copper surface during processing without impacting the ILD K. The caps can be formed with or without a silicon-nitride film (Fig. 7.46).

7.5.1.5 Fundamental Issues with Copper

As the copper interconnect width is reduced to nanoscale dimensions, two fundamental issues become more important: The liner occupies an increasingly larger portion of the damascene feature, and the resistance increases non-linearly as the line-width is reduced.

The liner accounts for about 15% of the feature volume at a minimum half-pitch of about 90 nm. This fraction must be maintained as an upper limit of trench or via width. The lower limit in barrier-liner thickness is the effectiveness in preventing the diffusion of copper and oxygen. Replacing *PVD TaN/Ta* liner with, for example, an *ALD TaN* (1 nm)/*ALD Ru* (2 nm) bi-layer, can significantly reduce the Cu volume displaced by the liner [325, 326].

Figure 7.47 shows how the copper specific resistivity increases with decreasing Cu line-width [325]. The copper resistivity increases with decreasing feature size because of additional inelastic scattering of electrons at the line sidewalls [325, 327–329]. The electron mean-free path in copper, aluminum, and tungsten is, respectively, about 39.3 nm, 14.9 nm, and 14.2 nm [329, 330]. As the line-width approaches the mean-free path, there is an increased contribution of inelastic electron scattering on interfaces and rough porous sidewalls [327, 331], and on grain boundaries [331]. The result is an increase in resistivity and RC delay as the line-width is scaled down [332] (Fig. 7.48).

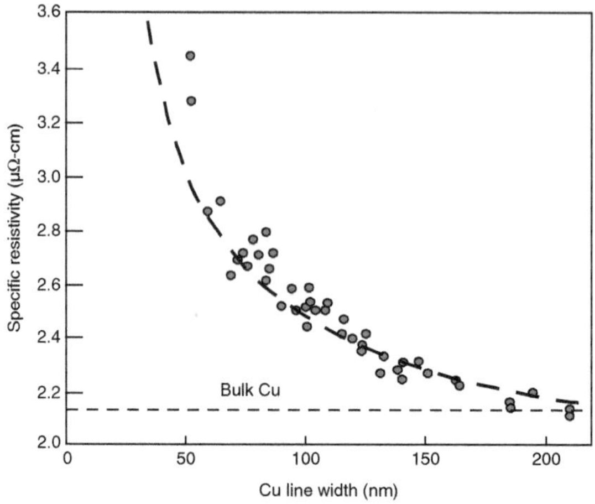

Fig. 7.47 Cu resistivity increase with decreasing line-width (Adapted from [325])

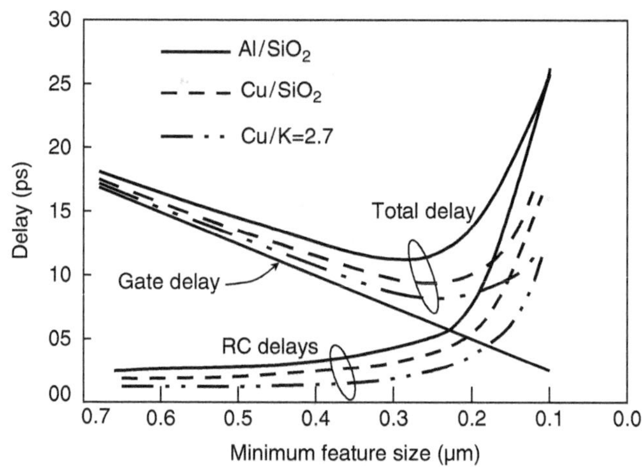

Fig. 7.48 Reduction in *RC* delay with low-*K* dielectric. As dimensions shrink below ~0.2 mm, RC delay exceeds gate delay. Low-*K* dielectric is necessary to improve performance (Adapted from [333, 334])

7.5.2 Low-K Dielectrics

The motivation for developing low dielectric-constant (low-*K*) inter-level and intra-level dielectrics is to reduce the capacitance between interconnects lines. This not only reduces the *RC* delay but also the cross-talk noise between adjacent signal-lines. Figure 7.48 shows how the *RC* delay is reduced by combining copper interconnects with low-*K* dielectric [333, 334].

Among the key requirements on low-*K* materials are the chemical and mechanical stability. The dielectric must maintain the low *K*-value during patterning, surface cleaning, and plasma processes, with minimal moisture absorption and gas

Table 7.2 Properties of some low-*K* materials

Material	K	References
SiOF	3.4	[335]
SiOC-base	2.9	[336]
SiOCH no pores	3.1	[337]
SiOCH, porous, ∼0.6 nm Ø pores	2.6	[337–343]
SiOCH, porous, 0.6–0.8 nm Ø pore	2.2–2.4	[337–343]
Air-gap	1.0	

permeation. It must have a Young's modulus larger than 8 GPa to avoid delamination and cracking during *CMP*, and must be thermally stable during subsequent heat treatments at temperatures up to about 500 °C. It must exhibit a minimum breakdown field of ∼1 MV/cm and have low leakage between conductors.

Several *PECVD* low-*K* dielectrics have been demonstrated on an integrated copper damascene process. Among them are materials that belong to the Organo-Silicate Glass (*OSG*) family, as summarized in Table 7.2 [335–343]. The *OSG* materials can be made silicon-rich, rigid without pores with $K = 3.1$. The dielectric constant can be extended to ultra-low-*K* values of about 2.0 by introducing pores of varying diameter. This can be achieved by, for example, adding an organic precursor, called porogens, to the tetramethylcyclotetrasiloxane (TMCTS) used for the preparation of *SiCOH* dielectrics and annealing to remove the thermally less stable organic CH_x radicals from the film, thus adding porosity [337].

The porosity reduces K but also degrades the mechanical strength, resistance to impurity penetration, and electric breakdown of the film. Because of porosity, etching and CMP processes can degrade K. In some cases, it is necessary to seal the pores with a special plasma treatment [344]. Figure 7.49 shows the capacitance as a function of reciprocal line-to-line space for different *ILD* combinations [339]. The capacitance increases from the value in point **a** to the value in point **b** in Fig. 7.49 as the space is reduced from 100 nm to 70 nm. For full-pore (both films with pores) and a space of 70 nm, the capacitance drops to the value point **c**. The capacitance drops further to the value in point **d** if the *Cu* thickness is scaled to about 2/3 of an aluminum line of same width and sheet resistance.

The values are given for the center line in the inset at a minimum distance from surrounding lines [339]. The value in **a** is for a combination of intra-level dielectric with pores, inter-level dielectric without pores, and a line-to-line space of 100 nm. The capacitance increases as the line-to-line space decreases.

The dielectric constant can further be reduced by forming air-gaps between metal lines [345–349]. An effective dielectric constant of K $\approx$ 1.9 has been achieved with an air-gap scheme shown in Fig. 7.50 [349]. The first *Cu* metal is formed on a low-*K* dielectric (*SiOC*) (Fig. 7.50a). A special *SiC* film is deposited and patterned (7.50b). The *SiC* film is etched and the insulator selectively removed from the air-gap regions (Fig. 7.50c). This step is referred to as gap-etch. Another *SiOC* layer is deposited under a condition to close the window at the top of the trench and form an encapsulated air-gap. The film is then planarized by *CMP* (Fig. 7.50d), followed by a second metallization level.

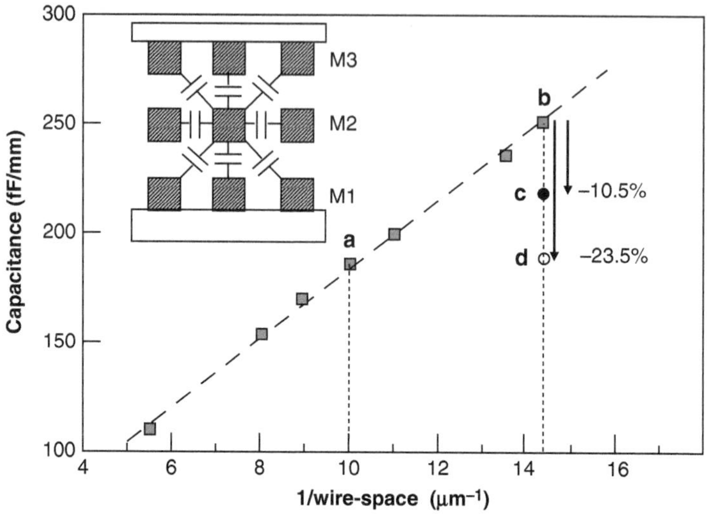

Fig. 7.49 Wire capacitance as a function of inverse wire-space (configuration in inset). **a** Hybrid: inter-level dielectric without pores, intra-level with pores, 100-nm space; **b** Hybrid, 70-nm space; **c** Full-pore, 70-nm space; **d** Full-pore, 70-nm space, scaled Cu thickness (Adapted from [339])

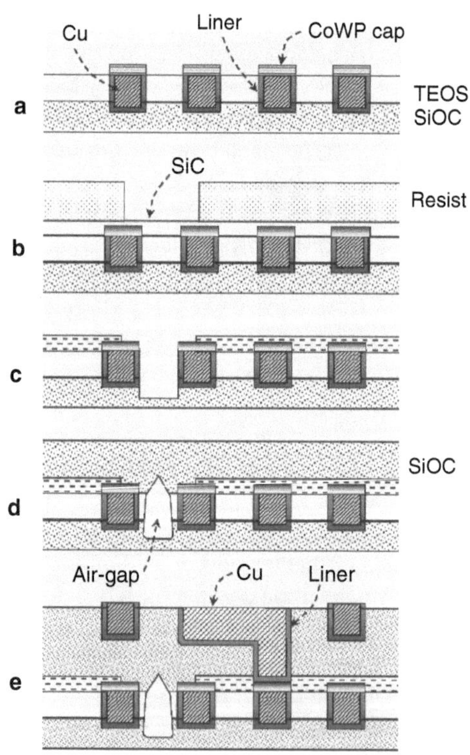

Fig. 7.50 Schematic of air-gap process flow [339]. **a** First metal level; **b** Deposition of *SiC* film and patterning; **c** Etch of air-gap region; **d** *SiOC* deposition and planarization; **e** Second metal level

7.6 Problems

The temperature is 300 K unless otherwise stated.

1. An NMOS p-well is implanted through a 10-nm screen oxide with boron at a maximum energy of 400 keV and a dose of $2\times10^{14}\,cm^{-2}$. At this energy, the range and straggle of boron in the patterning resist are, respectively, 1.7 μm and 0.1 μm.

(a) Estimate the resist thickness necessary to ensure that the surface boron concentration in the blocked areas does not increase by more than $10^{15}\,cm^{-3}$.
(b) What experiment and test would you implement to verify your estimate?

2. Vias are patterned with resist and etched in TEOS oxide to a depth of 1.0 μm with over-etch of 30%. The ratio of TEOS etch-rate to resist etch-rate is 1.5:1. What is the minimum resist thickness required to protect the covered TEOS regions from etching?

3. Calculate the required thickness of thermal oxide required to reduce the top silicon thickness in an SOI wafer by 100 nm.

4. A 500-nm deep and 250-nm wide trench for copper wiring is lined with 20-nm tantalum of resistivity 500 μΩ-cm. Copper is deposited and polished to the top of the trench. A 100-nm thick silicon-nitride cap layer is deposited, followed by 800-nm low-K dielectric ($K = 2.9$). Find:

(a) The effective interconnect sheet resistance.
(b) The effective dielectric constant of the dielectric stack above copper.

5. The nominal NMOS threshold voltage is measured lower than expected.

(a) Identify five possible process-induced mechanisms that would cause this behavior.
(b) What experiments and tests would you implement to determine the actual cause(s)?

6. The following electrical test-results were obtained on a PMOSFET:

Parameter	**Unit**	**Measured**	**3σ-low**	**Target**	**3σ-high**
$V_{T\text{-lin}}$	V	−0.45	−0.54	−0.62	−0.70
g_{msat}	μS/μm	190	225	250	275
C_{max}	fF/μm^2	4.5	4.4	4.6	4.8
ΔW	nm	320	270	300	330
ΔL	nm	240	225	250	275

Identify the most probable cause for the low threshold voltage and low transconductance.

7. The following electrical test-results were obtained on a PMOSFET:

Parameter	Unit	Measured	3σ-low	Target	3σ-high
$V_{T\text{-lin}}$	V	0	−0.54	−0.62	−0.70
g_{msat}	μS/μm	305	225	250	275
C_{max}	fF/μm^2	4.5	4.4	4.6	4.8
ΔW	nm	320	270	300	330
ΔL	nm	240	225	250	275

Identify the most probable cause for the low threshold voltage and high transconductance.

8. How can the NPN collector-base breakdown voltage be 25 V in a stand-alone collector-base capacitor, and 10 V in a transistor?

9. The leakage current between two adjacent collectors is in the mA range for an array of bipolar transistors of the type shown in Fig. 7.3, but in the sub-fA range between two isolated transistors. What is the most probable cause for the difference? How can you confirm this?

10. The gain β and gain-bandwidth product f_T of a bipolar transistor are 80% higher than expected. Identify process parameters that can cause this behavior. What independent electrical test would you implement to verify the cause?

11. Why should the etched polysilicon gate of a MOSFET always overlap the field oxide? How far should the drawn polysilicon image overlap the drawn field oxide image to ensure adequate overlap of the on-chip patterns? Assume: Polysilicon linewidth 200 nm; radius of corner rounding 100 nm; line-end foreshortening 60 nm; worst-case alignment tolerance of polysilicon pattern to field-oxide pattern 100 nm.

12. The space between two 400-nm high copper interconnect lines consists of 50-nm thick dielectrics of K = 3.0 on each side and a 100-nm wide air-gap in the middle. Disregard fringe effects and calculate the line-to-line capacitance per mm interconnect length.

13. The vertical and horizontal impurity profiles of the channel, source-drain, source-drain extensions, and halos are optimized for a 100-nm MOSFET channel length. The threshold voltage measured on a 1 μm long channel MOSFET fabricated in the same process is 0.42 V in the linear mode and 0.39 V in the saturation mode. Suggest a mechanism that causes this behavior.

14. The diffusion coefficient of Cu in SiO_2 in the temperature range 300 °C–500 °C is estimated as $D_{Cu} = 57.8\,e^{-1.82(eV)/kT}$, where $k = 8.62\times10^{-5}\,\text{eV/K}$ is the Boltzmann constant and T is the absolute temperature. Assume an 800-nm thick SiO_2 ILD under a copper interconnect line, no barrier at the bottom of the line, and an initial Cu concentration of $10^{16}\,\text{cm}^{-3}$ in the oxide near the Cu – SiO_2 interface at an anneal temperature of 480 °C. Estimate the time required for Cu to diffuse down to the bottom of the ILD.

15. The source and drain pn junctions are contaminated with copper. Assume one-sided abrupt junctions with a uniform background concentration of $10^{17}\,cm^{-3}$, and a uniform Cu concentration of $5 \times 10^{12}\,cm^{-3}$ within the depletion region. Estimate the reverse junction leakage per unit junction area at 85 °C and 2.5 V reverse voltage.

16. In a BiCMOS process, the epitaxial film was deposited 30% thicker than specified. What electrical parameters would be affected by this change?

17. Compare the equivalent oxide thickness (EOT) of 9-nm hafnium oxide ($K = 25$) to that of a dual layer 7-nm HfO_2 and a 2-nm protective thermal SiO_2.

References

1. B. El-Kareh, *Fundamentals of Semiconductor Processing Technologies*, Kluwer Academic Publishers, Boston, 1995.
2. W. B. De Boer, M. J. J. Theunissen, and R. H. J. Van der Linden, "The necessity of RTCVD in advanced epitaxial growth of Si and SiGe," Rapid Thermal and Integrated Processing IV. Symposium, 287–298, Mater. Res. Soc., 1995.
3. B. El-Kareh, S. Balster, W. Leitz, P. Steinmann, H. Yasuda, M. Corsi, K. Dawoodi, C. Dirnecker, P. Foglietti, A. Haeusler, P. Menz, M. Ramin, T. Scharnagl, M. Schiekofer, M. Schober, U. Schulz, L. Swanson, D. Tatman, M. Waitschull, J. W. Weijtmans, and C. Willis, "A 5V complementary-SiGe BiCMOS technology for high-speed precision analog circuits," Proceedings Bipolar/BiCMOS Circuits and Technology Meeting, 211–214, 2003.
4. J. Crochalski, "Ein neues Verfahren zur Messung der Kristallisationsgeschwindigkeit der Metalle," Z. Phys. Chem., 92, 219–221, 1918.
5. W. C. Dash, "Growth of silicon crystals free from dislocations," J. Appl. Phys., 30 (4), 459–474, 1959.
6. R. B. Swaroop, "Advances in silicon technology," Solid-State Technol., 26, 111–114, 1983.
7. B. Bergholz, *Grown-in and Process-Induced Defects, Semiconductors and Semimetals*, Vol. 42, 513–574, Academic Press, New York, 1994.
8. (a) K. Hoshi, N. Isawa, T. Suzuki, and Y. Ohkubo, "Czochralski silicon crystals grown in a transverse magnetic field," J. Electrochem. Soc., 132 (3), 693–700, 1985. (b) Th. Wetzel, A. Muiznieks, A. Muhlbauer, Y. Gelfgat, L. Gorbunovc, J. Virbulisd, E. Tomzigd, and W. v. Ammond, "Numerical model of turbulent CZ melt flow in the presence of AC and CUSP magnetic fields and its verification in a laboratory facility," J. Cryst. Growth, 230, 81–91, 2001.
9. M. Watanabe, M. Eguchi, T. Hibiya, "Silicon crystal growth by electromagnetic Czochralski (EMCZ) method," Jpn. J. Appl. Phys., 38, L10–L13, 1999.
10. T. Y. Tan, E. E. Gardner, and W. K. Tice, "Intrinsic gettering by oxide precipitate induced dislocations in Czochralski Si," Appl. Phys. Lett., 30 (4), 175–176, 1977.
11. C.-O. Lee and P. J. Tobin, "The effect of CMOS processing on oxygen precipitation, wafer warpage, and flatness," J. Electrochem. Soc., 133 (10), 2147–2152, 1986.
12. H.-D. Chiou, "Criteria for choosing initial oxygen concentration in CZ wafers," Proceeding of the 2nd Symposium on defects in silicon II, Electrochem. Soc., W. M. Bullis and U. Gosele, Eds., 577–588, 1991.
13. H. Shimizu, T. Watanabe, and Y. Kakui, "Warpage of Czochralski-grown silicon wafers as affected by oxygen precipitation," Jpn. J. Appl. Phys., 24 (7), 815–821, 1985.
14. H. Shimizu and T. Aoshima, "Thermal warpage of large diameter Czochralski-grown silicon wafers," Jpn. J. Appl. Phys., 27 (12), 2315–2323, 1988.
15. H.-D. Chiou, Y. Chen, R. W. Carpenter, and J. Jeong, "Warpage and oxide precipitate distributions in CZ silicon wafers," J. Electrochem. Soc., 141 (7), 1856–1862, 1994.

16. H. Lu, D. Yang, L. Li, Z. Ye, and D. Que, "Thermal warpage of Czochralski silicon wafers grown under a nitrogen ambience," Phys. Stat. Sol., 169, 193–198, 1998.
17. D. Yang, G. Wang, J. Xu, D. Li, D. Que, C. Funke, and H. J. Moeller, "Influence of oxygen precipitates on the warpage of annealed silicon wafers," Microelectron. Eng., 66, 345–351, 2003.
18. J. Chen, D. Yang, X. Ma, H. Li, and D. Que, "Intrinsic gettering based on rapid-thermal annealing in germanium-doped Czochralski silicon," J. Appl. Phys., 101 (033526), 1–4, 2007.
19. V. Savolainena, J. Heikonena, J. Ruokolainena, O. Anttilab, M. Laaksob, and J. Paloheimob, "Simulation of large-scale silicon melt flow in magnetic Czochralski growth," J. Cryst. Growth 243, 243–260, 2002.
20. W. Kaiser, "Electrical and optical properties of heat-treated silicon," Phys. Rev., 105 (6), 1751–1757, 1957.
21. C. A. Londos, M. J. Binns, A. R. Brown, S. A. McQuaid, and R. C. Newman, "Effect of oxygen concentration on the kinetics of thermal donor formation in silicon at temperatures between 350 and 500 °C," Appl. Phys. Lett., 62 (13), 1525–1526, 1993.
22. M. Pesola, Y. J. Lee, J. vom Boehm, M. Kaukonen, and R. M. Nieminen, "Structures of thermal double donors in silicon," Phys. Rev. Lett., 84 (23), 5343–5346, 2000.
23. B. A. Andreev, V. V. Emstev, D. I. Kryzhkov, and V. B. Shmagin, "Study of IR absorption and photoconductivity spectra of thermal double donors in silicon," Phys. Stat. Sol., 325 (1), 79–84, 2003.
24. M. Bruzzi, D. Menichelli, M. Scaringella, J. Härkönen, E. Tuovinen, and Z. Li, "Thermal donors formation via isothermal annealing of magnetic Czochralski high resistivity silicon," J. Appl. Phys., 99 (093706), 1–8, 2006.
25. Y. Yamadaa, H. Yamamoto, H. Ohbaa, M. Sasaseb, F. Esakac, K. Yamaguchia, H. Udonod, S.-I. Shamotoa, A. Yokoyama, and K. Hojoue, "Local neutron transmutation doping using isotopically enriched silicon film," J. Phys. Chem. Sol., 68 (11), 2204–2208, 2007.
26. J. Meese, *Neutron Transmutation Doping*, Plenum Press, New York, 1979.
27. R. D. Larrabee, *Neutron Transmutation Doping of Semiconductor Materials*, Plenum press, New York, 1984.
28. A. C. Reyes, S. M. El-Ghazaly, S. J. Dorn, M. Dydyk, D. K. Schroder, and H. Patterson, "Coplanar waveguides and microwave inductors on silicon substrates," IEEE Trans. Microwave Theory Tech., 43 (9), 2016–2022, 1995.
29. M. Yoshimi, A. Nishiyama, O. Arisumi, M. Terauchi, K. Matsuzawa, and N. Shigyo, "Reduction of the floating-body effect in SOI MOSFETs by the bandgap engineering method," Proc. 7th International. Symposium on SOI Technology and Devices, P. L.F. Hemment and S. Cristoloveanu, Eds., 231–236, 1996.
30. A. Nishiyama, O. Arisumi, M. Terauchi, S. Takeno, K. Suzuki, C. Takakuwa, and M. Yoshimi, "Formation of SiGe source/drain using Ge implantation for floating-body effect resistant SOI MOSFETs," Jpn. J. Appl. Phys., 35 (2B), 954–959, 1996.
31. Y. Domae, N. Miura, T. Okumura, A. Kumar, and J. Ida, "Suppression of floating body effect in low leakage FD-SOI with fluorine implantation technology," Proceedings of the IEEE International SOI Conference, 97–98, 2006.
32. R. M. Huang, T. F. Chen, S. F. Hong, Y. H. Lin, T. L. Tsai, E. C. Liu, C. W. Yang, Y. S. Hsieh, Y. T. Huang, J.-L. Pelloi, C. T. Tsai, and G. H. Ma, "Optimizing floating body effect & AC performance in 65 nm PD-SOI CMOS," Proceedings of the IEEE International SOI Conference, 107–108, 2007.
33. M. Watanabe and A. Tooi, "Formation of SiO_2 films by oxygen-ion bombardment," Jpn. J. Appl. Phys., 5, 737–738, 1966.
34. J. Dylewski and M. C. Joshi, "Thin SiO_2 films formed by oxygen ion implantation in silicon: electron microscope investigation of the Si-SiO_2 interface structures and their CV characteristics," Thin Solid Films, 37, 241–248, 1976.
35. M. H. Badawi and K. V. Anand, "A study of silicon oxides prepared by oxygen implantation into silicon," J. Phys. D, 10, 1931–1942, 1977.
36. K. Izumi, M. Doken, and H. Ariyoshi, "C.M.O.S. devices fabricated on buried SiO_2 layers formed by oxygen implantation into silicon," Electron. Lett., 14 (18), 593–594, 1978.

37. O. W. Holland, D. Fathy, and D. K. Sadana, "Formation of ultrathin, buried oxides in Si by O^+ ion implantation," Appl. Phys. Lett., 69 (5), 474–476, 1996.
38. Y. Dong, J. Chen, X. Wang, M. Chen, and X. Wang, "Optimized implant dose and energy to fabricate high-quality patterned SIMOX SOI materials," Solid State Commun., 130 (3–4), 275–279, 2004.
39. M. Kimura, K. Egami, and M. Kanamori, "Epitaxial film transfer technique for producing single crystal Si film on an insulating substrate," Appl. Phys. Lett., 43 (3), 263–265, 1983.
40. T. R. Anthony, "Dielectric isolation of silicon by anodic bonding," J. Appl. Phys., 58 (3), 1240–1247, 1985.
41. J. B. Lasky, S. R. Stiffler, F. R. White, J. R. Abernathey, "Silicon on insulator (SOI) by bonding and etch-back," IEEE IEDM Tech. Digest, 684–687, 1985.
42. H. Muraoka, T. Ohhashi, and Y. Sumitomo, "Controlled preferential etching technology" Semiconductor Silicon, H. R. Huff and R. R. Burgess, Eds., 327–329, 1973.
43. M. Bruel, "Application of hydrogen ion beams to silicon on insulator material technology," Nucl. Instrum. Meth. Phys. Res. B, 108, 313–319, 1996.
44. A. J. Auberton-Hervé, M. Bruel, B. Aspar, C. Maleville, and H. Moriceau, "Smart-cut®: The basic fabrication process for Unibond® SOI wafers," IEICE Trans. Electron., E80-C (3), 358–363, 1997.
45. T. O. Sedgwick, "Short time annealing," J. Electrochem. Soc., 130 (2), 484–493, 1983.
46. J. Kato and S. Iwamatsu, "Rapid annealing using halogen lamps," J. Electrochem. Soc., 131 (5), 1145–1152, 1984.
47. R. Singh, "Rapid isothermal processing," J. Appl. Phys., 63 (8), R59–R114, 1988.
48. J. M. Ranish, "Design of halogen lamps for rapid thermal processing," 11th IEEE Conf. on Advanced Thermal Processing of Semiconductors, 195–202, 2003.
49. C. M. Osburn, D. F. Downey, S. B. Felch, and B. S. Lee, "Ultrashallow junction formation using very low energy B and BF_2^+ sources," Proc. 11th Intl. Conf. on ion implantation technology, 607–610, 1997.
50. T. E. Seidel, "Rapid thermal annealing of BF_2^+ implanted preamorphized silicon," IEEE Electron Dev. Lett., EDL-4 (10), 353–355, 1983.
51. F. Simard-Normandin, "Electrical characteristics and contact resistance of B^+- and BF_2^+-implanted silicon diodes with furnace and rapid thermal annealing," IEEE Trans. Electron Dev., ED-32 (7), 1354–1357, 1985.
52. S. D. Hossain, M. F. Pas, G. Miner, and C. R. Cleavelin, "Rapid thermal processing (RTP) applied to ion implant anneal for 0.25 μm technology," IEEE/SEMI Adv. Semicon. Manuf. Conference, 5–7, 1995.
53. E. Vandenbossche, H. Jaouen, and B. Baccus, "Modeling arsenic activation and diffusion during furnace and rapid thermal annealing," IEEE IEDM Tech. Digest, 81–85, 1995.
54. H. Takemura, t. Makiya, S. Ohi, M. Sugiyama, T. Tshiro, and M. Nakamae, "Submicron epitaxial layer and RTA technology for extremely high speed bipolar transistors," IEEE IEDM Tech. Digest, 424–427, 1986.
55. T. Hashimoto, M. Tanemura, H. Fujii, F. Sato, T. Aoyama, H. Suzuki, H. Yoshida, and T. Yamazaki, "A CMOS-based RF SiGe BiCMOS technology featuring over-100 GHz fmax SiGe HBTs and 0.13 mm CMOS," IEEE BCTM Tech. Digest, 189–192, 2002.
56. B. El-Kareh, S. Balster, P. Steinmann, H. Yasuda, "Integration of a complementary-SiGe BiCMOS process for high-speed analog application", The Silicon Heterostructure Handbook: Materials, Fabrication, Devices, Circuits, and Applications of SiGe and Si Strained-Layer Epitaxy, John D. Cressler, Editor, CRC Press, July 2005.
57. S. Moffatt, A. Murrell, G. de Cock, D. Armour, M. Foad, and E. Collart, "Electron-volt, high-current implant into silicon SDR (surface damage region) and the effect of anneal time to form 200 to 700 Angstrom, low leakage junctions," IEEE Int. Conf. Ion Implant Tech., 682–685, 1999.
58. M. A. Foad, G. de Cock, D. Jenings, T.-S. Wang, and T. Cullis, "Uniform spike anneals of ultra low energy boron implants using xR LEAP and RTP Centura XEplus: ramp rate effects up to 150 °C/sec," IEEE Int. Conf. Ion Implant Tech., 732–735, 1999.

59. A. Agarwal, A. T. Fiory, H.-J. Gossmann, C. Rafferty, S. P. Frisella, J. Hebb, and J. Jackson, "Ultra-shallow junctions and the effect of ramp-up rate during spike anneals in lamp-based and hot-walled RTP systems," IEEE Int. Conf. Ion Implant Tech., 22–25, 1999.
60. T. Kubo, M. Hori, and M. Kase, "Formation of ultra-shallow junction by BF_2^+ implantation and spike annealing," IEEE Int. Conf. Ion Implant Tech., 195–198, 2000.
61. Y. Bykov, A. Eremeev, V. Holoptsev, I. Plotnikov, and N. Zharova, "Spike annealing of silicon wafers using millimeter wave power," 9th Int. Conf. Advanced Therm. Proc. of Semiconductors-RTP, 232–239, 2001.
62. S. Abo, S. Ichihara, T. Lohner, J. Gyulai, F. Wakaya, and M. Takai, "Ultra shallow As profiling before and after spike annealing using medium energy ion scattering," Ext. Abs. the 5th Int. Workshop on Jct. Tech., 49–50, 2005.
63. C. I. Li, C. C. Chien, K. T. Huang, P. Y. Chen, H. Y. Wang, S. T. Tzou, S. Chen, J. Lin, T. Fu, R. Tandjaja, S. Ramamurthy, E. Chung, J. Chuang, and W.-S. Chen, "Superior spike annealing performance in 65 nm source/drain extension engineering," 13th Int. Conf. Advanced Therm. Proc. of Semiconductors-RTP, 163–167, 2005.
64. K. S. Jones, S. P. Crane, C. E. Ross, T. Malborg, D. Downey, E. Arevalo, S. McCoy, and J. Gelpey, "The role of pre-anneal conditions on the microstructure of Ge^+ implanted Si after high temperature millisecond flash annealing," IEEE 14th Int. Conf. Ion Implant Tech., 76–78, 2002.
65. T. Ito, T. Inuma, A. Murakoshi, H. Akutso, K. Suguro, T. Arikado, K. Okumura, M. Yoshioka, T. Owada, Y. Imaoka, H. Murayama, and T. Kusuda, "10–15 nm ultrashallow junction formation by flash-lamp annealing," Jpn. J. Appl. Phys., 41 (4B), 2394–2398, 2002.
66. T. Ito, K. Suguro, T. Itani, K. Nishinohara, K. Matsuo, and T. Saito, "Improvement of threshold voltage roll-off by ultra-shallow junction formed by flash annealing," Symp. VLSI Tech. Digest, 53–54, 2003.
67. K. Adachi, K. Ohuchi, N. Aoki, H. Tsujii, T. Ito, H. Itokawa, K. Matsuo, K. Suguro, Y. Honguh, N. Tamaoki, K. Ishimaru, and H. Shiuchi, "Issues and optimization of millisecond anneal process for 45 nm node and beyond," Symp. VLSI Tech. Digest, 142–143, 2005.
68. W. Skorupa, T. Gebel, R. A. Yankov, S. Paul, W. Lerch, D. F. Downey, and E. A. Aravelo, "Advanced thermal processing of ultrashallow implanted junctions using flash lamp annealing," J. Electrochem. Soc., 152 (6), G436–G440, 2005.
69. S. H. Jain, P. B. Griffin, J. D. Plummer, S. McCoy, J. Gelpey, T. Selinger, and D. F. Downey, "Low resistance, low-leakage ultrashallow p+-junction formation using millisecond flash anneals," IEEE Trans. Electron Dev., 52 (7), 1610–1615, 2005.
70. C. F. Nieh, K. C. Ku, C. H. Chen, L. T. Wang, L. P. Huang, Y. M. Sheu, C. C. Wang, T. L. Leee, S. C. Chen, M. S. Liang, and J. Gong, "Millisecond anneal and short-channel effect control in Si CMOS transistor performance," IEEE Electron Dev. Lett., 27 (12), 969–971, 2006.
71. W. Lerch, S. Paul, J. Niess, J. Chan, S. McCoy, J. Gelpey, F. Cristiano, F. Severac, P. F. Fazzini, D. Bolze, P. Pichler, A. Martinez, A, Mineji, and S. Shishiguchi, "Experimental and theoretical results of dopant activation by a combination of spike and flash annealing," Ext. Abs. Int. Workshop on Jct. Tech., 129–134, 2007.
72. W. Lerch, S. Paul, J. Niess, S. McCoy, J. Gelpey, D. Bolze, F. Cristiano, F. Severac, P. F. Fazzini, A. Marinez, and P. Pichler, "Advanced activation and deactivation of arsenic-implanted ultra-shallow junctions using flash and spike + flash annealing," 15th Int. Conf. on Advanced Thermal Processing of Semiconductors, 215–220, 2007.
73. T. Feudel, M. Horstmann, L. Herrmann, M. Herden, M. Berhardt, D. Greenlaw, P. Fisher, and J. Kluth, "Process integration issues with spike, flash and laser anneal implementation for 90 and 65 nm technologies," 14th Int. Conf. Advanced Thermal Processing of Semiconductors, 73–78–2006.
74. S. Talwar, G. Varma, K. Weiner, and C. Gelatos, "Laser thermal processing for shallow junction and silicide formation," SPIE Conf. Microelectron. Dev. Tech., 3506, 74–81, 1988.
75. K. Tsuji, K. Takeuchi, and T. Moagami, "High performance 50-nm physical gatelength pMOSFETs by using low temperature activation by re-crystallization scheme," Proc. VLSI Tech. Digest, 9–10, 1999.

76. S. Talwar, G. Verma, and K. H. Weiner, "Ultra-shallow, abrupt, and highly-activated junctions by low-energy ion implantation and laser annealing," IEEE 11th Int. Conf. Ion Implant Tech., 1171–1173, 1999.
77. K.-i. Goto, T. Yamamoto, T. Kubo, M. Kase, Y. Wang, T. Lin, S. Talwar, and T. Sugii, "Ultra-low contact resistance for deca-nm MOSFETs by laser annealing," IEEE IEDM Tech. Digest, 931–933, 1999.
78. B. Yu, Y. Wang, H. Wang, Q. Xiang, C. Riccobene, S. Talwar, and M.-R. Lin, "70nm MOSFET with ultra-shallow, abrupt, and super-doped S/D extension implemented by laser thermal process (LTP)," IEEE IEDM Tech. Digest, 509–512, 1999.
79. R. Murto, K. Jones, M. Rendon, and S. Talwar, "An investigation of species dependence in germanium pre-amorphized and laser thermal annealed ultra-shallow abrupt junctions," IEEE 12th Int. Conf. Ion Implant Tech., 182–185, 2000.
80. C. Park, S.-D. Kim, Y. Wang, S. Talwar, and J. C. S. Woo, "50nm SOI CMOS transistors with ultra shallow junction using laser annealing and pre-amorphization implantation," Proc. VLSI Tech. Digest, 69–70, 2001.
81. S.-D. Kim, C.-M. Park, and J. C. S. Woo, "Advanced source/drain engineering for box-shaped ultrashallow junction formation using laser annealing and pre-amorphization implantation in sub-100-nm SOI CMOS," IEEE Trans. Electron Dev., 49 (10), 1748–1754, 2002.
82. Y. F. Chong, K. L. Pey, A. T. S. Wee, T. Osipowicz, A. See, and L. Chan, "Control of transient enhanced diffusion of boron after laser thermal processing of preamorphized silicon," J. Appl. Phys., 92 (3), 1344–1350, 2002.
83. A. Matsuno and K. Shibahara, "Effect of pulse duration on formation of ultrashallow junction by excimer laser annealing," Jpn. J. Appl. Phys., 45 (11), 8537–8541, 2006.
84. T. Yamamoto, K.-i. Goto, T. Kubo, Y. Wang, T. Lin, S. Talwar, M. Kase, and T. Sugii, "Drive current enhancement in sub-50 nm CMOS by reduction of SDE resistance with laser thermal process," J. Electrochem. Soc., 152 (12), G895–G899, 2004.
85. S. Earles, M. Law, R. Brindos, K. Jones, S. Talwar, and S. Corcoran, "Nonmelt laser annealing of 5-KeV and 1-KeV boron-implanted silicon," IEEE Trans. Electron Dev., 49 (7), 1118–1123, 2002.
86. S. Earles, M. E. Law, K. S. Jones, J. Frazer, S. Talwar, D. Downey, and E. Arevalo, "Formation of ultrashallow junctions in 500 eV boron implanted silicon using nonmelt laser anneal," 12th Int. Conf. on Advanced Thermal Processing of Semiconductors, 143–145, 2004.
87. S. K. H. Fung, H. T. Huang, S. M. Cheng, S. W. Wang, Y. P. Wang, Y. Y. Yao, C. M. Chu, S. J. Yang, W. J. Liang, Y. K. Leung, C. C. Wu, C. Y. Lin, S. J. Chang, S. Y. Wu, C. F. Nieh, C. C. Chen, T. L. Lee, Y. Jin, S. C. Chen, L. T. Lin, Y. H. Chiu, J. T. Tao, C. Y. Fu, S. M. Jang, K. F. Yu, C. H. Wang, T. C. Ong, Y. C. See, C. H. Diaz, M. S. Liang, and Y. C. Sun, "65nm CMOS high speed, general purpose and low power transistor technology for high volume foundry applications," Symp. VLSI Tech. Digest, 92–93, 2004.
88. A. Shima, Y. Wang, S. Talwar, and A. Hiraiwa, "Ultra-shallow junction formation by non-melt laser spike annealing for 50-nm gate CMOS," Symp. VLSI Tech. Digest, 174–175, 2004.
89. T. Yamamoto, T. Kubo, T. Sukegawa, A. Katakami, Y. Shimamune, N. Tamura, H. Ohta, T. Miyashita, S. Sato, M. Kase, and T. Sugii, "Advantages of new scheme of junction profile engineering with laser spike annealing and its integration into a 45-nm node high performance CMOS technology," Symp. VLSI Tech. Digest, 122–123, 2007.
90. T. Yamamoto, T. Kubo, T. Sukegawa, E. Takii, Y. Shimamune, N. Tamura, T. Sakoda, M. Nakamura, H. Ohta, T. Miyashita, H. Kurata, S. Satoh, M. Kase, and T. Sugii, "Junction profile engineering with novel multiple laser spike annealing scheme for 45-nm node high performance and low leakage CMOS technology," IEEE IEDM Tech. Digest, 143–146, 2007.
91. A. Shima, T. Mine, K. Torii, and A. Hiraiwa, "Enhancement of drain current in planar MOSFETs by dopant profile engineering using nonmelt laser spike annealing," IEEE Trans. Electron Dev., 54 (11), 2953–2959, 2007.
92. M. O. Thompson, G. J. Galvin, J. W. Mayer, P. S. Peercy, J. M. Poate, D. C. Jacobson, A. Cullis, and N. Chew, "Melting temperature and explosive crystallization of amorphous silicon during pulsed laser irradiation," Phys. Rev. Lett., 52 (26), 2360–2363, 1984.

93. P. S. Peercy and M. O. Thompson, "Kinetic and thermodynamic studies of pulsed laser irradiation," SPIE, 668, 72–81, 1986.
94. J. M. Poate, "High speed crystal growth and solidification using laser heating," J. Cryst. Growth, 79 (1–3), 549–561, 1986.
95. R. Murto, K. Jones, M. Rendon, and S. Talwar, "Activation and deactivation studies of laser thermal annealed boron, arsenic, phosphorus, and antimony ultra-shallow abrupt junctions," Proc. Int. Conf. Ion Implantation Technology, 155–158, 2000.
96. H. Wakabayashi, M. Ueki, M. Narihiro, T. Fukai, N. Ikezawa, T. Matsuda, K. Yoshida, K. Takeuchi, Y. Ochiai, T. Mogami, and T. Kunio, "45-nm gate length CMOS technology and beyond using steep halo," IEEE IEDM Tech. Digest, 49–52, 2000.
97. J. Nulmnan, J. P. Krusius, and A. Gat, "Rapid thermal processing of thin gate dielectrics. Oxidation of silicon," IEEE Electron Dev. Lett., EDL-6 (5), 205–207, 1985.
98. S. T. Ang, J. J. Wortman, "Rapid thermal oxidation of silicon," J. Electrochem. Soc., 133 (11), 2361–2362, 1986.
99. J. Nulman, J. Scarpulla, T. Mele, and J. P. Krusius, "Electrical characteristics of thin gate implanted MOS channels grown by rapid thermal processing," IEEE IEDM Tech. Digest, 376–379, 1985.
100. J. Nulman, J. P. Krusius, N. Shah, A. Gat, and A. Baldwin "Ultrathin gate dielectrics on 150 mm Si wafers via rapid thermal processing," J. Vac. Sci. Technol. A, 4 (3), 1005–1008, 1986.
101. K. X. Zhang and C. M. Osburn, "Reliability of in-situ rapid thermal gate dielectrics in deep submicrometer MOSFETs," IEEE Trans. Electron Dev., 412 (12), 2181–2188, 1995.
102. R. Das, M. K. Bera, S. Chakraborty, S. Saha, J. F. Woitok, and C. Maiti, "Physico-chemical and electrical properties of rapid thermal oxides on Ge-rich SiGe heterolayers," Appl. Surf. Sci., 253 (3) 1323–1329, 2006.
103. M. K. Bera, S. Chakraborty, R. Das, G. K. Dalapati, S. Chattopadhyay, S. K. Sumanta, W. J. Yoo, A. K. Chakraborty, Y. Butrnko, L. Šiller, M. R. C. Hunt, S. Saha, and C. K. Maiti, "Rapid thermal oxidation of Ge-rich $Si_{1-x}Ge_x$ heterolayers," J. Vac. Sci. Technol. A, 24 (1), 84–90, 2006.
104. A. Terrasi, S. Scalese, R. Adorno, E. Ferlito, M. Spadafora, and E. Rimini, "Rapid thermal oxidation of epitaxial SiGe thin films," Mater. Sci. Eng. B, 89 (1–3), 269–273, 2002.
105. M.-J. Chen and C.-S. Hou, "A novel cross-coupled inter-poly-oxide capacitor for mixed-mode CMOS processes," IEEE Electron Dev. Lett., 20 (7), 360–362, 1999.
106. S. Itoh, G. Q. Lo, D. L. Kwong, V. K. Mathews, and P. C. Fazan, "Formation of high-quality oxide/nitride stacked layers on rugged polysilicon electrodes by rapid thermal oxidation," IEEE Trans. Electron Dev., 40 (6), 1176–1177, 1993.
107. Z. L. Chun, J. H. Jaqn, Y. H. Fei, G. Y. Zhi, N. B. Jun, and M. B. Xian, "Improvement of RCA transistor using RTA annealing after formation of interfacial oxide," IEEE Trans. Electron Dev., 49 (6), 1075–1076, 2002.
108. A. Tilke, M. Förster, K. Schupke, A. Freigofas, C. Wagner, and C. Dahl, "As-doped polysilicon emitters with interfacial oxides and correlation to bipolar device characteristics," J. Vac. Sci. Technol. B, 23 (5), 1877–1882, 2005.
109. R. J. Kriegler, Y. C. Cheng, and D. R. Colton, "The effect of HCl and Cl_2 on the thermal oxidation of silicon," J. Electrochem. Soc., 119 (3), 388–392, 1972.
110. R. S. Ronen and P. H. Robinson, "Hydrogen chloride and chlorine gettering: An effective technique for improving performance of silicon devices," J. Electrochem. Soc., 119 (6), 747–752, 1972.
111. C. M. Osburn, "Dielectric breakdown properties of SiO_2 films grown in halogen and hydrogen-containing environments," J. Electrochem. Soc., 121 (6), 809–814, 1974.
112. M. Uematsu, H. Kaheshima, and K. Shiraishi, "The effect of chlorine on silicon oxidation: simulation based on the interfacial silicon emission model," Jpn. J. Appl. Phys., 40 (4A), 2217–2218, 2001.
113. C.-C. Hao, M.-h. Chi, C.-C. Chen, H.-J. Lin, Y. F. Lin, C. H. Hsieh, C. H. Lee, K. H. Chyang, H. T. Wu, and C.-H. Shen, "NBTI improvement for pMOS by Cl-contained 1st

oxidation in 25A/65Å dual nitrided gate-oxide of 0.1 mm CMOS technology," Proc. SPIE, 5042, 180–187, 2003.
114. M. M. Moslehi, C. Y. Fu, and K. C. Saraswat, "Thermal and microwave nitrogen plasma nitridation techniques for ultrathin gate insulators of MOS VLSI," Proc. VLSI Tech. Digest, 14–15, 1985.
115. M. M. Moslehi, C. J. Han, K. C. Saraswat, C. R. Helms, and S. Shatas, "Compositional studies of thermally nitrided silicon dioxide (Nitroxide)," J. Electrochem. Soc., 152 (9), 2189–2197, 1985.
116. T. Hori, H. Iwasaki, and K. Tsuji, "Electrical and physical properties of ultrathin reoxidized nitrided oxides prepared by rapid thermal processing," IEEE Trans. Electron Dev., 36 (2), 340–350, 1989.
117. H. Hwang, W. Ting, D.-L. Kwong, and J. Lee, "Electrical reliability characteristics of submicrometer nMOSFETs with oxynitride gate dielectric prepared by rapid thermal oxidation in N_2O," IEEE Trans. Electron Dev., 2712–2713, 1991.
118. H. Hwang, W. Ring, D.-L. Kwong, and J. Lee, "Improved reliability characteristics of submicrometer nMOSFETs with oxynitride gate dielectric prepared by rapid thermal oxidation in N_2O," IEEE Electron Dev. Lett., 12 (9), 495–497, 1991.
119. Z. Liu, H.-J. Wann, P. K. Ko, C. Hu, and Y. C. Cheng, "Effects of N_2O anneal and reoxidation on thermal oxide characteristics," IEEE Electron Dev. Lett., 13 (8), 402–404, 1992.
120. E. C. Carr and R. A. Buhrman, "Role of interfacial nitrogen in improving thin silicon oxides grown in N_2O," Appl. Phys. Lett., 63 (1), 54–56, 1993.
121. H. S. Momose, T. Morimoto, Y. Ozawa, K. Yamabe, and H. Iwai, "Electrical characteristics of rapid thermal nitrided-oxide gate n- and p-MOSFETs with less than 1 atom% nitrogen concentration," IEEE Trans. Electron Dev., 41 (4), 546–551, 1994.
122. Z.-Q. Yao, H. B. Harrison, S. Dimitrijevm, D. Sweatman, and Y. T. Yeow, "High quality ultrathin dielectric films grown on silicon in a nitric oxide ambient," Appl. Phys. Lett., 64 (26), 3584–3586, 1994.
123. Y. Okada, P. J. Tobin, K. G. Reid, R. I. Hedge, B. Maiti, and S. A. Ajuria, "Gate oxynitride grown in nitric oxide (NO)," Proc. VLSI Tech. Digest, 105–106, 1994.
124. L. K. Han, G. W. Yoon, J. Kim, J. Yan, and D. L. Kwong, "Formation of high quality ultrathin oxide/nitride (ON) stacked capacitors by in situ multiple rapid thermal processing," IEEE Electron Dev. Lett., 16 (8), 348–350, 1995.
125. K. A. Ellis and R. A. Buhrman, "Nitrous oxide (N_2O) processing for silicon oxynitride gate dielectrics," IBM J. Res. Dev., 413 (3), 287–300, 1999.
126. H. Wong, V. M. C. Poon, C. W. Kok, P. J. Chan, and V. A. Gritsenko, "Interface structure of ultrathin oxide prepared by N_2O oxidation," IEEE Trans. Electron Dev., 50 (9) 1941–1945, 2003
127. D. Matsushita, "Novel fabrication process to realize ultra-thin (EOT = 0.7 nm) and ultra-low-leakage SiON gate dielectrics," 13th Int. Conf. Advanced Thermal Processing of Semiconductors, 23–30, 2005.
128. J. L. Everaert, T. Conrad, and M. Schaekers, "SiON gate dielectric formation by rapid thermal oxidation of nitrided Si," 13th Int. Conf. Advanced Thermal Processing of Semiconductors, 135–138, 2005.
129. C. H. Kao, W. H. Sung, and C. S. Chen, "Investigation of the doping and thickness effects of polysilicon oxide by rapid thermal N_2O oxidation," Microelectron. Eng., 85 (2), 408–413, 2008.
130. S. Matsuda, T. Sato, H. Yoshimura, Y. Takegawa, A Sudo, I. Mizushima, Y. Tsunashima, and Y. Toyoshima, "Novel corner rounding process for shallow trench isolation utilizing MSTS (Micro-structure transformation of silicon)," IEEE IEDM Tech. Digest, 137–140, 1998.
131. J.-H. Lee, S.-H. Park, K.-M. Lee, K.-S. Youn, Y.-J. Park, C.-J. Choi, T.-Y. Seong, and H.-D. Lee, "A study of stress-induced p^+/n salicided junction leakage failure and optimized process conditions for sub-0.15-μm CMOS technology," IEEE Trans. Electron Dev., 49 (11), 1985–1992, 2002.

132. S. Y. Mun, C. Shin, K. C. Yoon, J. S. Kwak, H. H. Ryu, and Y. H. Jeong, "Shallow trench isolation top corner rounding using Si soft etching following diluted hydrofluorine solution," Jpn. J. Appl. Phys., 43 (11A), 7701–7704, 2004.
133. C. Lee, D. Park, B. Jo, C. Hwang, H. J. Kim, and W. Lee, "Deep submicron CMOS technology using top-edge round STI and dual gate oxide for low power 256M-bit mobile DRAM," Jpn. J. Appl. Phys., 42 (4B), 1892–1896, 2003.
134. T. Ohashi, T. Kubota, and A. Nakajima, "Ar annealing for suppression of gate oxide thinning at shallow trench isolation edge," IEEE Electron Dev. Lett., 28 (7) 562–564, 2007.
135. T. Park, J. Y. Kim, K. W. Park, H. S. Lee, H. B. Shin, Y. H. Kim, M. H. Park, H. K. Kang, and M. Y. Lee, "A novel simple shallow trench isolation (SSTI) technology using high selective CeO_2 slurry and liner SiN as a CMP stopper," Proc. VLSI Tech., 159–160, 1999.
136. J. H. Park, S.-W. Shin, S.-W. Park, Y.-T. Dong, D.-J. Kim, M.-S. Suh, S.-C. Lee, N.-Y. Kwak, C.-D. Dong, D.-W. Kim, G.-I. Lee, O.-J. Kwon, and H. S. Yang, "Effect of liner oxide densification on stress-induced leakage current characteristics in shallow-trench isolation processing," J. Electrochem. Soc., 150 (7) G359–G364, 2003.
137. T. Suntola and J. Antson, "Method for producing compound thin films," United States Patent #4,058,430, 1977.
138. T. Suntola and J. Hyvärinen, "Atomic layer epitaxy," Ann. Rev. Mater. Sci., 15, 177–195, 1985.
139. C. H. L. Goodman and M. V. Pessa, "Atomic layer epitaxy," J. Appl. Phys. 60 (3), R65–R81, 1986.
140. S. M. George, A. W. Ott, and J. W. Klaus, "Surface chemistry for atomic layer growth," J. Phys. Chem., 100(31), 13121–13131, 1996.
141. O. Sneh, R. B. Clark-Phelps, A. R. Londergan, J. Winkler, and T. E. Seidel, "Thin film atomic layer deposition equipment for semiconductor processing," Thin Solid Films, 402, 248–261, 2002.
142. M. Ritala and M. Leskelä, "Atomic layer epitaxy – a valuable tool for nanotechnology?" Nanotechnology, 10, 19–24, 1999.
143. R. L. Puurunen and W. Vandervorst, "Island grown as a growth mode in atomic layer deposition: a phenomenological model," J. Appl. Phys., 96 (12), 7686–7695, 2004.
144. M. Cho, R. Degraeve, G. Pourtois, A. Delabie, L.-Å. Ragnarsson, T. Kauerauf, G.o Groeseneken, S. De Gendt, M. Heyns, and C. S. Hwang, "Study of the reliability impact of chlorine precursor residues in thin atomic-layer-deposited HfO_2 layers," IEEE Trans. Electron Dev., 54 (4), 752–758, 2007.
145. M. Youm, H. S. Sim, H. Jeon, S.-I. Kim, and Y. T. Kim, "Metal oxide semiconductor field effect transistor characteristics with iridium gate electrode on atomic layer deposited ZrO_2 high-K dielectrics," Jpn. J. Appl. Phys., 42 (8), 5010–5013, 2003.
146. R. L. Puurunen, "Analysis of hydroxyl group controlled atomic layer deposition of hafnium dioxide from hafnium tetrachloride and water," J. Appl. Phys., 95 (9), 4777–4786, 2004.
147. D. R. Burgess, Jr., J. E. Maslar, W. S. Hurst, E. F. Moore, W. A. Kimes, R. R. Fink, and N. V. Nguyen, "Atomic layer deposition – process models and metrologies," Characterization and Metrology for ULSI Tech., AIP, 141–145, 2005.
148. Y. Senzaki, K. Choi, P. D. Kirsch, P. Majhi, and B. H. Lee, "Atomic layer deposition of high-K dielectric and metal gate stacks for MOS devices," Characterization and Metrology for ULSI Tech., AIP, 69–72, 2005.
149. I. C. Kizilyalli, R. Y. S. Huang, and P. K. Roy, "MOS transistors with stacked SiO_2-Ta_2O_5-SiO_2 gate dielectrics for giga-scale integration of CMOS technologies," IEEE Electron Dev. Lett., 19 (11) 423–425, 1998.
150. B. He, T. Ma, S. A. Campbell, and W. L. Gladfelter, "A 1.1 nm oxide equivalent gate insulator formed using TiO_2 on nitrided silicon," IEEE IEDM Tech. Digest, 1038–1040, 1998.
151. R. A. McKee, F. J. Walker, and M. F. Chisholm, "Crystalline oxides on silicon: The first five monolayers," Phys. Rev. Lett., 81 (14), 3014–3017, 1998.
152. E. P. Gusev, D. A. Buchanan, E. Cartier, A. Kumar, D. DiMaria, S. Guha, A. Callegari, S. Zafar, P. C. Jamison, D. A. Nemayer, M. Copel, M. A. Gribelyuk, H. Okorn-Schmidt,

C. D'Emic, P. Kozlowski, K. Chan, N. Bojarczuk, L.-Å. Ragnarsson, P. Rosenheim, K. Rim, R. J. Fleming, A. Mocuta, and A. Ajmera, "Ultrathin high-K gate stack for advanced CMOS devices," IEEE IEDM Tech. Digest, 451–454, 2001.
153. C. M. Perkins, B. B. Triplett, P. C. McIntyre, K. C. Saraswat, S. Hauka, and M. Tuominen, "Electrical and material properties of ZrO_2 gate dielectrics grown by atomic layer chemical vapor deposition," Appl. Phys. Lett., 78 (16), 2357–2359, 2001.
154. J. Pan, C. Woo, C.-Y. Yang, U. Bhandary, S. Guggilla, N. Kriswhna, H. Chung, A. Hui, B. Yu, Q. Xiang, and M.-R. Lin, "Replacement metal-gate NMOSFETs with ALD TaN/EP-Cu, PVD Ta, and PVD TaN electrode," IEEE Electron Dev. Lett., 24 (5), 304–306, 2003.
155. A. Nakajima, T. Ohashi, S. Zhu, S. Yokoyama, S. Michimata, and H. Miyake, "Atomic layer-deposited Si-Nitride/SiO_2 stack gate dielectrics for future high-speed DRAM with enhanced reliability," IEEE Electron Dev. Lett., 26 (8), 538–540, 2005.
156. S.-J. Ding, H. Hu, S. J. Kim, X. F. Yu, C. Zhu, M. F. Li, B. J. Cho, D. S. H. Chan, S. C. Rustagi, M. B. Yu, A. Chin, and D.-L. Kwong, "High-performance MIM capacitor using ALD high-K HfO_2-Al_2O_3 laminate dielectrics," IEEE Electron Dev. Lett., 24 (12), 730–732, 2003.
157. A. Satta, J. Schuhmacher, C. M. Whelan, W. Vandervorst, S. H. Brongersma, G. P. Beyer, K. Maex, A. Vantomme, M. M. Vitanen, H. H. Brongersma, and W. F. A. Besling, "Growth mechanism and continuity of atomic layer deposited TiN films on thermal SiO_2," J. Appl. Phys., 92 (12), 7641–7646, 2002.
158. N.-J. Bae, K.-I. Na, H.-I. Cho, K.-Y. Park, S.-E. Boo, J.-H. Bae, and J.-H. Lee, "Thermal and electrical properties of 5-nm thick TaN film prepared by atomic layer deposition using pentakis(ethylmethylamino)tantalum precursor for copper metallization," Jpn. J. Appl. Phys., 45 (12) 9072–9074, 2006.
159. R. Solanki and B. Pathangey, "Atomic layer deposition of copper seed layers," Electrochem. Solid-State Lett., 3 (10) 479–480, 2000.
160. H. H. Andersen, B. Stenum, T. Sørensen, and H. Whitlow, "Angular distribution of particles sputtered from Cu, Pt and Ge targets by keV Ar + ion bombardment," Nucl. Instrum. Methods B, 6 (3), 459–465, 1985.
161. S. M. Rossnagel and J. Hopwood, "Metal ion deposition from ionized magnetron sputtering discharge," J. Vac. Sci. Technol. B, 12 (1), 449–453, 1994.
162. S. M. Rossnagel, D. Mikalsen, H. Kinoshita, and J. J. Cuomo, "Collimated magnetron sputter deposition," J. Vac. Sci. Technol. A, 9 (2), 261–265, 1991.
163. J. C. S. Kools, A. P. Paranjpe, D. H. Heimanson, P. V. Schwartz, K. Song, B. Bergner, and S. McAllister, "Novel approach to collimated physical vapor deposition," J. Vac. Sci. Technol. A, 17 (4), 1941–1945, 1999.
164. M. S. Barnes, J. C. Forster, and J. H. Keller, "Apparatus for depositing material into high aspect ratio holes," United States Patent #5,178,739.
165. J. Hopwood, "Ionized physical vapor deposition of integrated circuit interconnects," Phys. Plasma, 5 (2), 1624–1631, 1998.
166. G. Zhong and J. P. Hopwood, "Ionized titanium deposition into high-aspect ratio vias and trenches," J. Vac. Sci. Technol. B, 17 (2), 405–409, 1999.
167. S. Hamaguchi and S. M. Rossnagel, "Liner conformality in ionized sputter metal deposition processes," J. Vac. Sci. Technol. B, 14 (4), 2603–2608, 1996.
168. K. C. Park, I.-R. Kim, B.-S. Suh, S.-M. Choi, W.-S. Song, Y.-J. Wee, S.-G. Lee, J.-S. Chung, J.-H. Chung, S.-R. Hah, J.-H. Ahn, K.-T. Lee, K.-K. Kang, and K.-P. Suh, "Advanced i-PVD barrier metal deposition technology for 90nm Cu interconnects," IEEE IITC Tech. Digest, 165–167, 2003.
169. D. Mao and J. Hopwood, "Ionized physical vapor deposition of titanium nitride: a deposition model," J. Appl. Phys., 96 (1), 820–827, 2004.
170. N. Li, D. M. Ruzic, and R. A. Powell, "Chemically enhanced physical vapor deposition of tantalum nitride-based films for ultra-large scale integrated devices," J. Vac. Sci. Technol. B, 22 (6), 2734–2742, 2004.
171. J. Brcka and R. L. Robison, "Wafer redeposition impact on etch rate uniformity in IPVD system," IEEE Trans. Plasma Sci., 35 (1), 74–81, 2007.

172. B. J Lin, "The optimum numerical aperture for optical projection microlithography," SPIE, 1463, 42–63, 1991.
173. C. A. Mack, "Optical proximity effects," Microlithography World, 22–23, 1996.
174. J. F. Chen, T. Laidig, K. E. Wampler, and R. Caldwell, K. H. Nakagawa, and A. Liebchen, "Practical method for full-chip optical proximity correction," SPIE, 3051, 790–803, 1997.
175. J. F. Chen, T. Laidig, K. E. Wampler, R. Caldwell, K. H. Nakagawa, and A. Liebchen, "A practical technology path to sub-0.10 micron process generations via enhanced optical lithography," SPIE, 3873, 995–1016, 1999.
176. H. Chuang, P. Gilbert, W. Grobman, M. Kling, K. Lucas, A. Reich, B. Roman, E. Travis, P. Tsui, T. Vuong, and J. West, "Practical applications of 2-D optical proximity corrections for enhanced performance of 0.25 μm random logic devices," IEEE IEDM Tech. Digest, 483–456, 1997.
177. S. Roy, D. Van Deb Broeke, J. F. Chen, A. Liebchen, T. Chen, S. Hsu, X. Shi, and R. Socha, "Extending aggressive low-k_1 design rule requirements for 90 and 65 nm nodes via simultaneous optimization of numerical aperture, illumination and optical proximity correction," J. Microlith., Microfab., Microsyst., 4 (2), 023003 1–10, 2005.
178. T. Winkler, W. Dettmann, M. Hennign, W. Koestler, M. Moukara, J. Thiele, and K. Zeiler, "AOPC for double exposure lithography," SPIE, 5754, 1169–1178, 2005.
179. S. Lee, G. Chen, and R. Lee, "Application of reverse scattering bar for memory device, combined with model-based OPC," Int. Symp. Semicond. Manuf. (ISSM), 446–449, 2005.
180. T. Ebihara, M. D. Levenson, W. Liu, J. He, W. Yeh, S. Ahn, T. Oga, M. Shen, and H. M'saad, "Beyond $k_1 = 0.25$ lithography: 70nm L/S patterning using KrF scanners," SPIE 5256, 985–994, 2003.
181. S. Hsu, J. F. Chen, N. Cororan, W. Knose, D. J. Van Den Broeke, T. Laidig, K. E. Wampler, X. Shi, M. Hsu, M. Eurlings, J. Finders, T.-B. Chiou, R. Socha, W. Conley, Y. W. Hsieh, S. Tuan, F. Hsieh, "65nm full-chip implementation using double dipole lithography," Proc. SPIE, 5040, 215–231, 2003.
182. S.-Y. Oh, W.-H. Kim, H.-S. Yune, H.-B. Kim, S.-M. Kim, C.-N. Ahn, and K.-S. Shin, "The double exposure strategy using OPC & simulation and the performance on wafer with sub-0.1mm design rule in ArF lithography," Digest. SPIE, 4591, 1537–1543, 2002.
183. W.-K. Ma, C.-M. Lim, S.-Y. Oh, S.-M. Kim, B.-H. Nam, S.-C. Moon, and K.-S. Shin, "Double exposure to reduce overall line-width variation of 80nm DRAM gate," Digest. SPIE, 5377, 939–946, 2004.
184. M. Dusa, B. Arnold, and A. Fumar-Pici, "Prospects and initial exploratory results for double exposure/double pitch technique," IEEE Int. Symp. Semicond. Manuf. (ISSM), 177–180, 2005.
185. S.-K. Kim, "Double exposure and double patterning studies with inverse lithography," IEEE 20th Int. Microprocesses and Nanotechnology Conf., 80–1, 2007.
186. W. Y. Jung, C.-D. Kim, J.-D. Eom, S.-Y. Cho, S.-M. Jeon, J.-H. Kim, J. I. Moon, B.-S. Lee, and S.-K. Park, "Patterning with spacer for expanding the resolution limit of current lithography tool," Digest, SPIE 6165, J1–J9, 2006.
187. J. Finders, M. Dusa, and S. Hsu, "Double patterning lithography: The bridge between low k_1 ArF and EUV," Microlithography World, February 2008.
188. M. D. Feuer and D. E. Prober, "Projection photolithography-liftoff technique for production of 0.2-μm metal patterns," IEEE Trans. Electron Dev., ED-28 (11), 1375–1378, 1981.
189. B. J. Lin, "Immersion lithography and its impact on semiconductor manufacturing," Proc. SPIE, 5377, 46–67, 2004.
190. M. McCullum, M. Kameyama, and S. Owa, "Practical development and implementation of 193 nm immersion lithography," Microelectron. Eng., 83 (4–9), 640–642, 2006.
191. H. Kawata, J. M. Carter, A. Yen, and H. I. Smith, "Optical projection lithography using lenses with numerical apertures greater than unity," Microelectron. Eng., 9 (1–4), 31–36, 1989.
192. G. Owen, R. F. W. Pease, D. A. Markle, A. Grenville, R. L. Hsieh, R. von Bünau, and N. I. Maluf, "1/8 μm optical lithography," J. Vac. Sci. Technol. B, 10 (6), 3032–3036, 1992.

193. H. Kawata, I. Matsumura, H. Yoshida, and K. Murata, "Fabrication of 0.2 μm fine patterns using optical projection lithography with an oil immersion lens," Jpn. J. Appl. Phys., 31 (12B), 4174–4177, 1992.
194. J. A. Hoffnagle, W. D. Hinsberg, M. Sanchez, and F. A. Houle, "Liquid immersion deep-ultraviolet interferometric lithography," J. Vac. Sci. Technol. B, 17 (6), 3306–3309, 1999.
195. M. Switkes and M. Rothschild, "Resolution enhancement of 157 nm lithography by liquid immersion," Proc. SPIE, 4691, 459–465, 2002.
196. B. W. Smith, A. Bourov, H. Kang, F. Cropanese, Y. Fan, N. Lafferty, and L. Zavyalova, "Water immersion optical lithography at 193 nm," J. Microlith., Microfab., Microsyst., 3 (1), 46–51, 2004.
197. Th. Zell, "Present and future of 193 nm lithography," Microelectron. Eng., 83 (4–9), 624–633, 2006.
198. G. O'Sillivan, A. Cummings, P. Dunne, K. Fahy, P. Hayden, L. McKinney, N. Murphy, E. Sokkel, and J. White, "Recent progress in the development of sources for EUV lithography," American Institute of Physics (AIP) Conf. Proc., 108–116, 2005.
199. Th. Kruecken, "Plasma and radiation modeling of EUV sources for micro lithography," American Institute of Physics (AIP) Conf. Proc., 181–190, 2007.
200. Th. Kruecken, "Discharge plasmas as EUV sources for future micro lithography," American Institute of Physics (AIP) Conf. Proc., 259–269, 2007.
201. S. Y. Chou, P. R. Krauss, and P. J. Renstrom, "Nanoimprint lithography," J. Vac. Sci. Technol. B, 14 (6), 4129–4133, 1996.
202. S. Y. Chou, P. R. Krauss, W. Zhang, L. Guo, and L. Zhuang, "Sub-10 nm imprint lithography and applications," J. Vac. Sci. Technol. B, 15 (6), 2897–2904, 1997.
203. K. Kincade, "Imprint lithography challenges EUV for next-generation chip manufacturing," Laser Focus World, 43 (7), 97–104, 2007.
204. S. N. Hong, G. A. Ruggles, J. J. Wortman, E. R. Myers, and J. J. Hren, "Characterization of ultra-shallow p^+ -n junction diodes fabricated by 500-eV boron-ion implantation," IEEE Trans. Electron Dev., 38 (1), 28–31, 1991.
205. A. Bousetta, J. A. van den Berg, and D. G. Armour, "Formation of 0.05-nm p^+-n and n^+ -p junctions by very low (<500 eV) ion implantation," IEEE Electron Dev. Lett., 13 (5), 250–252, 1992.
206. A. Al-Bayati, S. Tandon, A. May, M. Foad, and D. Wagner, "Exploring the limits of pre-amorphization implants on controlling channeling and diffusion of low energy B implants and ultra shallow junction formation," IEEE Proc. Conf. Ion Implantation Tech., 54–57, 2000.
207. J. Liu, U. Jeong, M. Meloni, and S. Mehta, "Effects of pre-amorphization on junction characteristics and damage behavior in low energy boron implantation," IEEE Proc. Conf. Ion Implantation Tech., 191–194, 2000.
208. N. Natsuaki, A. Shima, M. Honda, S. Nagayama, H. Sato and T. Hashimoto, "Surface sensitive redistribution of low energy implanted B in Si substrate," IEEE Proc. Conf. Ion Implantation Tech., 474–477, 1998.
209. N. Variam, S. Falk, S. Mehta, T. Miranda, and J. Luke, "Challenges and solutions in the process integration of ultra-shallow junctions in advanced CMOS technology," IEEE Proc. Conf. Ion Implantation Tech., 77–80, 2000.
210. H. C.-H. Wang, C.-C. Wang, C.-S. Chang, T. Wang, P. B. Griffin, and C. H. Diaz, "Interface induced uphill diffusion of boron: An effective approach for ultrashallow junction," IEEE Proc. Conf. Ion Implantation Tech., 65–67, 2001.
211. C. Laviron, F. Milesi, and G. Mathieu, "Ultrashallow P^+/N junctions using BCl_2^+ implantations for sub 0.1 μm CMOS devices," IEEE Proc. Conf. Ion Implantation Tech., 100–102, 2002.
212. K. Goto, J. Matsuo, Y. Tada, T. Sugii, I. Yamada, "Decaborane ($B_{10}H_{14}$) ion implantation technology for sub 0.1 μm PMOSFET's", IEEE Trans. Electron Dev., 46 (4), 683–689, 1999.
213. D. C. Jacobson, K. Bourdelle, H-J. Gossmann, M. Sosnowski, M. A. Albano, V. Babaram, J. M. Poste A. Aganval, A. Perel, T. Horsky, "Decaborane, an alternative approach to ultra low energy ion implantation," IEEE Proc. Conf. Ion Implantation Tech., 300–303, 2000.

214. D. Lenoble, A. Grouillet, F. Arnaud, M. Haond, S. B. Felch, Z. Fang, S. Walther, and R. B. Liebert, "Direct comparison of electrical performance of 0.1-pm pMOSFETs doped by plasma doping or low energy ion implantation," IEEE Proc. Conf. Ion Implantation Tech., 468–471, 2000.
215. B. Mizuno, I. Nakayama, N. Aoi, M. Kubota, and T. Komeda, "New doping method for subhalf micron trench sidewalls by using an electron cyclotron resonance plasma," Appl. Phys. Lett., 53 (21), 2059–2061, 1988.
216. N. W. Cheung, "Plasma immersion ion implantation for ULSI processing," Nucl. Instrum. Meth. Phys. Res. B, 55 (1–4), 811–820, 1991.
217. X. Y. Qian, N. W. Cheung, and M. A. Lieberman, "Plasma immersion ion implantation of SiF_4 and BF_3 for sub-100 nm p^+ n junction fabrication," Appl. Phys. Lett., 59 (3), 348–350, 1991.
218. P. K. Chu, A. Qin, C. Chan, N. W. Cheung, and L. A. Larson, "Plasma immersion ion implantation – a fledging technique for semiconductor processing," Mater. Sci. Eng. R, 17, 207–280, 1996.
219. R. B. Liebert, S. W. Walther, S. B. Felch, Z. Fang, B. O. Pederson, and D. Hacker, "Plasma doping system for 200 and 300mm wafers," IEEE Proc. Conf. Ion Implantation Tech., 472–475, 2000.
220. J. T. Sheuer, D. Lenoble, J.-P. Reynard, F. Lallement, A. Grouillet, A. Arevalo, D. Distaso, Z. Fang, L. Godet, B. W. Koo, T. Miller, and J. Weeman, "USJ formation using pulsed plasma doping,"Surf. Coat. Tech., 186 (1–2), 57–61, 2004.
221. S. Walther, D. Lenoble, F. Lallement, A. Grouillet, Y. Erokhin, V. Singh, and A. Testoni, "Advanced 65 nm CMOSW devices fabricated using ultra-low energy plasma doping," Nucl. Instrum. Meth. Phys. Res. B, 237 (1–2), 126–130, 2005.
222. D. Lee, S. Baek, S. Heo, C. Cho, G. Buh, T. Park, Y. Shin, and H. Hwang, "Ultrashallow p^+/n junction prepared by low energy BF_3 plasma doping and KrF excimer laser annealing," Electrochem. Soc. Solid-State Lett., 9 (1), G19–G21, 2006.
223. S. Heo, S. Baek, D. Lee, G. Buh, Y. Shin, and H. Hwang, "Ultrashallow arsenic n^+/p junction formed by AsH_3 plasma doping," Jpn. J. Appl. Phys., 45 (13), L373–L375, 2006.
224. S. Qin and A. NcTeer, "Device performance improvement of PMOS devices fabricated by B_2H_6 PIII/PLAD processing," IEEE Trans. Electron Dev., 54 (9), 2497–2501, 2007.
225. A. Agarwal and M. J. Kushner, "Characteristics of pulsed plasma doping sources for ultra-shallow junction formation," J. Appl. Phys., 101 063305, 1–16, 2007.
226. H. Strack, "Ion bombardment of silicon in a glow discharge," J. Appl. Phys., 34 (8), 2405–2409, 1963.
227. H. Ruecker, B. Heinemann, D. Bolze, D. Knoll, D. Kruger, R. Kurps, H. J. Osten, P. Schley, B. Tillack, and P. Zaumseil, "Dopant diffusion in C-doped Si and SiGe: Physical model and experimental verification," IEEE IEDM Tech. Digest, 345–348, 1999.
228. H. J. Osten, D. Knoll, B., Heinemann, H. Rücker, and B. Tillack, "Carbon doped SiGe heterojunction bipolar transistors for high frequency applications," IEEE BCTM Tech. Digest, 109–116, 1999.
229. K. E. Ehwald, D. Knoll, B. Heinemann, K. Chang, J. Kirchgessner, R. Mauntel, I. S. Lim, J. Steele, P. Schley, B. Tillack, A. Wolff, K. Blum, W. Winkler, M. Perschel, U. Jadghold, R. Barth, T. Gabolla, H. J. Erzgräber, B. Hunger, and H. J. Osten, "Modular integration of high-performance SiGe:C HBTs in a deep submicron, epi-free CMOS process," IEEE IEDM Tech. Digest, 561–564, 1999.
230. H.. Baudry, B. Martinet, C. Fellous, O. Kermarrec, M. Laurens, M. Marty, J. Mourier, G. Troillard, A. Monroy, D. Dutartre, D. Bernshel, G. Vincent, and A. Chantre, "High performance 0.25 μm SiGe and SiGe:C HBTs using non selective epitaxy," IEEE BCTM Tech. Digest, 52–55, 2001.
231. T. Tominari, S. Wada, K. Tokunaga, K. Koyu, M. Kubo, T. Udo, M. Seto, K. Ohhata, H. Hosoe, Y. Kiyota, K. Washio, and T. Hashimoto, "Study on extremely thin base SiGe:C featuring sub 5-ps ECL gate delay," IEEE BCTM Tech. Digest, 107 110, 2003.

232. M. W. Xu, S. Decoutere, A. Sibaja-Hernandez, K. van Wichelen, L. Witters, R. Loo, E. Kunnen, C. Knorr, A. Sadovnikov, and C. Bulucea, "Ultra low power SiGe:C HBT for 0.18 mm RF-BiCMOS," IEDM Tech. Digest, 125–128, 2003.
233. L. S. Lai, C. S. Liang, P. S. Chen, Y. M. H. Y. H. Liu, Y. T. Tseng, S. C. Ly, M.-J. Tsai, C. W. Liu, C. Rosenblad, T. Buschbaum, M. Buschbeck, and J. Ram, "Optimal SiGe:C HBT module for BiCMOS applications," VLSI Tech. Digest, 113–116, 2003.
234. F. Ducroquet, T. Ernst, J.-M. Hartmann, O. Weber. F. Andrieu, P. Holliger, P. Laugier, P. Rivallin, G. Guégan, D. Lafond, V. Laviron, V.; Carron, L. Brévard, C. Tabone, D. Bouchu, A. Toffili, J. Cluzel, and S. Deleonibus, "Double SiGe:C diffusion barrier channel 40nm CMOS with improved short-channel performances," IEEE IEDM Tech. Digest, 437–440, 2004.
235. P. H. C. Magnée, A. L. A. M. Kemmeren, N. E. B. Cowern, J. W. Slotboom, R. J. Havens, and H. G. A. Huizing, "Ultra shallow boron base profile with carbon implantation," IEEE BCTM, 64–57, 2001.
236. C. I. Li, R. Liu, M. Chan, T. F. Hsiao, C. L. Yang, and S. F. Tzou, "Control of source and drain extension phosphorus profile by using carbon co-implant," 15th IEEE International Conference on Advanced Thermal Processing of Semiconductors, 127–130, 2007.
237. Y. Momiyama, K. Okabe, H. Nakao,M. Kojama, M. Kase, and T. Sugii,"Extension engineering using carbon co-implantation technology for low power CMOS design with phosphorus- and boron-extension," Ext. Abs. 7th International Workshop on Junction Technology, 63–64, 2007.
238. A. Mineji and S. Shishiguchi, "Ultra shallow junction and super steep halo formation using carbon co-implantation for 65nm high performance CMOS devices," IEEE International Workshop on Junction Technology, 84–87, 2006.
239. B. Colombeau, A. J. Smith, N. E. B. Cowern, W. Lerch, S.-Paul, B. J. Pawlak, F.Cristiano, X.Hebras, D.Bolze,C.Ortiz, and P. Pichler, "Electrical deactivation and diffusion of boron in preamorphized ultrashallow junctions: Interstitial transport and F co-implant control," IEEE IEDM Tech. Digest, 971–974, 2004.
240. W. Kang, J. Kim, K. Lee, Y. Shin, T. Kim, Y. Park, and J. Park, "The leakage current improvement in an ultra shallow junction NMOS with Co silicided source and drain," IEEE Trans. Electron Dev. Lett., 21 (1), 9–11, 2000.
241. J. B. Lasky, J. S. Nakos, O. J. Cain, and P. J. Geiss, "Comparison of transformation to low-resistivity phase and agglomeration of $TiSi_2$ and $CoSi_2$," IEEE Trans. Electron Dev., 38 (2), 262–269, 1991.
242. R. A. Roy, L. A. Clevenger, C. Cabral, Jr., K. L. Saenger, S. Brauer, J. Jordan-Sweet, J. Bucchignano, G. B. Stephenson, G. Morales, and K. F. Ludwig, Jr., "In situ x-ray diffraction analysis of the C49–C54 titanium silicide phase transformation in narrow lines," J. Appl. Phys., 66 (14), 1732–1734, 1995.
243. E. G. Colgan, J. P. Gambino, and Q. Z. Hong, "Formation and stability of silicides on polycrystalline silicon," Mater. Sci. Eng. Reps., R16, 43–96, 1996.
244. S. P. Murarka, D. B. Fraser, A. K. Sinha, H. J. Levinstein, E. J. Lloyd, R. Liu, D. S. Williams, and S. J. Hillenius, "Self-aligned cobalt disilicide for gate and interconnection and contacts to shallow junctions," J. Appl. Phys., 58 (2). 971–973, 1985.
245. T. Ymazaki, K. Goto, T. Fukano, Y. Nara, T. Sugii, and T. Ito, "21 psec switching 0.1mm-CMOS at room temperature using high performance Co silicide process," IEEE IEDM Tech. Digest, 906–908, 1993.
246. T. Morimoto, H. S. Momose, T. Inuma, I. Kunishima, K. Suguro, H. Okano, I. Katakabe, H. Nakajima, M. Ono, Y. Katsumata, and H. Iwai, "A NiSi silicide technology for advanced logic devices," IEEE IEDM Tech. Digest, 653–656, 1991.
247. J. P. Lu, D. Miles, J. Zhao, A. Gurba, Y. Xu, C. Lin, M. Hewson, J. Ruan, L. Tsung, R. Kuan, T. Grider, D. Mercer, and C. Montgomery, "A novel nickel salicide process technology for CMOS devices with sub-40nm physical gate length," IEEE IEDM Tech. Digest, 371–373, 2002.

248. R. Chau, J. Kavalieros, B. Roberds, R. Schenker, D. Lionberger, D. Barlage, B. Doyle, R. Arghavani, A. Murthy, and G. Dewey, "30nm physical gate length CMOS transistors with 1.0 ps n-MOS and 1.7 ps p-MOS gate delays," IEEE IEDM Tech. Digest, 45–48, 2000.
249. Q. Xiang, C. Woo, E. Paton, J. Foster, B. Yu, and M.-R. Lin, "Deep sub-100nm CMOS with ultra low gate sheet resistance by NiSi," Symp. VLSI Tech. Digest, 76–77, 2000.
250. B. Froment, M. Muller, H. Brut, R. Pantel, V. Carron, H. Achard, A. Halimaoui, F. Boeuf, F.Wacquant, C. Regnier, D. Ceccarelli, R. Palla, A. Beverina, V. DeJonghe, P. Spinelli, O. Leborgne, K. Bard, S. Lis, V. Tirard, P.Morin, F. Trentesaux, V. Gravey, T. Mandrekai, D. Rabilloud, S.Van, E. Olson, J. Diedrick, "Nickel vs. Cobalt silicide integration for sub-50nm CMOS," Europ. Solid-State Dev. Res. Conf., 215–218, 2003.
251. T.-H. Hou, T.-F Lei, and T.-S. Chao, "Improvement of junction leakage of nickel silicided junction by a Ti-capping layer," IEEE Electron Dev. Lett., 20 (11), 572–573, 1999.
252. D. Mangelinck, P. Gas, J. M. Gay, B. Pichaud, and O. Thomas, "Effect of Co, Pt, and Au additions on the stability and epitaxy of $NiSi_2$ films on (111) Si," J. Appl. Phys., 84 (5), 2583–2590, 1998.
253. P. S. Lee, K. L. Pey, D. Mangelinck, J. Ding, D. Z. Chi, and L. Chan, "New salicidation technology with Ni(Pt) alloy for MOSFETs," IEEE Electron Dev. Lett., 22 (12), 568–570, 2001.
254. T. Jarmar, J. Seger, F. Ericson, D. Mangelinck, U. Smith, S.-L. Zhang, "Morphological and phase stability of nickel–germanosilicide on $Si_{1-x}Ge_x$ under thermal stress," J. Appl. Phys., 92 (12), 7193–7199, 2002.
255. L. J. Jin, K. L. Peya, W. K. Choi, E. A. Fitzgerald, D. A. Antoniadis, A. J. Pitera, and M. L. Lee, D. Z. Chi, Md. A. Rahman, T. Osipowicz, and C. H. Tung, "Effect of Pt on agglomeration and Ge out diffusion in Ni(Pt) germanosilicide," J. Appl. Phys., 98 (033520–1–6), 2005.
256. K. Ohuchi, C. Lavoie, C. Murray, C. D'Emic, I. Lauer, J. O. Chu, B. Yang, P. Besser, L. Gignac, J. Bruley, G. U. Singcol, F. Pagettel, A. W. Topoll, M. J. Rooks, J. J. Bucchignano,V. Narayanan, M. Khare, M. Takayanagi, K. Ishimaru, D.-G. Park, G. Shahidi, and P. Solomon, "Extendibility of NiPt silicide contacts for CMOS technology demonstrated to the 22-nm node," IEEE IEDM Tech. Digest, 1029–1031, 2007.
257. R. T.-P. Lee, K.-M. Tan, A. E.-J. Lim, T.-Y. Liow, G. S. Samudra, D.-Z. Chi, and Y.-C. Yeo, "P-Channel tri-gate FinFETs featuring $Ni_{1-y}Pt_ySiGe$ source/drain contacts for enhanced drive current performance," IEEE Electron Dev. Lett., 29 (5), 438–441, 2008.
258. B. H. Lee, L. Kang, W.-J. Qi, R. Nieh, Y. Jeon, K. Onishi, and J. C. Lee, "Ultrathin hafnium oxide with low leakage and excellent reliability for alternative gate dielectric application," IEEE IEDM Tech. Digest, 133–136, 1999.
259. Y.-Y. Chen, W.-Y. Fu, and C.-F. Yeh, "Electrical characteristics of the HfAlON gate dielectric with interfacial UV-ozone oxide," IEEE Electron Dev. Lett., 29 (1), 96–98, 2008.
260. G. D. Wilk, R. W. Wallace, and J. M. Anthony, "Hafnium and zirconium silicates for advanced gate dielectrics," J. Appl. Phys., 87 (1), 484–492, 2000.
261. S. J. Lee, C. H. Lee, Y. H. Kim, H. F. Luan, W. P. Bai, T. S. Jeon, and D. L. Kwong, "High-K gate dielectrics for sub-100 nm CMOS technology," Int. Conf. Solid-State and Integrated Tech., 1, 303–308, 2001.
262. C. Hobbs, H. Tseng, K. Reid, B. Taylor, L. Dip, L. Hebert, R. Garcia, R. Hedge, J. Grant, D. Gilmer, A. Franke, V. Dhandapani, M. Azrak, L. Prabhu, R. Rai, S. Bagchi, J. Conner, S. Backer, F. Dumbuya, B. Nguyen, and P. Tobin, "80 nm poly-Si gate CMOS with HfO_2 dielectric," IEEE IEDM Tech. Digest, 651–653, 2001.
263. Q. Lu, R. Lin, H. Takeuchi, T.-J. King, C. Hu, K. Onishi, R. Choi, C.-S. Kang, and J. C. Lee, "Deep-submicron CMOS process integration of HfO_2 gate dielectric with poly-Si gate," Semicon. Dev. Res. Symp. Digest., 377–380, 2001.
264. M. L. Green, M.-Y. Ho, B. W. Busch, G. D. Wilk, T. Sorsch, T. Conrad, B. Brijs, W. Vandervorst, P. I. Räisänen, D. Muller, M. Bude, and J. Grazul, "Nucleation and growth of atomic layer deposited HfO_2 gate dielectric layers on chemical oxide (Si-O-H) and thermal oxide (SiO_2 or Si-O-N) underlayers," J. Appl. Phys., 7168–7174, 2002.
265. M.-Y. Ho, H. Gong, G. D. Wilk, B. W. Busch, M. L. Green, P. M. Voyles, D. A. Muller, M. Bude, W. H. Lin, A. See, M. E. Loomans, S. K. Lahiri, and P. I. Räisänen, "Morphology

and crystallization kinetics in HfO_2 thin films grown by atomic layer deposition," J. Appl. Phys., 93 (3), 1477–1481, 2003.

266. M. Koike, T. Ino, Y. Kamimuta, M. Koyama, Y. Kamata, N. Susuki, Y. Mitani, A. Nishiyama, and Y. Tsunashima, "Effect of Hf-N bond properties of thermally stable amorphous HfSiON and applicability of this material to sub-50 nm technology node LSIs," IEEE IEDM Tech. Digest, 107–110, 2003.
267. T. Y. Luo, M. Laughery, G. A. Brown, H. N. Al-Shareef, V. H. C. Watt, A. Karamcheti, M. D. Jackson, and H. R. Huff, "Effect of H_2 content on reliability of ultrathin In-Situ Steam Generated (ISSG) SiO_2," IEEE Electron Dev. Lett., 21 (9), 430–432, 2000.
268. Y. Ma, Y. Ono, L. Stecker, D. R. Evans, and S. T. Hsu, "Zirconium oxide based gate dielectrics with equivalent oxide thickness of less than 1.0 nm and performance of sub-micron MOSFET using a nitride gate replacement process," IEEE IEDM Tech. Digest, 149–152, 1999.
269. A. Chin, C. C. Liao, C. H. Lu, W. J. Chen, and C. Tsai, "Device and reliability of high-K Al_2O_3 gate dielectric with good mobility and low Dit," Symp. on VLSI Tech., 135–136, 1999.
270. D. A. Buchanan, E. P. Gusev, E. Carter, H. Okorn-Schmidt, K. Rim, M. A. Gribelyuk, A. Mocuta, A. Ajmera, M. Copel, S. Guha, N. Bojarczuk, A. Callegari, C. D'Emic, P. Kozlowski, K. Chan, R. J. Fleming, P. C. Jamison, J. Brown, and R. Arndt, "80 nm poly-silicon gated n-FET with ultra-thin Al_2O_3 gate dielectric for ULSI applications," IEEE IEDM Tech. Digest, 223–226, 2000.
271. R. A. B. Devine, L. Vallier, J. L. Autran, P. Paillet, and J. L. Leray, "Electrical properties of Ta_2O_5 films obtained by plasma enhanced chemical vapor deposition using a TaF_5 source," Appl. Phys. Lett., 68 (13), 1775–1777, 1996.
272. S. Iwata, N. Yamamoto, N. Kobayashi, T. Terada, and T. Mizutani, "A new tungsten gate process for VLSI applications," ED-31(9), 1174–1179, 1984.
273. B. Doris, M. Ieong, H. Zhu, Y. Zhang, M. Steen, W. Natzle, S. Callegari, V. Narayanan, J. Cai, S. H. Ku, P. Jamison, Y. Li, Z. Ren, V. Ku, D. Boyd, T. Kanarski, C. D'Emic, M. Newport, D. Dobuzinsky, S. Seshpante, J. Petrus, R. Jammy, and W. Haensch, "Device design considerations for ultra-thin SOI MOSFETs," IEEE IEDM Tech. Digest, 631–634, 2003.
274. J. Widiez, M. Vinet, B. Guillaumot, T. Poiroux, D. Lafond, P. Holliger, O. Weber, V. Barral, B. Previtali, F. Martin, M. Mouis, and S. Deleonibus, "Fully depleted SOI MOSFETs with WSi_x, metal gate on HfO_2 gate dielectric, "IEEE Int. SOI Conf. Proc., 161–162, 2006.
275. D. Aimé, C. Fenouillet-Beranger, P. Perreau, S. Denorme, J. Coignus, A. Cros, D. Fleury, O. Faynot, A. Vandooren, R. Gassilloud, F. Martin, S. Barnola, T. Salvetat, G. Chabanne, L. Brevard, M. Aminpur, F. Leverd, R. Gwoziecki, F. Boeuf, C. Hobbs, A. Zauner, M. Müller, V. Cosnier, S. Minoret, D. Bensahel, M. Orlowski, H. Mingam, A. Wild, S. Deleonibus, and T. Skotnicki, "Fully-Depleted SOI CMOS technology using W_xN metal gate and $HfSi_xO_yN_z$ high-k dielectric," Eur. Solid-State Dev. Res. Conf. (ESSDERC), 255–258, 2007.
276. A. Yagishita, T. Saito, K. Nakajima, S. Inumiya, K. Matsuo, T. Shibata, Y. Tsunashima, K. Suguro, and T. Arikado, "Improvement of threshold voltage deviation in damascene metal gate transistors," IEEE Trans. Electron Dev., 48 (8), 1604–1611, 2001.
277. S. S. Suryagandh, M. Garg, and J. C. S. Woo, "A device design methodology for sub-100-nm SOC applications using bulk and SOI MOSFETs," IEEE Trans. Electron Dev., 51 (7), 1122–1128, 2004.
278. A. Vandooren, C. Hobbs, O. Faynot, P. Perreau, S. Denorme, C. Fenouillet-Beranger, C. Gallon, C. Morin, A. Zauner, G. Imbert, H. Bernard, P. Gamier, L. Gabette, M. Broekaart, M. Aminpur, S. Barnola, N. Loubet, D. Dutartre, T. Korman, G. Chabanne, F. Martin, Y. Le Tiec, N. Gierczynski, S. Smith, C. Laviron, M. Bidaud, I. Pouilloux, D. Bensahel, T. Skotnicki, H. Mingam, and A. Wild, "0.525 μm^2 6T-SRAM bit cell using 45nm fully-depleted SOI CMOS technology with metal gate, high K dielectric and elevated source/drain on 300mm wafers," IEEE Int. SOI Conf., 221–222, 2005.
279. A. Vandooren, A. Barr, L. Mathew, T. R. White, S. Egley, D. Pham, M. Zavala, S. Samavedam, J. Schaeffer, J. Conner, B.-Y. Nguyen, Bruce E. White, Jr., M. K. Orlowski,

and J. Mogab, "Fully-depleted SOI devices with TaSiN gate, HfO_2 gate dielectric, and elevated source/drain extensions," IEEE Electron. Dev. Lett., 24 (5), 342–344, 2003.
280. A. Chatterjee, R. A. Chapman, K. Joyner, M. Otobe, S. Hattangady, M. Bevan, G. A. Brown, H. Yang, Q. He, D. Rogers, S. J. Fang, R. Kraft, A. L. P. Rotondaro, M. Terry, K. Brennan, S.-W. Aur, J. C. Hu, H.-L. Tsai, P. Jones, G. Wilk, M. Aoki, M. Rodder, and I.-C. Chen, "CMOS metal replacement gate transistors using tantalum pentoxide gate insulator," IEEE IEDM Tech. Digest, 777–780, 1998.
281. Y. Abe, T. Oishi, K. Shiozawa, Y. Tokuda, and S. Satoh, "Simulation study on comparison between metal gate and polysilicon gate for sub-quarter-micron MOSFETs," IEEE Electron Dev. Lett., 20 (12), 632–634, 1999.
282. K. Maitra and V. Misra, "A simulation study to evaluate the feasibility of midgap workfunction metal gates in 25 nm bulk CMOS," IEEE Electron Dev. Lett., 24 (11), 707–709, 2003.
283. I. De, D. Johri, A. Srivastava, and C. M. Osburn, "Impact of gate workfunction on device performance at the 50nm technology node," Solid-State Electron., 44, 1077–1080, 2000.
284. E. Jossel and T. Skotnicki, "Polysilicon gate with depletion – or – metallic gate with buried channel: What evil worse?" IEEE IEDM Tech. Digest, 661–664, 1999.
285. M. Masahara, S.-i. O'uchi, Y. Liu, K. Sakamoto, K. Endo, T. Matsukawa, T. Sekigawa, H. Koike, and E. Suzuki, "Optimum gate workfunction for V_{th}-controllable four-terminal-driven double-gate MOSFETs (4T-XMOSFETs) - Band-edge workfunction versus midgap workfunction," IEEE Trans. Nanotech., 5 (6), 716–722, 2006.
286. Y.-C. Yeo, Q. Lu, P. Ranade, H. Takeuchi, K. J. Yang, I. Polishchuk, T.-J. King, C. Hu, S. C. Song, H. F. Luan, and D.-L. Kwong, "Dual-metal gate CMOS technology with ultrathin silicon nitride gate dielectric," IEEE Electron Dev. Lett., 22 (5), 227–229, 2001.
287. S. B. Samavedam, L. B. La, J. Smith, S. Dakshina-Murthy, E. Luckowski, J. Schaefer, M. Zavala, R. Martin, V. Dhanapani, D. Triyoso, H. H. Tseng, P. J. Tobin, D. C. Gilmer, C. Hobbs, W. J. Taylor, J. M. Grant, R. I. Hedge, J. Mogab, C. Thomas, P. Abramowitz, M. Moosa, J. Conner, J. Jiang, V. Arunachalam, M. Saad, B.-Y. Nguyen, and B., White, "Dual-metal gate CMOS with HfO_2 gate dielectric," IEEE IEDM Tech. Digest, 433–436, 2002.
288. Z. B. Zhang, S. C. Song, C. Huffman, J. Barnett, N. Moumen, H. Alshareef, P. Majhi, M. Hussain, M. S. Akbar, J. H. Sim, S. H. Bae, B. Sassman, and B. H. Lee, "Integration of dual metal gate CMOS with TaSiN (NMOS) and Ru (PMOS) gate electrodes on HfO_2 gate dielectric," VLSI Symp. Tech. Dig., 50–51, 2005.
289. B. Tavel, T. Skotnicki, G. Pares, N. Carrière, M. Rivoire, F. Leverd, C. Julien, J. Torres, and R. Pantel, "Totally silicided ($CoSi_2$) polysilicon: a novel approach to very low-resistive gate ($\sim 2\Omega/\square$) without metal CMP nor etching," IEEE IEDM Tech. Digest, 825–828, 2001.
290. T. Hoffmann, A. Veloso, A. Lauwers, H. Yu, M. Van Dal, H. Tigelaar, T. Chiarella, C. Kerner, R. Mitsuhashi, I. Satoru, M. Niwa, A. Rothschild, B. Froment, J. Ramos, A. Nackaerts, S. Brus, C. Vrancken, P. P. Absil, M. Jurczak, J. A. Kittl, and S. Biesemans, "Low power CMOS featuring dual work function FUSI on HfSiON and 17ps inverter delay," VLSI Symp. Tech. Dig., 154–155, 2006.
291. A. Lauwers, A. Veloso, T. Hoffmann, M. J. H. van Dal, C. Vrancken, S. Brus, S. Locorotondo, J.-F. de Marneffe, B. Sijmus, S. Kubicek, T. Chiarella, M. A. Pawlak, K. Opsomer, M. Niwa, R. Mitsuhashi, K. G. Anil, H. Y. Yu, C. Demeurisse, R. Verbeeck, M. de Potter, P. Absil, K. Maex, M. Jurczak, S. Biesemans, and J. A. Kittl, "CMOS integration of dual work function phase controlled Ni FUSI with simultaneous silicidation of NMOS (NiSi) and PMOS (Ni-rich silicide) gates on HfSiON,"IEEE IEDM Tech. Digest, 661–664, 2005.
292. A. Lauwers, A. Veloso, S.-Z. Chang, H. Y. Yu, T. Hoffmann, C. Kerner, M. Demand, A. Rothschild, M. Niwa, I. Satoru, R. Mitsuhashi, M. Ameen, G. Whittemore, M. A. Pawlak, C. Vrancken, C. Demeurisse, S. Mertens, W. Vandervorst, P. Absil, S. Biesemans, and J. A. Kittl, "Cost-effective low Vt Ni-FUSI CMOS on SiON by means of Al implant (pMOS) and Yb + P coimplant (nMOS)," IEEE Electron Dev. Lett., 29 (1), 34–37, 2008.
293. J. Yuan and J. C. S. Woo, "Tunable work function in fully nickel-silicided polysilicon gates for metal gate MOSFET applications," IEEE Electron Dev. Lett., 26 (2), 87–89, 2005.

294. N. Kumar, P. Pourrezaei, M. Fissel, T. Begley, B. Lee, and E. C. Douglas, "Growth and properties of radio frequency reactively sputtered titanium nitride thin films", J. Vac. Sci. Technol. A, 5 (4), 1778–1782, 1987.
295. J. F. Creemer, W. van der Vlist, C. R. de Boer, H. W. Zandbergen, P. M. Sarro, D. Briand, and N. F. de Rooij, "MEMS hotplates with TiN as a heater material," IEEE Sensors, 330–333, 2005.
296. C. W. Kaanta, S. G. Bombardier, W. J. Cote, W. R. Hill, G. Kerszykowski, H. S. Landis, D. J. Poindexter, C. W. Pollard, G. H. Ross, J. G. Ryan, S. Wolff, and J. E. Cronin, "Dual Damascene: A ULSI wiring technology," Proc. VLSI Multilevel Interconnect Conference (VMIC), 144–152, 1991.
297. J. G. Ryan, R. M. Geffken, N. R. Poulin, and J. R. Paraszczak, "The evolution of interconnection technology at IBM," IBM J. Res. Dev., 39 (4), 371–381, 1995.
298. D. Edelstein, J. Heidemreich, R. Goldblatt, W. Cote, C. Uzioh, N. Lustig, P. Roper, T. McDevitt, W. Motsiff, A. Simon, J. Dubovic, R. Wachnik, H. Rathore, R. Schulz, L. Su, S. Luce, and J. Slattery, "Full copper wiring in sub-0.25 mm CMOS ULSI technology," IEEE IEDM Tech. Digest, 273–276, 1997.
299. K. Mosig, H. Cox, E. Klawuhn, T. S. de Filipe, and A. Shiota, "Integration of porous low-k dielectric with CVD barriers," IEEE IEDM Tech. Digest, 88–91, 2001.
300. D. Edelstein, C. Uzoh, C. Cabral, Jr., P. DeHaven, P. Buchwalter, A. Simon, E. Cooney, S. Malhotra, D. Klaus, H. Rathore, B. Agarwala, and D. Nguyen, "A high performance liner for copper damascene interconnects," IEEE IITC Tech. Digest, 9–11, 2001.
301. J. C. Lin, R. Augur, S. L. Shue, C. H. Yu, M. S. Liang, A. Vijayendran, T. S. d Filipe, and M. Danek, "CVD barriers for Cu with nanoporous ultra-low-k: Integration and reliability," IEEE IITC Tech. Digest, 21–23, 2002.
302. P. Moon, V. Dubin, S. Johnston, J. Leu, K. Raol, and C. Wu, "Process roadmap and challenges for metal barriers," IEEE IEDM Tech. Digest, 841–844, 2003.
303. H. Chung, M. Chang, S. Chu, N. Kumar, K. Goto, N. Maity, S. Sankaranarayanan, H. Okamura, N. Ohtsuka, and S. Ogawa, "An ultra-thin ALD TaN barrier for high-performance Cu interconnects," IEEE Int. Symp. Semicond. Manuf., 454–456, 2003.
304. J. W. Hong, K. I. Choi, Y. K. Lee, S. G. Park, S. W. Lee, J. M. Lee, S. B. Kang, G. H. Choi, S. T. Kim, U.-I. Chung, and J. Moon, "Characteristics of PAALD-TaN thin fims derived from TAIMATA precursor for copper metallization," IEEE IITC Tech. Digest, 9–11, 2004.
305. C.-C. Yang, D. Edelstein. L. Clevenger, A. Cowley, J. Gill, K. Chanda, A. Simon, T. Dalton, B. Agarwala, E. Cooney III, D. Nguyen, T. Spooner, and A. Stamper, "Extendibility of PVD barrier/seed for BEOL Cu metallization," IEEE IITC Tech. Digest, 135–137, 2005.
306. T. Usui, H. Nasu, S. Takahashi, N. Shimizu, T. Nishikawa, M. Yoshimaru, H. Shibata, M. Wada, and J. Koike "Highly reliable copper dual-damascene interconnects with self-formed $MnSi_xO_y$ barrier layer," IEEE Trans. Electron Dev., 53 (10), 2492–2498, 2006.
307. Y. Ohoka, Y. Ohba, A. Isobayashi, T. Hayashi, N. Komai, S. Arakawa, R. Kanamura, and S. Kadomura, "Integration of High Performance and Low Cost Cu/Ultra Low-k SiOC($k = 2.0$) "Interconnects with self-formed barrier technology for 32nm-node and beyond," IEEE IITC Tech. Digest, 67–69, 2007.
308. H. Kudo, M. Haneda, H. Ochimizu, A. Tsukune, S. Okano, N. Ohtsuka, M. Sunayama, H. Sakai, T. Suzuki, H. Kitada, S. Amari, T. Tabira, H. Matsuyama, N. Shimizu, T. Futatsugi, and T. Sugii, "Copper wiring encapsulation with ultra-thin barriers to enhance wiring and dielectric reliabilities for 32-nm nodes and beyond," IEEE IEDM Tech. Digest, 513–516, 2007.
309. J.-J. Tan, Q. Xie, M. Zhou, T. Chen, Y.-L. Jiang, and X.-P. Qu, "Investigation of Ru/TaN on low dielectric constant material with $k = 2.7$," Int. Conf. Solid-State and Integrated Circuit tech., 339–341, 2006.
310. S. Smith, G. Book, W. M. Li, Y. M. Sun, P. Gillespie, M. Tuominen, and K. Pfeifer, "The application of ALD WN_xC_y as a copper diffusion barrier," IEEE IITC Tech. Digest, 135–137, 2003.
311. S.-M. Choi, K.-C. Park, B.-S. Suh, I.-R. Kim, K.-K. Kang, K.-P. Suh, H.-S. Park, H.-S. Ha, and D.-K. Joo, "Process integration of CVD Cu seed using ALD Ru glue layer for sub-65nm Cu interconnect," IEEE Symp. VLSI Tech., 64–65, 2004.

312. S. Lin, C. Jin, L. Lui, M. Tsai, M. Daniels, A. Gonzalez, J. T. Wetzel, K. A. Monnig, P. A. Winebarger, S. Jang, D. Yu, and M. S. Liang, "Low-k dielectrics characterization for damascene integration," IEEE IITC Tech. Digest, 146–148, 2001.
313. S. Kondo, B. U. Moon, S. Tokitoh, K. Misawa, S. Sone, H. J. Shin, N. Ohashi, and N. Kobayashi, "Low-pressure CMP for 300-mm ultra low-k (k = 1.6–1.8)/Cu integration," IEEE IEDM Tech Digest, 151–154, 2003.
314. N. Chandrasekaran, S. Ramarajan, W. Lee, G. M. Sabde, and S. Meikle, "Effects of CMP process conditions on defect generation in low-k materials. An atomic force microscopy study," J. Electrochem. Soc., 151 (12), G882–G889, 2004.
315. R. Chang and C. J. Spanos, "Dishing-radius model of copper CMP dishing effects," IEEE Trans. Semicond. Manuf., 18 (2), 297–303, 2005.
316. L. Economikos, X. Wang, A. Sakamoto, P. Ong, M. Naujok, R. Knarr, L. Chen, Y. Moon, S. Neo, J. Salfelder, A. Duboust, A. Manems, W. Lu, S. Shrauti, F. Liu, S. Tsai, and W. Swaert, "Integrated electro-chemical mechanical planarization (Ecmp) for future generation device technology," IEEE IITC Tech. Digest, 233–235, 2004.
317. F. Q. Liu, T. Du, A. Duboust, S. Tsai, and W.-Y. Hsu, "Cu planarization in electrochemical mechanical planarization," J. Electrochem. Soc., 153 (6), C377-C381, 2006.
318. M. Mellier, T. Berger, R. Duru, M. Zaleski, M. C. Luche, M. Rivoire, C. Goldberg, G. Wyborn, K.-L. Chang, Y. Wang, V. Ripoche, S. Tsai, M. Thothadri, W.-Y. Hsu, L. Chen, "Full copper electrochemical mechanical planarization (Ecmp) as a technology enabler for the 45 and 32nm nodes," IEEE IITC Tech. Digest, 70–72, 2007.
319. P. Besser, A. Marathe, L. Zhao, M. Herrick, C. Caspasso, and H. Kawasaki, "Optimizing the electromigration performance if copper interconnects," IEEE IEDM Tech. Digest, 119–122, 2000.
320. C.-K. Hu, L. Gignac, R. Rosenberg, E. Liniger, J. Rubino, C. Sambucetti, A. Domenicucci, X. Chen, and A. K. Stamper, "Reduced electromigration of Cu wires by surface coating," Appl. Phys. Lett., 81 (10), 1782–1784, 2000.
321. T. Ko, C. L. Chang, S. W. Chou, M. W. Lin, C. I. Lin, C. H. Shih, H. W. Su, M. H. Tsai, W. S. Shue, and M. S. Lian, "High performance/reliability Cu interconnect with selective CoWP cap," IEEE VLSI Tech. Digest, 109–110, 2003.
322. T. Ishagami, T. Kurokawa, Y. Kakuhara, B. Withers, J. Jacobs, A. Kolics, I. Ivanov, M. Sekine, and K. Ueno, "High reliability Cu interconnection utilizing a low contamination CoWP capping layer," IEEE IITC Tech. Digest, 75–77, 2004.
323. J. Gambino, J. Wynne, S. Smith, Y. Kakuhara, B. Withers, J. Jacobs, A. Kolics, I. Ivanonv, M. Sekine, and K. Ueno, "Effect of CoWP cap thickness on via yield and reliability for Cu interconnects with CoWP-only cap process," IEEE IITC Tech. Digest, 111–113, 2005.
324. T. Itabashi, H. Nakano, and H. Akahoshi, "Electroless deposited CoWB for copper diffusion barrier metal," IEEE IITC Tech. Digest, 285–287, 2002.
325. O. Hinsinger, R. Fox, E. Sabouret, C. Goldberg, C. Verovel, W. Besling, P. Brun, E. Jossel, C. Monget, O. Belmont, J. Van Hasse, B. G. Sharma, J. P. Jacquemin, P. Vannier, A. Humbert, D. Bune, R. Gonella, E. Mastromatteo, D. Reber, A. Farcy, J. Mueller, P. Christie, V. H. Nguyen, C. Cregut, and T. Berger, "Demonstration of an extendable and industrial 300" BEOL integration for the 65-nm technology node," IEEE IEDM Tech. Digest, 317–320, 2004.
326. S. M. Rossnagel, R. Wisnieff, D. Edelstein, and T. S. Kuan, "Interconnect issues post 45nm," IEEE IEDM Tech. Digest, 95–98, 2005.
327. I. E. H. Sondheimer, "The mean free path of electrons in metals," Adv. Phys., 1, 1–42, 1952.
328. A. F. Mayadas and M. Shatzkes, "Electrical-resistivity model for polycrystalline films: the case of arbitrary reflection at external surfaces," Phys. Rev. B 1, 1382–1389, 1970.
329. C. Reale, "Thickness dependence of the electrical conductivity in vacuum deposited copper films," Proceedings of the IEEE, 57 (11), 2073–2075, 1969.
330. S. M. Rossnagel and T. S. Kuan, "Alteration of Cu conductivity in the size effect regime," J. Vac. Sci. Tech. B 22 (1), 240–247, 2004.

331. W. Steinhögl, G. Schindler, G. Steinlesberger, M. Traving, and M. Engelhardt, "Comprehensive study of the resistivity of copper wires with lateral dimensions of 100 nm and smaller," J. Appl. Phys., 97, 023706 1–3, 2005.
332. W. Steinhoegl, G. Schindler, G. Steinlesberger, M. Traving, and M. Engelhardt, "Scaling laws for the resistivity increase of sub-100 nm interconnects," International Conference on Simulation of Semiconductor Processes and Devices (SISPAD), 27–30, 2003.
333. M. T. Bohr, "Interconnect scaling – the real limiter to high performance ULSI," IEEE IEDM Tech. Digest, 241–244, 1995.
334. S. C. Sun, "Process technology for advanced metallization and interconnect systems," IEEE IEDM Tech. Digest, 765–768, 1997.
335. M. Igarashi, A. Harada, H. Amishiro, H. Kawashima, N. Morimoto, Y. Kusumi, T. Saito, A. Ohsaki, T. Mori, T. Fukuda, Y. Toyoda, K. Higashitani, and H. Arima, "The best combination of aluminum and copper interconnects for high performance 0.18mm CMOS logic device," IEEE IEDM Tech. Digest, 829–892, 1998.
336. K. Tokashiki, T. Maruyama, and A. Nishizawa, "Influence of process chamber ambient on SiOC (k = 2:9) ILD Cu damascene ashing," IEEE Trans. Semicond. Manuf., 17 (3), 305–310, 2004.
337. A. Grill, V. Patel, K. P. Rodbell, E. Huang, S. Christiansen, M. R. Baklanov, J. L. Veteran, D. L. O'Meara, V. Misra, and P. S. Ho, "Characteristics of low-k and ultralow-k PECVD deposited SiCOH films," Mater. Res. Soc. Proc., 716, 569–574, 2002.
338. T. Nakamura and A. Nakashima, "Robust multilevel interconnects with a nano-clustering porous low-k (k-33)," IEEE IITC Tech. Digest, 175–177, 2004.
339. I. Sugiura, Y. Nakata, N. Misawa, S. Otsuka, N. Nishikawa, Y. Iba, E. Sugirnoto, Y. Setta, H. Sakai, Y. Mizushima, Y. Kotaka, C. Uchibori, T. Suzuki, H. Kitada, Y. Koura, K. Nakano, T. Karasawa, Y. Ohkura, H. Watatani, M. Sato, S. Nakai, M. Nakaishi, N. Shimizu, S. Fukuyama, M. Miyajima, T. Nakamura, E. Yano, and K. Watanabe, "45nm-Node BEOL integration featuring porous-ultra-low-K/Cu multilevel interconnects," IEEE IITC Tech. Digest, 15–17, 2005.
340. M. Aimadeddine, V. Jousseaume, V. Arnal, L. Favennec, A. Farcy, A. Zenasni, M. Assous, M. Vilmay, S. Jullian, P. Maury, V. Delaye, N. Jourdan, T. Vanypre, P. Brun, G. Imbert, Y. LeFriec, M. Mellier, H. Chaabouni, L. L. Chapelon, K. Hamioud, F. Volpi, D. Louis, G. Passemard, and J. Torres, "Robust integration of an ULK SiOCH dielectric (k = 2.3) for high performance 32 nm node BEOL," IEEE IITC Tech. Digest, 175–177, 2007.
341. N. Inoue, M. Tagami, F. Itoh, H. Yamamoto, T. Takeuchi, S. Saito, N. Furutake, M. Ueki, M. Tada, T. Suzuki, and Y. Hayashi, "45nm-node interconnects with porous SiOCH-stacks, tolerant of low-cost packaging applications," IEEE IITC Tech. Digest, 181–183, 2007.
342. D. Ryuzaki, H. Sakurai,' K. Abe, K. Takeda, and H. Fuhda, "Enhanced dielectric-constant reliability of low-k porous organosilicate glass (k = 2.3) for 45-nm-generation Cu interconnects," IEEE IEDM Tech. Digest, 949–952, 2004.
343. Y. Hayashi, F. Itoh, Y. Harada, T. Takeuchi, M. Tada, M. Tagami, H. Ohtake, K. Hijioka, S. Saito, T. Onodera, D. Hara, and K. Tokudome, "Novel molecular-structure design for PECVD porous SiOCH films toward 45nm-node, ASICs with k = 2.3," IEEE IITC Tech. Digest, 225–227, 2004.
344. S. Arakawa, I. Mizuno, Y. Ohoka, K Nagahata, K. Tabuchi, R. Kanamura and S. Kadomura, "Breakthrough integration of 32 nm-node Cu/ultra-low-k SiOC (k = 2.0) interconnects by using advanced pore-sealing and low-k hard mask technologies," IEEE IITC Tech. Digest, 21–212, 2006.
345. T. Ueda, E. Tamaoka, K. Yamashita, N. Aoi, and S. Mayumi, "A novel air gap integration scheme for multi-level interconnects using self-aligned via plugs," IEEE VLSI Tech. Digest, 46–47, 1998.
346. V. Amal, J. Torres, P. Gayet, R. Gonella, P. Spinelli, M. Guillermet, J.-P. Reynard, C. Vérove, "Integration of a 3 Level Cu - SiO_2 Air Gap Interconnect for Sub 0.1 micron CMOS Technologies," IEEE IITC Tech. Digest, 298–300, 2001.
347. S. E. Schulz and K. Schulze, "Achieving ultra low k dielectric constant for nanoelectronics interconnect systems," Solid-State Integr. Circuit Tech., 298–301, 2006.

348. L. G. Gosset, F. Gaillard, D. Bouchu, R. Gras, J. de Pontcharra, S. Orain1, O. Cueto, Ph. Lyan, O. Louveau, G. Passemard, and J. Torres, "Multi-level Cu interconnects integration and characterization with air gap as ultra-low K material formed using a hybrid sacrificial oxide/polymer stack," IEEE IITC Tech. Digest, 58–60, 2997.
349. T. Harada, A. Ueki, K. Tomita, K. Hashimoto, J. Shibata, H. Okamura, K. Yoshikawa, T. Iseki, M. Higashi, S. Maejima, K. Nomura, K. Goto, T. Shono, S. Muranaka, N. Torazawa, S. Hirao, M. Matsumoto, T. Sasaki, S. Matsumoto, S. Ogawa, M. Fujisawa, A. Ishii, M. Matsuura, T. Ueda, "Extremely low K_{eff} ($\approx$1.9) Cu interconnects with air gap formed using SiOC," IEEE IITC Tech. Digest, 141–143, 2007.

Chapter 8
Applications

8.1 Introduction

The first part of this chapter describes examples of *NMOS, CMOS* and *BiCMOS* logic units. The presence of both *NMOS* and *PMOS* in an integrated bulk[1] *CMOS* process makes the circuit susceptible to a parasitic effect known as latch-up. A *CMOS* inverter will be used to describe the latch-up mechanism and methods to prevent it. The second part of the chapter covers different types of memory cells, including dynamic random-access memory, *DRAM*; static random-access memory, *SRAM*; and nonvolatile memory, *NVM*. The chapter concludes with a summary of BiCMOS features that are important for analog/RF applications.

8.2 Logic Units

Four basic logic units are discussed in this section: the inverter, the NAND gate, the NOR gate, and the transmission gate. A more detailed discussion of logic designs can be found in Ref. [1].

8.2.1 The Inverter

This basic building block of digital elements is the inverter. It consists of a driver transistor and a "load" as shown in Fig. 8.1.

The driver is always an enhancement mode device because this type of *MOSFET* requires that the drain and gate voltages be of the same polarity which allows direct coupling between stages. For this discussion, it is assumed that the driver is

[1] The term "bulk" is used to distinguish the wafer on which CMOS is constructed from silicon on insulator, SOI.

B. El-Kareh, *Silicon Devices and Process Integration: Deep Submicron and Nano-Scale Technologies,*

© Springer Science+Business Media, LLC 2009

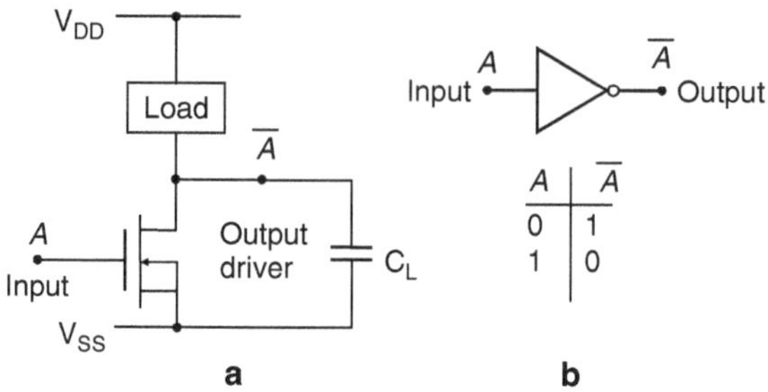

Fig. 8.1 **a** Inverter circuit diagram. **b** Symbol and truth table. A "1" corresponds to V_{DD} and a "0" to V_{SS}

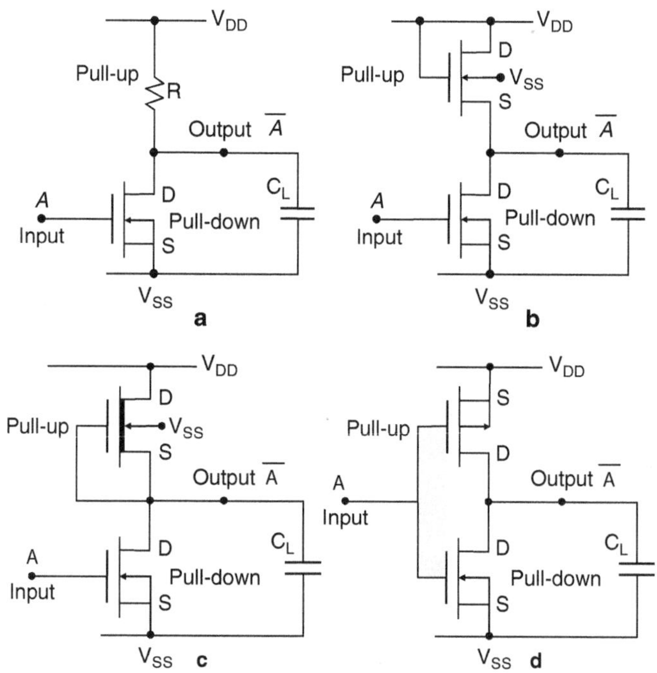

Fig. 8.2 Inverters with different types of pull-up devices. **a** Linear resistance. **b** Enhancement mode *NMOS* in saturation. **c** Depletion-mode NMOS. **d** Enhancement mode *PMOS (CMOS)*

an enhancement mode NMOS of threshold voltage V_{Tn}. C_L is the load capacitance that is the equivalent to the sum of all capacitors seen by the output. V_{DD} is the power-supply voltage and V_{SS} is typically at ground. Four types of load devices are shown in Fig. 8.2. The load can be a resistor (Fig. 8.2a), an enhancement-mode *NMOS* connected in saturation with the gate tied to the drain (Fig. 8.2b), a depletion-mode *NMOS* (Fig. 8.2c), or an enhancement mode *PMOS* in a *CMOS* technology (Fig. 8.2d).

Assume initially that the pull-up device is a simple resistor of resistance R. If the input voltage is "Low," at or near $V_{SS} < V_{Tn}$ ("0" in the truth table), the *NMOS* driver is off and the current through the resistor charges C_L to "*high*," at or near V_{DD}, with a time constant RC_L. The load is hence called a "*pull-up*" device since the current through the load brings the output to "*high*," that is, V_{DD}. If the input is rapidly switched to "*high*," the *NMOS* conducts and the capacitor discharges through the driver. The output drops to a low potential, at or near V_{SS}. The driver is hence called the "*pull-down*" device. Thus, the output voltage signal is the "inverse" of the input signal.

A plot of V_{out} versus V_{in} transfer characteristic of an inverter with a passive pull-up resistor is shown in Fig. 8.3a [1].

Ideally, the output voltage V_{OH} should be at V_{DD} when the input voltage is at V_{SS} (0 V), and V_{OL} should be at V_{SS} (0 V) when the input voltage is at V_{DD}. This

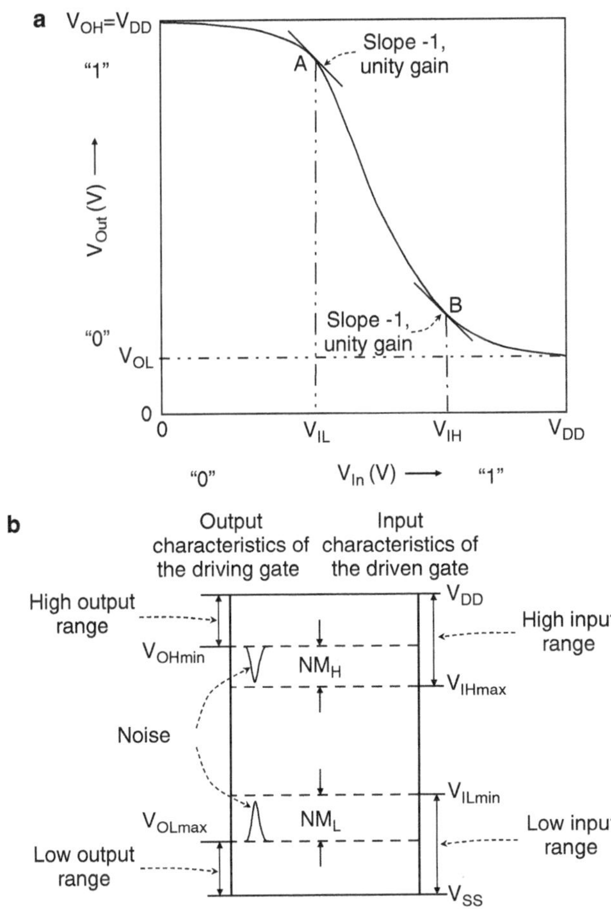

Fig. 8.3 **a** Inverter transfer characteristic and unity-gain points (Adapted from [1]). **b** Definition of "low" and "high" noise margins, NM_L and NM_H (Adapted from [1])

condition is approached in a CMOS inverter, as will be shown in the next section. It can be seen in Fig. 8.3a that for an inverter with pull-up resistor, the output does not drop completely to ground (0) but to a value $V_{Out\text{-}Low}$ (V_{OL}) that depends on R and the NMOS on-resistance R_{on} which includes the *NMOS* channel and extrinsic resistances

$$V_{Out-Low} = \frac{V_{DD}R_{On}}{R+R_{On}}. \tag{8.1}$$

With the pull-down transistor in the conducting state, current passes through both the resistor and the *NMOS*, dissipating power. To reduce power dissipation and decrease V_{OL}, R must be large. But as R increases, the RC_L delay increases. Thus, the value of R is a trade-off between speed, power, and the maximum V_{OL} that can be allowed.

In the transfer characteristic of Fig. 8.3a, there are two points A and B, called unity-gain points, where $dV_{Out}/dV_{In} = -1$. The input voltages at these points are denoted by V_{IL} at point A and V_{IH} at point B.

V_{OH} and V_{OL} can deviate from their nominal values and can also shift from circuit to circuit due to variability in component and process parameters. Transient voltage excursions superimposed on V_{OH} and V_{OL}, referred to as noise, can occur from external influences, such as line to line coupling from adjacent signal lines. An inverter must therefore be designed with adequate margin to prevent false switching in the presence of noise.

A figure of merit of an inverter is its static noise-margin, *NM*, defined as the absolute difference between the applied voltage, that is, from the driving gate and the input voltage corresponding to the nearest *unity-gain* point. Since there are two *unity gain* points on the transfer characteristic, there are two noise margins denoted by the *low* noise margin, NM_L, and the *high* noise margin NM_H, as shown in Fig. 8.3b [1–5].

NM_L is defined as the difference in magnitude between the maximum *low* output voltage, V_{OLmax} of the driving inverter, and the minimum *low* input voltage, V_{ILmin} (at the unity-gain point) that can be recognized by the driven inverter

$$NM_L = |V_{ILmn} - V_{OLmax}|. \tag{8.2a}$$

Similarly, N_{MH} is the difference in magnitude between the minimum *high* output voltage, V_{OHmin} of the driving gate and the maximum *high* input voltage V_{IHmax} (at the unity-gain point) that can be recognized by the driven inverter

$$NM_H = |V_{OHmin} - V_{IHman}|. \tag{8.2b}$$

For example, if the minimum high output, V_{OHmin}, of the driving inverter in Fig. 8.4 is less than the sum of maximum high input, V_{IHmin}, and noise margin NM_H of the driven inverter, there is a possibility that noise will flip the driven inverter to the opposite state.

A performance figure of merit is given in Chap. 5, (5.77), as

$$\tau \approx \frac{C_L V_{DD}}{I_{Dsat}} \quad \text{s}, \tag{8.3}$$

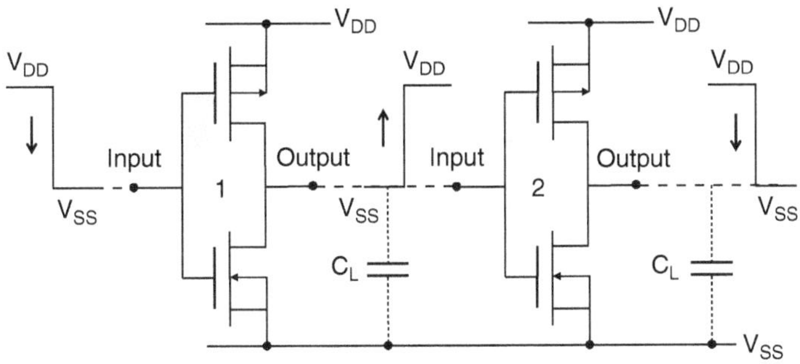

Fig. 8.4 Part of an inverter chain. The output of the driving inverter-1 is the input of the driven inverter-2

where C_L is the load capacitance, V_{DD} the total voltage swing, and I_{Dsat} the saturation drain current of a pull-up or pull-down transistor. For minimum physical space between inverters in Fig. 8.4, the wiring capacitance is typically negligible and C_L is just the sum of the *MOSFET* drain capacitances in inverter-1 and the *MOSFET* gate capacitances in inverter-2.

Pull-up MOSFETs have replaced passive resistors because of power, speed and the considerably smaller area occupied by a *MOSFET* of the same equivalent resistance. One option is an enhancement-mode *NMOS* connected in saturation with $V_G = V_D$, as shown in Fig. 8.2b. The main disadvantage of this option is that, as for the passive resistor load, the "*high*" output is lower than V_{DD}. From Chap. 5, it can be determined that when the pull-up drain and gate are connected and V_{In} is switched to "*low*," the output will be at approximately one threshold voltage V_T below V_{DD} rather than at V_{DD}. Actually, the output voltage will be lower by more than V_T because, as the output goes "*high*," the source becomes reversed biased with respect to body. This is known as the body-effect discussed in Chap. 5. The output low voltage V_{OL} decreases as the "gain" ratio β_R of pull-down to pull-up resistance increases. This ratio is given as (Chap. 5)

$$\beta_R = \frac{W/L \text{ of } \textit{pull-down}}{W/L \text{ of } \textit{pull-up}}. \tag{8.4}$$

This is why the *pull-down NMOS* is designed with a considerably higher W/L ratio than the *pull-up* transistor. If the gate of the *pull-up NMOS* is separately connected to a higher supply voltage than V_{DD}, the transistor can be made to operate in the linear region and the "*high*" output can again approach V_{DD}.

An alternative is to reduce V_T of the *pull-up* transistor below zero by using a depletion-mode *NMOS*. This can be achieved by forming a thin "n-skin" at the channel surface, for example, by ion implantation, so that the transistor is normally-on (Chap. 5). The channel is turned-off by applying a negative voltage on the gate with respect to source. In the configuration of Fig. 8.2c, the depletion-mode *NMOS* pull-up is always on since the gate is connected to source. One advantage of this two-terminal configuration is that the "*high*" output can go all the way to V_{DD} without necessitating a separate pull-up gate connection. Another advantage is the faster switching transition because the depletion mode *pull-up NMOS* provides more current to charge C_L than other *NMOS* pull-up structures throughout the output voltage-swing from zero to V_{DD} [6].

8.2.2 The CMOS Inverter

A schematic cross-section of a *CMOS* inverter formed on a p-type substrate is shown in Fig. 8.5. It consists of a complementary pair of enhancement-mode *NMOSFETs* in series with an enhancement-mode *PMOS*, with their gates connected to a common input, V_{in}, and drains connected to a common output, V_{out}. The *NMOS* source is typically connected to the p-well at ground potential $V_{SS} = 0\,\text{V}$. The *PMOS* source is connected to the n-well at the power-supply voltage V_{DD}. Under steady-state conditions, the output is either at V_{DD} or V_{SS} (ground). The *NMOS* is characterized by a threshold voltage V_{Tn}, and β_n defined as

$$\beta_n = \mu_n C_{eq} W_n / L_n, \tag{8.5a}$$

while the *PMOS* is characterized by a threshold voltage, V_{Tp} and β_p defined as

$$\beta_p = \mu_p C_{eq} W_p / L_p, \tag{8.5b}$$

where μ, C, W, and L are, respectively, the effective mobility, gate-oxide equivalent capacitance, effective channel width, and effective channel length.

The basic operation of the *CMOS* inverter can be described by examining the circuit diagram in Figs. 8.2d and 8.5. When the input voltage is "*high*," at $V_{DD} > V_{Tn}$, the *NMOS* is conducting, while the *PMOS* is not conducting because the *PMOS* source is at the same potential as the gate; the output voltage is "*low*," at V_{SS}. When

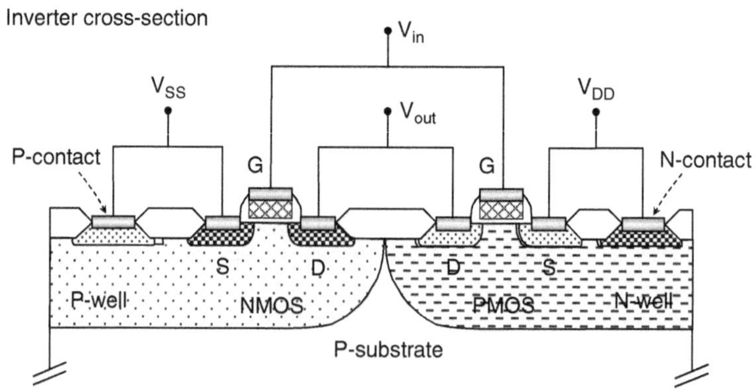

Fig. 8.5 Schematic cross-section of a *CMOS* inverter

the input voltage is switched from V_{DD} to V_{SS}, the *NMOS* turns-off, the *PMOS* turns-on, and the output voltage is "*high*," at V_{DD}. The main advantage of a *CMOS* over an *NMOS* inverter is that, under all *DC* (steady-state, stand-by) conditions, one of the transistors is off and there is no direct current through the inverter from V_{DD} to V_{SS}. This is the key attribute that has made *CMOS* the technology of choice for low-power applications. The voltage conditions for three regions of *MOSFET* operation in a *CMOS* inverter are given in Table 8.1.

The CMOS regions of operation are shown in Fig. 8.6. Since the effective inversion-hole mobility is about 1/3 of the electron mobility, $(W/L)_{PMOS}$ is often made $3 \times (W/L)_{NMOS}$ and V_{Tn} and V_{Tp} are made the same magnitude so that the *NMOS* and the *PMOS* currents are equal. One important attribute of a *CMOS* inverter is the high noise-margin because, as can be seen from Fig. 8.6, the inverter switches all the way between V_{DD} and ground ("rail-to-rail").

There is only a narrow region between the center dotted lines where both transistors are in saturation. Ideally, the output impedance of both transistors is infinite in this region and a small current causes a large instantaneous change in voltage.

Table 8.1 Voltage conditions of three regions of operation of *MOSFET*s in a *CMOS* inverter

MOSFET	Off	Linear	Saturated
PMOS	$V_{In} > V_{Tp} + V_{DD}$	$V_{In} < V_{Tp} + V_{DD}$ $V_{Out} > V_{In} - V_{Tp}$	$V_{In} < V_{Tp} + V_{DD}$ $V_{Out} < V_{In} - V_{Tp}$
NMOS	$V_{In} < V_{Tn}$	$V_{In} > V_{Tn}$ $V_{Out} < V_{In} - V_{Tn}$	$V_{In} > V_{Tn}$ $V_{Out} > V_{In} - V_{Tn}$

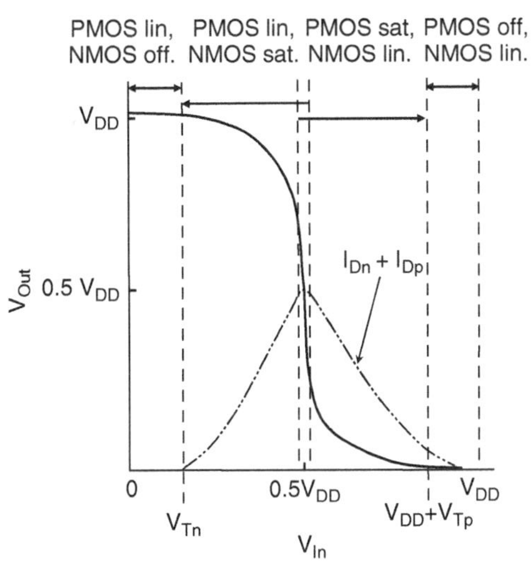

Fig. 8.6 Static transfer and current characteristics of a *CMOS* inverter and operating regions

Because of channel length modulation, however, the output impedance is finite and there is a finite slope in the transition in this region. Compared to other inverter configurations, the transition is still very sharp.

8.2.2.1 Latch-Up

Latch-up is a phenomenon which forms a low resistance path between V_{DD} and ground. It is triggered by an electric or radiation pulse; however, the path stays in a low resistance state after the pulse is removed [7–9]. This can cause loss of information or, if the current is not limited by series resistances, the destruction of the circuit. The *CMOS* inverter can be susceptible to latch-up because of the presence of four-layer *PNPN* (or *NPNP*) structures, such as the sequence of *p-source, n-well, p-well, n-source* shown in Fig. 8.7 [10,11]. Under normal operating conditions, the *n-well* to *p-well* junction is reverse-biased while the other junctions are either reverse-biased or at the same potential. Thus, the impedance between V_{DD} and ground (V_{SS}) is high. If a junction becomes forward-biased, even for a very short duration, internal gain can cause the structure to switch from high-impedance to low-impedance, resulting in high current between V_{DD} and ground. This effect, known as *latch-up*, is best explained by examining the inverter cross-section in Fig. 8.7, where a circuit consisting of an *NPN-PNP* transistor pair and two resistors is overlaid. A simplified lumped circuit model is shown in Fig. 8.8 [8, 12–16]. A more accurate model requires the resistors, transistors, and junction capacitors to be treated as distributed elements.

In the example shown in Fig. 8.7, the p^+*-source* of the *PMOS* acts as the emitter, the *n-well* as the base, and the *p-substrate/p-well* as the collector of the *PNP* transistor. Similarly, the n^+*-source* of the *NMOS* acts as the emitter,

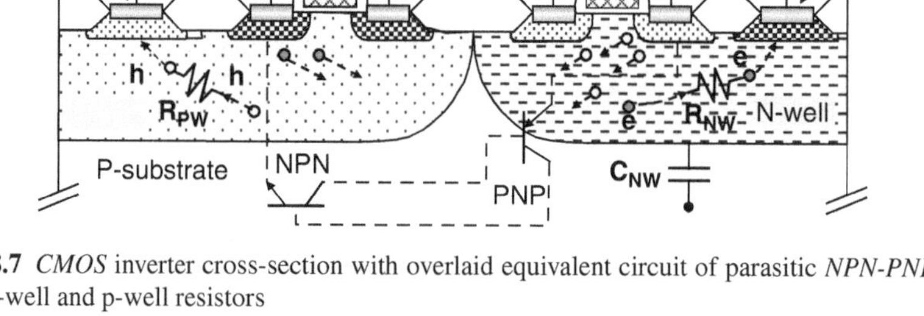

Fig. 8.7 *CMOS* inverter cross-section with overlaid equivalent circuit of parasitic *NPN-PNP* pair and n-well and p-well resistors

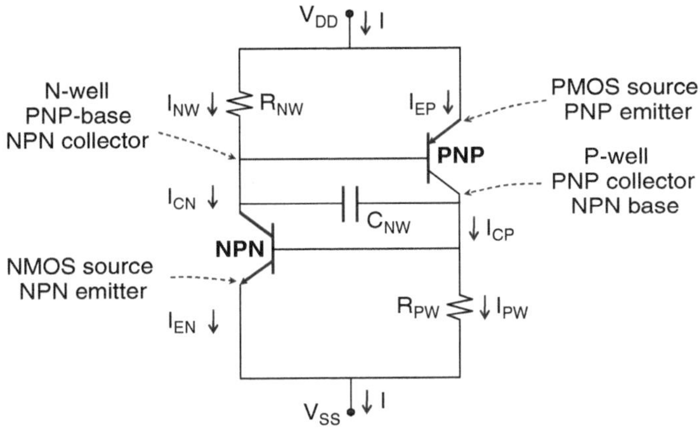

Fig. 8.8 Lumped circuit model of *CMOS* inverter latch-up structure. Also identified are the *NPN* and *PNP* emitter, base and collector along a *PNPN* current path in Fig. 8.7

the *p-substrate/p-well* as the base, and the *n-well* as the collector of the *NPN* transistor. Thus, the base of the *PNP* transistor is the collector of the *NPN* transistor and the base of the *NPN* transistor is the collector of the *PNP* transistor. Assume, for example, that a voltage-spike of amplitude above V_{DD} is applied to the *PMOS* source. Since the *n-well* is at V_{DD}, the voltage-spike can transiently forward-bias the *source* to *n-well* junction, although the *source* and *n-well* are shorted externally. Minority-carrier holes are injected from the *source* into the *n-well* (electrons are simultaneously injected from *n-well* into the source). Due to the large minority-carrier diffusion length in the well region, nearly all injected holes diffuse through the *n-well* and drift in the *p-substrate/p-well* to be collected at the *p-well contact*. The hole current I_{PW} passing through the *p-well* resistance, R_{PW}, creates a positive voltage drop $I_{PW}.R_{PW}$ that can *locally* forward-bias the *NMOS* source to p-well junction. Minority-carrier electrons are then injected from the *NMOS* source into the *p-well* and diffuse toward the *n-well* where they drift to the *n-well* contact (holes are simultaneously injected from the *p-well* into the *NMOS* source). The flow of electrons in the *n-well* causes a negative voltage-drop $I_{NW}.R_{NW}$ across the *n-well* resistance, R_{NW}, that can locally forward-bias the *PMOS source* to *n-well* junction This is a regenerative feed-back that quickly causes the structure to latch to a low-impedance path between V_{DD} and V_{SS}. Lateral currents can also result from forward-biasing the n^+*-source* to *p-well* junction, from avalanche multiplication current from the *n-well* to *p-well/p-substrate*, or from $C_{NW}(dv/dt)$ displacement current caused by a voltage-spike across the *n-well* to the *p-well/p-substrate* junction. Most susceptible to latch-up are circuits where large voltage transients and high currents are present, such as input-output circuits. An example of the current-voltage characteristic in a latch-up structure is shown in Fig. 8.9.

The current-voltage plot is illustrated for latch-up triggered by impact ionization at the junction between *n-well* and *p-well/p-substrate*. Assume initially that R_{NW}

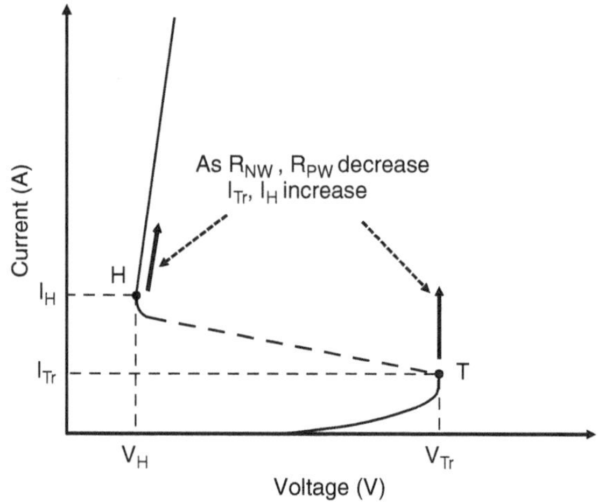

Fig. 8.9 Example of latch-up *I–V* characteristic

and R_{PW} are infinite. The *PNP* and *NPN* structures would then each exhibit the characteristics of an open-base bipolar transistor. For this case, the current is [10–12]

$$I = \frac{I_0}{1-(\alpha_n+\alpha_p)} \quad \text{A}, \tag{8.6}$$

where I_0 is the n-well junction reverse current and is a function of junction reverse bias, and α_n, α_p are, respectively, the current gains of the parasitic *NPN* and *PNP* transistors (Chap. 3). It follows that when $\alpha_n + \alpha_p$ approaches 1, the current goes to infinity. This condition is triggered at the point T that initiates a region of unstable negative resistance (shown as a dashed line) where α_n and α_p increase with increasing current and the voltage necessary to sustain the current decreases. Point H defines the holding voltage V_H and holding current I_H where $\alpha_n + \alpha_p = 1$ (or $\beta_n \cdot \beta_p = 1$) and $\delta I/\delta I_0 = \infty$. V_H is less than approximately $2V_{BE}$, (<1.4 V), where V_{BE} is the emitter-base forward voltage (Chap. 3). The current beyond point H is only limited by series resistances. As the shunt resistances R_{NW} and R_{PW} decrease, I_H and I_{Tr} increase, indicating a higher immunity to latch-up. This is the main reason for implementing retrograde wells in bulk *CMOS*. Three conditions must be met to sustain the latch-up state [17]:

(a) For given R_{NW} and R_{PW}, sufficient lateral current through the *p-well* and *n-well* must be present to forward-bias the emitter-base junctions of the parasitic transistors; (b) The sum $\alpha_n + \alpha_p$ (or product $\beta_n \cdot \beta_p$) must be a minimum of 1 and exceed the value necessary for regeneration and (c) the bias supply must be capable of "sinking" a current greater that the holding current I_H.

Several process and circuit techniques are available to suppress latch-up in bulk CMOS. Among them are:

1. Lowering the well resistance by increasing the retrograde well concentration and adding multiple well-contacts at short intervals. A higher well-concentration also increases the effective base Gummel-number of the parasitic transistors, thus reducing their current gain.[2] Lowering the well resistance is, however, limited by junction breakdown and parasitic capacitances. One method to efficiently lower the well resistance and increase the Gummel number is to form a heavily-doped buried layer under the well. The buried layers can be formed epitaxially without added complexity in a *BiCMOS* process [17], or implanted at high energy beneath the wells [18].
2. Placing p^+ guard rings, for example, *PMOS* source-drain in p-well, around the *n-well*, and n^+ guard rings, for example, *NMOS* source-drain in *n-well*, around the *p-well*. The guard rings act as collectors that intercept most of the injected minority carriers before they reach the circuit within the well [19,20]. This option will, however, increase circuit size and is only considered for circuits where large voltage transients and high currents are present, such as at input/output (I/O) terminals.
3. Reducing the distance between silicide and source-drain metallurgical junctions to increase minority-carrier injection into the source, thus degrading the parasitic bipolar current gain [21].

Latch-up can be eliminated by encapsulating the *NMOS* and *PMOS* with dielectrically-filled deep trenches merging with the buried oxide in *SOI* substrates.

8.2.3 The BiCMOS Inverter

BiCMOS allows the combination of both bipolar and *CMOS* circuits on the same die, taking advantage of high bipolar current drive capability, low quiescent power and the high packing density of *CMOS*. Thus, *BiCMOS* improves switching speed by driving high capacitive loads with bipolar transistors while maintaining the low *CMOS* static current for digital circuits. One variant of a *BiCMOS* inverter is shown in Fig. 8.10 [22]. When the input transitions from high (at V_{DD}) to low (at V_{SS}), the *PMOS* turns on and the *NMOS* turns off. A high current through the *PMOS* biases *NPN* transistor T_1 transiently to the on-state, thus charging-up the load capacitance C_L to approximately $V_{DD} - V_{BE}$ through T_1 at a higher speed than with *CMOS* alone. V_{BE} is the forward voltage on the *NPN* base-emitter junction (Chap. 3). Final charging to V_{DD} is through the *PMOS* and resistor R_1. Since *NMOS* is off, there is no base current in transistor T_2, and T_2 remains off. Similarly, when the input is switched to V_{DD}, *NMOS* turns-on and *PMOS* turns-off. T_2 turns-on and discharges the load capacitor at a higher speed than with *CMOS* alone.

[2] The effective base Gummel number is the integral of base concentration along the active path of the bipolar transistor (Chap. 3).

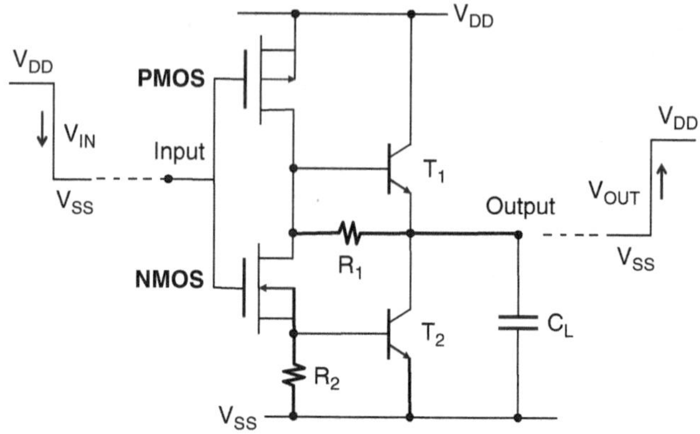

Fig. 8.10 Example of *BiCMOS* inverter (Adapted from [22])

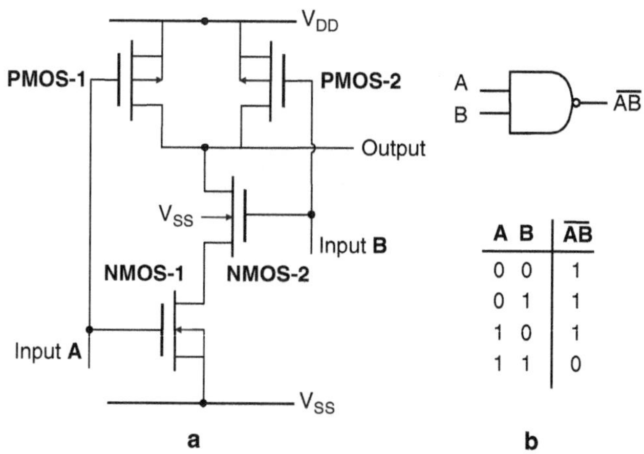

A	B	$\overline{AB}$
0	0	1
0	1	1
1	0	1
1	1	0

Fig. 8.11 **a** *CMOS NAND* gate. **b** Circuit symbol, truth table

8.2.4 CMOS NAND and NOR Gates

A *CMOS* two-input *NAND* gate is shown in Fig. 8.11a. The corresponding logic symbol and truth table are shown in Fig. 8.11b. As for the inverter, each input consists of a pair of complementary *MOSFETs*. When input *A* is low (0) and input *B* is low (0), both *PMOS* devices are on and both *NMOS* devices are off, and the output is high (1). When *A* is low and *B* is high, *PMOS-1* is on, *NMOS-1* is off, *PMOS-2* is off and *NMOS-2* is on. The path from output to V_{SS} is blocked and the output is high (1). Similarly, when *A* is high and *B* is low, the output is high (1). The only condition where the output is low (0) is when both *A and B* are high. Therefore, the output is high (1) only when *A and/or B* are *not* high.

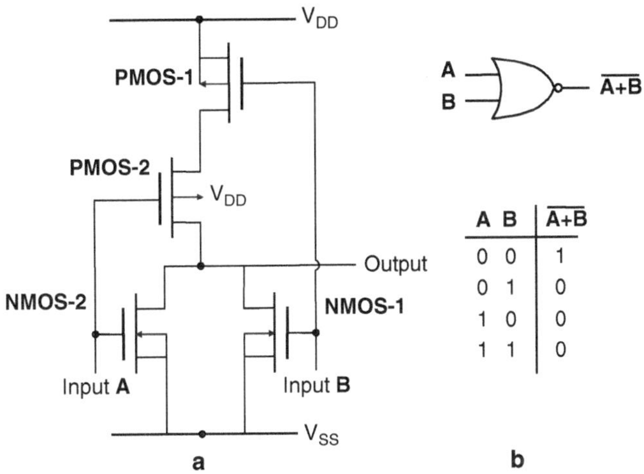

Fig. 8.12 **a** *CMOS NOR* gate. **b** Circuit symbol, truth table

A *CMOS* two-input *NOR* is shown in Fig. 8.12a and its logic symbol and truth table in Fig. 8.12b. When input *A* is low and input *B* is low, both *PMOS* devices are on and both *NMOS* devices are off and the output is high. When *A* is low and *B* is high, *PMOS-1* is off, *NMOS-1* is on, *PMOS-2* is on and *NMOS-2* is off. The output is low. Similarly, when *A* is high and *B* is low, or both *A* and *B* are high, the output is low. The only condition for the output to be high is when *neither* A *nor* B is high.

8.2.5 BiCMOS Two-Input NAND

The *NAND* output current drive capability can be enhanced with bipolar transistors, as shown in Fig. 8.13. The basic operation of a BiCMOS *NAND* gate is similar to that of the *BiCMOS* inverter shown in Fig. 8.10. When at least one input is low, at least one of the *NMOS* transistors in series is off and one of the parallel PMOS transistors is on. Base current is thus supplied to NPN transistor T_1 but not to NPN transistor T_2. T_1 is on and provides most of the current to charge C_L to approximately $V_{DD} - V_{BE}$. Final charging to V_{DD} is through *PMOS-2* and R_2. Only when both inputs *A* and *B* are high does T_2 turn on and provide most of the current to discharge the load capacitor to V_{SS}.

The performance advantages of *BiCMOS* over *CMOS* alone are achieved at higher processing complexity and lower circuit density (Chap. 7). Thus, applications of *BiCMOS* are limited to digital circuits where high drive currents are needed, such as input/output drivers, and analog and mixed analog/digital designs. It should be noted that as the *CMOS* channel length is reduced below ≈200 nm and the power supply voltage is reduced below ≈2V, *CMOS* performance improves while bipolar performance degrades, and digital BiCMOS loses its advantage [23].

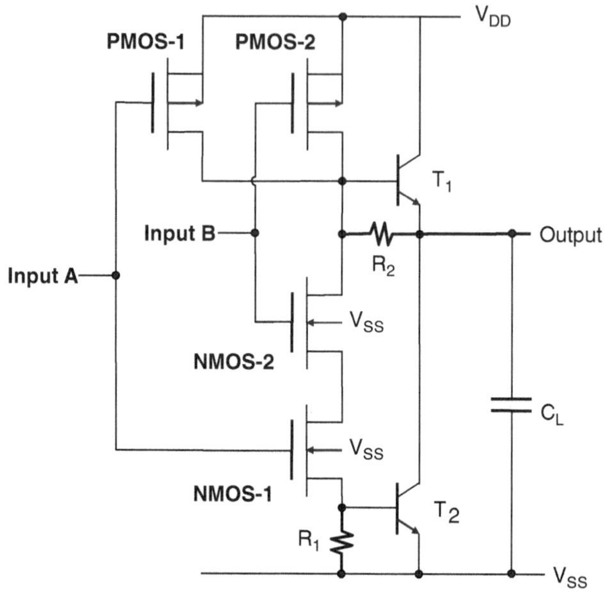

Fig. 8.13 *BiCMOS NAND* gate (Adapted from [22])

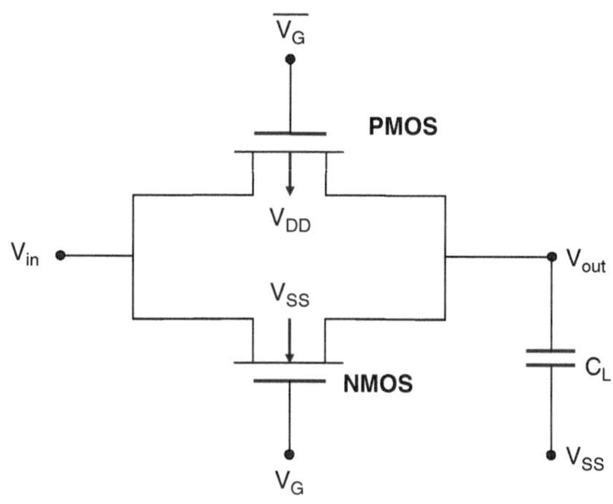

Fig. 8.14 *CMOS* transmission gate (or pass gate)

8.2.6 The Transmission Gate

The transmission gate (or pass gate) is shown in Fig. 8.14. It consists of a parallel arrangement of an *NMOS* and a *PMOS* transistor that form a complementary switch. Its main objective is to transfer logic levels without degradation from one node to another. For this purpose, a control signal is applied to the gate of the *NMOS*, and its complement applied to the gate of the *PMOS*.

Assume initially that the *PMOS* is not present and load capacitor C_L is at V_{SS} (0). When V_G is at V_{DD} (1), *NMOS* is on. For $V_{in} = V_{SS}$, there is no current between V_{in} and V_{out} and C_L remains uncharged at V_{SS}. For $V_{in} = V_{DD}$ (1), electron current flows through the *NMOS* from V_{in} (source) to V_{out} (drain), charging the capacitor toward V_{DD}. As V_{out} approaches the *NMOS* threshold voltage V_{Tn}, *NMOS* begins to turn-off. V_{out} reaches its maximum value of $V_G - V_{Tn}$. This means that the transmission of a "1" degrades to $V_{DD} - V_{Tn}$ when the signal passes through the *NMOS* alone. Note that the body effect is included in V_{Tn}. If C_L is initially charged to $V_{DD}(1)$, $V_{in} = V_{SS}$ (0) and V_G again switches to V_{DD}, the capacitor completely discharges to V_{SS} through the *NMOS*. Thus, the *NMOS* degrades a "1" but not a "0."

Similarly, consider the *PMOS* alone. For $V_G = V_{DD}(1)$, the complement $\overline{V_G} = VSS(0)$ and the *PMOS* is on. Assume C_L to be initially discharged at $V_{SS}(0)$. If the input signal to be transmitted is $V_{in} = V_{SS}(0)$, there is no current between V_{in} and V_{out} and the capacitor remains uncharged. If the input signal is a "1" (V_{DD}), hole current flows from V_{in} (source) to V_{out} (drain) until the capacitor fully charges to V_{DD}. If, however, the load capacitor is initially charged to V_{DD} and the signal to be transmitted is a "0," hole current flows from $V_{out}(V_{DD})$ to $V_{in}(V_{SS})$ until $V_{out} = V_{Tp}$, the *PMOS* threshold voltage. Thus, the *PMOS* degrades a "0" but not a "1." By combining both transistors as in Fig. 8.14, a gate is constructed that transmits both a logic "1" and a logic "0" without degradation [1].

8.3 Memories

Memories store information in the form of bits (binary digits). If they lose their information when the power is turned-off they are said to be volatile. The most important volatile memory types are Dynamic-Random Access Memory (*DRAM*) and Static Random Access Memory (*SRAM*). "Dynamic" refers to the fact that the cell information must be refreshed periodically (or "dynamically"), even if the power is on. Otherwise, the information would be lost because of leakage, as discussed below. "Static" means that the cell retains its information as long as power is on. Nonvolatile memories (*NVM*) retain the information even when the power is tuned off.

Memories can be designed as "stand alone," occupying a full chip, or embedded as part of a chip that incorporates other functions, such as in a system on a chip (SoC).

8.3.1 Dynamic Random-Access Memories, DRAM

The *DRAM* cell was briefly described in Chap. 5. It consists of a storage capacitor, of capacitance C_S, and one transfer device, typically an *NMOS* for stand-alone *DRAM* (Fig. 8.15). Hence, the cell is referred to as a one-device, or 1T-cell [24, 25]. The small cell-size allows the design of high-packing density memories with low-cost

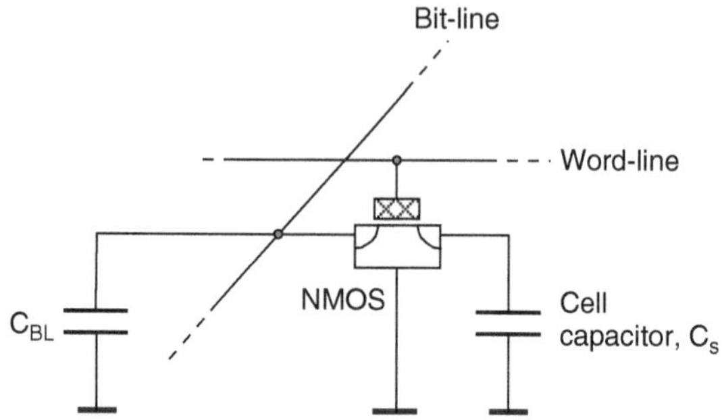

Fig. 8.15 *DRAM* cell circuit

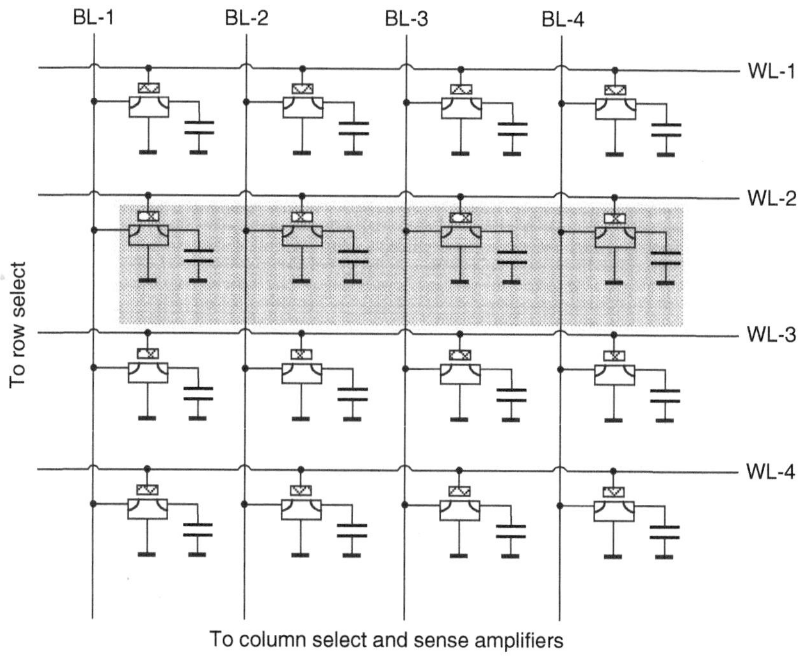

Fig. 8.16 Schematic of *DRAM* 4 × 4 array-section

per bit. A density figure of merit is the area occupied by the cell, measured in F^2, where F is the minimum lithographic feature size.

A charged capacitor may represent a logic "1" and a discharged capacitor a logic "0." The cells are arranged in rectangular arrays of rows and columns, as illustrated in Fig. 8.16 where only 4 × 4 cells are chosen for illustration. The control lines that connect the *MOSFET* gates are called word lines (*WL*), and the lines orthogonal to the word-lines are called bit-lines (*BL*). For example, a "256 Mega-bit" (Mb) array, that actually consists of 2^{28} bits, can be arranged as multiple sub-arrays of 256 (2^8) bit-lines by 64 (2^6) word lines.

The full array can be thought of as two half-arrays mirror-imaged with respect to the column-select and sense amplifier circuits. A sense amplifier is essentially a pair of cross-coupled inverters between the bit lines, with one bit-line of one half of the array connected to one node of the amplifier and the bit-line of the second half of the array to the second node of the amplifier.

8.3.1.1 Read Operation

To read the information of the highlighted cells in Fig. 8.16, for example, the sense amplifiers are first de-activated and all bit-lines along *WL-2* and their "mirror" bit-lines are pre-charged to exactly matching voltages that are intermediate between high and low logic levels. The pre-charge circuit is then turned-off and the transfer devices in the shaded cells turned-on by applying a voltage on *WL-2* gates sufficiently above the *NMOS* threshold voltage. This connects the storage capacitors in the shaded cells to their corresponding bit-lines. The charge in each cell is now re-distributed between C_S and the corresponding C_{BL}, where $C_S \ll C_{BL}$. A charged cell causes the voltage on the bit-line to rise by a value approximately proportional to the ratio $\mathrm{C_S/C_{BL}}$ with respect to the "mirror" bit-line. A discharged cell causes the voltage on the bit-line to dip below the level of the "mirror" bit-line. The charge transfer ratio is a measure of the voltage division that occurs when the cell capacitor is connected to the corresponding bit-line, defined as

$$T = \frac{C_S}{C_S + C_{BL}} \ll 1. \tag{8.7}$$

The capacitance of each bit-line is sufficiently large to hold the pre-charge voltage for a time long enough to complete the read cycle. The ratio C_S/C_{BL}, however, must be sufficiently high to result in a voltage signal typically in the range of 50–100 mV at the sense amplifier to allow signal amplification so that the row can be read at the output terminal.

8.3.1.2 Restore Operation

Since the cell capacitance is considerably smaller than the bit-line capacitance, the information is lost after each read access. This is referred to as a destructive readout. Thus, at the end of a read cycle, the values must be restored to the cell capacitors immediately after the read operation. The sense-amplifier detects the bit-line information and writes it back into the cell. This is done simultaneously for all cells that are addressed by the word-line.

8.3.1.3 Access and Cycle Times

The access time t_{Access} is the delay between the rising edge at the clock-pulse and the time the information becomes available at the output terminal. The cycle time t_{Cycle} is the minimum time between two accesses. Because of the need to refresh the cells after a "read," $t_{Cycle} > t_{Access}$. Typically, $t_{Cycle} \approx 2 \times t_{Access}$.

8.3.1.4 Write Operation

To write new information in a particular cell, the word-line is activated and a read operation is first performed on the entire row. The voltage on the bit-line corresponding to the cell is then forced by the sense-amplifier to the desired "1" or a "0." The amplifiers then drive the bit-lines, charging the cells with the new information.

8.3.1.5 Retention Time and Refresh Operation

Since the inner plate of the cell capacitor is connected to a junction, the cell charge can gradually leak-out because of several junction leakage mechanisms, even when the power is not interrupted. Among the leakage mechanisms are thermal generation, sub-threshold leakage, gate-induced drain leakage (*GIDL*), or direct tunneling through gate oxide (Chap. 5). Assume, for example, that a cell is charged to 2.5 V, and that detection of a "1" by the sense amplifier is only possible if the cell voltage remains above 2.3 V. For a cell capacitance of 30 fF and junction leakage of 1 fA, the cell voltage drops to 2.3 V in 6 s. In this case, the retention time is roughly 6 s. Because of loss of charge, the cells must be refreshed periodically by reading all cells in a row and writing-back the information before losing the ability to distinguish between a "1" and a "0." Long retention time is required to reduce power and delays associated with refreshing the data. Since material and process cause leakage to vary from cell to cell, the retention time is specified for the leakiest bit within a chip. The minimum retention time is typically specified as 64 ms although most of the cells exhibit longer retention times. In this case, each row must be refreshed every 64 ms. Assuming 8192 rows, this requires a refresh rate of one row every 7.8 μs.

8.3.1.6 Variable Retention Time, VRT

A single DRAM cell exhibits variable retention time (*VRT*) when its leakage current decreases or increases abruptly with time. *VRT* was first observed during testing of 1 Mb *DRAM* arrays [26]. Multiple-state and two-state *VRT* were observed on a very small fraction ($\sim$0.01%) of cells. The retention time of an individual cell varies abruptly from one storage cycle to another as shown in Fig. 8.17 [26,27]. The duration of one leakage state ranged from seconds to hours and decreased with increasing temperature. Also, the frequency of transitions between states increased with

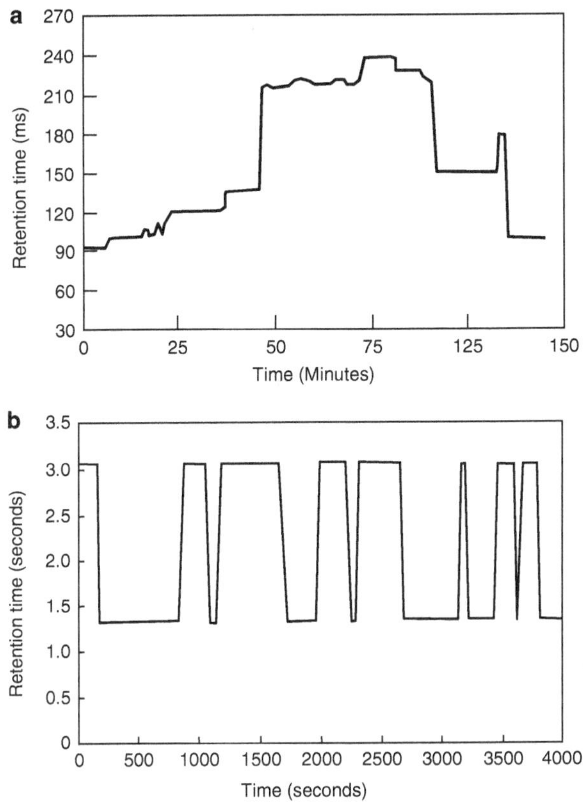

Fig. 8.17 **a** Multiple-state retention times measured on one cell as a function of time (Adapted from [26] **b** Two-state retention times measured on one cell as a function of time (Adapted from [27])

temperature. The observed multiple-state or two-state *VRT* can result in a failure if the shortest retention time measured is smaller than specified for all operating temperatures. As a result, the minimum retention time should be specified with a large "safety margin" to protect against failure of *VRT* cells. This results in an increase in power dissipation and cycle time. Thus, understanding the mechanism behind *VRT* and implementing process techniques to reduce the occurrence of *VRT* become important to the development of low-power, high-capacity *DRAM*.

The *VRT* pattern is analogous to that of the random telegraph signal (*RTS*) described in Chap. 6. The origin of *VRT* has been correlated to the fluctuation in the generation-recombination leakage current of the pn junction connecting the cell to the transfer device [28]. This was determined by separately monitoring substrate and bit-line leakage currents on a specially-designed structure that accesses the nodes connecting the *NMOS* junction to the cell-capacitors. Both the bit-line and node junction leakage currents are measured on the substrate but, with the word-line "low," only the sub-threshold current can be measured on the bit-line. *RTS*-like fluctuations were observed on the substrate current but not on the bit-line current, indicating that fluctuations are caused by generation-recombination and not sub-threshold current. The fluctuation is attributed to an intermediate, near mid-gap energy state that can switch from a low-leakage, stable level to a high-leakage meta-stable level. Its power spectral density is found to follow the Lorentzian noise spectrum described in Chap. 6 [28].

8.3.1.7 Single-Event Upsets, SEU

The information in the cell can be lost by generation of electron–hole pairs in silicon along the path of incident highly energetic particles, such as alpha particles [29], or cosmic rays in the form of neutrons and protons [30], striking sensitive regions of the die. An alpha particle is a double-ionized helium atom (2 protons, 2 neutrons) that can emanate from trace amounts of radioactive contamination in materials used for manufacturing. Alpha-particles can possess energy typically in the range of 4 MeV to 8 MeV and penetrate 25–50 μm in silicon. Cosmic rays collide with atoms in the atmosphere and create cascades or showers of predominantly protons and neutrons which can penetrate deep in silicon and collide with silicon nuclei, generating additional energetic particles [30–32]. The particles create a large amount of electron–hole pairs along their path. Minority carriers can be collected by a reverse-biased cell node within or in the vicinity of the particle path. If the collected charge is sufficiently large, the state of the cell can be upset. The resulting error is recoverable. It is thus called a single-event upset, *SEU*, or a "soft-error" to distinguish it from a "hard error" which causes permanent damage to the circuit and is nonrecoverable. Single-event upsets are observed in both memory and digital circuits. The minimum collected charge that upsets the cell is referred to as the critical charge, Q_{crit}. The soft-error rate (*SER*) caused by alpha particles can be reduced by designing the node with a higher capacitance to increase Q_{crit}, by the use of purified materials with reduced concentration of radioactive contamination, and by shielding the cell with a buried well as shown in Fig. 8.18. The entire chip may also be shielded with a particle-absorbent film. In case of cosmic rays, however, process fixes are more difficult to implement.

8.3.1.8 Cell Structures

High-density *DRAM* memory cell-capacitors are constructed three-dimensionally to reduce the horizontal area occupied by the cell. The cell capacitor can be formed in a deep trench below the silicon surface, or stacked above the silicon surface.

Trench-Capacitor Cell

One variant of a trench capacitor cell is shown in Fig. 8.18 [33]. Its unique feature is the self-aligned buried strap that connects the inner plate to one junction of the transfer device.

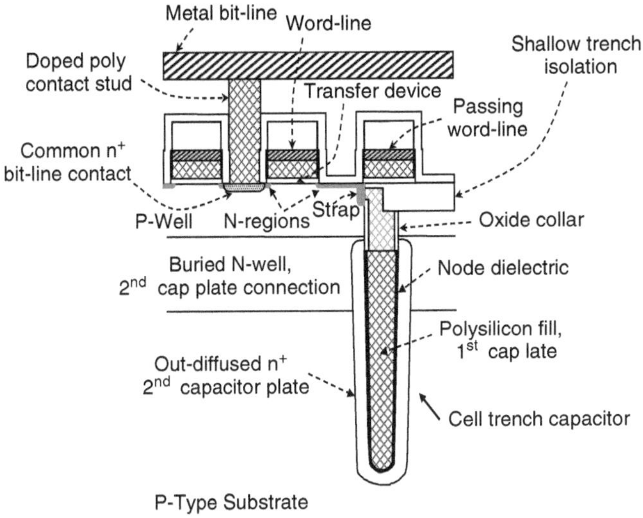

Fig. 8.18 Trench-capacitor DRAM cell

The cell in Fig. 8.18 is constructed on a p-type substrate. A buried n-well is first implanted beneath the p-well to connect all outer capacitor plates in the array. A trench is directionally etched into the substrate, about 8 µm deep for a 180-nm minimum feature size on the chip. The trench is filled with arsenic-doped polysilicon that is recessed down to the top level of the cell-capacitor. Out-diffusion of arsenic from polysilicon into the trench sidewalls connects to the *n-well* to form the common outer (2nd) capacitor plate. The arsenic-doped polysilicon is then etched away and a thin oxidized nitride capacitor dielectric grown on the trench sidewalls. This is followed by depositing a first n^+-polysilicon trench-fill and recessing it to a controlled depth to expose the trench-top and grow a thick isolating oxide collar. A second n^+-polysilicon trench-fill is deposited and etched-back deep enough to expose and etch the thick oxide collar where a buried "strap" will be formed. A third n^+-doped polysilicon fill is deposited and recessed below the silicon surface. The three n^+-polysilicon fills constitute the inner plate of the trench capacitor. Out-diffusion from the third (trench-top) polysilicon forms the buried "strap" that connects the inner plate to one junction of the *NMOS* transfer device. The shallow-trench isolation is patterned in a U-shape over the trench such that it cuts through the buried strap on three sides and leaves the strap only where the strap makes connection between inner plate and transfer device (Fig. 8.19).

Defects and heavy metal contamination increase leakage and reduce retention time. To reduce defects and metal contamination in the array, the transfer-device junctions are implanted with low-dose arsenic and blocked from silicidation. Also, the bit-line contact is formed by diffusion from n^+-polysilicon rather than with a conventional metal stud on silicide (Fig. 8.18). Although the low junction concentration increases extrinsic resistances, the impact on *DRAM* access- and cycle-time is negligible.

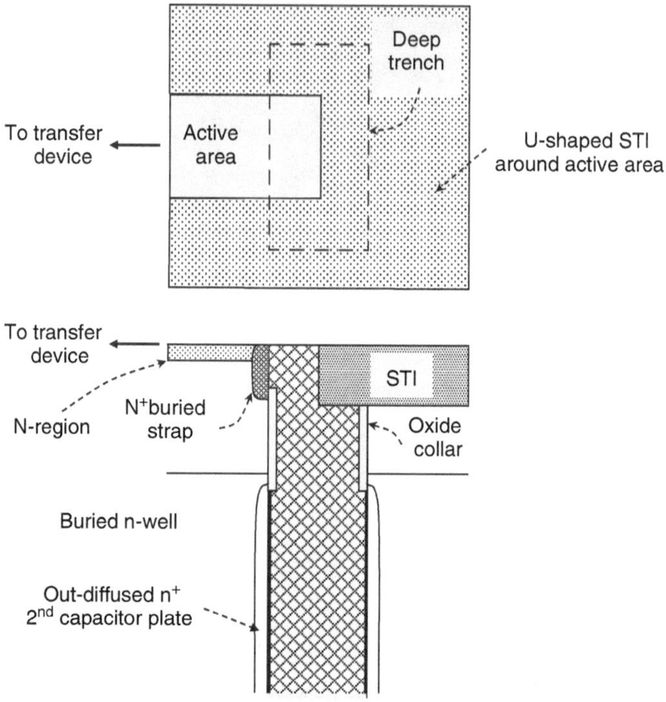

Fig. 8.19 Details of the buried strap

Scaling of the horizontal dimensions of the transfer device and cell-capacitor is limited. The threshold voltage and channel length of the transfer device must be sufficiently large to ensure that the off-current does not exceed the minimum specification for retention time, typically less than 0.1 fA at 85 °C for sub-100 nm technologies. As the channel length is decreased, the dopant concentration must be increased to reduce short-channel effects on threshold voltage discussed in Chap. 5. Also, as the channel width is reduced, the drive current decreases. This is exacerbated by a decrease in mobility because of an increase in channel doping concentration and electric field (Chap. 5). Several variants of the trench-cell in Fig. 8.18 have been reported [34–36, 47]. They are mostly aimed at reducing the cell size while maintaining acceptable cell-capacitance and drive current of the transfer device.

To reduce the off-state leakage, the physical channel length of the transfer device can be increased by recessing the channel region and forming a curved channel area below the silicon surface. This technique, called recessed-array channel transistor (*RCAT*), achieves long-channel characteristics without increasing the horizontal channel dimensions [39, 40]. Also, the drive current of the transfer device can be increased while maintaining very low cell leakage by forming the transfer channel on a fin (Chap. 5). One method for integrating an array FinFET is described in [41]. The *RCAT* and *FinFET* features can be combined to optimize the transfer device with respect to size, leakage, and drive current [42].

Constructing three-dimensional transfer devices directly above the trench-cell is shown to reduce the cell area to $8F^2$ (F: minimum feature size) in 0.175-μm rules, resulting in a cell area of $0.245\,\mu m^2$ [34–37], and an area of $6F^2$ ($0.135\,\mu m^2$ cell area) in 0.150-μm design rules [38].

The cell capacitor area can be enhanced while keeping minimum dimensions of the trench-top by forming silicon nodules, similar to hemispherical silicon grains (HSG), before filling the deep trench with the inner plate [43, 44]. HSG utilizes special surface treatment of polysilicon and was first introduced in a stacked-cell capacitor, as will be described in the next section [45]. Further improvement of capacitor surface area can be achieved by etching the trench in a "bottled" form rather than directionally [43, 46, 48].

The trench-capacitor dielectric must sustain thermal cycles subsequent to dielectric growth or deposition. This limits the choices of cell-dielectric and is the main reason why oxynitride has prevailed as the dielectric of choice for high-density *DRAM* cells [47]. Recent reports show, however, that aluminum oxide (Al_2O_3) can be a viable enhancement of the trench-cell capacitance [44, 48, 49]. A thin carbon film was deposited as an outer capacitor plate of high conductivity and thermal stability, and shown to be compatible with high-K dielectrics [49].

Stacked-Capacitor Cell

The principle of operation of the stacked-capacitor cell is identical to that of a trench-capacitor cell. The difference between the two cells is in the construction of the capacitor. An example of a method of forming a stacked capacitor is shown in Fig. 8.20 [45, 50].

Contacts to the inner capacitor plate are patterned in an inter-level oxide film (Fig. 8.20a). A first amorphous silicon (a-Si) film is deposited, heavily doped with phosphorus (P) at a concentration of about $3 \times 10^{20}\,cm^{-3}$, filling the contact hole. This is followed by the deposition of a sacrificial borophosphosilicate (*BPSG*) layer. The *BPSG* and P-doped a-Si films are simultaneously patterned and directionally etched to form a capacitor "mold" (Fig. 8.20b). Another phosphorus-doped a-Si film is deposited and directionally etched to form cylindrical side-walls around the *BPSG*, contacting the first P-doped a-Si on the sides (Fig. 8.20c). The *BPSG* "mold" is selectively etched in HF vapor (Fig. 8.20d). Hemispherical-grained silicon (*HSG-Si*) is formed on the exposed surfaces of the P-doped *a-Si* cylinder [45, 51, 52]. The surfaces are first cleaned by removing the native oxide in a dilute HF solution. Hemispherical grains are then formed by, for example, the "seeding" method described in [45, 51], resulting in an increase in as much as 30% in effective capacitor-plate area (Fig. 8.20e). The structure is annealed at approximately 800 °C to recrystallize the *a-Si* and activate phosphorus. A capacitor dielectric, such as oxynitride or higher-K material, is then formed or deposited, for example, by atomic layer deposition (ALD), then covered by a doped polysilicon film that constitutes the outer capacitor plate (Fig. 8.20f).

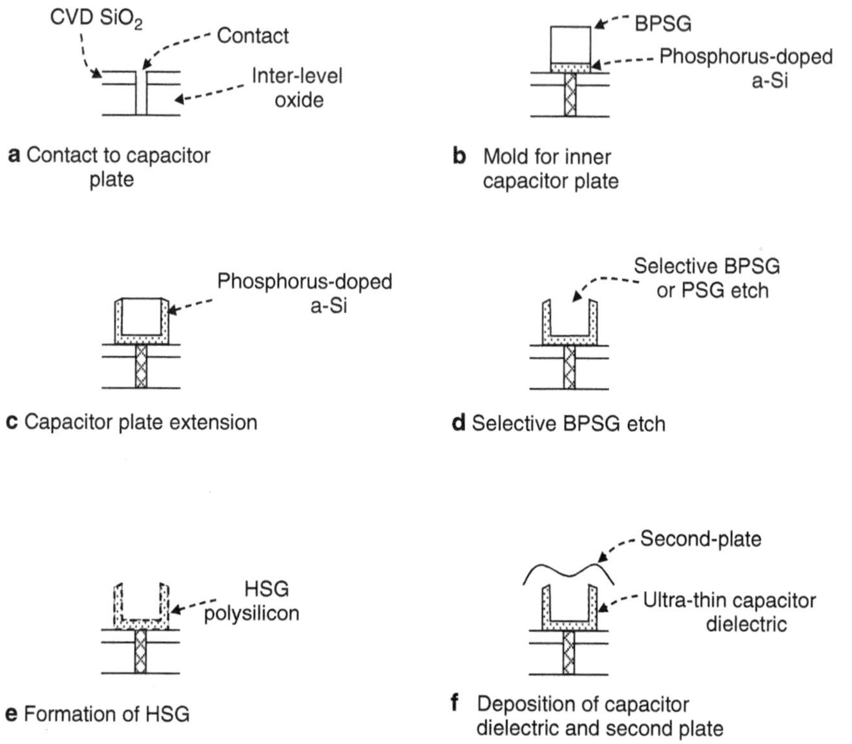

Fig. 8.20 Example of stack-capacitor cell (Adapted from [45])

One of the main advantages of the stacked capacitor cell is the flexibility in the choice of capacitor dielectric. This is because the capacitor is formed after all high-temperature steps are completed, enabling the use of temperature-sensitive high-*K* cell-dielectrics (Chap. 7), such as Barium-Strontium Titanate, *BST* $[(Ba,Sr)TiO_3]$, and Ruthenium Oxide (RuO_2) [53], Aluminum Oxide (Al_2O_3) [54], Aluminum Oxide and HfO_2 dual dielectric (*AHO*) [55], and Zirconium oxide (ZrO_2) [56]. The main disadvantage of the cell is the topography created by the stack, compared to a planar top-surface in a trench-capacitor process.

8.3.2 Static Random Access Memories, SRAM

A circuit schematic of the most common *SRAM* design is a 6-device CMOS cell shown in Fig. 8.21. A typical layout of the cell is shown in Fig. 8.22 [57, 58]. The cell can be viewed as consisting of two cross-coupled inverters, forming a "latch," and two access devices for read-write operations. The *NMOS-PMOS* drains of each inverter are connected to the gates of the other inverter. When "Node-1" is high, T_3

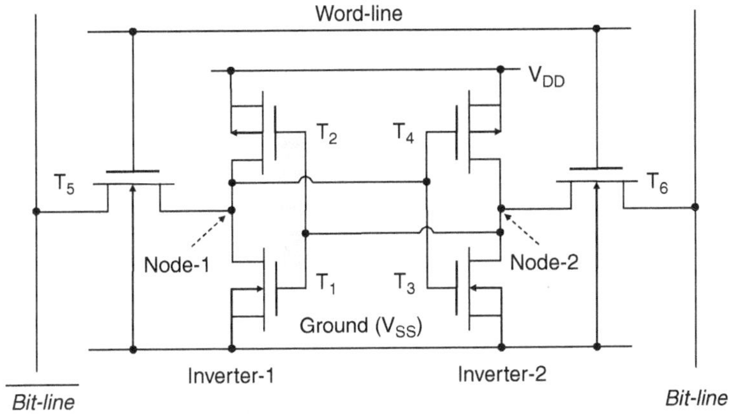

Fig. 8.21 Circuit diagram of 6-device *CMOS SRAM* cell

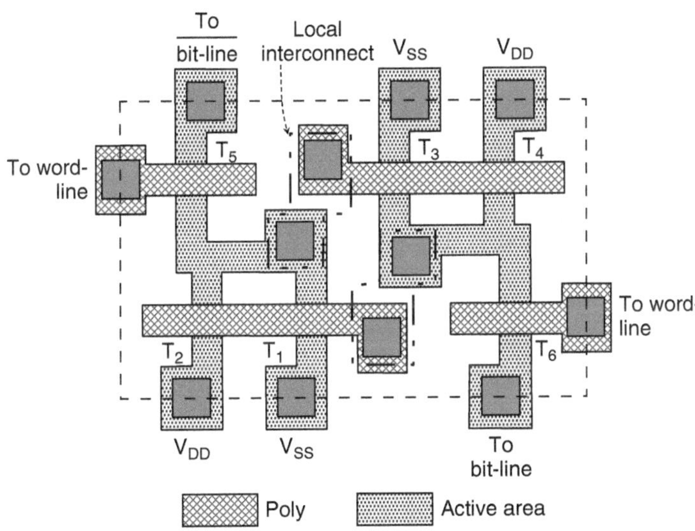

Fig. 8.22 Typical layout of 6-device CMOS SRAM cell. The dotted outline indicates the cell boundaries. V_{DD}, V_{SS}, word-line and bit-lines are connected with metal at different levels (Adapted from [57, 58])

is on and "Node-2" goes low, T_2 is on and T_1 is "off." Let this state define a "0." When "Node-2" is high, T_1 is "on" and node-1 goes low, T_4 is "on" and T_3 is "off." This state defines a "1." The feedback in the cell ensures the data is retained when the power is on. Thus, the cell has two stable states that define a "0" and a "1."

The word-line controls access to the cell through the two access transistors T_5 and T_6. When the word-line is high, T_5 and T_6 are turned on and connect the nodes of the cell to the bit-lines (Fig. 8.21).

8.3.2.1 Stand-By

In standby, the word-line is low, T_5 and T_6 are off and the cell is idle. The cell is designed such that the feedback (reinforcement) between the two inverters ensures that the cell retains its information as long as the power is on, so there is no need for periodic "refresh" as in *DRAM*. The only current in this state is leakage current consisting of junction and *MOSFET* leakage components described in Chap. 5. For low-power applications, the leakage currents must be kept very low.

8.3.2.2 Read Operation

The read operation is nondestructive, that is, there is no need for restoring the data after a read. The two capacitive bit-lines are first pre-charged to V_{DD} and then disconnected from the pre-charge circuit (not shown). Assume that a "1" is stored in the cell and node-2 is high. When the *word-line* is brought to a high voltage, turning-on access transistors T_5 and T_6, the cell information is transferred to the bit-lines (Fig. 8.21). This leaves $Bit-line$ at its pre-charged level and discharges the complementary $\overline{Bit-line}$ to $V_{DD}-100\,\text{mV}$ through T_1 and T_5. If a "0" were stored in the cell, $\overline{Bit-line}$ would remain at a high level and $Bit-line$ would go to $V_{DD}-100\,\text{mV}$. In both cases, the established differential signal is typically about 100 mV and sufficiently high to be detected by the sense-amplifier (not shown). The direct differential sensing without the need to refresh after a read is one of the reasons an *SRAM* memory is faster and consumes less power than a *DRAM*.

8.3.2.3 Write Operation

The voltages to be written are first applied to the bit-lines. To write a "1," for example, $Bit-line$ is set high ("1"), and $\overline{Bit-line}$ is low ("0"). Access transistors T_5 and T_6 are turned on by addressing the word-line and the voltages on the bit-lines are written to the cell, regardless of previous information in the cell.

8.3.2.4 Cell Stability

A measure of cell stability is the static noise margin (*SNM*) [2–5, 59–62]. This is the maximum magnitude of noise voltage that can be tolerated at the internal cell nodes before the cell flips its state. The most common way of representing the *SNM* graphically is shown in Fig. 8.23 in which the voltage transfer characteristic (*VTC*) of inverter-2 and the inverse *VTC* of inverter-1 are plotted [59–62]. For symmetrical inverters, the inverse plot for inverter-1 is obtained by rotating the *VTC* of inverter-2 about the axis of unity slope (45°) through the origin. The noise margins for a given inverter characteristic are defined as the lengths of the sides of the largest rectangles that can be fit in the upper and lower half of the curve. These are shown

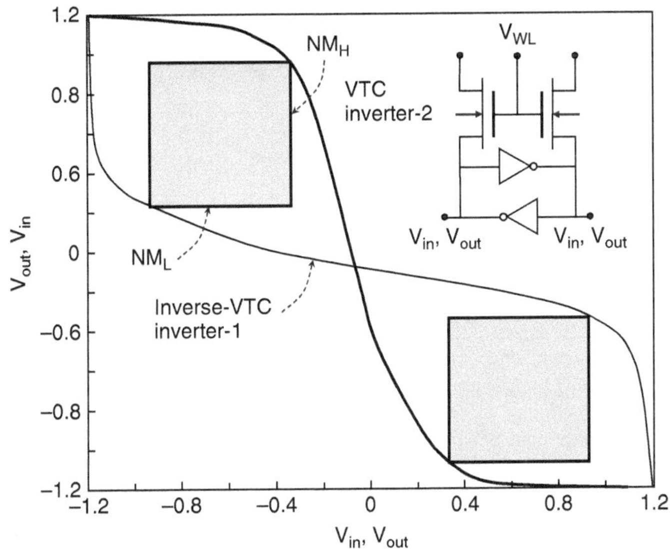

Fig. 8.23 Measured "butterfly curve" on a 3-D SRAM cell fabricated in a 1.2 V CMOS technology (Adapted from [62])

in Fig. 8.23 for the upper half characteristic as NM_L and NM_H. To illustrate this definition assume, for example, that in the presence of noise the *VTC* of inverter-2 shifts down and the *VTC* of inverter-1 shifts to the right. Once they both shift by the *NM* values, they meet at only one point. Any increase in noise would then flip the cell [59]. For $NM_H = NM_L$, the rectangle becomes a square as shown in the figure.

The noise margin is degraded during an active operation of the cell because an additional noise component is induced by the active operation itself. Thus, the worst-case noise margin should be specified for active operation, such as "read."

8.3.2.5 Scaling to Smaller Dimensions

When designing an *SRAM* cell for high-density, low power applications, the primary objective is to reduce the cell-area and operating voltage while maintaining cell stability and speed [57, 58, 63–70]. As the operating voltage is reduced, however, the noise margins decrease and it becomes more difficult to achieve adequate cell stability [64]. Also, the susceptibility of the cell to single-event upsets increases with decreasing operating voltage because the critical charge Q_{crit} that upsets the cell decreases with decreasing voltage [31, 32, 65, 71]. For example, an *SRAM* array of 6-device cells designed in a 90-nm technology shows an average increase in soft error rate by 18% for a reduction of 10% in operating voltage [32].

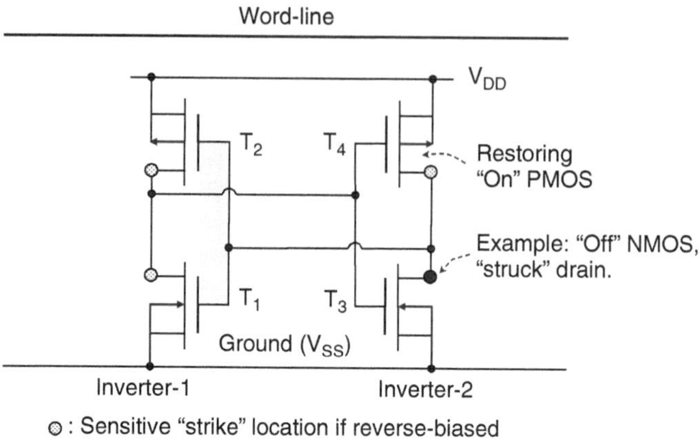

Fig. 8.24 *SRAM* regions sensitive to "strike" by energetic particle. Access devices are not shown (Adapted from [72])

The region of the *SRAM* cell that is most sensitive to a "strike" by an energetic particle is a reverse-biased drain junction of an *NMOS* or *PMOS* in the cross-coupled inverters. There are four possible sensitive "strike" locations, namely the two *PMOS* drains and the two *NMOS* drains (Fig. 8.24) [72]. Assume, for example, that a "1" is stored, that is, the *NMOS* of inverter-2 is "Off" and its *PMOS* is "On." In this case, the *NMOS* drain is reverse-biased at V_{DD}. A "strike" through the drain or within its vicinity creates electron–hole pairs causing a transient flow of minority-carrier electrons to the drain while the holes flow to the substrate/p-well contact. The flow of electrons tends to discharge the drain to ground while the On-*PMOS* tends to restore the drain to its high potential. Since the PMOS has a finite conductance, the transient current can drop the drain potential sufficiently to cause the cell to flip.

Since collection by an *NMOS* drain causes a transition from high to low and collection by the *PMOS* causes a transition from low to high, it is possible to measure the *SER* either on *NMOS* or *PMOS* drains by programming the nodes to either low or high voltage [32]. In an n-well technology the *PMOS* drain is located inside the well and the well-substrate junction provides a potential barrier that prevents minority carriers generated outside the well from diffusing to the "struck" drain. Consequently, the *SER* for low to high transitions is smaller than for high to low transitions [32, 72].

The *SER* is found to be proportional to the sensitive volume defined by the cell's drain area and the minority-carrier collection depth [32, 73]. Scaling the drain to smaller dimensions has opposing effects on the *SER*: The sensitive volume for collection of generated minority carriers is reduced, but the drain capacitance also decreases, making the cell more susceptible to soft errors. One method to avoid the conflict is to place a capacitor plate at a fixed potential above the drain, increasing the drain capacitance without increasing its area [67, 68]. This was achieved, for example, by optimizing a metal insulator-metal (*MIM*) capacitor between the metal contacting the drain and a titanium-nitride (*TiN*) plate separated from the drain by silicon nitride (Si_3N_4) or tantalum-oxide (Ta_2O_5) as the insulator. A minimum-size cell of area 0.46 μm^2 was obtained while considerably reducing the *SER* [67].

An important consideration when scaling the cell to smaller dimensions is the increased impact of fluctuations in transistor characteristics. As the channel reaches nanoscale dimensions, fluctuations in the number of dopant atoms within the channel become significant. For example, a channel of 50-nm length and width, doped at a concentration of around $10^{18}\,cm^{-3}$, will contain about 100 dopant atoms. A small fluctuation in the number of dopant atoms can have a significant effect on threshold voltage. One approach to solve the problem is to construct a near intrinsic channel on *SOI* [74], or on a double-gated fin-structure [58]. Since no dopant atoms are needed, fluctuations are no longer a problem.

8.3.3 Nonvolatile Memory, NVM

A semiconductor memory is said to be nonvolatile if it retains its information after its power supply is turned-off. Nonvolatile semiconductor memories can be categorized as Read-Only-Memory (*ROM*), Electrically-Programmable *ROM* (*EPROM*), and Electrically Erasable and Programmable *ROM* (*EEPROM* or E^2PROM). Programming is the operation of setting (or writing) the desired bit-state of each cell. A *ROM* has permanent storage of information that can be read but not readily programmed. An *EPROM* can be erased only by exposing the entire chip to strong ultraviolet light, typically of 235-nm wavelength for about 20 min. After erasure, the memory array can be electrically re-programmed. *EEPROM* allows electrical programming and erasing of individual bit or byte, whereas in flash *EEPROM* large blocks of cells, ranging from 512 bytes to entire chip, are erased simultaneously "in a flash" [75]. This section describes four types of *EEPROM* cells: the floating-gate, the trapped-charge, the phase-change, and the magneto-resistive.

8.3.3.1 Floating-Gate Cell

A floating gate cell is shown in Fig. 8.25 [76]. It consists of a *MOSFET*, typically an *NMOS*, with an additional gate that is encapsulated in a dielectric and referred to as the floating gate, FG. The gate oxide thickness ranges typically from 5 nm to 12 nm while the inter-poly dielectric equivalent oxide thickness ranges from 25 nm to about 50 nm.

Basic Concept

Charge trapped in the floating gate (FG) modulates the *MOSFET* threshold voltage V_T as seen by the control gate (CG). Absence of electron charge in the floating gate results in low V_T. The *MOSFET* turns on when the control gate voltage is increased

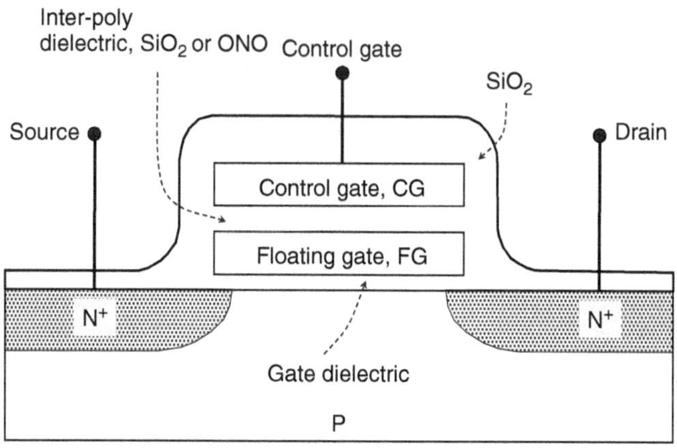

Fig. 8.25 A typical floating-gate memory cell

to a specified level above V_T and a large drain to source current can be measured. This state represents a logical "1." Presence of electron charge in the floating gate shifts V_T to a higher positive value. The *MOSFET* remains off when the control gate voltage is increased to the specified level. This state defines a logical "0."

Read Operation

The band-diagram of a floating gate structure is shown schematically in Fig. 8.26. Figure 8.26a shows the band diagram for an erased cell, with the control gate at ground potential and no charge in the floating gate or dielectrics. This state represents a logical "1," The small field created in the gate oxide is due to the work function difference between the polysilicon and the p-type body. To read the cell, a positive voltage is applied to the control gate and capacitively coupled to the floating gate, increasing the field in the gate oxide so that the field at the silicon-oxide interface is high enough to invert the surface (Fig. 8.26b).

In a programmed cell, the floating gate is negatively charged by ΔQ_{FG} causing a positive shift in threshold voltage, ΔV_T, proportional to ΔQ_{FG}, expressed as

$$\Delta V_T = V_T - V_{T0} = \frac{\Delta Q_{FG}}{C_{FG-CG}}, \tag{8.8}$$

where V_{TO} is the threshold voltage with zero floating-gate charge, and $C_{FG\text{-}CG}$ is the floating-gate to control-gate capacitance.

The $I_D - V_{CG}$ characteristics for an erased and a programmed cell are shown schematically in Fig. 8.27. The plots for programmed and erased sates are parallel to each other. The read voltage V_{RD} is chosen between the two curves, ensuring that it is above V_{TO} for the erased state but below V_T for the programmed state. It should be noted that erasure does not always result in zero charge on the floating gate, and programming does not always place the same amount of charge on the floating gate. Thus, safe margins should be allowed when defining the read voltage, V_{RD}.

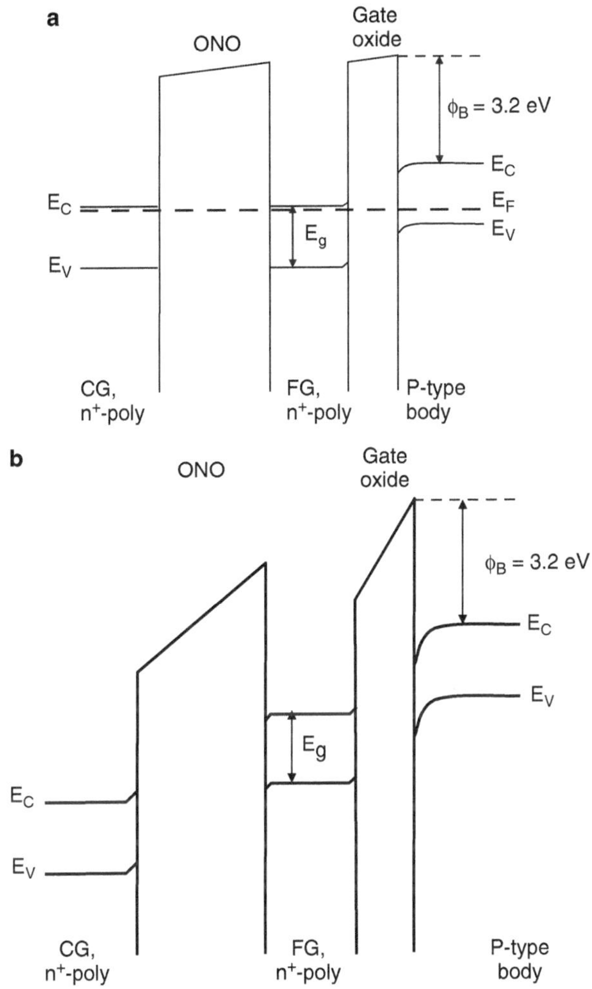

Fig. 8.26 Energy band diagram of floating gate cell. **a** Zero charge in floating gate (FG), zero bias on control gate (CG). **b** Positive voltage on control gate

Programming (Write) Operation

The charge needed to program the cell must be injected into the floating gate, either by Fowler-Nordheim (*FN*) tunneling [77], or by channel hot-electron (*CHE*) injection. [78].

FN tunneling was discussed in Chaps. 4 and 5. Tunneling through the thin oxide can be induced by applying a high voltage on the control gate, creating a field of about 10^7 V/cm in the oxide (Fig. 8.28). The tunneling current has approximately an exponential dependence on the induced electric field in the gate oxide. When a

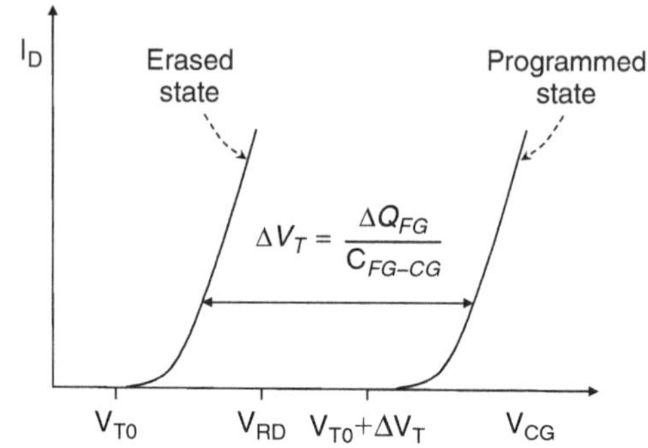

Fig. 8.27 $I_D - V_{CG}$ characteristics for erased and programmed cell

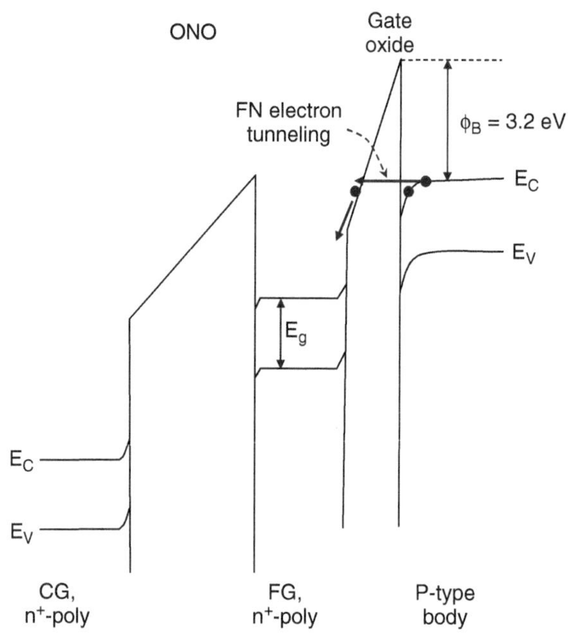

Fig. 8.28 Energy-band diagram of floating-gate cell during programming by Fowler-Nordheim tunneling

positive voltage is applied to the control gate while source, drain and body are kept at ground, tunneling occurs across the entire gate-oxide region.

Programming can also be achieved by hot-electron injection from the high field region near the drain, as shown schematically in Fig. 8.29. To enhance hot-electron injection, the *MOSFET* is biased in the pinch-off mode by applying a high voltage on the drain, for example, $V_D = 5\,\text{V}$, and a high voltage on the control gate, for example, 10 V, while the source and body are at ground. Channel electrons are accelerated by the lateral field between pinch-off point and source. They gain additional kinetic energy as they enter the high-field region near the drain and become "hot" (Chap. 5). A fraction of hot electrons can possess energy in excess of the $Si\text{-}SiO_2$

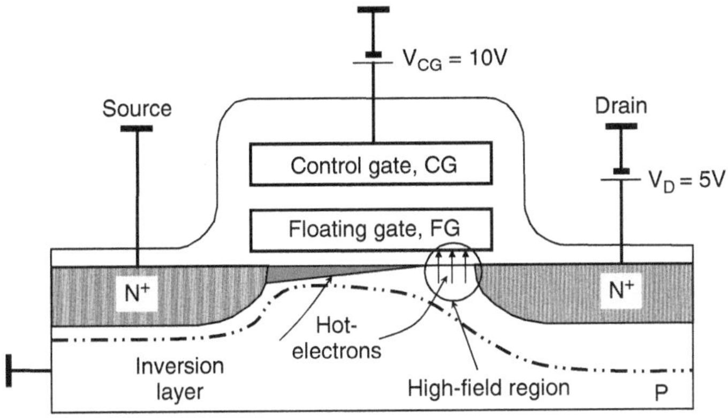

Fig. 8.29 Illustration of programming by hot-electron injection

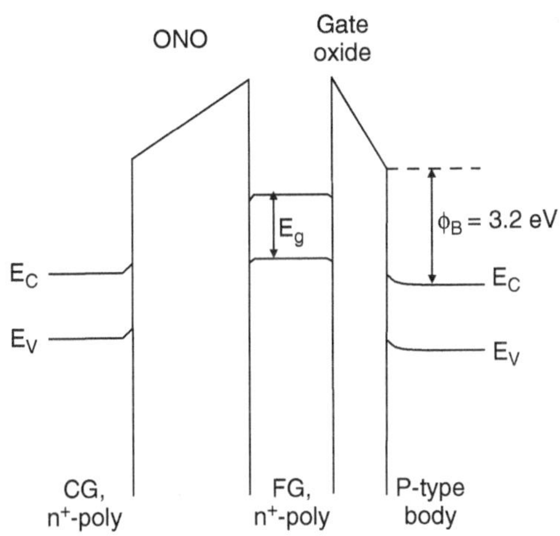

Fig. 8.30 Energy band diagram for a charged (logical "0") floating gate structure

barrier of 3.2 eV and, if directed toward the interface, surmount the barrier and travel to the floating gate. The high voltage on the control gate ensures that an adequate fraction of electrons are directed toward the floating gate and get trapped in it. The energy band diagram for a charged floating gate structure is shown schematically in Fig. 8.30. As can be seen, the floating-gate charge is in a potential well with the gate and inter-poly insulators acting as potential barriers. The probability for electrons to spontaneously overcome the barriers is extremely low. This is why the cell is nonvolatile and has a long retention time.

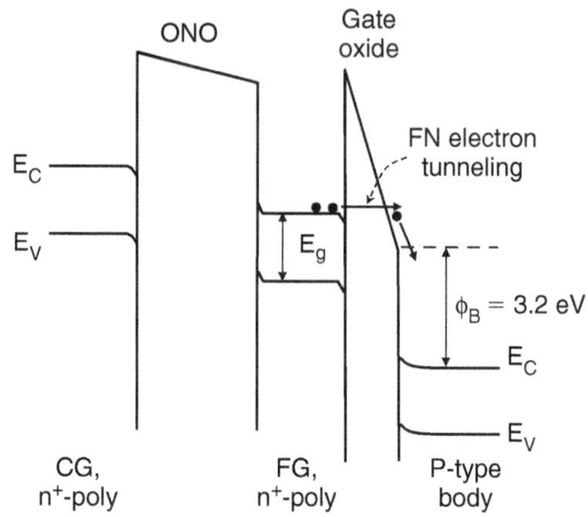

Fig. 8.31 Energy band diagram for a floating gate structure under FN "erase" conditions

Erase Operation

In flash *EEPROM*, all cells in a block must be erased, that is, set to "1," before reprogramming the memory. The most common method to erase the cell is by *FN* tunneling. A large negative voltage is applied to the floating gate with respect to drain or source while keeping the other terminals at ground. The energy band diagram under "erase" conditions is shown in Fig. 8.31. The negative voltage on the control gate creates a sufficiently high field in the gate oxide, allowing trapped electrons to tunnel through the barrier to the drain or source.

NAND Structure

An 8-bit *NAND* structure is illustrated in Fig. 8.32 [79]. The structure can be formed in its own p-well on an n-type substrate for ease of isolation [80]. Its main advantage over an array of individual cells is the elimination of contacts between the cells, considerably reducing the memory size.

To read a particular bit, a low voltage, for example, 1 V is applied to the selected bit-line and a high voltage, for example, 5 V applied to all control-gates except the selected cell. The high voltage ensures that all unselected cells are on, whether programmed or not. The voltage on the control gate of the selected cell is chosen above the threshold voltage of an erased cell, but below the threshold voltage of a programmed cell. It is assumed here that the *MOSFET* is on when $V_{CG} = 0$, that is, V_T is negative. Thus, current is measured between the bit-line and source if a "1" is stored in the selected cell, and no current measured if a "0" is programmed in the cell.

The cells can be block-erased, typically by *FN* tunneling from the floating gates to the drains or p-well. To program a selected cell, a high voltage is applied to the

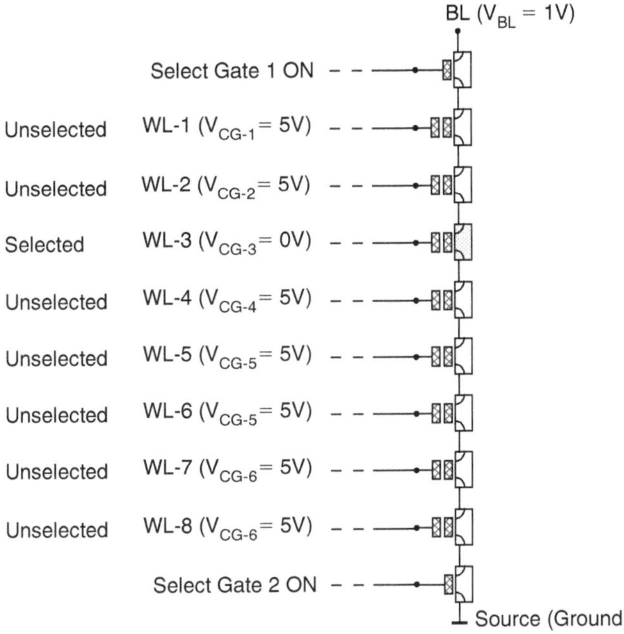

Fig. 8.32 Schematic representation of a *NAND* flash structure (Adapted from [79, 81])

selected bit-line and to all control gates except that of the selected cell where a low voltage is applied. The applied voltages are such that all unselected cells operate in the linear mode while the selected cell operates above pinch-off, creating a high field in the pinch-off region of the selected cell. Thus, programming by hot-electron injection is achieved in the selected cell only.

NOR Structure

In the *NOR* flash structure, the cells on a control-gate line are arranged in parallel between the respective bit-line and common source at ground, as illustrated in Fig. 8.33.

To read a cell, a positive voltage above threshold is applied to the selected control-gate line (word-line, *WL*) while all other word-lines are grounded. Thus, the state of the cell is sensed by the presence or absence of current measured on the bit-line, provided that all unselected (grounded control gates) connected to the same bit-line have a sufficiently high threshold voltage. It is possible, however, that "over-erasing" may result in a positive charge on the floating gate of one of the unselected cells. This would reduce its threshold voltage below zero. In this case, the read operation would be seriously disturbed by leakage through the unselected cell. The problem can be avoided by implementing a split-gate structure that merges a select transistor with the *EEPROM* structure as shown schematically in Fig. 8.34. This

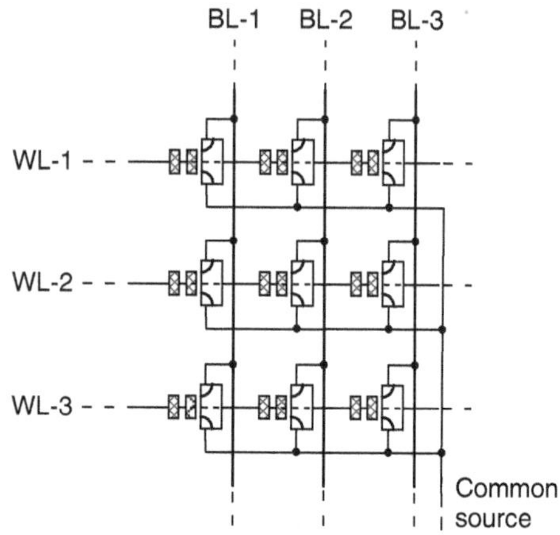

Fig. 8.33 Schematic representation of a *NOR* flash structure (Adapted from [81, 82])

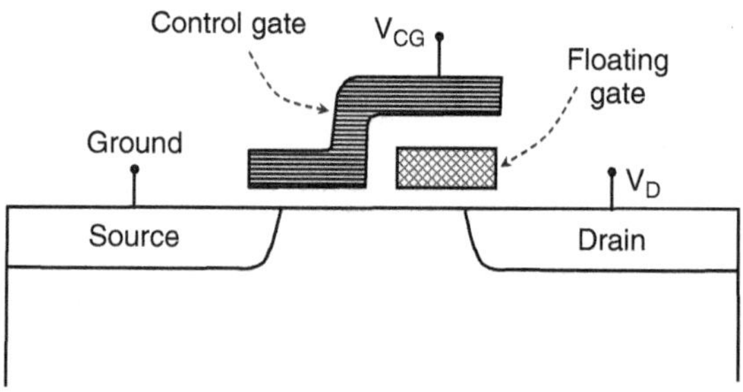

Fig. 8.34 Schematic of split-gate NOR cell (Adapted from [83])

ensures that unselected cells remain turned off regardless of the threshold voltage of the storage device since current is measured only when both parts of the channel surface are inverted [82–84].

In the *NOR* architecture, the cell can be sensed at a higher speed than in the *NAND* architecture because the selected cell is directly connected between bit-line and ground. It can be programmed by *FN* tunneling or channel hot-electron injection at the drain side. Block erase can be accomplished by *FN* tunneling from the floating gate to the silicon surface. The main disadvantage of the *NOR* architecture is the larger memory size because a bit-line contact is needed for every two cells and larger cell-size if the structure in Fig. 8.34 is chosen [82].

Data Retention

Nonvolatile memories are specified to retain data for at least 10 years. Thus, leakage from the floating gate in a programmed cell through the gate oxide or inter-poly dielectric (*IPD*) must be very low at all operating temperatures. Depending on the size of the structure, a threshold voltage shift of 2 V in a programmed cell corresponds to about 10^3–10^4 electrons stored in the floating gate. A loss of 20% of this stored charge (about 2–20 electrons per month) can cause a wrong "read" of the cell and data loss [85]. Triple-layer stacks composed of oxide-nitride-oxide (*ONO*) have been shown to exhibit very low leakage and are commonly used as inter-poly dielectrics [86, 87]. Both FN tunneling and channel hot-electron (*CHE*) injection that are used to program and erase the cell are known to stress the gate dielectric and create traps in the oxide and at its interface with silicon. Traps generated in the oxide by *FN* tunneling and *CHE* can create low-field leakage paths from the floating gate to the silicon surface. This leakage component, referred to as stress-induced leakage current, *SILC*, becomes more important as the oxide thickness is reduced. Charge loss due to *SILC* may prevent the tunnel oxide from being thinned below 8 nm [88, 89].

Program/Erase Endurance

The difference in threshold voltage between a programmed and erased state, referred to as the threshold-voltage window, decreases with time due to drift in V_T caused by programming and erasing (Fig. 8.35) [90]. Program/erase (P/E) cycling can significantly increase the trap density in the gate oxide and oxide-silicon interface. The trapped charge causes the threshold voltage to shift and the threshold-voltage window to decrease. Traps in the oxide also increase leakage from the floating gate to silicon, reducing data retention time. Each P/E cycle can introduce an incremental decrease in the threshold-voltage window. The maximum number of cycles that the cell can withstand before the window closes to such a level that a "1" and a "0" can no longer be distinguished is called "endurance." This number is specified as 10^5–10^6 cycles. The plot in Fig. 8.35 is for programming with *CHE* and erasing by

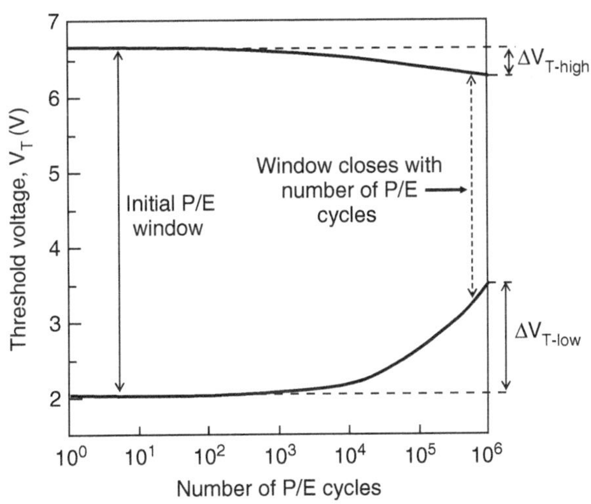

Fig. 8.35 Threshold voltage window closure as a function of program/erase cycles (Adapted from [90])

FN tunneling [90]. The V_T shift is more pronounced for erasing ($\Delta V_{T\text{-}low}$) than programming ($\Delta V_{T\text{-}high}$) because stressing by channel hot-electrons causes less damage than *FN* tunneling [91].

Multi-Level Cell, MLC

The discussion so far focused on a memory element that can store one bit of data in a cell in which two states can be distinguished, a "1" and a "0," by measuring the presence or absence of drain current associated with the absence or presence of electron charge in the floating gate. Multi-level cells (*MLC*) are capable of storing more than one bit per cell by placing controlled amounts of electron charge in the floating gate and sensing the associated magnitude of drain current rather than merely its presence or absence. For example, if four threshold voltages can be distinguished by injecting controlled amounts of charge into the floating gate, two bits of data can be stored in each cell, doubling the memory density [92–94]. The threshold voltage distributions are compared in Fig. 8.36 for a two-state and four-state cell. A current sensing algorithm is used to distinguish between the four states in the two-bit cell, requiring a tight V_T distribution to maintain an acceptable threshold voltage window. The lowest V_T corresponds to unselectively erasing the entire block or sector. Its distribution is highest because of the dispersion in the cell characteristics. Tighter V_T distributions can be obtained for the higher states by programming the cell in controlled voltage pulse increments [85]. The measured current will be different for a "11," "10," and "01" measured at fixed read voltage, V_{RD}, while no current is measured for the "00" level.

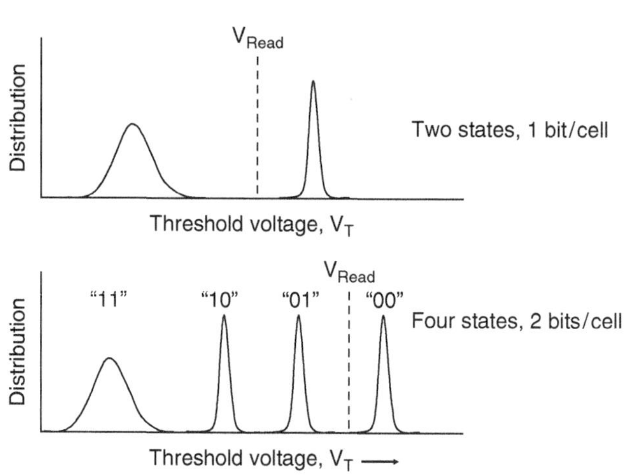

Fig. 8.36 Comparison of threshold voltage distribution of a two-state to a four-state *EEPROM* cell (Adapted from [85])

8.3.3.2 Trapped-Charge Memory

Scaling floating-gate cells to thinner gate oxide and smaller cell-size is limited by the increased charge loss due to *SILC* [95,96] and capacitive coupling between cells causing disturbs [96, 97]. Trapped-charge memory cells, such as polysilicon-oxide-nitride-oxide-silicon (*SONOS*) stacks [98], utilize uniform or localized charge trapping in the nitride rather than charging the floating gate to store information. *SONOS* cells are attractive for high density memories and embedded applications because of the absence of *SILC*, reduced coupling between cells, ease of integration in a *CMOS*-base process, and ease of extension to multi-level cells [100, 101]. The first trapped-charged device was demonstrated in 1967 [99]. A schematic cross-section of a *SONOS* cell is shown in Fig. 8.37. It is simply a *MOSFET* with a composite dielectric that consists of approximately 2–5 nm oxide in contact with silicon, 10-nm silicon-nitride, 5–6 nm blocking oxide, and a polysilicon gate. The blocking oxide is thermally grown on silicon-nitride to a thickness above 5 nm to prevent undesirable charge injection from the polysilicon gate, improving data retention [98].

The cell can be programmed, for example, by drain-sided channel hot electron injection with the source and body grounded, the gate at about +6 V, and the drain at +5 V [102]. Electrons are then trapped in the nitride above a channel region within a distance of about 100-nm from the drain [102]. The average threshold voltage in this region can be increased by about 3 V in 10 μs programming time. The structure thus consists of two channels, one near the drain with the high V_T and the other on the source-side with a "charge-neutral" low V_T. Erasing can be accomplished by applying a negative voltage on the gate with respect to the drain, resulting in hole tunneling, neutralizing the trapped charge. To read the cell, the gate is brought to an intermediate positive voltage V_{RD} above the "natural" V_T and the drain current is measured. A programmed cell will not conduct, indicating a "0" and an erased cell conducts, indicating a "1."

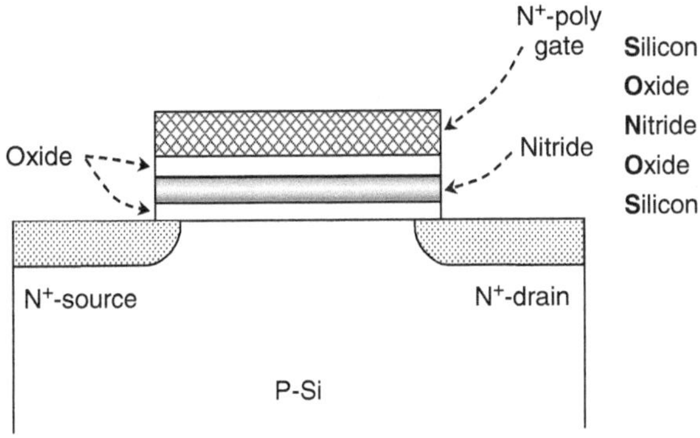

Fig. 8.37 Schematic cross-section of a SONOS structure [103]

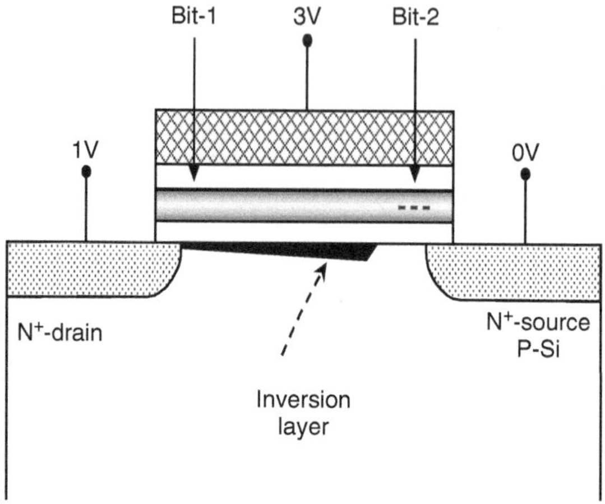

Fig. 8.38 Example of a "read" in reverse direction after right side was programmed. Inversion layer "pinched" by trapped charge, low current measured between source and drain

Multiple-Bit Cells

A two-bit trapped-charge memory cell relies on highly localized trapping and de-trapping of electrons at the drain and source sides in a structure similar to that in Fig. 8.38. Thus, each side will have two states, a charged, high V_T state and a discharged low-V_T state. Programming is achieved by channel hot-electron injection [101, 102]. The left side is programmed by applying, for example, 9 V on the gate and 4.5 V on the right junction that becomes the drain, while grounding the left junction that becomes the source. The cell is read by applying a low read bias of about 3 V on the gate and switching the roles of the junctions with 1–2 V applied to the left junction and the right junction grounded [101]. This ensures that the full effect of the trapped charge is sensed (Fig. 8.38). To program the right junction, the procedure is reversed. The erase operation consists of applying a high voltage on the junction with the gate at ground or at a negative bias. Holes are generated by band-to-band tunneling in silicon in the gate-induced depletion region at the junction boundary. Holes are accelerated by the two-dimensional field and injected into the trapped charge region where they neutralize trapped electrons or are trapped. Techniques to extend the cell capacity to four bits per cell have been reported [104].

8.3.3.3 Phase-Change Memory, PCM

In phase-change memory, *PCM* (also called *PRAM* or Ovonic Unified Memory, *OUM*), information is permanently stored by reversibly changing the resistance of a thin film with current pulses. A small region of a chalcogenide-based material,[3] such as $Ge_2Sb_2Te_2$ (called *GST*), is rapidly switched between a crystalline phase having a low resistance, referred to as the "set-state" and representing a "0," and an amorphous phase with several orders of magnitude higher resistance, referred to as the "reset-state" and representing a "1" [105–115].

The transition of one phase to the other is accomplished by appropriate heating and cooling of the material. Locally heating the material above the melting point with a current pulse, typically above 600°C, causes it to lose all crystalline structure. Cooling it rapidly, at a faster rate than nucleation and crystal growth, prevents it from re-crystallizing [106]. The material can be switched to the crystalline state by heating it to a temperature between the glass transition and the melting point. In this range, the rate of nucleation and growth of crystallites increases rapidly as the melting temperature is approached [107]. The write/read performance is in the nanosecond range and comparable to that of *DRAM* [108, 109]. It improves with decreasing volume of the material to be heated and cooled [108–111]. Figure 8.39 shows a schematic cross-section of a phase-change memory cell [111]. To achieve a high current density, the size of the bottom electrode contact is reduced to the limit allowed by lithography [107–112], or to sub-lithographic dimensions [113],

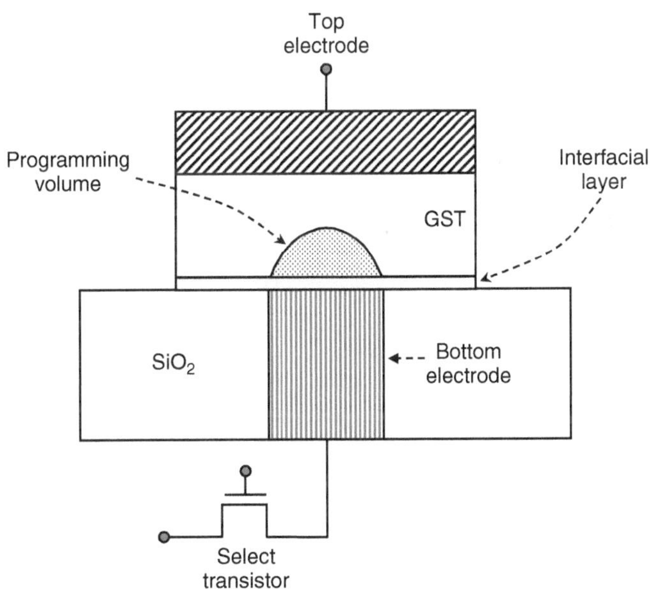

Fig. 8.39 Schematic cross-section of a phase-change cell with tantalum-oxide interfacial layer (Adapted from [111])

[3] Chalcogenides are compounds of chalcogens, such as Selenium (Se) and Tellurium (Te). This is similar to the material used in re-writable optical media (such as CD-RW and DVD-RW). In this case, the optical properties of the material are modified with a laser beam rather than its resistivity [106].

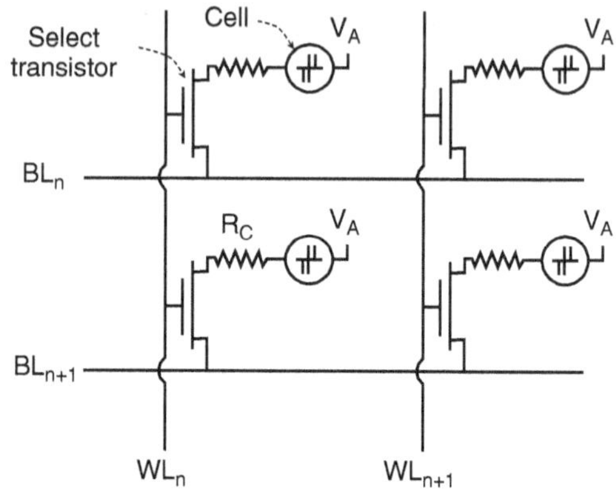

Fig. 8.40 Illustration of possible array of cells connected in rows (*BL*) and columns (*WL*) (Adapted from [107])

reducing the reset current to less than 50μA. In the cell of Fig. 8.39, a very thin tantalum-oxide interfacial layer is deposited to improve adhesion and reduce heat dissipation from the heated *GST* volume to the bottom electrode. Current for heating passes through the interfacial layer by direct tunneling [111].

A section of a memory array is shown schematically in Fig. 8.40 [107]. Every cell is connected to a select transistor, a *MOSFET* or bipolar transistor, which also provides the write and read current. V_A is the applied write or read voltage. The "contact" resistance R_C limits the current when the cell is above the glass transition temperature and highly conductive. In the amorphous state, the material is highly resistive and does not conduct significant current. Thus, to change the cell-state, it is necessary to increase the voltage above a "threshold level" to create a field in the material of approximately 3×10^5 V/cm that allows conduction and Joule heating by a trap-to-trap hopping mechanism referred to as the Poole-Frenkel conduction [107].

Scaling to smaller dimensions will affect the relative attractiveness of the different cell types. As the minimum feature size is scaled beyond the 65-nm "node," for example, the design of floating-gate *EEPROM* memory appears to encounter serious limitations [109]. One important issue with the floating-gate design is the gate-to-gate proximity disturb. Phase-change nonvolatile memory has emerged as a viable alternative to floating-gate flash *EEPROM* and is predicted to scale-down to 22 nm without programming thermal-disturbs between adjacent cells [109]. It is predicted to exhibit better performance and longer endurance than the floating gate flash memory [107–109, 114–117]. Also, the extension of storage capability of a *GST*-based phase-change memory from 1 bit per cell to 2 bits per cell has been demonstrated in [117] by determining two distinct partially crystalline states in addition to the amorphous and "fully" crystalline states.

8.3.3.4 Magneto-Resistive Cell

A memory is universal if it is nonvolatile, exhibits the performance of *SRAM*, and has the density and cost of *DRAM*. Magnetic tunnel junction magnetoresistive

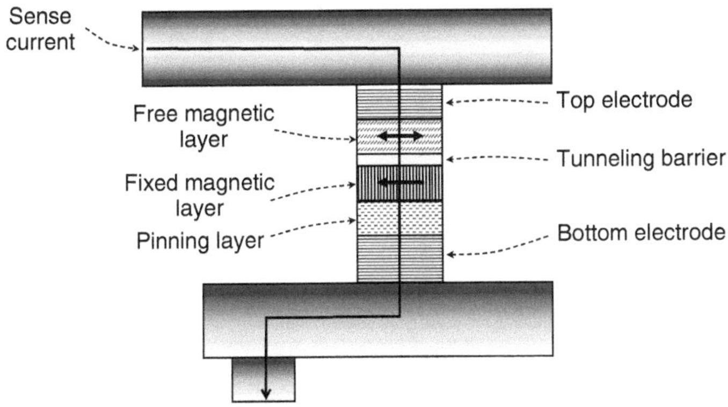

Fig. 8.41 Schematic of an *MTJ-MRAM* cell. Arrows indicate possible direction of polarization (Adapted from [121])

random-access memory (*MRAM* or *MTJ-MRAM*) is found to have the potential of becoming a universal memory technology with the speed of *SRAM*, the nonvolatility of flash memory and storage densities that approach those of *DRAM* [118–124]. *MRAM* appears to exhibit much greater write-erase endurance than flash memory [122]. An *MTJ-MRAM* cell cross-section is shown schematically in Fig. 8.41 [121]. It consists of a stack of two magnetic layers separated by an ultra-thin insulator that acts as a tunneling barrier. The pinning layer keeps the polarization of one of the magnetic layers in a fixed direction. The polarization direction of the "free" magnetic layer can be switched from parallel or anti-parallel to that of the fixed magnetic layer. This is where the information is stored. The resistance of the memory bit can be low or high, depending on the direction of spin-polarization of the free layer with respect to that of the fixed layer. The resistance to tunneling is lowest if the magnetization directions of the two layers are parallel and highest if they are aligned anti-parallel.

The stored information is detected by selecting the cell with, for example, a *MOSFET* [123] (not shown) and sensing the current between the two electrodes. Switching of the magnetic field is produced by forcing current through orthogonal conductors under and above the bit, to create an external magnetic field in the desired direction. The cell is designed such that switching occurs only if current passes through both lines crossing the selected cell but not if current passes through only one line [121].

MRAM has been shown to exhibit a speed similar to that of *SRAM* and a density comparable to that of *DRAM*. It is considerably faster than flash memory and is predicted to have "infinite" retention time and endurance when compared to flash memory. These features make it a potential choice for a "Universal Memory," replacing *SRAM, DRAM*, and flash memory in the future.

8.3.4 BiCMOS for Analog/RF Applications

In Chap. 6, it was shown that the trend to deep submicron and nanoscale *CMOS* results in conflicting technology requirements for digital and analog applications. As the channel length is reduced to improve *CMOS* performance for digital circuits, the power-supply voltage and output resistance (Early voltage) decreases. This conflicts with the analog need for high signal-to-noise ratio, high output resistance to maintain high amplifier gain, and for maximum oscillation frequency, f_{max}, derived for short-channels as [125]

$$f_{\max} = \frac{f_T}{2[\sqrt{(r_G + r_S/r_0)} + 2\pi f_T r_G C_{GD}]}, \tag{8.9}$$

where f_T is the cut-off frequency (gain-bandwidth product) given by (5.73), r_G is the gate resistance, r_S is the source resistance, r_0 is the output resistance, and C_{GD} is the gate to drain capacitance. Equation (8.9) may be considered as an extension of (5.74) derived for long channel *MOSFETs*. It can be seen that as r_0 decreases so does f_{max}. To increase r_0 the channel length must be increased. Thus, there is a trade-off between *CMOS* Early voltage and speed which makes deep submicron and nanoscale *CMOS* not suitable for many analog applications. Also, as the gate oxide thickness is reduced below ~1.5 nm to improve the drive current (Chap. 5), the tunneling gate current increases to a level that becomes prohibitive for low-noise and low power analog applications, such as low-noise amplifiers (*LNA*).

Because of the *CMOS* limitations, many analog designs utilize bipolar transistors to improve performance and reduce power and noise. It was shown in Chap. 3 that introducing *Ge* in the base and an ultra-thin interface oxide (*IFO*) between polysilicon and the single-crystal silicon emitter provides the flexibility of simultaneously optimizing the transistor gain, Early voltage and frequency response. This flexibility and the ease of integrating complementary bipolar transistors with *CMOS* have made *BiCMOS* very attractive for combining digital and analog/RF circuits on the same die. Among high-performance analog/RF designs are low-noise amplifiers, analog-to-digital, and digital-to-analog converters. In addition, isolation between the "quiet" analog and "noisy" digital circuits can be easily and efficiently achieved with silicon-on-insulator (*SOI*) and dielectrically-filled deep-trench merging into the buried oxide (*BOX*) of the *SOI*, as illustrated in Fig. 7.3.

Thus, there is increased interest in *SiGe*-base *BiCMOS* that utilize *CMOS* primarily for digital functions, and complementary bipolar transistors for analog and *RF* functions [126].

8.4 Problems

The temperature is 300 K unless otherwise stated.

1. The PMOS pull-up transistor in a CMOS inverter has the following parameters: $V_T = -0.5\,V$, $W_{eff} = 3\,\mu m$, $L_{eff} = 0.3\,\mu m$, $t_{eq} = 6.5\,nm$, $V_{DD} = +2.5\,V$, $\mu_{eff} = 100\,cm^2/Vs$. Assume the gate to be switched instantaneously to 0 V and estimate the time to charge a load capacitance of 1 pF?

2. An enhancement NMOS pull-up transistor is connected as in Fig. 8.2b. Given: Linear $V_T = +0.4\,V$, $V_{in} = 0\,V$, $V_{DD} = 2.5\,V$, Maximum output on load capacitor $V_{out} = 1.9\,V$. Explain the difference between the magnitudes of V_{DD} and V_{out}.

3. An inverter with a depletion-mode pull-up device has the following characteristics: $V_{DD} = 5\,V$, $V_{SS} = 0\,V$. $V_{T(Pull\text{-}up)} = -3.5\,V$, $V_{T(Pull\text{-}down)} = 0.5\,V$, $\beta_{Pull\text{-}down}/\beta_{Pull\text{-}up} = 10$. The inverter is driven by an identical inverter. Approximate graphically its noise margin NM_H when the input is high, and NM_L when the input is low.

4. What process features should be added to eliminate latch-up in a CMOS constructed on an SOI wafer?

5. A 30-fF storage capacitor of a DRAM cell is charged to 2 V. The bit-line of capacitance 0.4 fF per cell is pre-charged to 1 V. What is the maximum number of cells that can be connected to one bit-line if the minimum read signal on the sense latch should be 50mV?

6. At what maximum wavelength does light begin to erase electron charge in a floating gate?

7. A floating-gate NMOS EEPROM cell has the following parameters: Gate oxide thickness: 10 nm, effective channel dopant concentration: $10^{17}\,cm^{-3}$, inter-poly oxide thickness: 20 nm, n^+-poly floating gate (FG) width and length: 1 μm.

(a) Assume zero oxide, interface, and FG charge. Calculate the voltage that must be applied to the n^+-poly control gate (CG) to turn-on the NMOS.
(b) How many electrons must be injected into the floating gate to raise the turn-on voltage by 4 V?
(c) If the turn-on voltage is allowed to drop by only 2 V in 10 years, how many electrons may be lost by leakage every month?

References

1. N. H. E. Weste and K. Eshraghian, *Principles of CMOS VLSI Design*, Addison Wesley Publishing Company, 1993.
2. C. F. Hill, "Definitions of noise margins in logic systems," Mullard Tech. Commun. 89, 239–245, Sept. 1967.

3. C. F. Hill, "Noise margin and noise immunity in logic circuits," Microelectronics 1 (5), 16–21, 1968.
4. J. R. Hauser, "Noise margin criteria for digital logic circuits," IEEE Trans Edu. 36 (4), 363–368, 1993.
5. J. S. Yuan and L. Yang, "Teaching digital noise and noise margin issues in engineering education," IEEE Trans Edu. 48 (1), 162–168, 2005.
6. DeWitt G. Ong, *Modern MOS Technology, Process, Devices, & Design*, McGraw-Hill Book Company, 1984.
7. W. J. Dennehy, A. G. Holmes-Siedle, and W. F. Leopold, "Transient radiation response of complementary-symmetry MOS integrated circuits," IEEE Trans. Nucl. Sci. NS-16 (6), 114–119, 1969.
8. B. L. Gregory and B. D. Shafer, "Latchup in CMOS integrated circuits," IEEE Trans. Nucl. Sci. NS-20 (6), 293–299, 1973.
9. D. B. Estreich, "The physics and modeling of latch-up and CMOS integrated circuits," Tech. Rept. No. G-201-9, Nov. 1980, Stanford University, Stanford, California.
10. J. J. Ebers, "Four-terminal p-n-p-n transistors," Proc. IRE 40 (11), 1361–1364, 1952.
11. J. L. Moll, M. Tanenbaum, J. M. Goldley, and N. Holonyak, "P-N-P-N transistor switches," Proce. IRE 44 (9), 1174–1182, 1956.
12. W. D. Raburn, "A model for the parasitic SCR in bulk CMOS," IEEE IEDM Tech. Digest. 252–255, 1980.
13. R. D. Rung and H. Momose, "DC holding and dynamic triggering characteristics of bulk CMOS latchup," IEEE Trans. Electron Dev. ED-30 (12), 1647–1655, 1983.
14. R. C.-Y. Fang and J. L. Moll, "Latch-up model for the parasitic p-n-p-n path in bulk CMOS," IEEE Trans. Electron Dev. ED-31 (1), 113–120, 1984.
15. G. J. Hu, "A better understanding of CMOS latch-up," IEEE Trans. Electron Dev. ED-31 (1), 62–67, 1984.
16. R. R. Troutman, *Latch-up in CMOS Technology*, Kluwer Academic Publishers, 1986.
17. D. B. Estreich, A. Ochoa, Jr., and R. W. Dutton, "An analysis of latch-up prevention in CMOS IC's using an epitaxial-buried layer process," IEEE IEDM Tech. Digest. 230–234, 1978.
18. H.-Y. Lin and C. H. Ting, "Improvement of CMOS latch-up immunity using a high energy implanted buried layer," Nucl. Instrum. Methods Phys. Res., B37/38, 960–964, 1989.
19. M.-J. Chen and C.-Y. Wu, "A simplified computer analysis for n-well guard ring efficiency in CMOS circuits," Solid-State Electron. 30 (8), 879–882, 1987.
20. D. Tremouilles, M. I. Natarajan, M. Scholz, N. Azilah, M. Bafleur, M. Sawada, T. Hasebe, and G. Groeseneken, "A novel method for guard ring efficiency assessment and its applications for ESD protection, design and optimization," IEEE IRPS 606–607, 2007.
21. L. J. McDaid, S. Hall, W. Eccleston, and J. C. Alderman, "Suppression of latch up in SOI MOSFETs by silicidation of source," Electron. Lett. 27 (11), 1003–1005, 1991.
22. R. Alvarez, *BiCMOS Technology and Applications*, Kluwer Academic Publishers, 1993.
23. S. H. K. Embabi, A. Bellaouar, and M. I. Elmasry, *Digital BiCMOS Integrated Circuits*, Kluwer Academic Publishers, 1993.
24. R. H. Dennard, "Field-effect transistor memory," US Patent 3, 387, 286, June 4, 1968.
25. V. L. Rideout, "One-device cells for dynamic random-access memories: a tutorial," IEEE Trans. Electron. Dev. ED-26 (6), 839–852, 1979.
26. D. S. Yaney, C. Y. Lu, R. A. Kohler, M. J. Kelly, and J. T. Nelson, "A meta-stable leakage phenomenon in DRAM charge storage: variable hold time," IEEE IEDM Tech. Digest. 336–339, 1987.
27. P. J. Restle, J. W. Park, and B. F. Lloyd, "DRAM variable retention time," IEEE IEDM Tech. Digest. 807–810, 1992.
28. Y. Mori, K. Ohyu, K. Okonogi, and R.-I. Yamada, "The origin of variable retention time in DRAM," IEEE IEDM Tech. Digest. 1034–1037, 2005.
29. T. C. May and M. H. Woods, "Alpha-particle-induced soft errors in dynamic memories," IEEE Trans. Electron. Dev. ED-26 (1), 2–9, 1979.
30. J. F. Ziegler and W. A. Lanford, "The effect of sea-level cosmic rays on electronic devices," J. Appl. Phys. 62 (6), 4205–4215, 1981.

31. Y. Tosaka, S. Satoh, T. Itakura, H. Ehara, T. Ueda, G. A. Woffinden, and S. A. Wender, "Measurement and analysis of neutron-induced soft errors in sub-half-micron CMOS circuits," IEEE Trans. Electron. Dev. 45 (7), 1453–1458, 1998.
32. P. Hazucha, T. Kamik, J. Maiz, S. Walstra, B. Bloechel, J. Tschanz, G. Dermer, S. Hareland, P. Armstrong, and S. Borkar, "Neutron soft error rate measurements in a 90-nm CMOS process and scaling trends in SRAM from 0.25-μm to 90-nm generation," IEEE IEDM Tech. Digest. 523–526, 2003.
33. L. Nebit, J. Alsmeier, B. Chen, J. DeBrosse, P. Fahey, M. Gall, J. Gambino, S. Gernhardt, H. Ishiuchit, R. Kleinhenz, J. Mandelman, T. Mii, M. Morikado, A. Nitayama, S. Parke, H. Wong, and G. Bronner, "A 0.6 μm^2 256Mb trench DRAM cell with self-aligned buried strap (BEST)," IEEE IEDM Tech. Digest. 627–630, 1993.
34. W. F. Richardson, D. M Bordelon, G. P. Pollack, A. H. Shah., S. D. S. Malhi, H. Shichijo, S. K. Brlnerjee, M. Elahy, R. H. Womack, C.-P. Wang, J. Gallia, H. E. Davis, and P.K. Chattarjee, " A Trench transistor cross-point DRAM cell," IEEE IEDM Tech. Digest. 714–717, 1985.
35. U. Gruening, C. J. Radens. J. A. Mandelman, A. Michaelis, M. Seitz, N. Arnold, D. Lea, D. Casarotto, A. Knorr, S. Halle, T. H. Ivers, L. Economikos, S. Kudelka, S. Rahn, H. Tews, H. Lee, R. Divakaruni, J. J. Welser, T. Furukawa, T. S. Kanarsky, J. Alsmeier, and G. B. Bronner, "A novel trench DRAM cell with a VERtIcal access transistor and BuriEd STrap (VERI BEST) for 4Gb/l6Gb," IEEE IEDM Tech. Digest. 25–28, 1999.
36. R. Weis, K. Hummler, H. Akatsu, S. Kudelka, T. Dyer, M. Seitz, A. Scholtz, B. Kim, M. Wise, R. Malik, J. Strane, Th. Goebel, K. McStay, J. Beintner, N. Arnold, R. Gerber, B. Liegl, A. Knorr, L. Economikos, A. Simpson, W. Yan, D. Dobuzinski, J. Mandelman, L. Nesbit, C. J. Radens, R. Divakaruni, W. Bergner, G. Bronner, and W. Mueller, "A highly cost efficient 8F2 DRAM cell with a double gate vertical transistor device for 100 nm and beyond," IEEE IEDM Tech. Digest. 415–418, 2001.
37. T. Schlosser, D. Manger, R. Weis, S. Slesazeck, F. Lau, S. Tegen, M. Sesterhenn, K. Meummler, J. Nuetzel, D. Temmler, B. Kowalski, U. Scheler, M. Stavrev, and D. Koehler, "Highly scalable sub-50 nm vertical double gate trench DRAM cell," IEEE IEDM Tech. Digest. 57–60, 2004.
38. C. J. Radens, U. Gruening, J. A. Mandelman, M. Seitz, T. Dyer, D. Lea, D. Casarotto, L. Clevenger, L. Nesbit, R. Malik, S. Halle, S. Kudelka, H. Tews, R. Divakaruni, J. Sim, A. Strong, D. Tibbel, N. Arnold, S. Bukofsky, J. Preuninger, G. Kunkel, and G. Bronner, "A 0.135 μm^2 $6F^2$ trench-sidewall vertical device cell for 4Gb/16Gb DRAM," IEEE Symp. VLSI Tech. Digest. 81–82, 2000.
39. J. Y. Kim, C. S. Lee, S. E. Kim, I. B. Chung, Y. M. Choi, B. J. Park, J. W. Lee, D. I. Kim, Y. S. Hwang, D. S. Hwang, H. K. Hwang, J. M. Park, D. H. Kim, N. J. Kang, M. H. Cho, M. Y. Jeong, H. J. Kim, J. N. Han, S. Y. Kim, B.Y. Nam, H.S. Park, S.H. Chung, J. H. Lee, J. S. Park, H. S. Kim, Y. J. Park, and K. Kim, "The breakthrough in data retention time of DRAM using recess-channel-array transistor (RCAT) for 88 nm feature size and beyond," IEEE Symp. VLSI Tech. Digest. 11–12, 2003.
40. I.-G. Kim, S.-H. Park, J.-S. Yoon, D.-J. Kim, J.-Y. Noh, J.-H. Lee, Y.-S. Kim, M.-W. Hwang, K.-H. Yang, J. Park, and K. Oh, "Overcoming DRAM scaling limitations by employing straight recessed channel array transistors with <100> uni-axial and {100} uni-plane channels," IEEE IEDM Tech. Digest. 319–322, 2005.
41. D.-H. Lee, B.-C. Lee, I.-S. Jung, T.-J. Kim, Y.-H. Son, S.-G. Lee, Y.-P. Kim, S. Choi, U.-I. Chung, and J.-T. Moon, Fin-channel-array transistor (FCAT) featuring sub-70 nm low power and high performance DRAM," IEEE IEDM Tech. Digest. 407–410, 2003.
42. M. J. Lee, S. Jin, C.-K. Baek, S.-M. Hong, S.-Y. Park, H.-H. Park, S.-D. Lee, S.-W. Chung, J.-G. Jeong, S.-J. Hong, S.-W. Park, I.-Y. Chung, Y. J. Park, and H. S. Min, "A proposal on an optimized device structure with experimental studies on recent devices for the DRAM cell transistor," IEEE Trans. Electron Dev. 54 (12) 3325–3335, 2007.
43. P. S. Parkinson, K. Settlemyer, L. McStay, D.-G. Park, R. Ramachandran, M. Chudzik, K. Cheng, C.-Y. Sung, F. Chen, A. Strong, P. Papworth, and R. Jammy, "Novel techniques

for scaling deep trench DRAM capacitor technology to 0.11 μm and beyond," IEEE Symp. VLSI Tech. Digest. 21–22, 2003.
44. J. Amon, A Kieslich, L. Heineck, T. Schuster, J. Faul, J. Luetzen, C. Fan, C.-C. Huang, B. Fischer, G. Enders, S. Kudelka, U. Schroeder, K.-H. Kuesters, G. Lange, and J. Alsmeier, "A highly manufacturable deep trench based DRAM cell layout with a planar array device in a 70 nm technology," IEEE IEDM Tech. Digest. 73–76, 2004.
45. H. Watanabe, T. Tatsumi, S. Ohnisihi, T. Hamada, I. Honma, and T. Kikkawa, "A new cylindrical capacitor using hemispherical grained Si (HSG-Si) for 256 Mb DRAMS," IEEE IEDM Tech. Digest. 259–262, 1992.
46. T. Sanuki, Y. Sogo, A. Oishi, Y. Okayama, R. Hasumi, Y. Morimasa, T. Kinoshita, T. Komoda, H. Tanaka, K. Hiyama, T. Komoguchi, T. Matsumoto, K. Oota, T. Yokoyama, K. Fukasaku, R. Katsumata1, M. Kido1, M. Tamura, Y. Takegawa, H. Yoshimura, K. Kasai, K. Ohno, M. Saito, H. Aochi, M. Iwai, N. Nagashima, F. Matsuoka, Y. Okamoto, and T. Noguchi, "High density and fully compatible embedded DRAM cell with 45 nm CMOS Technology (CMOS6)," IEEE VLSI Tech. Digest. 14–15, 2005.
47. Y.-H. Wu, C.-M. Chang, C.-Y. Wang, C.-K. Kao, C.-M. Kuo, A. Ku, and T. Huang, "Augmented cell performance of NO-based storage dielectric by N_2O-treated nitride film for trench DRAM," IEEE Electron Dev. Lett. 29 (2), 149–152, 2008.
48. J. Lützen, A. Birner, M. Goldbach, M. Gutsche, T. Hecht, S. Jakschik, A. Orth, A. Sänger, U. Schröder, H. Seidl, B. Sell, and D. Schumann, "Integration of capacitor for sub-100-nm DRAM trench technology," IEEE VLSI Tech. Digest. 178–179, 2002.
49. G. Aichmayr, A. Avellán, G. S. Duesberg, F. Kreupl, S. Kudelka, M. Liebau, A. Orth, A. Sänger, J. Schumann, and O. Storbeck, "Carbon/high-k trench capacitor for the 40nm DRAM generation," IEEE VLSI Tech. Digest. 186–187, 2007.
50. M. Koyanagi, H. Sunami, N. Hashimoto, and M. Ashikawa, "Novel high density, stacked capacitor MOS RAM," IEEE IEDM Tech. Digest. 348–351, 1978.
51. Sakai and T. Tatsumi, "Novel seeding method for the growth of polycrystalline Si films with hemispherical grains," Appl. Phys. Lett. 61 (2), 159–161, 1992.
52. H. Watanabe, N. Aoto, S. Adachi, and T. Kikkawa, "Device application and structure observation for hemispherical-grained Si," J. Appl. Phys. 71, 3538–3543, 1992.
53. S. Yamamichi, P.-Y. Lesaicherre, H. Yamaguchi, K. Takemura, S. Sone, H. Yabuta, K. Sato, T. Tamura, K. Nakajima, S. Ohnishi, K. Tokashiki, Y. Hayashi, Y. Kato, Y. Miyasaka, M. Yoshida, and H. Ono, "A stacked capacitor technology with ECR plasma MOCVD (Ba,Sr)TiO and RuO/Ru/TiN/TiSi storage nodes for Gb-scale DRAMs," IEEE Trans. Electron. Dev. 44 (7) 1076–1083, 1997.
54. K. N. Kim, H. S. Jeong, W. S. Yang, Y. S. Hwang, C. H. Cho, M. M. Jeong, S. Park, S. J. Ahn, Y. S. Chun, S. H. Shin, J. S. Park, S. H. Song, J. Y. Lee, S. M. Jang, C. H. Lee, J. H. Jeong, M. H. Cho, H. I. Yoon, and J. S. Jeon, "Highly manufacturable and high performance SDR/DDR 4 Gb DRAM," IEEE VLSI Tech. Digest 7–8, 2001.
55. J. M. Park, Y. S. Hwang, H. K. Hwang, S. H. Lee, G. Y. Kim, M. Y. Jeong, B. J. Park, S. E. Kim, M. H. Cho, D. I. Kim, J.-H. Chung, I. S. Park, C.-Y. Yoo, J. H. Lee, B. Y. Nam, Y. R. Park, C.-S. Kim, M.-C. Sun, J.-H. Ku, S. Choi, H. S. Kim, Y. G. Park, and K. Kim, "A novel robust TiN/AHO/TiN capacitor CoSi2 cell pad structure for 70 nm stand-alone and embedded DRAM technology and beyond," IEEE IEDM Tech. Digest. 823–836, 2002.
56. Berthelot, C. Caillat, V. Huard, S. Barnola, B. Boeck, H. Del-Puppo, N. Emonet, and F. Lalanne, "Highly reliable TiN/ZrO_2/TiN 3D stacked capacitors for 45 nm embedded DRAM technologies," Device Res. Conf. (DRC) 343–346, 2006.
57. Y. Fukaura, K. Kasai, Y. Okayama, H. Kawasaki, K. Isobe, M. Kanda, K. Ishimaru, and H. Ishiuchi, "A highly manufacturable high density embedded SRAM technology for 90 nm CMOS," IEEE IEDM Tech. Digest. 515–418, 2002.
58. Z. Guo, S. Balasubramanian, R. Zlatanovici, T.-J. King, and B. Nikolić, "FinFET-based SRAM design," International Symposium on Low Power Electronics and Design (ISLPED) 2–7, 2005.
59. B. H. Calhoun and A. P. Chandrakasan, "Static noise margin variation for sub-threshold SRAM in 65-nm CMOS," IEEE J. Solid-State Circuits 41 (7), 1673–1679, 2006.

60. E. Seevinck, F. List, and J. Lohstroh, "Static noise margin analysis of MOS SRAM cells," IEEE J. Solid-State Circuits SC-22 (5), 748–754, 1987.
61. J. Lohstroh, E. Seevinck, and J. De Groot, "Worst-case static noise margin criteria for logic circuits and their mathematical equivalence," IEEE J. Solid-State Circuits SC-18 (6), 803–807, 1983.
62. X. Wu, P. C. H. Chan, S. Zhang, C. Feng, and M. Chan, "A three-dimensional stacked Fin-CMOS technology for high-density ULSI circuits," IEEE Trans. Electron Dev. 52 (9), 1998–2003, 2005.
63. K.-L. Cheng, C. C. Wu, Y. P. Wang, D. W. Lin, C. M. Chu, Y. Y. Tamg, S. Y. Lu, S. J. Yang, M. H. Hsieh, C. M. Liu, S. P. Fu, J. H. Chen, C. T. Lin, W. Y. Lien, H. Y. Huang, P. W. Wang, H. H. Lin, D. Y. Lee, M. J. Huang, C. F. Nieh, L. T. Lin, C. C. Chen, W. Chang, Y. H. Chiu, M. Y. Wang, C. H. Yeh, F. C. Chen, C. M. Wu, Y. H. Chang, S. C. Wang, H. C. Hsieh, M. D. Lei, K. Goto H. J. Tao, M. Cao, H. C. Tuan, C. H. Diaz, and Y. J. Mii "A highly scaled, high performance 45 nm bulk logic CMOS technology with $0.242\,\mu m^2$ SRAM cell," IEEE IEDM Tech. Digest. 243–246, 2007.
64. K. Zhang, F. Hamzaoglu, and Y. Wang, "Low-power SRAMs in nanoscale CMOS technologies," IEEE Trans. Electron Dev. 55 (1), 145–151, 2008.
65. T. Miyashita, K. Ikeda, Y S. Kim, T. Yamamoto, Y. Sambonsugi, H. Ochimizu, T. Sakoda, M. Okuno, H. Minakata, H. Ohta, Y Hayami, K. Ookoshi, Y Shimamune, M. Fukuda, A. Hatada, K. Okabe, T. Kubo, M. Tajima, T. Yamamoto, E. Motoh, T. Owada, M. Nakamura, H. Kudo, T. Sawada, J. Nagayama, A. Satoh, T. Mori, A. Hasegawa, H. Kurata, K. Sukegawa, A. Tsukune, S. Yamaguchi, K. Ikeda, M. Kase, T. Futatsugi, S. Satoh, and T. Sugii, "High-performance and low-power bulk logic platform utilizing FET specific multiple-stressors with highly enhanced strain and full-porous low-k interconnects for 45-nm CMOS technology," IEEE VLSI Tech. Digest. 251–252, 2007.
66. S. Inaba, H. Kawasaki, K. Okano, T. Izumida, A. Yagishita, A. Kaneko, K. Ishimaru, N. Aoki, and Y. Toyoshima, "Direct evaluation of DC characteristic variability in FinFET SRAM cell for 32 nm node and beyond," IEEE IEDM Tech. Digest. 487–490, 2007.
67. S.-M. Jung, H. Lim, W. Cho, H. Cho, H. Hong, J. Jeong, S. Jung H. Park, B. Son, Y. Jang, and K. Kim, "Soft error immune $0.46\,\mu m^2$ SRAM cell with MIM node capacitor by 65 nm CMOS," IEEE IEDM Tech. Digest. 280–292, 2003.
68. E. Ootsuka, M. Nakamura, T. Miyake, S. Iwahashi, Y. Ohira, T. Tamaru, K. Kikushima, and K. Yamaguchi, "A novel $0.20\,\mu m$ full CMOS SRAM cell using stacked cross couple with enhanced soft error immunity," IEEE IEDM Tech. Digest. 205–208, 1998.
69. M. Hashimoto, N. Nagashima, Y. Miyazawa, M. Shimanoe, H. Satoh, and T. Matsushita "Small geometry SO1 CMOS cell technology for high density SRAMs," IEEE IEDM Tech. Digest. 973–976, 1991.
70. T.-S. Park, H. J. Cho, J. D. Choe, S. Y. Han, D. Park, K. Kim, E. Yoon, and J.-H. Lee, "Characteristics of the full CMOS SRAM cell using body-tied TG MOSFETs (Bulk FinFETs)," IEEE Trans. Electron Dev. 53 (3), 481–487, 2006.
71. R. Baumann, "The impact of technology scaling on soft error rate performance and limits to the efficacy of error correction," IEEE IEDM Tech. Digest. 329–332, 2002.
72. P. E. Dodd and L. W. Massengill, "Basic mechanisms and modeling of single-event upset in digital microelectronics," IEEE Trans. Nucl. Sci. 50 (3), 583–602, 2003.
73. Y. Tosaka, H. Kanata, S. Satoh, and T. Itakura, "Simple method for estimating neutron-soft error rates based on modified BGR model," IEEE Electron Dev. Lett. 20 (2), 89–91, 1999.
74. K. Takeuchi, R. Koh, and T. Mogami, "A study of the threshold voltage variation for ultra-small bulk and SOI CMOS," IEEE Trans. Electron. Dev. 48 (9), 1995–2000, 2001.
75. F. Masuoka, M. Asano, H. Iwahashi, T. Komuro, and S. Tanaka, "A new flash E^2PROM cell using triple polysilicon technology," IEEE IEDM Tech. Digest. 464–467, 1984.
76. D. Khang and S. M. Sze, "A floating gate and its application to memory devices," Bell Syst. Tech. J. 46, 1283–1286, 1967.
77. J. R. Yeargain and C. Kuo, "High density floating-gate EEPROM cell," IEEE IEDM Tech. Digest. 24–27, 1981.

78. D. C. Guterman, I. H. Rimawi, T.-L. Chiu, R. D. Halvorson, and D. J. McElroy, "An electrically alterable nonvolatile memory cell using a floating-gate structure," IEEE Trans. Electron. Dev. ED-26 (4), 576–586, 1979.
79. F. Masuoka, M. Momodomi, Y. Iwata, and R. Shirota, "New high-density EPROM and flash EEPROM cell with NAND structure cell," IEEE IEDM Tech. Dig. 552–555, 1987.
80. R. Kirisawa, S. Aritome, R. Nakayama, T. Endoh, R. Shirota, and F. Masuoka, "ANAND structures cell with new programming technology for highly reliable 5 V only flash EEPROM," IEEE VLSI Tech. Digest. 129–130, 1990.
81. W. D. Brown and J. E. Brewer, Eds., *Nonvolatile Semiconductor Memory Technology*, IEEE Press, New York, 1998.
82. B. Riccò. G. Torelli, M. Lanzoni, A. Manstretta, H. E. Maes, D. Montarani, and A. Modelli, "Nonvolatile memories for digital applications," Proc. IEEE 86 (12), 2399–2420, 1998.
83. F. Masuoka, M. Asano, H. Iwahashi, T. Komuro, and S. Tanaka, "A new flash E^2PROM cell using triple polysilicon technology," IEEE IEDM Tech. Digest. 464–467, 1984.
84. G. Samachisa, C.-S. Su, Y.-S. Kao, G. Samarandoiu, C.-Y. M. Wong, and C. Hu, "A 128 K flash EEPROM using double-polysilicon technology," IEEE J. Solid-State Technol. 22 (5) 676–683, 1987.
85. R. Bez, E. Camerlenghi, A. Modelli, and A. Visconti, "Introduction to flash memory," Proc. IEEE 91 (4), 489–502, 2003.
86. S. Mori, Y. Kaneko, N. Arai, Y. Ohshima, H. Araki, K. Narita, E. Sakagami, and K. Yoshikawa, "Reliability study of thin inter-poly dielectric for nonvolatile memory applications," IEEE IRPS 132–144, 1990.
87. S. Mori, E. Sakagami, H. Araki, Y. Kaneko, K. Narita, Y. Ohshima, N. Arai, and K. Yoshikawa, "ONO inter-poly dielectric scaling for nonvolatile memory applications," IEEE Trans. Electron. Dev. 38 (2), 386–391, 1991.
88. K. Naruke, S. Taguchi, and M. Wada, "Stress induced leakage current limiting to scale down EEPROM tunnel oxide thickness," IEEE IEDM Tech. Digest. 424–427, 1988.
89. S. Takagi, N. Yasuda, and A. Toriumi, "Experimental evidence of inelastic tunneling and new I-V model for stress-induces leakage current," IEEE IEDM Tech. Digest. 323–326, 1996.
90. N.-K. Zous, Y.-J. Chen, C.-Y. Chin, W.-J. Tsai, T.-C. Lu, M.-S. Chen, W.-P. Lu, T. Wang, S. C. Pan, and C.-Y. Lu, "An endurance evaluation method for flash EEPROM," IEEE Trans. Electron. Dev. 51 (5), 720–725, 2004.
91. M. Suhail, T. Harp, J. Bridwell, and P. J. Kuhn, "Effects of Fowler-Nordheim tunneling stress vs. channel hot electron stress on data retention characteristics of floating gate non-volatile EEPROM," IEEE IRPS Proc. 439–440, 2002.
92. C. Bleiker and H. Melchior, "A four-state EEPROM using floating-gate memory cells," IEEE J. Solid-State Circuits SC22 (3), 460–463, 1987.
93. M. Bauer, R. Alexis, G. Atwood, B. Baltar, A. Fazio, K. Frary, M. Hensel, M. Ishac, J. Javanifard, M. Landgraf, D. Leak, K. Loe, D. Mills, P. Ruby, R. Rozman, S. Sweha, S. Talreja, and K. Wojciechowski, "A multilevel-cell 32 Mb flash memory" IEEE ISSCC 132–133, 1995.
94. J.-H. Park, S.-H. Hur, J.-H. Lee, J.-T. Park, J.-S. Sel, J.-W. Kim, S.-B. Song, J.-Y. Lee, J.-H. Lee, S.-J. Son, Y.-S. Kim, M.-C. Park, S.-J. Chai, J.-D. Choi, U.-I. Chung, J.-T. Moon, K.-T. Kim, K. Kim, and B.-I. Ryu, "8 Gb MLC (Multi-Level Cell) NAND flash memory using 63 nm process technology," IEEE IEDM Tech. Digest. 873–876, 2004.
95. J. De Blauwe, J. Van Houdt, D. Wellekens, R. Degraeve, Ph. Roussel, L. Haspeslagh, L. Deferm, G. Groeseneken, and H.E. Maes, "A new quantitative model to predict SILC-related disturb characteristics in Flash E^2PROM devices," IEEE IEDM Tech. Digest. 343–346, 1996.
96. S. Lai, "Flash memories: where we were and where we are going", IEEE IEDM Tech. Digest. 971–973, 1998.
97. Ghetti, L. Bortesi, and L. Vendrame, "3D simulation study of gate coupling and gate cross-interference in advanced floating-gate non-volatile memories," Solid-State Electron. 49 (11), 1805–1812, 2005.
98. P. C. Y. Chen, "Threshold-alterable Si-gate devices," IEEE Trans. Electron. Dev. ED-24 (5), 584–585, 1977.

99. H. A. R. Wegener, A. J. Lincoln, H. C. Pao, M. R. O'Connell, R. E. Oleksiak, and H. Law, "The variable threshold transistor, a new electrically alterable, non-destructive read-only storage device," IEEE IEDM Tech. Digest. 70, 1967.
100. K.-T. Chang, W.-M. Chen, C. Swift, J. M. Higman, W. M. Paulson, and KM Chang, "A New SONOS memory using source-side injection for programming," IEEE Electron. Dev. Lett. 19 (7), 253–255, 1998.
101. B. Eitan, P. Pavan, I. Bloom, E. Aloni, A. Frommer, and D. Finzi, "NROM: a novel localized trapping, 2-bit nonvolatile memory cell," IEEE Electron. Dev. Lett. 21 (11), 543–545, 2000.
102. C. T. Swift, G. L. Chindalore, K. Harber, T. S. Harp, A. Hoefler, C. M. Hong, P. A. Ingersoll, C. B. Li, E. J. Prinz, and J. A. Yater, "An embedded 90 nm SONOS nonvolatile memory utilizing hot electron programming and uniform tunnel erase," IEEE IEDM Tech. Digest. 927–930, 2002.
103. T. Y. Chan, K. K. Young, and C. Hu, "A true single-transistor oxide-nitride-oxide EEPROM device," IEEE Electron. Dev. Lett. EDL-8 (3), 93–95, 1987.
104. C. W. Oh, S. H. Kim, N. Y. Kim, Y. L. Choi, K. H. Lee, B. S. Kim, N. M. Cho, S. B. Kim, D. W. Kim, D. Park, and B. I. Ryu, "A 4-Bit double SONOS memory (DSM) with 4 storage nodes per cell for ultimate multi-bit operation," IEEE VLSI Tech. Digest. 40–41, 2006.
105. S. R. Ovshinski, "Reversible electrical switching phenomena in disordered structures," Phys. Rev. Lett. 21 (20), 1450–1453, 1968.
106. N. Yamada, E. Ohno, K. Nishiuchi, N. Akahira, and M. Takao, "Rapid-phase transitions of GeTe-Sb_2Te_3 pseudobinary amorphous thin films for an optical disk memory," J. Appl. Phys. 69 (5), 2849–2856, 1991.
107. G. Wicker, "Nonvolatile, high density, high performance phase change memory," SPIE 3891, 2–9, 1999.
108. S. Lai and T. Lowrey, "OUM – a 180 nm nonvolatile memory cell element technology for stand alone and embedded applications," IEEE IEDM Tech. Digest. 803–806, 2001.
109. S. Lai, "Current status of the phase change memory and its future," IEEE IEDM Tech. Digest. 255–258, 2003.
110. Y. N. Hwang, S. H. Lee, S. J. Ahn, S. Y. Ryoo, H. S. Hoong, H. C. Koo, F. Yeung, J. H. Oh, H. J. Kim, W. C. Jeong, J. H. Park, H. Horii, Y. Ha, J. H. Yi, G. H. Koh, G. T. Jeong, H. S. Jeong, and K. Kim, "Writing current reduction for high-density phase-change RAM," IEEE IEDM Tech. Digest. 893–896, 2003.
111. Y. Matsui, K. Kurotsuchi, O. Tonomura, T. Morikawa, M. Kinoshita, Y. Fujisaki, N. Matsuzaki, S. Hansawa, M. Terao, N. Takaura, H. Moriya, T. Iwasaki, M. Moniwa, and T. Koga, "Ta_2O_3 interfacial layer between GST and W plug enabling low power operation of phase change memories," IEEE IEDM Tech. Digest. 1–4, 2006.
112. L. Lacaita, A. Radaelli, D. Ielmini, F. Pellizzer, A. Pirovano, A. Benvenuti, and R. Bez, "Electrothermal and phase-change dynamics in chalcogenide-based memories," IEEE IEDM Tech. Digest. 911–914, 2004.
113. D. L. Kencke, I. V. Karpov, B. G. Johnson, S. J. Lee, D.C. Kau, S. J. Hudgens, J. P. Reifenberg, S. D. Savransky, J. Zhang, M. T. Giles, and G. Spadini, "The role of interfaces in damascene phase-change memory," IEEE IEDM Tech. Digest. 323–326, 2007.
114. Pirovano, A. L. Lacaita, A. Benvenuti, F. Pellizzer, S. Hudgens, and R. Bez, "Scaling analysis of phase-change memory technology," IEEE IEDM Tech. Digest. 699–672, 2003.
115. F. Pellizzer, A. Benvenuti, B. Gleixner, Y. Kim, B. Johnson, M. Magistretti, T. Marangon, A. Pirovano, R. Bez, and G. Atwood, "A 90 nm phase change memory technology for stand-alone non-volatile memory applications," IEEE Symp. VLSI Tech. Digest. 122–123, 2006.
116. G. Mueller, T. Harp, M. Kund, G. Y. Lee, N. Nagel, and R. Sezi, "Status and outlook of emerging memory technologies," IEEE IEDM Tech. Digest. 567–570, 2004.
117. F. Bedeschi, R. Fackenthal, C. Resta, E. M. Donze, M. Jagasivamani, E. Buda, F. Pellizzer, D. Chow, A. Cabrini, G. M. A. Calvi, R. Faravelli, A. Fantini, G. Torelli, D. Mills, R. Gastaldi, and G. Casagrande, "A multi-level-cell bipolar-selected phase-change memory," IEEE ISSCC 428–429, 625, 2008.

118. K. Nordquist, S. Pendharkar, M. Durlam, D. Resnick, S. Tehrani, D. Mancini, T. Zhu, and J. Shi, "Process development of sub-0.5 mm nonvolatile magnetoresistive random access memory arrays," J. Vac. Sci. Technol. B, 15 (6), 2274–2278, 1997.
119. J.-G. Zhu and Y. Zheng, "Ultrahigh density vertical magnetoresistive random access memory," J. Appl. Phys. 87 (9), 6668–6673, 2000.
120. N. Nishimura, T. Hirai, A. Koganei, T. Ikeda, K. Okano, Y. Sekiguchi, and Y. Osada, "Magnetic tunnel junction device with perpendicular magnetization films for high-density magnetic random access memory," J. Appl. Phys. 91 (8), 5246–5249, 2002.
121. S. Tehrani, J. M. Slaughter, M. Deherrera, B. N. Engel, N. D. Rizoo, J. Salter, M. Durlam, R. W. Dave, J. Janesky, B. Butcher, K. Smith, and G. Grynkewich, "Magnetoresistive random access memory using magnetic tunnel junctions," Proc. IEEE 91 (5), 703–712, 2003.
122. B. F. Cockburn, "Tutorial on magnetic tunnel junction magnetoresistive random-access memory," Memory Technol Design. Test 46–51, 2004.
123. J. DeBrosse, D. Gogl, A. Bette, H. Hoenigschmid, R. Robertazzi, C. Arndt, D. Braun, D. Casarotto, R. Havreluk, S. Lammers, W. Obermaier, W. R. Reohr, H. Viehmann, W. J. Gallanger, and G. Müller, "A high-speed 128-kb MRAM core for future universal memory applications," IEEE J. Solid-State Circuits 39 (4), 678–682, 2004.
124. C.-C. Hung, M.-J. Kao, Y.-S. Chen, Y.-H. Wang, Y.-J. Lee, W.-C. Chen, W.-C. Lin, K.-H. Shen, K.-L. Chen, S. Chao, D.-L. Tang, and M.-J. Tsai, A 6-F^2 bit cell design based on one transistor and two uneven magnetic tunnel junctions structure and low power design for MRAM," IEEE Trans. Electron. Dev. 53 (7), 1530–1538, 2006.
125. M. A. Hollis and R. A. Murphy, "Homogeneous Field-Effect Transistors," *High-Speed Semiconductor Devices*, S. M. Sze, Ed., John Wiley and Sons, 1990.
126. T. H. Ning, "Why BiCMOS and SOI BiCMOS," IBM J. Res. Dev. 46 (2/3), 181–186, 2002.

Appendix A: Universal Physical Constants

Quantity	Symbol	Value	Unit
Avogadro's constant	A_0	6.0225×10^{23}	Mole^{-1}
Boltzmann constant	k	1.3806×10^{-23}	J/K
		8.6181×10^{-5}	eV/K
Electron rest mass	m_0	9.1091×10^{-31}	kg
Energy equivalent of m_0		0.511	MeV
Electron charge	q or e	1.6021×10^{-19}	C
Planck constant	h	6.6256×10^{-34}	$\text{J} \cdot \text{s}$
		4.1356×19^{-15}	$\text{eV} \cdot \text{s}$
Planck constant/2π	$\hbar$	1.0546×10^{-34}	$\text{J} \cdot \text{s}$
Permittivity in vacuum	ε_0	8.8542×10^{-12}	F/m
Speed of light in vacuum	c	2.9979×10^{8}	m/s
Thermal energy	kT	0.02586 (300 K)	eV
Thermal voltage	kT/q	0.02586 (300 K)	V

Appendix B: International System of Units, SI

Quantity	Unit name	Symbol	Dimension
Capacitance	*farad*	F	C/V
Conductance	*siemens*	S	A/V
Energy	*joule*	J	$N \cdot m$
Electric charge	*coulomb*	C	$A \cdot s$
Force	*newton*	N	$kg \cdot m/s^2$
Frequency	*hertz*	Hz	s^{-1}
Inductance	*henry*	H	Wb/A
Length	*meter*	m	m
Magnetic flux	*weber*	Wb	$V \cdot s$
Magnetic flux density	*tesla*	T	Wb/m^2
Mass	*kilogram*	kg	kg
Power	*watt*	W	J/s
Potential	*volt*	V	J/C
Pressure	*pascal*	Pa	N/m^2
Resistance	*ohm*	Ω	V/A
Temperature	*kelvin*	K	K
Time	*second*	s	s

Appendix C: The Greek Alphabet

Upper case	Lower case		Upper case	Lower case	
Α	α	alpha	Ν	ν	Nu
Β	β	beta	Ξ	ξ	xi
Γ	γ	gamma	Ο	ο	omicron
Δ	δ	delta	Π	π	pi
Ε	ε	epsilon	Ρ	ρ	rho
Ζ	ζ	zêta	Σ	σ	sigma
Η	η	êta	Τ	τ	tau
Θ	θ	thêta	γ	υ	upsilon
Ι	ι	iota	Φ	φ	phi
Κ	κ	kappa	Χ	χ	chi
Λ	λ	lambda	Ψ	ψ	psi
Μ	μ	mu	Ω	ω	omega

Appendix D: Properties of Silicon and Germanium (300 K, Intrinsic Semiconductor Unless Otherwise Stated)

Property	Unit	Si	Ge
Atomic number	–	14	32
Atomic weight	g/mole	28.09	72.59
Atomic density	cm^{-3}	5.0×10^{22}	4.42×10^{22}
Crystal structure	–	Diamond	Diamond
Density	g/cm^3	2.328	5.323
Density of surface atoms	cm^{-2}		
(100)		6.78×10^{14}	6.27×10^{14}
(110)		9.59×10^{14}	8.87×10^{14}
(111)		7.83×10^{14}	7.24×10^{14}
Dielectric constant	ε_s	11.7	16.0
Effective electron mass	kg		
Longitudinal (4.2 K)		$0.9163m_0$	$1.58m_0$
Transverse (4.2 K)		$0.1905m_0$	$0.082m_0$
Density-of-states (4.2 K)		$1.062m_0$	
Density of states (300 K)		$1.090m_0$	
Effective hole mass	kg		
Longitudinal (4.2 K)		$0.537m_0$	$0.28m_0$
Transverse (4.2 K)		$0.153m_0$	$0.044m_0$
Density-of-states (4.2 K)		$0.059m_0$	
Density of states (300 K)		$1.15m_0$	
Elastic constants	Pa		
c_{11}		1.656×10^{11}	1.26×10^{11}
c_{22}		0.639×10^{11}	0.44×10^{11}
c_{44}		0.796×10^{11}	0.68×10^{11}
Electron affinity	eV	4.15	4.00
Energy gap	eV	1.124	0.67
Index of refraction	–	3.44	3.97
Lattice constant	nm	0.543095	0.564613

(Continued)

(Continued)

Property	Unit	Si	Ge
Mobility	cm^2/Vs		
Electron		1,450	3,900
Hole		500	1,900
Melting point	°C	1,412	937
N_C	cm^{-3}	2.80×10^{19}	1.04×10^{19}
N_V	cm^{-3}	1.04×10^{19}	6.00×10^{18}
n_i	cm^{-3}	1.4×10^{10}	2.4×10^{13}
Phonon mean-free path	nm	7.6	10.5
Poisson ratio	–	0.28	0.26
Thermal conductivity	$W/(cm \cdot K)$	1.412	0.606
Young's modulus, < 100 >	Pa	1.3×10^{11}	1.03×10^{11}
α	K^{-1}	2.5×10^{-6}	5.7×10^{-6}
ΔE_p	eV	0.063	0.037
λ_{ph}	nm		
Electrons		6.2	6.5
Holes		4.5	6.5

ΔE_p: Average energy loss per phonon scattering
α: Coefficient of linear expansion
n_i: Intrinsic carrier concentration
N_C: Effective density of states, conduction band
N_V: Effective density of states, valence band
m_0: Free electron mass $= 9.1091 \times 10^{31}$ kg
λ_{ph}: Optical phonon mean-free path

Sources

W. E. Beade, J. C. C. Tsai, R. D. Plummer, *Quick Reference Manual for Silicon Integrated Circuit Technology*, Wiley-Interscience, New York, 1985.

R. Hull, Editor, *Properties of Crystalline Silicon*, EMIS Datareviews Series, Nr. 20, INSPEC, London, 1999.

M. Shur, *Physics of Semiconductor Devices*, Prentice-Hall, Englewood Cliffs, NJ, 1990.

S. M. Sze, *Physics of Semiconductor Devices*, John Wiley and Sons, 1969.

H. Wolf, *Semiconductors*, John Wiley and Sons, 1971.

Appendix E: Conversion Factors

Length

nm		Å	μm	cm	m	mil	in
nm	1	10	10^{-3}	10^{-7}	10^{-9}	$C' \times 10^{-5}$	$C' \times 10^{-8}$
Å	10^{-1}	1	10^{-4}	10^{-8}	10^{-10}	$C' \times 10^{-6}$	$C' \times 10^{-9}$
μm	10^{3}	10^{4}	1	10^{-4}	10^{-6}	$C' \times 10^{-2}$	$C' \times 10^{-5}$
cm	10^{-7}	10^{8}	10^{4}	1	10^{-2}	$C' \times 10^{2}$	$C' \times 10^{-1}$
m	10^{-9}	10^{10}	10^{6}	10^{2}	1	$C' \times 10^{4}$	$C' \times 10$
mil	$C \times 10^{4}$	$C \times 10^{5}$	C	$C \times 10^{-3}$	$C \times 10^{-5}$	1	10^{-3}
in	$C \times 10^{7}$	$C \times 10^{8}$	$C \times 10^{4}$	C	$C \times 10^{-2}$	10^{3}	1

$C = 2.54$, $C' = 1/C = 3.937$

Energy

	eV	J	W · s	N · m	Erg
eV	1	K	K	K	$K \times 10^{7}$
J	K′	1	1	1	10^{7}
W · s	K′	1	1	1	10^{7}
N · m	K′	1	1	1	10^{7}
Erg	$K' \times 10^{-7}$	$K' \times 10^{-7}$	$K' \times 10^{-7}$	$K' \times 10^{-7}$	1

$K = 1.6022 \times 10^{-19}$, $K' = 1/K = 6.2415 \times 10^{18}$
1 eV corresponds to $\lambda = 1.240\,\mu\text{m}$ ($\nu = 2.418 \times 10^{14}\,\text{s}^{-1}$)

Index

E

FSC
www.fsc.org
MIX
Papier | Fördert gute Waldnutzung
FSC® C083411

Kolibri 360 GmbH
Ferdinand-Jühlke-Straße 7
99095 Erfurt, Deutschland
produktsicherheit@kolibri360.de